Geophysical Monograph Series
American Geophysical Union

*Geophysical Monograph Series*
A. F. Spilhaus, Jr., Managing Editor

1  **Antarctica in the International Geophysical Year,** A. P. Crary, L. M. Gould, E. O. Hulburt, Hugh Odishaw, and Waldo E. Smith (editors)

2  **Geophysics and the IGY,** Hugh Odishaw and Stanley Ruttenburg (editors)

3  **Atmospheric Chemistry of Chlorine and Sulfur Compounds,** James P. Lodge, Jr. (editor)

4  **Contemporary Geodesy,** Charles A. Whitten and Kenneth H. Drummond (editors)

5  **Physics of Precipitation,** Helmut Wieckmann (editor)

6  **The Crust of the Pacific Basin,** Gordon A. Macdonald and Hisashi Kuno (editors)

7  **Antarctic Research: The Matthew Fontaine Maury Memorial Symposium,** H. Wexler, M. J. Rubin, and J. E. Caskey, Jr. (editors)

8  **Terrestrial Heat Flow,** William H. K. Lee (editor)

9  **Gravity Anomalies: Unsurveyed Areas,** Hyman Orlin (editor)

10  **The Earth Beneath the Continents: A Volume in Honor of Merle A. Tuve,** John S. Steinhart and T. Jefferson Smith (editors)

11  **Isotope Techniques in the Hydrologic Cycle,** Glenn E. Stout (editor)

12  **The Crust and Upper Mantle of the Pacific Area,** Leon Knopoff, Charles L. Drake, and Pembroke J. Hart (editors)

13  **The Earth's Crust and Upper Mantle,** Pembroke J. Hart (editor)

14  **The Structure and Physical Properties of the Earth's Crust,** John G. Heacock (editor)

15  **The Use of Artificial Satellites for Geodesy,** Soren W. Henriksen, Armando Mancini, and Bernard H. Chovitz (editors)

16  **Flow and Fracture of Rocks,** H. C. Heard, I. Y. Borg, N. L. Carter, and C. B. Raleigh (editors)

17  **Man-Made Lakes: Their Problems and Environmental Effects,** William C. Ackermann, Gilbert F. White, and E. B. Worthington (editors)

18  **The Upper Atmosphere in Motion: A Selection of Papers With Annotation,** C. O. Hines and Colleagues

19  **The Geophysics of the Pacific Ocean Basin and Its Margin: A Volume in Honor of George P. Woollard,** George H. Sutton, Murli H. Manghnani, and Ralph Moberly (editors)

20  **The Earth's Crust,** John G. Heacock (editor)

21  **Quantitative Modeling of Magnetospheric Processes,** W. P. Olson (editor)

geophysical monograph 21

# Quantitative Modeling of Magnetospheric Processes

W. P. OLSON
editor

American Geophysical Union
Washington, D. C.
1979

Published under the aegis of the AGU Geophysical Monograph Board; Bruce Bolt, Chairman; Thomas E. Graedel, Rolland L. Hardy, Barry E. Parsons, George R. Tilton, and William R. Winkler, members.

ISBN 0-87590-021-6

Copyright 1979 by the American Geophysical Union
1909 K Street, N.W.
Washington, D. C. 20006

Library of Congress Card Number 79-84163

LITHOCRAFTERS, CHELSEA, MICHIGAN

CONTENTS

OVERVIEW     1
   W. P. Olson

SECTION 1   MAGNETIC FIELD SESSION

   OVERVIEW     7
      K. S. Pfitzer

   QUANTITATIVE MODELING OF PLANETARY MAGNETOSPHERIC MAGNETIC FIELDS     9
      R. J. Walker

   THE SYNCHRONOUS ORBIT MAGNETIC FIELD DATA SET     35
      R. L. McPherron

   A METHOD OF EVALUATING QUANTITATIVE MAGNETOSPHERIC FIELD MODELS BY AN ANGULAR PARAMETER $\alpha$     48
      M. Sugiura and D. J. Poros

   MAGNETIC FIELD DISTORTIONS AT L = 3-4 INFERRED FROM SIMULTANEOUS LOW AND HIGH ALTITUDE OBSERVATIONS OF RING CURRENT IONS     64
      L. R. Lyons

   MODELING THE MAGNETOSPHERIC MAGNETIC FIELD     77
      W. P. Olson, K. A. Pfitzer, and G. J. Mroz

   SELF-CONSISTENT THEORY OF A MAGNETOSPHERIC $B$-FIELD MODEL     86
      K. Fuchs and G.-H. Voigt

   CONJUGATE LOW ENERGY ELECTRON OBSERVATIONS MADE BY ATS-6 AND DMSP-32 SATELLITES     96
      C.-I. Meng

   QUANTITATIVE MODELS OF THE EARTH'S INTERNAL FIELD -- STATUS AND FUTURE TRENDS     110
      G. D. Mead

SECTION 2   HIGH ENERGY PARTICLES SESSION

   OVERVIEW     118
      J. I. Vette

   MODELING OF ELECTRON TIME VARIATIONS IN THE RADIATION BELTS     121
      K. Chan, M. J. Teague, N. Schofield, and J. I. Vette

   THE SIGNATURES OF THE VARIOUS REGIONS OF THE OUTER MAGNETOSPHERE IN THE PITCH ANGLE DISTRIBUTIONS OF ENERGETIC PARTICLES     150
      H. I. West, Jr.

EFFECTS OF THE SOLAR WIND ON MAGNETOSPHERIC DYNAMICS:    180
ENERGETIC ELECTRONS AT THE SYNCHRONOUS ORBIT
*G. A. Paulikas and J. B. Blake*

PITCH ANGLE DISTRIBUTIONS OF >30 keV ELECTRONS AT GEO-    203
STATIONARY ALTITUDES
*P. R. Higbie, D. N. Baker, E. W. Hones, Jr., and*
*R. D. Belian*

SOLAR ENERGETIC PARTICLES: FROM THE CORONA TO THE    220
MAGNETOTAIL
*E. C. Roelof*

THE EFFECT OF MAGNETIC FIELD MODELS ON COSMIC RAY    242
CUTOFF CALCULATIONS
*K. A. Pfitzer*

SECTION 3   ELECTRIC FIELDS SESSION

OVERVIEW    253
*D. P. Cauffman*

SEMIEMPIRICAL MODELS OF MAGNETOSPHERIC ELECTRIC    261
FIELDS
*H. Volland*

QUASISTATIC ELECTRIC FIELD MEASUREMENTS ON THE GEOS-1    281
AND GEOS-2 SATELLITES
*A. Pedersen and R. Grard*

GENERATION OF THE MAGNETOSPHERIC ELECTRIC FIELD    297
*T. W. Hill*

INDUCED MAGNETOSPHERIC ELECTRIC FIELDS    316
*G. J. Mroz, W. P. Olson, and K. A. Pfitzer*

COUPLING OF MAGNETOSPHERIC SUBSTORM ELECTRIC EFFECTS    326
INTO THE GLOBAL ATMOSPHERIC ELECTRICAL CIRCUIT
*P. B. Hays and R. G. Roble*

SECTION 4   LOW ENERGY PARTICLES SESSION

OVERVIEW    338
*E. C. Whipple, Jr. and S. E. DeForest*

ION COMPOSITION MEASUREMENTS IN MAGNETOSPHERIC MODELING    340
*D. T. Young*

QUANTITATIVE MODELS OF THE 0 TO 100 keV MID-MAGNETOSPHERIC    364
PARTICLE ENVIRONMENT
*H. B. Garrett*

ENERGY BUDGET FOR SOLAR WIND-MAGNETOSPHERIC INTER-    389

ACTIONS
  W. J. Heikkila

THE MAGNETOPAUSE LAYER AND PLASMA BOUNDARY LAYER OF    401
THE MAGNETOSPHERE
  T. E. Eastman and E. W. Hones, Jr.

THE MAGNETOSPHERIC BOUNDARY LAYER: A STOPPER REGION FOR    412
A GUSTY SOLAR WIND
  J. Lemaire

EMPIRICAL RELATIONSHIPS BETWEEN INTERPLANETARY CONDITIONS,    423
MAGNETOSPHERIC FLUX TRANSFER, AND THE AL INDEX
  J. A. Slavin and R. E. Holzer

STATISTICAL STUDY OF THE DEPENDENCE OF GEOMAGNETIC    436
ACTIVITY ON SOLAR WIND PARAMETERS
  K. Maezawa

A MACROSCOPIC MODEL FOR FIELD LINE INTERCONNECTION    448
BETWEEN THE MAGNETOSPHERE AND THE INTERPLANETARY SPACE
  G. H. Voigt and K. Fuchs

SECTION 5  MODELING TECHNIQUES SESSION

OVERVIEW    460
  W. P. Olson and R. A. Wolf

A KINETIC APPROACH TO MAGNETOSPHERIC MODELING    462
  E. C. Whipple, Jr.

ON CALCULATING MAGNETIC AND VECTOR POTENTIAL FIELDS DUE    473
TO LARGE-SCALE MAGNETOSPHERIC CURRENT SYSTEMS AND IN-
DUCED CURRENTS IN AN INFINITELY CONDUCTING EARTH
  J. L. Kisabeth

COMPUTER MODELING OF EVENTS IN THE INNER MAGNETOSPHERE    499
  M. Harel, R. A. Wolf, P. H. Reiff, and M. Smiddy

MOTIONS OF CHARGED PARTICLES IN THE MAGNETOSPHERE UNDER    513
THE INFLUENCE OF A TIME-VARYING LARGE SCALE CONVECTION
ELECTRIC FIELD
  P. H. Smith, N. K. Bewtra, and R. A. Hoffman

GLOBAL MAGNETOHYDRODYNAMIC SIMULATION OF THE TWO-    536
DIMENSIONAL MAGNETOSPHERE
  J. N. Leboeuf, T. Tajima, C. F. Kennel, and J. M. Dawson

IONOSPHERIC ELECTRIC FIELDS DRIVEN BY FIELD-ALIGNED    557
CURRENTS
  R. W. Nopper, Jr. and R. L. Carovillano

INFLUENCE OF ELECTRIC FIELDS ON CHARGED PARTICLE MOTION    569

AND ELECTRON FLUXES AT SYNCHRONOUS ALTITUDES
    J. C. Kosik

   MAGNETIC SHELL TRACING: A SIMPLIFIED APPROACH                582
       J. G. Luhmann and M. Schulz

   SHAPE INTEGRAL METHOD FOR MAGNETOSPHERIC SHAPES              592
       F. C. Michel

SECTION 6  MODELING APPLICATIONS SESSION

   OVERVIEW                                                     599
       H. B. Garrett

   ENVIRONMENTAL MODEL NEEDS FOR SPACECRAFT INTERACTIONS        602
       S. E. DeForest

   IONOSPHERIC AND MAGNETOSPHERIC MODELING FOR AIR FORCE        609
   APPLICATIONS
       A. L. Snyder, Jr.

   CURRENT LEAKAGE FOR LOW ALTITUDE SATELLITES: MODELING        617
   APPLICATIONS
       A. Konradi, J. E. McCoy, and O. K. Garriott

   THE RADIATION ENVIRONMENT AND ITS EFFECTS ON SPACECRAFT      634
       J. F. Janni and G. E. Radke

SECTION 7  REVIEW OF JULY 29, 1977 SUBSTORM                     644
       R. H. Manka

SECTION 8  AN ANNOTATED LIST OF EXISTING QUANTITATIVE MODELS    647
OF MAGNETOSPHERIC PARTICLE AND FIELD FEATURES

   MAGNETIC FIELD MODELS                                        647

   ENERGETIC PARTICLE RADIATION MODELS                          649

   ELECTRIC - FIELD MODELS                                      650

KEY WORD LIST INDEX                                             653

## Introductory Remarks

There is some confusion concerning what is meant by quantitative modeling. This is caused primarily by the existence of a wide variety of model users. Quantitative models may be purely empirical, that is based on data sets, semi-empirical, that is relying on physics whenever appropriate and available, but the "holes" being filled in with data as necessary, and finally purely physical models. It was apparent at the meeting that most current quantitative modeling efforts at this point in time are semi-empirical. This is partly because the physics of the magnetosphere is not yet entirely understood. At the same time, however, there is a large user community that needs models that can accurately describe magnetospheric features. The preponderance of semi-empirical models results because current data sets describing various magnetospheric features and processes are limited. There is some additional discussion of this topic in the section summaries. In particular, Cauffman in his introduction to the electric field section makes the useful distinction between "explanatory" and "representative" models. It is believed that all kinds of modeling will continue because of the various needs of the model user community.

The general purpose of the meeting was to bring together the authors of newly developed quantitative models of magnetospheric features and processes and have them exchange ideas and concerns with the model user community, and with the collectors and analyzers of data. It was then planned that by having these groups of people interact through the presentation of their papers and in discussion periods, the meeting output would be an up-to-the-minute assessment of the effects to quantitatively model magnetospheric features and processes. As was done at the 1975 conference on quantitative modeling, it was planned that this meeting provide a "snap shot" of the progress being made in quantitative modeling. While these conferences provide an up-to-date look at modeling activities, there is within IAGA (International Association of Geomagnetism and Aeronomy) a working group on magnetospheric models that continuously seeks to monitor the same subject. Thus, the outputs of this meeting are closely linked to the work on the IAGA group, and it was attempted to have an international representation at the meeting. The meeting was also timed to take place so that these proceedings would be helpful to workers participating in the IMS (International Magnetospheric Study). This volume therefore, in addition to containing papers from the meeting, contains discussion on each of the meeting sessions and provides a summary of currently available quantitative models of several magnetospheric features and processes. There were over 90 meeting participants with 14 foreign attendees. The general areas covered in the meeting were: magnetospheric magnetic and electric fields, low and high energy particle populations, model users, modeling techniques, and attempts at modeling several magnetospheric features or processes simultaneously.

The meeting subject was restricted to <u>quantitative</u> models. These models are distinct from the conjectural models ("cartoon approximations") that abound in the literature. A quantitative model has, as its output, numbers which are meant to describe a magnetospheric feature or process. These numbers may be compared with observations. The quality of the model can then be tested by comparing model output with observation. Thus, the development and testing of quantitative models is a natural step in the process of physically describing the magnetosphere.

Quantitative models are also necessary tools for users with practical applications. In fact, it is these practical needs that have been responsible for the development of many quantitative models. For example, in the past decade quantitative descriptions of the high-energy particle environment were required in the inner magnetosphere since this environment posed a radiation damage threat to vehicles operating in "near earth orbital space." More recently attention has been given to modeling the low-energy particle distribution, especially at geosynchronous orbit, because of the threat that they pose to satellite hardware in the form of charging and electrical arcing. Quantitative magnetospheric models have in some instances been extrapolated from the setting of their development (the earth's magnetosphere) and extended to planetary, stellar and astronomical applications and thus are valuable to other sciences.

It is interesting to compare this conference with the 1975 AGU conference on magnetospheric models. In 1975 the emphasis in magnetospheric modeling was clearly the magnetic field. In the intervening years, the emphasis has been shifted to the magnetospheric electric field. That is not to say that interesting work is no longer being done on the magnetospheric magnetic field, but that good satellite DC electric field data are now becoming available, finally making it possible to discuss and describe its properties quantitatively. There has also been a trend away from the development of static models and in their place an increased emphasis has been placed on dynamic models. For example, in 1975 all of the models dealt with static magnetospheric features, like the average configuration of the magnetic field. It is now agreed that the average magnetic field configuration is well described, if not quite so well understood. Of more interest are the time variations in the magnetic field, electric field, and corresponding changes in the low-energy particle populations. There now appears to be a trend away from the description of magnetospheric features (e.g., the magnetospheric magnetic field) and toward the description of magnetospheric processes (e.g., the self-consistent particle and field description of the magnetospheric substorm). It is clear when these two meetings are compared, that much progress has been made in the intervening three years. Hopefully, with all of the data that is becoming available during the IMS, even more rapid progress will be made in the next few years in the quantitative modeling and understanding of magnetospheric processes.

The meeting sessions were initially confined to descriptions of the magnetic and electric fields, low and high energy particles, and papers by model users. However, it was found during the planning stages that another very important subject was emerging. This resulted in a session on "modeling techniques." In it several papers were presented in which the interaction of charged particles with their electric and magnetic field environment were quantitatively described. Also, several papers concerned with the simplification of computer procedures necessary for the understanding and quantitative descriptions of various magnetospheric processes were presented. These included new techniques for determining magnetopause shape, simplified descriptions of drift shell topology, the analytical determination of low energy particle motions in time varying electric and magnetic fields, the kinetic description of the low energy particle and field interactions, and the use of the magnetic charge concept in the description of scalar and vector magnetic potentials as they apply to the representation of field-aligned currents.

In addition to the formal daytime sessions, an evening session was held in which the ability of available models to describe detailed observations of a magnetospheric substorm event was discussed. The need for "dynamic" models was apparent. A summary of this session is provided at the end of the volume. Work on the testing of models is progressing and related "working group" meetings are now being planned on this topic in conjunction with IMS activities.

At least two problems of importance to this community were not discussed in detail at the meeting. Both should be important topics in future modeling meetings. First, the subject of where to break into the "interactive particle-field" loop was not considered in detail. The problem may be described as follows. The magnetospheric magnetic field is largely determined by several magnetospheric current systems, which are in turn the result of motions of low energy charged particles. These motions are influenced by the magnetic field and sometimes, more importantly, by the electric field that resides in the magnetosphere. The currents produced by these particle motions contribute to the magnetic field and are in turn determined by the magnetic field. Thus, there is the need to solve the particle-field interaction problem in a self-consistent way. Other self-consistency problems were mentioned at the meeting. For example, it was suggested that it is not sufficient to simply calculate the magnetopause currents but instead it is necessary to accurately determine the entire bow-shock, magnetosheath magnetopause interactions with the solar wind in order to quantitatively represent the flow of energy between the magnetosphere and solar wind. However, again, the magnetopause shape and current and the solar wind must be considered simultaneously in some self-consistent way in order to accurately determine the problem. There are many such "loops" that must be considered before the quantitative descriptions of many magnetospheric processes can be obtained. Papers were presented which considered magnetic and electric fields and charged particles simultaneously; but there does not at present exist a self-consistent model that quantitatively determines the interaction of the magnetosphere with the solar wind and interplanetary magnetic fields to the extent that it can be used to describe such magnetospheric processes as the substorm and magnetic storm.

The other subject that was omitted in the formal presentation, but arose in several contexts, concerned the use of indices and input parameters to the models. In the past decade or two, it has been popular to use magnetic indices as inputs to various magnetospheric models. The reason for this centers around the positive correlation between such indices as $K_p$ and various observed magnetospheric parameters. It is not surprising that the correlations exist, but it is also clear (based on our current physical understanding of the magnetosphere) that there is not necessarily any direct casual relationship between the observed magnetospheric feature and variations in these magnetic indices. Rather, these indices describe roughly the variation in some magnetic parameter at the earth's surface with time. Obviously these observations are produced by many changes in magnetospheric current systems and other magnetospheric processes. What is required for the input to these quantitative magnetospheric models is the proper set of physical parameters. It is hoped that data from the ISEE-C satellite may be used to describe the solar wind and interplanetary conditions prior to their interactions with the magnetospheric parameters will lead to more physical input parameters which can be used with models. With such inputs, it should be possible to <u>predict</u>, with a 15-minute to half-hour lead time, the variability in several observed magnetospheric features and processes. In the next few years, the use of magnetic indices may thus be supplanted by the development and availability of such physical parameters which can be used as more appropriate inputs to quantitative models of magnetospheric processes.

This volume is divided into eight sections. The Magnetic Field is discussed first (Section 1) as it is a relatively stable feature of the magnetosphere and controls the motions of plasmas and energetic particles in the magnetosphere. The discussion of High Energy Particles follows in Section 2. The behavior of these particles can be discussed solely in terms of the magnetic field. The Electric Field session papers are presented in Section 3. Note that Cauffman's introduction to this section contains an up-to-date description of past, present, and future electric field experiments. Descriptions of magnetic field and particle experiments are given in papers in other sections. Low Energy Particles are discussed in Section 4. Their behavior in the magnetosphere is influenced by both the magnetic and electric fields which have been described earlier in the text. Section 5 deals with Modeling Techniques. It includes descriptions of current efforts to quantitatively describe particle-field interactions self consistently and new techniques to make the computer solution of various magnetospheric problems more accurate and rapid. Model uses are discussed in Section 6. Several examples of instances where quantitative models of magnetospheric features were used to solve "practical" problems are given.

There was one evening session during which an assessment of the ability of currently available models to describe an observed event was made. The well-observed July 29, 1977 substorm was used as an example. A review of the session is provided in Section 7.

An annotated list of existing quantitative models of magnetospheric particle and field features is given in Section 8. It provides the reader with an up-to-date reference on available electric and magnetic field models and low and high energy particle distribution representations. Information on the region of validity and availability of each model is included.

<u>Acknowledgements</u>. It is hoped that this volume will be useful to many workers in the magnetospheric physics community and to model users whose applications are in several fields. Both the conference and this volume are the result of encouragement and help from several people. Financial support for the meeting was received from the Office of Naval Research (which coordinated the other grants), Headquarters of the National Aeronautics and Space Administration, the Air Force Office of Scientific Research, and McDonnell Douglas Astronautics Company. The

publication of this volume was supported by a grant from the National Science Foundation. The Program Committee, consisting of David P. Cauffman, Sherman E. DeForest, Captain Henry B. Garrett, Willard P. Olson, George A. Paulikas, Karl A. Pfitzer, Masahisa Sugiura, James I. Vette, Elden C. Whipple, and Richard A. Wolf, was largely responsible for the structure of each of the sessions. Drs. George Paulikas, David Cauffman, Richard Wolf, Karl Pfitzer, Robert Manka, and Raymond Walker provided additional help on the Editorial Review Committee for this volume and on other specialized tasks. Ms. Susan Poling and Ms. Cindy Beadling from the AGU provided excellent help with administrative matters concerning the conference. Likewise, Ms. Linda Masnik and Ms. Judy Holoviak were responsible for the timely handling of the format and production of the volume.

This final paragraph concerns the phone calls (probably hundreds) and letters (many dozens) and myriad other chores required for the conference and the timely assembly of this volume that were cheerfully done by Mrs. Dawn Ustick. A special note of thanks is extended to her.

           W. P. Olson
           Editor

OVERVIEW - MAGNETIC FIELDS

K. A. Pfitzer

McDonnell Douglas Astronautics Company, 5301 Bolsa Avenue,
Huntington Beach, California 92647

Considerable progress has been made in the modeling of the magnetospheric magnetic field since the last modeling symposium three years ago. The precision of the magnetic field models as well as the quality of the data sets has improved significantly. Whereas the last modeling symposium primarily treated static magnetic field models, in this symposium we noted considerable progress in modeling quantitatively the dynamic structure of the magnetic fields. Walker's review article provides a concise summary of the current state of magnetic field modeling.

Almost a decade of magnetic field data is now available at synchronour orbit. McPherron lists the currently available data sets. He points out that only the more recent measurements, such as ATS-6 and GEOS have the precision necessary for much of today's work. The early data at synchronous orbit has large offset errors and may not be very useful.

The synchronous orbit data is used by model makers as an independent check on the models validity. The synchronous orbit data exhibits variations from all the current sources and thus a model must correctly predict the effects of all the currents in order to agree with the synchronous orbit measurement. As dynamic models become available, the synchronous orbit data will become even more important in testing the validity of the dynamic models, since most of the magnetic field fluctuations created by the various magnetospheric processes are observable from synchronous orbit.

The most important data sets for model makers to date have been the $\Delta B$ contours of Sugiura. The scalar $\Delta B$ contours organized the data over a large region space and thus provided the required input to design magnetic field models. At this symposium Sugiura has introduced $\Delta \alpha$ contours which will supplement the $\Delta B$ contours. The $\Delta \alpha$ contours contain vector information and will provide a more stringent test for model verification and construction.

The static magnetic field models are not very precise. For many applications such as field line mapping, the external models must be used with accurate internal field models (the dipole field approximations is not sufficiently accurate). Mead describes four new internal field models in his review. Users are advised to use any three of the recently released internal magnetic field models (IGS/75; AWC/75; POGO 8/71) instead of the 20 year old models still used by many investigators. These three internal field models are clearly more accurate than the IGRF or the older models. The internal field models must be combined with the external field model by coordinate transformation programs.

Many such programs now exist using various algorithms and approximations. The user is urged to fully understand these approximations. Once an accurate internal field model is combined with an accurate external field model, exceptional accuracies can be obtained. Meng details the results of a synchronous orbit, low latitude correlation experiment in which mapping errors of the order of a few tenths of a degree can be observed. (Note: When using synchronous orbit data such as ATS-6 data, the 0.5° inclination of the ATS-6 orbit to the geographic equator <u>cannot</u> be ignored.)

Dynamic field models are now becoming available. Olson, et al., presented a magnetic field model in which the various current strengths and thus the associated magnetic fields are permitted to fluctuate independently. These types of models will be useful in testing the effect the various current systems have on the acceleration and deceleration of charged particles. Along with the time varying field the new model described by Olson, et al., also makes available the time varying vector potential. The time varying vector potential will permit the study of induced electric field effects. In the Electric Field Session of this symposium large electric field fluctuation at synchronous orbit were observed along with the magnetic field fluctuations. Correlated magnetic/electric field models may assist us in developing quantitative models of magnetospheric acceleration processes.

Lyons shows variation in the magnetic field that required large displacement of a field line for L's as low as 3 or 4. Variation predicted by the dynamic model described by Olson, et al., can only produce small changes. Changes of the order observed by Lyons must be created by local (perhaps parallel) currents which are not yet included in the dynamic model.

A paper by Fuchs and Voigt attempted to present a more self-consistent treatment of the magnetic field models. Most of the presently available models rely very heavily on magnetic field data to produce accurate models. Although considerable physics is used in the development of these semiempirical models, a complete understanding of the physics of the magnetosphere requires the development of completely self-consistent models.

The papers in this section present a coordinated set describing the current state of magnetic field modeling. Much has been learned since the last conference and the emergence of dynamic models should introduce the era of a quantitative modeling of particle energization processes.

# QUANTITATIVE MODELING OF PLANETARY MAGNETOSPHERIC MAGNETIC FIELDS

Raymond J. Walker

Institute of Geophysics and Planetary Physics, University of California
Los Angeles, California 90024

*Abstract.* Considerable progress has been made in the development of quantitative magnetospheric magnetic field models in recent years. Several models include the contributions from the major current systems thought to be responsible for the quiet time magnetic configuration. These models include currents in the inner magnetosphere in addition to the boundary currents and the magnetotail current system. The currents in the inner magnetosphere evolve smoothly into the tail current system thereby eliminating the artificial neutral lines found at the inner edge of the current sheet in earlier models. Several models contain realistic tail current systems of finite thickness which close on the magnetopause surface. All of the models include the dipole tilt angle as an input parameter. A recent model incorporates the effects of the interplanetary magnetic field on the magnetospheric configuration. The newest models reproduce the observed field magnitude well and model the field direction on the dayside reasonably well, also. In the near tail region there are considerable differences in the field direction between the theoretical models and between the theoretical models and an empirical model. Much recent effort has been directed toward modeling the magnetospheres of Mercury and Jupiter. Mercury's magnetosphere is sufficiently earth-like that simple models have been constructed by scaling models of the earth's field. There are no models of the entire Jovian magnetosphere. However, considerable progress has been made in modeling the field adjacent to the equatorial current sheet.

## Introduction

The first quantitative magnetospheric magnetic field models included the field contribution from magnetopause currents in addition to the earth's main field. Two types of magnetopause models have been widely used: the boundary surface model of Mead [1964] and the image dipole calculation of Hones [1963]. Only in the boundary surface models was the shape of the magnetopause calculated self-consistently. The boundary was determined by a pressure balance between the magnetic field and a steady solar wind with the requirement that the normal component of the magnetospheric field vanish at the boundary. Since a calculation of the boundary requires a knowledge of currents on the boundary, the method for determining the pressure balance surface was iterative [Beard, 1960]. In this method, an initial boundary was assumed and the currents flowing on this first surface were calculated.

Then, the field calculated from these currents was used to find an improved surface. This procedure was repeated until a self-consistent solution was obtained. The only magnetospheric source included in the boundary surface calculations was the main dipole field. The calculation of the image dipole surface was much simpler. A large image dipole was placed between the earth and the sun to approximate the magnetopause currents.

Following the first observations of the magnetotail [Ness et al., 1964; Ness, 1965], a semi-infinite current sheet located in the equatorial plane on the night side was added to the models [Williams and Mead, 1965; Taylor and Hones, 1965]. The current sheet in the Mead-Williams model was infinitesimally thin while that in the Hones-Taylor model was of finite thickness. In both models the current was from east to west.

The boundary surface models have physical input parameters such as the magnetopause stand off distance, the location of the near and far edges of the current sheet and the tail field intensity. The image dipole model has unphysical input parameters including the image dipole position and size. However, an image dipole model can be formulated as an analytic function and is therefore simpler and less expensive to use on a computer than a boundary surface model which is expressed as an expansion in Legendre polynomials. Both models are only valid for zero dipole tilt and neither model includes the effects of currents in the inner magnetosphere. These early models are reviewed in greater detail by Roederer [1969].

The intersection of several model magnetopause surfaces with the equatorial plane is plotted in Figure 1. The thin solid line is the average observed magnetopause shape determined by Fairfield [1971] and the dashed line with large dots is the average uneroded magnetopause surface of Holzer and Slavin [1978]. The self-consistent boundary surface calculations (heavy solid line) agree quite well with the observed surfaces for longitudes within about 60° of noon, however on the night side the gas dynamic surfaces are a poor approximation to the observed magnetopause. The most recent image dipole surface [Willis and Pratt, 1972] is a much better approximation to the observations although the day side magnetopause appears to be slightly more blunted than is observed.

Models based on these two types of magnetopause surfaces have been gradually improved in recent years. Models have been presented in which the tail current sheet closes on the magnetopause surface [Choe and Beard, 1974b] and in which the tail current is a function of distance down the tail [Willis and Pratt, 1972; Choe and Beard, 1974b]. Calculations also have included tail current regions of finite thickness [Willis and Pratt, 1972; Beard, 1978]. The resulting magnetospheric configuration is qualitatively quite reasonable as is demonstrated in Figure 2. Field lines calculated by using the model of Choe and Beard [Choe and Beard, 1974a,b; Halderson et al., 1975; Kosik, 1977] demonstrate the expected day side compressed field and extended night side field lines. However, the models are unable to reproduce quantitatively the observed field magnitude [Walker, 1976].

The $\Delta B$ contours of Suguira and Poros [1973] have proved to be a very useful data set for evaluating magnetospheric models (Figure 3). The $\Delta B$ contours were calculated by subtracting the earth's main field

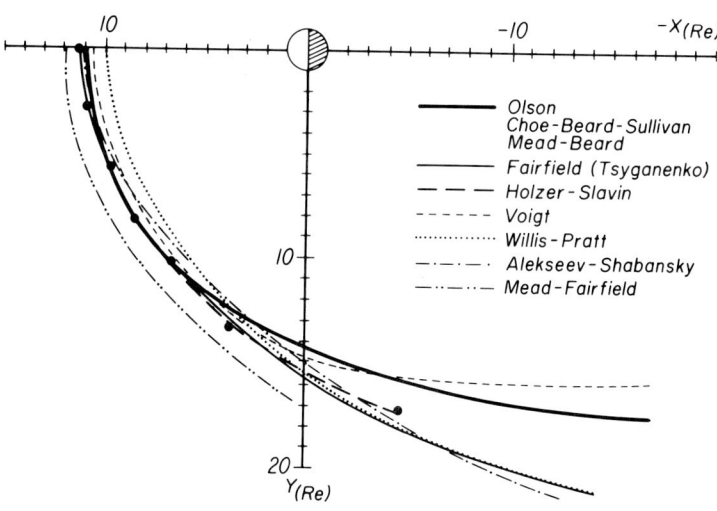

Fig. 1. The intersections of magnetopause surfaces with the equatorial plane. The curves calculated by using the boundary surface models of Olson (1969), Choe, Beard and Sullivan (1973) and Mead and Beard (1964) were so similar that one curve was used for all three. The curves from Fairfield (1971) and Holzer and Slavin (1978) are empirical surfaces. The Mead and Fairfield (1975) curve is the boundary of their empirical model. The remaining curves denote the surfaces used in various magnetospheric magnetic field models (Tsyganenko, (1976); Voigt, (1972); Willis and Pratt, (1972); Alekseev and Shabansky (1972)).

from the fields measured on Ogo-3 and Ogo-5. The jagged curves at the top and bottom in Figure 3 delimit the region of data coverage. Only periods for which $K_p = 0-1$ were included in the data set. All data in a 90° wide wedge in longitude about the noon-midnight meridian were included in the average. Because of the wide longitude range over which the data were averaged and because all values of dipole tilt were included in the average, exact agreement should not be expected between the model and observed $\Delta B$ contours. However, such a comparison should give a good indication of the role played by the current systems included in the model.

$\Delta B$ contours calculated by using the Choe-Beard model are plotted in Figure 4. Near the day side magnetopause, the model agrees rather well with the observed field. At higher latitudes, both the model and the observed $\Delta B$ are negative but the predictions are smaller ($-30\gamma$) than the observations ($-10\gamma$). The largest differences occur near the equator where large positive values are predicted but large negative values are observed. Similar results are obtained with all models which contain only the contributions from magnetopause currents and tail currents [Walker, 1976]. That these models are unable to reproduce the observed $\Delta B$ contours argues strongly for the inclusion of currents distributed in the inner magnetosphere.

The earliest models which included the field from currents in the

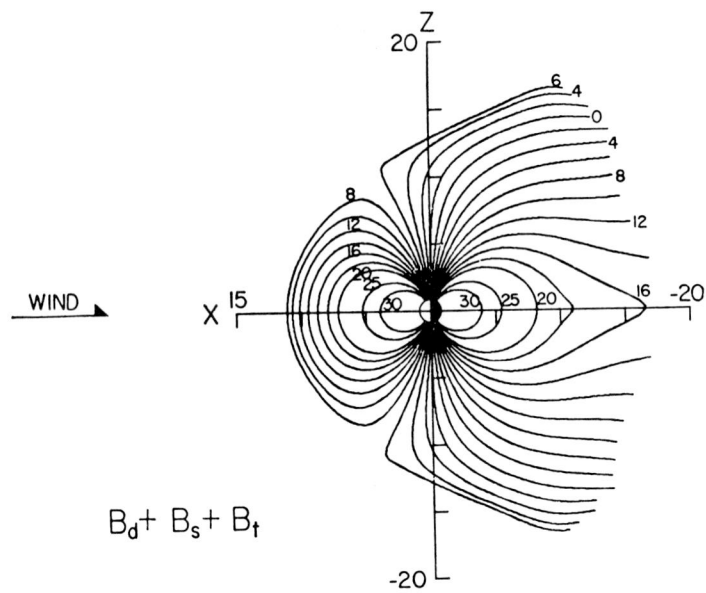

Fig. 2. Magnetic field lines in the noon-midnight meridian calculated by using the magnetospheric model of Choe and Beard (1974b).

inner magnetosphere were by Suguira and Poros [1973] and Olson and Pfitzer [1974]. Both of these models gave a much better fit to the $\Delta B$ contours than the models without currents distributed in the inner magnetosphere.

Unfortunately even models that reproduce the average observed field reasonably well may not give physically reasonable field gradients. In particular the Olson-Pfitzer model, which modeled the average field best, contained a region in the near tail where the field gradient reversed sign. Unphysical neutral lines or unphysical changes in the field gradient occurred at the inner edge of the tail current or at the intersection between current systems in most other models as well [Walker, 1976].

Recently, three new models have been presented. In this paper these new models will be reviewed in some detail with emphasis on the extent to which they have succeeded in improving on the earlier models. The models will be compared with the observed field in both magnitude and direction. Finally, the application to other planetary magnetospheres of the techniques used to model the earth's magnetospheric magnetic field will be briefly discussed.

### The Olson-Pfitzer (1977) Model

A computer plot of the magnetospheric current system used in the latest version of the model of Olson and Pfitzer [1977] is presented in Figure 5. The currents were assumed to flow along wire loops placed throughout the magnetosphere. Three sets of nested loops were used to represent currents distributed in the inner magnetosphere. The tail currents were required to flow almost linearly across the

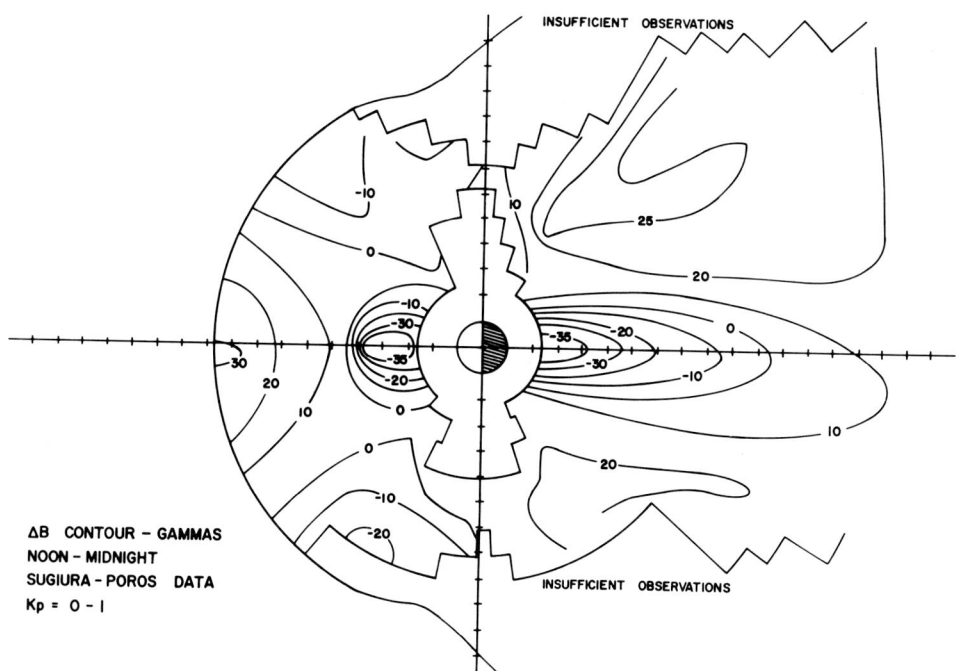

Fig. 3. Observed ΔB contours for Kp=0-1 in the noon-midnight meridian (from Sugiura and Poros, 1973).

upper and lower boundaries of the plasma sheet in order to approximate a current region of finite thickness. The currents were closed on the magnetopause surface. Near the earth the tail current loops were curved toward the ring current loops. The plane of symmetry in the far tail was raised and lowered as a function of dipole tilt. The orientation of the nested loops and near earth tail loops was rotated as a function of tilt so that the current system remained symmetric with respect to the magnetic equator. The position and current in the loops were varied to fit the observed ΔB contours.

Instead of using a self-consistent magnetopause shape determined by the pressure balance condition, as was done in the earlier Olson-Pfitzer model [Olson and Pfitzer, 1974], a more realistic shape based on observations was used. The magnetopause shape in Figure 6 was chosen to approximate the average observed surface of Fairfield [1971]. The tilt dependent magnetopause currents were calculated by using the method developed by Olson [1969].

The Olson-Pfitzer model is available in a form suitable for general use in magnetospheric physics. The field components calculated from the current loops in Figure 5 and the magnetopause currents were expressed as a polynomial expansion including terms to fourth order in space and second order in tilt. The field lines in Figure 6 are limited to R<15 Re because the fit rapidly diverges at larger radial distances.

ΔB contours calculated by using this version of the Olson-Pfitzer model are presented in Figure 7. The agreement between the observa-

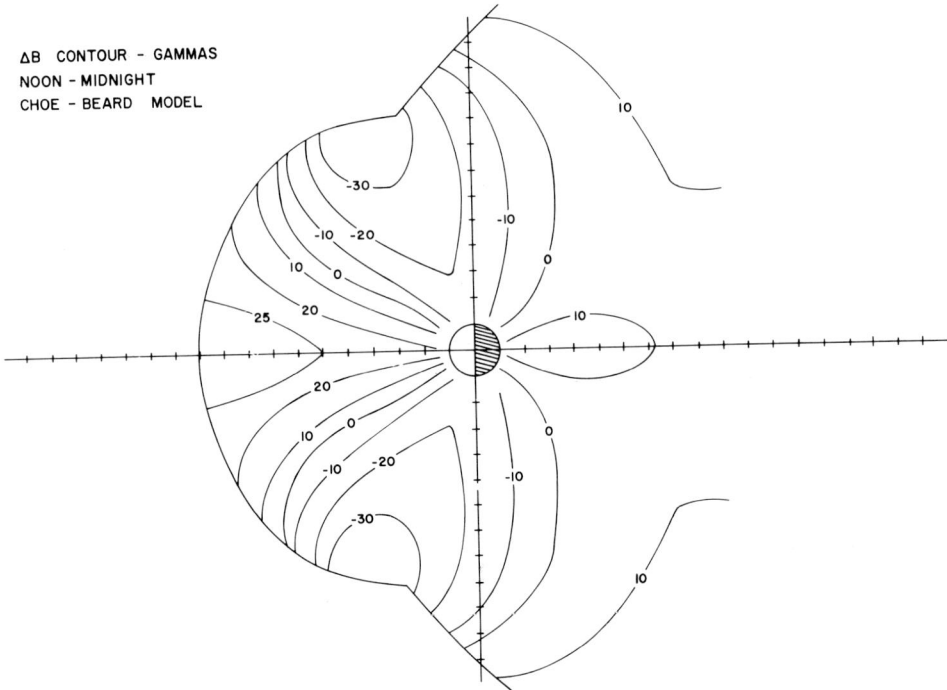

Fig. 4. The ΔB contours in the noon-midnight meridian calculated by using the magnetospheric model of Choe and Beard (1974a,b) (from Walker, 1976).

tions and the model is very good, much better than with the original Olson-Pfitzer model. The only region of disagreement on the day side is near the polar cusp where the model ΔB is still too small. However, the largest difference is near the limit of the observations so that this difference may not be too significant. The model reproduces the negative values near the equator well. Note that on the night side the contours extend deeper into the tail than is observed. For instance, the zero contour from the model stretches about 2 $R_e$ tailward of the observed contour. This difference can be understood if we recall that the observed ΔB contours are averages over all tilt angles while the calculation was made for zero tilt. If the model is averaged over all tilts this difference vanishes. The difference between the ΔB contours at zero tilt and the tilt averaged ΔB contours caused the unphysical field gradient reversal which occurred in this region in the original Olson-Pfitzer calculation. In that calculation, the model parameters for zero tilt were adjusted to fit the tilt averaged ΔB contours. This effect does not occur in the present version which gives physically reasonable gradients.

## The Tsyganenko Model

Recently a new approach has been proposed for calculating the field from the magnetopause currents [Alekseev and Shabansky, 1972; Voigt, 1972]. In these calculations a magnetopause of given geometry was

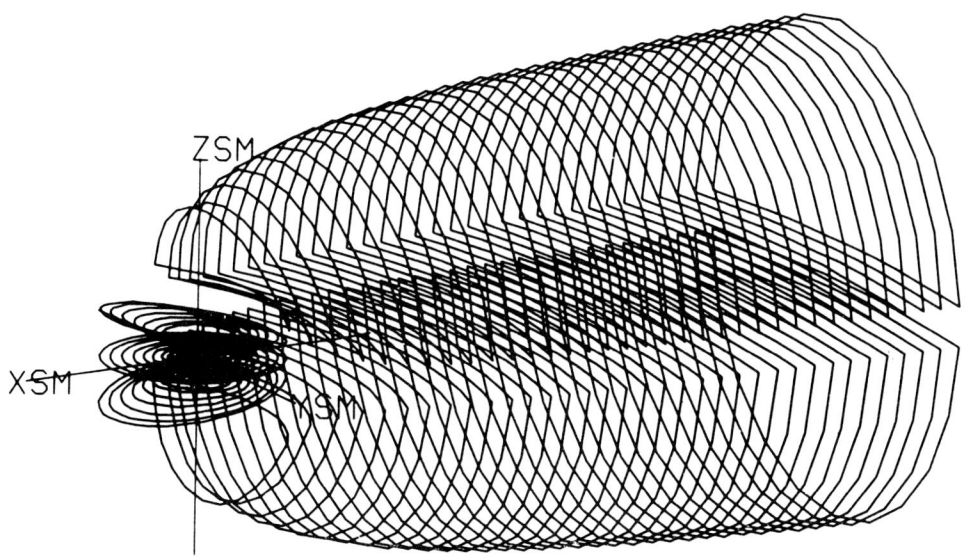

Fig. 5. The wire current loops used in the Olson-Pfitzer magnetospheric model (from Olson and Pfitzer, 1977).

assumed. The field from magnetospheric sources was assumed to be completely shielded outside the magnetopause and the field from magnetopause currents was assumed to be derivable from a scalar potential. The scalar potential was thus found by solving Laplace's equation with Neumann's boundary condition. This approach has the advantage of providing a relatively simple means of finding a solution for the earth's main field plus any additional current system. In the initial models, simple magnetopause geometries were chosen to simplify the calculation. Alekseev and Shabansky [1972] chose a paraboloid of revolution while

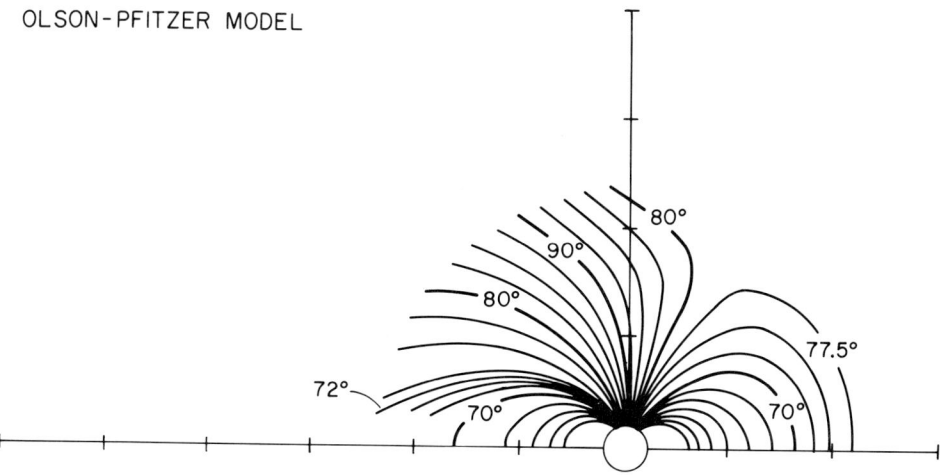

Fig. 6. Magnetic field lines in the noon-midnight meridian calculated by using the magnetospheric model of Olson and Pfitzer (1977).

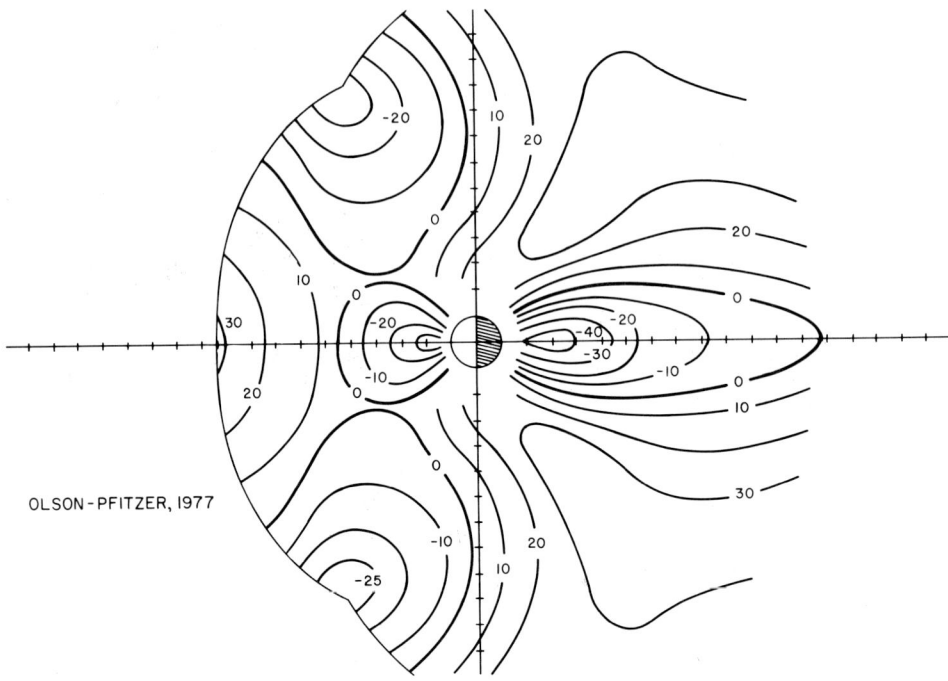

Fig. 7. The ΔB contours in the noon-midnight meridian calculated by using the magnetospheric model of Olson and Pfitzer (1977).

Voigt used a hemisphere on the day side and a cylinder on the night side (Figure 1). The only magnetospheric source in these calculations other than the dipole was the tail current.

The current system used in the new model of Tsyganenko [1976] is presented in Figure 8. The tail current was an infinitesimally thin sheet which closed on the magnetopause surface. The height of the current sheet above the equatorial plane was determined using a tilt dependent formula similar to that of Russell and Brody [1967]. The tail current changes continuously from the tail region to the ring current thereby eliminating the neutral line found in previous models at the intersection of the two current systems.

A much more complex magnetopause surface was used in this calculation than in the previous models (Figure 9). The boundary was assumed to have approximately the same shape as Fairfield's [1971] average magnetopause (Figure 1). The position of the tail magnetopause was displaced along the Z-axis as a function of tilt. In the noon-midnight meridian there is an indentation in the surface near the polar cusp which is similar to the neutral points obtained in the gas dynamic models. The field contributions from all of the currents in Figure 8 were used to calculate the magnetopause field.

On the night side the field lines calculated by using this model are much more tail-like than those from any other model (Figure 9). The model reproduces the observed field magnitude very well (Figure 10). The ΔB values from the model are slightly too large on the day

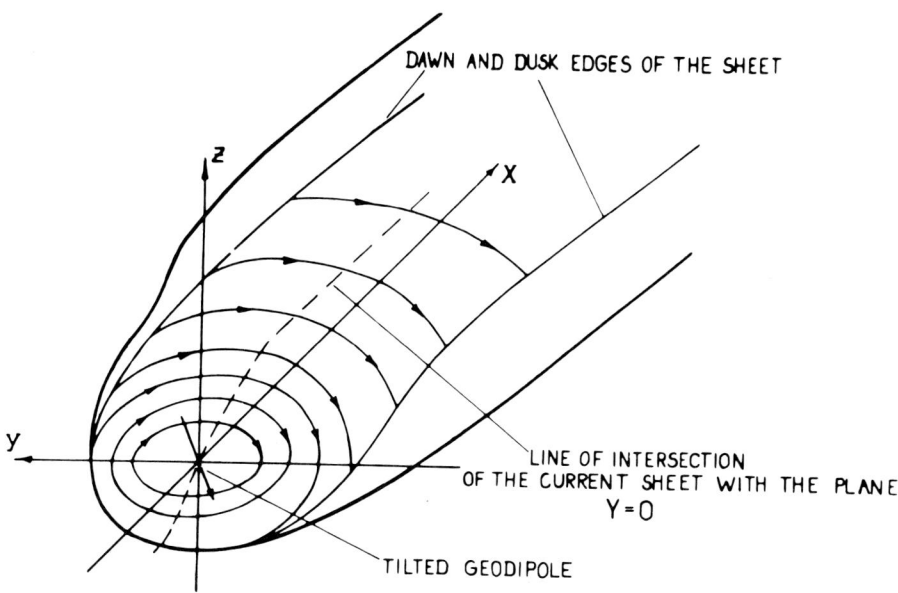

Fig. 8. The current sheet used in the magnetospheric model of Tsyganenko (from Tsyganenko, 1976).

side near the subsolar point. On the night side the $\Delta B$ contours like the corresponding field lines extend deeper into the tail than in the Olson-Pfitzer calculation.

## The Voigt Model

The versatility of the defined magnetopause approach to magnetospheric modeling has been demonstrated by the recent calculations of Voigt [1978; Voigt and Fuchs, 1979]. Unlike the Tsyganenko model and all previous models of this type, the normal component of the magnetospheric field was not required to be zero at the boundary in this calculation. Instead $B_{normal}$ was allowed to penetrate the boundary continuously. This enabled Voigt to model the macroscopic properties of the interaction of the magnetosphere with the interplanetary magnetic field. The magnetospheric configuration obtained for both a northward and southward IMF is given in Figure 11. $C_0$ is an input parameter that gives the fraction of the normal component of the IMF which penetrates the magnetopause. The same fraction of $B_{normal}$ from sources within the magnetosphere was assumed to penetrate the boundary to outside the magnetosphere. Note the neutral line which forms on the night side when the IMF is southward and the inter-connection coefficient is sufficiently large. The model has been used to successfully reproduce the observed dependance of the polar cusp position on the north-south component of the IMF [Voigt and Fuchs, 1979]. However, it should be emphasized that this is a static model and that the interaction of the magnetosphere with the IMF has not been calculated self-consistently since the magnetopause position and shape were fixed in the calculation.

TSYGANENKO MODEL

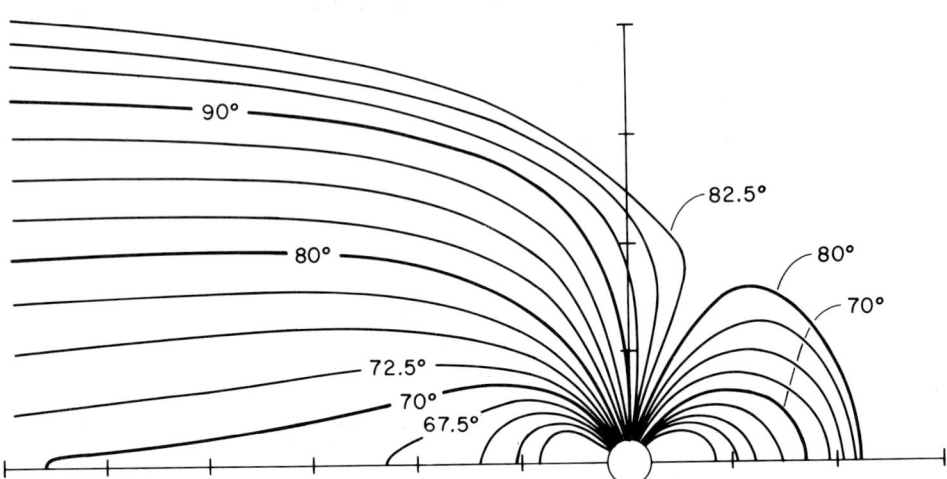

Fig. 9. Magnetic field lines in the noon-midnight meridian calculated by using the magnetospheric model of Tsyganenko (1976).

Comparison with Vector Observations

A vector standard field is necessary to adequately evaluate the model calculations. Unfortunately, an adequate standard requires a large set of observations distributed with sufficient uniformity over the entire magnetosphere. Furthermore, the data should represent all levels of magnetospheric activity and in order to include seasonal and diurnal variations in the shape of the magnetosphere, the data set should include all tilt angles. This presents a difficult organizational problem (for a more complete discussion see Suguira and Poros [1979]). Hedgecock and Thomas [1975, 1979] have developed a new coordinate system which helps solve this problem by assigning equivalent positions and directions in a "zero tilt magnetosphere." In their coordinate system, the distance in the north-south direction in geocentric solar magnetospheric (GSM) coordinates is calculated from a complex equatorial surface which was derived from spacecraft measurements. The properties of this equatorial surface are demonstrated in Figure 12. In the far tail regions (top panel), the surface is similar to neutral sheet models currently in use. The equator moves north and south along the $Z_{GSM}$ direction in response to changes in the dipole tilt. This tilt dependence is similar to that found in the earlier models of Murayama [1966]; Speiser and Ness [1967]; Russell and Brody [1967]; Fairfield and Ness [1970] and Bowling and Russell [1976]. The intercept of the equatorial surface in $Y_{GSM}$ - $Z_{GSM}$ plane is more complex than in the earlier models. The surface is an ellipse near the center of the tail which decreases exponentially to the Z=0 plane near the magnetopause. In the near tail, the equatorial surface joins smoothly to the magnetic equator (central panel). The "hinge" point where the surface begins to bend toward the $Z_{GSM}$ = 0 plane is Y de-

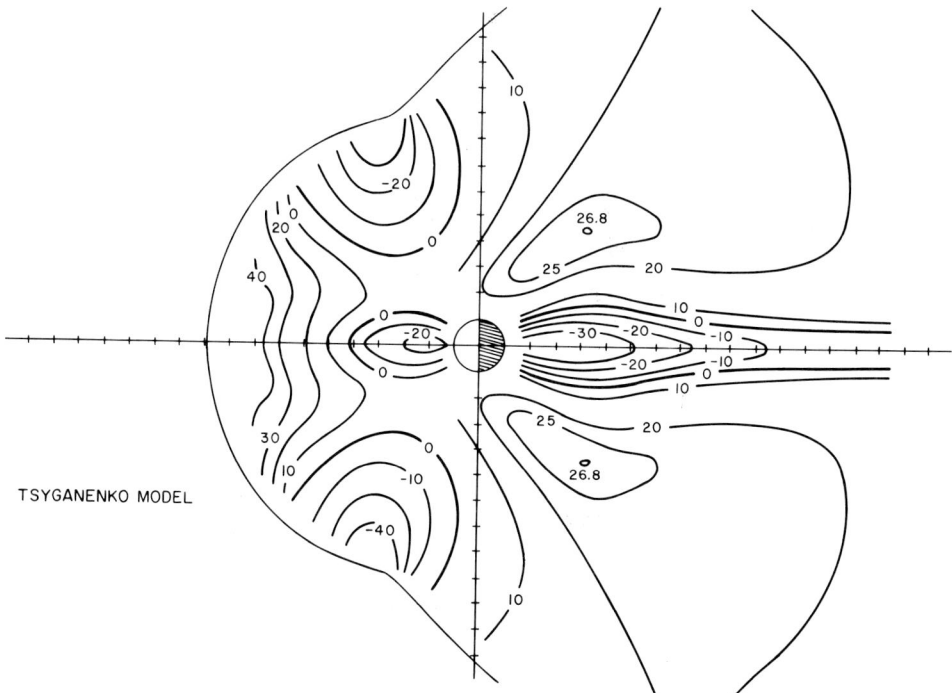

Fig. 10. The ΔB contours in the noon-midnight meridian calculated by using the magnetospheric model of Tsyganenko (1976).

pendent (bottom panel). The exact shape and position of the equatorial surface was determined by fitting IMP-1, -3, -4, -5, -6, OGO-5 and HEOS-1 neutral sheet crossings.

In addition to using the equatorial surface described above, Hedgecock and Thomas used an "effective tilt" in place of the dipole tilt. They noted that the distant polar magnetosphere was poorly organized in both GSM and SM (solar magnetic) coordinates. They discovered, however, that the data was better organized if an "effective tilt" which is a function of both X and Z was used. Therefore, the resulting transformation to geocentric magnetospheric equatorial coordinates (GME) is a two step process. To transform a position in GSM coordinates into GME coordinates, one first calculates the distance (ZN) between the given $Z_{GSM}$ value and the $Z_{GSM}$ position of the surface in Figure 12 by using the actual dipole tilt angle. ZN and $X_{GSM}$ are then used to calculate an "effective" tilt angle. $Z_{GME}$ is the difference between the original $Z_{GSM}$ value and the $Z_{GSM}$ position of the equatorial surface calculated by using this "effective" tilt angle in place of the actual dipole tilt angle. The $X_{GME}$ and $Y_{GME}$ values are the same as $X_{GSM}$ and $Y_{GSM}$.

Observations organized in GME coordinates are presented in Figure 13. The arrows are projections into the noon-midnight meridian of unit vectors derived from 1 hour averages of observations for $|Y|<4Re$. Data from all tilt angles and from all levels of activity were includ-

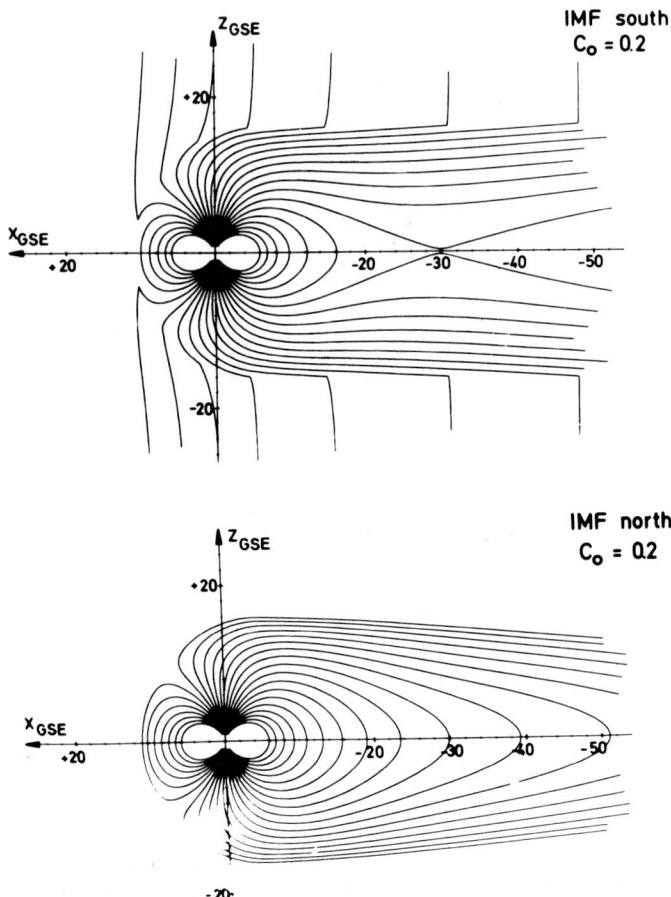

Fig. 11. Magnetic field lines in the noon-midnight meridian calculated by using the magnetospheric model of Voigt (from Voigt and Fuchs, 1979). $C_o$ is the fraction of the normal component of the IMF which penetrates the magnetopause. The IMF was assumed to be 5γ directed southward (top) and northward (bottom).

ed. The observations are extremely well organized by this coordinate system. The data are better organized in this system than in GSM coordinates [Hedgecock and Thomas, 1979]. The field direction is least well organized in the polar cusp region. This probably results from motion of the cusp due to variable solar wind and IMF conditions.

The field lines in Figure 13 are from an extended Mead-Fairfield type empirical model [Mead and Fairfield, 1975] generated by using observations for $K_p < 2_o$ and DST > - 10γ [Hedgecock et al., 1979]. There is surprisingly good agreement between the model and observations even though the observations cover all levels of activity. One area of difference is in the tail, near the equator where the arrows indicate a more tail-like configuration than does the model. The overall good fit to the observations indicates that the HEOS-Q model can serve

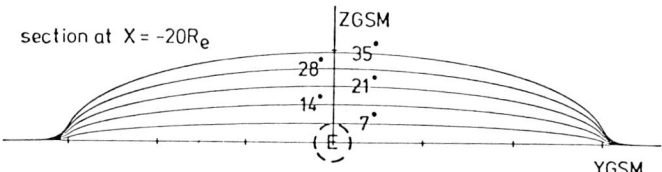
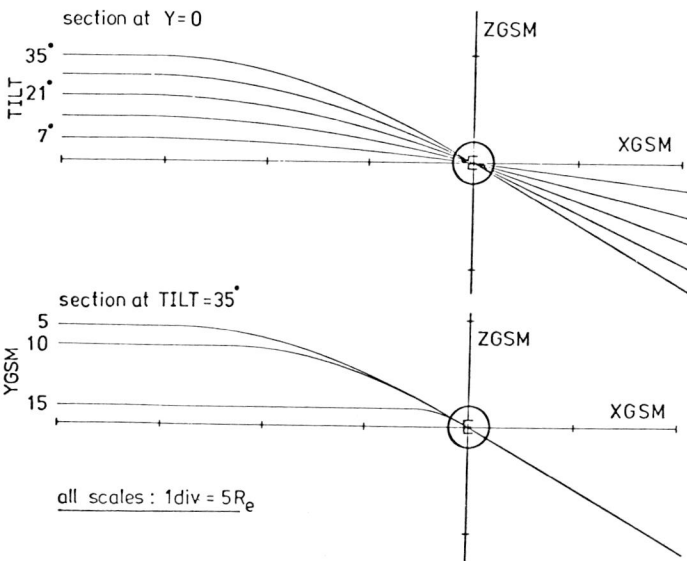

Fig. 12. Cross sections of the equatorial surface used in the GME coordinate system (from Hedgecock and Thomas, 1979). The top panel contains the cross section in the YGSM-ZGSM plane at $X_{GSM}$ = -20 $R_e$ for selected dipole tilt values. The center panel contains the cross section in the noon-midnight meridian for selected dipole tilt values. The bottom panel contains the cross section in the $X_{GSM}$-$Z_{GSM}$ plane for selected YGSM values and dipole tilt of 35°.

as a basis with which to compare the vector field of the theoretical models.

Field lines calculated by using the Voigt model (solid lines) are superposed on the field lines from the HEOS-Q model (dashed lines) in Figure 14. The theoretical field lines were calculated by using the closed magnetosphere version of the model. The field lines with an earth intercept of <70° are quite similar in the two models. At higher latitudes the differences become larger possibly because a slightly smaller sub-solar distance was used in the Voigt calculation. The last closed field line calculated from the model differs by about 1° from the observations. On the night side, the field lines are not even qualitatively similar. At high latitudes the large differences may have resulted from Voigt's use of a cylindrical magnetopause

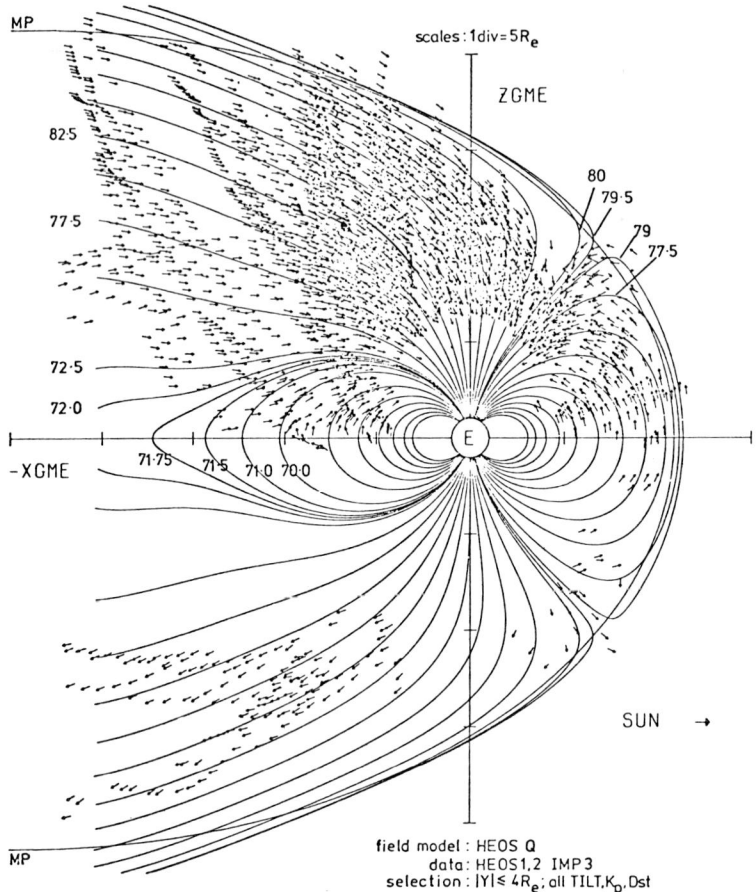

Fig. 13. Magnetic field lines in the noon-midnight meridian calculated using the HEOS-Q magnetic field model (from Hedgecock and Thomas, 1979). The arrows are projections into the noon-midnight meridian of unit vectors derived from 1 hour averages of HEOS 1, 2, and IMP 3 magnetic field observations. The data were taken for $|Y| < 4\ R_e$.

(Figure 1). Near the equator the HEOS-Q field lines are stretched much farther into the tail than the theoretical field lines.

A similar display containing field lines calculated by using the Olson-Pfitzer model is given in Figure 15. Again the theoretical field lines are very similar to the field lines from the empirical model for field lines originating at latitudes <70°. The sub-solar point is even closer to the earth in this calculation than in Voigt's with correspondingly larger differences in the high latitude configuration. The high latitude tail lobe field lines show some of the flaring seen in the HEOS-Q model. Again the differences in the high latitude field lines may be due to the differences in the magnetopause shape and position. Nearer the equator, the Olson-Pfitzer field lines are not as tail-like as those determined from the observations but they are more

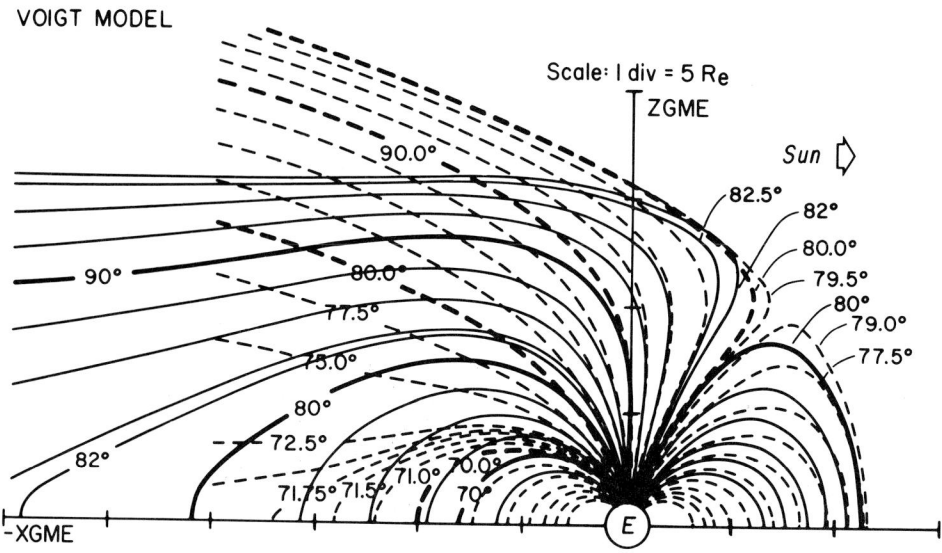

Fig. 14. A comparison of magnetic field lines in the noon-midnight meridian calculated by using the Voigt magnetospheric model (Voigt and Fuchs, 1979) (solid lines) and those from the HEOS-Q model (Hedgecock and Thomas, 1979) (dashed lines).

tail-like than those from the Voigt model. Recently, West et al., [1978] have argued, using electron pitch angle distributions observed on OGO-5, that the tail field even when it is in its most dipole-like state is much more tail-like than the original Olson-Pfitzer model. The comparison with the HEOS-Q model indicates that this is still true with the current version of the model.

Sugiura and Poros [1979] have assembled a data set which is also useful for evaluating the direction of the model field. Their observations are presented as contours of $\Delta I$ and $\Delta \alpha$ where $\Delta I$ is the difference in inclination between the observed field and the earth's main field and $\Delta \alpha$ is the difference in total angle. They compared these observations with the original Olson-Pfitzer model and found significant differences especially on field lines originating at high latitudes.

Like the other two models the Tsyganenko model agrees very well with the observed field direction for initial latitudes <70° on the day side (Figure 16). Nearer the magnetopause larger differences are evident. The high latitude tail field lines are flared but not as flared as those from the empirical model. Nearer the equator the field lines from the model are even more tail-like than the observations suggest. This may be due, in part, to the infinitesimally thin current sheet used in the model. However, it should be emphasized that while the observations are limited very close to the equator tailward of about $X_{GME} = -10$ Re, these limited observations suggest that the field is more tail-like than the HEOS-Q model indicates. Recently a revised version of the HEOS-Q model has been developed in which the field is

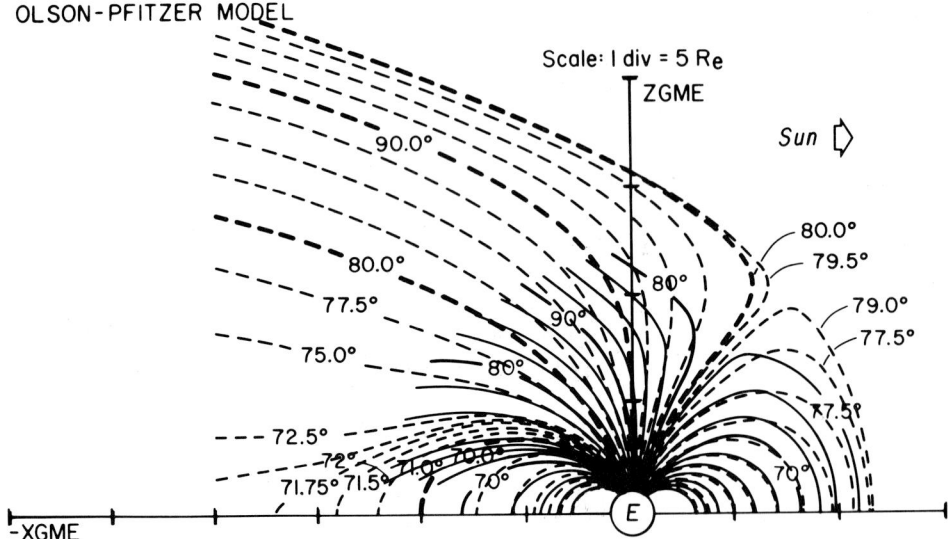

Fig. 15. A comparison of magnetic field lines in the noon-midnight meridian calculated by using the Olson-Pfitzer magnetospheric model (Olson and Pfitzer, 1977) (solid lines) and those from the HEOS-Q model (Hedgecock and Thomas, 1979).

more tail-like in the near tail region near the equator (B.T. Thomas, personal communication).

## Modeling Planetary Magnetospheres

In recent years considerable effort by magnetospheric model makers has been directed toward modeling the planetary magnetospheres of Mercury and Jupiter. In the next two sections these calculations will be briefly outlined with emphasis on how techniques used in modeling the earth's magnetosphere can be applied to modeling planetary magnetospheres.

### Mercury

Following the surprising discovery on Mariner 10 that Mercury has a well defined magnetosphere [Ness et al., 1974], two models of Mercury's magnetosphere have been presented. The magnetic field observations in Mercury's magnetosphere closely resemble those in the earth's magnetosphere. Jackson and Beard [1977] argued that the Hermian observations were sufficiently similar to those at earth that the field could be modeled by scaling a model of the earth's field. They assumed that, at a given point in the magnetosphere, the contributions from the magnetopause and magnetotail currents had the same fixed ratios to one another as those at the earth when distances are scaled in terms of the distance to the subsolar point. Thus their model is

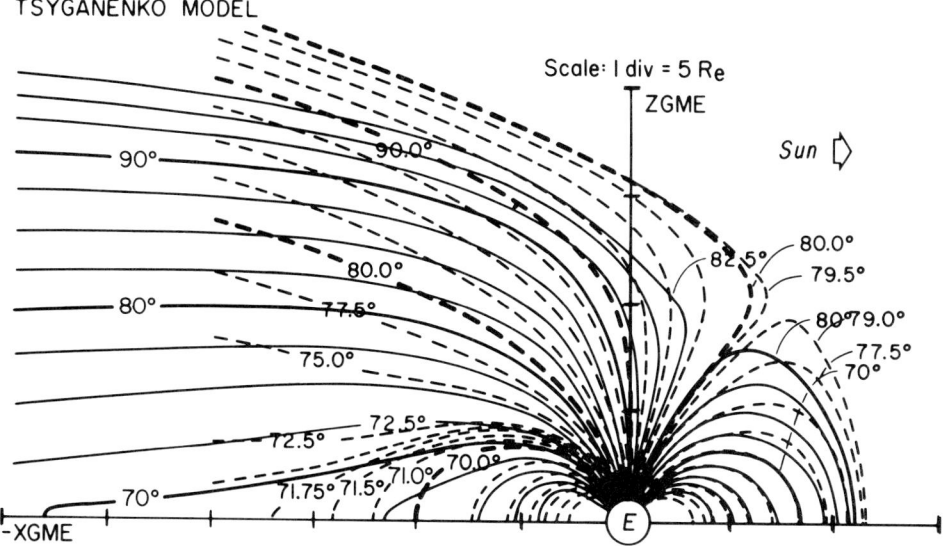

Fig. 16. A comparison of magnetic field lines in the noon-midnight meridian calculated by using the Tsyganenko magnetospheric model (Tsyganenko, 1976) (solid lines) and those from the HEOS-Q model (Hedgecock and Thomas, 1979).

a scaled down version of the Choe-Beard model of the earth's magnetosphere (Figure 2). Field lines calculated by using this model are presented in Figure 17. The latest version of this model uses a dipole offset along the dipole axis as the main field [Ng and Beard, 1978]. The distance to the subsolar point is only 1.4 Rm in this calculation. The most outstanding feature of Figure 17 is the large fraction of the magnetosphere occupied by the planet.

The field observed during the third Mercury flyby is given by the dots in Figure 18. This pass was chosen since this was the quieter of the two magnetospheric encounters [Slavin and Holzer, 1978]. The model field based on data from both magnetospheric encounters is shown as a solid line. The agreement is good. A couple of additional points should be kept in mind concerning this calculation. Both encounters were on the night side of the planet. As a result the distance to the subsolar point was determined by scaling the terrestrial magnetopause surface calculated by Choe, Beard and Sullivan [1973]. As noted above the stand off distance was found to be 1.4 Rm. As can be seen in Figure 1, this model would give an erroneous value for the stand off distance if applied to the earth. Scaling their empirical model for the earth's magnetopause shape to Mercury, Slavin and Holzer [1978] have calculated a stand off distance of 1.9 Rm. A change in the stand off distance of this size can make a large difference in the configuration in Figure 17. In addition, Hood and Schubert [1978] have argued that currents induced in the planet may also make important contributions to the magnetic configuration especially during times when the solar wind compresses the magnetosphere.

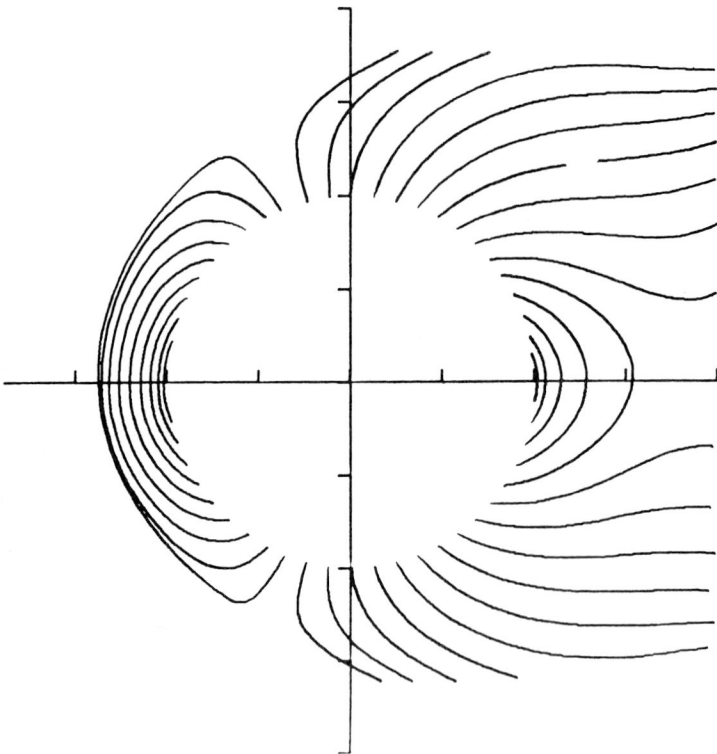

Fig. 17. Magnetic field lines in the noon-midnight meridian of Mercury's magnetosphere calculated by using the Jackson-Beard magnetospheric model (from Jackson and Beard, 1977).

The other Hermian magnetospheric model [Whang, 1977] is basically a version of the Hones-Taylor type image dipole model. This model uses a semi-infinite current sheet of finite thickness to model the tail current. The resulting configuration is qualitatively similar to that of Jackson and Beard and also fits the observations well.

### Jupiter

There are no models of the entire Jovian magnetosphere since Jupiter's rapidly rotating magnetosphere differs greatly from the earth's and since the observations are limited. For discussion purposes, the magnetosphere of Jupiter can be divided into three regions, the magnetopause region, the current sheet region, and the inner region.

The Jovian magnetopause appears to be similar in shape and size to that which would be expected from a simple scaling of the earth's magnetopause [Smith et al., 1976]. Beard and Jackson [1976] have used this information to create a zeroth order model of the Jovian magnetopause. They assumed that the magnetospheric field could be approximated by the main field of Jupiter plus a current sheet located in the equatorial plane on the day side. They calculated the shape which would result if this field balanced the solar wind pressure.

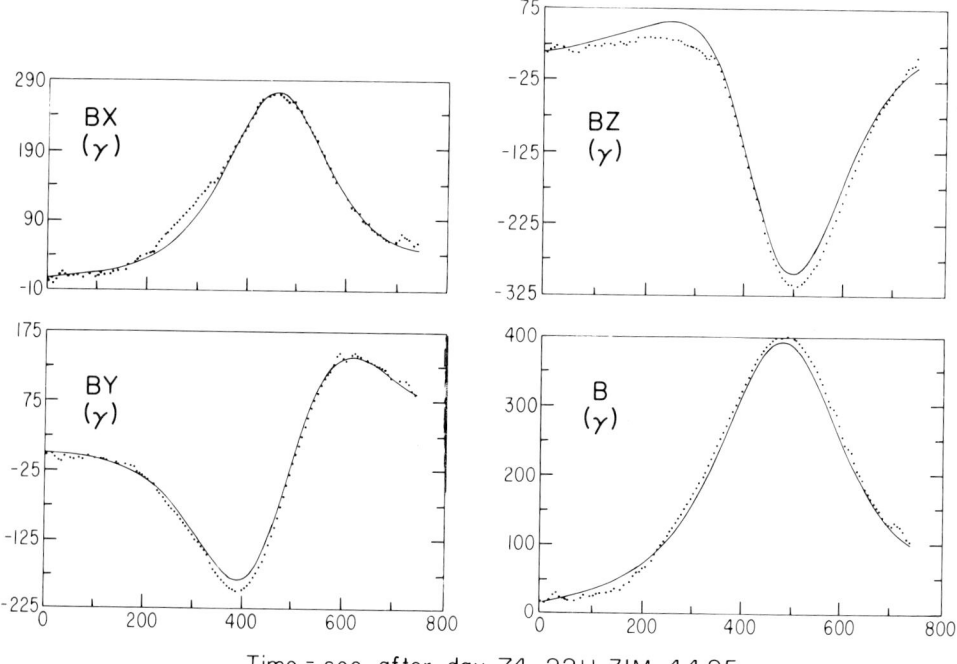

Fig. 18. Magnetic field observations during the third Mercury encounter of Mariner 10 (dotted lines) (after Jackson and Beard, 1977). The solid lines give the field calculated by using the Jackson-Beard model.

Their surface is only slightly more disk like in shape than the pressure balance models give for the earth's field. Although the Jovian magnetopause is similar in shape to the earth's, it is subject to much larger and more frequent fluctuations as evidenced by the multiple magnetopause crossings observed by Pioneers 10 and 11 on both inbound and outbound trajectories.

The second region, the equatorial current sheet region, extends from near the planet at about 15 $R_J$ to 20 $R_J$ to near the magnetopause at 80 $R_J$ to 100 $R_J$. This region is very similar to the earth's plasma sheet but was observed on the dayside by Pioneer 10 and 11 inbound as well as on the pre-dawn track of Pioneer 10 outbound.

The final region, the inner magnetosphere, is dominated by Jupiter's main field. The measurements indicate that within about 6 $R_J$ models of the Jovian internal field reasonably fit the data if terms up to octupole are included [Smith et al., 1976, Acuna and Ness, 1976].

In Figure 19 observations of $|B|$ from the outbound (pre-dawn) portion of the Pioneer 10 trajectory are plotted. This is an excellent interval for studying the Jovian current sheet since the data in this interval is very well organized. That the decrease is less rapid than expected for a dipole indicates the presence of the current sheet. The magnitude exhibits rapid decreases with a 10 hour periodicity, the rotation period of Jupiter. These decreases are believed to occur as the high $\beta$ plasma associated with the current sheet is swept over the spacecraft [Goertz, 1976; Kennel and Coroniti, 1977; Walker et al., 1978].

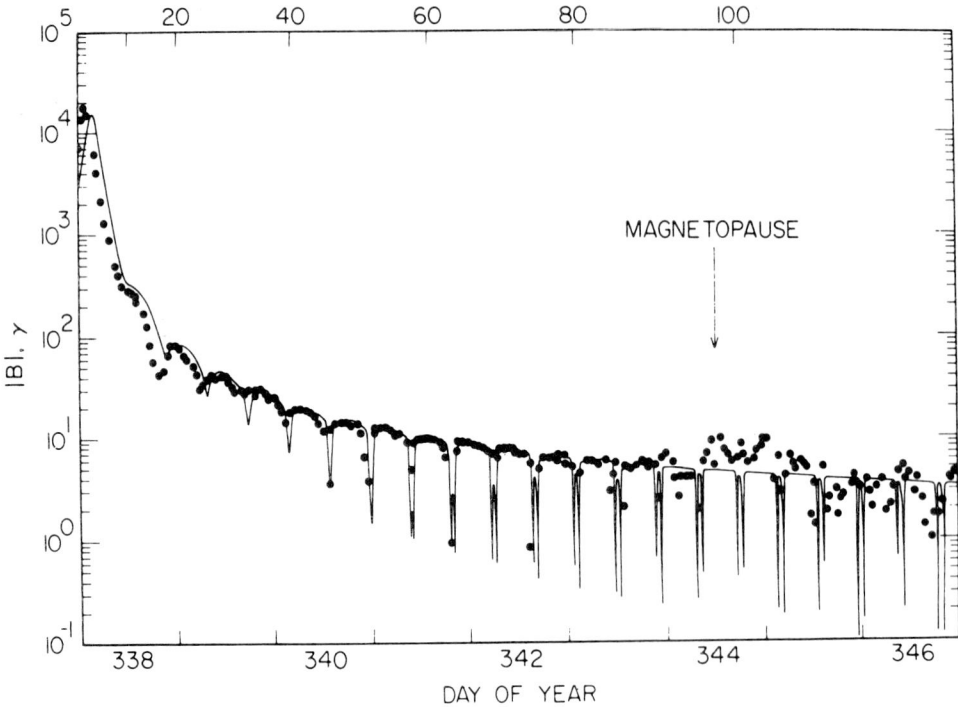

Fig. 19. The magnitude of $\vec{B}$ observed on the outbound portion of the Pioneer 10 trajectory (dots). The solid line is the field calculated by using the magnetospheric model of Goertz et. al. (from Goertz et. al., 1976).

There have been several calculations of the magnetic field in a rapidly rotating magnetosphere which have application to Jupiter [Gleeson and Axford, 1976; Goertz et al., 1976; Sozzou, 1978; Carbary and Hill, 1978]. However only the models of Goertz et al., and Carbary and Hill, have been used directly to calculate the field geometry. Carbary and Hill assumed an ionospheric plasma source and found that the resulting field lines were much more dipolar than was observed. They concluded that an equatorial plasma source was required to model the field. Field lines calculated by using the Goertz et al. model are plotted in Figure 20. Goertz et al. assumed that the decreases in Figure 19 represented motion from a region with open field lines to a region with closed field lines. They, also, assumed a thin current sheet and took into account the observed sweep back of the field lines. The resulting field lines are stretched out toward dawn and resemble tail field lines on earth. The current sheet is near the magnetic equator in this model and therefore oscillates with respect to the spacecraft. The heavy line gives the path of Pioneer 10 with respect to the oscillating current sheet. The solid line in Figure 19 is the calculated field magnitude. The model gives reasonable agreement with the observations. This model is only valid in the pre-dawn region near the plasma sheet.

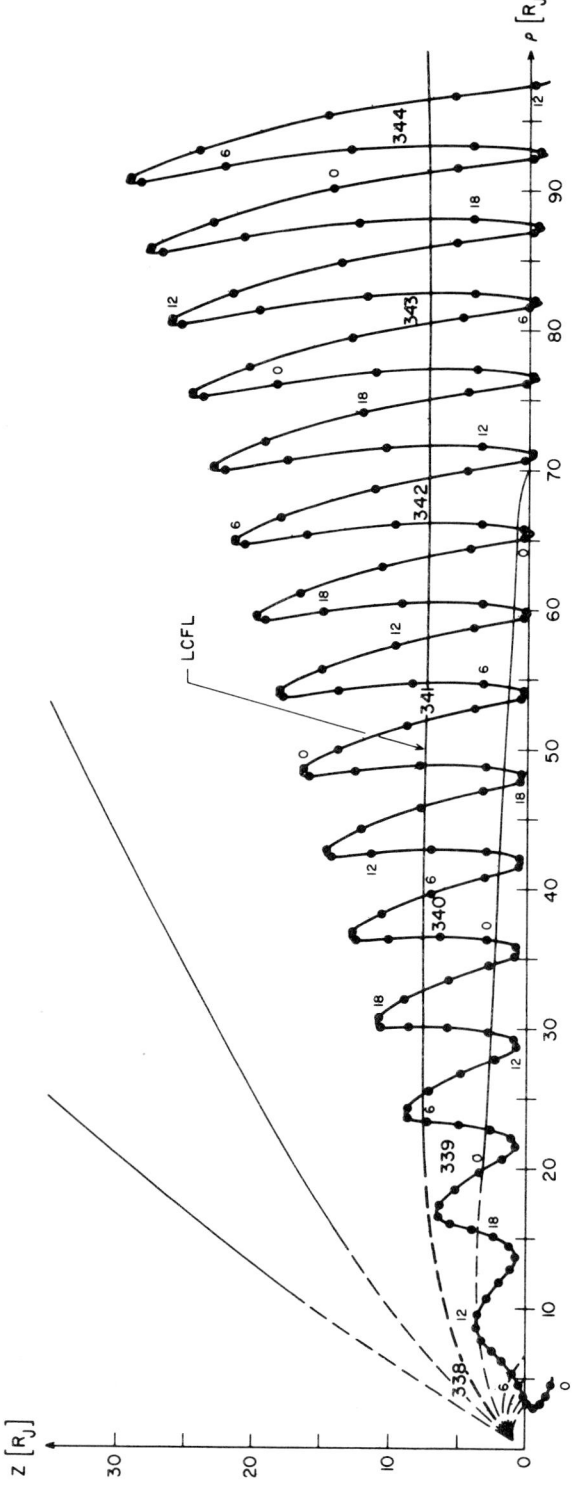

Fig. 20. Magnetic field lines in the pre-dawn Jovian magnetosphere calculated by using the model of Goertz et. al. The heavy line gives the orbit of Pioneer 10 with respect to the oscillating magnetospheric configuration (from Goertz et. al., 1976).

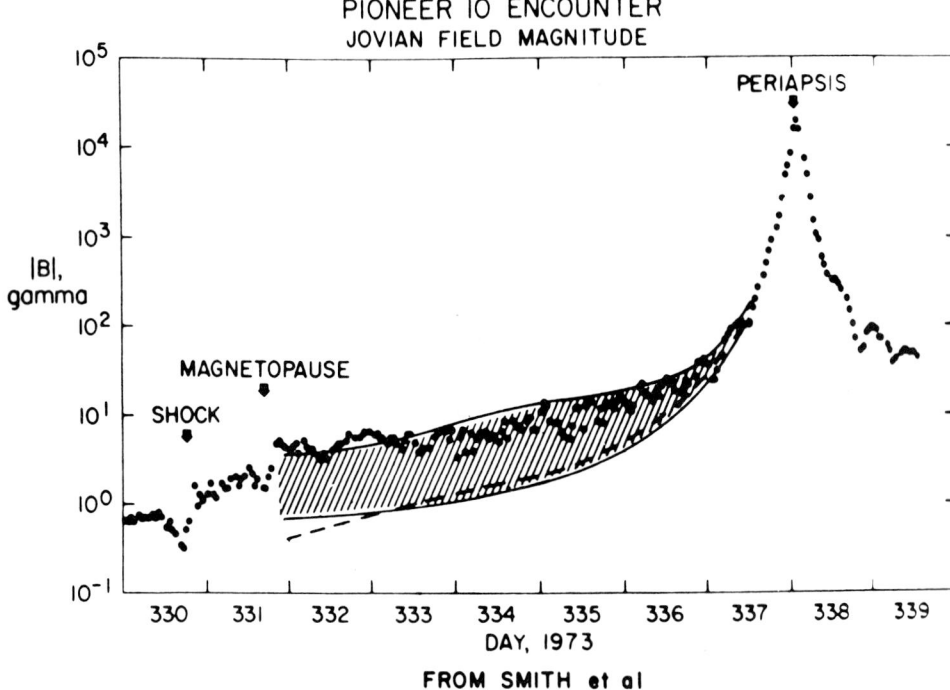

Fig. 21. The magnitude of $\vec{B}$ observed on the inbound portion of the Pioneer 10 trajectory (dots). The dashed line gives the field expected from a dipole. The shaded region delimits the range of values calculated by using the magnetospheric model of Barish and Smith (Barish and Smith, 1975).

The day side observations are not as well organized as the pre-dawn data. The data in Figure 21 are from Pioneer 10 inbound between dawn and noon. Again the field in the outer magnetosphere is enhanced over the dipole field. Barish and Smith [1975] have modeled the field in this region. They assumed an equatorially confined current sheet and a blunt earth-like magnetopause configuration. Barish and Smith also assumed azimuthal symmetry, however the model is valid only on the dayside. In this model, as in the Goertz et al. calculation, the field magnitude along the Pioneer 10 trajectory oscillates as the planet rotates. The shaded region in Figure 21 delimits the field magnitudes predicted by the model. Although the structure predicted by the model is not observed, the observations do fall within the predicted range of magnitudes.

Field lines from this model (Figure 22) are less tail-like than those in the Goertz et al. model but are much more tail-like than are dayside field lines in the earth's magnetosphere. The field at higher latitudes also differs greatly from the terrestrial field. No attempt was made to model the field at high latitudes (>50°) because of insufficient observations. Therefore, the results at high latitudes may be spurious.

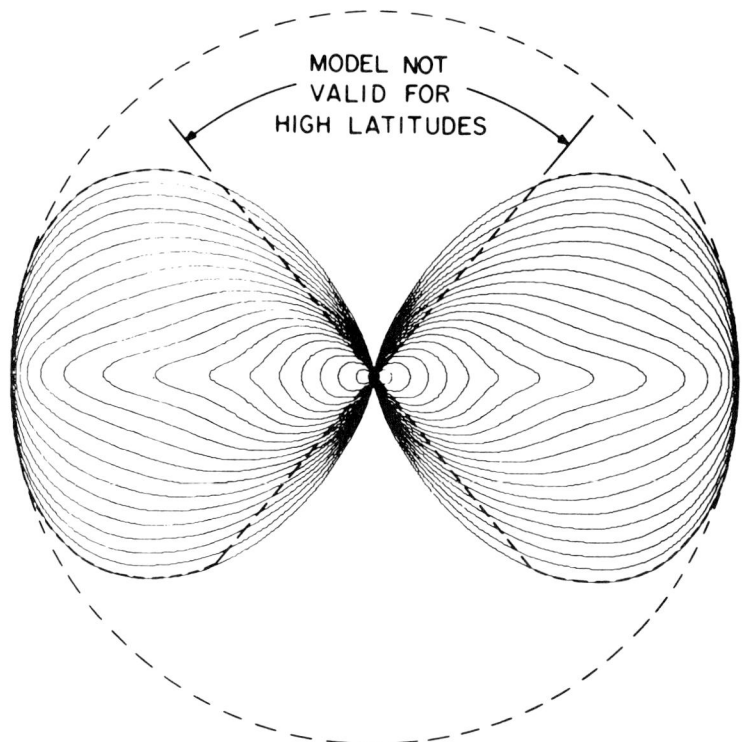

Fig. 22. Magnetic field lines in the day side Jovian magnetosphere calculated by using the Barish-Smith model (from Barish and Smith, 1975).

Summary and Conclusions

Considerable progress has been made in the development of quantitative models of the earth's magnetospheric magnetic field in recent years. Several models now include the contributions from the current systems thought most important in determining the magnetospheric configuration. These models include currents distributed in the inner magnetosphere in addition to boundary currents and magnetotail currents. Some of the latest models use more realistic magnetopause shapes based on observations.

The newest models of the earth's magnetosphere reproduce the field magnitude very well and model the field direction on the day side reasonably well, also. In the near tail region, there are considerable differences in field direction between the theoretical models and between the theoretical models and the latest empirical model. The models with the most realistic magnetopause shapes reproduce the observed field direction best in the tail lobes. Near the equator the observations suggest field lines which stretch far into the tail. The field lines from the Voigt and Olson-Pfitzer models are insufficiently tail-like while those from the Tsyganenko model may be too tail-like.

Clearly much additional work needs to be done in the tail region, both on the theoretical models and on establishing a vector standard for use near the equator.

Of the newer models, only the Olson-Pfitzer model is currently available in a form suitable for general use in magnetospheric physics.

Considering the limited observations provided by flyby missions, considerable progress has been made in modeling Mercury's magnetosphere and the region around the equatorial current sheet on Jupiter. The techniques used to model the earth's magnetosphere have proved useful in these calculations particularly in modeling Mercury's magnetic field. Considerable addition data is necessary before a global magnetospheric magnetic field model for Jupiter can be attempted.

Acknowledgments. This work was supported by NASA under contract NSG 7295 and grant NGL 05-007-004. UCLA Institute Geophysics and Planetary Physics Publication No. 1871.

## References

Acuna, M.H., and N.F. Ness, Results from the GSFC fluxgate magnetometer on Pioneer 11, in Jupiter, edited by T. Gehrels, Univ. of Arizona Tucson, 730, 1976.

Alekseev, I. I., and V.P. Shabansky, A model of a magnetic field in the geomagnetosphere, Planet. Space Sci., 20, 117, 1972.

Barish, F.D., and R.A. Smith, An analytical model of the Jovian magnetosphere, Geophys. Res. Lett., 2, 269, 1975.

Beard, D.B., The interaction of the terrestrial magnetic field with the solar corpuscular radiation, J. Geophys. Res., 65, 3559, 1960.

Beard, D.B., The magnetotail magnetic field, the Univ. of Kansas preprint, submitted to J. Geophys. Res., 1978.

Beard, D.B., and D.L. Jackson, The Jovian magnetic field and the magnetosphere shape, J. Geophys. Res., 81, 3399, 1976.

Bowling, S.B., and C.T. Russell, The position and shape of the neutral sheet at 30-Re distance, J. Geophys. Res., 81, 270, 1976.

Carbary, J.F., and T.W. Hill, A self-consistent model of a corotating Jovian magnetosphere, J. Geophys. Res., 83, 2603, 1978.

Choe, J.Y., and D.B. Beard, The compressed geomagnetic field as a function of dipole tilt, Planet. Space Sci., 22, 595, 1974a.

Choe, J.Y., and D.B. Beard, The near earth magnetic field of the magnetotail current, Planet. Space Sci., 22, 609, 1974b.

Choe, J.Y., D.B. Beard, and E.E. Sullivan, Precise calculation of the magnetosphere surface for a tilted dipole, Planet. Space Sci., 21, 485, 1973.

Fairfield, D.H., Average and unusual locations of the earth's magnetopause and bow shock, J. Geophys. Res., 76, 6700, 1971.

Fairfield, D.H. and N.F. Ness, Configuration of the geomagnetic tail during substorms, J. Geophys. Res., 75, 7032, 1970.

Gleeson, L.J. and W.I. Axford, An analytic mode illustrating the effects of rotation on a magnetosphere containing low-energy plasma, J. Geophys. Res., 81, 3403, 1976.

Goertz, C.K., The current sheet in Jupiter's magnetosphere, J. Geophys. Res., 81, 3393, 1976.

Halderson, D.W., D.B. Beard and J.Y. Choe, Corrections to the compressed geomagnetic field as a function of dipole tilt, Planet. Space Sci., 23, 887, 1975.

Hedgecock, P.C. and B.T. Thomas, HEOS observations of the configuration of the magnetosphere, Geophys. J.R. Astro. Soc., 41, 391, 1975.

Hedgecock, P.C. and B.T. Thomas, a New tilt-independent magnetospheric coordinate representation, Imperial College Preprint, in press, Planet. Space Sci., 1979.

Hedgecock, P.C., B.T. Thomas, A.M. Cornwall, and C.W. Davis, A dipole-tilt dependent model of the earth's magnetosphere, Imperial College Preprint, submitted to Planet. Space Sci., 1979.

Holzer, R.E. and J.A. Slavin, Magnetic flux transfer associated with expansions and contractions of the day side magnetosphere, J. Geophys. Res., 83, 3831, 1978.

Hones, E.W. Jr., Motion of charged particles trapped in the earth's magnetosphere, J. Geophys. Res., 68, 1209, 1963.

Hood, L.C. and G. Schubert, Inhibition of solar wind impingement on Mercury by planetary induction currents, submitted to J. Geophys. Res., 1978.

Jackson, D.J. and D.B. Beard, The magnetic field of Mercury, J. Geophys. Res., 82, 2828, 1977.

Kennel, C.F. and F.V. Coroniti, Jupiter's magnetosphere, in Annual Reviews of Astronomy and Astrophysics, 15, edited by Geoffrey R. Burbridge, Palo Alto, CA, 1977.

Kosik, J. CL., An analytical approach to the Choe-Beard magnetosphere, Planet. Space Sci., 25, 457, 1977.

Mead, G.D., Deformation of the geomagnetic field by the solar wind, J. Geophys. Res., 69, 1169, 1964.

Mead, G.D. and D.B. Beard, Shape of the geomagnetic field-solar wind boundary, J. Geophys. Res., 69, 1169, 1964.

Mead, G.D. and D.H. Fairfield, A quantitative magnetospheric model derived from spacecraft magnetometer data, J. Geophys. Res., 80, 523, 1975.

Murayama, T.J., Spatial distribution of energetic electrons in the geomagnetic tail, J. Geophys. Res., 71, 5547, 1966.

Ness, N.G., The earth's magnetotail, J. Geophys. Res., 70, 2989, 1965.

Ness, N.G., C.S. Searce and J.B. Seek, Initial results of Imp 1 magnetic field experiment, J. Geophys. Res., 69, 3531, 1964.

Ness, N.G., K.W. Behannon, R.P. Lepping, Y.C. Whang and K.H. Schatten, Magnetic field observations near Mercury: Preliminary results from Mariner 10, Science, 185, 150, 1974.

Ng, K.H. and D.B. Beard, Mercury's magnetic field, Univ. of Kansas Preprint, submitted to J. Geophys. Res., 1978.

Olson, W.P., The shape of the tilted magnetopause, J. Geophys. Res., 74, 5642, 1969.

Olson, W.P. and K.A. Pfitzer, A quantitative model of the magnetospheric magnetic field, J. Geophys. Res., 79, 3739, 1974.

Olson, W.P. and K.A. Pfitzer, Magnetospheric magnetic field modeling, McDonnell Douglas Astronautics Co., preprint, 1977.

Roederer, J.G., Quantitative models of the magnetosphere, Rev. Geophys. Space Phys., 7, 77, 1969.

Russell, C.T. and K.I. Brody, Some remarks on the position and shape of the neutral sheet, J. Geophys. Res., 72, 6104, 1967.

Slavin, J.A. and R.E. Holzer, The effect of erosionon the solar wind stand-off distance at Mercury, UCLA Institute of Geophysics and Planetary Physics Preprint. No. 1803, in press, J. Geophys. Res., 1978.

Smith, E.J., L. Davis, Jr., and D.E. Jones, Jupiter's magnetic field and magnetosphere, in Jupiter, edited by T. Gehrels, Univ. of Arizona Press, Tucson, AZ, 783, 1976.

Sozzou, C., A similarity model illustrating the effect of rotation on an inflated magnetosphere, Planet. Space Sci., 26, 311, 1978.

Speiser, T.W. and N.G. Ness, The neutral sheet in the geomagnetic tail: Its motion, equivalent currents, and field line connection through it, J. Geophys. Res., 72, 131, 1967.

Sugiura, M. and D.J. Poros, A magnetospheric field model incorporating the Ogo 3 and 5 magnetic field observations, Planet. Space Sci., 21, 1763, 1973.

Sugiura, M. and D.J. Poros, Evaluation of the quantitative magnetospheric field models by means of $\Delta I$ Contours, in Quantitative Modeling of Magnetospheric Processes, Geophys. Monogr. Ser., Vol. 21, edited by W.P. Olson, AGU, Washington, D.C., 1979.

Taylor, H.E. and E.W. Hones, Adiabatic motion of auroral particles on a model of the electric and magnetic fields surrounding the earth, J. Geophys. Res., 70, 3605, 1965.

Tsyganenko, H.A., A model of the cis-lunar magnetospheric field, Ann. Geophys., 32, 1, 1976.

Voigt, G.H., A three dimensional analytical magnetospheric model with defined magnetopause, Z. Geophys., 38, 319, 1972.

Voigt, G.H., A static-state field-line reconnection model for the earth's magnetosphere, J. Atmos. Terr. Phys., 34, 1955, 1978.

Voigt, G.H. and K. Fuchs, A macroscopic model for field line interconnection between the magnetosphere and the interplanetary space, in Quantitative Modeling of Magnetospheric Processes, Geophys. Monogr. Ser., Vol. 21, edited by W.P. Olson, AGU, Washington, D.C., 1979.

Walker, R.J., An evaluation of recent quantitative magnetospheric magnetic field models, Rev. Geophys. Space Phys., 14, 411, 1976.

Walker, R.J., M.G. Kivelson and A.W. Schardt, High $\beta$ plasma in the dynamic Jovian current sheet, Geophys. Res. Lett., 5, 799, 1978.

West, H.I., Jr., R.M. Buck and M.G. Kivelson, On the configuration of the magnetotail during quiet and weakly disturbed periods: Magnetic field modeling, J. Geophys. Res., 83, 3819, 1978.

Whang, Y.C., Magnetospheric magnetic field of Mercury, J. Geophys. Res., 82, 1024, 1977.

Williams D.J. and G.D. Mead, Night side magnetosphere configuration as obtained from trapped electrons at 1100 kilometers, J. Geophys. Res., 70, 3017, 1965.

Willis, D.M. and R.J. Pratt, A quantitative model of the geomagnetic tail, J. Atmos. Terr. Phys., 34, 1955, 1972.

THE SYNCHRONOUS ORBIT MAGNETIC FIELD DATA SET

Robert L. McPherron

Department of Earth and Space Sciences
Institute of Geophysics and Planetary Physics
University of California at Los Angeles
Los Angeles, California 90024

Abstract. Synchronous orbit magnetic field data provide a valuable resource for use in checking and parameterizing quantitative models of the earth's magnetic field. The field observed at this location depends on the earth's main field, magnetopause currents, the tail current, the ring current and field aligned currents. Each of the external currents constitutes as much as one third the total field. All are local time and seasonally varying. Magnetic field data has been acquired by 11 synchronous spacecraft during the last 10 years. The data are of varying quality depending on the instrument, spacecraft and efforts devoted to correcting spacecraft fields. A considerable quantity of graphical data is available from data centers and the original investigators. Digital data is less accessible and has not yet been organized for use in quantitative modelling. Some digital data are presently available, however, and should be used to check current models.

## Introduction

The magnetic field at synchronous orbit is the result of superposition of fields from many sources; the earth, the magnetopause, the geomagnetic tail, the ring current and field aligned currents. The field of the earth alone is approximately $110\gamma$ while the perturbations each produce fields of order $10 - 50\gamma$ depending on geomagnetic activity. Since the perturbations are not spatially uniform the observed field exhibits local time dependence. In addition, seasonal changes in the orientation of the earth's dipole axis causes significant changes in each of the external sources.

The synchronous orbit magnetic field data set is a potentially valuable resource for several reasons. First, these data have not been used in the creation of existing models so they provide an independent check of the models. Second, synchronous satellites make a complete circuit of the magnetosphere daily making a separation of local time and seasonal effects somewhat easier than is the case for eccentric satellites. Third, because the effects of each external source dominate in different local time regions the data provide a means for determining parameters characterizing each source separately.

The primary reason synchronous magnetic field data have not been used more extensively in magnetic field modelling is absolute errors in

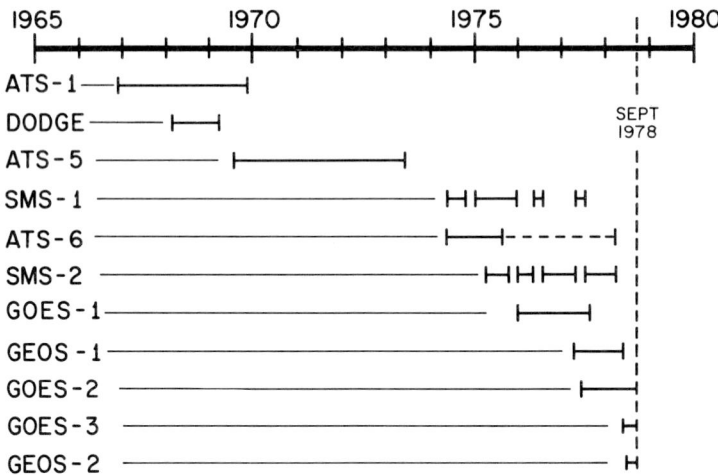

Fig. 1. Time intervals during which some magnetic field data are available for all synchronous spacecraft carrying magnetometers.

the measured fields. These errors are almost entirely a result of highly variable spacecraft fields present at the magnetometer. To date only the ESA GEOS satellite has been devoted entirely to scientific measurements. All other synchronous satellites carrying magnetometers have had as their primary mission communications or meterological applications. On these satellites short magnetometer booms and satellite design considerations have resulted in variable magnetic fields at the sensor location.

Despite such problems there exists a reasonably large collection of synchronous orbit magnetic field data. Some of these data can be of use in quantitative modelling of the earth's magnetic field. The purpose of this paper is characterization of the synchronous orbit magnetic field data set. In separate sections we briefly describe the spacecraft, the magnetometers, the standard graphical data displays, and the digital data files.

Magnetometers at Synchronous Orbit

At the present time, eleven satellites have carried out extensive measurements at or near synchronous orbit. These satellites are identified in Figure 1 in a graph showing the time intervals during which they were operational. Additional information including launch date, location in synchronous orbit, stabiliztion and availability of magnetometer data are summarized in Table 1.

The time span during which the synchronous magnetic field has been monitored is roughly 11 years. The first measurements were made by the ATS-1 (Application Technology Satellite) and the most recent by GEOS-2 (Geocentric Earth Orbiting Satellite). Although the field has

TABLE 1. SYNCHRONOUS SPACECRAFT CARRYING MANGETOMETERS

| SPACECRAFT & LAUNCH DATE | GEOSTATIONARY LONGITUDE | DATA COVERAGE |
|---|---|---|
| ATS-1 Dec 7, 1966 | 150°W Dec 66 - Jun 70 | Dec 6, 66 to Nov 2, 69 |
| DODGE July 1, 1967 | Near Synchronous 6.25 $R_e$, period 22 hr. | Mar 7, 68 to Mar 29, 68 |
| ATS-5 Aug 12, 1969 | 105°W Aug 69 - June 73 | Aug 12, 69 to Jun 1, 73 |
| SMS-1 May 16, 1974 | 45°W Jun-Sep 74<br>75°W Nov 74 - Dec 75<br>105°W Feb 76 --- | May 23, 74 to Jan 8, 76<br>May 19, 76 to Aug 16, 76<br>May 9, 77 to Aug 11, 77 |
| ATS-6 May 30, 1974 | 96°W May 74 - May 75<br>36°E Jun 75 - Aug 75<br>140°W Oct 75 --- | May 30 74 to Sep 9, 75<br>Sep 9, 75 to Mar 31, 78<br>-2 axes only |
| SMS-2 Feb 6, 1975 | 115°W Mar-Nov 75<br>135°W Jan-Apr 78<br>118°W Apr 78 --- | Feb 10, 75 to May 18, 76<br>Aug 16, 76 to May 9, 77<br>Aug 18, 77 to Apr 4, 78<br>Jun 1, 78 to Jun 21, 78 |
| GOES-1 Oct 16, 1975 | 55°W Nov-Dec 75<br>75°W Dec 75 - Aug 77 | None currently available |
| GEOS-1 Apr 20, 1977 | Eccentric Orbit 12-hour period with apogee at 7 $R_e$ | None currently available |
| GOES-2 Jun 16, 1977 | 75°W Aug 77 --- | July 18, 77 --- |
| GOES-3 June 13, 1978 | 135°W Jun 78 --- | Jun 31, 78 --- |
| GEOS-2 July 14, 1978 | 37.5°E Sep 78 --- | None currently available. |

been monitored nearly continuously for the entire time, observations directly useable in magnetic field modelling are currently available from only two spacecraft, ATS-1 and ATS-6. Observations acquired by the European Space Agency on GEOS-1, 2 will eventually be of great value for this purpose but are not yet routinely available. Data from the remaining spacecraft are badly contaminated by spacecraft fields and except in a few cases have not been corrected sufficiently for modelling purposes.

Two of the spacecraft shown in Table 1 were unsuccessful in some respect. The satellite ATS-5 was designed for three axis stabili-

TABLE - 2a. CHARACTERISTICS OF SYNCHRONOUS MAGNETOMETERS

| MAGNETOMETER CHARACTERISTICS | ATS-1 | DODGE |
|---|---|---|
| RESOLUTION | $0.125\gamma$ | $0.22\gamma$ |
| DYNAMIC RANGE | Transverse: $\pm 50\gamma$, $\pm 100\gamma$, $\pm 200\gamma$<br>Parallel: $-675\gamma$ to $+925\gamma$ by $25\gamma$ | $\pm 250\gamma$ |
| BANDWIDTH | 0 to 1.0 Hz | 20 Hz low pass with 3 db between 0.5 and 10.0 Hz |
| SAMPLE RATE | 0.32 sec | 1.26 sec and analog records |
| INSTRUMENT NOISE | $P(f) = \dfrac{4 \times 10^{-3}}{f} (\gamma^2/HZ)$ | Unknown |
| SPACECRAFT NOISE | Highly variable with multiple spectral lines. | Unknown |
| OFFSETS AND/OR SPACECRAFT FIELDS | Large and variable. Corrected data accurate $\approx 20\gamma$ parallel, $5\gamma$ transverse. | Continuously changing offsets of unknown magnitude |
| LINEARITY | Better than resolution over $\pm 50\gamma$ dynamic range | Unknown |
| TEMPERATURE DEPENDENCE | Sensitivity independent of T, offsets change but small compared to spacecraft field | Sensitivity independent of temperature. Offsets drift steadily in morning hours |
| SENSOR ORTHOGONALITY | Manufacturing tolerance, not measured | Unknown |
| WEIGHT<br>  SENSOR<br>  ELECTRONICS | <br>0.70 lbs<br>3.05 lbs | <br>Unknown |
| SIZE<br>  SENSOR<br>  ELECTRONICS | <br>3.5 x 3.125 x 3.25 in.<br>9.15 x 3.80 x 3.70 in. | <br>Unknown |
| POWER | 3.0 watts | Unknown |

zation, but, because a boom failed to deploy properly it had to be spin stabilized. Since the magnetometer filtering and sampling scheme had not been designed to operate on a spinning spacecraft it was virtually impossible to convert the observations to physical units in a non-rotating reference frame.

TABLE - 2b. CHARACTERISTICS OF SYNCHRONOUS MAGNETOMETERS

| MAGNETOMETER CHARACTERISTICS | ATS-5 | ATS-6 |
|---|---|---|
| RESOLUTION | $0.12\gamma$ | $0.0625\gamma$ |
| DYNAMIC RANGE | Basic magnetometer: $\pm 25\gamma$<br>Offset: $-495\gamma$ to $+525$ by $\approx 32\gamma$ | Basic magnetometer: $\pm 16\gamma$<br>Offset: $-512\gamma$ to $+496\gamma$ by $40\gamma$ |
| BANDWIDTH | 0 to 0.125 Hz | 0 to 2.25 Hz |
| SAMPLE RATE | Basic magnetometer: 2.97 or 5.12 sec<br>Offset: 5.12 or 95 sec | 0.125 sec |
| INSTRUMENT NOISE | $\sim 0.12$ | $P(f) = (1 \times 10^{-3}/f)(\gamma^2/Hz)$ |
| SPACECRAFT NOISE | Unknown | Depends on spacecraft state and local time. |
| OFFSETS AND/OR SPACECRAFT FIELDS | Large and variable. Calibrated prelaunch. | Transverse: $<2\gamma$<br>Parallel: $\leq 10$<br>(Relative to radius vector) |
| LINEARITY | Unknown | $\leq (2$ mv $\sim 5$ volts$)$ |
| TEMPERATURE DEPENDENCE | Calibrated over range $\pm 50°C$ to $+65°C$ | $\leq 1\gamma$ change in offset when cycled between $-10°$ and $+100°F$ in zero field. |
| SENSOR ORTHOGONALITY | Manufacturing tolerance of order $1.5°$ | $\leq 0.5°$ |
| WEIGHT<br>  SENSOR<br>  ELECTRONICS | <br>1.03 lbs<br>5.89 lbs | <br>0.97 lbs<br>7.18 lbs |
| SIZE<br>  SENSOR<br>  ELECTRONICS | <br>50 cubic inches<br>200 cubic inches | <br>4.1 x 4.2 x 3.8 in.<br>6.7 x 8.1 x 4.6 in. |
| POWER | 1.2 watts | 4.5 watts |

The satellite GEOS-1 could not be placed in geocentric orbit because of a booster malfunction. Instead it was placed in a 12 hour eccentric, equatorial orbit with apogee just beyond synchronous orbit. The unexpected orbit made it impossible to use programs and data displays prepared prior to launch and has delayed routine data processing. Eventually, however, the data should be of great value in studying radial gradients in the field around synchronous orbit.

TABLE - 2c. CHARACTERISTICS OF SYNCHRONOUS MAGNETOMETERS

| MAGNETOMETER CHARACTERISTICS | SMS/GOES | GEOS |
|---|---|---|
| RESOLUTION | $0.2\gamma$ | $0.25\gamma$ |
| DYNAMIC RANGE | Basic magnetometer: $\pm 50\gamma$<br>Offset: $-1170\gamma$ to $1450\gamma$<br>by $40\gamma$ | Transverse: $\pm 60\gamma$ and $\pm 180\gamma$<br>Parallel: Basic: $\pm 60\gamma$<br>Offset: $-420\gamma$ to $+480\gamma$<br>by $60\gamma$ |
| BANDWIDTH | 0 to 0.136 Hz | 0 to 5.0 Hz |
| SAMPLE INTERVAL | 0.74 sec | 0.08625 sec |
| INSTRUMENT NOISE | $\leq 0.1\gamma$ peak to peak | $P(f) = (1.6 \times 10^{-3}/\sqrt{f})(\gamma^2/Hz)$ |
| SPACECRAFT NOISE | Unknown | Unknown |
| OFFSETS AND SPACECRAFT FIELDS | Depend on spacecraft state and of order ambient field in parallel axis. | Ground calibration:<br>$X, Y < 2\gamma$; $Z < 1\gamma$ |
| LINEARITY | $\sim 0.5\gamma$ in $50\gamma$ | Better than measuring instruments. |
| TEMPERATURE DEPENDENCE | Parallel: $\leq 2\frac{1}{2}\gamma$ over normal temperature range<br>Transverse: $\leq \frac{1}{2}\gamma$ in normal range | Less than $0.25\gamma$ in range $-60°C$ to $+120°C$ |
| SENSOR ALIGNMENT | Manufacturing tolerance $\approx 1°$ | $\leq 1.0°$ measured on ground |
| WEIGHT<br>  SENSOR<br>  ELECTRONICS | <br>0.55 lbs<br>3.60 lbs | <br>0.65 kgm<br>2.15 kgm |
| SIZE<br>  SENSOR<br>  ELECTRONICS | <br>3.5 in. diameter, 5" tall<br>7.3" x 8" x 3.5" | <br>11 cm diameter, 14 cm tall<br>15 cm x 14 cm x 23 cm |
| POWER | $\leq 4.0$ watts | 0.8 watts |

Characteristics of Synchronous Orbit Magnetometers

Important characteristics of the magnetometers flown at synchronous orbit are summarized in Table 2. Since the field at this location is about $100\gamma$ most of the magnetometers have been designed with dynamic ranges of order $\pm 500\gamma$. Typical amplitude resolution for these instruments is 0.1 to $0.2\gamma$ and typical time resolution one second. Most of the instruments have noise spectra below those of the spacecraft enabling the instruments to observe a large variety of field fluctuations. In most cases measurement errors of instrumental origin such as non-linearity, sensor misalignment and temperature dependence of offset and sensitivity are negligible in comparison with errors due to spacecraft fields.

Large, highly variable spacecraft fields are present in data from all spacecraft except ATS-6 and GEOS 1 and 2. The satellite ATS-1 exemplifies the problem. This satellite was a spin stabilized cylinder with the magnetometer located on a short 15 cm boom close to the satellite electronics package. DC magnetic fields parallel to the spin axis were produced by spacecraft electronics. Typical parallel offsets were of order $100\gamma$. Changes in these offsets occurred several hundred times per day and were as large as $\pm 50\gamma$. AC magnetic fields at the satellite spin frequency were caused by modulation of currents in the solar cells on the surface of the spinning satellite. After spin demodulation these appeared as DC offsets in the transverse components. A change in spacecraft state altered the solar cell currents, hence the amplitude of the AC signal and finally the magnitude of the transverse offsets.

### Standard Synchronous Magnetometer Data Displays

Magnetic field data obtained at synchronous orbit are routinely processed to physical units and displayed graphically by pen and microfilm plotters. In some cases these plots are the only easily accessible data at the present time. This section briefly summarizes the graphical displays used for each spacecraft listed in Table 1.

The ATS-1 graphical data displays shown in Table 3 include three years of uncorrected data (1967-1969) and two years of corrected data (1967-1968). Uncorrected data are at high time resolution (0.32 seconds) and intermediate time resolution (15.0 seconds). Uncorrected data are despun to GSEE (Geocentric Solar Earth Equatorial) coordinates and corrected data are transformed to dipole VDH coordinates.

The DODGE (Department of Defense Gravity Experiment) graphical data displays are analog chart recording with various chart speeds and amplitude scales (See Table 3). Data were routinely obtained for about 4 hours per day for 4½ days in 11 day intervals. During these times the near synchronous satellite DODGE was in view of the John Hopkins Applied Physics Laboratory tracking station. Recordings were made between March 1968 and March 1969.

The ATS-5 graphical data displays as originally planned are listed in Table 3. Because the satellite was not despun these displays contain little information of scientific value. An extensive effort was carried out to despin and correct some PCM telemetry data. Low resolution microfilm plots were made from these data. These displays are of some value for correlative purposes.

The SMS (Synchronous Meterological Satellite) and GOES (Geocentric Orbiting Environmental Satellite) satellites are spacecraft operated by NOAA for meterological purposes. The satellites include a magnetometer, particle detector and x-ray detectors monitoring the synchronous environment. Graphical data have been produced for the SMS-1 and SMS-2 magnetometers for the interval from launch to December 31, 1975. For the later satellites in this series, renamed GOES, graphical data

TABLE 3. RESPONSIBLE INDIVIDUALS AND TYPES OF GRAPHICAL DISPLAYS

| SATELLITE | INDIVIDUALS AND GRAPHICAL DISPLAYS |
|---|---|
| ATS-1 | Prof. R.L. McPherron, Prof. P.J. Coleman, Jr., Mr. Neal Cline, Inst. of Geophys. and Planet. Physics, Univ. of California, Los Angeles 90024.<br>High Time Resolution: 163.84 second microfilm panels containing 0.32-second sample of uncorrected field components in GSEE coordinates.<br>Intermediate Time Resolution: One-hour microfilm panels containing 15.0-second averages of uncorrected field components in GSEE coordinates. Four-hour CalComp plots containing 15-second averages of corrected field components and magnitude in dipole VDH coordinates.<br>Low Resolution Plots: One-day CalComp plots containing 150-second averages of corrected field components and magnitude in dipole VDH coordinates. |
| DODGE | Dr. T. Potemera, John Hopkins University, Applied Physics Lab., Laurel, Maryland 20810. Analog chart records made by Brush recorder at speeds from 0.2 to 200 cm/sec, 2-20 gamma/division. No standard displays from digital data. |
| ATS-5 | Dr. M. Sugiura, Code 625, Goddard Space Flight Center, Greenbelt, Maryland 20771.<br>High Time Resolution: PLOTA, Ten-minute microfilm panels containing 5.12-second samples from basic magnetometer with fixed amplitude scale.<br>PLOTB: Ten-minute microfilm panels containing 5.12-second samples from basic magnetometer and offset generator with variable amplitude scale.<br>(Above for PFM telemetry data. For PCM plot lengths are 6 minutes and sample interval is 2.97 seconds.)<br>Intermediate Time Resolution: PLOTC, One-hour microfilm panels containing 30-second averages of basic magnetometer and offset generator with variable scale.<br>Low Time Resolution: PLOTD, one-day Gerber pen plots containing 30-second averages. Half-day microfilm panels of despun, corrected PCM data with 90-second averages. |
| SMS-1,2<br>GOES-1,2<br>and 3 | Dr. J.N. Barfield, Dr. D.J. Williams, Mr. Frank Cowley, Space Environment Laboratory, NOAA, Boulder, Colorado 80302<br>Intermediate Time Resolution: One-hour microfilm panels of automatically corrected 3-second averages of field components in PEN coordinates [Parallel, Earthward, Normal (west)].<br>Low Time Resolution: One-day microfilm panels of corrected 60-second averages of field components in PEN coordinates. |

TABLE 3   RESPONSIBLE INDIVIDUALS AND TYPES OF GRAPHICAL DISPLAYS

| SATELLITE | INDIVIDUALS AND GRAPHICAL DISPLAYS |
|---|---|
| ATS-6 | Prof. R.L. McPherron, Prof. P.J. Coleman, Jr., Mr. Neal Cline, Inst. of Geophys. and Planet. Physics, Univ. of California, Los Angeles 90024.<br>High Time Resolution: 128-second microfilm panels containing 0.25-second samples of field components and magnitude in spacecraft coordinates. Twelve-minute microfilm panels containing one-second samples of field components and magnitude in spacecraft coordinates.<br>Intermediate Time Resolution: One-hour microfilm panels containing 15-second averages of field components and magnitude in dipole VDH coordinates. CalComp plots of length four hours containing high pass filtered 5-second averages of fild components in spacecraft coordinates.<br>Low Resolution Plots: One-day CalComp plots containing 64-second averages of V, D, H, B, $\Theta$, $\phi$ in dipole coordinates.<br>Very Low Resolution Plots: 27-day Houston plotter plots containing one-hour averages of field components and magnitude in dipole VDH coordinates.<br>Pc1 Wave Index: One-day CalComp plots of 64-second averages of rms power in Pc1 band for each spacecraft field component.<br>Pc3 and Pc4 Wave Index: One-week CalComp plots containing five-minute averages or rms power in Pc3 and Pc4 band for each spacecraft field component. |
| GEOS-1<br>GEOS-2 | Dr. M. Candidi, Dr. F. Mariani, Laboratorio Di Recerca E. Tecnologia per lo Studio del Plasma Nello Spazio, Frascati, Italy.<br>Graphical data displays for these spacecraft have not been finalized as of November 1, 1978. |

are available from GOES 2 and 3 from September 1977 to the present time.

Data from all these spacecraft are automatically corrected for spacecraft fields. For SMS-1 and SMS-2 the corrections are not of high quality and require additional hand corrections when absolute accuracy is required. In the GOES spacecraft several sources of DC field changes were eliminated and the automatic correction procedure was improved. However, several problems remain and the data are not currently available as completely corrected field values.

The ATS-6 graphical data displays are summarized in Table 3. The spacecraft was operated at 96 W longitude from May 31, 1974 to May 20, 1975 and then moved to 35 east longitude arriving June 20, 1975. The magnetometer operated properly at this location until Sept. 9, 1975 when the Y axis (parallel to earth's axis) failed. Graphical data from Y axis failure until spacecraft shutdown on March 31, 1978 have been prepared on special request. Prior to Y axis failure graphical data is continuously available for all except the high time resolution plots and Pc3, 4 wave index plots. For highest resolution (¼ second) 1 month is available; for one second resolution, approximately 6 months and for Pc3, 4 wave index, 2 months.

Data from the GEOS 1 and 2 (Geosynchronous Earth Orbiting Satellites) are not presently available. Failure of GEOS 1 to achieve synchronous orbit seriously handicapped program development and delayed routine processing. These data are expected to be available in the near future on a routine basis.

## Digital Synchronous Magnetic Field Data

Synchronous orbit magnetic field data in digital form are much less accessible than the graphical data discussed in the previous section. The reasons for this include different data processing philosophies, different tape formats, different computer representations, the large size of many of the digital data files, lack of documentation and physical inaccessibility of the data files. At the present time it would be nearly impossible to bring together all of the existing data, whatever its accuracy, for use in quantitative modelling.

The only digital data currently available in the Space Science Data Center is from the ATS-1 spacecraft. Digital data from the ATS-6 spacecraft are in the process of submission and will become available in the near future. Data from the SMS/GOES series of spacecraft exist on digital archive tapes in the Space Environment Laboratory of NOAA in Boulder, Colorado, and will be submitted to WDC-A for geomagnetism in the near future. Data from GEOS 1 and 2 spacecraft have not yet been processed to final form. Data from DODGE and ATS-5 are of little value for modelling purposes and are not currently active data sets.

One of the primary requirements of a magnetic field data set used in modelling is accurate field values. At the present time none of the available data sets are particularly accurate. Both ATS-1 and the SMS/GOES spacecraft data have been computer corrected. The correction procedures involve a number of assumptions and intercomparisons with other spacecraft that limit the accuracy of the data. Since all these spacecraft were spinning about an axis parallel to the earth's rotation axis, errors in the transverse components are caused by AC signals at

the satellite spin frequency. These signals are usually small and the
resulting field values are accurate to a few gamma. The errors in the
parallel component are caused by DC spacecraft fields which on these
spacecraft are large. Corrections are usually based on comparisons
with data from other spacecraft. Consequently, errors in the parallel
component may be as large as 10 to 20 gamma.

Data from the ATS-6 spacecraft are somewhat more accurate than ATS-1
and SMS/GOES data and considerably less noisy. These data were acquired on a 3 axis stabilized spacecraft with the sensors at the end
of a 2 meter boom. Considerable care was taken in spacecraft design
and preflight calibrations. In addition, an in-flight roll maneuver
was used to check offsets for sensors transverse to the earth pointing
axis. For these two axes offsets are less than two gamma. Using
particle pitch angle data on the same spacecraft an upper limit to the
possible errors in the earth pointing sensor was determined to be 10
gamma. Model comparisons suggest that the error is about this large,
i.e., a 10 $\gamma$ corrections should be added to the spacecraft Z component
(Dipole V).

The second major requirement of a magnetic field data set for use in
modelling is convenient access. To provide such convenience the original observations must be averaged to lower time resolution. In our
work we find one minute values somewhat too high resolution and one
hour values too low. An appropriate sampling interval is 10 or 15
minutes. It is essential that these averages exist in a distinct
data file separated from data at higher time resolution. A common
programmer mistake is to write time averages as separate records interspersed with data records of higher time resolutions. Such formats
force the potential user to read as many as 100 tapes to acquire data
that will fit on a few tracks of a small disk pack.

## Discussion

Monitoring of the magnetic field at synchronous orbit has been in
progress for over 10 years. Much of the data acquired is available
in graphical form and some in digital form. The data is of great
value for studying variations in the field at synchronous orbit, particularly hydromagnetic waves. It is of less value for quantitative
modelling because of errors due to spacecraft fields. Considerable
effort has been spent on attempts to correct the field observations.
The resulting data are probably accurate to better than 10%.

Magnetic field data useful for modelling and conveniently accessible
at the present time exist only for the ATS-1 and ATS-6 and the SMS/GOES
spacecraft. Sometime in the near future data of greater accuracy from
the GEOS spacecraft should also become available.

Descriptions of the average field behavior at synchronous orbit are
available in several published reports. Cummings et al., 1971 and
Olson and Cummings, 1971, characterized the quiet field ATS-1 and
compared it to an early field model. Coleman and Cummings, 1971,
described the synchronous magnetic field during magnetic storms.
Review of this work and some extensions are presented in papers by
McPherron and Coleman, 1971; McPherron, et al., 1970; Coleman and
McPherron, 1970. Initial results of statistical study of ATS-1 data
are given in McPherron, 1975 and Coleman and McPherron, 1976. The

latter paper includes initial results from the ATS-6 spacecraft as well. The seasonal dependence of the quiet field observed by the ATS-5 spacecraft has been discussed by Skillman, 1972.

A number of publications or reports are available characterizing the instruments, data processing graphical displays and digital data sets for each synchronous magnetometers. These documents are listed in the bibliography under subheadings for each spacecraft. For many of the data sets for each synchronous magnetometer. These documents are listed in the bibliography under subheadings for each spacecraft. For many of the data sets any detailed study of the data would require extended discussions with the individuals currently responsible for the data sets. These individual are listed as part of Table 3.

Acknowledgments. The author would like to thank the following individuals for the information they provided the author while preparing this report; J. N. Barfield, M. Candidi, F. Cowley, F. Grubb, T. Potemra, M. Sugiura. This work was supported by NASA contract NAS 5-23702

## References

### Published Reports of Synchronous Field Observations

Coleman, P. J., Jr. and W. D. Cummings, Stormtime disturbance fields at ATS-1, J. Geophys Res., 76, 1, 1971.

Coleman, P. J., Jr. and R. L. McPherron, Fluctuations in the distance geomagnetic field during substorms: ATS-1, edited by B. M. McCormac, D. Reidel Pub. Co., Dordrecht, 1970.

Coleman, P. J., Jr. and R. L. McPherron, Substorm observations of magnetic perturbations and ULF waves at synchronous orbit by ATS-1 and ATS-6, Knott and Battrick (eds.), The Scientific Satellite Programme During the International Magnetospheric Study, 345-365, D. Reidel Publishing Co., Dordrecht-Holland, 1976.

Cummings, W. E., P. J. Coleman, Jr. and G. L. Siscoe, Quiet day magnetic field at ATS-1, J. Geophys. Res., 76(4), 1971.

McPherron, R. L., C. T. Russell and P. J. Coleman, Jr., Comparison of the Olson field model to quiet day magnetic field observations at ATS-1, (extended abstract), in Quantitative Magnetospheric Models (Proc ESSA Symp.), Boulder, Col., 1970.

McPherron, R. L. and P. J. Coleman, Jr., Magnetic field variations at ATS-1, in The ESRO Geostationary Magnetospheric Satellite, ESRO pub #SP-60 ESRO Colloquium, Lyngby, Denmark, Oct. 1969), 1971.

McPherron, R. L., The quiet time magnetic field at synchronous orbit, EOS Transactions, AGU, 56(9), 1975.

Olson, W. P. and W. D. Cummings, Comparison of the predicted and observed magnetic field at ATS-1, J. Geophys. Res., Space Physics, 75(34), 1970.

Skillman, T. L., Average daily variations in the magnetic field as observed by ATS-5, Goddard Space Flight Center Preprint X-645-72-301, Greenbelt, Md., 1972.

### Descriptive Publications for ATS-1

Barry, J. D. and R. C. Snare, A fluxgate magnetometer for the applications technology satellite, IEEE Trans. on Nuclear Science, NS-13 (6), 326-331.

McPherron, R. L., Correction and analysis of ATS-1 magnetometer data during 1971, IGPP Pub. #995, Oct. 1971.

McPherron, R. L., Coordinate transformations used in processing ATS-1 magnetic field data, IGPP Pub. #1053-05, Aug. 1972.

Russell, C. T. and R. L. McPherron, A bibliography for the ATS-1 and OGO-5 fluxgate magnetometers, IGPP Pub. #1065-07, Oct. 1972.

Snare, R. C. and G. N. Spellman, Digital offset field generator for spacecraft magnetometers, Proc. Symp. on Space Magnetic Exploration and Technology, Reno, Nevada, Aug. 30, 1967.

## Descriptive Publications for DODGE

Dwarkin, M. L., A. Z. Zmuda and W. E. Radford, Hydromagnetic waves at 6.25 earth radii with periods between 3 and 240 seconds, J. Geophys. Res., 76(16) 3668-3674, 1971.

Patel, C. L., R. J. Greaves, S. A. Wahob, T. A. Potemra, DODGE satellite observations of Pc3 and Pc4 magnetic pulsations and correlated effects in the ground observations, Preprint, Sept. 1978.

Space Development Department, The John Hopkins University, Applied Physics Laboratory, DODGE Satellite Performance July 1, 1967 - October 1, 1968, Tech. Mem. TG-1034A, December 1968; and DODGE Satellite Performance July 1, 1967 - October 1, 1968, Appendix Tech. Mem. TG-1034B, December 1968.

## Descriptive Publications for SMS/GOES

Grubb, R. N., The SMS/GOES space environment monitor subsystem, NOAA Tech. Mem. ERL SEL-42, Dec. 1975.

Wirth, R. J., Geostationary operational environmental satellite (GOES/synchronous meterological satellite (SMS) Operational Manual, NASS GSFC Report, unnumbered.

## Descriptive Publications for ATS-6

McPherron, R. L., Progress Report: UCLA fluxgate magnetometer on ATS-6 for the period April 1 - September 1, 1974, IGPP Pub. #1388-60, Oct. 1, 1974.

McPherron, R. L., Description of the UCLA fluxgate magnetometer on ATS-6: instrument data files, data displayes, preliminary observations, IGPP Pub. #1578, May 19, 1976.

McPherron, R. L., P. J. Coleman, Jr., R. C. Snare, ATS-6 UCLA fluxgate magnetometer IEEE Trans. on Aerospace and Elec., AES-11 (6), 1110-1117, 1975.

## Descriptive Publication for GEOS

Candidi, M., S. Cantarano, G. Ferri, F. Mariani, G. Martinelli, R. Orfei, C. Signorini, G. Vannarini, GEOS S 331 fluxgate magnetometer, Report #LPS-77-10, Lab. di Ricerca e Tech. per lo Studio del Plasma Nello Spazio, March 1977.

A METHOD OF EVALUATING QUANTITATIVE MAGNETOSPHERIC FIELD MODELS BY AN ANGULAR PARAMETER $\alpha$

Masahisa Sugiura

Laboratory for Planetary Atmospheres, Goddard Space Flight Center, Greenbelt, Maryland 20771

Demetrios J. Poros

Computer Sciences Corporation, Silver Spring, Maryland 20910

Abstract. Previously we presented a method in which the distribution of the scalar difference $\Delta B$ between the magnitude of the observed magnetospheric field, or of a theoretical model field, and the magnitude of the earth's internal field is used to study the magnetospheric field distortions. This technique has also been used widely as a standard test for evaluating quantitative magnetospheric field models. As an extension of the $\Delta B$ method an angular parameter, $\alpha$, is introduced which is defined as the angle between the observed (or model) magnetospheric field and the earth's internal field calculated from a spherical harmonic expansion. On the basis of the Ogo-5 GSFC fluxgate magnetometer data, the average $\alpha$ distribution is presented for the noon-midnight and dawn-dusk meridians either in the form of equal $\alpha$ contours or by numerical values. These representations are given for three magnetic activity groups: Kp = 0-1, 2-3, and 4-9. The parameter $\alpha$ is also calculated for the Olson-Pfitzer (1974) model, and in a more limited extent for the Mead-Fairfield (1975) model, and compared with the distribution of $\alpha$ deduced from observations. The advantages of the use of $\alpha$ over $\Delta I$ and $\Delta D$ are discussed. In certain areas $\alpha$ is more sensitive than $\Delta B$ in expressing magnetospheric field distortions. It is recommended to use both $\Delta B$ and $\alpha$ in comparing models with observations.

## Introduction

Large scale distortions of the earth's magnetic field in the magnetosphere are mainly produced by the magnetopause current and the equatorial current system that consists of the ring current and the cross tail current. The gross features of the magnetic field distortions are now fairly well known, observationally (e.g., Fairfield, 1968, 1971; Sugiura and Poros, 1973; Hedgecock and Thomas, 1975), and are reasonably well represented by several magnetospheric field models (e.g., Alekseev and Shabansky, 1972; Sugiura and Poros, 1973; Olson and Pfitzer, 1974; Mead and Fairfield, 1975).

To represent the deviations of the observed magnetospheric field from the earth's internal field a number of parameters can be used: for instance, $\Delta B$, defined as the difference between the magnitude of the

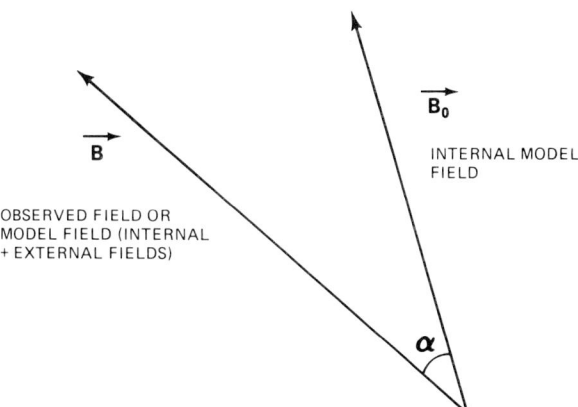

Fig. 1. Parameter $\alpha$ is defined as the angle that the observed magnetic field (or a model field), $\vec{B}$, makes with the magnetic field, $\vec{B}_o$, of an internal field model.

observed magnetic field and that of the earth's internal field calculated from a spherical harmonic expansion; $\Delta I$ and $\Delta D$, which are similarly defined for inclination, I, and declination, D, respectively. The scalar difference $\Delta B$ was initially used by Heppner et al. (1967) in their analysis of Ogo-1 magnetic field observations to classify regions of the magnetosphere by gross characteristics of the field behavior. Pursuing this method with the Ogo-3 and -5 data we arrived at the conclusion that the ring current is much more extensive than had generally been thought (Sugiura, 1972a, 1972b). This feature has since been incorporated in some of the magnetospheric field models, for instance in the models given by Sugiura and Poros (1973) and by Olson and Pfitzer (1974).

When a model field is constructed it is important to evaluate how well the model represents the observed features. Making such an evaluation on a quantitative basis is not necessarily easy. However, comparison of the observed $\Delta B$ and the $\Delta B$ calculated from a model has been found to be extremely useful in this regard (Sugiura and Poros, 1973; Olson and Pfitzer, 1974; a review paper by Walker, 1976). Nevertheless, $\Delta B$ is a scalar quantity and therefore it is desirable to expand the method and to incorporate the directional information of the vector difference field $\Delta\vec{B}$.

In this paper we introduce an angular parameter which is called $\alpha$ and which represents the angular difference between the observed, or model, field and the internal model field. We will discuss why this parameter is chosen and demonstrate its usefulness by applying the method to both observations and models.

## Angular Parameters

The most obvious extension of the $\Delta B$ analysis would be to analyze $\Delta I$ and $\Delta D$, using inclination and declination of the magnetic field which are the most widely used angular variables dealing with the earth's field; see Chapman and Bartels (1962) for the definition of these magnetic elements. A preliminary analysis of $\Delta I$ and $\Delta D$ was made

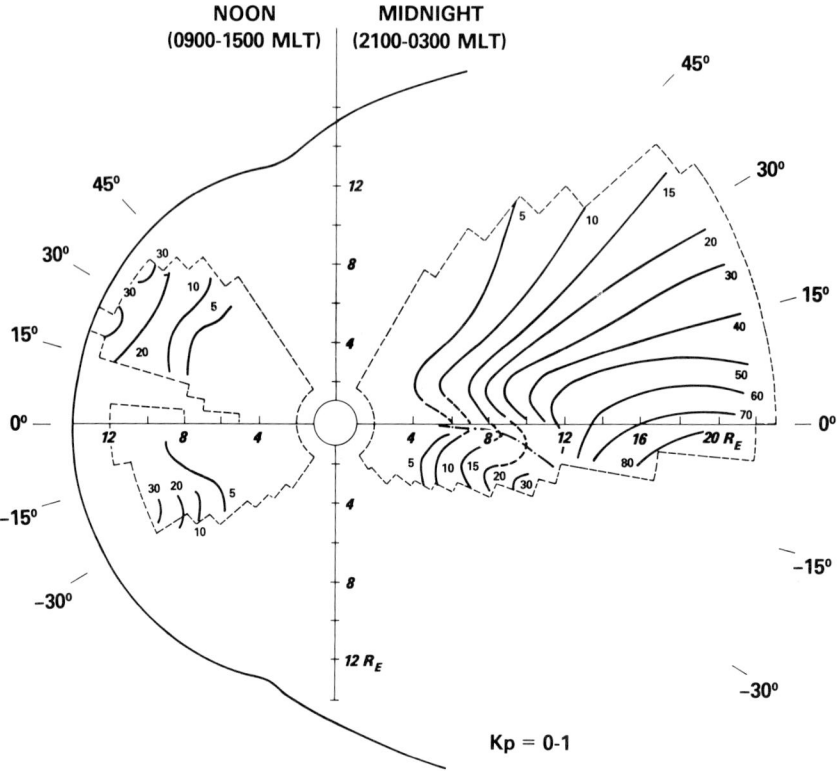

Fig. 2. Equal α contours (α in degrees) in the noon-midnight meridian, Kp = 0-1.

using the observed magnetic field data from Ogo-5 and for a few representative magnetospheric field models. However, several disadvantages have been found in the use of these parameters. In the case of ΔB, one parameter was equally useful for all local times; whereas with ΔI and ΔD their usefulness is dependent on local time. For instance, in the noon-midnight meridian, ΔD is nearly zero and ΔI alone is important, while in the dawn-dusk meridian ΔD as well as ΔI is equally important. In general, the two parameters have to be calculated and examined. There is an obvious advantage in dealing with only one angular parameter that is useful everywhere in the magnetosphere.

Thus we define a parameter α by the angle between the observed, or model, vector field and the vector reference field which is calculated usually from a spherical harmonic expansion of the earth's internal field, as shown in Figure 1. In calculating α for a magnetospheric model field, the model must include a representation of the internal field, which often is a simple dipole field. Although α does not give any azimuthal information on the field deviation relative to the direction of the reference field, it provides a convenient means of expressing the degree of distortion of the earth's field by the magnetospheric currents.

If in the noon-midnight meridian plane the magnetic field lies completely in this plane (as in the cases of theoretical models having

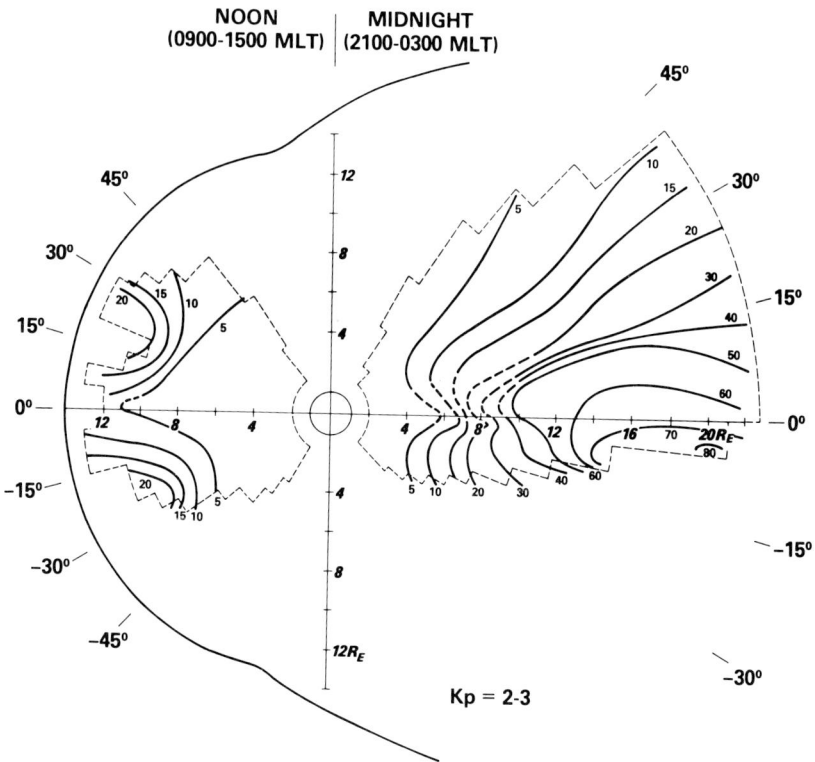

Fig. 3. Equal α contours (α in degrees) in the noon-midnight meridian, Kp = 2-3.

symmetry with respect to this meridian), α is identical with the magnitude of ΔI. A few remarks are made here concerning the differences between α and ΔI. Since ΔI changes sign in the vicinity of the dipole equator as the spacecraft crosses the equator from one hemisphere to the other, any averaging process applied to ΔI near the equator could lead to cancellation of positive and negative values. With α, being always a positive number, there is no such cancellation. The parameter α goes through a minimum near the equator, and the curved surface defined by the locus of α minimum can be regarded as the dividing surface between the northern and southern fields. In the tail this surface would coincide with the neutral sheet.

There is another, subtle but important, point about ΔI and α. Inclination, I, is measured (positively downward) from the local horizontal plane. Therefore, in a region where the magnetic field is nearly vertical, either downward or upward, the field measured from a spacecraft traversing such a region may cross the vertical direction. Then the quadrant in which I is calculated would change; I simply goes through a maximum ($90°$) and changes continuously. Such a condition is encountered near noon on the poleward side of the northern and southern polar cusp regions where the field lines are bent backward away from the solar direction. In calculating ΔI, the inclination, $I_o$, of a reference field is subtracted from the inclination, I, of the measured or a model

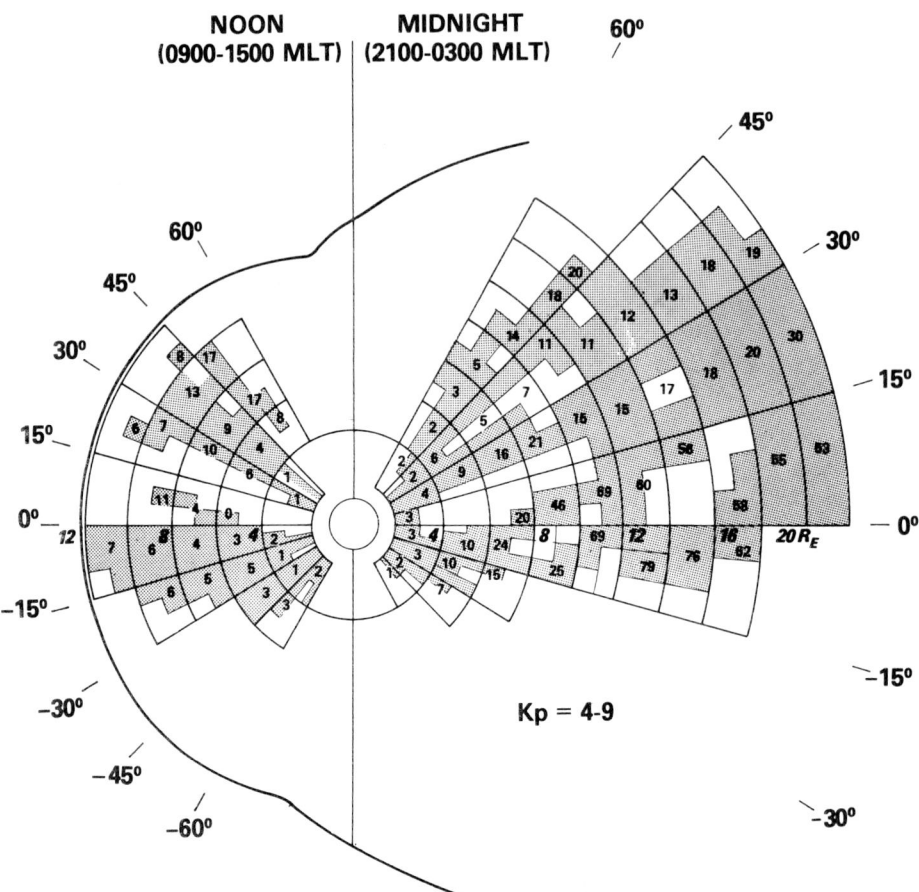

Fig. 4. Average values of α (in degrees) in the noon-midnight meridian, Kp = 4-9. The data coverage was not adequate to draw α contours as in Figures 2 and 3.

field. But $I_o$, being that of a dipole field or nearly so, remains in the same quadrant unless the pole or the equator is crossed, while I for the latter field may change quadrant. It might be thought that it would be more logical, for the sake of continuity, to measure I in the same quadrant as that for $I_o$ in calculating $\Delta I$, thus allowing I to exceed $90°$. However, this is not a meaningful procedure as will be explained in the Discussion section. The use of α avoids such a difficulty. This is another strong reason for recommending α over $\Delta I$.

## Data Used in the Analysis

The analysis presented in this paper uses data obtained by the Ogo-5 GSFC fluxgate magnetometer during the period March 1968 (the beginning of the operation) to July 1969. The data set consists of samples selected from the averages over 36.864 seconds (referred to as 37 second averages below) of individual measurements which were taken at

TABLE 1. Number of 37 second averages of vector magnetic field measurements used in the present analysis (March 1968 to July 1969).

| MLT Sector | Kp = 0-1 | Kp = 2-3 | Kp = 4-9 |
|---|---|---|---|
| 2100-0300 | 4,190 | 6,056 | 2,028 |
| 0300-0900 | 2,551 | 4,079 | 817 |
| 0900-1500 | 1,933 | 3,183 | 1,095 |
| 1500-2100 | 2,444 | 3,056 | 1,104 |
| TOTAL | 11,118 | 16,374 | 5,044 |
| | | TOTAL: | 32,536 |

one of the three data rates: 0.868, 6.94, or 55.55 times per second. The resolution of the measurements is ±1/8 nT and the total range of the instrument, 4000 nT (Ledley, 1970). The sampling of the 37 second averages was made, on the average, at the rate of about 4 times per earth-radius ($R_E$) of distance along the orbit. The sampling was not

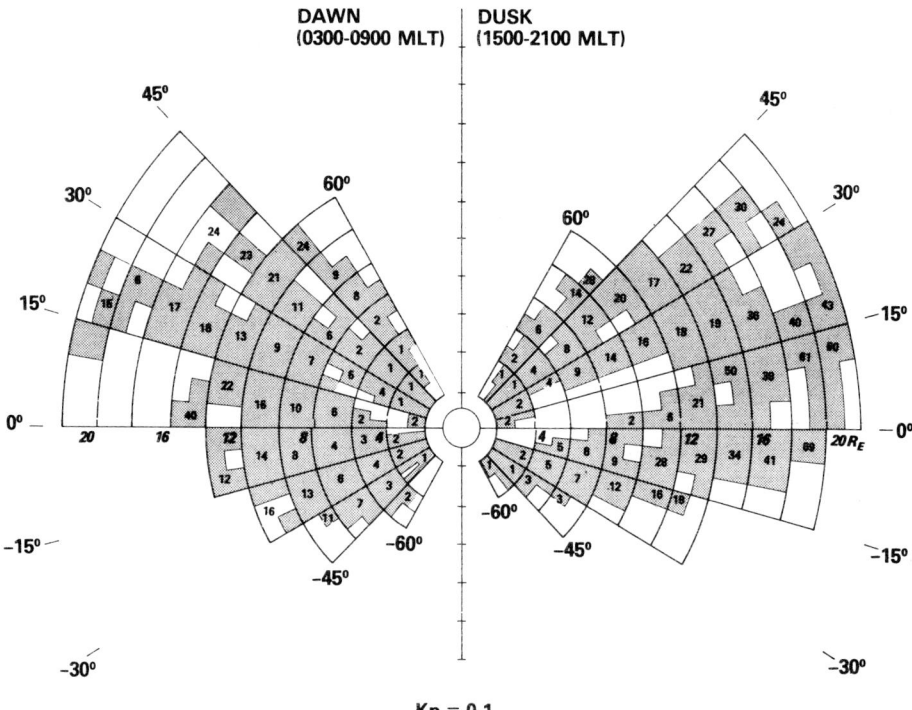

Fig. 5. Average values of α (in degrees) in the dawn-dusk meridian, Kp = 0-1.

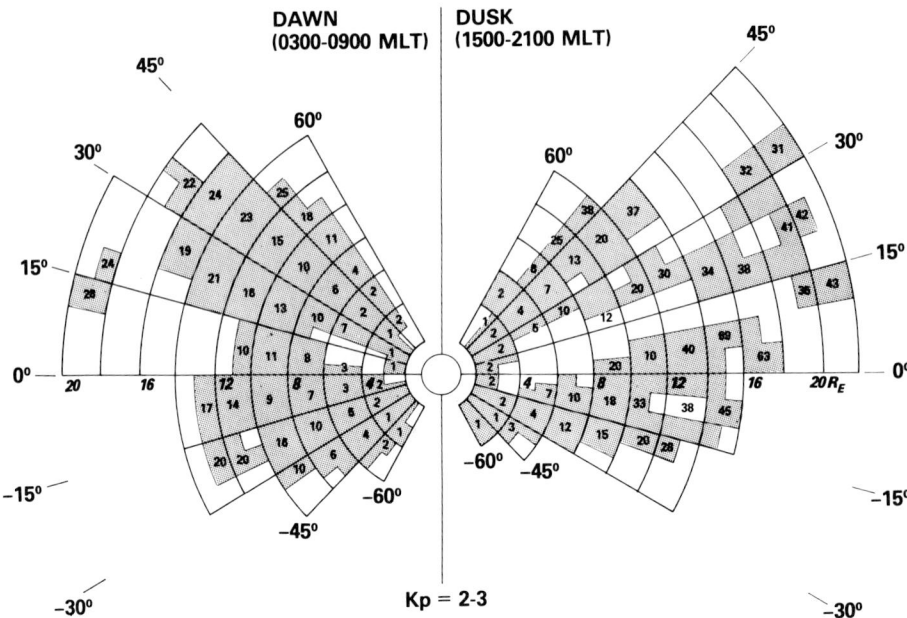

Fig. 6. Average values of α (in degrees) in the dawn-dusk meridian, Kp = 2-3.

made by any automatic procedure primarily because of data gaps. In our judgement this sampling method could not have introduced any bias in the selection of data. The total number of 37 second averages used in this analysis is about 32,000. These sampled averages are divided into three magnetic activity groups, Kp = 0-1, 2-3, and 4-9, and then into four magnetic local time (MLT) quadrants centered at 0000 (midnight), 0600 (dawn), 1200 (noon), and 1800 (dusk). The number of 37 second averages in each MLT sector and each activity group is given in Table 1.

Equal α Contours Deduced from Ogo-5 Data

Noon-Midnight Meridian

For each activity group, each of the MLT sectors was further divided into slices of volume element bounded by the dipole equator and the conical surfaces at dipole latitudes $15°$, $30°$, etc., and by geocentric spherical surfaces at 2, 3, ..., 23 $R_E$. The 37 second averages were then further averaged in each volume element. In the present analysis the network of averages obtained in this manner constitutes the data base for each MLT sector and for each activity group. The number of 37 second averages contributing to the final average value in each volume element was also recorded and was used as a weighting factor in subjective judgements in the analysis.

On the basis of these averages, contours of equal α were drawn for each MLT sector. Figures 2 and 3 give such α contours for the noon-midnight meridian for Kp = 0-1 and 2-3, respectively. The regions where the average values of α were available are indicated by the

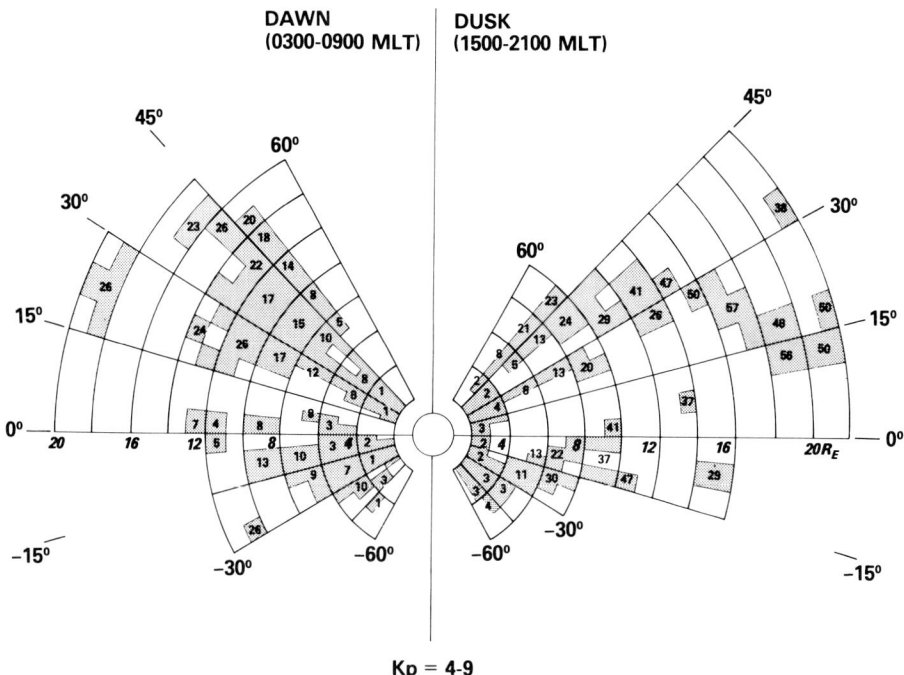

Fig. 7. Average values of α (in degrees) in the dawn-dusk meridian, Kp = 4-9.

broken lines. In both contour maps the dividing line between the northern and southern configurations hangs below the dipole equator on the night side. This is because there were more summer data than winter data in this region. It is remarked that this dividing line appears to begin deviating from the dipole equator at a geocentric distance less than 8 $R_E$, possibly from about 6 $R_E$.

When the contour maps for the two activity levels (which were drawn independently of each other) are superimposed, the night side contours are, on the whole, remarkably similar except in the equatorial region at approximately 8 to 12 $R_E$. This striking agreement shows that on the average the magnetic field directions change relatively little in the vast volume of the night side midlatitude magnetosphere under quiet and moderately disturbed conditions.

The α contours on the day side are less certain than on the night side because of an inadequate observational coverage. Therefore specific shapes of the day side α contours should not be taken seriously. However, a very crude distribution of average α can be inferred from Figures 2 and 3. In these figures a model magnetopause is drawn in only to indicate the overall shape of the magnetosphere relative to the contours; in no way it reflects the actual average position of the magnetopause at the times for which α was determined.

For the severely disturbed condition, Kp = 4-9, the data coverage was not enough to draw contours reliably. Therefore, values of α themselves are given in Figure 4. In the figure the area where data were available are shaded. On the basis of the distribution of α in

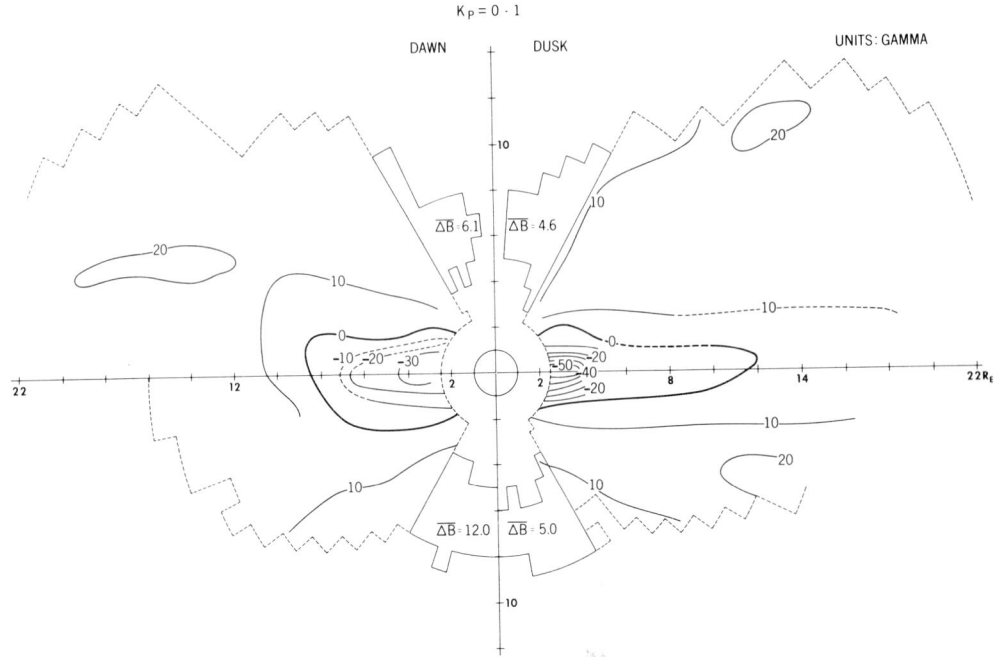

Fig. 8. ΔB contours in the dawn-dusk meridian, showing asymmetry (after Sugiura and Poros, 1973).

Figure 4, compared with those in Figures 2 and 3, it is not possible to characterize in general the changes in α from low activity level to high activity level. However, large angular deviations in the equatorial region at distances roughly 8 to 12 $R_E$ are again indicated. Since the magnetic field is expected to vary rapidly during a substorm, the classification of activity level by Kp would of course not be appropriate to derive substorm processes. However, the intent of this paper is not to investigate magnetic field processes in a substorm but to describe the average magnetospheric field distribution for different disturbance levels that are here specified by a parameter Kp.

On the front side in Figure 4, large α values than for lower Kp are seen beyond 5 $R_E$ and at latitudes above $45°$, but the limited observational coverage prevents us from knowing how α varies at higher latitudes. Relatively small values at large distances ($\gtrsim$ 8 $R_E$) and below $30°$ latitude, compared with corresponding values in Figures 2 and 3, are of interest and require a more detailed study for clarification.

Dawn-Dusk Meridian

For the dawn and dusk sectors the data coverage is less than that for the midnight sector (Table 1) and is not adequate to draw contours with confidence. Therefore we present average values of α for the three Kp groups in Figures 5, 6 and 7. As in Figure 4, the areas where data were available are indicated by shades. The general trend observed in these figures is that α becomes progressively greater, and large α values appear at closer distances, with increasing degree of disturbance.

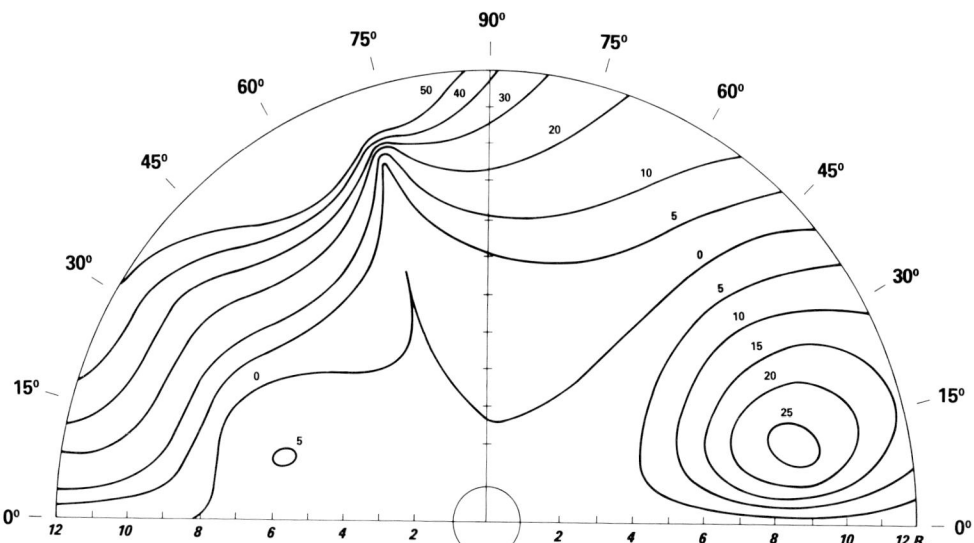

Fig. 9. α contours in the noon-midnight meridian, calculated from the Olson-Pfitzer (1974) model; α in degrees.

It is interesting to note that there is a considerable dawn-dusk asymmetry in the field distortions as expressed by α, namely that the field is more distorted on the dusk side than on the dawn side. An indication of such an asymmetry was found in our earlier analysis of ΔB, shown in Figure 8 which is based on the Ogo-3 and 5 Rb magnetometer observations (Sugiura and Poros, 1973).

### α Calculated from Models

As has been mentioned earlier the angle α can be calculated from any theoretical magnetospheric field model that includes the earth's internal field which is normally approximated by a dipole field. The primary purpose of the present paper is to present a technique of comparing theoretical models with observations on a quantitative basis and not to evaluate the existing models by this method. Therefore our emphasis is placed on the main topological features of α rather than on the details of the α distribution. It is recognized that averaging over $90°$ in MLT inevitably smoothes out various features. However, the limited extent of data gave us no other choice than taking such averages at this phase of investigation. The smoothing effect should be kept in mind in comparing the α distribution determined in this paper from the Ogo-5 observations with the α contours calculated for specific meridians from theoretical models.

Figure 9 shows α contours in the noon-midnight meridian calculated from the model of Olson and Pfitzer (1974). Since this model has no tilt, α is zero at the equator. This should be kept in mind in comparing Figure 9 with the α contours in Figures 2 and 3 or the α values in Figure 4, which are derived from the Ogo-5 observations. (Note the

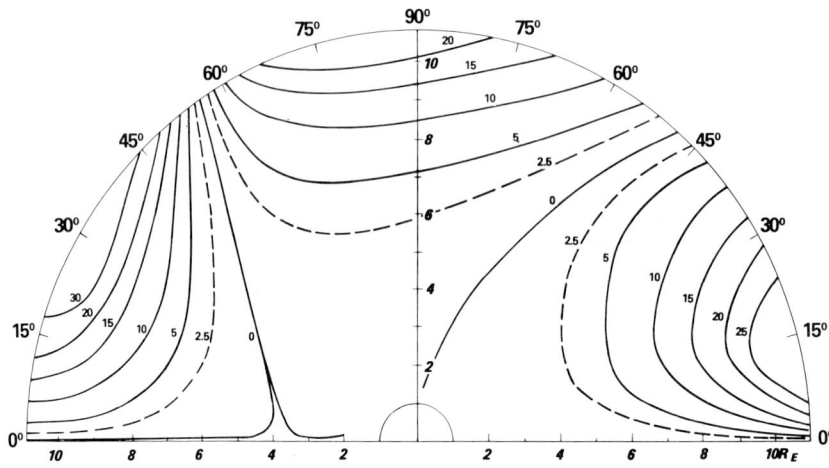

Fig. 10. α contours in the noon-midnight meridian, calculated from the Mead-Fairfield (1975) quiet model; α in degrees.

difference in scale; while Figure 9 extends to only 12 $R_E$, Figures 2, 3 and 4 give data to about 22 $R_E$. However, in making the following comparison, transparent graphs were prepared on the same scale and superimposed on each other.)

So far as the overall magnitudes of α are concerned there is a general agreement between the model and the observation. In the midnight meridian the contours for α = 5° and 10° in the Olson-Pfitzer model and the corresponding contours from the Ogo-5 observations for Kp = 0-1 agree almost perfectly between latitudes 5° and 25°. The most conspicuous difference is the existence of a focus in the midnight meridian at about 8.5 $R_E$ and 15° latitude in the model, while the observational data indicate that if there is such a focus it must be near or beyond 20 $R_E$.

The α contours in the polar cusp region have a configuration peculiar to that region. This characteristic could potentially be of value in comparing models with observations when high-latitude α contours are completed with data taken by spacecraft such as Imp's and Heos. Since the model is symmetric with respect to the noon-midnight meridian, α is the magnitude of the difference in inclination, I, between the model field and the internal dipole field when I is measured from the same horizontal direction for both fields.

As an example demonstrating the usefulness of the α contours we show, in Figure 10, α contours calculated from the model (tilt = 0°; Quiet) of Mead and Fairfield (1975). Since this model is based on the observational data from Imp satellites, the shapes of the α contours in the midnight meridian (Figure 10) are more similar to the Ogo-5 results (Figures 2 and 3) than to those in the Olson-Pfitzer model (Figure 9). There are substantial differences in the α contours between the two models in the polar cusp region.

Figure 11 gives α contours calculated from the Olson-Pfitzer model for the dawn-dusk meridian. Comparing these with the observed results

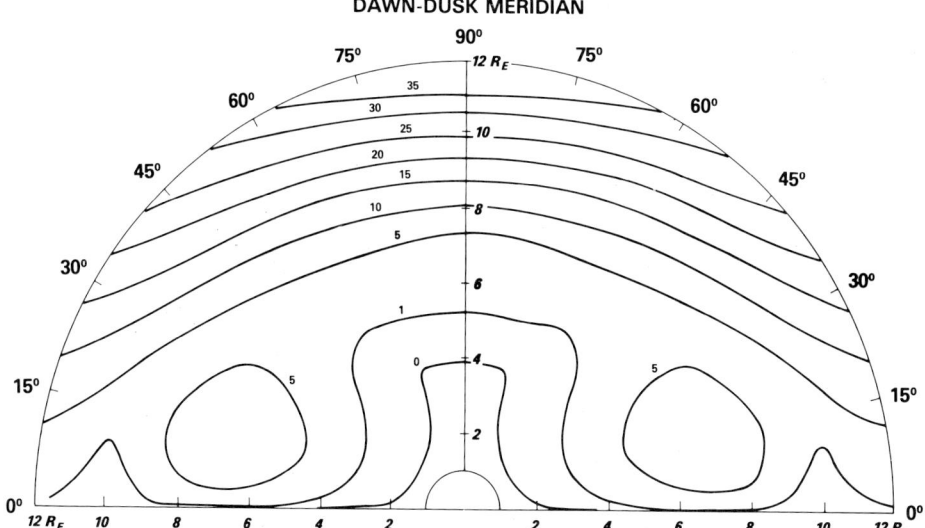

Fig. 11. α contours in the dawn-dusk meridian, calculated from the Olson-Pfitzer (1974) model; α in degrees.

(Figures 5, 6 and 7), the field distortions as represented by α are found to be greater in the observed field than in the model. The transition from the α distribution in the noon-midnight meridian to that in the dawn-dusk meridian is represented in Figure 12 for intermediate magnetic local times.

## Additional Remarks

### α and ΔI

In introducing the angular parameter α we gave reasons for preferring α over ΔI; in particular, we mentioned ambiguities in defining ΔI. This point is now discussed more fully because it is a problem of practical importance. For simplicity of argument, we take a magnetospheric field model which is symmetric with respect to the noon-midnight meridian plane, and we limit our discussion to this plane. If inclination, I, is calculated for the model field according to the standard definition of I (i.e., positively downward from the horizontal plane), this angle is not necessarily measured in the same quadrant as for a dipole field representing the earth's internal field. Therefore if the difference is taken between the I for the model and that for the dipole field, the magnitude of this difference is not necessarily the same as the angle α between the two field vectors. An example in which this discrepancy occurs is shown in Figure 13 which gives ΔI, calculated strictly in accordance with the usual definition of I, for the same Mead-Fairfield model as that used in deriving the α contours in Figure 10. The contours are grossly different in these two figures in the high latitude region between the midlatitude 0° contour on the front side and the vertical line through the north pole.

By measuring the inclination of the dipole field vector from the same horizontal direction as for the model field vector the magnitude

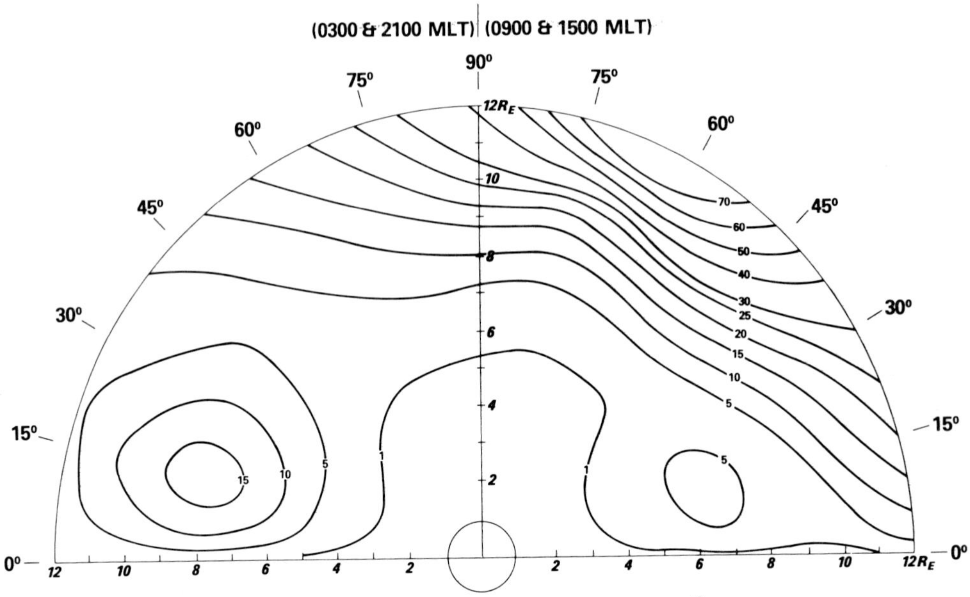

Fig. 12. α contours in intermediate meridians, 45° from the noon-midnight meridian, calculated from the Olson-Pfitzer (1974) model; α in degrees.

of $\Delta I$ can be made equal to $\alpha$. However, this revision of I for calculating $\Delta I$ is not recommended, because such a practice is more likely to introduce a confusion than removing the deficiency. For instance, if the vertical plane containing the field vector of a model does not coincide with the vertical plane containing the internal field (which need not be a dipole field) one has to revert to the conventional definition of declination for each vector separately. Therefore the above revision in the definition of I is not so meaningful as it might appear at first sight.

Use of Kp for activity classification.

In the present analysis, as in the previous work with $\Delta B$, we used the Kp index as a measure for magnetic activity level, and gave average $\alpha$ distributions for three levels of activity. However, Kp is too coarse in time resolution and also, being essentially an index representing high-latitude magnetic activity near the surface of the earth, Kp may not necessarily be a good index to use for classifying conditions of the magnetospheric field. Since the ring current is a major source of magnetic field distortion in the inner magnetosphere, the Dst index would be one of the indices that should be tried in the future. In describing the field distortions during substorms, specification of the substorm phase as well as the intensity may be necessary.

Dawn-dusk asymmetry.

The possibility exists that the dawn-dusk asymmetry discussed above is caused by the solar wind flow direction. To test this possibility the MLT quadrants were rotated by various angles up to 10°, and the

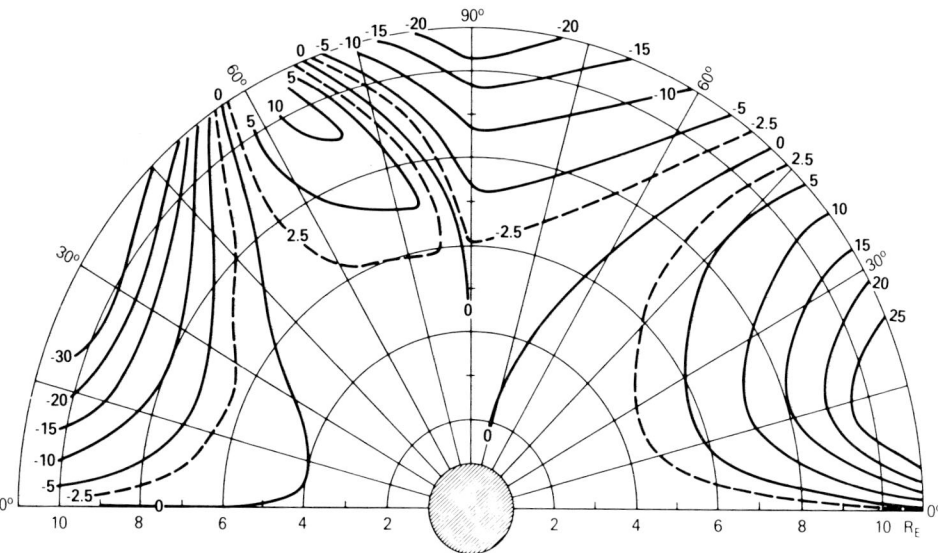

Fig. 13. ΔI contours in the noon-midnight meridian, calculated from the Mead-Fairfield (1975) quiet model; ΔI in degrees. Inclination, I, is calculated using the conventional definition of I both for the model and the internal (dipole) field. Compare with α contours for the same model in Figure 10.

same statistical averages as are presented in this paper were obtained in the newly defined quadrants. However, because of the inadequate coverage of data no definitive judgement could be made on the above question. But our somewhat subjective conclusion is that the dawn-dusk asymmetry found in this analysis is not a purely geometric feature resulting from the solar wind plasma flow direction.

In the Mead-Fairfield (1975) model the polynomial expansions used to represent the magnetic field contained a condition demanding a symmetry with respect to a plane which could be the noon-midnight meridian or any plane obtained by rotating it. Mead and Fairfield calculated the residuals in the least squares determination of the coefficients for various rotation angles. In this manner they found a minimum of the residuals at a rotation angle of from $3°$ to $5°$ depending on the data set. This rotation angle is much smaller than the values of $10°$ to $15°$ given by Cummings et al. (1971) based on their analysis of ATS-1 magnetometer data. The rotation angle for maximum symmetry determined by Hundhausen et al. (1969) from plasma flow directions observed by the Vela satellites in the magnetosheath was $8° \pm 1°$, and the value obtained by Roederer (1969) from the magnetopause positions detected also by the Vela satellites was $7°$. The dawn-dusk asymmetry found in α and also in ΔB appears to be related to the basic structure of the magnetosphere, and therefore it is of different nature from the offset of the plane of maximum symmetry relative to the noon-midnight meridian that has been discussed by other authors.

## Conclusion

The parameter, $\alpha$, defined as the angle between the observed magnetic field and the earth's internal field, is a useful parameter to describe the field distortions in the magnetosphere. Contours of equal $\alpha$ calculated for theoretical magnetospheric field models are also found to be useful in evaluating these models. It is recommended that both $\Delta B$ and $\alpha$ be used in evaluating and improving magnetospheric field models. The parameter $\alpha$ is found to be especially sensitive in the night side equatorial region and in the polar cusp region on the day side.

From the present analysis of $\alpha$ contours derived from the Ogo-5 observations the following aspects are considered to be of particular importance. First, while the gross features of the statistical average distribution of $\alpha$ remains, on the whole, relatively unchanged under disturbed conditions, localized large changes in $\alpha$ in the night side equatorial region at geocentric distances 8 to 12 $R_E$ noticeably stand out. This subject is being investigated relative to substorm variations in greater detail than discussed above. Secondly, the dawn-dusk asymmetry found in $\alpha$ (and previously in $\Delta B$) is likely to be a real asymmetry in the structure of the magnetosphere rather than a geometric feature related to the direction of the solar wind plasma flow.

Acknowledgements. We wish to thank Dr. B. G. Ledley for his assistance in the data analysis, Drs. W. P. Olson and K. A. Pfitzer for simullating and valuable discussions, and Dr. D. H. Fairfield for reading the manuscript and making helpful comments.

## References

Alekseev, I. I., and V. P. Shabansky, A model of a magnetic field in the geomagnetosphere, Planet. Space Sci., 20, 117, 1972.

Chapman, S., and J. Bartels, Geomagnetism, p. 2, Oxford University Press, London, 1962 (corrected edition).

Cummings, W. D., P. J. Coleman, Jr., and G. L. Siscoe, Quiet day magnetic field at ATS 1, J. Geophys. Res., 76, 926, 1971.

Fairfield, D. H., Average magnetic field configuration of the outer magnetosphere, J. Geophys. Res., 73, 7329, 1968.

Fairfield, D. H., Average and unusual locations of the earth's magnetopause and bow shock, J. Geophys. Res., 76, 6700, 1971.

Hedgecock, P. C., and B. T. Thomas, HEOS observations of the configuration of the magnetosphere, Geophys. J. R. astr. Soc., 41, 391, 1975.

Heppner, J. P., M. Sugiura, T. L. Skillman, B. G. Ledley, and M. Campbell, OGO-A magnetic field observations, J. Geophys. Res., 74, 5417, 1967.

Hundhausen, A. J., S. J. Bame, and J. R. Asbridge, Plasma flow pattern in the earth's magnetosheath, J. Geophys. Res., 74, 2799, 1969.

Ledley, B. G., Magnetometers for space measurements over a wide range of field intensities, Rev. Phys. appl., 5, 164, 1970.

Mead, G. D., and D. H. Fairfield, A quantitative magnetospheric model derived from spacecraft magnetometer data, J. Geophys. Res., 80, 523, 1975.

Olson, W. P., and K. A. Pfitzer, A quantitative model of the magnetospheric magnetic field, J. Geophys. Res., 79, 3739, 1974.

Roederer, J. G., Quantitative models of the magnetosphere, Rev. Geophys. Space Phys., 7, 77, 1969.

Sugiura, M., Equatorial current sheet in the magnetosphere, J. Geophys. Res., 77, 6093, 1972a.

Sugiura, M., The ring current, in Critical Problems of Magnetospheric Physics, edited by E. R. Dyer, pp. 195-210, Inter-Union Commission on Solar-Terrestrial Physics, c/o National Academy of Sciences, Washington, DC 1972b.

Sugiura, M., and D. J. Poros, A magnetospheric field model incorporating the OGO-3 and 5 magnetic field observations, Planet. Space Sci., 21, 1763, 1973.

Walker, R. J., An evaluation of recent quantitative magnetospheric magnetic field models, Rev. Geophys. Space Phys., 14, 411, 1976.

MAGNETIC FIELD DISTORTIONS AT L = 3-4 INFERRED FROM SIMULTANEOUS LOW
AND HIGH ALTITUDE OBSERVATIONS OF RING CURRENT IONS

Lawrence R. Lyons

Space Environment Laboratory, NOAA, ERL,
Boulder, Colorado 80303

Abstract. Simultaneous measurements of trapped particles near the equator and of precipitating particles at low altitudes can be used to map geomagnetic field lines from low altitudes to the equator when the particles are being pitch angle diffused into the loss cone by resonant wave-particle interactions. Using such measurements obtained during a storm recovery phase near local midnight, it is suggested that magnetic field lines labelled L = 3.0 at low altitudes by a magnetic field model were extended so as to map to field lines labeled L = 3.7 near the equator. If this suggested mapping is correct, simultaneous magnetic field measurements near the equator imply that a longitude slice at the surface mapped along field lines to a significantly smaller (e.g., 16% at L = 3 using an arbitrary model) longitude slice in the equatorial plane.

## Introduction

Satellite measurements of radiation belt particles obtained simultaneously at low altitudes and near the equator can at times be used to map geomagnetic field lines from low altitudes to the equatorial plane. This paper describes a technique for such mapping and an example where the technique has been partially applied to magnetic field lines near L = 3 to 4 during the recovery phase of a storm. The application yielded unexpected results which have yet to be tested by an independent analysis.

## Technique for Field-Line Mapping

The technique for mapping field lines from the equator to low altitudes utilizes the fact that plasma turbulence can pitch angle scatter trapped particles into the loss cone so that trapped particles are precipitated along field lines into the atmosphere (See Lyons, 1976 for a more detailed discussion than is presented here). Equatorial measurements can be used for identifying the characteristics of the precipitated particle flux expected from identified wave-particle interactions as a function of radial distance, and simultaneous low altitude observations as a function of latitude can directly measure the precipitated particle fluxes. By comparing the results from the two sets of measurements, the foot of the field lines at low altitudes as a function of latitude can be directly mapped to the equatorial plane as a function of radial distance.

For a spectrum of waves, with wave energy disturbed over a band of frequencies and a range of latitudes, cyclotron resonant interactions occur for all equatorial parallel particle energies $E_{||}$ greater than a minimum value $E_{||,min}$. This minimum interacting energy is determined by the wave distribution and the wave dispersion relation. In examining particle data obtained near the equator for the effects of resonant wave-particle energies, one should look for the effects of diffusion at values of $E_{||}$ greater than a minimum $E_{||,min}$. Examples where measured equatorial particle distributions clearly show the effect of cyclotron resonant diffusion occuring for $E_{||} \gtrsim E_{||,min}$ are shown by in Lyons (1974).

Since pitch-angle diffusion at a given L-shell only occurs for parallel energies $E_{||} \gtrsim E_{||,min}$, diffusion into the loss cone and precipitation to the atmosphere should occur only for energies $E \gtrsim E_{||,min}$. $E_{||,min}$ generally varies as a function of radial distance because of the radial dependence of the geomagnetic field, so that mapping of field lines is possible from the identification of $E_{||,min}$ versus radial distance near the equator and the identification of $E_{||,min}$ versus latitude from precipitation measurement of low altitudes.

The degree of simultaniety required in longitude and time between the two sets of particle measurements must be considered. It is necessary that the distribution of wave energy with frequency not change appreciably between the times and longitudes of the low-and high-altitude measurement, since it is necessary that $E_{||,min}$ be the same. However, the wave amplitude need not be the same in order to map $E_{||,min}$ from low to high altitudes. Since exact simultaneity of the measurement in time and longitude is unlikely, this mapping of the effects of wave-particle interactions should be most successful for widespread phenomena that do not vary rapidly with time.

Sufficiently good equatorial particle measurements to determine $E_{||,min}$ are only available from within 5.2 $R_e$, and identified wave-particle interactions within 5.2 $R_e$ are not sufficiently strong to drive strong pitch angle diffusion. Thus the precipitating particle fluxes are much less than the trapped particle fluxes, and unfortunately sufficiently sensitive low-altitude measurements of the low-level, precipitating particle fluxes as a function of particle energy and latitude are difficult to obtain. Thus field line mapping as described above has yet to be fully attempted. However, aspects of the above technique have been applied in one situation as described in the next section. In this case, we are aided by species identification at low altitude and species inferences at high altitudes.

## FIELD LINE MAPPING

### 1. *High Altitude Observations*

Williams and Lyons (1974) presented equatorial pitch angle distribution of 1 to 400 keV ring current ions obtained by Explorer 45 during the recovery phase of the large storm on Dec. 17, 1971 at radial distances of 2.6 to 5.2 $R_e$. An example of the pitch angle distribution for selected radial distances and energy channels for Orbit 103 on Dec. 18, approximately 16 hours after the minimum Dst of the storm main phase, is shown in Figure 1. At each radial distance, the pitch angle distributions are nearly isotropic at the lower energies. A transition to anisotropic distributions peaked at 90° pitch angle is observed at the higher energies, with the transition energy increasing with radial distance. The isotropic distributions show empty loss cones indicating a stably-trapped, isotropic particle distribution undergoing negligible losses from pitch angle diffusion. As the storm recovery progressed, distributions which were initially isotropic evolved to rounded distributions in such a manner that the transition energy decreased with time. This rounding of the pitch angle distributions was interpreted to result from pitch angle diffusion driven by resonant interactions with ion cyclotron waves.

Joselyn and Lyons (1976) analyzed the pitch angle distribution for effects of diffusion occurring for all $E_{||}$ greater than a minimum value. For each radial distance, the energy of the highest Explorer 45 energy channel showing a nearly isotropic pitch angle distribution was chosen to be $E_{||,min}$, and the pitch angles corresponding to this chosen value of $E_{||}$ were calculated for all higher energy channels. These pitch angles have been indicated by vertical ticks on the pitch angle distribution in Figure 1.

Notice that to within the accuracy of the pitch angle measurements, the pitch angle distributions are nearly isotropic between the ticks, i.e. for $E_{||} < E_{||,min}$. At larger $E_{||}$ (pitch angles approaching 0° and 180°), the pitch angle distributions are rounded. Such a separation between rounded and isotropic distribution was found for all cases analyzed by Joselyn and Lyons (1976), and we concluded that the chosen values for $E_{||,min}$ represented the minimum parallel particle energies that interacted with ion cyclotron waves for each L-value and time examined.

Lyons and Evans (1976) questioned how the lower energy ion distributions could remain isotropic, when charge exchange with neutral hydrogen of the earth's geocorona should cause the loss rates to increase markedly for equatorial pitch angles increasing or decreasing from 90°.

Figures 2 and 3 compare the observed pitch angle distributions at L = 3.5 and 4.0 with those expected to evolve from proton charge exchange with neutral hydrogen. Observations are shown from the inbound portion of orbits 101-106, the observations from orbit 101 being less than one hour following the minimum of the Dst index. The evolving pitch angle distributions from charge exchange were obtained by neglecting possible sources and assuming an isotropic proton distribution at the time of the orbit 101 observation, which was taken to be t = 0. Fluxes at subsequent times were obtained using the charge exchange lifetimes given by Tinsley (1976).

At both L values the charge exchange calculations predict that the pitch angle distribution for 2 and 10 keV protons will become greatly

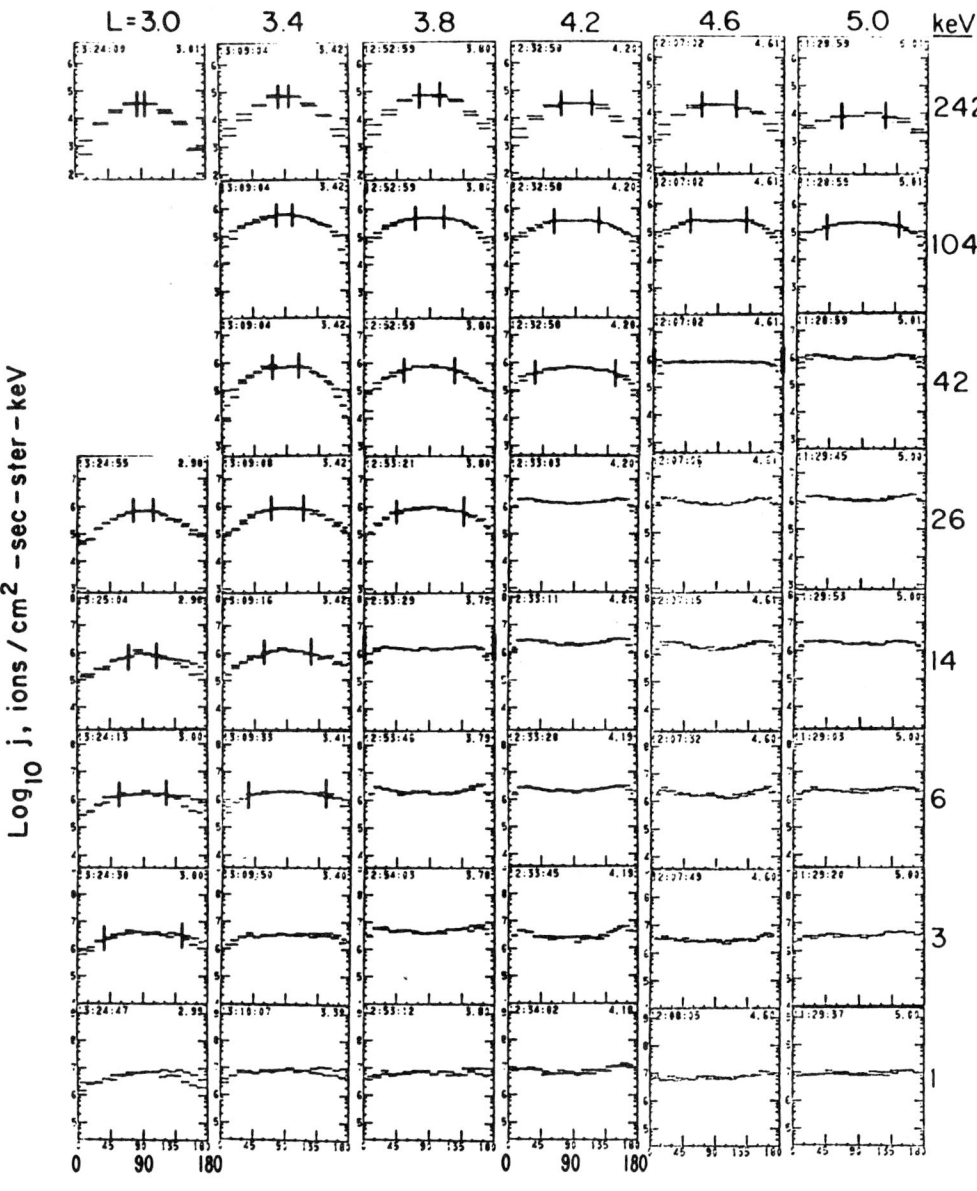

Fig. 1

anistropic. However, this dramatic rounding due to the loss of particles by charge exchange is not observed. In addition, charge exchange predicts that even the 90° pitch angle fluxes should decay much more rapidly than is observed, this discrepancy becoming particularly dramatic at lower L-values (see Lyons and Evans, 1976).

Lyons and Evans argue that the most likely explanation for the disagreement between the charge exchange predictions and the observations is that the ring current at particle energies $\lesssim$ 50 keV and radial distances $\lesssim$ 4 $R_e$ was dominated by some ion species other than protons during the recovery phase of the storm. Such ions must have much longer lifetimes for charge exchange with hydrogen than do protons. $O^+$ is a possibility, but the charge exchange lifetimes for $O^+$ are still somewhat shorter than what is necessary to account for the observation. The charge exchange lifetimes for $He^+$ are significantly longer than for $O^+$ at energies between 1 and 50 keV, so $He^+$ is an attractive candidate. Arguments that $He^+$ is dominant, based on a low altitude equatorial optical observations, are given by Tinsley (1976).

## 2. Low Altitude Observations

Precipitation of ring current ions due to the ion-cyclotron-wave interaction is expected to be of weak to moderate intensity and to occur from the outer region of the plasmasphere. No evidence of strong pitch angle diffusion was found in the Explorer 45 observations during the storm recovery phase, so that the strong auroral precipitation of ions known to exist at low altitudes must originate from outside the Explorer 45 apogee of 5.2 $R_e$. This led Williams and Lyons (1976) to predict a two-zone structure for precipitating ions during a storm recovery phase, consisting of an inner zone of weak to moderate precipitation from the outer region of the plasmaphere and an outer zone of strong precipitation outside the plasmasphere. Hultqvist (1975) has observed such a two-zone ion precipitation structure at 1 keV and 6 keV, and such a structure has been followed throughout a storm recovery phase for > 115 keV ions (Hauge and Søraas, 1975).

Sharp et al. (1976a, b) presented observations of precipitating ions obtained by an energetic ion mass spectrometer on the low-altitude polar-orbiting satellite 1971-089A during the recovery phase of the same storm discussed above. Throughout the day of Dec. 18, the first day of the storm recovery phase, measurable precipitating ion fluxes

---

Fig. 1   Equatorial proton pitch angle distributions observed on Explorer 45 orbit 103 inbound, $\sim$ 16 hours after the minimum Dst of the December 17, 1971, storm main phase. Distributions are shown every 0.4 in L from L = 3 to L = 5, and selected proton energy channels are stacked vertically at each L. Elevated fluxes at pitch angles of 90° - 180° for energies $\lesssim$ 14 keV are due to reflected sunlight. Ticks are at constant values of $E_{||}$ for each L, with the chosen value of $E_{||}$ at each L being equal to the energy of the highest Explorer 45 channel showing a nearly isotropic pitch angle distribution.

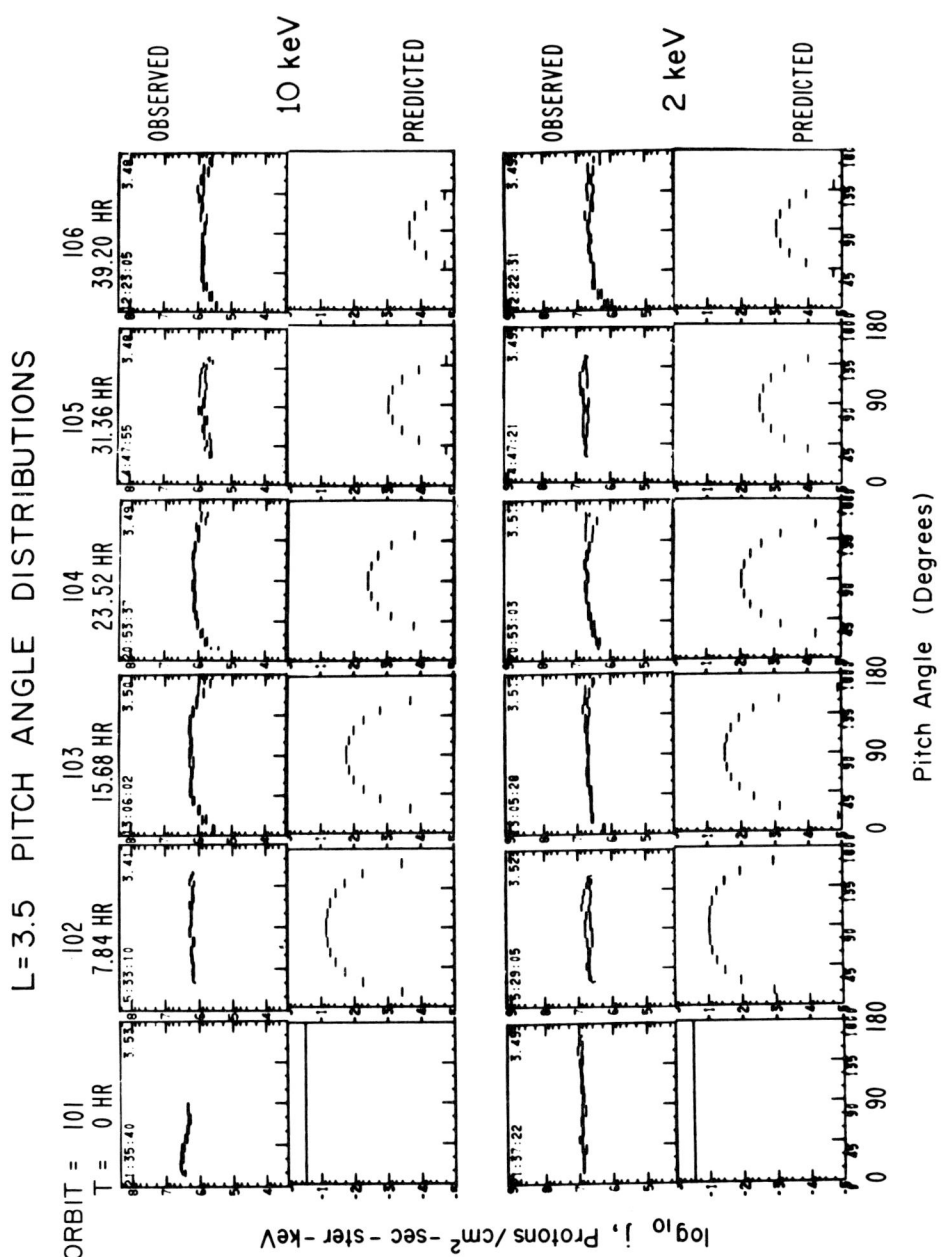

Fig. 2

were observed near 0300 LT predominantly in the outer precipitation zone at invariant latitudes $\stackrel{\sim}{<} 60°$ N (see the contour plots, Figures 7 and 9, of Sharp et al. [1976b]). A likely explanation for why they saw only one precipitation zone is that fluxes in the weaker, inner precipitation zone were below the sensitivity of their instrument (R. D. Sharp, private communication, 1976). However, that the strong precipitation zone was apparently detected down to 60° latitude (which corresponds to $L \stackrel{\sim}{\sim} 4$ in magnetic field models) while no evidence for the strong precipitation zone in the vicinity of local midnight was evident in the equatorial observations within 5.2 $R_e$, suggests that magnetic field lines near 60° latitude mapped to equatorial radial distances > 5.2 $R_e$ at the time of the observations.

Sharp et al. (1977) examined in detail the limited detectable ion fluxes near 0300 LT at latitudes below 60°. Detectable fluxes were found on four consecutive satellite passes from $\sim$ 0830 to 1330 UT on Dec. 18 within a narrow latitude range centered near 55° invariant latitude ($L \stackrel{\sim}{\sim} 3$ in magnetic field models) during a period of enhanced magnetic activity. This region may be a signature of the weaker, inner precipitation zone. Sharp et al. (1977) found barely detectable $H^+$ and $O^+$ fluxes at 8 and 12 keV in this region, but they found no detectable $He^+$. There is no reason for pitch angle diffusion driven by ion-cyclotron waves to strongly favor $H^+$ and $O^+$ over $He^+$, so that we can conclude that at least during the period of observable precipitation the ion population was not dominated by $He^+$ on those flux tubes from which the $H^+$ and $O^+$ were precipitating. In addition, by comparing the precipitation flux with the equatorial observations, Sharp, et al. concluded that a significant source of $H^+$ and $O^+$ was needed to account for the observed precipitating fluxes.

The existence of such an ion source to L = 3 would cast doubt on the conclusion of Lyons and Evans (1976), though a short-lived source during the period of observable precipitation would not by itself invalidate their conclusions.

## 3. *Field Line Mapping*

The last of the sequence of four orbits with observable, low-latitude precipitation was well coordinated with the inbound portion of

---

Fig. 2   Equatorial pitch angle distributions at L = 3.5 observed during the recovery phase of the December 17, 1971, storm are compared with those expected to evolve from porton charge exchange with neutral hydrogen. Observations are shown from the inbound portion of orbits 101-106, the observations from orbit 101 being less than 1 hour following the minimum of Dst. Elevated fluxes over the pitch angle range of 90°-180° resulted from reflected sunlight. The pitch angle distributions expected to evolve from charge exchange were obtained by neglecting possible sources, assuming an isotropic proton pitch angle distribution at the time of the orbit 101 observations (T = 0), and using the charge exchange lifetimes given by Tinsley, (1976). The initial fluxes for the calculations were arbitrarily normalized to approximately $3 \times 10^{-1}$.

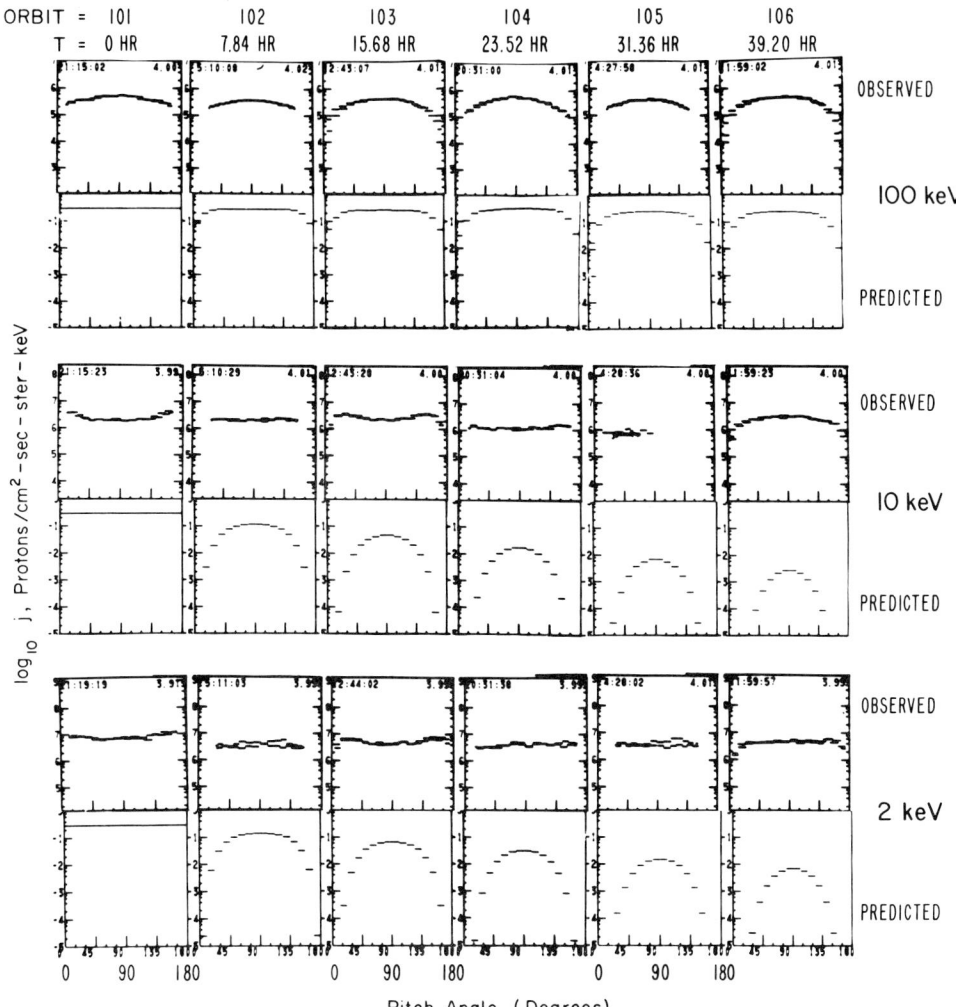

Fig. 3  Same as Figure 2, except for L = 4.0 and the inclusion of 100 keV ion measurements and charge exchange predictions.

Explorer 45 orbit 103. Explorer 45 crossed L = 3.2 nine minutes earlier in UT and ∼ two hours earlier in LT than 1971-0897. On this orbit, the precipitation ions at 55° latitude were observed at 12 keV. However, the analysis of Joselyn and Lyons (1976) (See Figure 1) shows that the minimum energy for pitch angle diffusion into the loss cone $E_{||,min}$ = 2 keV at L = 3.0, 9 keV at L = 3.6, and 14 keV at L = 3.8. In addition, the pitch angle diffusion coefficient for ions resonant with ion cyclotron waves varies as $B_w^2/E_{||}^{1/2}$ (Lyons and Thorne, 1972), where $B_w$ is the resonant wave magnetic field amplitude, so that pitch angle diffusion losses should maximize at energies near the lowest energy subject to such losses. Thus the low-altitude measurements at 55° latitude give $E_{||,min} \sim 12$ keV which maps to L = 3.7 in the equatorial plane.

This field line tracing yields the result: near local midnight during the period of observable precipitation, the magnetic field lines labeled L = 3 at low altitudes (L defined by using Goddard Space Flight Center magnetic field model 1266 with time coefficients for Dec. 1971) are extended so as to map to field lines labeled L $\sim$ 3.7 near the equatorial plane (L defined by using the POGO (8,69), epoch 1969.0 model; differences between the two magnetic field models should be negligible for L $\lesssim$ 4, and use of L throughout this paper refers to the values assigned by the models). If this field line tracing is correct, then the low altitude ion observations do not contradict the inferences of the equatorial trapped ion concentration of Lyons and Evans (1976) based on the equatorial observation.

This marked field line distortion is supported by two additional features of the Explorer 45 and observations:

1. The 2 and 10 keV equatorial pitch angle distributions (See Figures 2 and 3, and Lyons and Evans (1976)) for six consecutive Explorer 45 orbits (8 hours apart) during the storm recovery phase show perceptible flux increases for orbit 103 at L = 4 (12:43 UT) over the previous orbit but not at L = 3.5 (13:00 UT) or L = 3.0 (13:25 UT). Thus a high altitude ion source, as Sharp et al. (1977) inferred was necessary at the time of the observable precipitation, apparently was available at L = 4.0, but not at L $\lesssim$ 3.5.

2. Saturation of the dc electric field probe on Explorer 45 occurs at a plasma density of $\sim 60$ cm$^{-3}$ and is thus an indication of the plasmapause location (Morgan and Maynard, 1976). On orbit 103 inbound this probe saturated at L = 3.8, a few tenths of an earth radii further out than on the previous orbit. Direct injection to near L = 3.0 would be expected to be associated with displacement of the plasmapause inward towards L = 3.0 rather than with an expansion of the plasmasphere.

The above arguments are not beyond criticism. Direct injection to L = 3.0 could have occurred without being discernible in the equatorial pitch angle distributions, and our knowledge of plasmapause dynamics during magnetic activity is quite limited. In addition, a distribution of wave energy as a function of frequency could be devised that would precipitate 8- and 12-keV ions at detectable rates but result in lower energy precipitation at rates too small for detection. However, it may be more than a coincidence that all of the arguments are consistent with injection to equatorial L values of $\sim 3.7$, but not below, during the period of detectable low-altitude precipitation near L = 3.0.

Using the magnetic field magnitudes measured by Explorer 45 during orbit 103 inbound, we can relate the reference model L value to the equatorial crossing of a magnetic field line by using the technique described by Fairfield (1968). Integrating $\nabla \cdot B = 0$ over a longitude slice $\Delta\phi$ from the equator to a latitude $\theta_L$ at the earth's surface and mapping this longitude slice along field lines to an area A in the equatorial plane of longitudinal width $\Delta\phi_e(r)$ as a function of radial distance r, we have

$$\int_0^{\Theta_L} \int_{\Delta\phi} B_r(\Theta) \cos\Theta \, d\phi \, d\Theta = \int_1^{R(\Theta_L)} \int_{\Delta\phi_e(r)} B_e \, dA \qquad (1)$$

Here $B_r$ is the radial component of the earth's field at the surface, $B_e$ is the field through the equatorial plane, and $R(\Theta_L)$ gives the equatorial radial distance in earth radii of the field line which intersects the surface at latitude $\Theta_L$. The relation between the longitude slice $\Delta\phi_e(r)$ and $\Delta\phi$ at the surface is a major unknown in this equation.

The left-hand side of (1), which is evaluated at the earth's surface, should remain constant to within 0.1% throughout the period of a storm and should be accurately calculable by using the reference magnetic field. Thus for a given $\Theta_L$ we have

$$\int_1^{L_{ref}(\Theta_L)} B_{e,ref} L_{ref} \, dL_{ref} = \int_1^{L'_{ref}(\Theta_L)} \frac{\Delta\phi_e(r)}{\Delta\phi} B_e L_{ref} \, dL_{ref} = I(\Theta_L) \qquad (2)$$

Here "ref" refers to the reference magnetic field model. $L_{ref}(\Theta_L)$ is the reference model L value of the field line striking the surface at $\Theta_L$, while $L'_{ref}(\Theta_L)$ is the L value given by the reference model at the equatorial crossing of the real field line which strikes the surface at $\Theta_L$. The longitude slice is assumed to be sufficiently thin that $B_e$ does not vary over $\Delta\phi_e$, and we assume that $\Delta\phi_e = \Delta\phi$ in the model field. Thus a field line labeled $L_{ref}$ at low altitudes by the reference model would be labeled $L'_{ref}$ at high altitudes provided $\Delta\phi_e(r)$ and $B_e(r)$ are correctly specified.

Figure 4 shows $I(\Theta_L)$ calculated in three different ways as a function of $L_{ref}(\Theta_L)$. The curve labeled reference model is the left-hand side of (2) calculated by using the reference model magnetic field. The other two curves were calculated by using the Explorer 45 magnetic field measurements for orbit 103 inbound (see Lyons and Williams, 1976), assuming the measurements were at the equator. This assumption introduces some errors, since the satellite was at a reference model latitude of $-6°$. The curve labeled "without area reduction" was obtained by assuming $\Delta\phi_e(r) = \Delta\phi$. By using the horizontal dashed lines as a guide it can be seen that without area reduction the low-altitude $L_{ref} = 3$ maps only to $L_{ref} = 3.1$ in the equatorial plane.

However, $\Delta\phi_e(r)$ need not equal $\Delta\phi$. When a distortion toward a more tail-like magnetic field geometry is assumed, it is more likely that near midnight, $\Delta\phi_e(r) < \Delta\phi$ than that $\Delta\phi_e(r) > \Delta\phi$, as was found by Fairfield (1968) for distances further from the earth for average geomagnetic conditions. From Figure 1 it can be seen that $I(\Theta_L)$ for orbit 103 inbound must be reduced by 29% at $L_{ref} = 3.7$ for the low-altitude $L_{ref} = 3$ field line to map to $L_{ref} = 3.7$ at the equator. The only way that $I(\Theta_L)$ can be reduced is by taking $\Delta\phi_e(r) < \Delta\phi$. The curve labeled

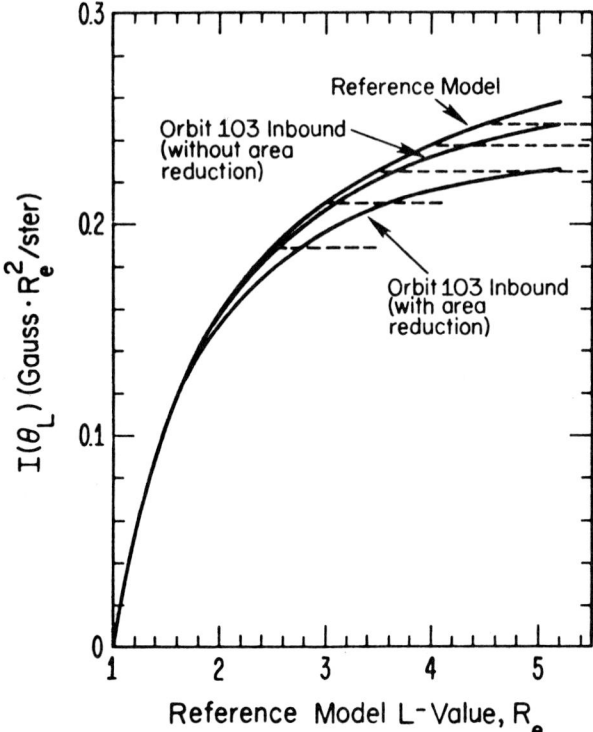

Fig. 4  $I(\Theta_L)$ as a function of the reference magnetic field model L value, $L_{ref}$. The reference model curve was obtained by using the reference magnetic field model, while the orbit 103 inbound curves were obtained by using the magnetic field measured on Explorer 45. The "without area reduction" curve was calculated by assuming that a longitude slice at the surface maps along field lines to the same size longitude slice in the equatorial plane. The curve "with area reduction" assumes that the equatorial longitude slice is smaller than the surface longitude slices as described in the text. The horizontal dashed lines are a guide for mapping every 0.5 in $L_{ref}$ (from 2.5 to 4.5) from low altitudes (by using the reference model curve) to the L value that the reference model would give to the same field line in the equatorial plane.

"with area reduction" maps the low-altitude $L_{ref} = 3$ to $L_{ref} = 3.7$ at the equator. This curve was obtained by assuming an area reduction increasing linearly from 0% at $L_{ref} = 1$ to 32% at $L_{ref} = 5$, so that $\Delta\phi_e(r)/\Delta\phi = 1 - 0.08(L_{ref} - 1)$. This form for $\Delta\phi_e(r)$ is arbitrary, but it was taken as an illustration of the area reduction that is required for the field line mapping suggested here.

Notice that the orbit 103 curve with area reduction maps the low-altitude $L_{ref} = 4$ field line to $L_{ref} > 5.2$ in the equatorial plane.

This may explain why Sharp et al. (1976a, b) see the outer precipitation zone inward to $L_{ref} = 4$, while Williams and Lyons (1974) contend that the outer precipitation zone at the equator was beyond the Explorer 45 apogee of $5.2\ R_E$.

While magnetic field measurements within radial distances of $4\ R_e$ are available (e.g. Sugiura, 1972), it appears that valid vector measurements within $4\ R_e$ are not currently available. Thus the proposal that the magnetic field lines labeled $L = 3$ at the earth's surface near local midnight mapped to $\sim 3.7\ R_E$ in the equatorial plane during a period of enhanced magnetic activity during a storm recovery phase cannot yet be substantiated or rejected by available magnetic field measurements. However, if the proposed mapping is correct, the Explorer 45 magnetic field measurements imply that a longitude slice at the surface mapped along field lines to a significantly smaller (i.e. 16% at $L = 3$ when the arbitrary model presented here is used) longitude slice in the equatorial plane. Also, if the proposed mapping is correct, then the observations of Sharp et al. do not contradict the suggestion of Lyons and Evans that the recovery phase $\lesssim 50$ keV ions were dominated by $He^+$ for $L \lesssim 3.7$, even during the period of observable precipitation.

## References

Fairfield, D. H., Average magnetic field configuration of the outer magnetosphere, J. Geophys. Res., 73, 7329, 1968.

Hauge, R., and F. Søraas, Precipitation of >115 keV protons in the evening and forenoon sectors in relation to the magnetic activity, Planet. Space Sci., 23, 1141, 1975.

Hultqvist, B., The ring current and particle precipitation near the plasmapause, Ann. Geophys., 31, 111, 1975.

Joselyn, J. A., and L. R. Lyons, Ion cyclotron wave growth calculated from observations of the proton ring current during storm recovery, J. Geophys. Res., 81, 2275, 1976.

Lyons, L. R., Trapped particles and waves and what can be learned from multisatellite experiments, The Satellite Programme during the International Magnetospheric Study, ed. by K. Knott and B. Battrick, D. Reidel Publ. Co., Dordrecht-Holland, 237, 1976.

Lyons, L. R. and D. S. Evans, The inconsistency between proton charge exchange and the observed ring current decay, J. Geophys. Res., 81, 1967.

Lyons, L. R., and R. M. Thorne, Parasitic pitch angle diffusion of radiation belt particles by ion cyclotron waves, J. Geophys. Res., 77, 5608, 1972.

Lyons, L. R., and D. J. Williams, Storm associated variations of equatorially mirroring ring current protons, 1-800 keV, at constant first adiabatic invariant, J. Geophys. Res., 81, 216, 1976.

Morgan, M. G. and N. C. Maynard, Evidence of dayside plasmaspheric structure through comparisons of ground-based whistler data and Explorer 45 plasmapause data, J. Geophys. Res., 81, 3992, 1976.

Sharp, R. D., R. G. Johnson, and E. G. Shelley, The morphology of energetic $O^+$ ions during two magnetic storms: temporal variations, J. Geophys. Res., 81, 3283, 1976a.

Sharp, R. D., R. G. Johnson, and E. G. Shelley, The morphology of energetic $O^+$ ions during two magnetic storms: latitudinal variations, J. Geophys. Res., 81, 3292, 1976b.

Sharp, R. D., E. G. Shelley, and R. G. Johnson, A search for helium ions in the recovery phase of a magnetic storm, J. Geophys. Res., 82, 2361, 1977.

Sugiura, M., The ring current and associated phenomena, in Critical Problems of Magnetospheric Physics, ed. by E. R. Dyer, IUCSTP Secretariat, % National Academy of Sciences, Washington, D. C., 195, 1972.

Tinsley, B. A., Evidence that the recovery phase ring current consists of helium ions, J. Geophys. Res., 81, 6193, 1976.

Williams, D. J. and L. R. Lyons, The proton ring current and its interaction with the plasmapause: storm recovery phase, J. Geophys. Res., 79, 4195, 1974.

# MODELING THE MAGNETOSPHERIC MAGNETIC FIELD

W. P. Olson, K. A. Pfitzer, and G. J. Mroz

McDonnell Douglas Astronautics Company
5301 Bolsa Avenue
Huntington Beach, California 92647

Abstract. The magnetic fields resulting from each of the three major magnetospheric current systems have been described quantitatively. This new model of the total magnetospheric magnetic field is defined for perpendicular incidence of the solar wind on the dipole axes. In it the strengths of the three currents systems can be varied separately. It is thus possible to describe the time varying magnetic field associated with dynamic processes in the magnetosphere such as the magnetic storm and the magnetospheric substorm. The magnetic vector potential is also determined and can be used to describe quantitatively the induced electric field associated with time variations in the magnetic field. This model should be useful in the description and study of charged particle motions and their acceleration associated with storm and substorm processes

## Introduction

In another conference paper we have described the vector magnetic potential, $\vec{A}$, the induced electric field, $\vec{E}_I$, produced by the daily variation in the earth's dipole field and the resulting response in the magnetopause ring and tail current systems (Mroz, et al., 1978). In that paper it is pointed out that even this daily change in the magnetic field produces an electric field large enough to be of consequence to the motion and energization of low energy particles in the magnetosphere.

It is of interest to consider the more radical variations in the magnetic field associated with storm and substorm processes. In order to describe such variations it is necessary to represent each of the current systems and their associated magnetic fields separately. In this paper we describe only the changes in magnetic field topology resulting from the variation in the strengths of the three current systems but with their shapes and locations fixed as described by our previous model which is valid only for quiet magnetic conditions. Eventually, we will have the capability of varying not only the strengths but also the location and geometry of each of the current systems.

## Procedure

Our procedure for determining the strengths and locations of the current systems has been described previously (e.g., see Olson and

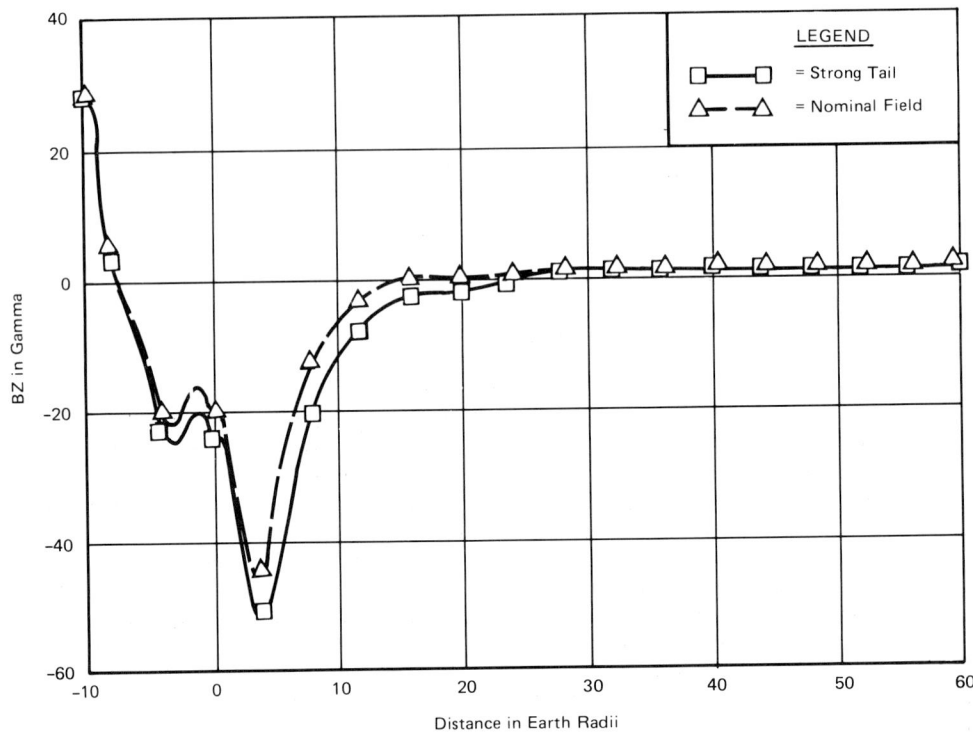

Fig. 1. The combined field of the boundary ring and tail currents along the sun-earth line (with only a $B_Z$ component because of symmetry) is shown for nominal conditions and for a strong tail current system.

Pfitzer, 1974). Here we have integrated over each of the current systems using the Biot-Savart law to determine the resultant vector magnetic field, $\vec{B}$, at a set of points distributed throughout the magnetosphere produced by that current system. This model is valid from the nose to the magnetosphere to lunar orbit in the magnetotail. The coefficients for the terms in a series representation for the magnetic field are then determined using a generalized least squares fitting program. This is done for the three components of $\vec{B}$ for each of the three current systems with terms in the series chosen to produce the best possible fit to the input data.

Along with the integration used to determine $\vec{B}$, we also integrate over the currents to determine the magnetic vector potential, $\vec{A}$. It is then possible (once the variations in the current systems are known) to use the series representations for $\vec{B}$ and $\vec{A}$ to describe the time varying electromagnetic environment throughout the model magnetosphere.

It is appropriate to make some comparison between this "disturbed model" and our quiet time tilt dependent model. In the tilt dependent model the change in the tilt angle with time is important. In order to accurately represent the total magnetic field it is required that the magnetospheric contribution be expressed in solar magnetospheric coordinates and that it be added vectorially to the representation of

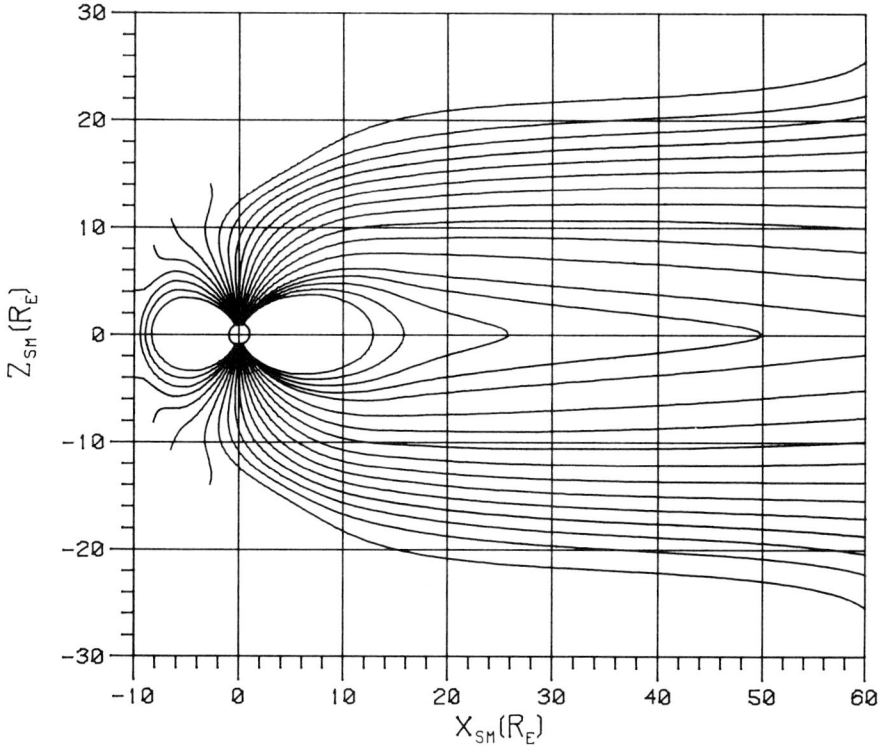

Fig. 2. Nominal field line topology in the noon midnight meridian plane.

the main field as represented in geographic coordinates. Such a procedure requires the use of several coordinate transformations. The tilt dependent model is also complicated in that the topology of the magnetic equatorial plane changes with time. In such a model, equatorial particles (with 90° pitch angles) do not remain on the euqator and it is difficult to study their motion and energization. In the present model the lack of tilt dependence may be considered a shortcoming. However, we feel that the study of magnetospheric dynamics is difficult enough without including this complication. Therefore, we take the rotation and dipole axes to be coparallel and the solar wind direction perpendicular to the dipole. This produces a high degree of symmetry. Thus, in the present model equatorial particles will remain on the equator even though $\vec{B}$ changes with time.

## Results

In this section the changes in magnetic field topology resulting from variation in the strengths of the three major magnetospheric current systems are described. We plan in the future to develop "driver programs" which describe the time history for each of the current systems (including strength, location, and extent) during a storm and substorm process. Eventually, other "drivers" may be used to represent actual observed events.

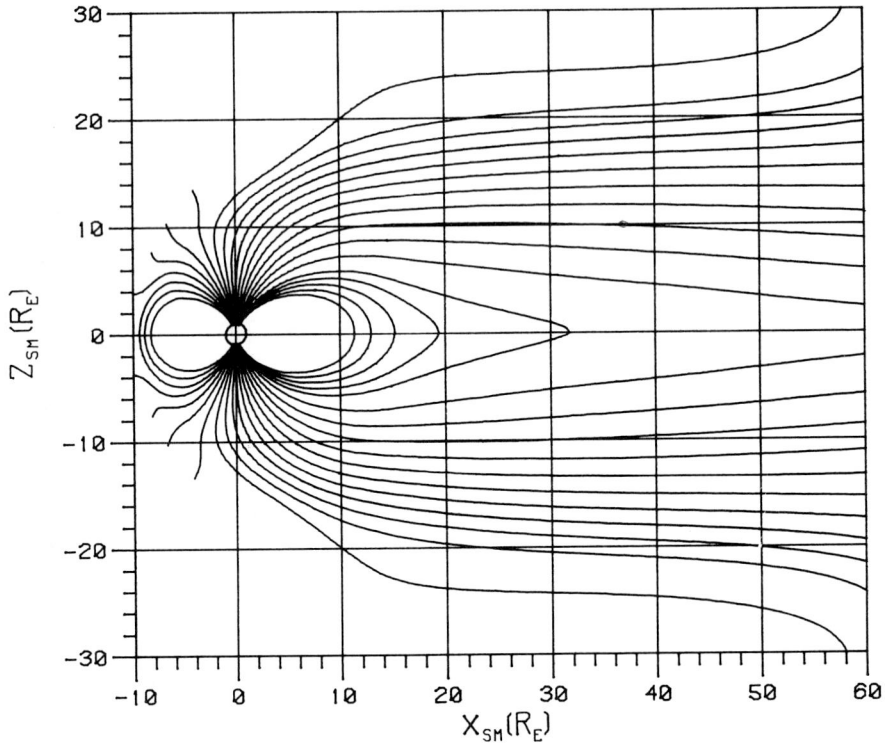

Fig. 3. Total magnetic field with a weak tail field and nominal ring and boundary current strengths.

It is useful in the testing of magnetic field models to compare their outputs with the observed $\Delta B$ contours. $\Delta B$ is the magnitude of the contribution of the magnetospheric field to the total observed field. It is typically determined by subtracting a model representation of the main field magnitude from the observed total field magnitude. By symmetry in the present model $\Delta B$ along the sun-earth line is the same as the $B_Z$ component of the field (in the coordinates used, the X axis points toward the sun, the Z axis is coparallel with the dipole and rotation axes and points north, while the Y axis is directed into the dusk meridian plane). Sugiura (1973) has published observed $\Delta B$ contours. However, his data are averaged over all tilt angles. We have found that this is not the same as determining the zero-tilt value separately from the current system. With this caution in mind, a comparison is made between the observed $\Delta B$ values along the sun-earth line and those obtained from the model for various strengths of the three magnetospheric current systems. Generally all of these $\Delta B$ plots share the following features: the value of $B_Z$ is largest as the nose of the magnetosphere is approached. This is because of the proximity to the magnetopause currents flowing in that region which adds to the existing field. The largest depression in the field occurs in the vicinity of the ring current and exhibits minima at about $\pm 4 R_E$ with the largest on the night side. In the tail of the magnetosphere,

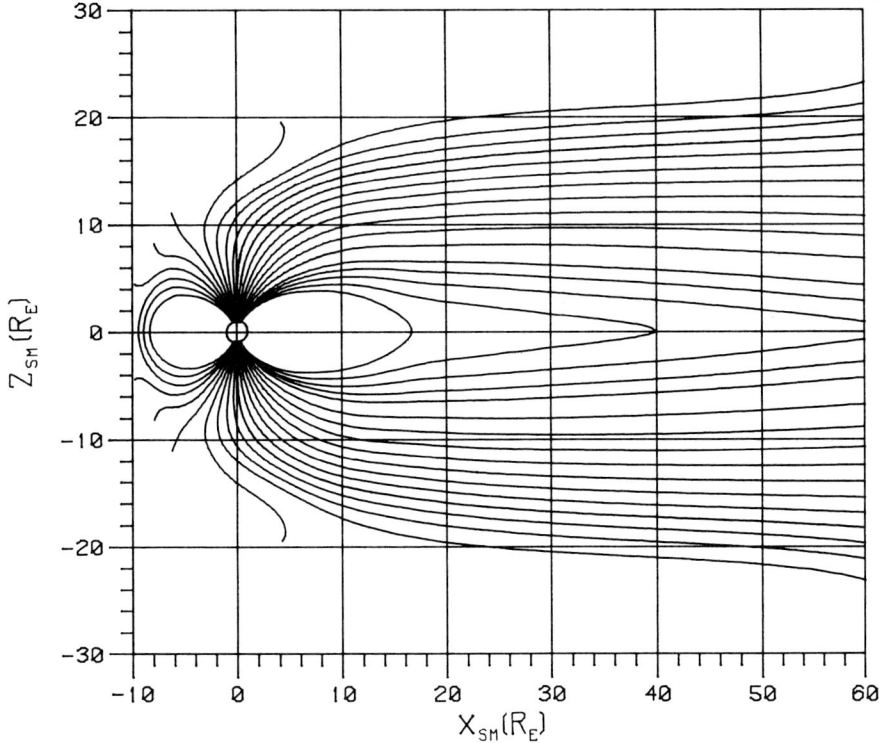

Fig. 4. Total magnetic field with a strong tail field and nominal ring and boundary current strengths.

$B_Z$ is a small positive number. It is small because the tail is extended. It is positive because the field lines close through the equatorial plane in the north direction. In Figure 1 three curves are given for varying strengths in the tail - weak, nominal and strong with the weak and strong values represented by currents 25 percent below and above the nominal value. It is seen that the $B_Z$ value in the tail along the sun-earth line does not vary appreciably, the largest effects occurring in the near earth region (there are appreciable changes in the strength of the tail lobe field also which are not shown in the figure).

A picture of the field lines in the noon-midnight meridian more graphically illustrates the effect of chaning the strength of the tail current system. The nominal field configuration is shown in Figure 2. The field lines are depicted in 1° intervals from latitude 70 to 74 and then continuing at 2° intervals to the poles in the midnight meridian. The lines are given in 2° intervals from 70 to 90° in the noon meridian. It is seen that the 72° field line goes back to 25 $R_E$ and the 74° line closes beyond lunar orbit. Thus, the nominal field in this model is much less dipolar in structure than previous models. No attempt has been made to force all of the field lines to be closed. Also the model is self-consistent only in that the final magnetic field values have been compared with observations and found to be in good agreement.

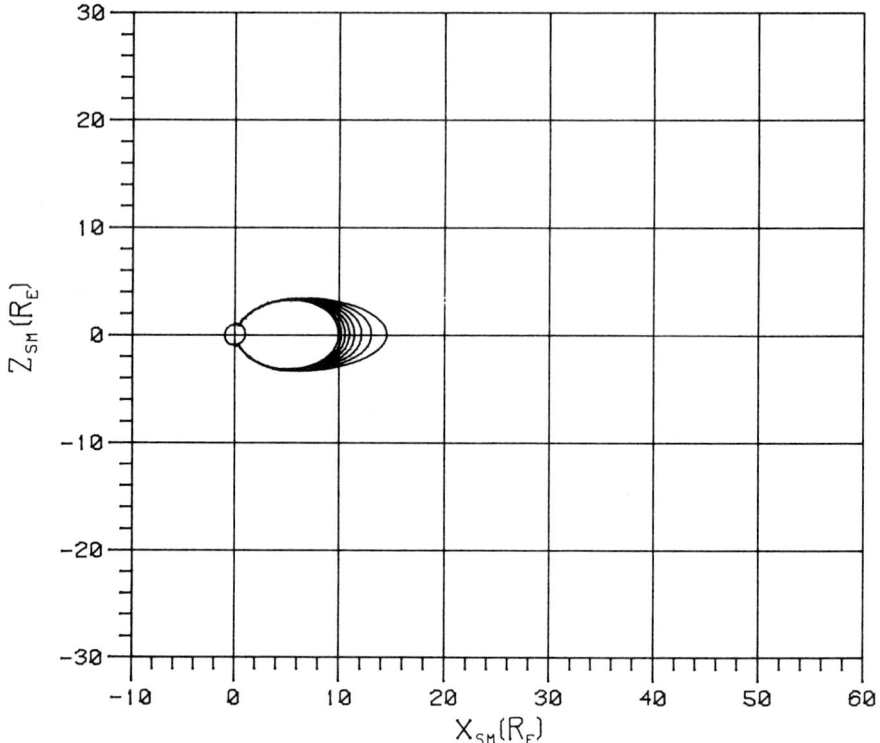

Fig. 5. Field lines on the midnight meridian at 69° magnetic latitude for varying tail field strengths.

The magnetopause currents and shape, however, have been determined with the usual internal dipole field geometries and not self-consistently with the contributions from the magnetospheric currents included. The bell-shaped field lines at lunar orbit for large $|Z|$ values are caused simply by the fitting procedure. The field in that region is not accurately represented.

In Figures 3 and 4 the field topology for weak and strong tail fields are given. In each the nominal values for the ring and magnetopause currents have been used. These configurations we believe come close to describing the state of the magnetotail during the extremes of the substorm process. Notice that when the tail field is large, the 70° field line already extends to about 16 $R_E$ and the 72° line closes beyond lunar orbit. Thus, a change in tail current strength from minus to plus 25 percent of the nominal value changes the extent of the 72° line by about 48 $R_E$. When such a change occurs during a substorm in an interval of less than one hour, the associated induced electric field is several millivolts per meter and can act to move and accelerate the magnetotail plasma.

In Figures 5 through 7 field lines at 69, 72, 75° magnetic latitude are shown for varying strengths of the tail magnetic field (0.7 to 1.4 of normal in steps of 0.1). The equatorial crossing of the 69° latitude

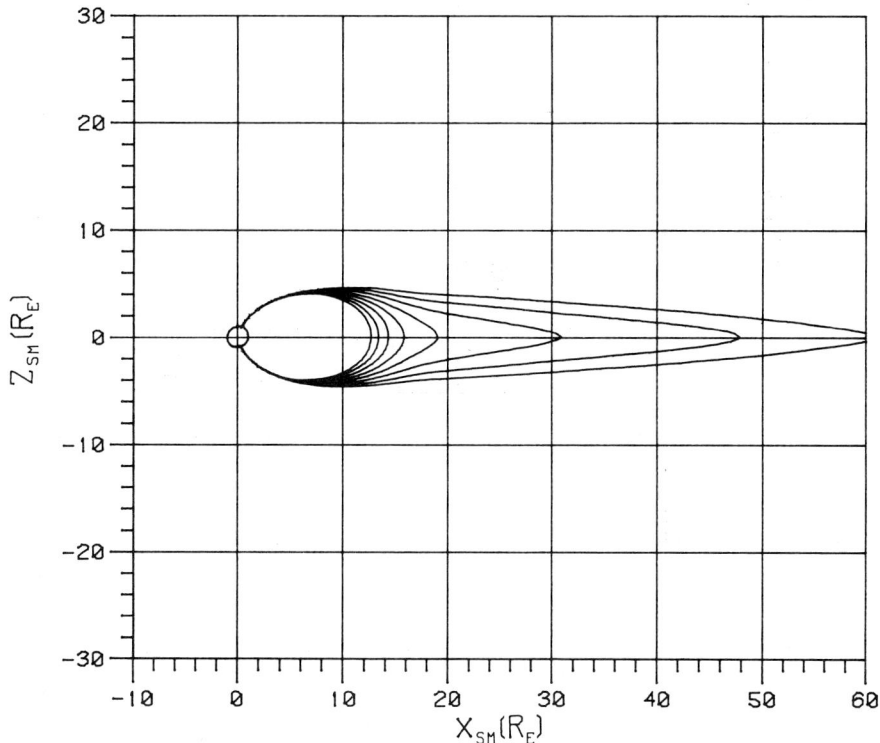

Fig. 6. Field lines on the midnight meridian at 72° magnetic latitude for varying tail field strengths.

field lines varies nominally but does not move the line into the tail region where the dense plasma sheet particles persist. The 72° line, however, has its equatorial crossing position always in the plasma sheet region. At the other extreme the 75° line for most strengths of the tail field closes beyond lunar orbit in a region where the particle densities are low. This study of field line extent confirms what we already know from observations concerning the extent of the auroral region (from about 70 to 74° magnetic latitude). Below 70° the field lines are quite rigid while above 75° the lines extend to the distant tail where the plasma density is quite low. In-between (in the auroral region) the field lines are influenced strongly by changes in the strengths of the tail currents and the resultant electric fields can act more readily to influence the behavior of the plasma on the lines.

In Figure 8 the changes in $B_Z$ along the sun-earth line are again shown - this time for a strong ring current. The increased depression in the field just within geosynchronous orbit is apparent. Such a signature should persist during the first day or two of the main phase decrease associated with a magnetic storm. These variations produce an electric field which is large enough to influence charged particle motions in the inner magnetosphere.

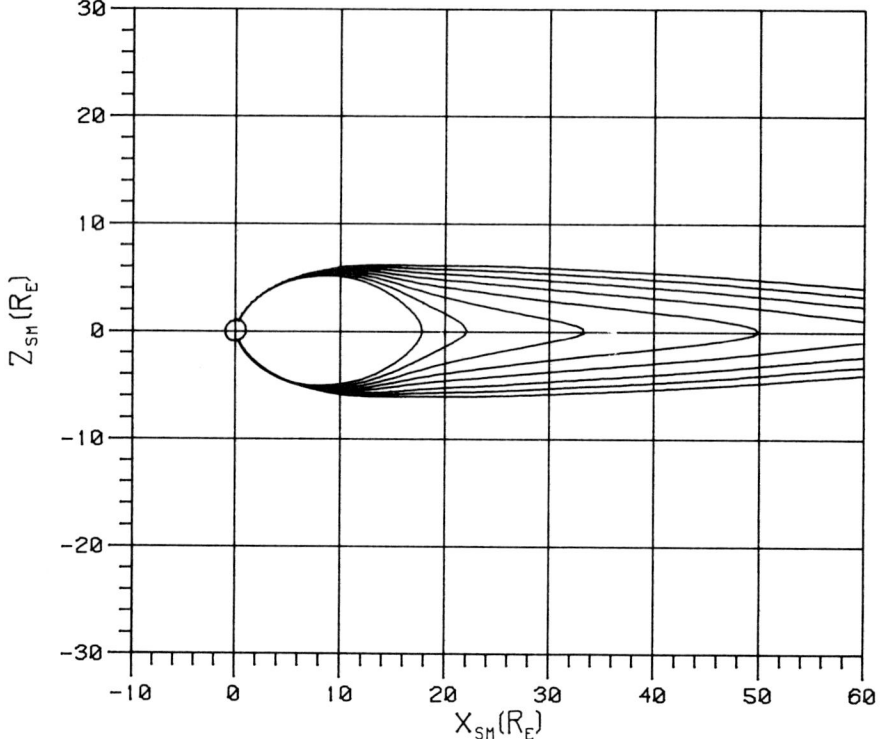

Fig. 7. Field lines on the midnight meridian at 75° magnetic latitude for varying tail field strengths

Conclusions

A model of the three major magnetospheric current systems has been described. It can be used in the study of the dynamics of magnetic storm and magnetospheric substorm processes. In this model the magnetic field from each of the three major magnetospheric current systems can be varied individually. Such a model is necessary in the quantitative study of magnetospheric dynamics. The model has been used here to examine the behavior and extent of the night time auroral region and the behavior of the field in the inner magnetosphere associated with magnetic storms. We have looked only at the changes in the strengths of the current systems and not changes in their patterns or locations. It is our intention in the near future to examine also changes in the location and morphology of the currents and to determine quantitatively the associated induced electric fields. The resultant models will then be used to study particle motions and acceleration in specified time varying electromagnetic environments. These studies are aided by making use of "zero-tilt" symmetries. The procedures for determining total electric fields in the presence of plasma, given an induced electric field associated with a time varying magnetic field, are given in an accompanying proceedings paper (see Mroz, et al.).

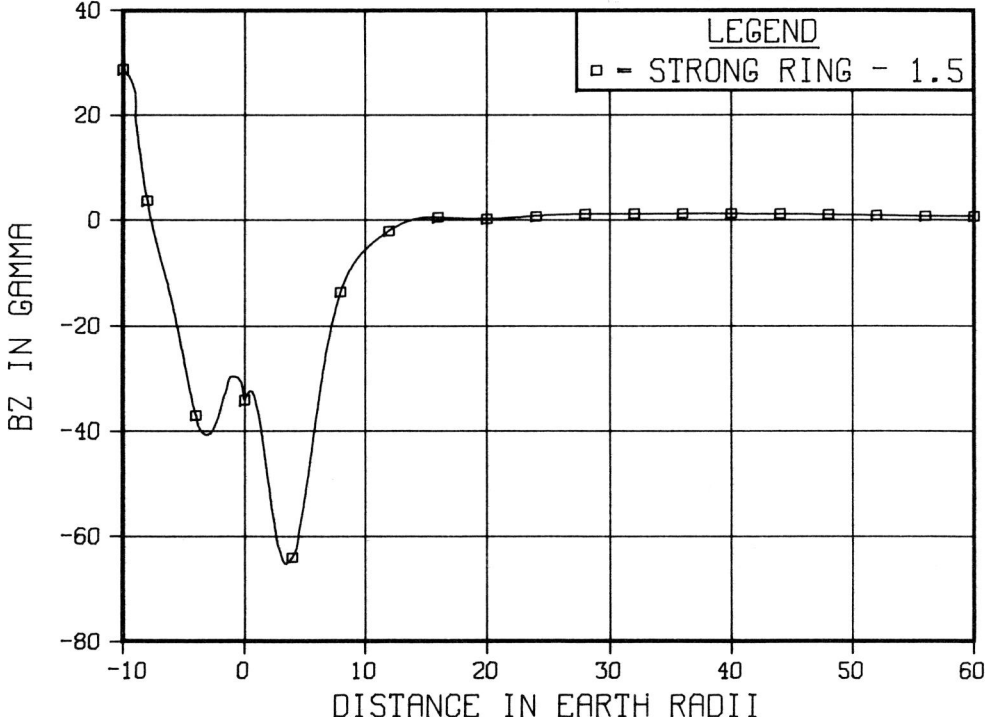

Fig. 8. The combined field of the boundary ring and tail currents along the sun-earth line (with on a Bz component because of symmetry) is shown for a strong ring current system.

Acknowledgements. Ths work was supported by the Air Force Office of Scientific Research under contract F44620-75-C-0033 and also by the Office of Naval Research under contract N00014-78-C-0215.

References

Mroz, G. J., et al., Induced magnetospheric electric fields, This Volume, 1979.
Olson, W. P., and K. A. Pfitzer, A quantitative model of the magnetospheric magnetic field, J. Geophys. Res., 79, 3739, 1974.
Sugiura, M., and D. J. Poros, A magnetospheric field model incorporating the Ogo 3 and 5 magnetic field observations, Planet. Space Sci., 21, 1763, 1973.

SELF-CONSISTENT THEORY OF A MAGNETOSPHERIC B-FIELD MODEL

K. Fuchs and G.H. Voigt

Angewandte Geophysik, Technische Hochschule Darmstadt,
Alexanderstrasse 35, D-6100 Darmstadt, West Germany

Abstract. A major problem of magnetospheric modeling is the consideration of the plasma in models which describe the magnetic field. The existing magnetospheric vacuum B-field models are incapable of considering the interaction between the plasma and the magnetic field. In these models, the tail field lines are stretched out by artificially inserted infinitesimally thin current layers in order to fit the field line picture to experimental data. Therefore, it is the purpose of this paper to discuss a method to calculate a self-consistent model which satiesfies the static MHD force balance equation within the whole magnetosphere. That is, the well known self-consistent theory of the distant tail is extended to the day-side magnetosphere. The total magnetic field consists of the earth's dipole field, the field produced by the Chapman-Ferraro currents, and the field of currents distributed in the magnetosphere. The assumption of an isotropic particle pressure leads to a restrictive condition for possible magnetic field topologies. One can find a two-dimensional equilibrium solution which describes the whole magnetosphere under quiet conditions. With a simplified geometry of the magnetopause, an analytical solution for the magnetic field components has been found. The final result will be a magnetospheric model whose tail field configuration depends on standard parameters, for example the stand-off distance and the dipole tilt angle.

## 1. Introduction

A lot of questions gave rise to the development of magnetospheric models which attempt to reproduce a realistic picture of the magnetic field lines. To the present day, however, all of these models have been vacuum-models [Williams and Mead, 1965; Voigt, 1972; Olson and Pfitzer, 1974; Choe and Beard, 1974]. These models are not based on a self-consistent force balance between plasma particles and the magnetic field. Instead, the field lines are reproduced more or less realistically by assuming singular sources which contribute all together to the total field within the magnetospheric cavity: The vacuum field of a given dipole is shielded against the interplanetary space by boundary currents on the magnetopause. In the inside, the picture of the field lines is corrected by assuming infinitesimally thin current sheets, for example in the neutral sheet of the tail. It should be mentioned that Olson and Pfitzer [1974] use a great maze of wires as a covenient numerical approximation to a volume distribution of current. Nevertheless, they do not examine the force balance between magnetic field and particles, and the corresponding questions of self-consistency.

On the other hand, there exists the self-consistent theory of the distant magnetotail [Schindler, 1972; Bird and Beard, 1972; Siscoe, 1972; Rich et al., 1972; Birn et al., 1975; Schindler, 1976; Birn et al., 1977]. In this case, continuously distributed currents meet the force balance equation. These models succeed in describing the physics of the distant tail. But to the present day, no connection has been made to the main geomagnetic field.

Therefore, we intend to describe a synthesis of these two ways of magnetospheric modeling. In other words, we extend the self-consistent theory to a model of the whole magnetosphere. We attempt to calculate a model in the presence of the earth's dipole field under the assumption of a continuously distributed plasma. The plasma is assumed to be in a static equilibrium from the day-side up to the distant tail. This enables us to examine the influence of the dipole field on the tail field configuration. Moreover, we are interested in calculating the current distribution in the tail and its dependence on day-side parameters. The extended field lines of the night-side are due to the consideration of the plasma. They are not extended due to artificially inserted current layers, as was the case in older vacuum-models.

## 2. Two-dimensional Theory

Our synthesis of the two different descriptions leads to the magnetohydrostatic momentum equation which must fulfill boundary conditions on the magnetopause. Assuming that we can describe magnetospheric physics by static equations for quiet times, we have a balance of forces between the $\underline{j} \times \underline{B}$ - force and the divergence of a pressure tensor

$$\underline{j} \times \underline{B} = \nabla \cdot \underline{\underline{P}} . \qquad (1)$$

We must consider that the magnetic field $\underline{B}$ is composed of three parts:

$$\underline{B} = \underline{B}_j + \underline{B}_d + \underline{B}_{cf} .$$

$\underline{B}_j$ : field of the continuously distributed current densities,

$\underline{B}_d$ : the earth's dipole field,

$\underline{B}_{cf}$ : field produced by the shielding Chapman-Ferraro currents.

Neither variations in time, nor the directed plasma flow and its connected electric fields are considered. Furthermore, our model is restricted to a two-dimensional description. It is also restricted to the assumption of pressure isotropy in the whole magnetosphere. Therefore, only plasma processes related to diffusion or scattering are implicitly reflected in the scalar function p; processes which are due to plasma convection cannot be considered on the level of our theory.

Thus, in the case of isotropic particle pressure, our task is reduced to the solution of the following set of MHD equations

$$\underline{j} \times \underline{B} = \nabla p \qquad (2)$$

$$\nabla \times \underline{B} = \mu_o \underline{j} \qquad (3)$$

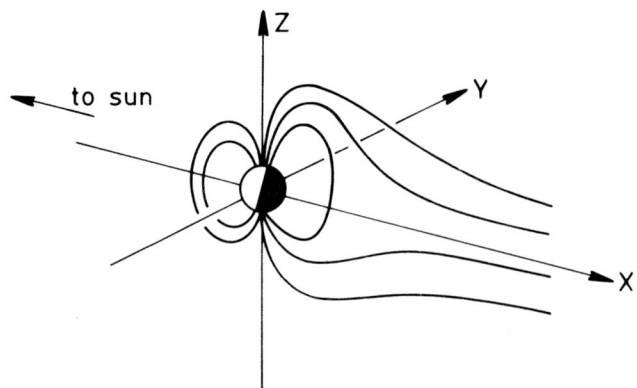

Fig. 1. Sketch of the coordinate system.

$$\nabla \cdot \underline{B} = 0 \quad . \tag{4}$$

As long as the noon-midnight-meridian plane is represented by the x- and z-coordinates, we can describe the vector potential by one component,

$$\underline{A} = A \cdot \underline{\hat{e}}_y \quad .$$

Thus, the momentum equation (2) takes the following form

$$\nabla A \times \nabla \Delta A = 0 \quad . \tag{5}$$

This corresponds to the relation $\nabla \times (\underline{j} \times \underline{B}) = 0$ which we obtain from the assumption of isotropic particle pressure in equation (2). All flux functions A are solutions of (5) as long as they satisfy the following relation:

$$\Delta A = f(A) \quad . \tag{6}$$

In other words, equation (5) is solved when the current density j can be described as any function of A,

$$j = j(A) \quad .$$

Both, the y-component of the vector potential and the electric current density are constant along the magnetic field lines.

A linear relation between the current density and the flux function allows for analytic methods. Therefore, we use a linear connection between j and A. Thus, equation (6) will be reduced to a Helmholtz equation

$$\Delta A + k^2 A = - m \frac{\partial}{\partial x} \delta(x) \delta(z) \quad , \tag{7}$$

which must be solved for the unknown function $A(x,z)$. We shall do that for a special case in Section 4. The inhomogeneity of Helmholtz' equation describes the earth's dipole two-dimensionally. (For the defini-

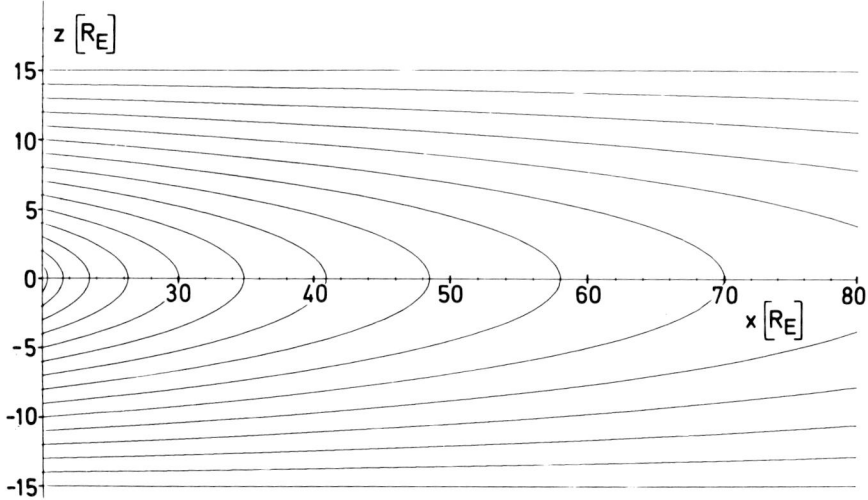

Fig. 2. Magnetic field lines for the example of the distant tail.

tion of the coordinate system see Fig. 1). Let m be a modified magnetic moment of the two-dimensional dipole. The constant k, which links the current distribution to the flux function A, is a measure for the density of the particles within the cavity (see Section 4).

Solutions of (7) describe stable magnetospheric configurations. All other relations between the current density and the flux function require numerical methods to solve non-linear second order differential equations.

### 3. General Solution

Let us describe a general solution of (7) for any kind of magnetopause geometry. Later on we shall restrict ourselves to a special simplified geometry.

The solution of the <u>inhomogeneous</u> Helmholtz equation which we can find by using Green's function, gives us the field of a dipole in the plasma.

The solution of the <u>homogeneous</u> equation fulfills the boundary conditions on the magnetopause. It results in an additional current distribution in the self-consistent equilibrium of forces and in additional Chapman-Ferraro currents at the magnetopause which shield the magnetosphere against the interplanetary space.

It should be noted that the normal component of the magnetic field at the magnetopause need not necessarily disappear. This model even allows us to link field lines with the interplanetary magnetic field, in analogy to the vacuum-model of Voigt and Fuchs [1979].

The Helmholtz equation allows the examination of two special cases:

i. In the case of $k = 0$ which is suited to describe magnetospheric vacuum-models, we obtain

$$\Delta A = - m \frac{\partial}{\partial x} \delta(x) \delta(z) \quad . \tag{8}$$

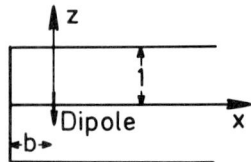

Fig. 3. Rectangular magnetopause geometry.
x and z in units of the half of tail diameter.

ii. The case of negligible inhomogeneity stands for models of the distant tail, where the influence of the main geomagnetic field is of no considerable importance. The distant tail is therefore determined mainly by k and the boundary values of the tail magnetopause,

$$\Delta A + k^2 A = 0 \quad . \tag{9}$$

According to a vanishing normal component of the magnetic field at the magnetopause assumed to be a field line (A = 0), we find a separated solution of the function A for the distant tail:

$$A = F(x) \cdot G(z) = -B_o \cos(\alpha z) \cdot e^{-\lambda x} \quad . \tag{10}$$

The x- and z-coordinates are dimensionless quantities in units of the half of the tail diameter. According to the boundary condition $A = 0$ at $z = \pm 1$, and to the usual curvature of the field lines, we have to choose

$$\alpha = \pi/2 \quad . \tag{11}$$

By a qualified value of $\lambda$, the exponential x-dependence is fitted to the known experimental data of the field strength falloff in the distant tail [Mihalov and Sonett, 1968]. We have chosen $\lambda = 0.176$. From (9) we conclude that the value of k is determined by these two quantities $\alpha$ and $\lambda$:

$$k^2 = \alpha^2 - \lambda^2 \quad . \tag{12}$$

The field lines corresponding to (10) are plotted in Fig. 2.

## 4. Simplified Magnetopause Geometry

We are able to describe our model analytically by assuming an especially simplified magnetopause geometry. In this case, we shall use a rectangular magnetopause (see Fig. 3) which allows us to shield the day-side by using the image dipole method. Note that the stand-off distance is specified by the parameter b.

Let us begin by assuming an infinite plane geometry which is symmetrical in the x-direction. This allows us to calculate A as well as the inhomogeneity in the Helmholtz equation according to Fourier. On the finite interval $-1 \leq z \leq 1$, the functions can be expanded into Fourier series. On the infinite interval, $-\infty < x < \infty$, the series are replaced by integrals, so that the expansions assume the form

$$A = \sum_{n=0}^{\infty} \int_{-\infty}^{+\infty} A_n(s) \cos(\alpha_n z) \cdot e^{-isx} ds \quad , \tag{13}$$

and

$$\frac{\partial}{\partial x} \delta(x) \delta(z) = -\frac{i}{2\pi} \int_{-\infty}^{+\infty} s e^{-isx} ds \sum_{n=0}^{\infty} \cos(\alpha_n z) \quad , \tag{14}$$

with $\alpha_n = (2n+1) \cdot \pi/2$, in analogy to (11).

The expansion coefficients $A_n$ follow immediately by comparing (13) to (14). Subsequent integration yields to

$$A = -\frac{m}{2} \sum_{n=0}^{\infty} \cos(\alpha_n z) \, e^{-|\lambda_n x|} \, \text{sign}(x) \tag{15}$$

with

$$\lambda_n^2 = \alpha_n^2 - k^2 > 0 \quad . \tag{16}$$

We obtain the shielding of the day-side at $x = -b$ by superimposing on (15) the function A of an image-dipole at $x = -2b$. Thus, the analytical solution of the total flux function A is given by:

$$A = -\frac{m}{2} \sum_{n=0}^{\infty} \left( e^{-|\lambda_n x|} \text{sign}(x) + e^{-|\lambda_n(x+2b)|} \text{sign}(x+2b) \right) \cdot \cos(\alpha_n z) \tag{17}$$

Figure 4 shows how the usual field line picture of the tail gradually appears as the plasma density increases. The extended field lines of the tail exist without introducing any additional current sheets, as was the case in all former magnetospheric models. Furthermore, the cusp regions incline more equatorwards as k increases. Note, that the case k = 0 refers to the field lines of the earth's dipole in a vacuum-magnetosphere. The fact that the field line picture near the earth always looks realistic regardless of k, is of course due to the fact that the dipole field itself is dominating.

Realistic values of k must be apparently orientated by those ones of the distant tail. Therefore, we choose such values as we successfully used for the description of the distant tail following relation (12).

According to the relation (16), k should not exceed $\pi/2$, because otherwise $\lambda_1$ would become imaginary. With increasing $k > \pi/2$, more and more quantities $\lambda_n$ take imaginary values. The exponential x-dependence of those terms of our expansion would be replaced by trigonometrical functions. This means, that the typical exponential falloff of the tail field would be gradually superimposed by trigonometrical x-dependences.

One can surely understand that the value chosen for k is coupled with the plasma population within the cavity of the magnetosphere. Assuming a linear connection between j and A (see equation (7) and Conclusion) one obtains from equation (2)

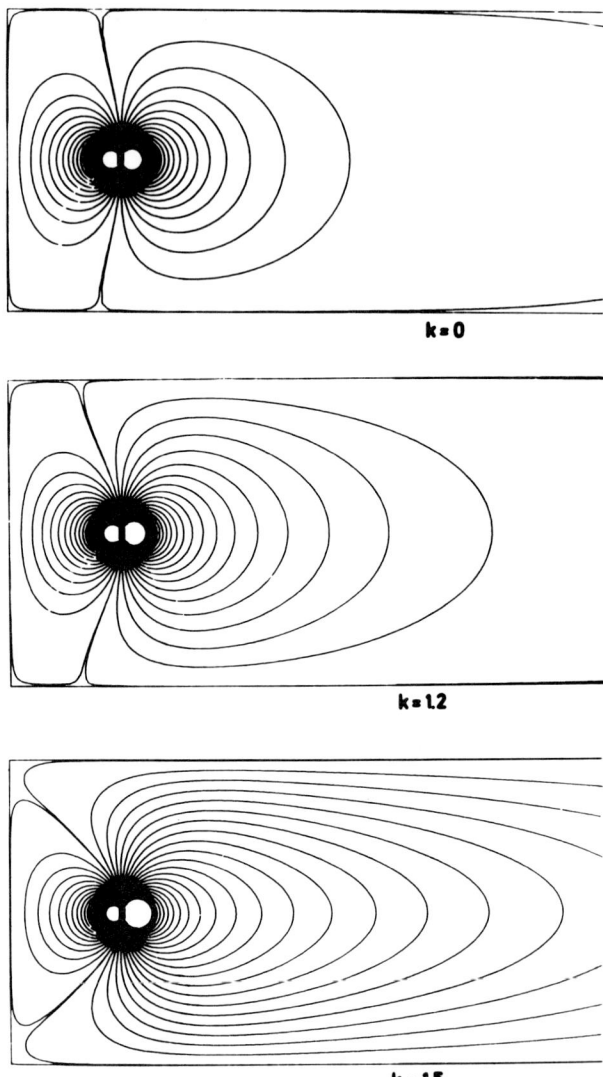

Fig. 4. Tail field lines including the earth's dipole. The first example (k = 0) shows the magnetic field lines of the earth's dipole in a vacuum-magnetosphere. With increasing values of k, the field lines are stretched out and become tail-like.

$$p(A) - p_o = \frac{k^2}{2\mu_o} A^2 , \qquad (18)$$

where $p_o$ is a reference particle pressure on the magnetopause field line (A = 0).

Using the magnetic flux function A given by the analytical expression (17) we can calculate p(x,z) from (18). If the ion and electron temper-

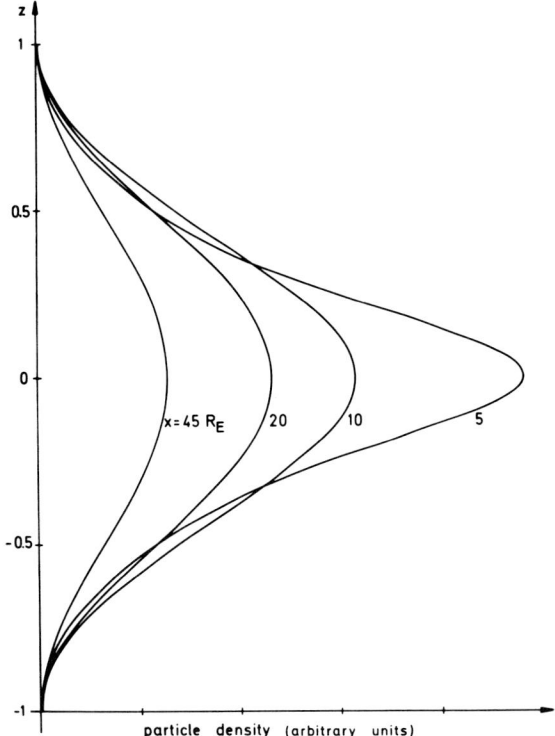

Fig. 5. Distribution of the plasma density across the tail for different distances from the earth (k = 1.5).

atures are assumed to be spatially constant, the number density follows immediately. For a reasonable value of k, the result is shown in Figure 5, where the particle density corresponding to different distances from the earth is plotted against the cross section of the magnetotail.

At present, we are also examining, how the neutral sheet's inner edge depends on the stand-off distance b. Moreover, the theory yields the displacement of the neutral sheet out of the ecliptic plane as function of the dipole tilt angle.

## 5. Conclusions

The previously described theory indicates the apparent importance of the concept of force balancing in the whole magnetosphere. This concept is basically new with respect to the vacuum concept of magnetospheric modeling. We have shown that the continuously distributed plasma, calculated from the static MHD force balance equation, is a dominant source of the magnetospheric field configuration. It creates the typical curvature of the field lines in the magnetotail without introducing infinitesimally thin current layers, as was the case in many vacuum models.

However, because of the restriction to two dimensions and the rectangular magnetopause geometry, our self-consistent model is incapable of reproducing magnetospheric situations which are of interest for model users.

The restrictions we made in our theory enabled us to find an analytical solution for the self-consistent $\underline{B}$-field plasma configuration. This solution is justified for the distant tail. Nevertheless, we have to consider in future work the characteristics of directed plasma flows and the effects of plasma anisotropy near the earth and in the region of the ring currents.

Moreover, we are beginning to compute a realistic geometry of the magnetopause by assuming an equilibrium of pressure with the solar wind from the day-side up to the tail. The numerical method envolved corresponds to that which Mead and Beard [1964] successfully used for their vacuum model.

In the special case of two dimensions, it is evident that the pressure p, the current density j, and the vector potential A are constant along magnetic field lines. Therefore, the function f(A) in equation (6) is related to the plasma pressure p by

$$f(A) = -\mu_o \, j(A) = -\mu_o \frac{dp}{dA} .$$

In order to find a physical argument for the linear connection between j and A, we must know the shape of the particle distribution functions which specify p. The investigation of the validity of the linear relation $j \propto A$ (see equation (7)) requires an examination of the microscopic behaviour of the plasma inside the magnetospheric current sheets. With other words, we have to examine the whole problem in the framework of the kinetic theory.

Although the model is simplified in the present stage, it could be regarded as a first step in considering the influence of the plasma on the global magnetospheric $\underline{B}$-field topology.

## References

Bird, M.K., and D.B. Beard, The self-consistent geomagnetic tail under static conditions, *Planet. Space Sci.*, 20, 2057-2072, 1972.

Birn, J., R. Sommer, and K. Schindler, Open and closed magnetospheric tail configurations and their stability, *Astrophys. Space Sci.*, 35, 389-402, 1975.

Birn, J., R. Sommer, and K. Schindler, Self-consistent theory of the quiet magnetotail in three dimensions, *J. Geophys. Res.*, 82, 147-154, 1977.

Choe, J.Y., and D.B. Beard, The compressed geomagnetic field as a function of dipole tilt, *Planet. Space Sci.*, 22, 595-608, 1974.

Mead, G.D., and D.B. Beard, Shape of the geomagnetic field solar wind boundary, *J. Geophys. Res.*, 69, 1169-1179, 1964.

Mihalov, J.D., and C.P. Sonett, The cislunar geomagnetic tail gradient in 1967, *J. Geophys. Res.*, 73, 6837-6841, 1968.

Olson, W.P., and K.A. Pfitzer, A quantitative model of the magnetospheric magnetic field, *J. Geophys. Res.*, 79, 3739-3748, 1974.

Rich, F.J., V.M. Vasyliunas, and R.A. Wolf, On the balance of stresses in the plasma sheet, *J. Geophys. Res.*, 77, 4670-4676, 1972.

Schindler, K., A self-consistent theory of the tail of the magnetosphere, in *Earth's Magnetospheric Processes, Astrophysics and Space Science Library*, vol. 32, edited by B.M. McCormac, pp. 200-209, D. Reidel, Publishing Company, Dordrecht-Holland, 1972.

Schindler, K., Magnetotail model, in *Magnetospheric Particles and Fields, Astrophysics and Space Science Library,* vol. 58, edited by B.M. McCormac, pp. 79-88, D. Reidel Publishing Company, Dordrecht-Holland, 1976.

Siscoe, G.L., Consequences of an isotropic static plasma sheet in models of the geomagnetic tail, *Planet. Space Sci.,* 20, 937-953, 1972.

Voigt, G.H., A three dimensional, analytical magnetospheric model with defined magnetopause, *Z. Geophys.,* 38, 319-346, 1972.

Voigt, G.H., and K. Fuchs, A macroscopic model for field line interconnection between the magnetosphere and the interplanetary space, in *Quantitative Modeling of the Magnetospheric Processes, Geophys. Monogr. Ser.,* vol. 21, edited by W.P. Olson, this volume, AGU, Washington, D.C., 1979.

Williams, D.J., and G.D. Mead, Nightside magnetosphere configuration as obtained from trapped electrons at 1100 kilometers, *J. Geophys. Res.,* 70, 3017-3029, 1965.

CONJUGATE LOW ENERGY ELECTRON OBSERVATIONS MADE BY
ATS-6 AND DMSP-32 SATELLITES

Ching-I. Meng

The Johns Hopkins University Applied Physics Laboratory
Laurel, Maryland 20810

**Abstract.** Coordinated simultaneous observations of low energy electrons at conjugate locations by using two satellites are presented. The geosynchronous ATS-6 measures the plasma sheet electrons near the magnetospheric equator, and the polar-orbiting DMSP-32 measures the auroral electron precipitation above the ionosphere. It is found that the particle characteristics of precipitated auroral electrons of diffuse auroras in the evening sector are nearly identical to those of the trapped plasma sheet electrons in the conjugate magnetospheric equator. This property of conjugate electrons reveals that the conjugate latitude of ATS-6 in the evening sector is at about 65.7° corrected geomagnetic latitude.

## Introduction

It is believed that the ionospheric projection of the magnetospheric plasma sheet is the auroral oval, and the auroral electrons are likely originated from the plasma sheet. However, the precise and definitive relation between auroral electrons and the electrons in the plasma sheet has not been determined. Purposes of this coordinated simultaneous conjugate observation are: (1) to examine the relation between the precipitated auroral electrons of the diffuse aurora and the trapped plasma sheet electrons near the magnetospheric equator and (2) to determine the precise conjugate location of the ATS-6 satellite in the ionosphere based on the particle characteristics.

The ATS-6 satellite was launched into geosynchronous orbit near 94° west longitude on May 30, 1974. Particle data of the plasma sheet are obtained from the University of California, San Diego (UCSD) auroral particle experiment consisting of five electrostatic analyzers; but only those of trapped fluxes from the two north-south oriented detectors are used in this report. The detector provides the plasma data of both ions and electrons over an energy range of 1 eV to 80 keV with an energy resolution of about $0.2E$, and the complete energy scan (64 steps) is covered in 16 seconds (Mauk and McIlwain, 1975). The DMSP-32 satellite was an Air Force, triaxially stabilized, meteorological satellite placed in a sun-synchronous circular polar orbit along the dawn-dusk meridian at $\sim$ 840 km altitude. A scanning radiometer which was sensitive to radiation in the 0.4 to 11 μm wavelength range provides auroral images covering a 2500 km width transverse to the satellite motion. In addition to the auroral picture, auroral electron precipitations were also monitored by a curved plate electrostatic analyzer measuring electrons between 0.2 and 20 keV in six logarithmic steps once every second, corresponding to a satellite motion of about 6 km (Meng, 1976).

Three magnetospheric data sets consisting of the ATS plasma, the DMSP auroral picture, and the DMSP electron precipitation, from October and November 1974 were used in this study.

## Observations

Similar to any other conjugate observations, the conjugacy is calculated based on a model geomagnetic field. The "foot" of the geo-

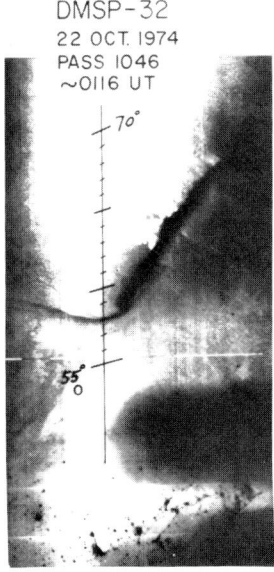

Fig. 1a

Fig. 1 A coordinated observation between the plasma sheet electrons at geosynchronous orbit (the ATS-6 satellites) and the auroral display and electron precipitation near its field line conjugate observed by the DMSP-32 satellite on October 22, 1974 (pass 1046). There are two parts of this Figure: (a) the DMSP auroral imagery (in negative) indicating the location of the calculated ATS-6 field line conjugate with respect to the auroral display, and (b) the ATS-6 E-t spectrogram representing the general particle characteristics at the geosynchronous altitude. The 'foot' of the ATS-6 field line, shown by an open circle, is located to the equatorward of the auroral display in a region of no aurora. While this coordinated observation was made at about ~ 0116 UT, the earthward edge of the plasma sheet was beyond the ATS-6 satellite. Neither auroral display nor the electron precipitation occurred at the 'foot' of the ATS-6 field line.

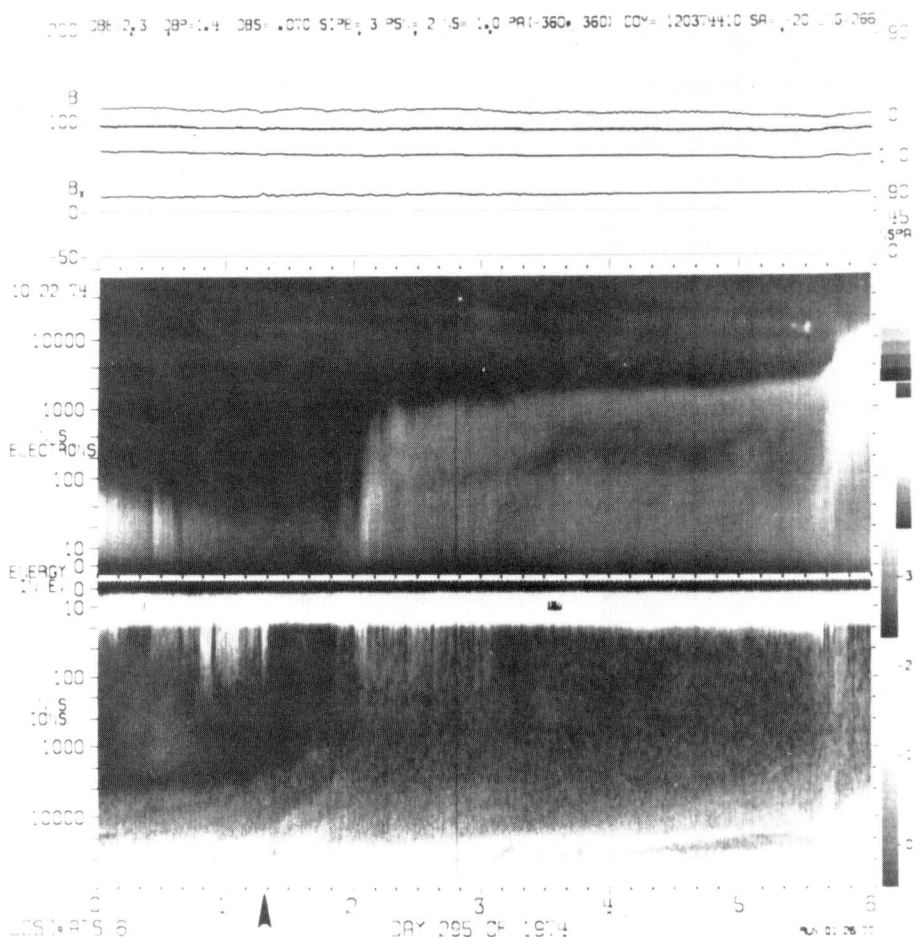

Fig. 1b

magnetic field line threading the ATS-6 satellite is determined from the field model of GSFC (9/65) with Olson and Pfitzer's (1974) modification for $K_p = 0$. The anchorage of the ATS-6 field line on the earth's surface is 54°N and 89.6°W, corresponding to a corrected geomagnetic coordinate (Gustafsson, 1970) of about 66.2°, 334°. Three representative examples of the simultaneous conjugate observations in the evening sector under different geophysical conditions are illustrated as follows.

The first example shows a very simple condition with the foot of the ATS-6 field line located to the south (equatorward) of the auroral oval under a $K_p = 2+$ condition observed on October 22, 1974. Figure 1a is the DMSP auroral photograph in negative, and the field line conjugate of ATS-6 is indicated by an open circle. Discrete evening auroral arcs were located at about 58°N near the satellite ground subtrack represented by a vertical line with the geographical latitudes. It is

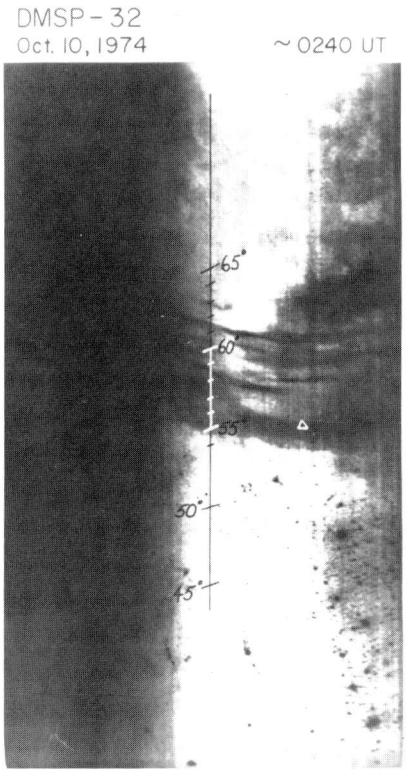

Fig. 2a

Fig. 2 Similar to Figure 1, but observed on October 10, 1974 (pass 877) when the 'foot' of the ATS-6 field line was inside the intense diffuse aurora. In addition to (a) and (b), other diagrams constituting the complete set of the Figure are: (c) the precipitation profiles of auroral electrons across the auroral region measured along the DMSP subtrack, (d) the spectral variation of precipitating electrons across the auroral region, and (e) the comparison of differential spectra between auroral electrons detected by DMSP-32 near the conjugate point of ATS-6 and trapped electrons in the plasma sheet detected by ATS-6 near the magnetospheric equator. While this coordinated observation was made at about 0241 UT, intense fluxes of electrons near and below 1 keV engulfed the ATS-6 satellite. Note the similarity between the spectrum of trapped plasma sheet electrons at ATS-6 and that of precipitating auroral electrons near the ATS-6 conjugate (see text).

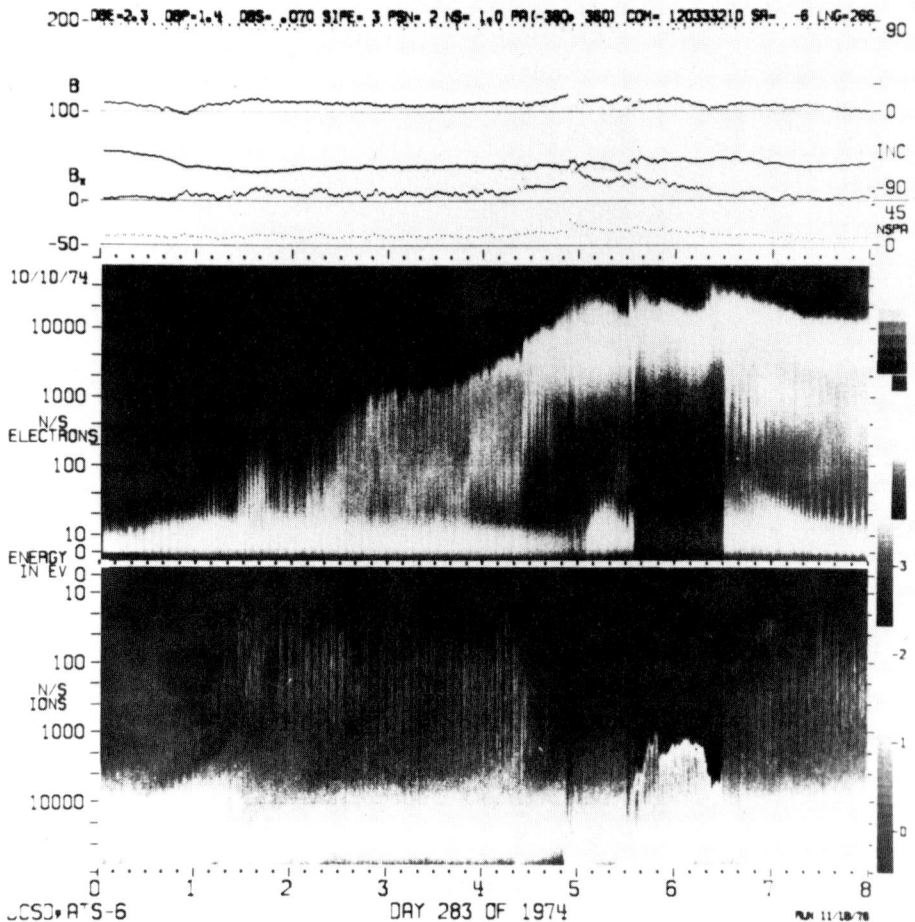

Fig. 2b

very clear that there was no auroral display adjacent to the foot of ATS-6 field line. Also, the particle measurement of the DMSP satellite recorded no auroral electron precipitation at this location. At the same time (∼ 0118 UT) near the conjugate magnetospheric equator, the ATS-6 satellite was not surrounded by any significant fluxes of plasma sheet electrons; and the first appearance of the soft plasma sheet particles occurred 40 minutes later at approximately 0200 UT as shown by the "E-t" spectrogram (Figure 1b). The format and interpretations of the "E-t" spectrogram can be seen in DeForest and McIlwain (1971). This observation reveals that at the time of the coordinated observation (∼ 0118 UT) near the 1930 magnetic local time meridian, the earthward edge of the plasma sheet was beyond the geosynchronous orbit (∼ 6.62 $R_e$), and the instantaneous ionospheric projection of the plasma sheet was at latitudes higher than the ATS-6 conjugate location of 54°N (66.2° gm lat).

The second example (October 10, 1974) illustrates the situation when

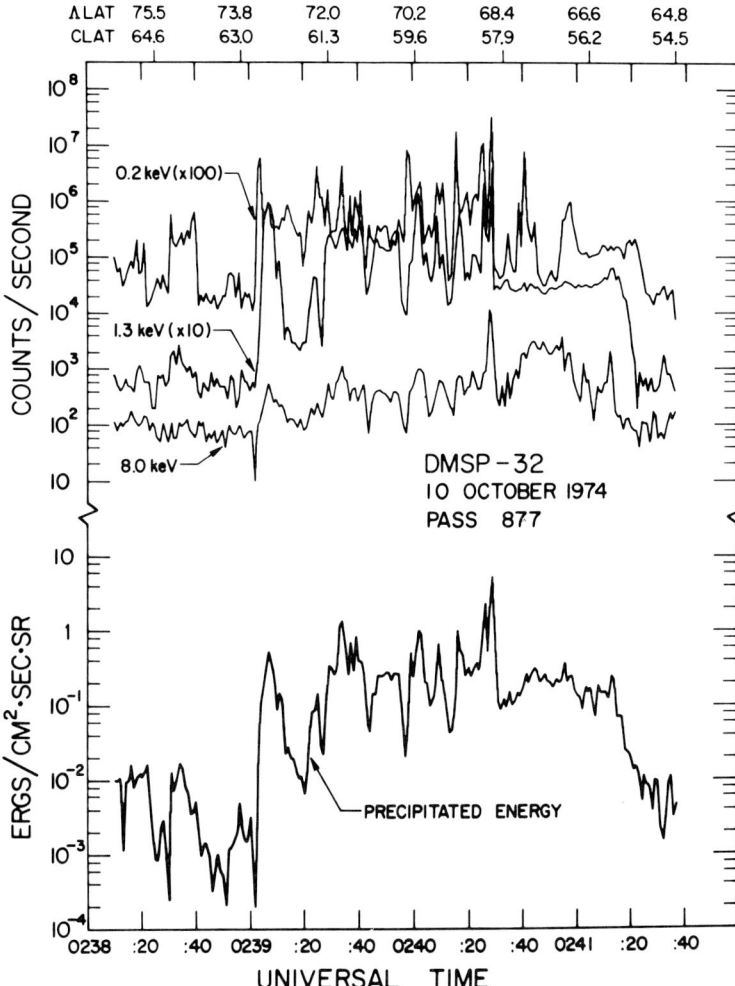

Fig. 2c

the foot of the ATS-6 field line was anchored into the bright diffuse aurora during a moderate geomagnetic activity ($K_p$ = 3+) as shown in Figure 2a. The subtrack was parallel to the terminator and it traversed the evening auroral oval along the ~ 19 MLT. At 0241:13 UT, DMSP-32 was at the calculated ATS-6 conjugate geomagnetic latitude (~ 66.2° gm lat), and the ATS-6 satellite was embedded in fluxes of soft plasma sheet electrons below about 1 keV (Figure 2b). The vertical striations in the ATS-6 "E-t" spectrogram resulted from the mechanical scan of the detectors revealing pitch angle anisotropies. This observation shows that when the ATS-6 was inside the plasma sheet, its field line conjugate was anchored into the auroral region. Thus, we can compare the auroral electron precipitations near this conjugate location with the trapped plasma sheet electrons in the magnetospheric equator. Figure 2c illustrates the precipitation profiles of three selected energies at 0.2, 1.3 and 8 keV, together with the integrated

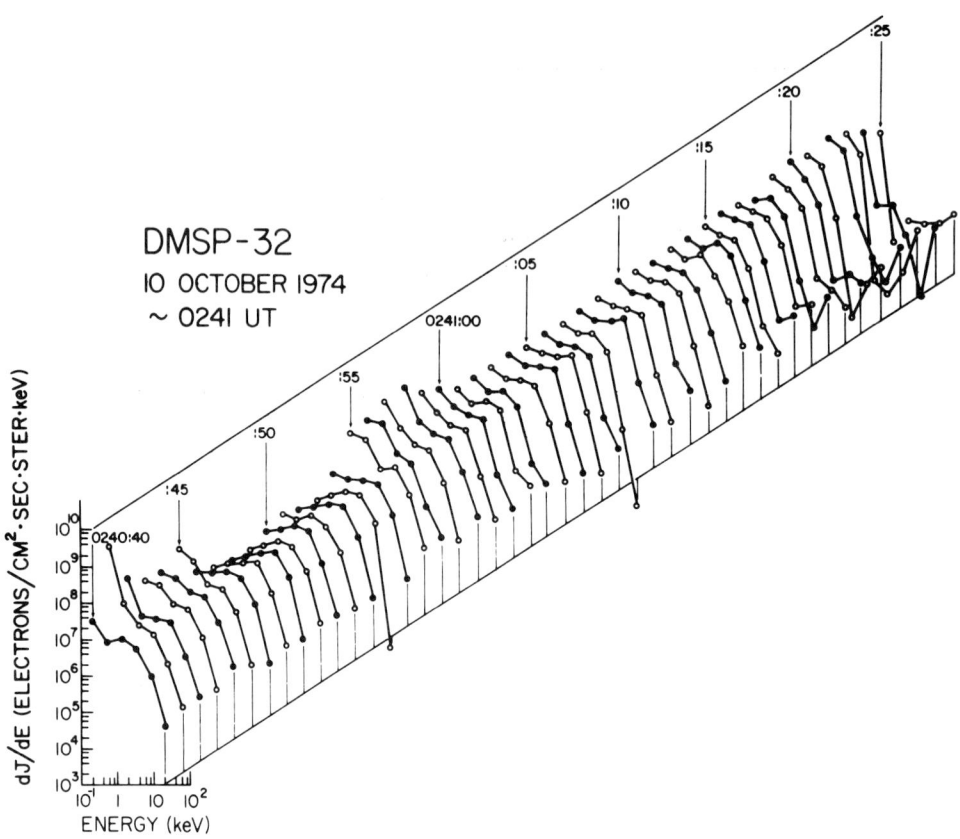

Fig. 2d

precipitating energy flux from 0.2 to 20 keV. DMSP-32 detected intense precipitating auroral electrons from ~ 73.5° gm lat (~ 63°N) to 65.7° gm lat (~ 55°N). Associated with the diffuse aurora from ~ 68.5° gm lat (58°N) to ~ 66° gm lat (55°N), electron precipitations were about 0.1 ergs/cm$^2$sec sr similar to the typical diffuse aurora (Meng, 1976). Differential electron spectra of this region are shown in Figure 2d from high to low latitudes; they are generally of the plasma sheet type. A detailed comparison of the electron spectra at two conjugate locations, namely above the polar ionosphere and near the magnetospheric equator, is shown in Figure 2e. Although the electron spectrum detected by DMSP-32 at the calculated ATS-6 conjugate latitude (0241:13 UT) revealed the existence of intense precipitations of plasma sheet electrons, the precipitating fluxes of electrons between 1 keV and 10 keV were more intense than the trapped equatorial plasma sheet fluxes. This indicates that the precipitation at the calculated conjugate latitude was different from the simultaneous plasma sheet spectrum at ATS-6. However, an auroral electron spectrum almost identical to that at the ATS-6 satellite was found slightly equatorward of the calculated conjugate latitude by about 0.5° at ~ 65.7° gm lat. Below a few keV the spectra of the trapped plasma sheet fluxes observed by ATS-6 and the precipitated fluxes at 65.7° gm lat detected by DMSP-32

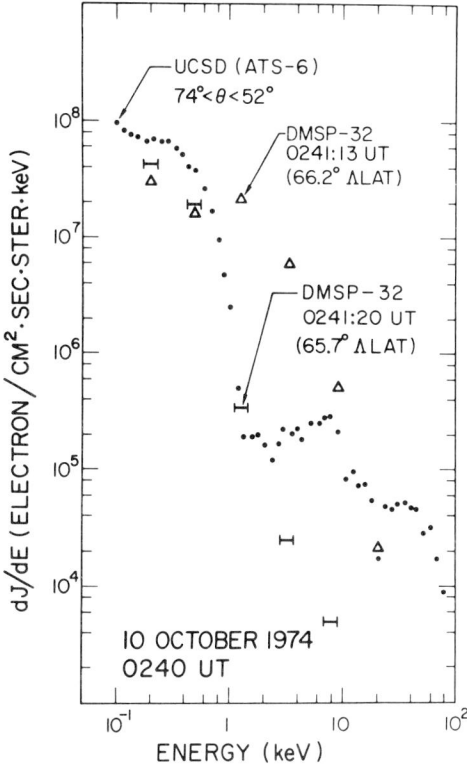

Fig. 2e

are almost identical. It is important to note that the ATS-6 spectrum was inappropriately sampled at fairly large pitch angles. At smaller pitch angles the (still trapped) ATS-6 fluxes fell below the DMSP fluxes. This low flux at small pitch angles precludes any simple, direct interpretation made on the basis of an assumption of exact conjugacy and time. However, the significance of the spectra comparison lies with the comparable fluxes and with the coincident observation by both satellites of our extremely sharp high energy cutoff near 1 keV. Above several keV the spectra do not match; the additional energetic electron fluxes observed by ATS-6 resulted from the past substorm enhancement, and they were stably trapped and circling around the earth in the inner magnetosphere due to the magnetic field gradient drift (DeForest and McIlwain, 1971). This example clearly demonstrates that the electron precipitation occurs near the foot of the ATS-6 field line when the plasma sheet engulfs the ATS-6 satellite. The spectrum and the intensity of the precipitation at 840 km above the earth's surface are almost identical to the spectrum of trapped plasma sheet electrons detected near the magnetospheric equator. The observed conjugate latitude of the ATS-6 is about 0.5° equatorward of the calculated one.

In the final example (November 9, 1974) the foot of the ATS-6 field line was embedded in the active auroral region under conditions of $K_p$ = 5+ and AE ~ 500γ (Figure 3a). Diffuse auroras and few discrete arcs occurred both poleward and equatorward of the ATS-6 field line conjugate point. Simultaneously (~ 0146:30 UT) the ATS-6 satellite was surrounded by hot plasma sheet electrons as shown by the "E-t" spectrogram (Figure 3b). The precipitated energy flux of auroral electrons was about 0.5 ergs/cm$^2$sec sr near the conjugate point (Figure 3c). Figure 3d illustrates the spectral variations of auroral electrons from ~ 68° gm lat (55.8°N) to 64.5° gm lat (51.9°N) in the active diffuse auroral region. Generally, the plasma sheet type of spectrum was observed over this region, with the exception of a few peaked spectra associated with discrete arcs (Meng, 1976). Comparison of the ATS-6 plasma sheet spectrum with the precipitated electrons (Figure 3e) reveals that the auroral electrons at the calculated ATS-6 conjugate latitude of 66.2° gm lat (traversed by DMSP-32 at 0146:28 UT) were similar to those of the plasma sheet at ATS-6; however, at about 0.4° equatorward (~ 65.8° gm lat) an almost identical spectrum was detected.

Fig. 3a

Fig. 3 Similar to Figure 2, but observed on November 9, 1974 (pass 1301) during more active condition. The precipitating electrons again have the same spectral shape and differential fluxes as those of the trapped plasma sheet electrons at ATS-6.

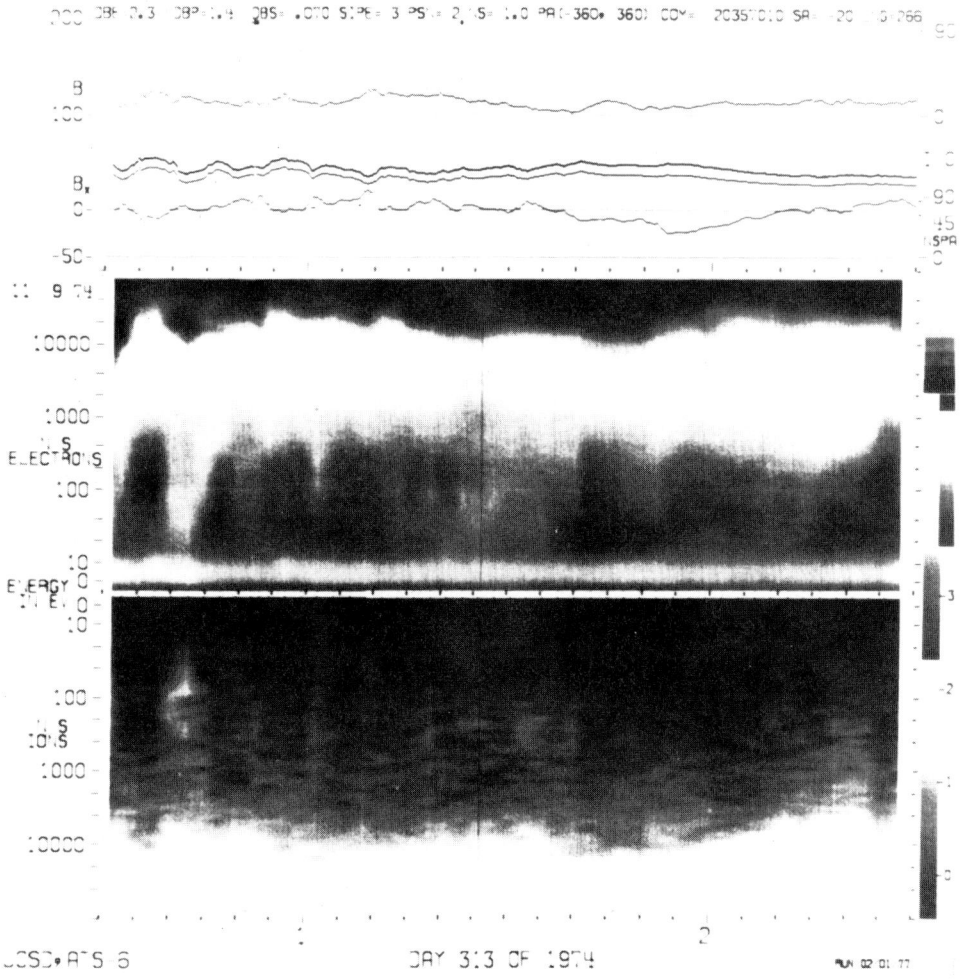

Fig. 3b

## Conclusion and Discussion

From the above examples and other coordinated observations between ATS-6 and its conjugate point in the auroral display, it is found that the spectral shape and differential fluxes of precipitated auroral electrons of diffuse auroras are very similar, and sometimes almost identical, to those of the trapped plasma sheet electrons simultaneously located in the conjugate magnetospheric equator. Thus, the characteristics of auroral electron precipitations are determined by the particle features in the conjugate magnetospheric equator, and it further indicates that the diffuse auroras are produced by the direct dumping of the plasma sheet electrons.

During the geomagnetically active period, a much closer likeness was observed between the ATS-6 plasma sheet and the precipitated elec-

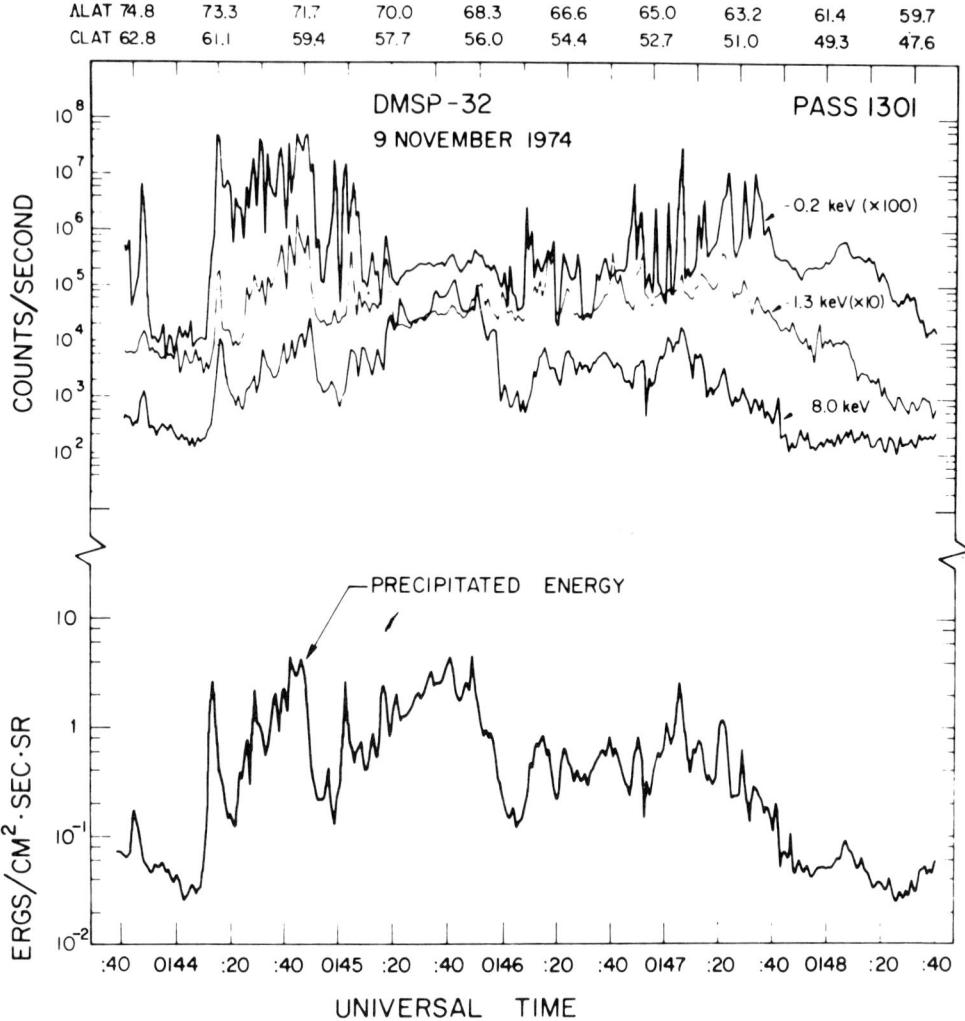

Fig. 3c

trons by shifting the calculated conjugate latitude by ~ 0.5° equatorward. One possible explanation of this ~ 0.5° equatorward shift is associated with the inflation of the inner magnetosphere. During the active geomagnetic period, the partial ring current is formed in association with each of the magnetospheric substorms; and as the consequence, the ATS-6 satellite at geosynchronous orbit is generally at lower L-shells than during the quiet condition. Quantitatively, this 0.5° shift is consistent with the difference of the geosynchronous conjugate latitude between the quiet and super disturbed conditions based on the Mead and Fairfield (1975) models.

Another possible cause of this latitudinal variation is the diurnal motion of the geosynchronous satellite in the magnetosphere. Olson and Pfitzer (1974) determined the variations of foot prints of the geomagnetic field line from the geosynchronous orbit for various field models

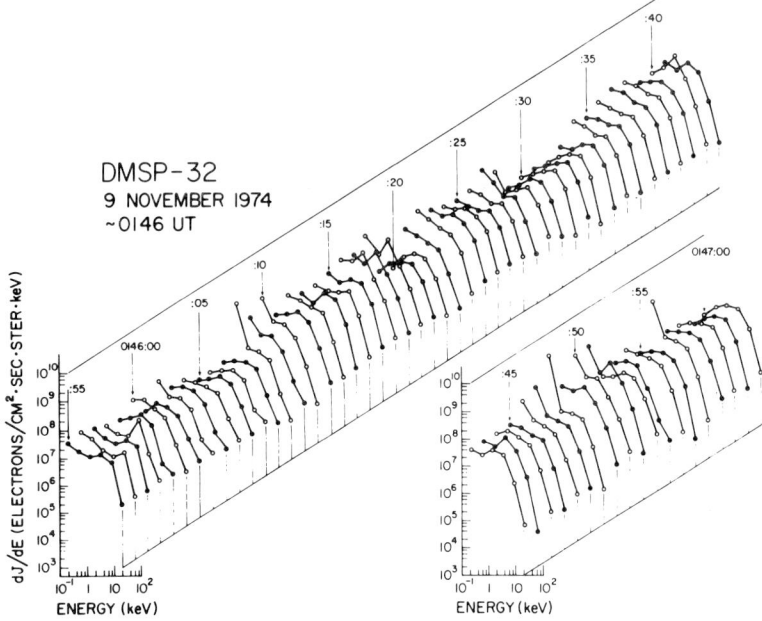

Fig. 3d

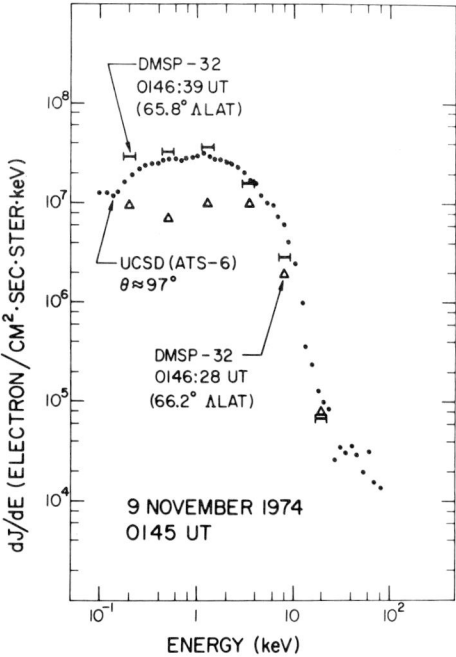

Fig. 3e

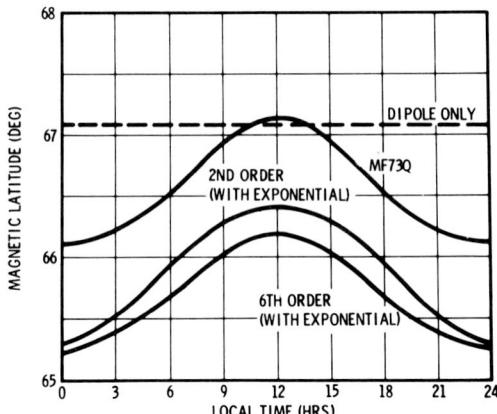

Fig. 4 Diurnal variations of the 'foot' of the geomagnetic field line at the geosynchronous orbit for different field models. Reproduced from Figure 11 of Olson and Pfitzer (1974).

and it is reproduced in Figure 4. Both the second-order and sixth-order expansions of Olson and Pfitzer (1974) model place the foot of the ATS-6 field line between about 65° to 66.5°, depending on the time of day. On the dayside near noon, the foot print of the geosynchronous orbit is at about 66.2° to 66.5° gm lat; whereas near the local midnight, it is located at about 65.3° gm lat. This approximately one degree difference of the geosynchronous conjugate latitudes is due to the solar wind compression of the dayside magnetosphere and the nightside inflation of the geomagnetic field. At about 19 local time meridian, the foot of the field line threading through the ATS-6 satellite is expected to be at about 65.7° gm lat which is in excellent agreement with the coordinate conjugate observations made by DMSP-32 and ATS-6 satellites.

It is clear that quantitative models of the magnetospheric magnetic field can be calibrated by the type of conjugate observations shown in this report. Unfortunately, the present coordinated measurements were limited to the local evening sector, and in this sector the plasma sheet engulfs the ATS-6 satellite only during the active geomagnetic condition. Thus, it is impossible from this set of observations to determine whether the observed 0.5° equatorward shift of the ATS-6 field line conjugate is attributed to the effect of the geomagnetic activity or to the diurnal variation of the conjugate latitude associated with the Olson and Pfitzer (1974) field line model. Such a distinction can be achieved by coordinated observations near midnight sector where the geosynchronous orbit is almost always adjacent to the earthward edge or inside the plasma sheet, except during the extremely quiet geomagnetic condition.

In summary, the coordinated conjugate measurements between precipitated auroral electrons and the trapped plasma sheet electrons reveal that (1) the diffuse auroras are produced by the direct dumping of the trapped plasma sheet electrons; (2) the coordinated conjugate particle observations, such as those presented in this report, can be used to "calibrate" the qualitative models of the magnetospheric magnetic field.

Acknowledgements. The author is grateful to Drs. C. E. McIlwain and B. Mauk for many helpful comments and suggestions as well as providing ATS-6 data for this project. The DMSP data were made available by the Air Force Weather Service through the National Geophysical and Solar-Terrestrial Data Center. This research was funded in part by the Air Force Office of Scientific Research under Grant AFOSR-79-0010, the Air Force Geophysics Laboratory under Contract F19628-76-C-0125. Some of the materials incorporated in this work have developed with financial support of the National Science Foundation Grant ATM 75-02621 through contract from the University of California at Berkeley to the Johns Hopkins University Applied Physics Laboratory.

## References

DeForest, S. E., and C. E. McIlwain, Plasma clouds in the magnetosphere, J. Geophys. Res., 76, 3587, 1971.

Gustafsson, G., A revised corrected geomagnetic coordinate system, Ark. Geofys., 5, 595, 1970.

Mauk, B. H., and C. E. McIlwain, ATS-6: UCSD auroral particles experiment, IEEE Trans. on Aerospace and Electronic Systems, AES-11, 1125, 1975.

Mead, G. D., and D. H. Fairfield, A quantitative magnetospheric model derived from spacecraft magnetometer data, J. Geophys. Res., 80, 523, 1975.

Meng, C.-I., Simultaneous observations of low energy electron precipitation and optical auroral arcs in the evening sector by the DMSP-32 satellite, J. Geophys. Res., 81, 2772, 1976.

Olson, W. D., and K. A. Pfitzer, A quantitative model of the magnetospheric magnetic field, J. Geophys. Res., 79, 3739, 1974.

AN EVALUATION OF RECENT INTERNAL FIELD MODELS

Gilbert D. Mead

Geophysics Branch, NASA Goddard Space Flight Center
Greenbelt, Maryland 20771

Abstract. Four recently published field models were evaluated by comparing their predictions with annual means of the magnetic field measured at 140 magnetic observatories from 1973 to 1977. Three of these, AWC/75, IGS/75 and Pogo 8/71 were nearly equal in their ability to predict the magnitude and direction of the current field. A fourth, IGRF 1975, was significantly poorer.

## Introduction

Models of the earth's internal magnetic field and its secular variation need to be periodically updated in order to accurately predict the current field. Since the secular change is not completely predictable, model predictions made for time periods well outside the time range of data used to construct the model are likely to have serious errors. Whether or not one should continue to use an older model depends upon the accuracy requirements. Most of the global models published over the last ten or 15 years can predict the current field with a directional accuracy of at least a degree or two. However, the best recent models are capable of predicting the average quiet-time field over most of the earth's surface with an accuracy of about 0.2 degrees in direction and 200 gammas in magnitude. Older models are simply incapable of predicting the current field with such accuracy. (External field models, whose magnitudes are typically of the order of 10-100 gammas, are generally not required to achieve these accuracies at the earth's surface; such models are frequently required, however, at satellite altitudes.)

The purpose of this study is to review the current status of internal field models and to evaluate several recently published models by comparing their predictions with the mean annual field measured recently at a number of magnetic observatories distributed aroung the globe. (A similar evaluation over Canada was made recently by Coles (1979).) Three of these models were found to be roughly equivalent in their ability to predict the current field. The fourth was significantly poorer.

## Field Models: Current Status

A large number of field models have been published in the literature since Gauss showed in 1839 that spherical harmonics could be conveniently used to represent the surface field. Barraclough (1978)

TABLE 1. Characteristics of the four models

| NAME | $N_{MAX}$ MF | $N_{MAX}$ SV | $N_{MAX}$ SA | DATA RANGE | DATA SOURCE |
|---|---|---|---|---|---|
| AWC/75 | 12 | 8 | 0 | 1939 to 1974.5 | Ground and Satellite |
| IGS/75 | 12 | 8 | 6 | 1955 to 1974 | Ground and Satellite |
| IGRF 1975 | 8* | 8 | 0 | * | * |
| POGO 8/71 | 10 | 10 | 0 | 1965.9 to 1970.3 | Satellite only (POGO 2, 4, and 6) Field magnitude (B) only |

$N_{MAX}$ = Maximum degree of spherical harmonic expansion (N = 1 is dipole)

MF = Main Field,   SV = Secular Variation,   SA = Secular Acceleration

*For IGRF 1975, the main field terms are determined solely by extrapolating the IGRF 1965 main field terms to 1975, using the IGRF secular variation terms, so as to make the field continuous in time. Thus the 1975 secular variation terms represent the only new analysis.

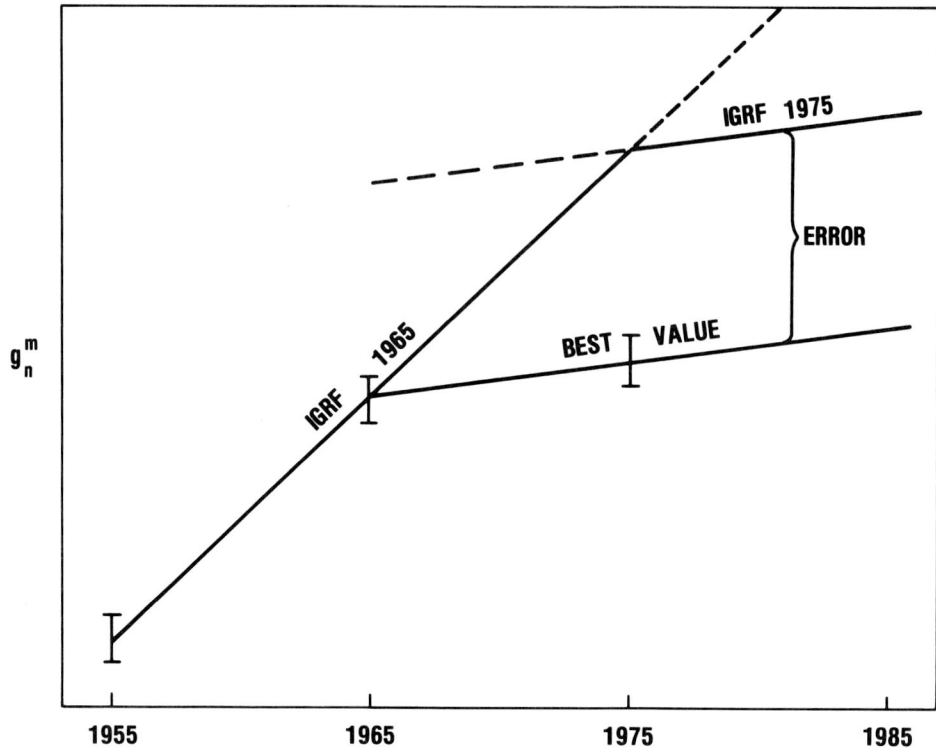

Fig. 1. Schematic illustrating how the IGRF 1975 model was derived. The solid line labeled IGRF 1965 shows the variation of a typical model coefficient, $g_n^m$ or $h_n^m$, with time. The IGRF 1975 main field coefficients were defined as the value of the corresponding IGRF 1965 coefficients at epoch 1975, so that the two models would predict the same field everywhere at 1975. Only the secular variation terms were adjusted to more accurate current values.

published an exhaustive survey of a total of 264 models that have appeared in the literature. There are at least two groups who have a continuing governmental mission to monitor changes in the earth's field and periodically publish up-to-date global field models for use in preparing magnetic charts for navigation. One of these is the USGS group located in Denver, which publishes an up-dated model of the main field and its secular variation every five years, the most recent of which is entitled AWC/75 (Peddie and Fabiano, 1976). (Navigational charts are published by the U.S. Naval Oceanographic Office.) The second is the British group located at the Institute of Geological Sciences in Edinburgh. The most recent British model is entitled IGS/75 (Barraclough et al., 1975). Recent publications from this group also include a definitive model for 1965 (Barraclough et al., 1978) and a review of methods used to determine the secular variation (Barraclough, 1976). In addition, a number of field models have been distributed which were based on the scalar magnetometer data obtained

## HISTOGRAMS FOR ΔB
## 425 OBSERVATORY ANNUAL MEANS: 1973.5 TO 1976.5

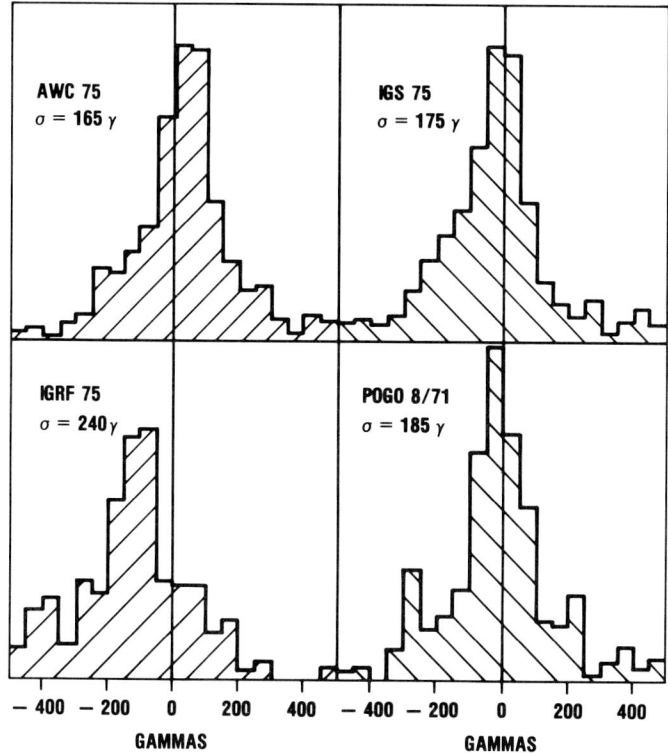

Fig. 2. Histograms of the ΔB residuals for the four models. The error, σ, was taken to be half the width of the median 68% of the residual values.

from the three Pogo satellites (Cain, 1971). One of these, Pogo 8/71 (Langel, 1974) is included in the current study. Finally, a model entitled IGRF 1975 has been recently published by the IAGA Division 1 Study Group (IAGA, 1976). The primary goal of this model, however, was to provide continuity in time with the IGRF 1965 model rather than to be an accurate current model. By vote of the IAGA General Assembly at Grenoble in 1975, the IGRF 1975 main field coefficients were arbitrarily set equal to the IGRF 1965 main field terms updated to 1975 by means of the IGRF 1965 secular variation (SV) terms (see Fig. 1). The IGRF 1975 SV coefficients were defined as the arithmetic mean of the SV terms from the AWC/75 and IGS/75 models. Errors in the IGRF 1965 SV model therefore contribute directly to errors in the IGRF 1975 main field model. Although continuity in time was achieved, the main field coefficients systematically depart from what one might determine to be the best value of the coefficients at epoch 1975. As can be seen in the next section, this seriously degrades the accuracy of its predictions. According to the IAGA publication, "the ref-

## HISTOGRAMS FOR ΔI
## 425 OBSERVATORY ANNUAL MEANS: 1973.5 TO 1976.5

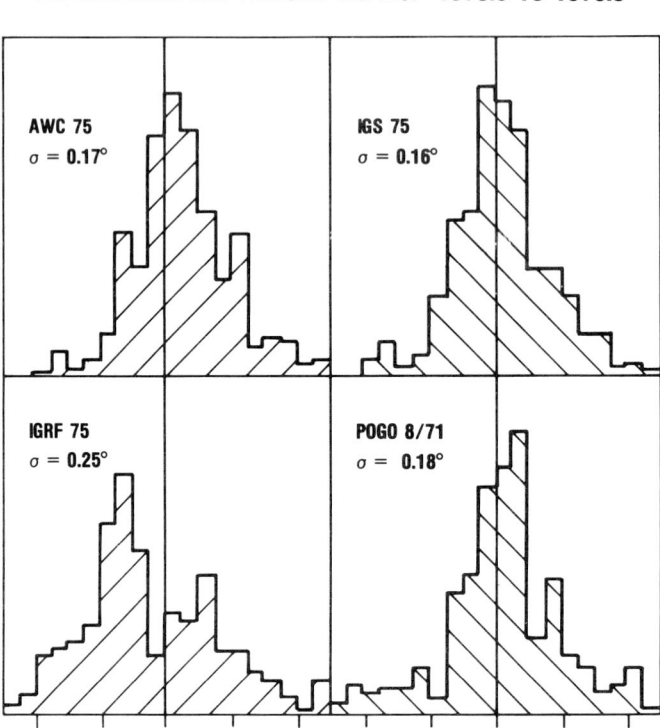

Fig. 3. Histograms of ΔI residuals.

erence field is not intended as a source of compass information for nautical and aeronautical charts." Regan and Cain (1975) have discussed the use of the IGRF and other models in magnetic surveys.

### Model Evaluations

Four models, AWC/75, IGS/75, IGRF 1975, and Pogo 8/71 were chosen for evaluation in the current study. Their characteristics are indicated in Table 1. All have secular variation terms, and one (IGS/75) also contains secular acceleration terms through degree and order six.

Values of the annual means of the three components of the magnetic field at about 140 magnetic observatories were obtain from NOAA. These observatories are not, of course, uniformly distributed around the globe. They show a rather high concentration in western Europe, Japan, and North and South America. Data were available from about 10 stations in Antarctica, four in Greeland and six in the Pacific. Comparatively little data was available in the South Atlantic, South Pacific, or Indian Oceans.

Each of the four models was used to calculate residuals representing the difference between the observatory annual means for each year in

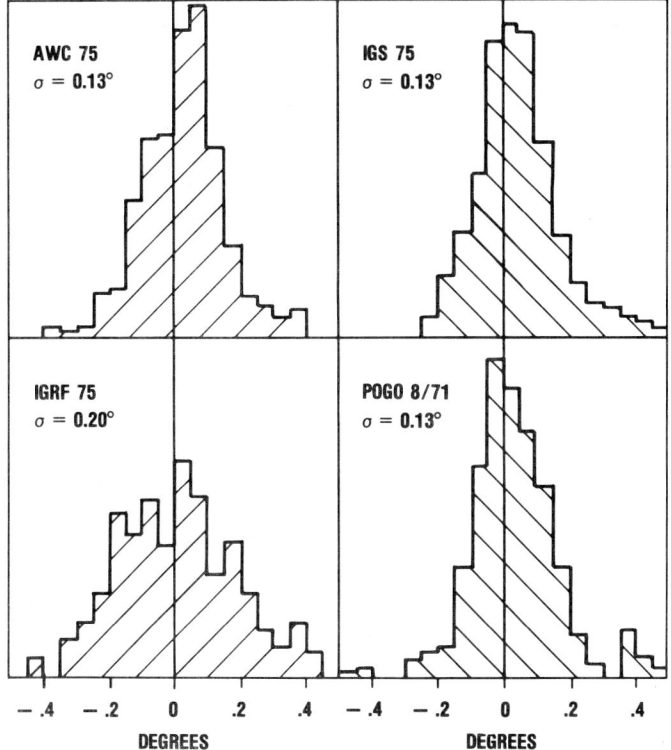

Fig. 4. Histograms of ΔD cosI.

which data was available and the model prediction. The components chosen for analysis were the field magnitude B, declination D and the inclination I. An initial examination of the difference between the measurements and models showed occasional large values of ΔD in regions near the magnetic dip pole, where D is undefined. In order to obtain realistic histograms, therefore, the quantity ΔD cosI, representing the true directional error in degrees of arc, was used instead of ΔD itself.

Residuals were calculated for each of the years 1970 through 1976 for which annual means were available. (Only a few values were available for 1977.) Since the sample size for any one year was rather small, histograms of the residuals were constructed for 425 available annual mean values for the years 1973 through 1976. These histograms are shown in Figures 2 through 4 for ΔB, ΔI, and ΔD cosI, respectively.

Root-mean-square values of the residuals were calculated, but were found not be reliable indicators of the accuracy of the predictions due to the non-normal nature of the distributions. A few very large residuals, indicating either possible errors in the data set or the presence of unusually large surface anomalies near some of the observatory sites, frequently dominated the calculation of rms values. To minimize the effect of the anomalous values, a different measure of the prediction accuracy was calculated, namely, half the width of the median 68% of the residual values. This would be equivalent to one standard error, $\sigma$, if the distributions were normal. These values of $\sigma$ are shown in the figures.

A glance at the histograms as well as the values of $\sigma$ for the various cases indicates that three of the models, vis., AWC/75, IGS/75 and Pogo 8/71, are nearly equal in their ability to estimate the true magnitude and direction of the field. Typical errors are 150-200 $\gamma$ in magnitude and 0.2 degrees of arc in direction, considering the combined effect of errors in declination and inclination. The AWC/75 seems to be slightly better in the magnitude estimations and IGS/75 is a trifle better in predictions of I. The IGRF 1975 model, however, is significantly poorer in its predictions of field magnitude and direction. In addition, the histogram for $\Delta B$ shows a significant offset in the median values of the residuals, indicating that this model consistently overestimates the true field strength; 329 predictions were too high and only 96 too low. This is probably a result of the way in which the main field coefficients were derived (see Fig. 1).

Separate histograms for each year were prepared and showed similar results; no one model was consistently superior than the others.

To check the ability of the field models to extrapolate the predictions well outside the data range of the measurements used to construct the model, residuals were calculated for 1975 annual means but with the models calculating field values for 1970.0, much closer to the period of the data range defining the models (see Table 1). These residuals were 25-50% larger than when the "proper" (i.e., 1975.5) time was used for model calculations. Thus it appears preferable to input the actual time for which the estimations are required, even though this means extrapolating several years outside the data range of the model.

Summary

1. Three of the four models studied, namely, AWC/75, IGS/75, and Pogo 8/71, were nearly equal in their ability to predict the magnitude and direction of the current field. This is particularly surprising in the case of Pogo 8/71, since only scalar data, taken over a restricted time span of 1966-1970, were used to construct this model.
2. A fourth model, IGRF 1975, was significantly poorer in its ability to predict the current field.
3. All models seemed to be able to extrapolate predictions quite well several years outside the data range used to construct the models.

A note of caution is needed. These comparisons were made with a very limited data set. No satellite data, repeat station data, project Magnet data, or survey data other than observatory annual means were used to make the comparisons. The relatively small differences noted, therefore, between the AWC/75, IGS/75 and Pogo 8/71 models are probably not significant.

Finally, a global set of vector data at altitudes of 300-500 km is expected from the Magsat satellite after its launch in late 1979 (Langel et al., 1977). Models based on this data are likely to provide excellent field predictions for the 1980's.

Acknowledgements. I wish to thank Robert Langel for several useful discussions and for critiquing the paper. The calculations were made by Ron Estes and George Maslyar of Business and Technical Systems, Inc.

## References

Barraclough, D.R., Spherical harmonic analysis of the geomagnetic secular variation--a review of methods, Phys. Earth Planet. Int., 12, 365-382, 1976.

Barraclough, D.R., Spherical harmonic models of the geomagnetic field, Institute of Geological Sciences, Geomagnetic Bull. No. 8, 1978.

Barraclough, D.R., J.M. Harwood, B.R. Leaton, and S.R.C. Malin, A model of the geomagnetic field at epoch 1975 [IGS/75], Geophys. J.R. Astr. Soc., 43, 645-659, 1975.

Barraclough, D.R., J.M. Harwood, B.R. Leaton, and S.R.C. Malin, A definitive model of the geomagnetic field and its secular variation for 1965--I. Derivation of model and comparison with the IGRF, Geophys. J.R. Astr. Soc., 55, 111-121, 1978.

Cain, Joseph C., Geomagnetic models from satellite surveys, Rev. Geophys. Space Phys., 9, 259-273, 1971.

Coles, R.L., Some comparisons among geomagnetic field models, observatory data and airborne magnetometer data: Implications for broad scale anomaly studies over Canada, submitted to Jour. Geomag. Geolectr., 1979.

IAGA Division 1 Study Group, International Geomagnetic Reference Field 1975 [IGRF 1975], E⊕S Trans. Amer. Geophys. Union, 57, 120-121, 1976.

Langel, R.A., Near-earth magnetic disturbance in total field at high latitudes, 1., Summary of data from OGO 2, 4, and 6 [Pogo 8/71], J. Geophys. Res., 79, 2363-2371, 1974.

Langel, R.A., R.D. Regan, and J.P. Murphy, Magsat: A satellite for measuring near earth magnetic fields, GSFC X-922-77-199, 1977.

Peddie, N.W., and E.B. Fabiano, A model of the geomagnetic field for 1975 [AWC/75], J. Geophys. Res., 81, 2539-2542, 1976.

Regan, R.D., and J.C. Cain, The use of geomagnetic field models in magnetic surveys, Geophys, 40, 621-629, 1975.

OVERVIEW - HIGH ENERGY PARTICLES

J. I. Vette

National Space Science Data Center, NASA/Goddard Space Flight Center, Greenbelt, Maryland 20771

The purpose of this overview is to provide some background to the reader of the role that high-energy particles have played in quantitative modeling, to place in that context the papers of this section, and to assess the progress represented by new work. By high-energy particles we mean electrons with energies between 0.05 and 6 MeV and protons between 0.05 and 600 MeV, although not all of these energies are explicitly discussed in the papers. The two main uses of these energetic particles in the modeling sense have been the production of flux maps and the testing of the accuracy of magnetic field models.

From the very beginning of the satellite era, starting with Explorer 1, individual satellites provided some incomplete mapping of the radiation belts and crude models were obtained by extending and closing flux contours to obtain a more extensive picture. Since late 1962, models of the flux distribution and energy spectrum throughout the inner magnetosphere have been produced by combining the results from several satellites. The results from single experiments are still used by some workers to form the basis for a model, but experience has shown that this should be treated with caution, paying particular attention to the range of validity. Although models of the radiation belts have provided a good picture of the general spatial distribution to guide theoretical ideas, the main quantitative use has been in computing radiation damage to solid state electronic and power components, determining shielding requirements, providing background rates for new experiments, and playing a role in mission analysis where orbit selection and operating lifetime are factors. Since the main use of the models for engineering design and mission analysis are predictive in nature, the modeling of the time behavior is of primary importance. For electrons the time behavior must be explicitly considered everywhere in the magnetosphere in construction of reasonably accurate average models.

The testing of magnetic field models by the use of high-energy particle measurements is based on the adiabatic motion of these particles. The energy density of the particles is always small in the cases of energetic particles. In addition the effects of large scale electric fields are negligible. Of course, inductive electric fields must play a role in particle acceleration and diffusion of these particles by both electric and magnetic fluctuations are possible.

The work presented by Teague et al. provides an update of the electron modeling effort carried out by NASA. In the inner zone, it is clear the effects of Starfish disappeared by 1971 and the model for this region is adequate. The high gradient regions of the outer zone, namely, at low altitude or high B, at the inner radial edge or 3 < L < 4 and at energies above 2 MeV can be represented better with the new data available from a number of satellites. In the transition region between the inner and outer belt, it is shown that even yearly averages are different because of the morphology of magnetic substorm effects. The depletion times obtained from various experiments in this region represent some significant new quantitative results and allow for building a model in which the user can take account of various levels of activity. In addition, the modeling of the lifetime of artificial enhancements in this region should be improved by these new results. The new electron model planned for completion in the near future will provide further improvement in the description of average and substorm enhanced flux levels.

The paper by Paulikas and Blake provides the most extensive information available, to date, on the time behavior of high-energy electrons for spatial regions occupied by geostationary satellites. These results demonstrate that the long term averages of 6 to 12 months show no significant changes with respect to the solar cycle. Comparing these results with those obtained by Singley and Vette (1972) who used the data from Winckler's experiment on OGO 1 and OGO 3, one can say with some confidence that from 5 earth radii outward there are no solar cycle effects expected in the average flux levels of these particles. Of course, the short-term behavior is very dynamic. The rough correlation of the most energetic electrons with the passage of fast solar wind streams is interesting and should stimulate work in understanding the physical process or at least determining the parameter which shows the best correlation with these flux levels.

The work by Paulikas and his co-workers has shown how important the omnidirectional electron fluxes observed at geostationary altitude with well calibrated detectors are to understanding the outer zone. New results presented by Higbie et al. and the review by West of the older but spatially more comprehensive OGO 5 data set show the value of the detailed directional measurements in this region. In particular, observations by Higbie et al. of the regular pattern of pitch angle variations preceding and during the course of nearly every substorm observed in the midnight sector provide an adequate one to two hour prediction method for magnetic substorm particle injection. This is an important advance in our knowledge and can be used in the future to coordinate measurements during such events.

The indication of the magnetic field topology, particularly its changes, by particle angular distribution functions was vividly illustrated by Higbie et al. West, in his survey using his OGO 5 results, showed the value of this technique throughout the outer magnetosphere. The mechanisms which can be discerned from these data are drift-shell splitting, magnetopause shadowing of drift paths on the dayside, motion through high-latitude minimum B regions, and an extensive view of the quiet time geomagnetic tail including the effects observed by Higbie et al.

Pfitzer demonstrated that the new Olson-Pfitzer model (unveiled at this meeting) is quite adequate to explain the entry of cosmic rays into the magnetosphere without invoking diffusion or some unknown process. This shows the importance of a good quantitative model in clarifying the important physical processes involved. Roelof also demonstrated that solar cosmic rays in traveling from the sun to the earth do not undergo significant diffusion perpendicular to the magnetic field lines; this ties in very nicely with Pfitzer's results in that no diffusion needs to be invoked. By using this fact, Roelof has been able to trace these solar particles back to the region of the corona where they were injected into the interplanetary medium. This work is significant in a number of respects -- it provides considerable insight into the coronal transport of accelerated particles, it defines a natural coordinate system for comparing particle fluxes on well separated spacecraft, it provides elements of a prediction technique for forecasting the arrival of solar particles at the earth, and it provides the increased use of solar flare particle distribution functions to study the bow shock and magnetosheath. It would be useful to apply the method of Roelof to the solar proton events of the 19th cycle to improve the statistical model of these events. Such models defining the radiation resulting from solar particle events have important influence on the design of extended manned missions.

Although much work remains to be done before a satisfactory explanation of magnetospheric processes can be given, the advances in our knowledge of energetic particles reflected in the work presented in this section signal a trend to modeling of magnetospheric dynamics. Fortunately, the energetic particle instruments presently operating and those planned to be launched in the near future will provide better distribution functions with the hopeful result that further knowledge will come more rapidly. There is a consistent trend to emphasize study and modeling of time dependent phenomena and to use measurement of pitch angle distribution to infer global properties of the magnetosphere. Changes in pitch angle distribution also show promise as potentially useful tools for developing predictive techniques. The correlation of magnetospheric changes with changes in the solar wind parameters, although still in their infancy, also show promise of leading to a better understanding of temporal effects, possibly, prediction capabilities.


References

Singley, G. W., and J. I. Vette, A model environment for outer zone electrons, National Space Science Data Center Report NSSDC-72-13, December 1973.

MODELING OF ELECTRON TIME VARIATIONS IN THE RADIATION BELTS

K. W. Chan, M. J. Teague, N. J. Schofield, and J. I. Vette

National Space Science Data Center, NASA/Goddard Space Flight Center
Greenbelt, Maryland 20771

Abstract. The temporal variations in the trapped electron population of the inner and outer radiation zones are reviewed. The different techniques presently used for modeling these zones are discussed and deficiencies in the present generation of models are identified. An intermediate region is identified between the zones in which the present modeling techniques are inadequate as a result of the magnitude of and frequency of occurrence of magnetic storms. Future trends in the modeling activity are discussed, and the suggestion is made that modeling of individual magnetic storms may be required in certain L bands. A simple analysis of seven magnetic storms is presented, and the independence of the depletion time of the storm flux and the storm magnitude is established. Provisional correlation between the storm magnitude and the Dst index is demonstrated.

## Introduction

The primary purpose of this paper is to review the temporal variations of electrons in the trapped radiation zones and to discuss how these variations are treated in the course of generating a model for the trapped electron population. Various generations of such models were produced since 1967 at the National Space Science Data Center, and earlier at the Aerospace Corporation (e.g., Vette et al., 1966; Singley and Vette, 1972; and Teague et al., 1976). The basic modeling technique used to date has changed very little and each generation of models has followed its predecessor in response to the availability of new and better data sets. In general, of course, the sophistication and information content of the observations has increased and the quality of the models has consequently improved. In the first section of this paper we will address the present techniques for modeling the temporal variations, and identify the deficiencies in the existing models. In the final sections we will discuss the future trends of the modeling activity.

## Outer Zone

### Temporal Variations

Figure 1 typifies the temporal variations observed in the outer radiation zone over recent years and shows the familiar 1.9-MeV ATS 1 data (Paulikas and Blake, 1971) obtained over the years 1966 through 1968. At high outer zone L values beyond the flux peak at L = 3.6 to 4.5, de-

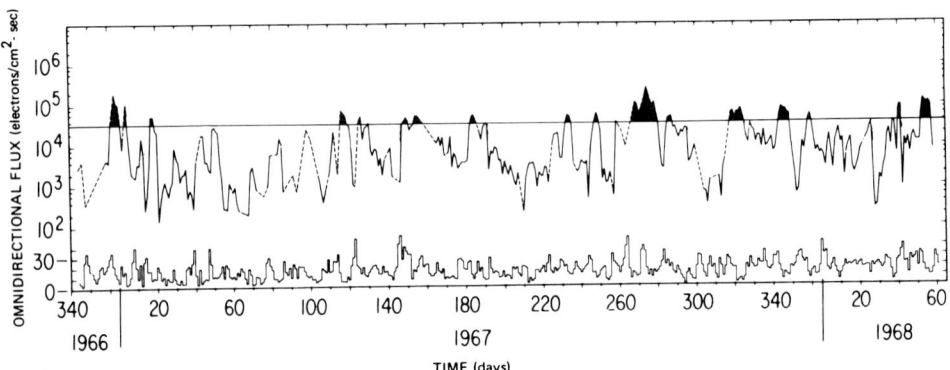

Fig. 1. Hourly averages of ATS 1 1.9 MeV omnidirectional electron flux at local midnight. Daily sum of Kp is plotted at the bottom of the figure. The horizontal line is the AEI-7 model equatorial flux at L = 6.6 and local midnight.

pending upon energy, there are large and frequent flux variations that are not necessarily correlated well with magnetic indices, such as Kp. At low outer zone L values inside the peak the data show less general activity and, at some energies, the flux changes exhibit a dominant event structure that begins to show correlation with magnetic indices. In modeling terms the outer zone (L $\geq$ 2.8) has classically been treated in a simple statistical manner (Vette and Lucero, 1967, and Singley and Vette, 1972). The particle flux was averaged because this relates directly to the dose rate. It is readily apparent from Figure 1 that the time during which the flux exceeds this average is small in relation to the time span of the data. Deviations from the average were treated statistically using a log-normal distribution for the data sample in a certain flux interval above or below the average. The implicit assumption of this technique is that the time scale of the temporal variations is small with respect to a typical mission duration, and that a spacecraft flying in the environment will, in a short time, encounter an average flux equal to the average of the observations, and therefore, to the model average. The data are treated separately at solar maximum and solar minimum, and the model average and log-normal distribution exhibit solar cycle effects for outer zone L values below 5 (Singley and Vette, 1972).

Outer Zone Models

The outer zone models, AE-4, issued in 1972 (Singley and Vette, 1972) have, in recent years, been compared by other workers with various new flux observations and dosimeter measurements obtained in the outer zone (e.g., Vampola et al., 1977). Various statements were published to the effect that the AE-4 model is 'low by one order of magnitude above 1 MeV,' that AE-4 'represents a quiet outer zone condition,' and that AE-4 'is in substantial error above 1.5 MeV.' For any of these statements to be completely accurate is to say that a substantial body of data included in this model is in error. Undeniably, however, there are

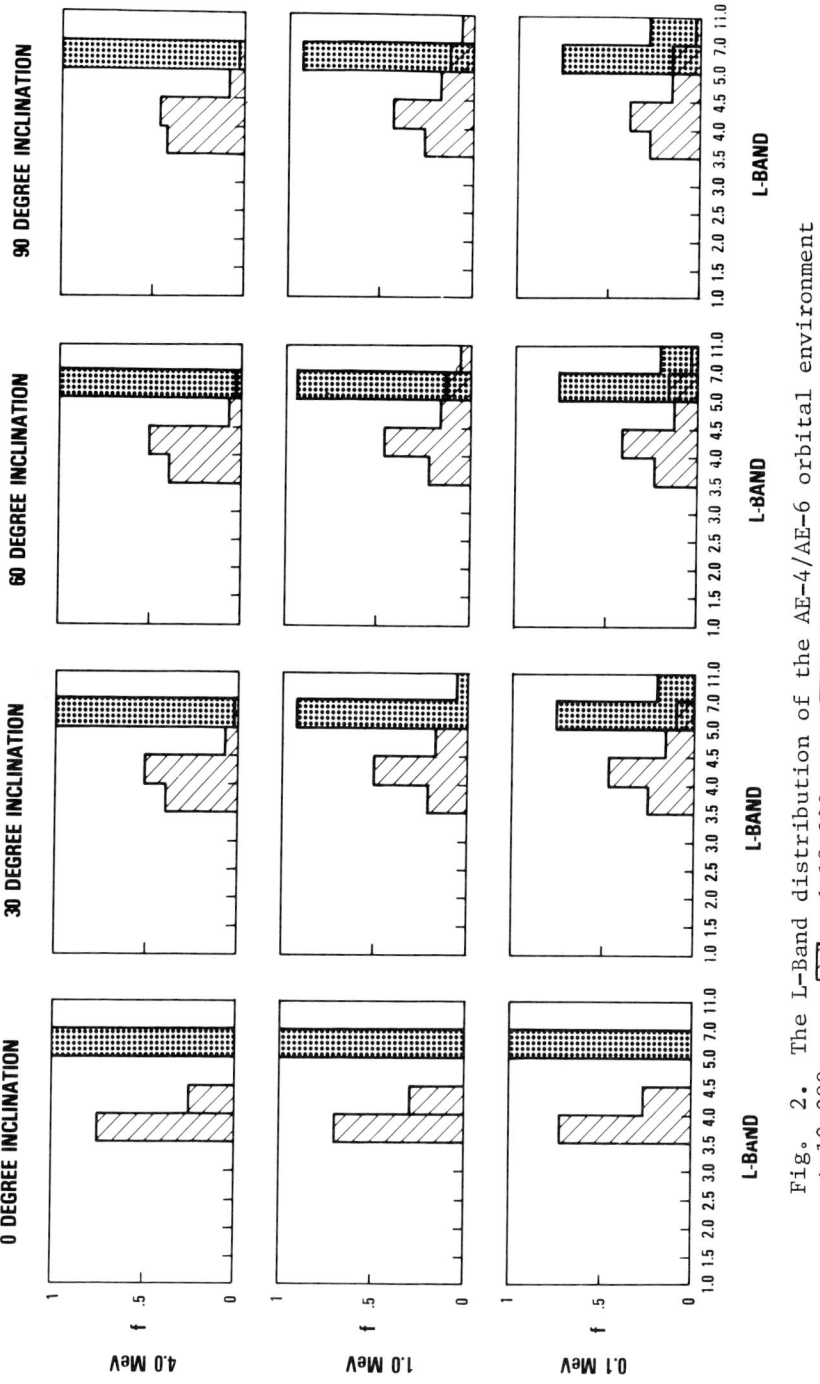

Fig. 2. The L-Band distribution of the AE-4/AE-6 orbital environment at 10,000 n.m. and 18,000 n.m. altitudes

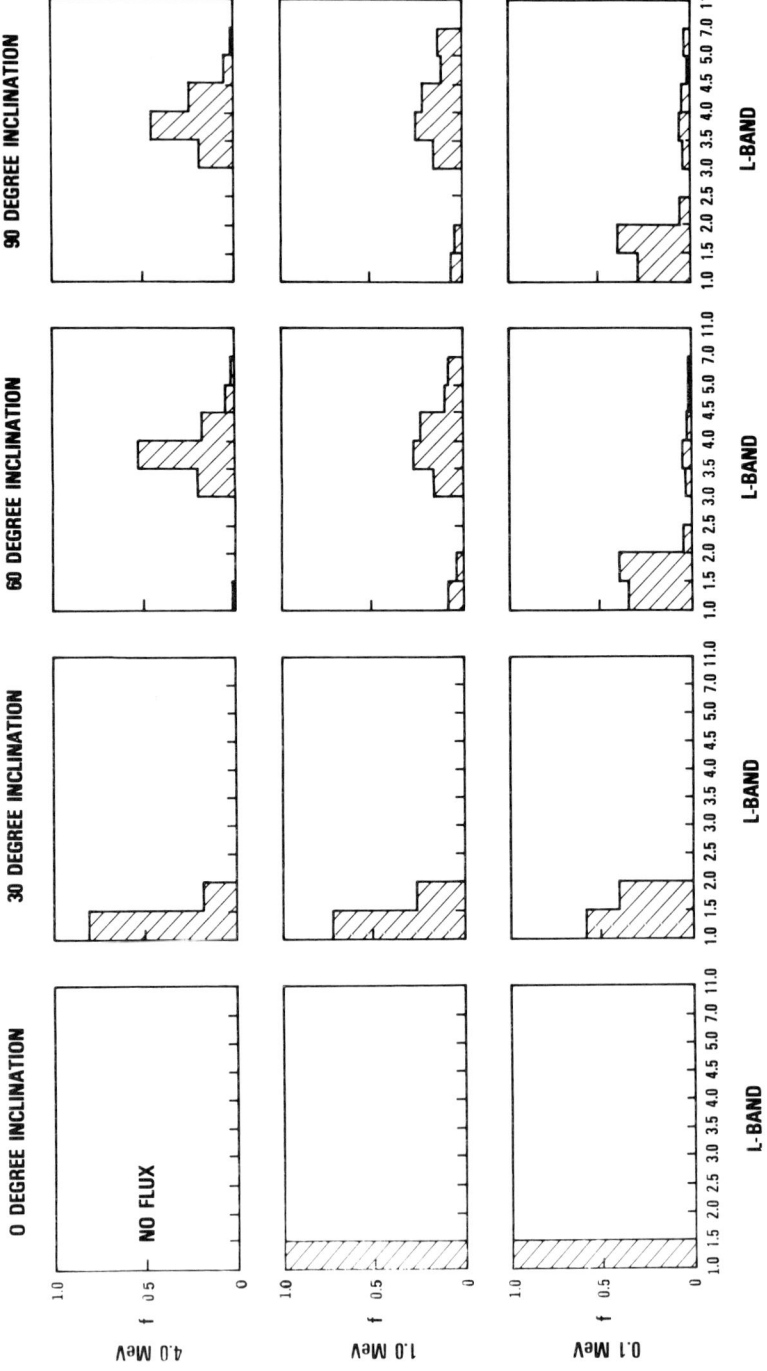

Fig. 3. The L-Band distribution of the AE-4/AE-6 orbital environment at 450 n.m. altitude

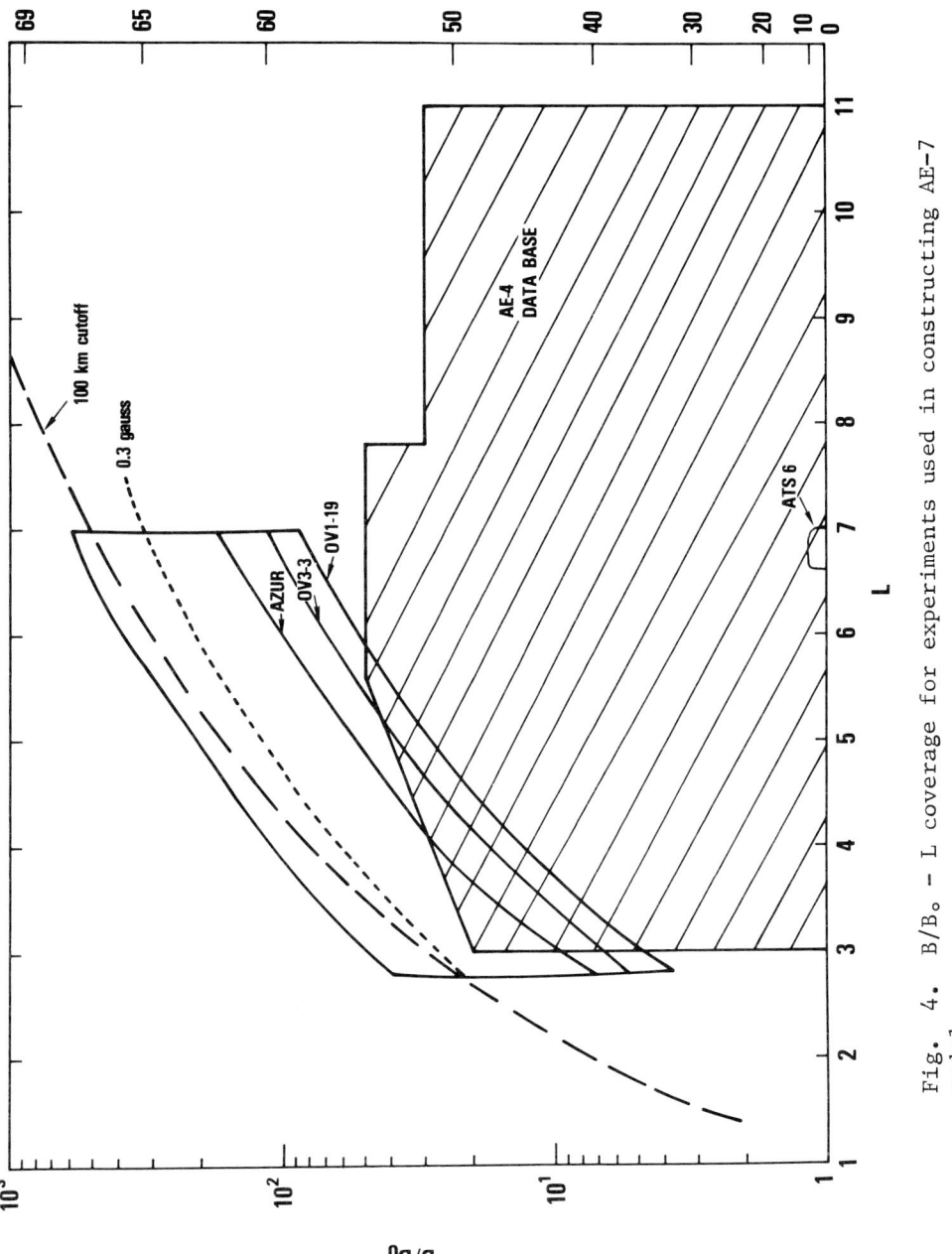

Fig. 4. $B/B_0$ - L coverage for experiments used in constructing AE-7 model

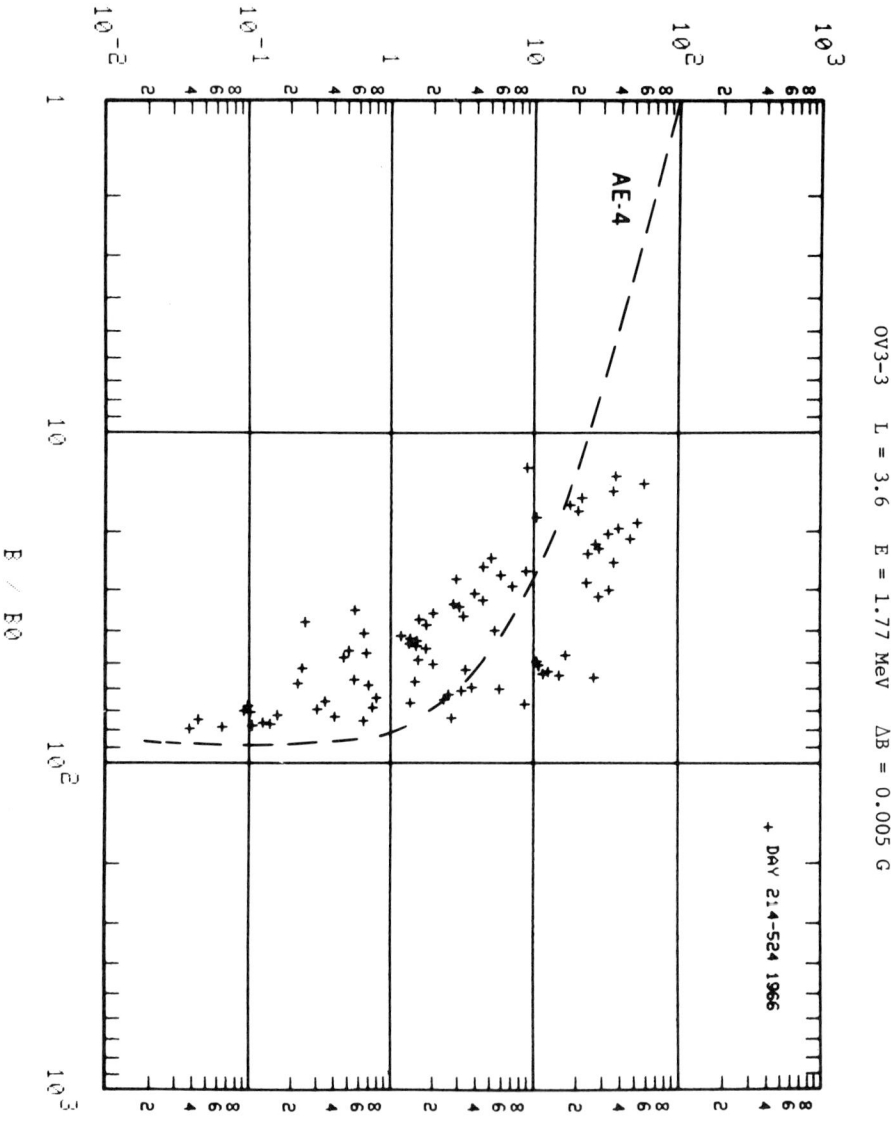

Fig. 5. Averaged locally mirroring flux

important problem areas in AE-4 that should, and will, be corrected when the AE-7 model is issued early in 1979. Basically, AE-4 shows disagreements with new data sets in regions of high gradients in energy and B. The most significant error occurs at high energies around the outer zone peak. This is an important region because an increasing number of missions are being flown with heavily shielded devices, which encounter the most significant fluences from this L range. This is illustrated in

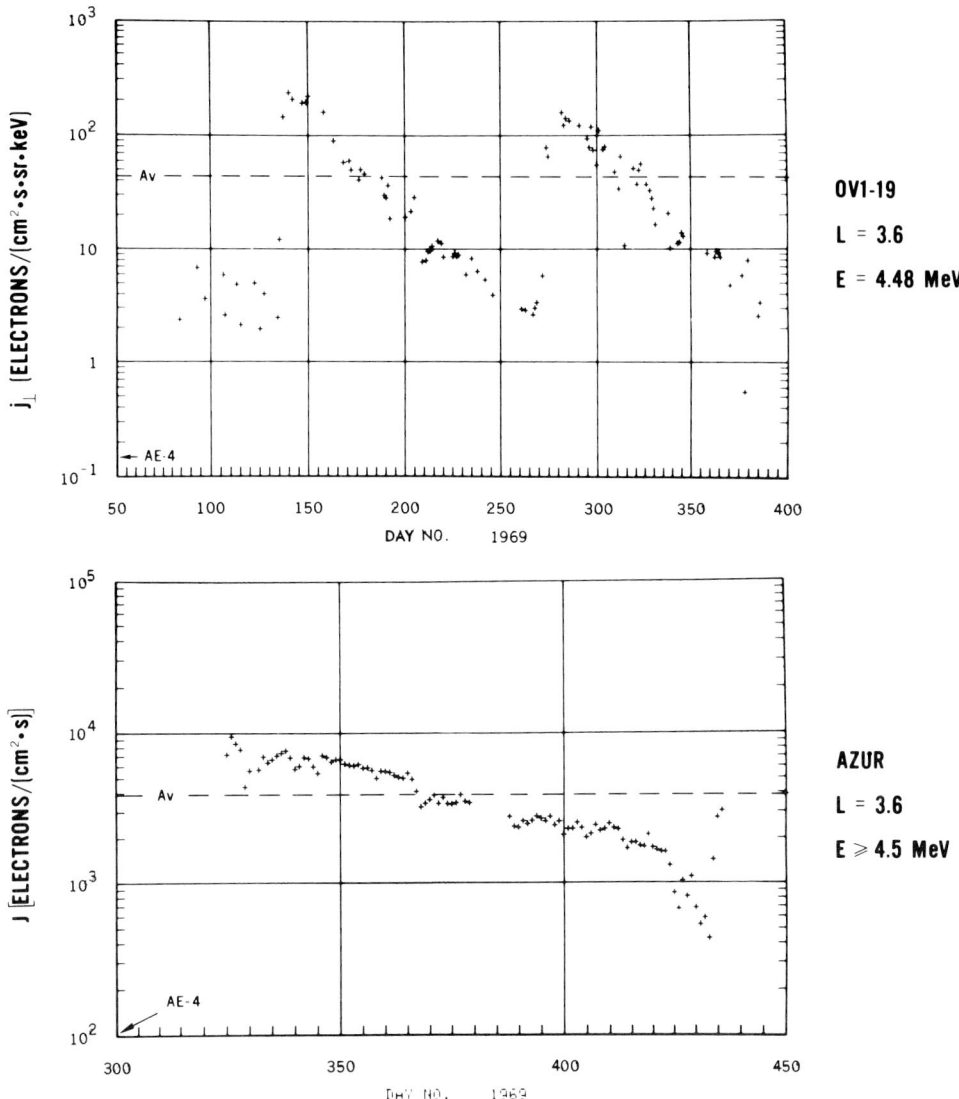

Fig. 6. Equatorial daily flux averages

Figures 2 and 3, in which the flux accumulated in various L bins is shown for various orbits and energies. For both the 450-n.m. and 10,000-n.m. orbits the most significant outer zone flux contribution arises from this suspect L range.

Figure 4 illustrates the B-L coverage of the data used for the generation of the AE-4 model. This is shown as a shaded area. It is clear that there are significant regions of B-L space near the atmospheric cutoff for which no data were available for inclusion in AE-4. This region is covered well by the OV3-3 (Vampola, 1969), OV1-19 (Vampola et al., 1977), and AZUR (Achtermann et al., 1970) satellites, as is shown

OV1-19 (UNIDIRECTIONAL, DIFFERENTIAL)   AZUR (OMNIDIRECTIONAL, INTEGRAL)
E = 4.48 MEV                            E ≥ 4.5 MEV

$$\frac{j \text{ (DAYS 77-387, 1969)}}{j \text{ (AE-4)}} \approx 320 \qquad \frac{J \text{ (DAYS 325-437, 1969)}}{J \text{ (AE-4)}} \approx 36$$

$$\frac{j \text{ (DAYS 325-387, 1969)}}{j \text{ (AE-4)}} \approx 100 \qquad \frac{J \text{ (DAYS 325-387, 1969)}}{J \text{ (AE-4)}} \approx 51$$

Fig. 7. Comparison of averaged fluxes with AE-4 model at L = 3.6

by the open box areas in Figure 4. Figure 5 compares the AE-4 model with 1.77-MeV OV3-3 data at L = 3.6. These data are not time selected and the data scatter results from the temporal variations occurring during the observations in 1966 and 1967. It is clear that the data show a faster cutoff than the model for $B/B_o \geq 45$. It should be noted

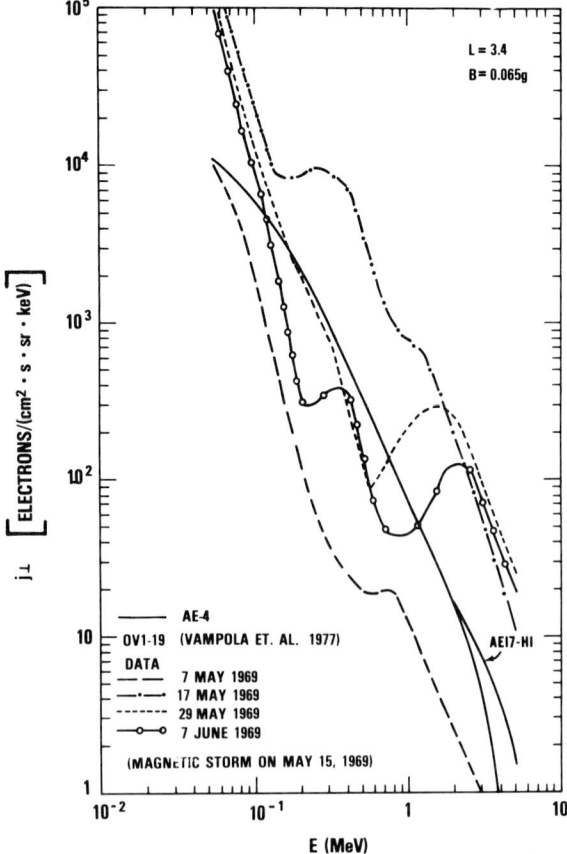

Fig. 8. Temporal variation of OV1-19 spectra and the AE-4 model

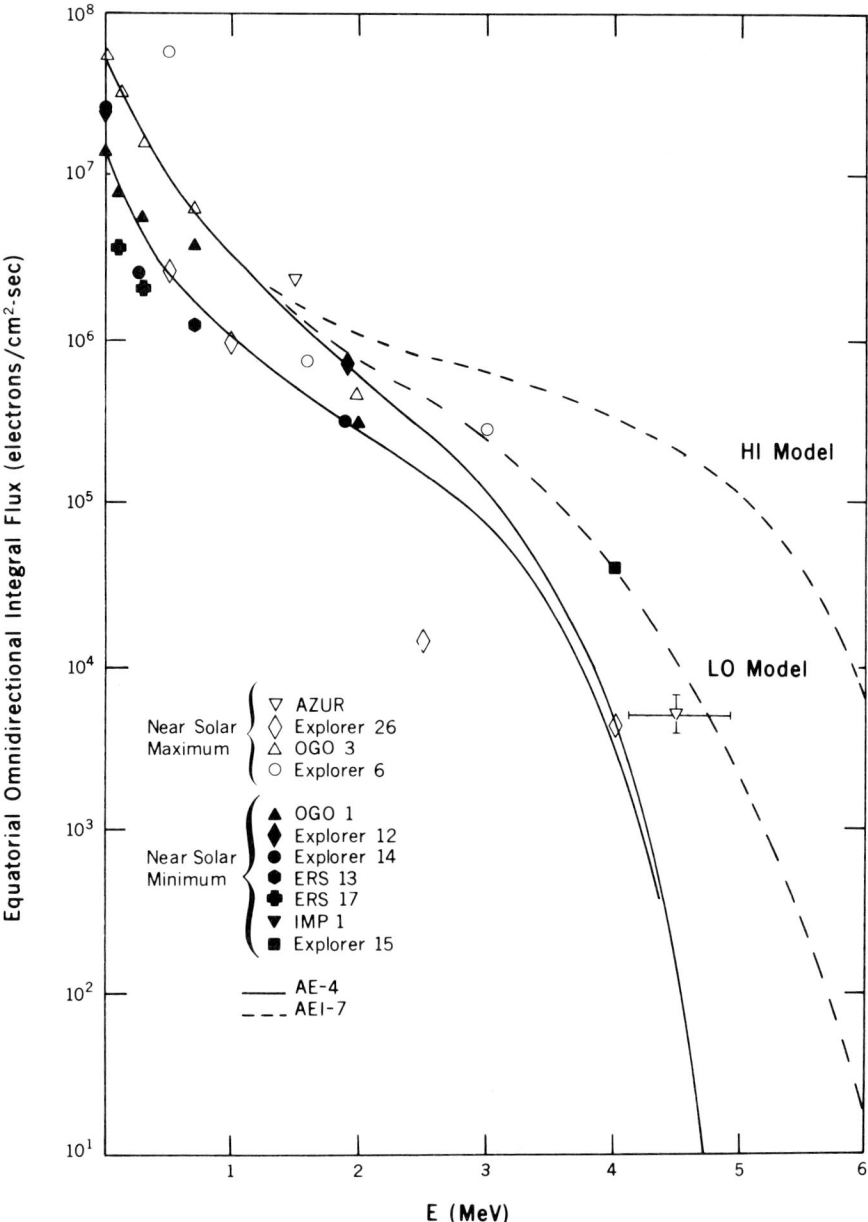

Fig. 9. Comparison of AEI-7 model spectra with various data at L = 4

that OGO 3 (Pfitzer, 1968) data are available to support the magnitude and the B-distribution of the model nearer the equator at this L value and energy (Singley and Vette, 1972). A similar situation is observed when AE-4 is compared to the B-distributions observed by the OV1-19 and AZUR satellites. The first problem with AE-4 is, therefore, in the region of steep gradients in B. In all probability this extends to all

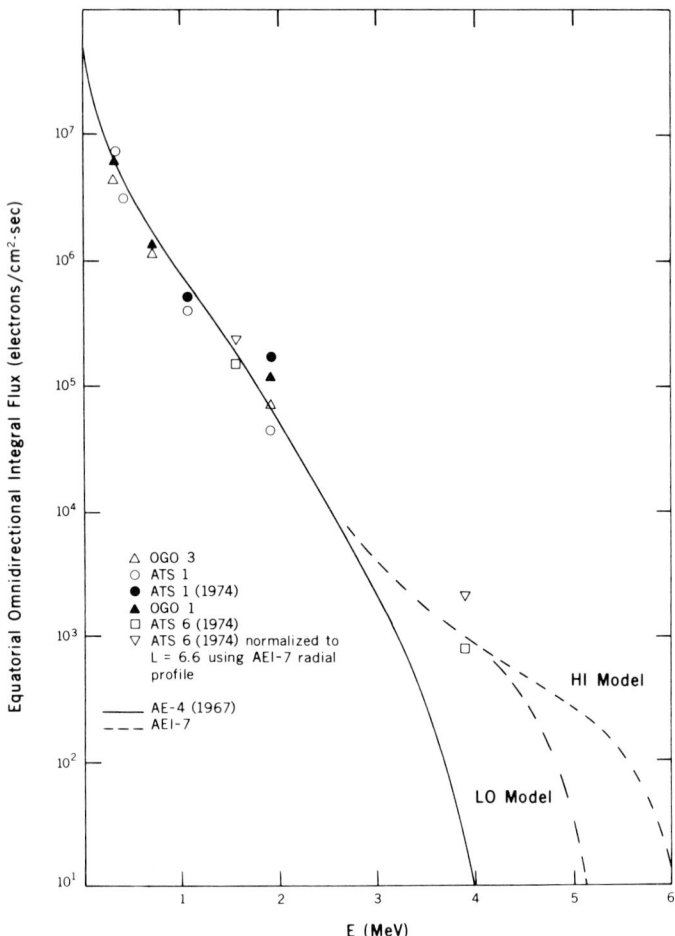

Fig. 10. Comparison of AEI-7 model spectra with various data at L = 6.6

energies and outer zone L values. In order to factor out this problem in the subsequent analysis, AZUR, OV3-3, and OV1-19 data are retrieved for $B < 0.3$ gauss (corresponding to $B/B_o = 45$ at $L = 3.6$). In addition, for these latter two data sets, only $J_{perp}$ data are used.

The second problem with AE-4 pertains to the model spectrum. Figure 6 shows the AZUR and OV1-19 data at L = 3.6 and ~4.5 MeV. The OV1-19 data are unidirectional differential and the AZUR data are omnidirectional integral for a subset of the time shown for OV1-19. The averages of the fluxes, which would have been used to determine the model had these data sets been available, are shown as broken lines. The AE-4 flux levels are shown and the large differences between model and data are self evident. Quantitatively, the differences are summarized in Figure 7. Averaging over the entire data set in each case, the model is lower than the data by a factor of 320 in one case and 36 in the other. The question arising is: why are these factors different by an order of magnitude? The answer is not that the data disagree, but that

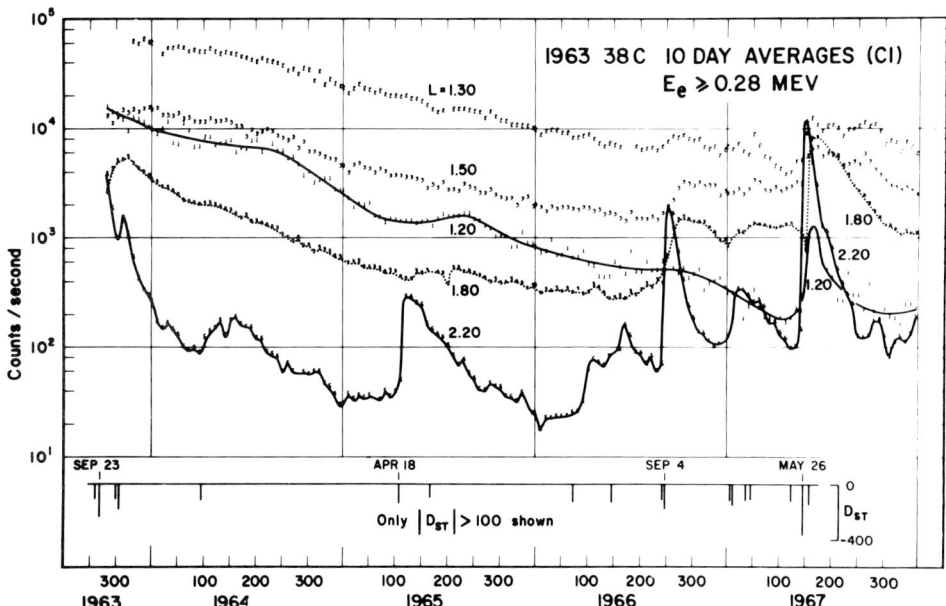

Fig. 11

the temporal variations are not of the type that the data can be averaged over different periods in excess of 100 days with the same average resulting. If the averaging is performed over an identical period of approximately 60 days, the ratio of the observed fluxes to the model is approximately the same. In Figure 6, the structure of the temporal variations shows little resemblance to the classical outer zone data that were assumed to exhibit continuous and large variations of characteristic time of a few days. Thus, Figures 6 and 7 show that not only is the AE-4 model low at certain energies and L values, but that the modeling technique used in these regions cannot adequately reflect the environment over reasonable mission durations. This is dramatized in Figure 8, taken from a paper by Vampola et al., which shows various OV1-19 spectra observed at L = 3.4 during the May 1969 storm. The important feature is that although all energies show an immediate enhancement, the high energy fluxes remain enhanced for considerable periods approaching 30 days. The AE-4 model is shown for comparison. In our opinion, a new technique of modeling is required for these high energy regions.

Various model energy spectra are summarized in Figures 9 and 10 at L = 4.0 and 6.6, respectively. At the lower L values, the AE-4 curve at high energies was based upon a single data set, Explorer 26 (McIlwain, 1963). Subsequently, AZUR, OV1-19, and Explorer 15 (McIlwain, 1963) data were added. The time-averaged OV1-19 data are represented by the upper broken line, labeled HI Model, which was derived using the published fits of Vampola et al., 1977, extrapolated from the 5-MeV limit of the observations to 7 MeV. Our present analysis of the OV1-19 data, which is provisional, results in a curve that is a factor of 2 to 3 lower than this line. The two broken lines represent the two versions

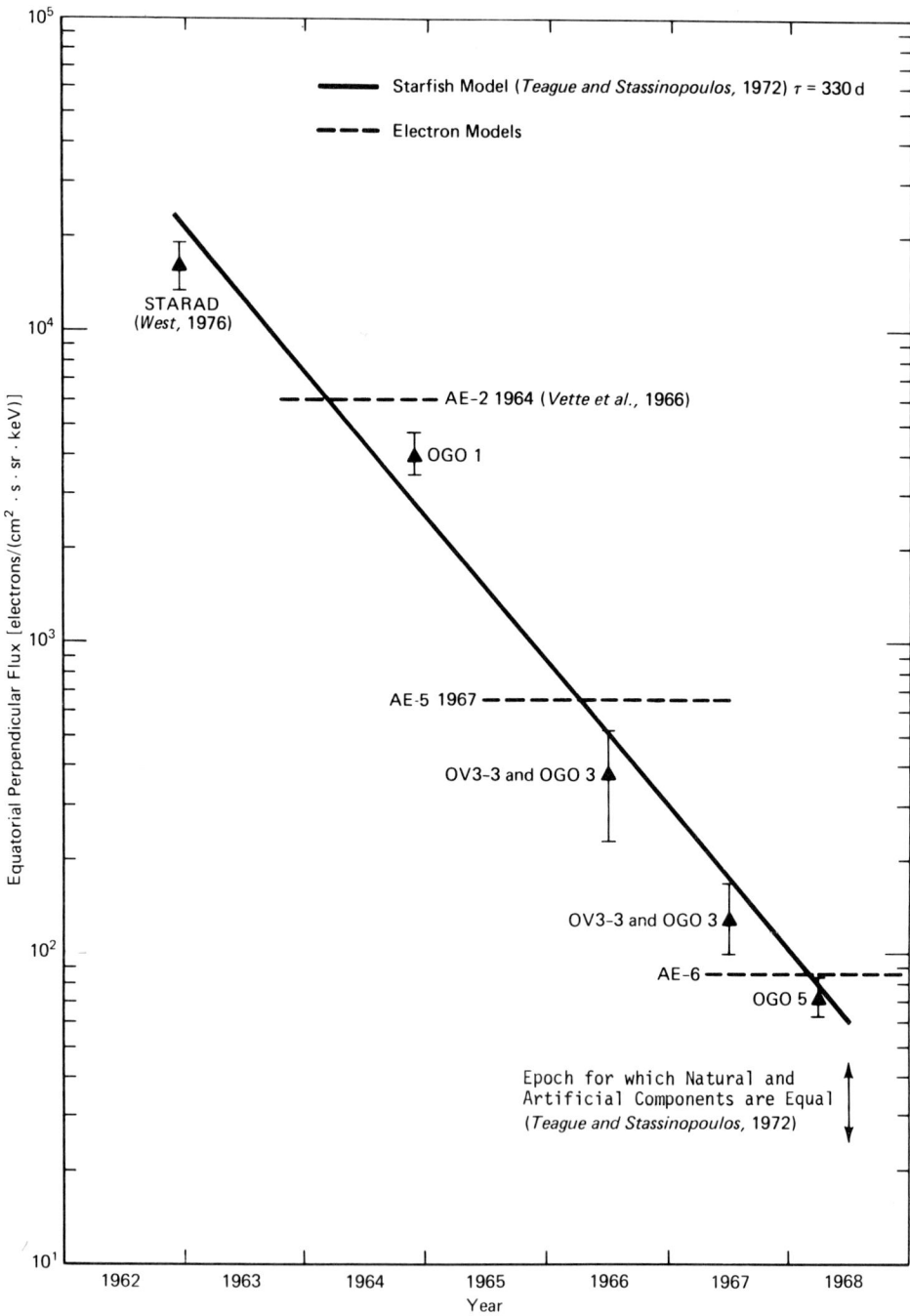

Fig. 12. Decay of Starfish electrons, L = 1.5, and E = 1.5 MeV

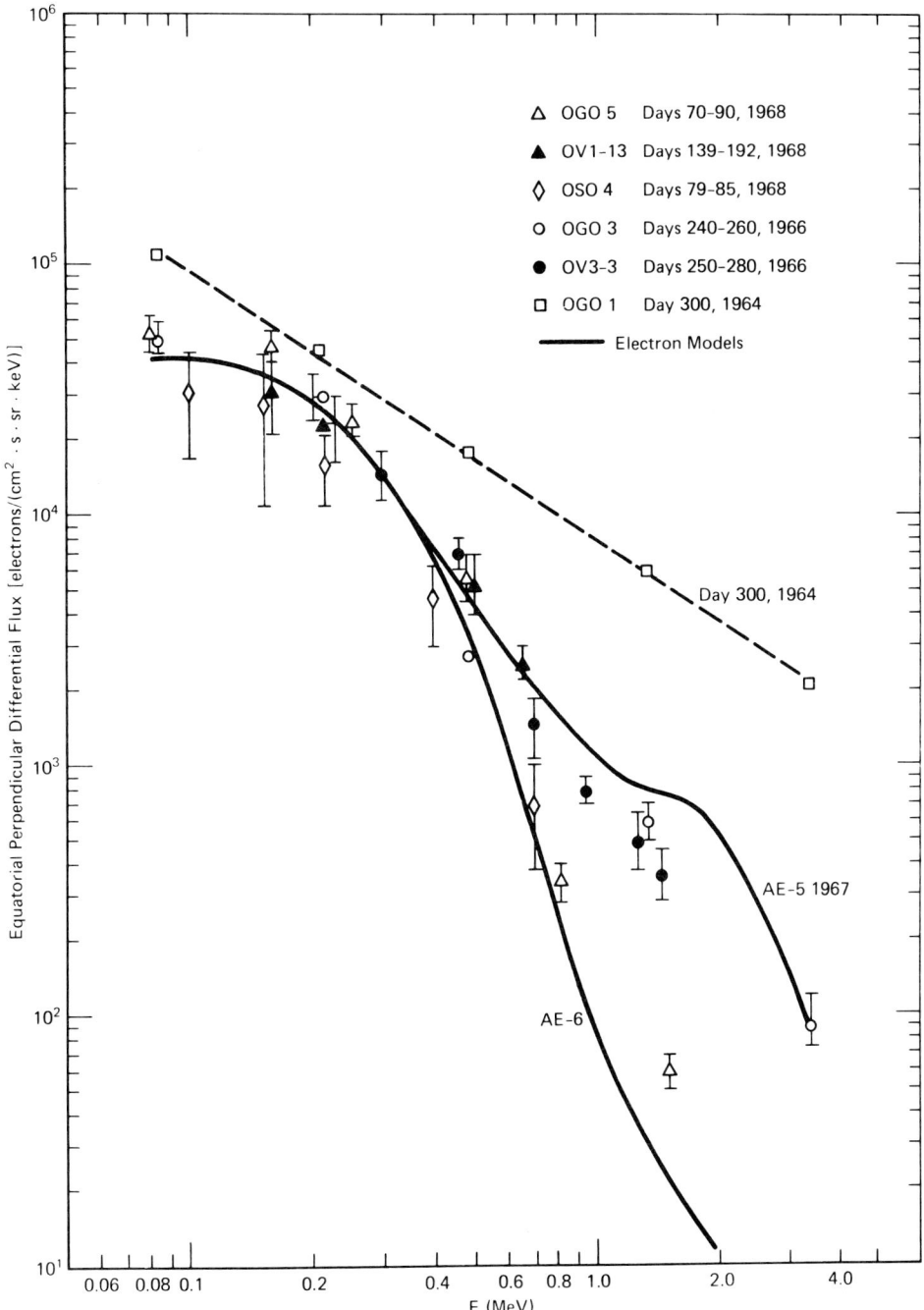

Fig. 13. Inner zone electron spectra, L = 1.3

| DATE OF COMMENCEMENT | PEAK HOURLY MEAN DST INDEX | L = 2.4, 500 KEV PEAK FLUX VALUE* | SATELLITES OBSERVING |
|---|---|---|---|
| APRIL 1965 | -185 | $2.9 \times 10^4$ | OGO 1 |
| SEPTEMBER 1966 | -229 | $5.0 \times 10^4$ | OGO 3, OV3-3 |
| MAY 1967 | -418 | $1.9 \times 10^5$ | OGO 3, OV3-3 |
| JUNE 1978 | -94 | $9.8 \times 10^3$ | OGO 5, OV1-13 |
| OCTOBER 1968 | -231 | $1.5 \times 10^5$ | OGO 5 |
| MAY 1969 | -133 | $1.6 \times 10^4$ | OV1-19 |
| SEPTEMBER/OCTOBER 1969 | -128 | $7.5 \times 10^3$ | OV1-19 |

*ELECTRONS/($CM^2 \cdot S \cdot SR \cdot KEV$)

Fig. 14. Geomagnetic storm data

of the interim model AEI-7, which was issued as a temporary response to the problems identified with the AE-4 spectrum. Our present recommendation is that the LO model should be used until the AE-7 model becomes available. At L = 6.6, in Figure 10, the problems with the AE-4 spectrum are significantly less severe. ATS-6 (Paulikas and Blake, 1978) data are available at this L value, and significant errors in the AE-4 spectrum are not observed below 3 MeV.

### Inner Zone

Temporal Variations

The inner zone temporal variations are markedly different from those observed in the outer zone for L > 5.0, as shown in Figure 11, which shows the well known 280-keV data from the 1963-038C satellite (Beall, 1969). At high inner zone L values, the dominant temporal effects are identifiable events correlated with large magnetic disturbances. The flux changes are, however, less frequent and of a smaller relative magnitude than the corresponding variations in the outer zone. The inner zone storm variations are quickly damped toward the lower L values, and below L = 1.7 such variations are unimportant in a modeling sense. The standard modeling technique in the inner zone was the simplest one of all: average the data regardless of the temporal variations. No effort was made to model these variations and it is clear that a statistical approach, such as that used in the outer zone, is inappropriate as the frequency of occurrence of significant inner zone events is small. Figure 11 shows only four significant events in the same number of years. The rationale for the modeling technique adopted for the inner zone is, firstly, there is no easy way to treat the storm variations in the inner zone on an average basis. Secondly, and more importantly, although the flux changes caused by storms at an L value toward the outer edge of the inner zone are significant in relation to a quiet time flux at the same L value, they are not significant in relation to the higher average

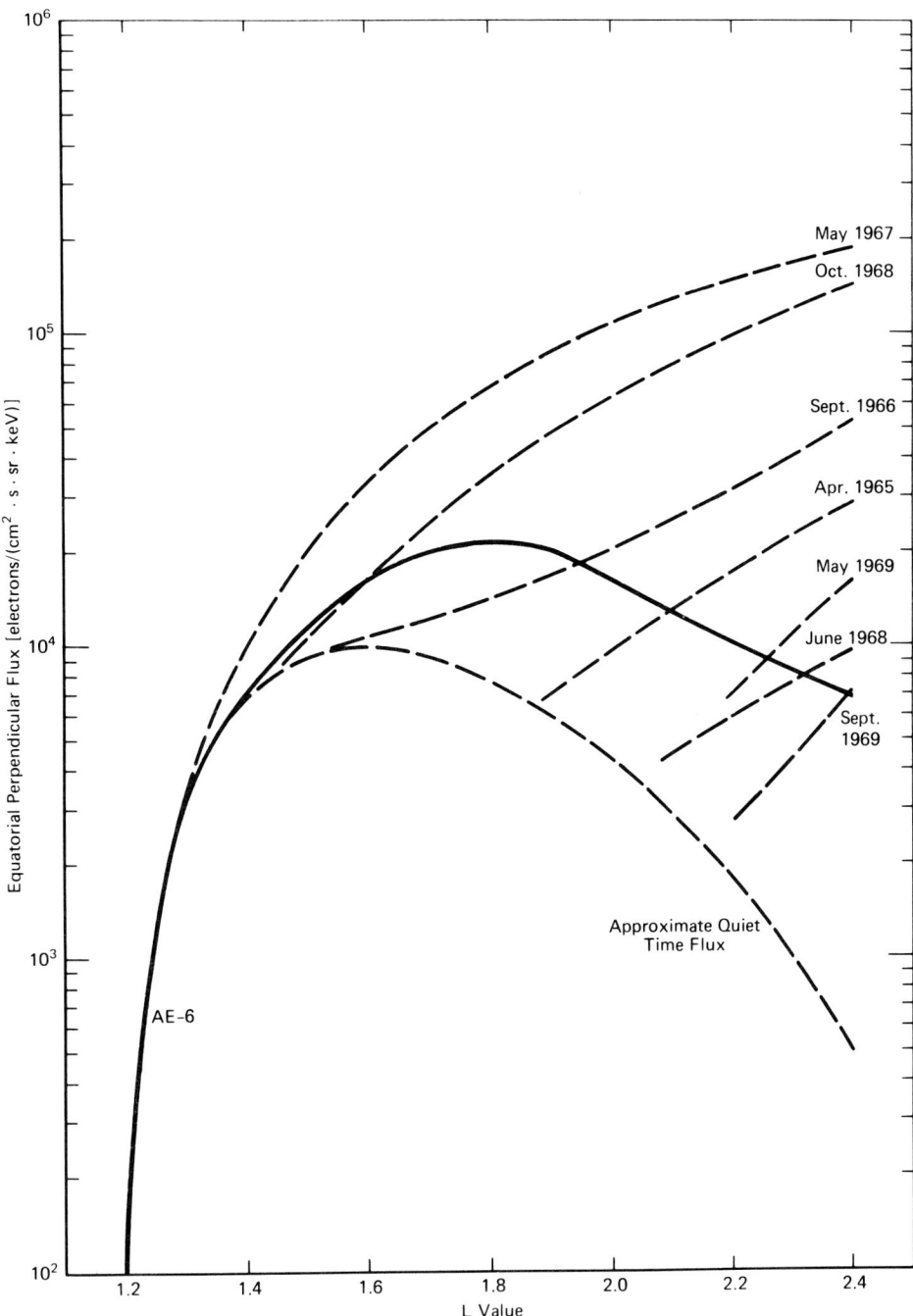

Fig. 15. Peak storm profiles for 500 keV electrons

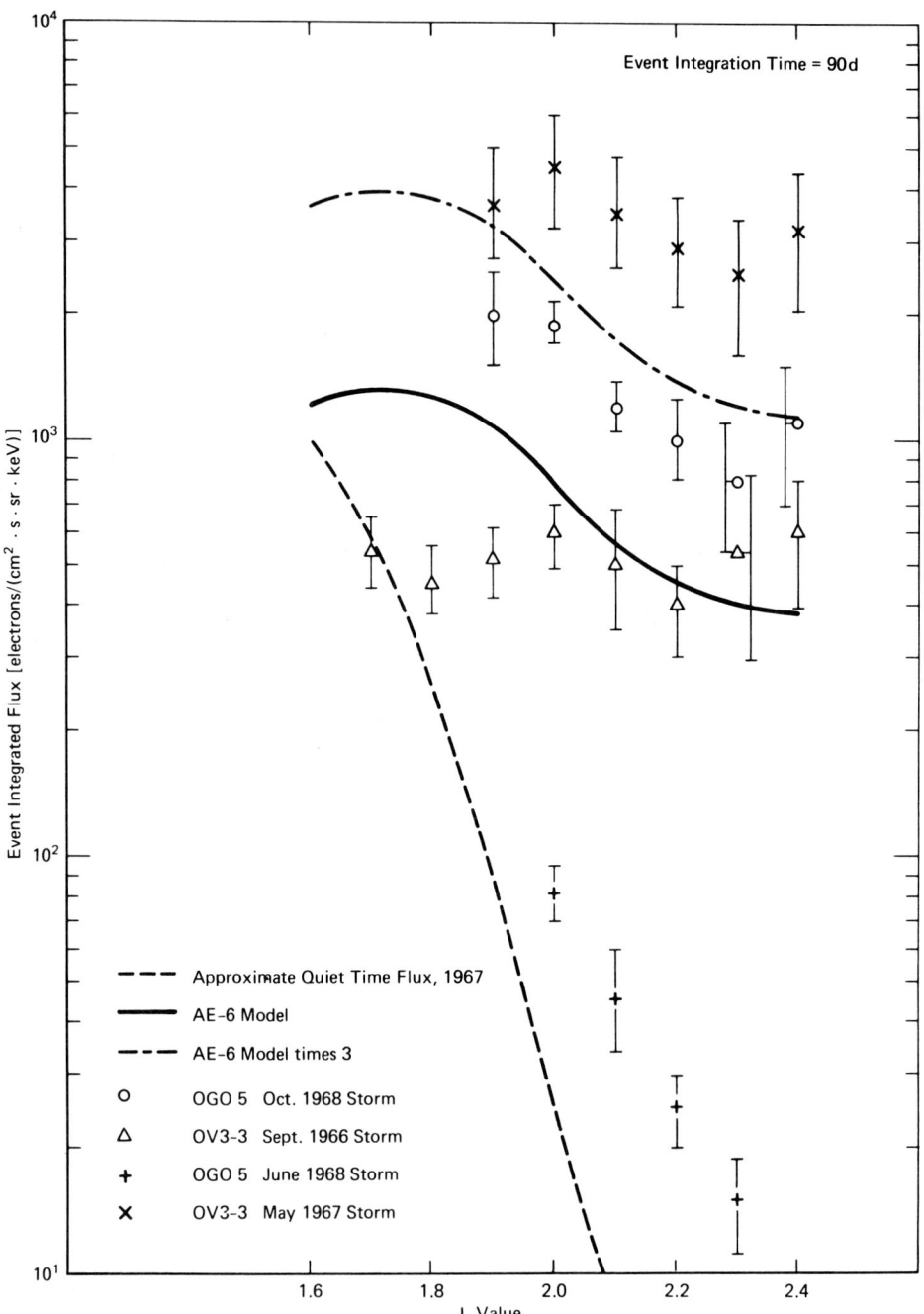

Fig. 16. Event integrated radial profiles, E = 800 keV

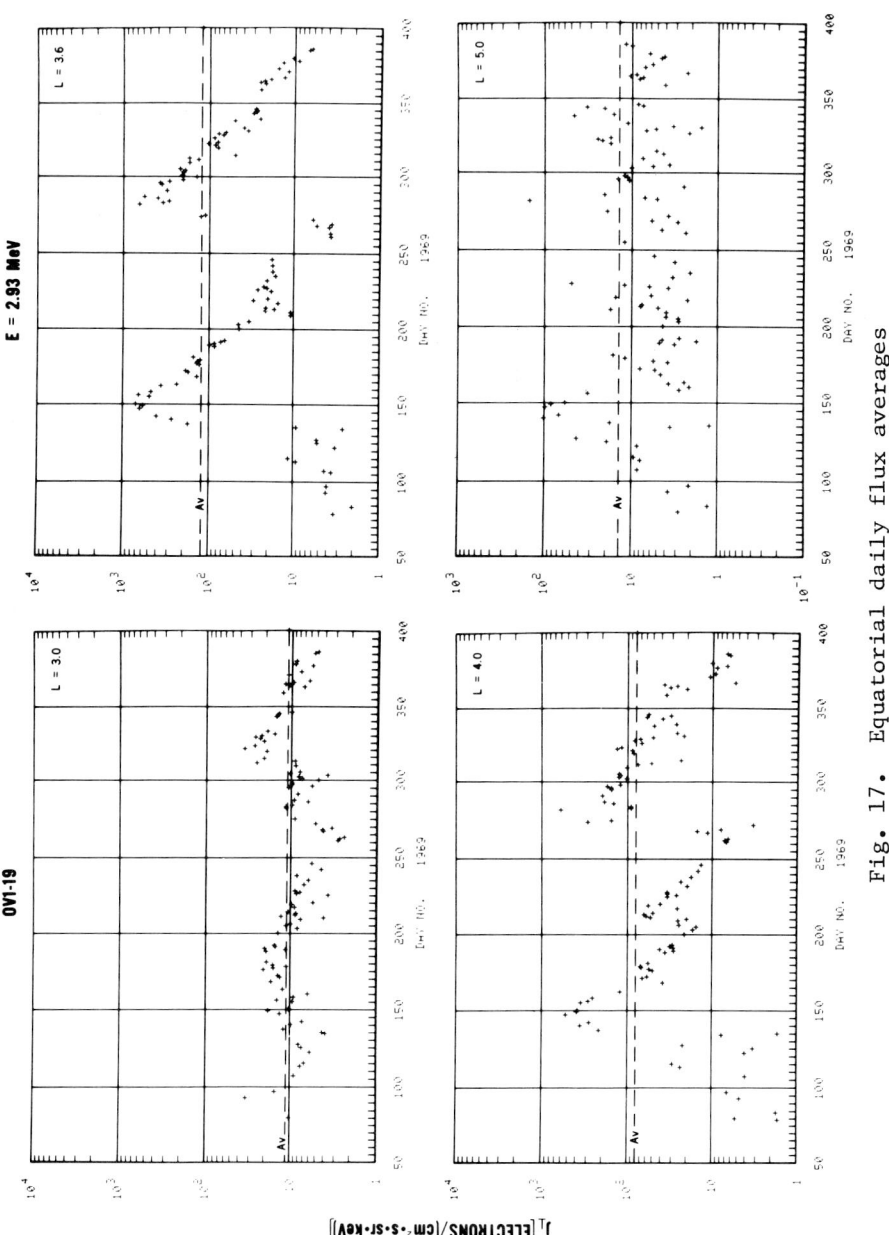

Fig. 17. Equatorial daily flux averages

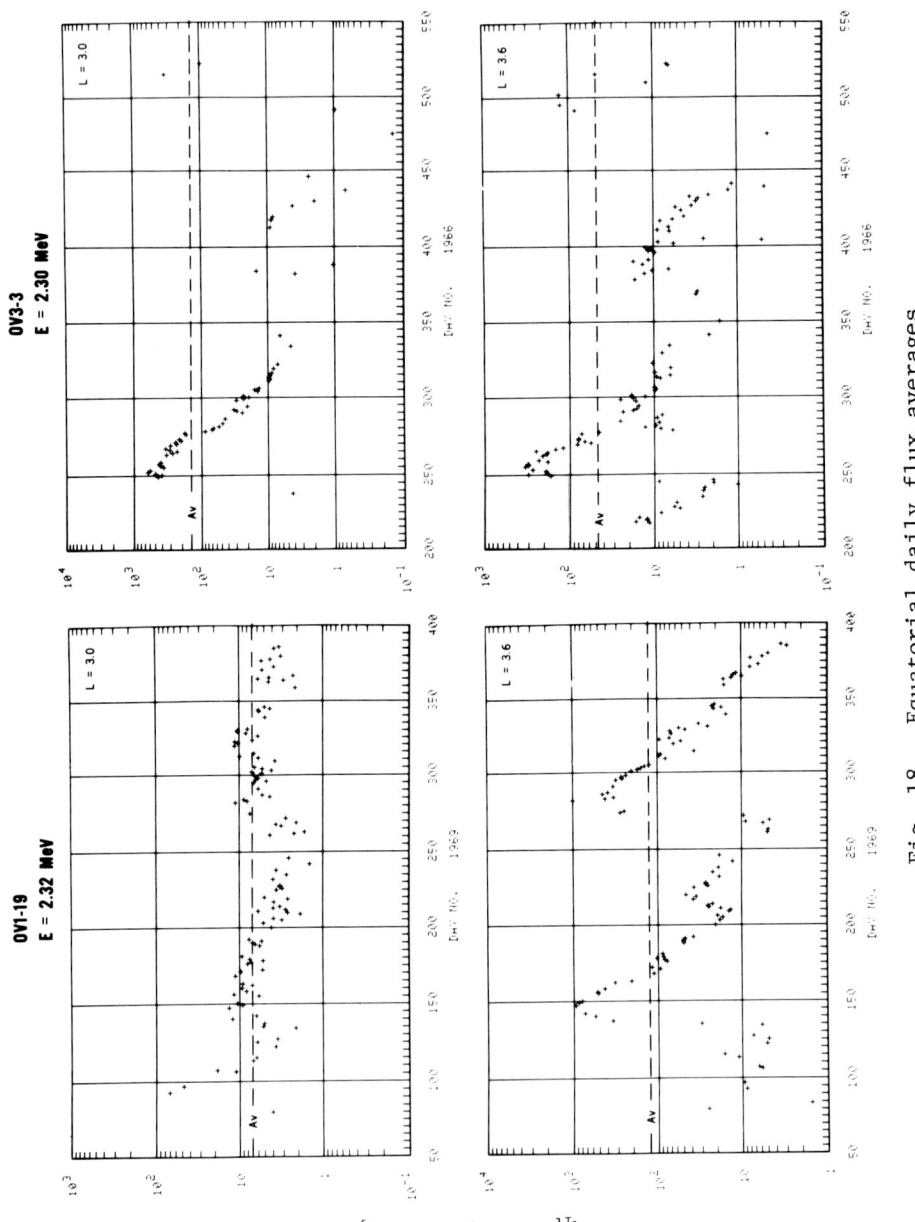

Fig. 18. Equatorial daily flux averages

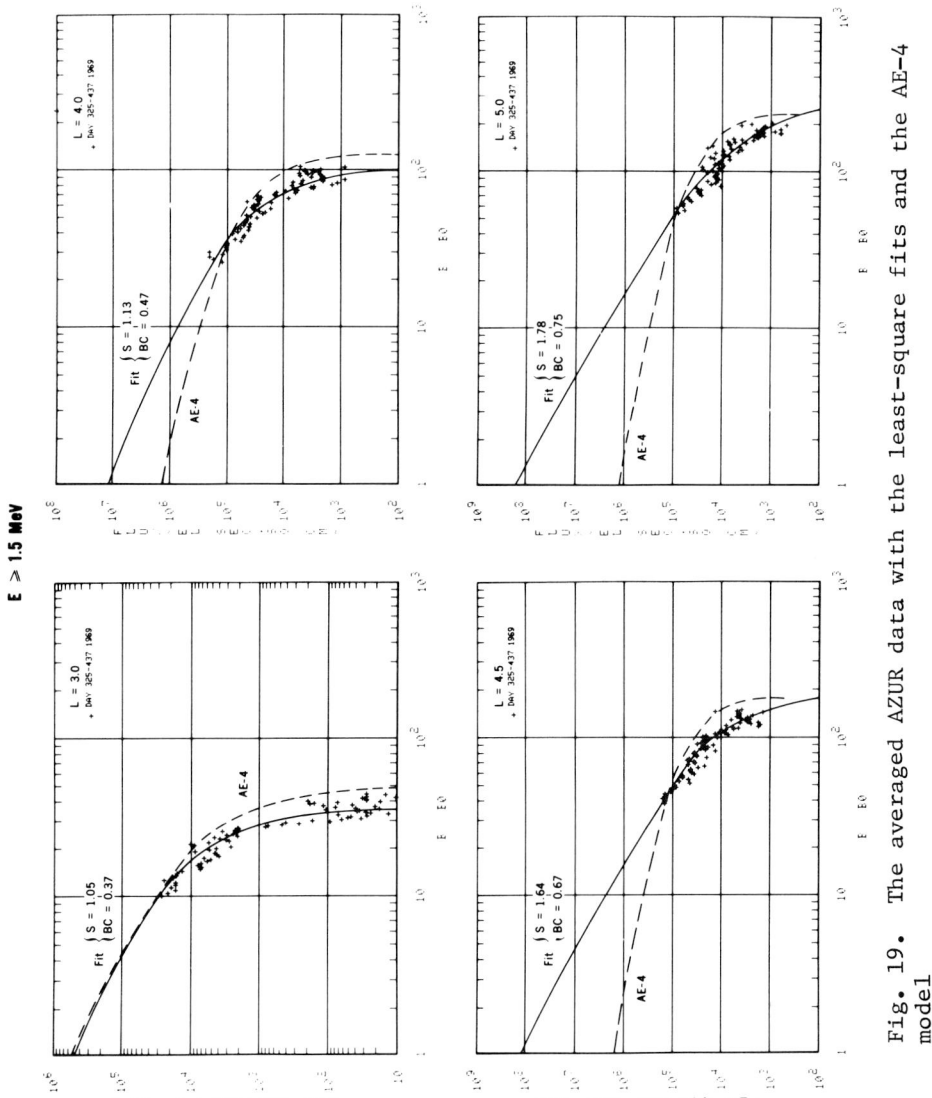

Fig. 19. The averaged AZUR data with the least-square fits and the AE-4 model

THE EMPIRICAL CUMULATIVE DISTRIBUTION FUNCTION

$$P(F > F_x) = \frac{\text{NUMBER OF DATA POINTS WITH } F > F_x}{\text{TOTAL NUMBER OF POINTS IN SAMPLE}}$$

THE GAUSSIAN DISTRIBUTION FUNCTION OF $Z \equiv LOG_{10} F$

$$P(Z > Z_x) = \int_{Z_x}^{\infty} \frac{1}{\sqrt{2\pi\sigma^2}} \; EXP\left[-\frac{(Z-\mu)^2}{2\sigma^2}\right] DZ$$

WHERE $\mu$ AND $\sigma$ ARE FITTING PARAMETERS

(AFTER *VETTE AND LUCERO* (1967), AND *PAULIKAS AND BLAKE* (1971))

Fig. 20

fluxes observed at lower inner zone and higher outer zone L values. On an orbit-integrated basis, therefore, the temporal variations in this region of the radiation zone were assumed to be relatively unimportant. This is illustrated in Figure 3 for various inclinations of a 450-n.m. circular orbit. Although in most cases a significant inner zone flux component is apparent, in no case does a significant contribution come from the L range 2.0 to 2.5. Strictly speaking, while the orbit integration rationale may be true, it does not result in an elegant model and certainly not in a model that provides a good description of the temporal variations at high inner zone L values.

In Figure 11, it is obvious that there is a very significant long-term variation at low inner zone L values. As is well known, this is

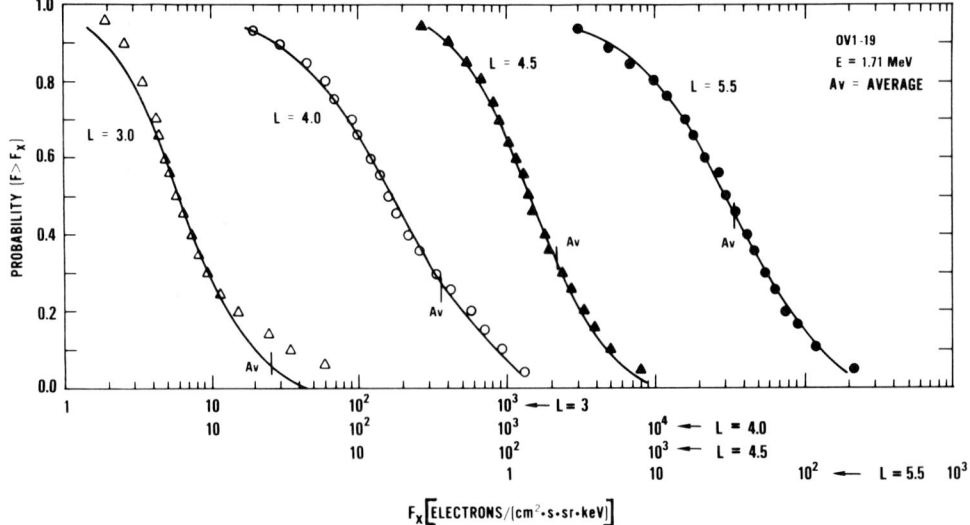

Fig. 21. Gaussian fit to the probability of observing flux $F > F_x$

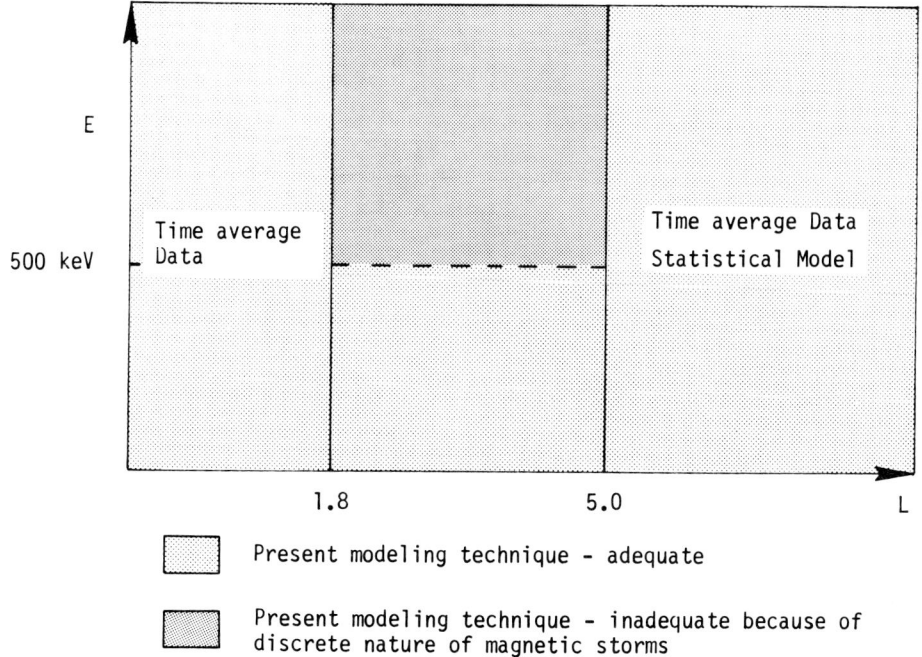

Fig. 22. Modeling technique for trapped electron models

the decay of the residue from the artificial Starfish injection event of July 1962. It is not appropriate to dwell on this to any great extent, because the effects of the injection no longer influence either the inner or the outer zone electron population. However, the event did have a very significant effect on the evolution of the inner zone electron models, as is shown in Figure 12. Data are shown from STARAD (West and Buck, 1976), OGO 1 (Pfitzer, 1968), OGO 3, OV3-3 and OGO 5 (West and Buck, 1976) for various epochs over the time period 1962-1968. Three generations of inner zone models are shown varying in flux levels over two orders of magnitude at L = 1.5 and 1.5 MeV. The models are AE-2 (Vette et al., 1966), AE-5 (Teague and Vette, 1972), and AE-6 (Teague et al., 1976). The solid line represents the Starfish model of Teague and Stassinopoulos, 1972, which was not available in machine-sensible form but was used in the derivation of the current inner zone model AE-6. The Starfish model included cutoff times or times for which the artificial flux component would have decayed to a level equal to the natural flux component. In many cases these cutoff times corresponded to epochs beyond those for which data were available, as in the case shown in Figure 12. The longest lived artificial particles were predicted to be at L = 1.3 and ~1.7 MeV, and to persist at significant levels until 90 to 95 months after the Starfish event.

Typical low L value, inner-zone spectra are shown in Figure 13. The OGO 1 data correspond approximately to the epoch of the AE-2 model while the OGO 3 and OV3-3 data correspond to the epoch of the AE-5 1967 model, and the OGO 5 and OV1-13 data (Rothwell and Moomey, 1972) correspond to

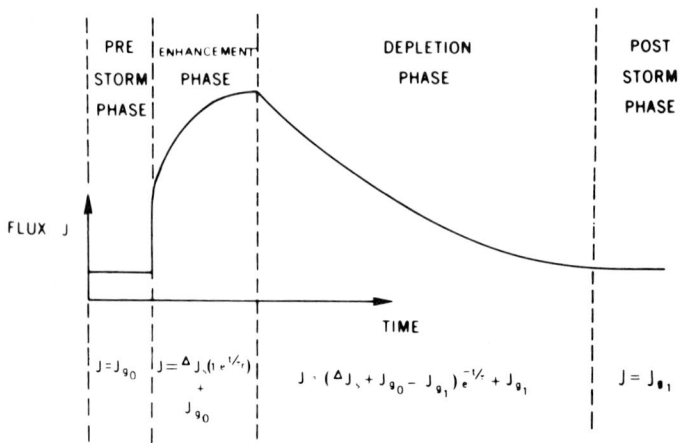

Fig. 23. Storm model

the AE-6 model (although these latter data sets were not available at the time AE-6 was generated). The relationship of this model to the 1.58-MeV OGO 5 data is interesting. Although the fact that the model flux is lower than the observation does not of itself establish that a Starfish decay process was still in effect at the epoch of the data; this was the prediction of the Starfish model based upon fitting a simple model of exponential decay to a constant natural background to the OGO 1, OGO 3, and OV3-3 observations. If this is correct, then there are still regions of the inner zone for which there are no observations of the post-Starfish natural high energy flux. Potentially, OV1-19 observations may help to resolve this question, although these observations suffer significant proton contamination in the relevent inner zone region.

The Inner Zone Model AE-6

The AE-6 model was derived from AE-5 1967 using the Starfish model to remove the artificial flux component. In addition, corrections were made for errors in the calibration coefficients used for the OV3-3 data set incorporated into AE-5 1967 (Teague et al., 1976). A study of the AE-6 model using data from OV1-19, OV1-13, OGO 5, and OSO 4 (Knox, 1972)

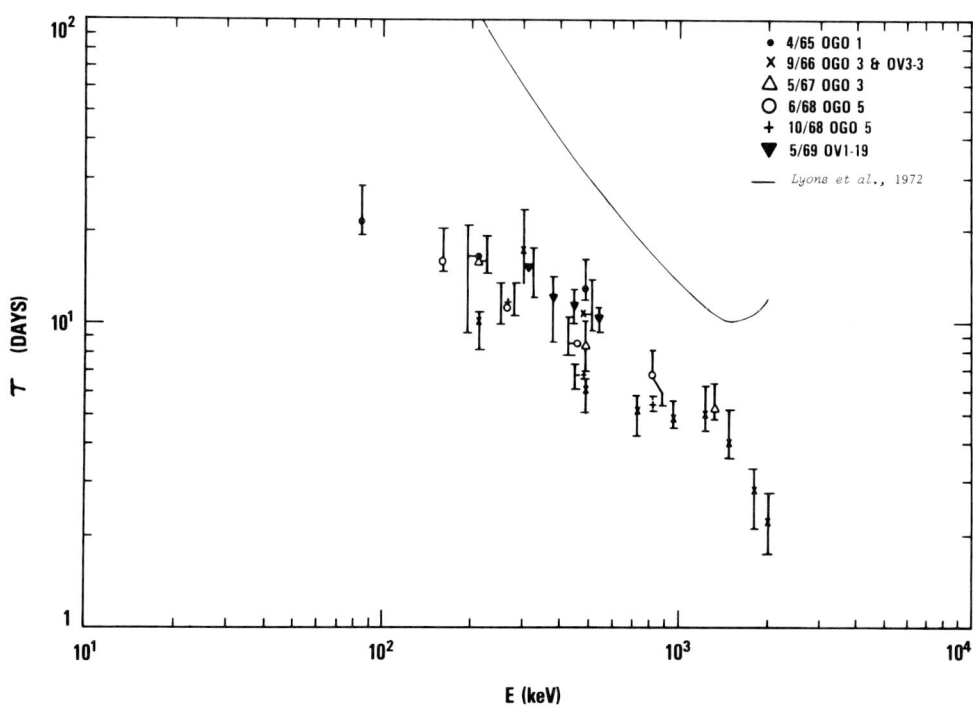

Fig. 24.  Storm depletion time, L = 2.4

was recently completed (Teague et al., 1978). None of these data sets were used in the derivation of AE-6. Figure 13 typifies the inner zone spectra over the L range 1.3 to 1.7. At the higher L values in this range, however, there is a smaller artificial component in the data taken in 1966 and 1967, and the OV3-3, OGO 3 and OGO 5 data show good agreement with the model.

At high inner zone L values, storm effects become significant and comparison of the model and data becomes less straightforward. A large body of inner zone storm data is now available and is summarized in Figure 14. Over a 4-year period, observations of seven storms were made covering a wide Dst range and showing peak fluxes at L = 2.4 and 500 keV, varying by a factor of 30. In three cases, multiple observations of the same storm were made. Figure 15 compares the peak flux observed during these storms with the AE-6 model at 500 keV. This energy is chosen because all of the satellites identified in Figure 14 made observations near this energy. An approximate quiet time or prestorm flux is shown, although this flux may vary by a factor of 2 to 3 from storm to storm. The largest flux changes exceed two orders of magnitude at L = 2.4, and peak fluxes an order of magnitude higher than the model are observed over the L range 2.1 to 2.4. At L = 1.7 and lower, however, the model is always within a factor of two of both the peak and quiet time fluxes.

A reasonable basis for comparing model and data over the L range 1.8 to 2.4 is to event-integrate the storm data. An event-integrated radial profile at 800 keV is shown in Figure 16, based upon an event integra-

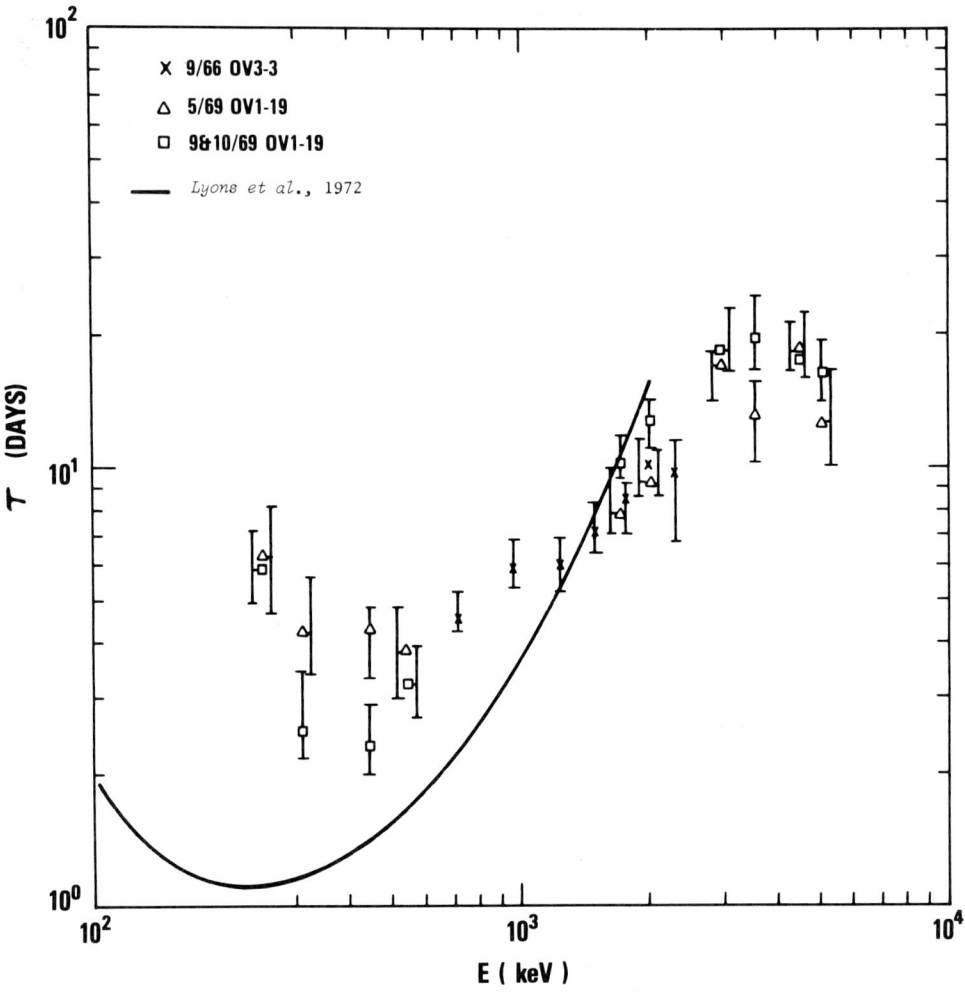

Fig. 25. Storm depletion time, L = 3.6

tion time of 90 days. The quiet time profile and the model profile, multiplied by three, are shown. The latter profile encompasses all but the largest storm. This multiplicative factor is recommended as the model accuracy factor at these L values and energies (Teague et al., 1976). Although the results of this kind of analysis are very dependent upon the integration time, the analysis does provide an indication of the ability of the model to predict the averaged effects of major inner zone storms when they occur. The corollary is that the model is particularly high if no storm occurs, or if only a small storm occurs. If an appropriate 6-month interval is chosen at L = 2.4, the AE-6 model may be higher than the average flux by an amount in excess of an order of magnitude at 1 to 2 MeV. Because the frequency of occurrence of storms in general, and large storms in particular, is low in the inner zone, the AE-6 model is a pessimistic one the majority of the time.

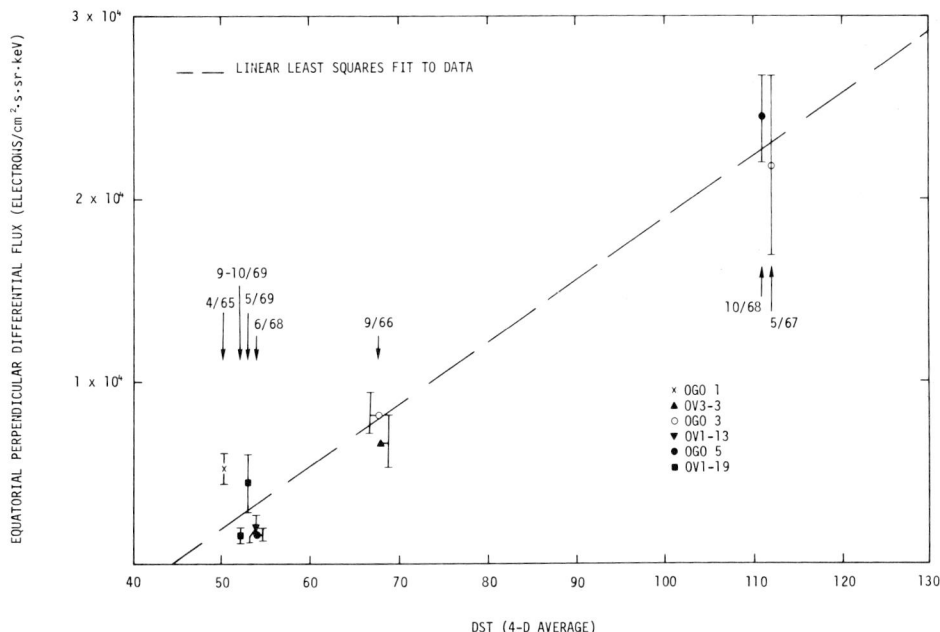

Fig. 26. Average event integrated flux vs Dst, L = 2.4, E = 500 keV

Over the L range 1.7 to 2.4 and for E > 0.5 MeV, the AE-6 model is lower than the observations integrated over an arbitrary 1-day period less than 10 percent of the time.

## The Future Modeling Activity

### Short Term

In the immediate future the problems with the AE-4 B-distribution and energy spectra need to, and will, be corrected when the new outer zone model, AE-7, comes out in early 1979. The form of this model and the modeling technique used will be similar to AE-4. Figures 17 and 18 show the two new data sets that will be incorporated into this model. Figure 17 shows 2.93-MeV OV1-19 data at four L values. Note the trend from little temporal variation at L = 3.0 to large event-oriented variation at L = 3.6 and 4.0, and finally to relatively high-frequency temporal variation at L = 5.0. Figure 18 compares 2.3-MeV OV3-3 and OV1-19 data at L = 3.0 and 3.6. At the lower L value, the average of the OV3-3 data is more than one order of magnitude higher than the OV1-19 data, whereas at the higher L value the OV3-3 average is a factor of two lower. Figure 19 shows B distributions for the 1.5-MeV AZUR data and compares fits of the Roberts function (Roberts, 1965) to the data with the AE-4 B distribution. The fits and the model become increasingly divergent toward the higher L values, indicating the danger of fitting to high latitude data alone. Because the AE-4 B distribution is supported by OGO 3 data near the equator, the AE-7 model will be the envelope of the AE-4 data base near the equator and data from the AZUR, OV1-19, and OV3-4 polar satellites in the high B region. The statistical treatment of the AE-7 data base will be similar to that used in the AE-3 (Vette and Lucero, 1971) and AE-4 models, and used by Paulikas and Blake (1971) for treating several synchronous data sets. Figure 20 shows that a gaussian log-normal distribution function is assumed for the probability that the particle flux F will exceed an arbitrary flux level $F_x$, which is determined from the data by observing the data sample size for $F > F_x$. Both $\mu$ and $\sigma$ are parameters determined by fitting the log-normal distribution to the data. An example of this distribution is shown in Figure 21, using the 1.75-MeV OV1-19 data. The fit in each case is good with the exception of the lowest L value shown, which lies in the region where the data exhibit the event-type structure, as opposed to the continuous high frequency behavior exhibited at high L values.

## Long Term

We have shown that the temporal variations of the electron flux in the radiation zone fall into three general categories. The influence of these on the modeling technique is summarized in Figure 22. For $L \leqslant 1.8$, storm-related temporal variations are small and not significant in a modeling sense. The present modeling technique is adequate, and, unless any Starfish-type artificial injection processes occur, the present AE-6 model and the associated solar minimum model AE-5 1975 (Teague and Vette, 1974) are likely to be the inner zone models for $L \leqslant 1.8$ for the forseeable future. For $L \geqslant 5.0$, the present statistical approach is adequate because of the short time scale, non-event dominated nature of the temporal variations; the AE-7 model should be adequate for the forseeable future. In the range $5.0 \geqslant L \geqslant 1.8$, the AE-6/AE-7 approach will be adequate for $E \leqslant 500$ keV. For higher energies the environment over a reasonable mission lifetime is determined almost exclusively by whether or not a storm does occur in the first place, and the magnitude of the storm in the second place. A future model should take into account these two parameters, thus an analysis activity is necessary on a per event basis.

A provisional move was made in this direction by examining the seven storms identified earlier. A simple storm model, shown in Figure 23, was assumed, which postulates a relatively sharp rise from a prestorm flux followed by an exponential decay to a constant post-storm flux. The depletion time, or $\tau$, of the exponential decay exhibits some interesting features. At L = 2.4, $\tau$ values were determined for six storms using five data sets, shown in Figure 24. The interesting feature is that although the peak flux for these storms varies by more than one order of magnitude, the depletion time appears to be independent of the magnitude of the event. The theoretical lifetimes given by Lyons et al. (1972) are compared with the data. At L = 3.6, shown in Figure 25, three storms were analyzed. However, $\tau$ appears to remain independent of the storm magnitude. The interesting feature is that $\tau$ exhibits the reverse trend with energy from that shown at L = 2.4. Both trends are supported by the analysis of Lyons et al. (1972). The other features of the storm model exhibit interesting characteristics. For instance, from storm to storm the post-storm flux is observed to be both higher and lower than the prestorm flux. Also, the rise time of the storm varies from approximately 10 days in the outer zone at some high energies to approximately 2 days in the slot, and back to a few tens of days at

L = 1.8. Considerable further analysis is required before the individual events can be modeled adequately.

The other question that we have addressed is the perennial one regarding the correlation between the magnitude of a particle event and geomagnetic indices. We have examined a variety of indices averaged over a variety of time periods, and both peak and event integrated fluxes. The best correlation observed was between the event integrated flux and a 4-day average of the Dst index beginning at the time of the peak hourly Dst index. The results are shown in Figure 26 for L = 2.4. The event integration time was determined by the time at which the decaying flux became 10 percent higher than the post-storm flux, and the event integrated flux can be regarded as a measure of the enhanced electron population on the L shell resulting from the magnetic storm. The results shown in Figure 26 are not entirely convincing, because we have examined only one L value and energy. Also, there are significant Dst changes for which no particle variations are observed. However, if a particle event does occur, there seems to be an approximate linear relationship between the two parameters.

It is not completely clear where this kind of analysis might lead in terms of a final model for the region for which the present modeling technique is not adequate. However, certain gross model features may be identified. Implicitly, time is now a parameter that must be included in the machine-sensible form of the model supplied to a user. A possible model may include a quiet time flux, plus a series of event-integrated storm fluxes that the user may invoke by dialing his own Dst value. The ramifications to the user are not trivial. Firstly, such a model is potentially much larger in a computer sense, because a new parameter was injected. Possibly this may be abbreviated by the use of analytic functions. Secondly, more responsibility would be placed on the user in determining the environment. For instance, a military application may dictate a more pessimistic environment than a commercial or scientific application. This judgement would lie with the user, as opposed to the modeler, as is the case at present.

## Conclusion

The temporal variations in the inner and outer radiation zones were discussed and the observed fluxes were compared with the present generation of electron models. In discussing the accuracy of the models, it is important to delineate between value judgements resulting from disagreements between observations, vagaries resulting from extrapolations into regions of B-L-E space for which no data are available, and limitations resulting from the modeling technique adopted. The first and second of these will probably always exist, although it is true that significantly better agreement between truly comparable data sets was observed in recent times. It was shown that the AE-4 model is in error in those regions of B-L-E space for which extrapolations were performed. The most significant error occurs at energies $E \geqslant 2$ MeV, near the outer zone flux peak.

Some inadequacies in the modeling technique were identified for $E \geqslant 500$ keV and $5.0 \geqslant L \geqslant 1.8$ that result from the event-dominated nature of the temporal variations in this region. Correction of these inadequacies will require extensive modifications to both the modeling technique and the final model distributed to the user community. This new generation of models will hopefully make the transition from appealing predominantly to an engineering community to, in addition, finding an application within the scientific community. It is likely that such models will be available on a 2- to 3-year time scale.

Typically, a model has in the past required 3 to 5 years of work before it is distributed, and may be based upon data that are 10 years old. The reasons are many. The dangers of basing models on single data sets, we believe, is well known by most people, and it takes time to accumulate a reasonably comprehensive data base. Also, principal investigators are very understandably reluctant to release data for modeling purposes until a substantial time has passed for their own scientific analyses. In recent years a number of data sets were made available to us on a proprietary basis, however. In addition, simply performing the data set processing preparatory to generating the model has in the past been very time consuming using batch processing computer techniques. We are now operating at the National Space Science Data Center with an interactive minicomputer system that should significantly reduce this processing time. We consider that it is most important for the user community that a given model be compared with a new data set, and the model possibly modified in response to that comparison on the shortest possible time scale. The National Space Science Data Center, as part of its role as the Satellite Situation Center, is hosting an International Magnetospheric Study-related workshop in December of this year. The principle behind this workshop is that if a large data base is maintained on a computer disk, and if the responsible principal investigators are offered an interactive computer system that will allow them to compare parameters in real time, not only will the time scale of scientific analysis be reduced but equally importantly the scope of the analysis will be greatly increased. Such a principle can be applied to trapped particle modeling using new data sets and the relevant models as a data base. The National Space Science Data Center is presently investigating the possibility of holding such a workshop in the course of the next year or so.

## References

Achtermann, E., B. Häusler, D. Hovestadt, G. Paschmann, E. Künneth, and P. Laeverenz, "The Experiments EI88 and EI93 for the Measurement of High-Energy Electrons, Protons, and Alpha Particles in the AZUR Satellite: Physical Properties and Test Measurements," The Max-Planck-Institut of Physics and Astrophysics, Research Report W70-67, Interplanetary Research, 1970.

Beall, D. S., "Graphs of Selected Data from Satellites 1963-038C," Applied Physics Laboratory TG 1050-1 through TG 1050-5, 1969.

Knox, R. J., OSO-4 Electron Observations on the Edge of the Inner Radiation Zone, UCRL-51185, 1972.

Lyons, L. R., R. M. Thorne, and C. F. Kennel, "Pitch Angle Diffusion of Radiation Belt Electrons Within the Plasmasphere," J. Geophys. Res., 77, 19, 3455, 1972.

McIlwain, C. E., "The Radiation Belts, Natural and Artificial," Science 142, 335-361, 1963.

Paulikas, G. A., and J. B. Blake, "Energetic Electrons at Synchronous

Altitude 1967-1977," Aerospace Corp. Report TR-0078-75, 1978.

Paulikas, G. A., and J. B. Blake, "The Particle Environment of the Synchronous Altitude," *Models of the Trapped Radiation Environment, Vol. VII, Long Term Time Variations*, NASA SP-3024, 1971.

Pfitzer, K. A., *An Experimental Study of Electrons Fluxes from 50 keV to 5 MeV in the Inner Radiation Belt*, University of Minnesota Technical Report No CR-123, 1968.

Roberts, C. S., "On the Relationship between the Unidirectional and Omnidirectional Flux of Trapped Particles on a Magnetic Line of Force," *J. Geophys. Res.*, 70, 11, 2517, 1965.

Rothwell, P. L., and W. R. Moomey, *Calibration of a Magnetic Spectrometer Designed to Measure 0.1-1.0 MeV Electrons in Space*, AFCRL-72-0710, 1972.

Singley, G. W., and J. I. Vette, *A Model Environment for Outer Zone Electrons*, NSSDC/WDC-A-R&S 72-13, 1972.

Teague, M. J., and E. G. Stassinopoulos, *A Model of the Starfish Flux in the Inner Radiation Zone*, NASA/GSFC X-601-72-487, 1972.

Teague, M. J., and J. I. Vette, *A Model of the Trapped Electron Population for Solar Minimum*, NSSDC/WDC-A-R&S 74-03, 1974.

Teague, M. J., and J. I. Vette, *The Inner Zone Electron Model AE-5*, NSSDC/WDC-A-R&S 72-10, 1972.

Teague, M. J., K. W. Chan, and J. I. Vette, *AE 6: A Model Environment of Trapped Electrons for Solar Maximum*, NSSDC/WDC-A-R&S 76-04, 1976.

Teague, M. J., N. J. Schofield, K. W. Chan, and J. I. Vette, *A Study of Inner Zone Electron Data and their Comparison with Trapped Radiation Models*, NSSDC/WDC-A-R&S 78-07, 1978.

Vampola, A. L., "Energetic Electrons and Latitudes Above the Outer Zone Cutoff," *J. Geophys. Res.*, 74(5), 1254-1269, 1969.

Vampola, A. L., J. B. Blake, and G. A. Paulikas, "A New Study of the Magnetospheric Electron Environment," *J. Spacecraft and Rockets*, 14, 11, 690-695, 1977.

Vette, J. I., and A. B. Lucero, "Electrons at Synchronous Altitudes," *Models of the Trapped Radiation Environment, Vol. III*, NASA SP-3024, 1967.

Vette, J. I., A. B. Lucero, and J. A. Wright, "Inner and Outer Zone Electrons," *Models of the Trapped Radiation Environment, Vol. II*, NASA SP-3024, 1966.

West, H. I., and R. M. Buck, "A Study of Electron Spectra in the Inner Belt," *J. Geophys. Res.*, 81, 4696, 1976.

THE SIGNATURES OF THE VARIOUS REGIONS OF THE OUTER MAGNETOSPHERE IN THE PITCH ANGLE DISTRIBUTIONS OF ENERGETIC PARTICLES

Harry I. West, Jr.

Lawrence Livermore Laboratory, University of California
Livermore, California  94550

Abstract. An account is given of the observations of the pitch angle distributions of energetic particles in the near equatorial regions of the earth's magnetosphere. The emphasis is on relating the observed distributions to the field configuration responsible for the observed effects. The observed effects relate to drift-shell splitting, to the breakdown of adiabatic guiding center motion in regions of sharp field curvature relative to partial gyro radii, to wave-particle interactions, and to moving field configurations.

## Introduction

The signature that the magnetosphere leaves in the pitch angle distributions (PAD's) of azimuthally-drifting energetic particles can be used as an important diagnostic tool in the understanding of the magnetic field configuration. This paper examines these signatures primarily emphasizing the PAD's of energetic electrons rather than protons. The reason for this is straightforward. For example, the rigidity $B\rho$ for 79-keV electrons, the lowest energy electrons that will usually be considered, is 0.154 $\gamma R_E$. In contrast, the $B\rho$ for the lowest energy protons that will be considered, 100-150 keV, is 8 $\gamma R_E$. The electrons are the more useful of the two for probing the fine structure of the magnetosphere since in the outer regions of the magnetosphere the protons are more subject to breakdown of the adiabatic invariants than are the electrons. Of course, in principle, one could use low-energy proton data but then the results are strongly affected by convection.

The effects presented fall mainly into three categories. The first has to do with shell splitting and how the electrons at various pitch angles drift through the distorted magnetosphere; in this case the particle motion is completely adiabatic. The second case pertains to the breakdown of adiabatic motion in those distant regions in which the gyroradius of the particle is no longer small relative to the field-line curvature. The third case pertains to scattering through wave-particle interactions, especially those periods of time when the particles are close to the magnetopause; this case is more difficult to treat theoretically and is less strongly emphasized in this review.

This review relies heavily upon Ogo-5 observations by the Lawrence Livermore Laboratory (LLL) experiment during 1968 and 1969 covering the equatorial regions out to 24 $R_E$ [West et al., 1973a; West and Buck, 1974]. The experiment consisted of a 7-channel magnetically-selected

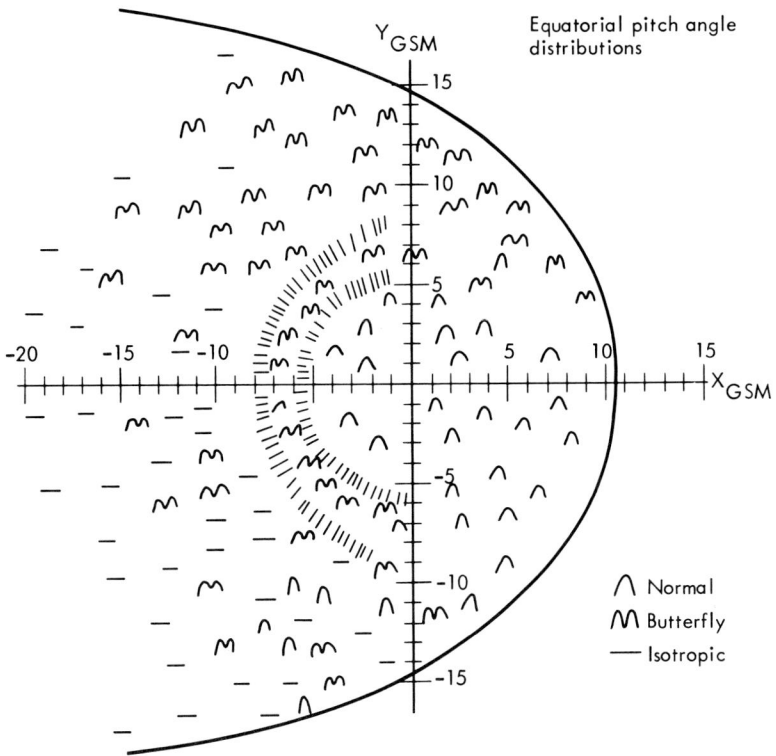

Fig. 1. Survey of energetic electron PAD's in the near equatorial magnetosphere as determined by the Lawrence Livermore Laboratory's experiment on Ogo 5.

electron spectrometer and proton telescope located on a scanning boom. The success of the Ogo-5 data analysis has depended greatly upon the ready availability of good magnetometer data from the UCLA experimenters covering all periods of data acquisition. When appropriate, work other than our Ogo-5 results are cited. Most of the results presented are from data acquired close to the geomagnetic equator. This leaves out the whole problem of what happens to the field and particle distributions at the edge of the polar cap near the earth and the ingenious methods that people have applied in the study of the associated field topology. For example, there is the use of the so called "trapping boundary" which has been widely used to define, ostensibly, the change from closed to open field lines. Also, during solar electron events the observation of transition from a double to a single loss cone as the satellite entered the polar caps has been used extensively to define the transition to open field lines; we will not have much further to say about these methods.

In the first part of this paper we present the studies of PAD's of energetic electrons in the equatorial regions of the magnetosphere from the magnetopause on the dayside of the earth to about 17 $R_E$ on the nightside of the earth. We first discuss how drift-shell splitting alters the PAD's of drifting electrons. We then follow the eastward

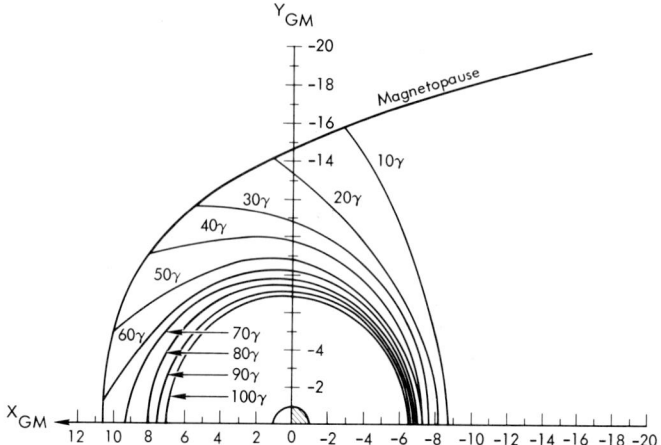

Fig. 2. Contours of constant B in the equatorial regions for an average magnetosphere [Fairfield, 1968].

azimuthal drift of the electrons, starting on the morning side of the earth, examining their encounter with the magnetopause in terms of drift-shell splitting, emphasizing the high-latitude regions near noon, and noting the resultant evolution of the PAD's in the extended afternoon magnetosphere. In the premidnight magnetosphere we examine PAD's during quiet times showing the marked effects of drift-shell splitting.

Near midnight we show spatially-dependent quiettime PAD data which have been analyzed in terms of field modeling and particle trajectories. Next we discuss the PAD changes observed during substorms and show how these changes are used to infer the field configurations during the various phases of the substorm. This first part of the paper ends with a discussion of the PAD effects observed post-midnight.

In the second part of the paper we present a potpourri of proton and electron observations, with the emphasis on the signatures that moving boundaries leave in the distributions of energetic particles. We first present a brief picture of our knowledge of the PAD's of energetic protons throughout the magnetosphere. We then discuss the observation of the spatial gradients of energetic protons by means of the east-west effect and how their temporal variations can be interpreted in terms of moving boundaries. This is followed by the use of energetic-particle directional distributions in the magnetotail to infer the motion of field-line structures through the use of the Compton-Getting effect.

Pitch Angle Distributions of Energetic Electrons

PAD Survey

We begin with a brief survey of the LLL Ogo-5 electron observations at all local times to provide a framework for what follows. Figure 1 shows the Ogo-5 PAD survey. A few words on nomenclature are in order. We have termed the bell-shaped distribution a normal loss-cone distribution because in the early history of space physics it was the

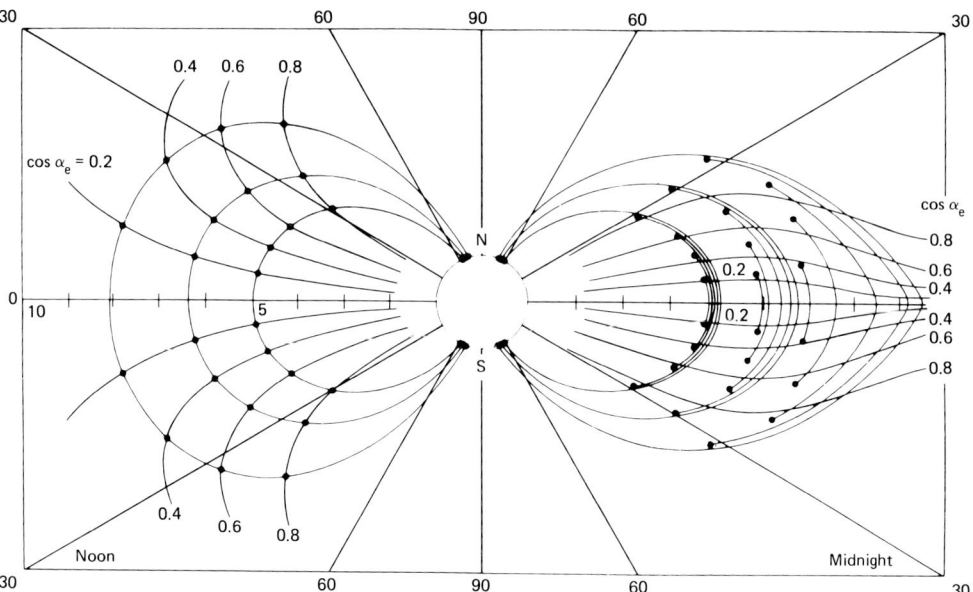

Fig. 3. Transformation of drift shells from the dayside to nightside magnetosphere as calculated by Roederer [1967]. The dots in each case represent the mirror points of each particle group as identified by the cosine of the equatorial pitch angle.

distribution normally expected, since the measurements at that time were mostly in the near-earth regions; also the distribution is shaped much like a normal frequency distribution (when presented in a Cartesian plot 0-180°) and is peaked normal to the field direction thus emphasizing the term normal distribution. Occasionally in the literature this has been called a pancake distribution. The distribution depleted near 90° pitch angles we call a butterfly loss-cone distribution, reflecting its shape (j($\phi$) vs $\phi$, 0-360°) in a polar plot. It has been called an anti-loss-cone distribution which is a misnomer since the loss cone is usually empty. The butterfly distribution has also been called cigar-shaped which seems particularly inappropriate. Finally, we have the isotropic distribution, a universally-accepted term. However, in the context of the work presented here, the term isotropic does not necessarily mean that the loss cone is filled with precipitating particles.

Let us now discuss Figure 1. We start with the morning sector and proceed eastward following the azimuthal drift of the electrons. In the morning sector we find the normal PAD out to the magnetopause. In the early afternoon we find the butterfly PAD at extended distances and by dusk we find the butterfly PAD extending inwards to 5.5 to 7.5 $R_E$ depending upon energy. The dashes at the more extended distances refer to the crossover from butterfly to normal for ~79-keV electrons, and the inner band of dashes refers to the crossover for ~822-keV electrons. Note that this band extends across the entire nighttime sector. You will also notice that in the extended early evening sector the butterfly PAD prevails whereas in the corresponding region past midnight we

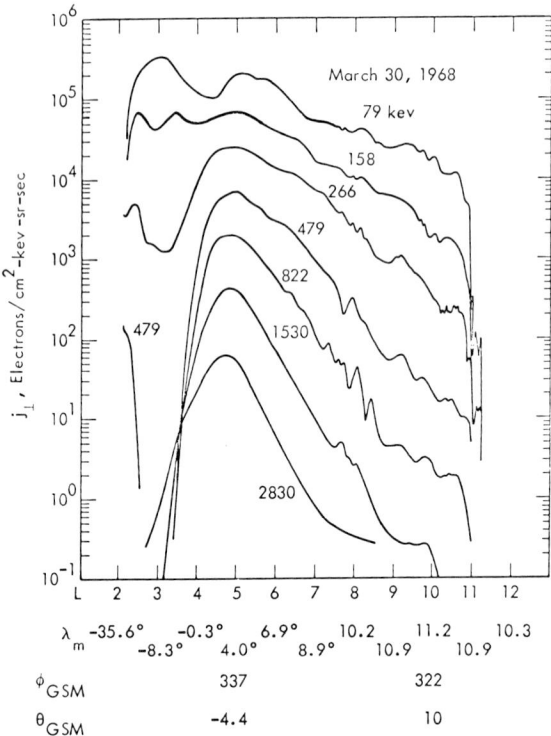

Fig. 4. Radial $j_\perp$ profiles of energetic electron fluxes on March 30, 1968, as determined on Ogo 5.

frequently find isotropy. Obviously the data show the marked effects of drift-shell splitting, and in the more distant nighttime magnetosphere we have the effects of substorms. We expand on this later. Some aspects of this survey were known prior to the Ogo-5 observation; for example, Serlimitsos [1966] and Haskell [1969] had observed the butterfly PAD in the nighttime magnetosphere. However, the observations in the afternoon magnetosphere are unique to our Ogo-5 experiment, and it took Ogo-5 observations [West et al., 1973a; West and Buck, 1974] to put together the picture for the total magnetosphere. We want to keep the survey in mind as we proceed with the rest of the presentation.

Drift-Shell Splitting — Basic Ideas

The idea of drift-shell splitting was probably put forward initially by Northrop and Teller [1960] but has received major impetus through the efforts of Roederer [e.g. 1967]. (Also see Schultz [1972] for later references.) The particle motion under discussion is completely adiabatic. One examines the drift shells of the particles of various equatorial pitch angles and finds that if he breaks the azimuthal symmetry, exemplified by the near-earth dipole field, that the degeneracy of the drift paths is removed.

The picture of separation of drift shells is usually explained by examining the drift paths of $90°$ particles along with particles of small equatorial pitch angle. In the first case the equatorially-mirroring particles must drift at constant B as shown, e.g., by the data of Fairfield [1968] in Figure 2, thus moving from 9 $R_E$ near noon to ~6.5 $R_E$ near midnight. For the second case we find that in evaluating the integral for the second invariant (I = $\oint \sqrt{1-B/B_m}$ ds) that I is close to being equal to twice the bounce path of the particle, that is twice the distance along the field line between mirror points. Of course, these particles must conserve the first invariant, also, and thus drift at constant B at their mirror point. Roederer [1967] has calculated these effects using the Mead [1964] model. The results are shown in Figure 3. Here e.g. we see $37^o$ equatorial pitch-angle particles (cos $\alpha_e$ = 0.8) moving from 8.1 $R_E$ to an equatorial crossing of 9.2 $R_E$ at midnight at an equatorial pitch angle $\leq 20^o$ (cos $\alpha_e \approx 0.95$), and $78^o$ particles (cos $\alpha_e$ = 0.2) initially, moving to 7.1 $R_E$ at $68^o$.

Lets proceed to see how the PAD's transform from the dayside to the nightside of the earth. We first examing PAD's on the dayside, then transform these data to midnight, and finally compare them with measurements. Figure 4 shows a radial profile of the $j_\perp$ (differential flux perpendicular to $\vec{B}$) fluxes on an inbound pass of Ogo 5 on March 30, 1968. First note the abrupt increase in particle flux at the magnetopause ~11 $R_E$. In the outer trapping regions the data are characterized by the negative radial gradients that we have come to associate with diffusion. In Figure 5 we see the corresponding PAD's. The very pronounced energy dependence evident for the lowest two L shells (3.06 and 3.32) and to a lesser extent L = 3.89, we now know is due to the special way that whistler-mode radiation interacts with electrons in the plasma sphere as shown by the theoretical results of Lyons, Thorne, and Kennel [1972] and experimental results of, e.g., Lyons and Williams [1975]. In the more extended magnetosphere (L = 8.38 and 9.18) a narrowing of the PAD's with increasing energy is evident. We have not attempted to determine the cause of the effect but expect that it is related to diffusion.

To make the transformation of the electron PAD data to midnight we use the calculated results of Roederer [1967] in Figure 6, making use of Liouville's theorem. We choose the shell crossing the equator at 7.4 $R_E$ at midnight as the place to generate the PAD and note that the particles that are to fill this shell came from shells crossing the equator at noon over the range 7.4 to 9.7 $R_E$. The corresponding fluxes at the various pitch angles can be obtained from the PAD's in Figure 5 transformed to the equator and interpolated. (Note that the negative radial Flux gradients, exemplified in Figure 4, Figure in the generation of the butterfly PAD.) The results are shown in Figure 7. For comparison with our transformed results we examine the PAD's for September 18 in Figure 8; these data were aquired near midnight on a quiet day. Note that the transformed results compare most favorably with the measurements at 8 to 9 $R_E$ rather than the 7.4 $R_E$ of the transformation. The discrepancy can be attributed to the temporal variations in the particle fluxes between the two days (the PAD's even on the noon meridian depend upon the degree of magnetic activity) and deficiencies in the Mead [1964] model used by Roederer in the calculations.

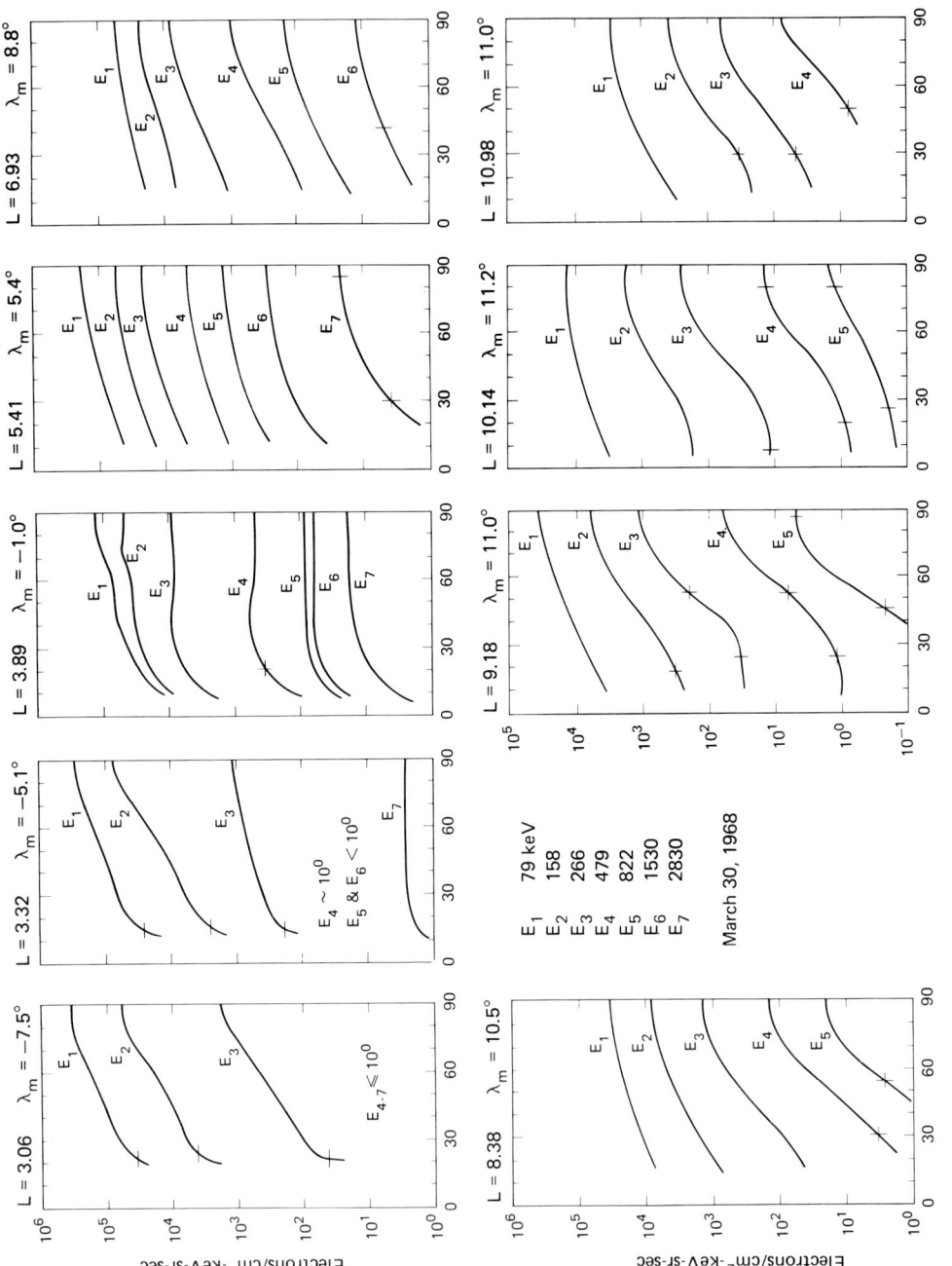

Fig. 5. Energetic electron PAD's measured by Ogo 5 on March 30, 1968.

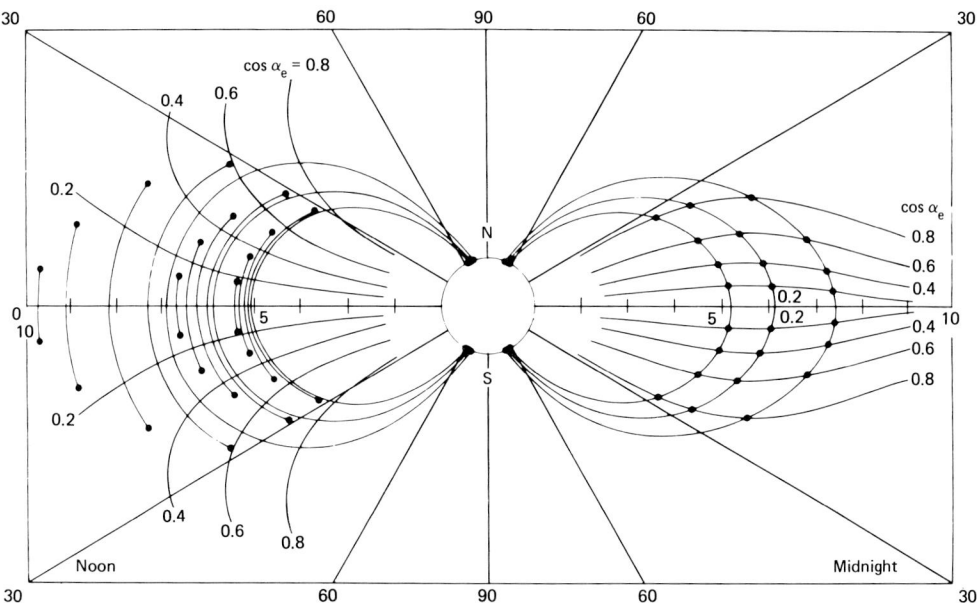

Fig. 6. Transformation of drift shells from the nightside to dayside magnetosphere as calculated by Roederer [1967]. Conversely, we may use this figure to determine the dayside origins of particles of a given equatorial pitch angle for a particular shell at midnight.

Magnetopause Shadowing

Magnetopause shadowing is an extreme variation of drift-shell splitting that occurs on the day side of the earth. In the region of the magnetosphere past noon, beyond that contour of constant equatorial B (Figure 2) that maps from the noon magnetopause to local midnight, we find that the drift paths for 90° pitch-angle electrons map back (westward) to the magnetopause. Unless there is a source of electrons at the magnetopause (scattering from other pitch angles can be a source) there will be no 90° fluxes for the PAD under question. It is very easy to find examples of this in Ogo-5 data, and in Figure 9, left panel, are shown PAD's acquired in the afternoon showing the effect. In actuality the picture presented, in terms of Fairfield [1968] contours of constant equatorial-B, are too simplistic for much of what occurs near noon; in this prespect, note the middle two panels of Figure 9. For this region we consider the electron drift paths in more detail.

Effect of the Dayside Minimum-B Regions

As the electrons drift eastward towards noon in the extended magnetosphere beyond ~9 $R_E$ they may encounter branch points, depending upon equatorial pitch angle, which moves them north or south away from the equator onto new minimum-B surfaces. (Note that in the undistorted magnetosphere the minimum-B surface in the field topology encompasses the equator.) Shabansky [1971] has analyzed the situation and points

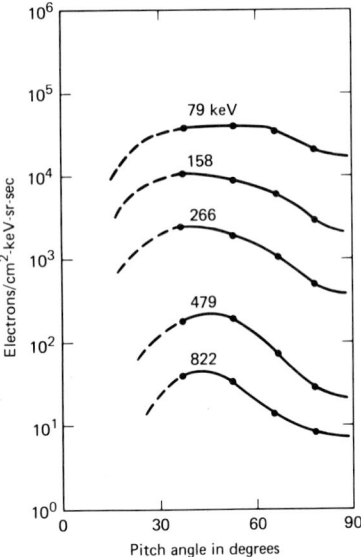

Fig. 7. Ogo-5 March-30 electron PAD data from Figure 5 transformed to midnight by means of Roederer's calculations of drift-shell splitting in Figure 6. The dashes at small angles represent both extrapolations based on the Roederer calculations and the author's expectations based on data.

out that the second invariant is halved at the branch points. The picture we have, then, is a particle bouncing back and forth on field lines threading the minimum-B surface, with one mirror point toward the geomagnetic equator and the other toward the earth. The minimum-B surface moves to its highest latitude near the noon meridian.

The author's colleague, R. M. Buck, has calculated some of these drift effects by means of invariant tracing in a model field after first following detailed particle motion to show that the equator and high-latitude regions are topologically connected; early results were presented at a previous meeting [Buck, 1975]. Figure 10 shows results for the Choe and Beard [1974] magnetic field model. The particles were started 70° away from noon on the equator at equatorial pitch angles of 70, 75, 80, and 85° at radial distances of 9.5, 10, 11 and 12 $R_E$. Particles near 90° pitch angles drift to the magnetopause to be lost as described under "magnetopause shadowing," and those with equatorial pitch angles less than 65° drift through noon well inside the magnetosphere.

Figure 10 shows only the near-magnetopause mirror points of the respective particle groups. Here we note that for 9.5 and 10 $R_E$, the 70° pitch-angle particles are mirroring close to but just inside the magnetopause, whereas for 11 and 12 $R_E$, the 70° particles do not reach this point in space having encountered the magnetopause earlier. This all means that these particles are scattered preferentially by encounters with the wave-rich region near the magnetopause and are lost from our distribution of drifting particles whereas the 85° pitch-angle particles are subject to less scatter.

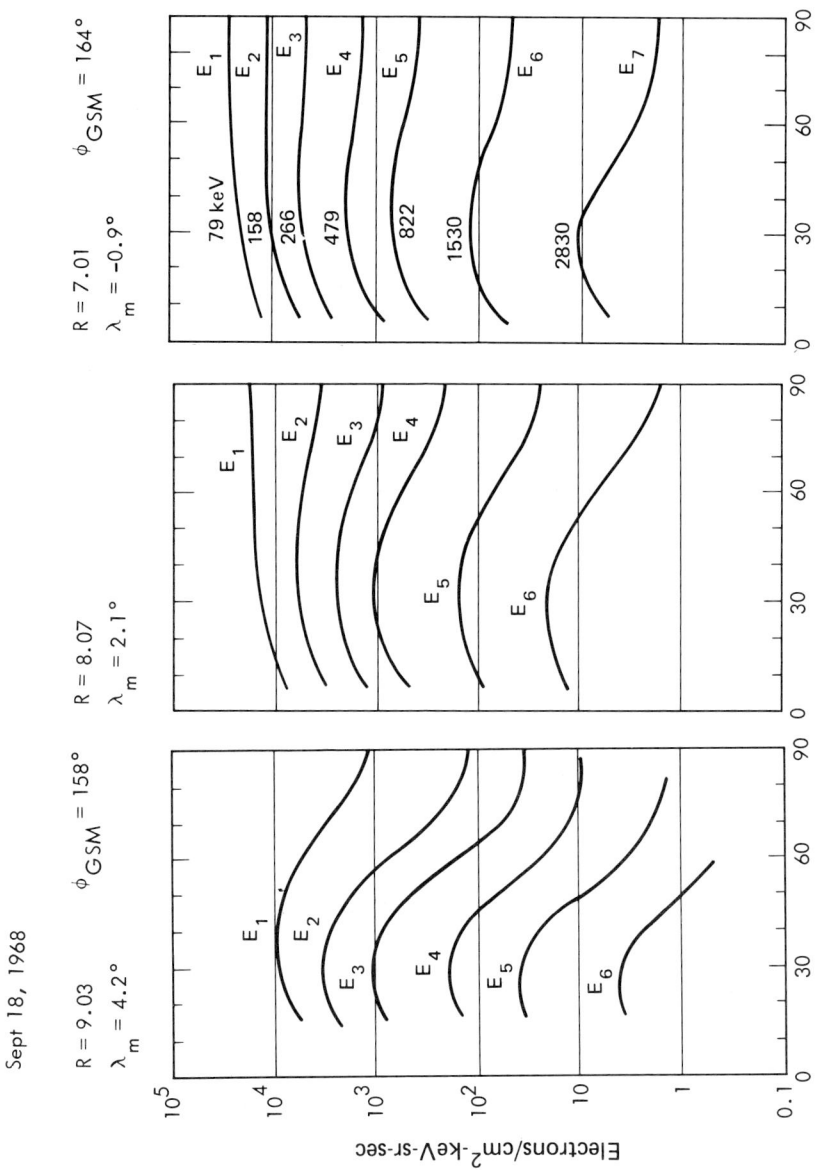

Fig. 8. Electron PAD's in the quiet nighttime magnetosphere on September 18, 1968, as measured on Ogo 5. These results are presented partially for comparison with the transformed results in Figure 7.

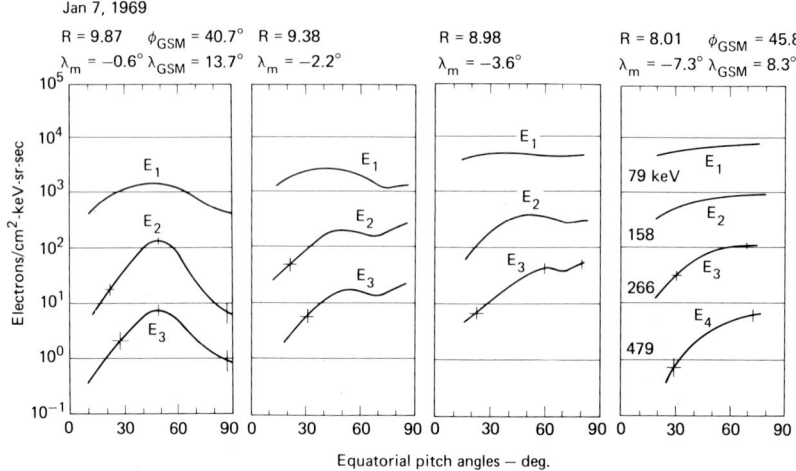

Fig. 9. PAD's of energetic electrons measured near the magnetopause in the early afternoon on January 7, 1969, by Ogo 5.

Confirmation of these ideas comes from examining equatorial PAD's in the afternoon magnetosphere after the various drift shells have reassembled. Some striking results are shown in Figure 9, center two panels. Near the magnetopause we see simple magnetopause shadowing as discussed earlier, but inward a bit we find PAD's with minima near 65° as predicted by our model. Figure 11 shows a radial profile of the different pitch-angle groups for the data on this day. The extent of the region in which $j_\parallel$ (shown here as $j_{50}$, the flux at 50° pitch angles) is greater than $j_\perp$ increases as we examine radial data acquired later and later in the afternoon magnetosphere. The region of crossover, $j_\perp = j_\parallel$, follows roughly that constant-B contour which maps from the noon magnetopause to midnight (see Figure 2).

PAD's Near Dusk

Figure 12 shows a radial profile of data acquired near dusk on November 29, 1968, a particularly quiet day magnetically (designated a QQ day). Here we have plots of the perpendicular flux $j_\perp$ and the $j_\parallel$ flux, which is the peak flux at small pitch angles (20 to 40°). One notes the gradual increase in $J_\perp$ relative to $j_\parallel$ as we approach the smaller radial distances including the energy-dependent crossover from $j_\parallel$ to $j_\perp$ at 6 to 8 $R_E$.

Figure 13 shows data in marked contrast to that of Figure 12. Here we see data acquired on November 3, 1968, two days after the most disturbed period of 1968. Kp at this time was 5⁻. Here each data point is plotted for 4.6-sec acquisition times as the experiment scanned back and forth at 3°/sec. The upper and lower envelopes of the data correspond to $j_\parallel$ (the peak flux at 20 to 40°) and $j_\perp$. To provide the right perspective, proton data, 100-150 keV, are included; the protons have drifted through the nighttime magnetosphere not subject to the violent magnetic activity that occurred for the electrons near the magnetopause. The outer boundary of the protons is at ~16.9 $R_E$ which

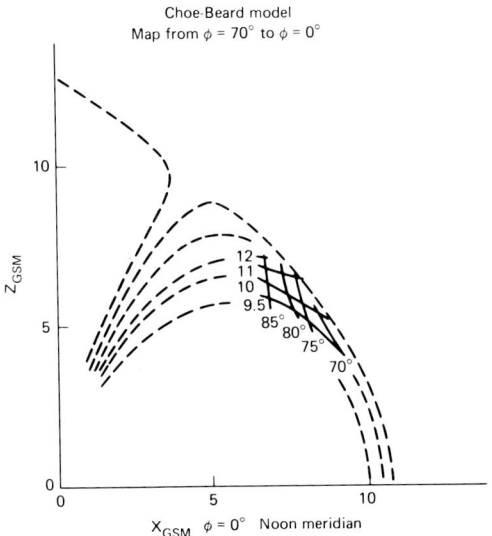

Fig. 10. Results of following the drift paths of particles of equator pitch angles of 70, 75, 80 and 85° at $\phi_{GSM}$ = 70°, started at radial distances of 9.5, 10, 11, and 12 $R_E$, into the high-latitude minimum-B regions in the noon meridian [Buck, 1975]. The calculations are based on the Choe-Beard field model. The intersections of the grid lines represent the mirror points of the respective particles on the sunward side of the configuration. The conjugate near-earth mirror points are not shown.

corresponds to the magnetopause in agreement with the magnetometer data.

In examining the electron data, in particular the lowest energy channel, we find that the $j_\perp$ fluxes show far greater temporal fluctuations than do the $j_\parallel$ fluxes. This is vivid proof that the $j_\perp$ fluxes were closer to the magnetopause during their drift through noon than were the $j_\parallel$ fluxes. Another interesting facet of the data is that beyond ~15.2 $R_E$, the $j_\parallel$ fluxes as well as $j_\perp$ fluxes, are wiped out. We expect the $j_\perp$ fluxes to be gone but it takes pronounced temporal changes in the field configuration to wipe out the $j_\parallel$ fluxes as well.

Premidnight Magnetosphere

Let us proceed azimuthally into the nighttime magnetosphere in our examination of data. Figure 14 shows a radial profile of the $j_\perp$ and $j_\parallel$ fluxes for October 21, 1968, and Figure 15 shows corresponding PAD's for the lowest energy channel (79 keV) of the LLL Ogo-5 spectrometer. Here we find that the effects of drift-shell splitting are very strong. This was an especially quiet day magnetically (Kp = $0^+$). In Figure 16 we show a second example of quiettime data. These data were acquired on September 18, 1968, when Ogo 5 was inbound 1 to 2 hours before midnight during a very quiet time magnetically (Kp = $1^-$). The corresponding PAD's were presented earlier in Figure 8. So far we have shown only quiettime data in the nighttime magnetosphere. However, even during substorms the butterfly PAD is seen in this region of space

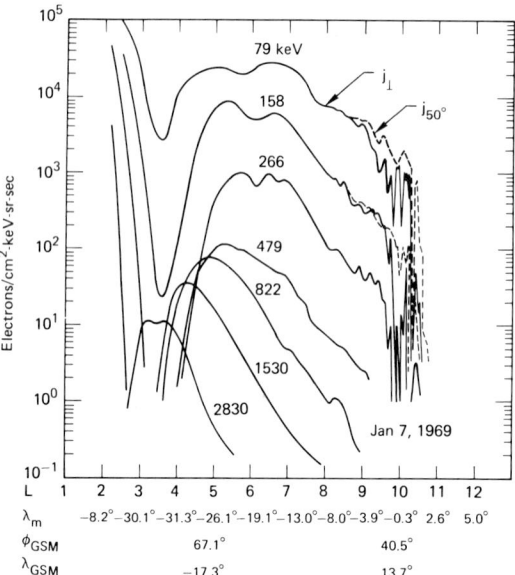

Fig. 11. Radial profiles of electron fluxes measured in the early afternoon on January 7, 1969, by Ogo 5.

for all but ~30 min before onset and ~10-15 min after onset. During these periods isotropy usually prevails. Such quiettime data in this part of the magnetosphere is not the norm, more often being punctuated by a substorm or two on such an inbound pass. However, we now examine the results of a modeling study of quiettime data acquired in the midnight meridian before proceeding to substorm effects.

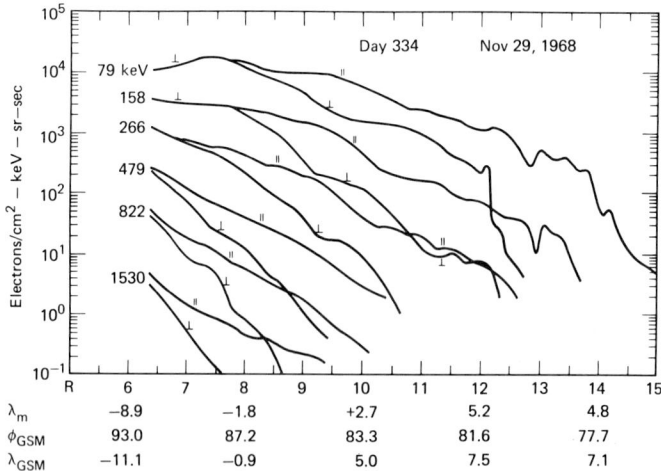

Fig. 12. Radial profiles of electron fluxes on November 29, 1968, measured on Ogo 5. This was an especially quiet day magnetically ($Kp \approx 1$).

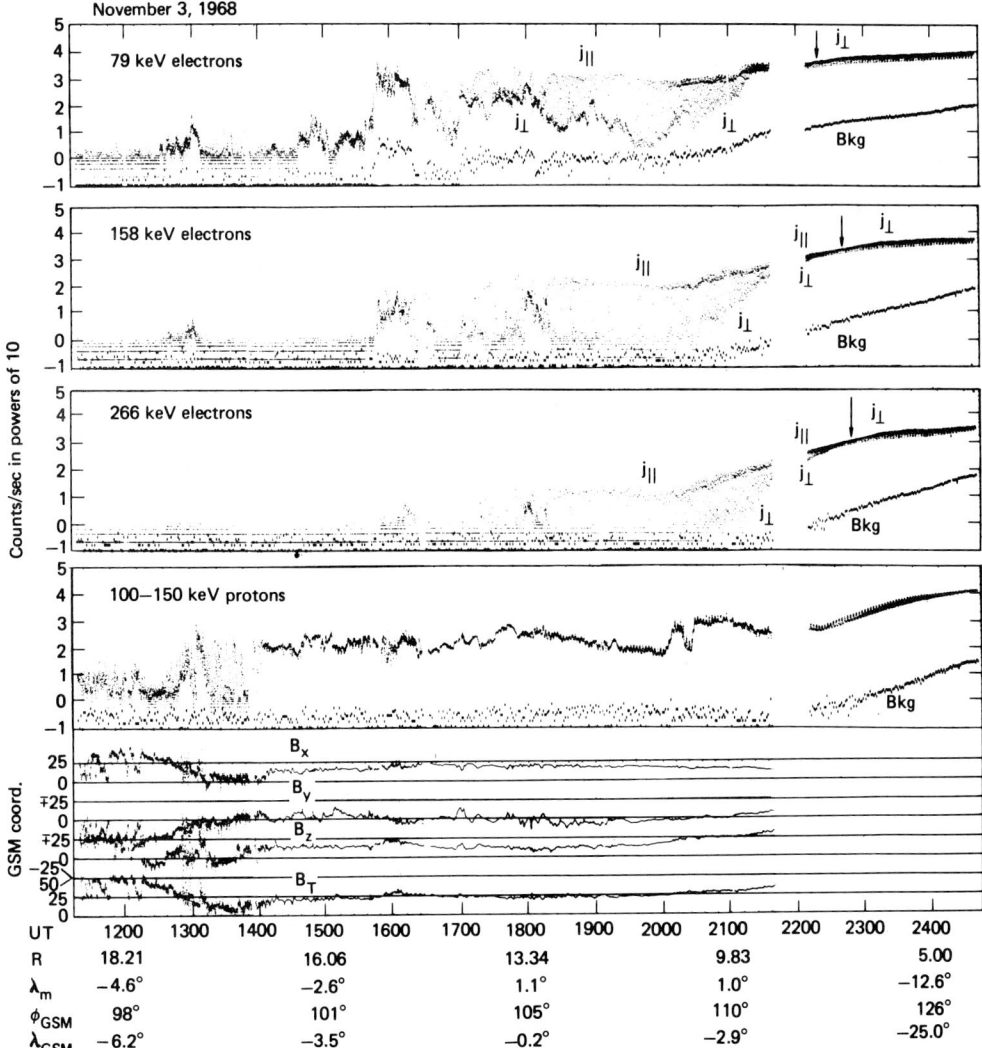

Fig. 13. Radial profiles of electron and proton fluxes on November 3, 1968, measured on Ogo 5. This was a very disturbed day magnetically (Kp = 5⁻). The electron pitch angle data clearly show the effect of their drift through noon near the fluctuating magnetopause.

PAD's Near Midnight — Quiettime Modeling Study

On a number of Ogo-5 inbound passes near midnight during quiettimes the PAD's of the particles when first observed were isotropic which were followed by the rigidity-dependent transition to the butterfly PAD nearer the earth. The low-rigidity particles made the transition first followed in turn by the higher rigidity particles. Figures 17 and 18 show two extremes in the data acquired near midnight during quiet times [West et al., 1978ab] (the August-2 data were acquired during a period

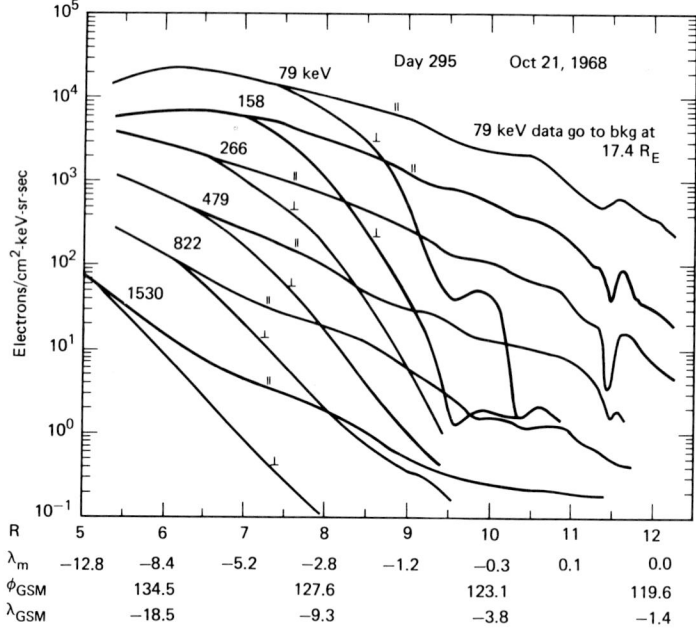

Fig. 14. Radial profiles of the electron fluxes on October 21, 1968. This was a very quiet day magnetically (Kp = 0⁺).

of enhanced dynamic pressure [West et al., 1978a] which may account for the difference between the August-2 and -25 data). For each figure the top panel shows the Ogo-5 magnetic field data in GSM coordinates, measured by the UCLA experiment; the middle panel shows the scan-modulated data (plotted every 4.6 sec), obtained by the LLL experiment on Ogo 5; and, the bottom panel shows a rough sketch of the field configuration inferred from the data. The August-2 data strikingly show the point to be made and that is in the region of sharpest curvature along the field line (neutral sheet) that for the particle motion to be adiabatic (conservation of first and second invariants, $\mu$ and J) we need the maximum gyroradius of the electrons to be less than ~1/10 the minimum radius of curvature of the guiding field line, or stated differently, less than 1/10 of $B_z$ divided by the gradient of B in the neutral sheet [Alfvén and Fälthammer, 1963]. The details of the particle motion in the region of the neutral sheet have been discussed by Speiser [1967], Shabansky [1971], Sonnerup [1971] and Eastwood [1972]. The transition in the scan modulation of the August-2 data at 0634 UT is the dividing line between $\mu$-J breakdown in the neutral sheet for trajectories on field lines farther from the earth and $\mu$-J conservation for trajectories on field lines closer to the earth. On this particular day we were able to identify the transition points for electrons in five channels, 79 to 822 keV, and protons in two channels, 100-570 keV. In contrast, we note that for the August-25 data the point of $\mu$-J violation for the 79-keV electrons was much farther from the earth than for August 2. The pitch angle changes in conjunction with the UCLA magnetic field data were used in a modeling study [West et al., 1978b].

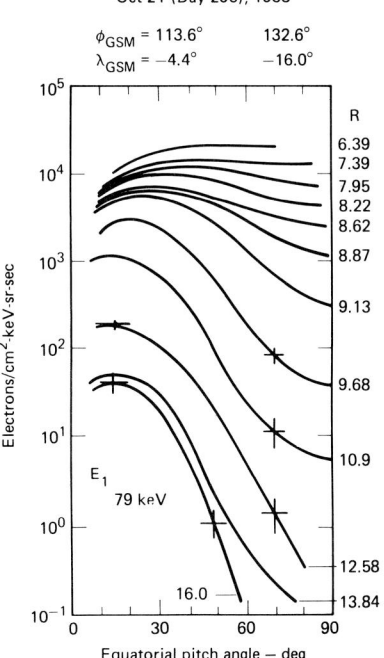

Fig. 15. PAD's of 79-keV electrons in the early evening on October 12, 1968, as measured on Ogo 5.

It was possible to find a model which fits both aspects of the data quite well, that is the particle pitch-angle changes and magnetic field values. Particle motion in the model field was then studied and the results give insight to the structure of the plasma sheet.

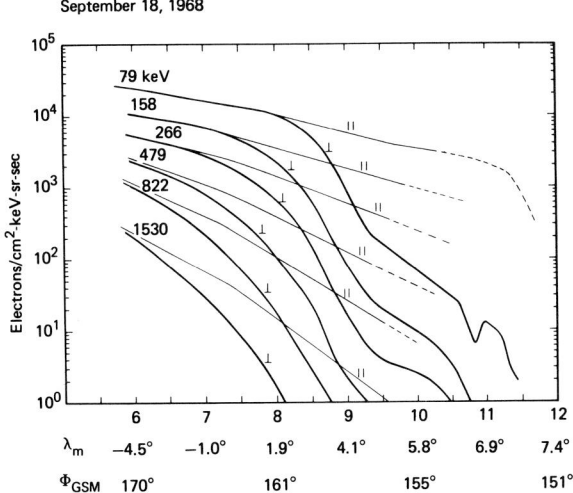

Fig. 16. Radial profile of the $j_\perp$ and $j_\parallel$ fluxes on September 18, 1968. The $j_\parallel$ fluxes are the peak fluxes at 20 to 40° pitch angles. See Figure 8 for the corresponding PAD's.

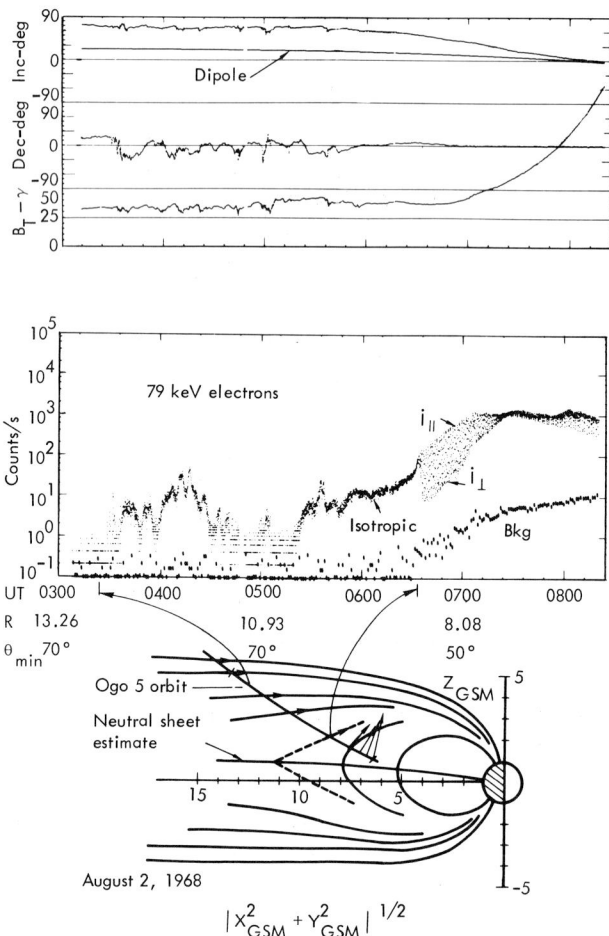

Fig. 17. Electron data acquired on an inbound pass of Ogo 5 near midnight on August 2, 1968. The abrupt change in the scan modulation of the 79-keV electrons at 0634 UT is indicative of a change from isotropy to the butterfly PAD. Computer modeling [West et al., 1978b] puts the transition point in the neutral sheet between μ-J breakdown farther and conservation nearer the earth at ~11 $R_E$. Magnetically August 2 was very quiet (Kp ≈ 1, a QQ day).

These findings enhance earlier results at low altitudes near the trapping boundary. For example, Fritz [1968] has noted a region of isotropy near midnight even during quiet times. Imhof et al. [1977] have made detailed PAD measurements at low altitudes near midnight showing profiles of rigidity-dependent transitions to isotropy, being highest in rigidity at lower L-shells as expected from our modeling study. The profiles were acquired at a variety of magnetic activity levels, Kp = 2 to $5^+$. In general, the higher the activity the lower in L-shell was the measured transition point, showing that the field was more taillike for these cases as expected.

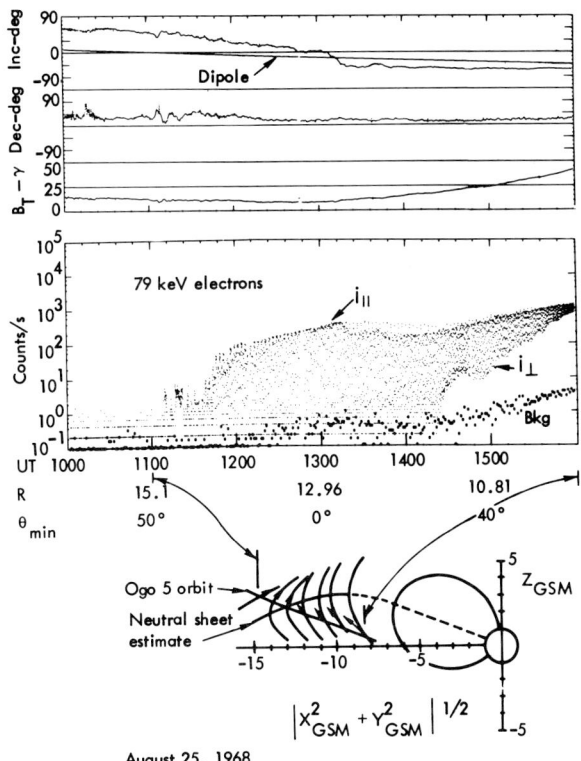

Fig. 18. Electron data acquired on an Ogo-5 inbound pass near midnight on August 25, 1968. Kp was 0⁺. Note the scan modulation in the count rates, almost from the first detection of the electrons. Computer modeling [West et al., 1978b] puts the crossover in the neutral sheet between μ-J breakdown farther and conservation nearer the earth at $17 \pm 1$ $R_E$. The most obvious difference between this day and that of August 2, Figure 17, is the much higher solar wind dynamic pressure for the latter day.

## PAD's at 6.6 to 15 $R_E$ During Substorms Near the Midnight Meridian

The PAD of energetic electrons drifting into the region at 9 to 19 $R_E$ (as observed on Ogo-5) near midnight during quiettimes ranges from isotropic to butterfly. At 9 $R_E$, almost without exception, we find the butterfly PAD whereas at the greater distances, depending upon just how taillike the field is, we may find that the isotropic PAD prevails. The concepts governing the changeover in observed PAD during quiettimes were discussed earlier.

At times, dynamic changes in the field configuration can dominate the azimuthal effects that the electrons are subject to during their drift. In plasma-sheet observations in the near-magnetotail, copious quantities of energetic electrons and protons are almost always found. Usually, in the period of a half hour or so before a substorm expansion or onset, the magnetic field is observed to become more taillike. Proof

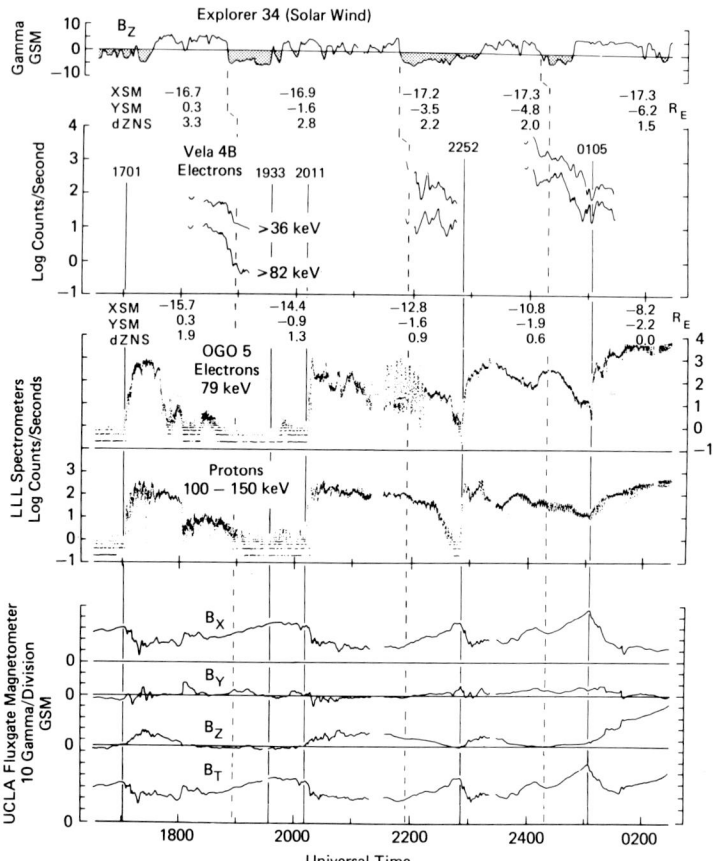

Fig. 19. Data from a substorm study [Pytte and West, 1978]. The scan modulation (or lack of it) of the Ogo-5 energetic electron data provides a key to the field configuration (see text). Substorm onset as determined by Pi2 micropulsations was at 1701, 1933, 2011, 2252, and 0105 UT as marked.

that the field rotation observed only at the satellite truly results in a more taillike field over a large region of the magnetosphere comes from the fact that Ogo-5 observations show repeated examples of the transition from the butterfly PAD to isotropy during these growth-phase periods. Some of these effects are shown in Figure 19 taken from [Pytte and West, 1978]. Here we see substorm expansion phases as marked at 1701, 1933, 2011, 2252, and 0105 UT. Onset was determined primarily by Pi 2 micropulsations, and all but the onset at 1933 UT are readily apparent in the Ogo-5 particle and field data. The effects between 2011 and 2252 UT are especially interesting. Following onset at 2011 UT the field became more dipolelike. Initially at onset, and this is always the case, the PAD's were isotropic but then on the time scale

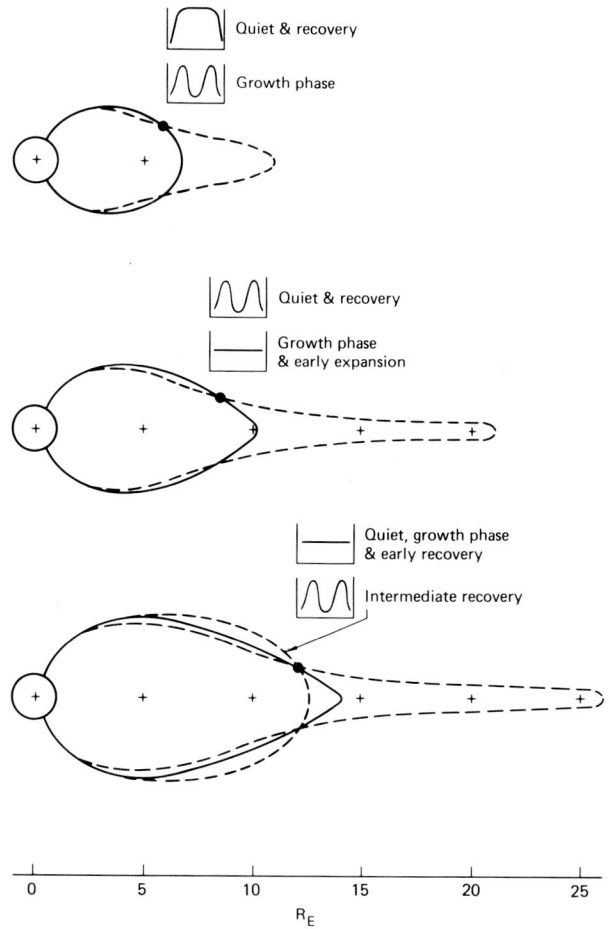

Fig. 20. Summary diagram showing the PAD's of energetic electrons that can be expected at various distances in the midnight magnetosphere during various phases of substorms. The dot on each Figure is the observation point. Note that field configuration at early expansion is generally believed to be more dipolar than the final field configuration.

of minutes the butterfly PAD gradually reemerged, first at the higher energies (not shown) and later the lower energies. Kivelson et al. [1973] attribute this time dispersion to the drift of the electrons from an undisturbed region of the magnetosphere. The reemergence of the butterfly PAD during the expansion is vivid proof that the field configuration, at least to the west, had become more dipolelike than that which prevailed earlier. A southward turning of the IMF reached the magnetopause at ~2155 UT signaling the start of a new growth phase. Possibly the normal recovery of the magnetosphere to a more taillike field may have meant the loss of the butterfly PAD at Ogo 5. In any respect, the new growth phase hastened the change of the PAD to isotropy which was accompanied by the observation of electron-precipitation

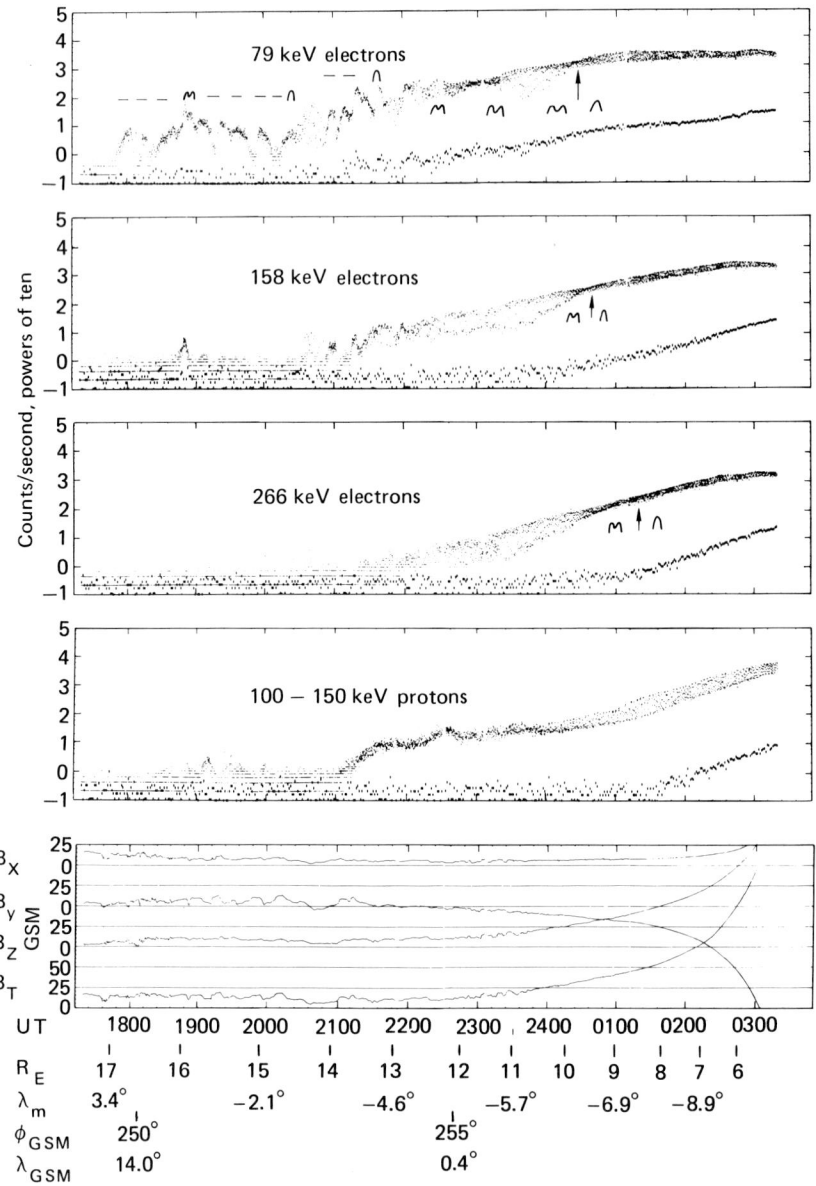

Fig. 21. Radial profiles of scan-modulated electron fluxes in the early dawn magnetosphere as measured on Ogo 5 June 5, 1968. Kp was 1+. The PAD forms are sketched for the various regions.

bremsstrahlung at balloon altitudes in Scandinavia. During the next expansion phase, after 2252 UT, the field never became dipolelike enough and/or long enough to allow the butterfly PAD to emerge. Although not too obvious in the data presented here (observational reasons), the butterfly PAD did reemerge after the 0105-UT expansion. Further examples of substorm related PAD changes observed on Ogo-5

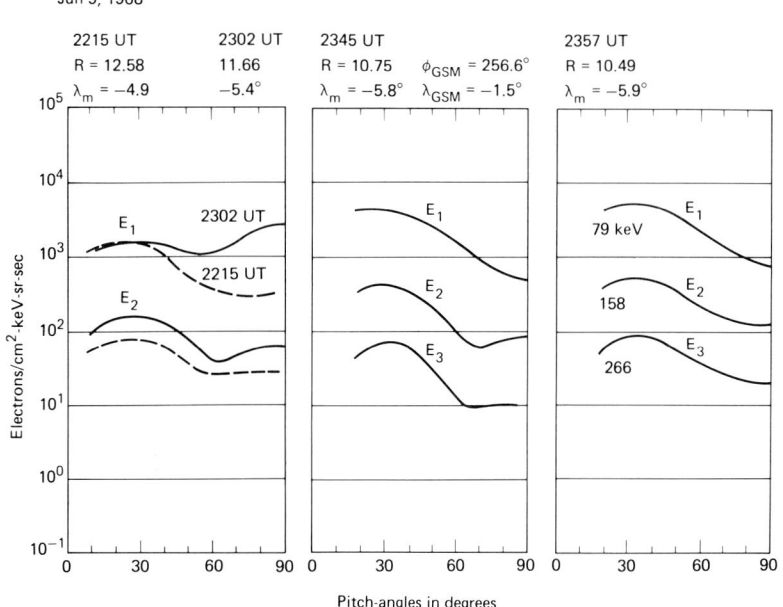

Fig. 22. PAD's of energetic electrons on June 5, 1968, in the early dawn magnetosphere. See Figure 21 for the corresponding radial profiles. See Figure 16 for PAD's acquired in a roughly equivalent region premidnight.

are reported in West et al. [1973b], Pytte and West [1978] and Pytte et al. [1976].

Baker et al. [1978] have made repeated observations of PAD's of electrons ~30 keV at synchronous orbit in the nighttime sector during substorms. During quiettimes the normal PAD was observed, but almost without exception the butterfly PAD was observed during the growth to a more taillike field configuration in the period prior to substorm onset. At onset, the normal PAD returned. Some aspects of these observations were reported previously by Bogott and Moser [1971]. The transition from the normal distribution to the butterfly PAD during the growth phase is marked evidence of the appearance of more taillike fields. The concepts have been discussed earlier. We need the concept of drift-shell splitting coupled with a negative radial gradients in the electron fluxes. Although relatively smooth negative radial gradients often exist, the flux change may be more drastic than that. Quite often, we expect, that during substorm growth phases the field distortions and inward motion of the magnetopause are such that magnetopause shadowing exists for the equatorially mirroring electrons as described in a previous section. In such cases the drift paths of these electrons map westward to the magnetopause along contours of constant B so that only electrons of smaller equatorial pitch angles are seen near midnight.

In Figure 20, top panel, we have provided a summary sketch for the ideas presented in this section. (The reader may wish to refer to

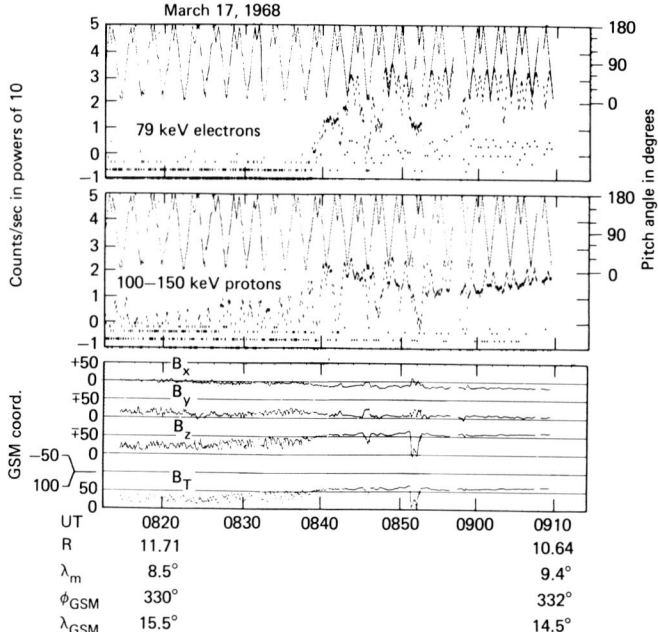

Fig. 23. Magnetopause crossing on March 17, 1968, into the dayside near-equatorial magnetosphere. The instantaneous pitch angle of the particles being detected is given by the zigzag pattern read from the scale to the right.

Pytte and West [1978] and West et al. [1978a] for further discussion of these ideas.) The top panel shows the situation that prevails for the data of Baker et al. [1978], the bottom two panels show the contrasting situation at Ogo-5. In the top panel we note that prior to substorm thinning the normal PAD prevails, which gives way to the butterfly PAD during substorm thinning. In the second panel we find the butterfly PAD during quiet times, which gives way to isotropy during thinning and early expansion. In the bottom panel, at the more extended distances, note that it is probable that the isotropic PAD prevails presubstorm because taillike fields usually prevail. Here the butterfly PAD is expected only for a period during recovery when the field is dipolelike enough to support adiabatic guiding center motion.

Post Midnight — 9 to 15 $R_E$
_____

The PAD found at roughly 9 to 15 $R_E$ past midnight to near dusk tends toward isotropy (however, in many cases the PAD we observe may have a loss cone but we cannot tell) but quite often the butterfly effect is observed. We have already discussed the effects that occur near midnight during both quiet and disturbed periods. During very quiet periods, probably accompanied by low dynamic-pressure of the solar wind with small or northward IMF, we can expect the electrons to drift through midnight still maintaining a marked butterfly PAD. However,

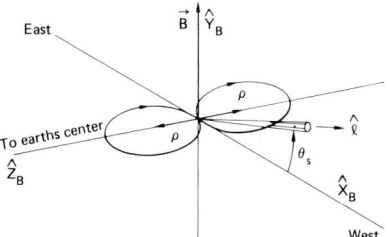

Fig. 24. Idealized geometry for observing the proton east-west effect. Here $\hat{\ell}$ refers to the look direction of the experiments aperture which is scanned through $\theta_S$ observing protons with gyro radii varying from $Z_B = -\rho$ to $+\rho$. In general the analysis of such scan data must allow for the scan plane being tilted at an appreciable angle relative to $\vec{B}$.

we noted that often during very quiet times, probably in association with enhanced dynamic pressure in the solar wind, that the field configuration could be taillike enough to cause isotropy. Conversely during substorms, beyond 8 to 9 $R_E$, we found isotropy during the growth and early expansion phases followed by the emergence of the butterfly PAD during the times of the more dipolar fields.

Although much is known about how the PAD's evolve as they drift through midnight, we do not have all of the answers. Figure 21 shows a radial profile of scan-modulated electron fluxes obtained on an inbound pass of Ogo-5 near dawn on June 5, 1968, a very quiet time magnetically (Kp = 1$^+$). The PAD forms are sketched on the figure. Figure 22 shows detailed PAD's obtained on this pass. Here we find periods of isotropy interspersed with the butterfly PAD. Such distributions of PAD's are common in this region of the magnetosphere. However, note that the flux at 90° is markedly enhanced relative to that observed just past dusk, Figure 15, in an equivalent region premidnight. It is very possible that other mechanisms not previously mentioned are operative. Assume for the moment that the electrons become isotropic due to the necked-down field configuration near midnight but only moments later in their azimuthal drift are back on field lines allowing μ-J conservation. At this point, assuming that the particle motion is still taking place in the plasma sheet, we find that we have electrons near 90° pitch angles drifting faster than those at low pitch angles [e.g., West et al., 1978b]. The production rate is proportional to the drift rate of electrons at low pitch angles into the isotropizing region, and here we have a situation which can lead to the evolution of the butterfly effect in an hour or so of azimuthal drift. However, this does not usually lead to the very low values of $j_\perp$ relative to $j_\parallel$ seen premidnight. In addition, for the lower energies we expect that electric fields in the nighttime magnetosphere are contributing somewhat to modification of the PAD's.

## Proton and Proton-Electron Associated Observations

### General

Most of the pitch-angle effects seen in the PAD's of energetic electrons in the equatorial magnetosphere have been seen also for energetic protons. For example, the butterfly effect has been seen at 6.6 $R_E$ by Stevens et al. [1970] and Bogott and Mozer [1971]. The LLL Ogo-5 PAD observations at 100-150 keV show the combined effects of spatial gradients, the butterfly effect, and breakdown of adiabatic guiding-center motion. There have been case studies of these PAD's but no systemmatic studies. However, it can be clearly stated that the butterfly PAD exists across the nighttime magnetosphere, but even on quiet days isotrophy exists (occurs when the gyro radii of the protons are greater than ~1/10 the minimum curvature of the field lines) much beyond 8 $R_E$. In the inner magnetosphere the normal PAD exists and the effects there are documented by, e.g. Williams and Lyons [1974]. In the morning magnetosphere at, say, 9 $R_E$ we might expect to see the effects of magnetopause shadowing. All too often the protons studied from Ogo-5 data had gyro radii that were too large to allow traversal of the minimum-B regions without scattering. Nevertheless, we have made observations of proton PAD's on disturbed days, that is with a contracted magnetopause, in which PAD's similar to the center panel of Figure 9 were observed. There was one major difference, however, and that is a strong component of isotropy (near 0 and 180°) existed in conjunction with the PAD. A strong component of isotropy appears to be the norm in the PAD's of energetic protons in the extended regions on the dayside of the earth.

Region of the Magnetopause

The transition from magnetosheath to magnetosphere is usually signaled by the change in the field orientation and noise and by changes in the plasma-flow pattern. Quite as specific though is the appearance of a double loss cone in the PAD's of the energetic electrons and protons, the electron data being most easily interpreted. Figure 23 shows an example of an Ogo-5 magnetopause crossing near noon. The electron and proton data from the LLL experiment are plotted every 4.6 sec as the experiment scanned back and forth at 3°/sec. The zigzag pattern at the top of the figure is the instantaneous pitch angle of the particles read from the scale to the right. For perspective note that peaks in the electron counting rate are at 90°.

From the magnetometer we note the magnetopause crossing at 0837 UT. The spikes in the proton data prior to that are the signatures of flowing sheath protons [West and Buck, 1976; Roelof et al., 1976]. The data are consistent with momentary return to the sheath at ~0846 and 0851 UT.

In this pass the signature of the transition from sheath to magnetosphere is relatively clear. However, at high latitudes and along the flanks of the magnetosphere, where the boundary layer is relatively thick, the transition is not always obvious. This is especially the case when appreciable fluxes of protons and electrons are observed flowing downstream in the magnetosheath. Particularly good examples of the situation under discussion are to be found in Figures 3 and 4 of West and Buck [1976] showing data acquired on satellite passes near dusk.

Energetic-Proton Spatial Gradients and the East-West Effect

The gyro radius of 100-150 keV protons in a 50-γ field is 0.16 $R_E$, a situation which pertains to the equatorial dayside magnetopause.

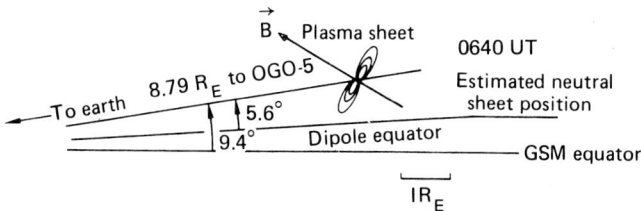

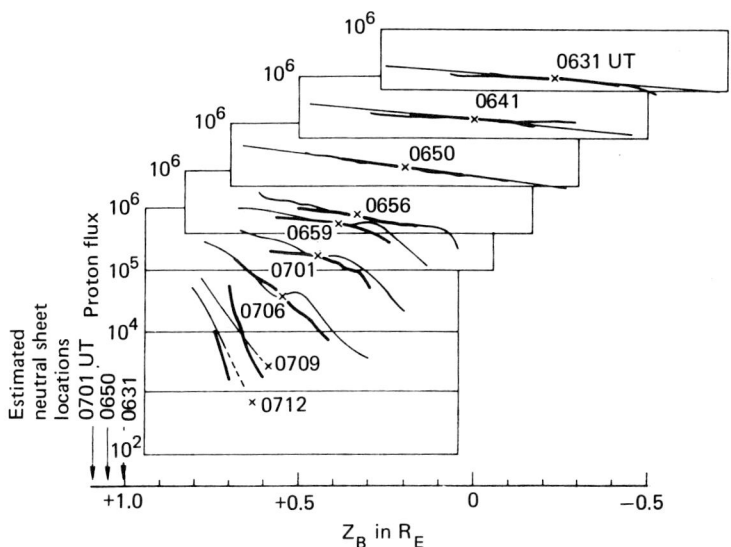

Fig. 25. Use of the proton east-west effect to determine the plasma sheet boundary motions during the 0714-UT August 15, 1968, substorm. The upper panel shows typical proton orbits for 100-150, 230-570, and 570-1350 keV at the edge of the plasma sheet at the position of Ogo 5. The lower panel shows profiles of the proton fluxes at different times inferred from data acquired by the LLL scanning proton spectrometer.

Measurements of $j_\perp$ from a scanning spectrometer, such as the LLL Ogo-5 experiment, show asymmetries in the data that are due to the fact that with the spectrometer looking westward it measures protons with gyro centers 0.16 $R_E$ farther from the earth than the radial position of the spacecraft and when looking eastward measures protons with gyro centers 0.16 $R_E$ closer to the earth, a range of 0.32 $R_E$. Appreciable proton flux gradients can occur over this distance, and thus from a single point in space one can obtain a snapshot of the particle distribution. The geometry of this situation is shown in Figure 24. Kaufman and Konradi [1973] have taken advantage of such effects in studying magnetopause boundary motions. Also, Kaufman et al. [1972] have studied field-line motions at L = 5 at high latitudes on the nightside of the earth during a magnetic storm. Recently Williams [1978] has analyzed early Isee results using three-dimensional data from his scanning proton spectrometer. His paper touches on aspects of the previous section as well as the east-west effect.

We have taken advantage of the east-west effect during the growth phase of a substorm [Buck et al., 1973] to study the motion of the plasma sheet in the near magnetotail. In this situation the partical fluxes fall off north and south of the plasma sheet and in a very real sense the particle intensities reflect the field configuration. Figure 25 provides the picture. The upper panel shows typical proton orbits at the edge of the plasma sheet. The Ogo-5 spectrometer measured these protons by scanning in the plane perpendicular to the earth's radius vector. (Note that the geometry of Figure 24 is idealized; for the case in hand the scan plan is at a marked angle with respect to $\vec{B}$.) Since the protons were intrinsically isotropic we were able to use data at all pitch angles which were then ordered in terms of the distance of the gyro centers from the neutral sheet. These results are shown in the lower panel as proton flux profiles at different times. Note that $Z_B$ is the perpendicular distance from $\vec{B}$ at the time of onset of substorm thinning, 0641 UT. The velocity of thinning of the edge of the plasma sheet was inferred from these data [cf. Buck et al., 1973].

Vela experimenters [Palmer et al., 1976] have employed a similar analysis in the lobes of the magnetotail. They used proton spatial gradients measured during a solar particle event to provide the first measurement of field line motions high in the lobes of the magnetotail. Obviously the proton east-west effect is a very powerful tool, especially when one is limited to only one satellite in making observaions.

## Energetic Proton and Electron Flow in the Magnetotail

In the last section we made use of the gradients of the particle fluxes to determine boundary motions. Here we discuss a different feature in the proton fluxes, the Compton-Getting effect, which is important when the center-of-flow motion is appreciable in respect to particle velocities. Here the observer sees particles increased in energy when looking upstream and decreased in energy when looking downstream. As a result, an anisotropy appears in the directional distributions if the energy spectra decrease sufficiently rapid with increasing energy. Historically, field line motion in the magnetotail has been measured by means of low-energy plasma observations, of course, taking advantage of the Compton-Getting effect. Interestingly though, energetic protons can provide such measurements. Recently Roelof et al. [1976] and Keath et al. [1976] have used data from the 16-sector 50-200-keV proton spectrometer on Imp 7 for such measurements at 35 $R_E$ down the tail. Well-defined anisotropies during substorms were observed which are readily interpreted as a flow. These data support the current idea of the formation of an X-type neutral line and its attendant motions.

Baker and Stone [1976] have made measurements of >200-keV electrons using their 8-sector scanning spectrometer on Imp 8 at ~30 $R_E$ in the magnetotail. Unlike protons the Compton-Getting effect causes very little anisotropy in the counting rates of energetic electrons since the velocities of the electrons are large compared to the flow velocities. Normally the PAD's are isotropic or on occasion show the butterfly PAD. However, Baker and Stone have observed asymmetries in the PAD's of the electrons which are associated with substorms.

During substorms they have observed streaming away from the earth suggestive of an X-type neutral line between the satellite and earth and that the observations were being made on open field lines.

## Conclusions

Clearly the PAD's of energetic particles that we have been discussing can be used as excellent diagnostic tools in the study of field configurations. Although much of what was presented was qualitative, in most cases the underlying theory is well known. Obviously the routine use of PAD data as discussed here, but on a more quantitative basis, is very important in our future studies of magnetospheric structure.

Acknowledgments. I would like to thank those colleagues who have worked the most closely with me in the Ogo-5 researches, my immediate colleague R. M. Buck, M. G. Kivelson from UCLA, and T. Pytte from the University of Bergen. I am especially grateful to the UCLA magnetometer experimenters, P. J. Coleman and C. T. Russell, for the ready availability of good magnetic field data. These field data were essential to the interpretation of the Ogo-5 particle data presented in this review. This work was performed under the auspices of the U. S. Department of Energy by the Lawrence Livermore Laboratory under contract No. W-7405-Eng-48.

## References

Alfvén, A., and C. G. Fälthammer, Cosmical Electrodynamics, Fundamental Principles, 2nd edition, Clarendon, Oxford, 1963.

Baker, D. N., and E. C. Stone, Energetic electron anisotropies in the magnetotail: Identification of open and closed field lines, Geophys. Res. Letters, 3, 557, 1976.

Baker, D. N., P. R. Higbie, E. W. Hones, and R. D. Belian, >30-keV electron anistropies at 6.6 $R_E$ as precursors to substorms, EOS Trans. Amer. Geophys. Union, 59, 357, 1978. Submitted to Geophys. Res. Letters, 1978.

Bogott, F. H., and F. S. Mozer, Equatorial electron angular distributions in the loss-cone and at large angles, J. Geophys. Res., 76, 6790, 1971.

Buck, R. M., H. I. West, Jr., and R. G. D'Arcy, Jr., Satellite studies of magnetospheric substorms on August 15, 1968: Ogo-5 energetic proton observations-spatial boundaries, J. Geophys. Res., 78, 3103, 1973.

Buck, R. M., Energetic electron drift motions in the outer dayside magnetosphere — observations and calculations, EOS Trans. Amer. Geophys. Union, 56, 628, 1975.

Choe, J. Y., and D. B. Beard, The compressed geomagnetic field as a function of dipole tilt, Planet. Space Sci., 22, 595, 1974.

Eastwood, J. W., Consistency of fields and particle motion in the 'Speiser' model of the current sheet, Planet. Space Sci., 20, 1555, 1972.

Fairfield, D. H., Average magnetic field configuration of the outer magnetosphere, J. Geophys. Res., 73, 7329, 1968.

Fritz, T. A., High-latitude outer-zone boundary region for ⩾40-keV electrons during geomagnetically quiet periods, J. Geophys. Res., 73, 7245, 1968.

Haskell, G. P., Anisotropic fluxes of energetic particles in the outer magnetosphere, J. Geophys. Res., 74, 1940, 1969.

Imhof, W. L., J. B. Reagan, and E. E. Gaines, Fine-scale spatial structure in the pitch-angle distributions of energetic particles near the midnight trapping boundary, J. Geophys. Res., 82, 5215, 1977.

Kaufman, R. L., and A. Konradi, Speed and thickness of the magnetopause, J. Geophys. Res., 78, 6549, 1973.

Kaufman, R. L., J. T. Horng, and A. Konradi, Trapping boundary and field-line motion during geomagnetic storms, J. Geophys. Res., 77, 2780, 1972.

Keath, E. P., E. C. Roelof, C. O. Bostrom, and D. J. Williams, Fluxes of ⩾50-keV protons and ⩾30-keV electrons at ~35 $R_E$: 2. Morphology and flow patterns in the magnetotail, J. Geophys. Res., 81, 2315, 1976.

Kivelson, M. G., M. P. Aubry, and T. A. Farley, Satellite studies of magnetospheric substorms on August 15, 1968, 5. Ogo-5 energetic electron observations: spatial boundaries and wave-particle interactions, J. Geophys. Res., 78, 3079, 1973.

Lyons, L. R., R. M. Thorne, and C. S. Kennel, Pitch angle diffusion of radiation belt electrons within the plasmasphere, J. Geophys. Res., 77, 3455, 1972.

Lyons, L. R., and D. J. Williams, The quiet time structure of energetic (35-560 keV) radiation belt electrons, J. Geophys. Res., 80, 943, 1975.

Mead, G. D., Deformation of the geomagnetic field by the solar wind, J. Geophys. Res., 69, 1184, 1964.

Northrop, T. G., and E. Teller, Stability of the adiabatic motion of charged particles in the earth's field, Phys. Rev., 117, 215, 1960.

Palmer, I. D., P. R. Higbie, and E. W. Hones, Jr., Gradients of solar protons in the high-latitude magnetotail and the magnetospheric electron field, J. Geophys. Res., 81, 562, 1976.

Pytte, T., R. L. McPherron, M. G. Kivelson, H. I. West, Jr., and E. W. Hones, Jr., Multiple-satellite studies of magnetospheric substorms: radial dynamics of the plasma sheet, J. Geophys. Res., 81, 5921, 1976.

Pytte, T., and H. I. West, Jr., Ground-satellite correlations during presubstorm magnetic field changes and plasma sheet thinning in the near-earth magnetotail, J. Geophys. Res., 83, 3791, 1978.

Roederer, J. G., On the adiabatic motion of energetic particles in a model magnetosphere, J. Geophys. Res., 72, 981, 1967.

Roelof, E. C., E. P. Keath, C. O. Bostrom, and D. J. Williams, Fluxes of ⩾50-keV protons and ⩾30-keV electrons at ~35 $R_E$: 1. Velocity anisotropies and plasma flow in the magnetotail, J. Geophys. Res., 81, 2304, 1976.

Schultz, M., Drift-shell splitting at arbitrary pitch angle, J. Geophys. Res., 77, 624, 1972.

Serlimitsos, P., Low energy electrons in the dark magnetosphere, J. Geophys. Res., 71, 61, 1966.

Shabansky, V. P., Some processes in the magnetosphere, Space Sci. Rev., 12, 299, 1971.

Sonnerup, B. U. O., Adiabatic particle orbits in a magnetic null sheet, J. Geophys. Res., 76, 8211, 1971.

Speiser, T. W., Particle motion in model current sheets, 2. Applications to auroras using a geomagnetic tail model, J. Geophys. Res., 72, 3919, 1967.

Stevens, J. R., E. F. Martina, and R. S. White, Proton energy distributions from 0.060 to 3.3 MeV at 6.6 earth radii, J. Geophys. Res., 75, 5373, 1970.

West, H. I., Jr., R. M. Buck, and J. R. Walton, Electron pitch angle distributions throughout the magnetosphere as observed on Ogo-5, J. Geophys. Res., 78, 1064, 1973a.

West, H. I., Jr., R. M. Buck, and J. R. Walton, Satellite studies of magnetospheric substorms on August 15, 1968: 6. Ogo 5 energetic electron observations — pitch angle distributions in the nighttime magnetosphere, J. Geophys. Res., 78, 3093, 1973b.

West, H. I., Jr., and R. M. Buck, Pitch angle distributions of energetic electrons in the equatorial regions of the outer magnetosphere — Ogo 5 observations, in Magnetospheric Physics, edited by B. M. McCormac and D. Reidel, p. 93, Dordrecht-Holland, 1974.

West, H. I., Jr., and R. M. Buck, Observations of >100-keV protons in the earth's magnetosheath, J. Geophys. Res., 81, 569, 1976.

West, H. I., Jr., R. M. Buck, and M. G. Kivelson, On the configuration of the magnetotail during quiet and weakly disturbed periods: State of the magnetosphere, J. Geophys. Res., 83, 3805, 1978a.

West, H. I., Jr., R. M. Buck, and M. G. Kivelson, On the configuration of the magnetotail near midnight during quiet and weakly disturbed periods: Magnetic field modeling, J. Geophys. Res., 83, 3819, 1978b.

Williams, D. J., and L. R. Lyons, Further aspects of the proton ring current interaction with the plasmapause: Main and recovery phases, J. Geophys. Res., 79, 4791, 1974.

Williams, D. J. (Northern Oceanic and Atmospheric Administration, Boulder, Colorado), unpublished manuscript, 1978.

NOTICE

"This report was prepared as an account of work sponsored by the United States Government. Neither the United States nor the United States Department of Energy, nor any of their employees, nor any of their contractors, subcontractors, or their employees, makes any warranty, express or implied, or assumes any legal liability or responsibility for the accuracy, completeness or usefulness of any information, apparatus, product or process disclosed, or represents that its use would not infringe privately-owned rights."

# EFFECTS OF THE SOLAR WIND ON MAGNETOSPHERIC DYNAMICS: ENERGETIC ELECTRONS AT THE SYNCHRONOUS ORBIT

G. A. Paulikas and J. B. Blake

Space Sciences Laboratory, The Aerospace Corporation
Los Angeles, California 90009

Abstract. Data on energetic electron fluxes at the synchronous orbit, covering the 1967-1978 time interval, obtained by experiments flown on the ATS-1, ATS-5 and ATS-6 spacecraft, have been analyzed. Long term (year) and short term (days) electron flux averages are found to correlate positively with corresponding averages of the solar wind velocity.

## Introduction

One of the fundamental questions in magnetospheric physics concerns the connection between the interior dynamics of the magnetosphere and the input into the magnetosphere of energy, momentum and mass by the solar wind. Viewing the solar wind-magnetosphere interaction in a macroscopic way, various workers have sought to establish the connections between the parameters which characterize the solar wind and the resulting geophysical effects. Knowledge of the variation of the strength of this interaction as a function of solar wind parameters is important in the testing of theories of magnetospheric dynamics. More practically, an unambiguous association of geomagnetic effects with the parameters of the solar wind could lead to long term and short term predictions of geomagnetic activity and the levels of space radiation.

The approach frequently used in studying the macroscopic interaction of the solar wind with the magnetosphere consists of correlating the parameters of the solar wind with direct or indirect measures of the energy content of the magnetosphere or the rate of dissipation of magnetospheric energy. In such an approach, which we shall follow, the details of the microscopic processes are ignored and global quantities describing the state of the magnetosphere are related to those parameters of the solar wind thought to be governing the efficiency of the coupling process. Russell et al. (1974) considered the conditions existing in interplanetary space associated with the development of the main phase of magnetic storms. Kivelson et al. (1973) studied the influence of the direction of the interplanetary field on the location of the polar cusp. Crooker et al. (1977) correlated long-term averages of the solar wind speed with geomagnetic activity. Arnoldy (1971) and Caan et al. (1977) studied the signature in the interplanetary medium associated with the occurrence of substorms. Akasofu (1975) and Burch (1973) examined the effect of the interplanetary magnetic field direction on auroral zone dynamics, and Burton et al. (1975) developed a relationship between the rate of injection of energy into the magnetospheric ring current and the interplanetary $E_y = -(V \times B)_y$ field. The citations above are meant to be a representative, not exhaustive, review of recent work in the field. These papers and the review by Russell (1974) should be consulted for additional details and references.

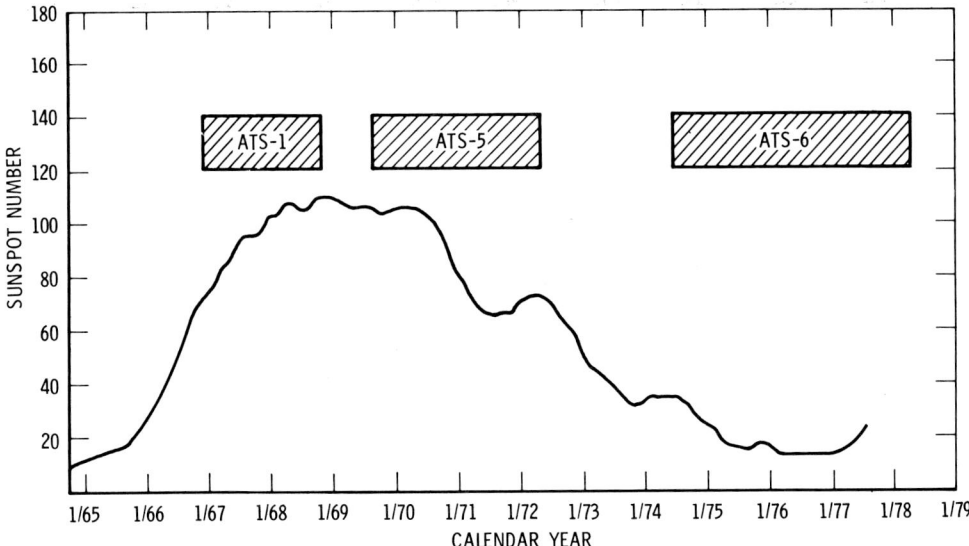

Fig. 1. Relationship between the intervals of data acquisition on energetic electrons at synchronous altitude by experiments on board ATS spacecraft and the 11-year solar activity cycle as defined by the Zurich sunspot number.

The picture which emerges from the work cited earlier is that the interplanetary magnetic field direction as well as the solar wind velocity determine the efficiency of coupling between the solar wind and magnetosphere. It has been known, for some time now, that during times when the magnetosphere "sees" a net southward interplanetary magnetic field, the interaction between the solar wind and the magnetosphere (presumed to be proceeding via magnetic merging) is enhanced, leading to enhanced geomagnetic activity and various effects resulting from such activity. It has been shown that, while the presence of a net southward field correlates very well with geomagnetic activity when this correlation is examined on a time scale of an hour, such a correlation may not hold true for longer averages of the parameters to be correlated. Crooker et al. (1977) have demonstrated that on time scales of the order of a year, correlation between geomagnetic activity, as expressed by the Ap index, and solar wind parameters scales as approximately as $B_z V^2$, where V is the average solar wind velocity and $B_z$ is the southward component of the interplanetary magnetic field in GSM coordinates. The work of Crooker et al. has been criticized by Dessler and Hill (1977), who point out that earlier work by Garrett et al. (1974) had shown that the positive derivative of the solar wind speed is the factor controlling geomagnetic activity. It is thus of interest to continue work along these lines, looking both at time scales intermediate to those discussed above as well as at correlations between the parameters of the solar wind and various other magnetospheric observables.

We have recently presented a preliminary report (Paulikas and Blake, 1976) in which we showed that the sector structure of the interplanetary magnetic field exerted a strong influence on the intensity of energetic electrons contained in the outer magnetosphere. Specifically, we found that in the fall, more energetic electrons are found when the earth is in a positive sector of the interplanetary field (with a consequent net southward $B_z$) than when a negative sector envelops the magnetosphere. The situation is reversed in the spring, and negative

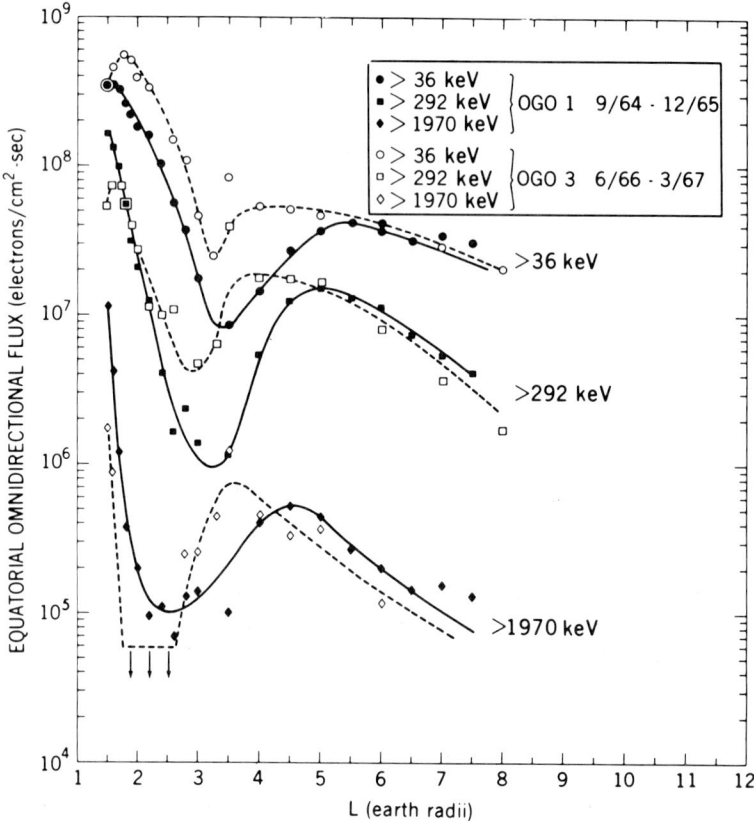

Fig. 2. Equatorial radial profiles of energetic electrons obtained from OGO data (from Singley and Vette, 1972).

interplanetary magnetic field sectors appear to be able to generate larger fluxes of energetic magnetospheric electrons. These findings, based on a relatively limited set of data obtained in late 1974-early 1975, were interpreted in the context of the work of Burton et al. (1975) and Russell and McPherron (1973) who viewed energetics of the magnetosphere as being influenced primarily by the north-south component of the interplanetary field.

These conclusions were based on limited data and no attempt was made to correlate the behavior of energetic electrons in the magnetosphere with the various parameters which describe the interplanetary medium. We have since extended our data base on energetic electrons substantially in time and have acquired data on the properties of the interplanetary medium during the period of interest. This paper presents the results of a study correlating the observations of energetic electrons at the synchronous orbit in the 1974-1977 interval with the state of the interplanetary medium. In addition, we also present here an overview of the behavior of energetic trapped radiation at the synchronous altitude since 1967.

## Description of the Data

Energetic electrons at the synchronous orbit were measured by the Aerospace Corporation experiment on ATS-6. Additional data, used to compile the 1967-

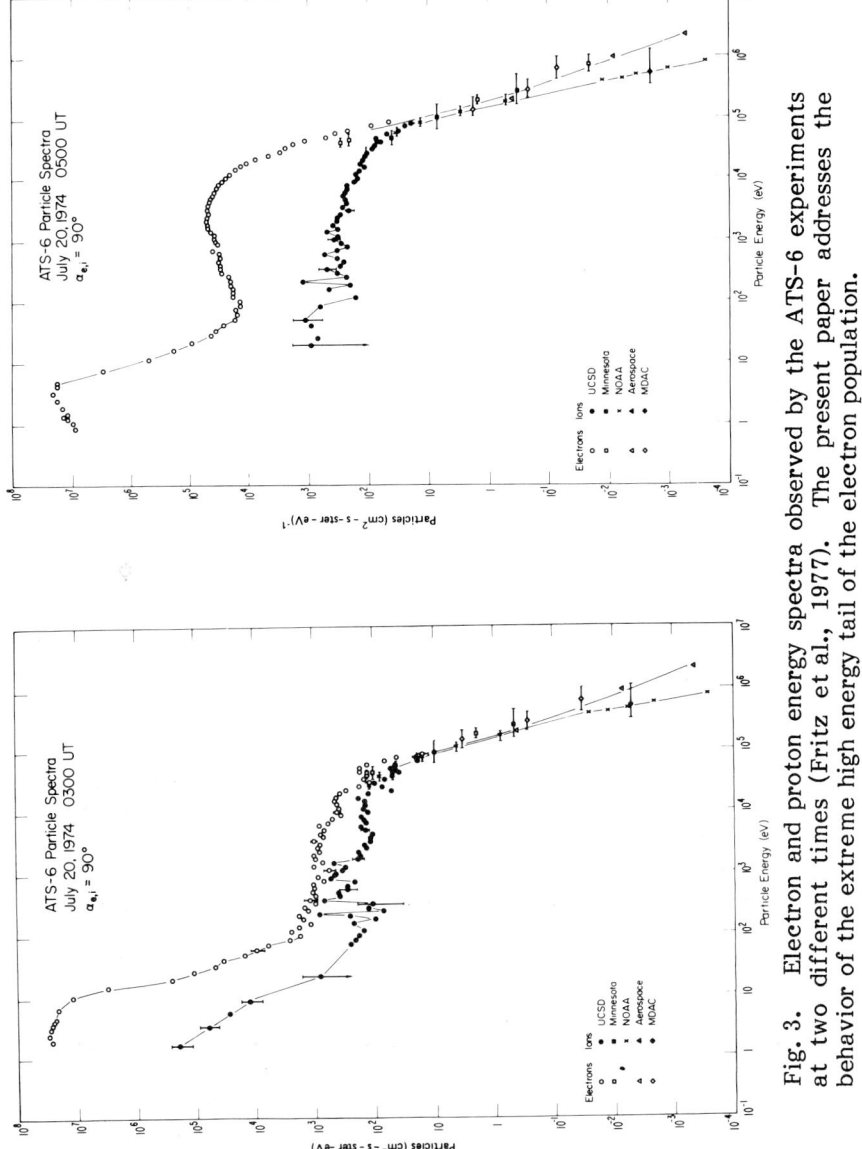

Fig. 3. Electron and proton energy spectra observed by the ATS-6 experiments at two different times (Fritz et al., 1977). The present paper addresses the behavior of the extreme high energy tail of the electron population.

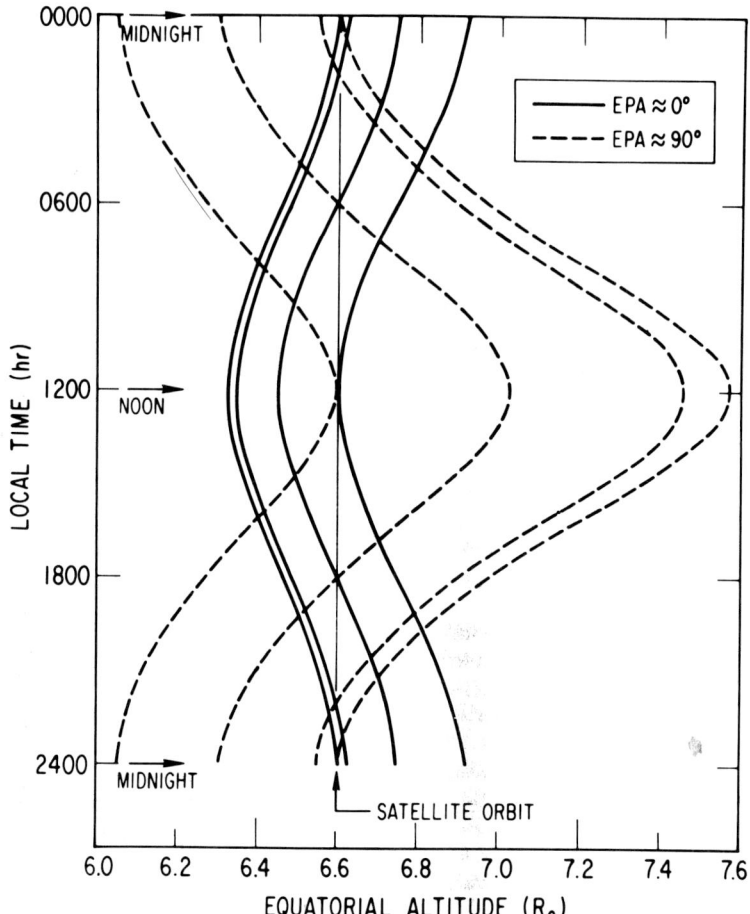

Fig. 4. Representative drift paths (equatorial crossing altitudes) of electrons for particles mirroring near the equator (EPA = 90°) and those mirroring at high latitudes (EPA = 0°).

1977 history of energetic electron radiation at the synchronous orbit, were obtained by our earlier experiment on ATS-1 and the UCSD instrument on ATS-5. The latter data were graciously provided to us by C. E. McIlwain. The periods of data coverage relative to the recent history of solar activity, as described by the Zurich sunspot number, are shown in Figure 1. The details of the instrumentation have been described in detail elsewhere (Paulikas et al., 1968, 1975). Briefly, the experiments measured the fluxes of energetic (mainly relativistic) electrons. We are using the flux of relativistic electrons as an indication of the efficiency of the solar wind – magnetosphere engine: these particles are a measure of the capability of magnetospheric processes to generate relativistic particles from thermal plasma.

It is useful to recall that the synchronous orbit is well beyond the maximum of energetic trapped radiation (Fig. 2) and that the particles of concern in this paper are the extreme high energy tail of the spectrum (Fig. 3). The particles of concern can be thought to be end products of magnetospheric dynamics, but not as

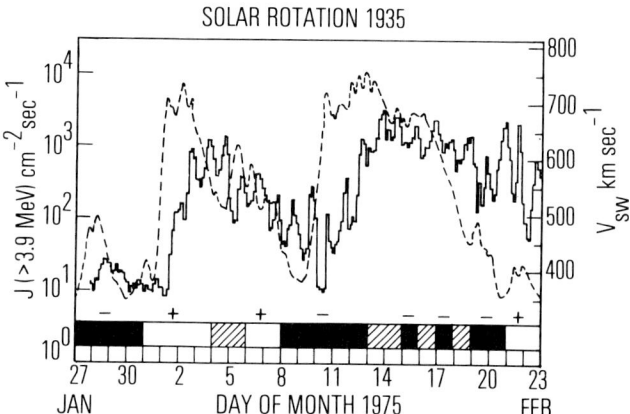

Fig. 5. Flux of energetic electrons (solid curve with logarithmic scale at left) as a function of time for solar rotation 1935. Three-hour averages of electron flux are plotted. Solar wind velocity (dashed curve with linear scale at right) is superimposed on these data. The pattern at the bottom of the figure is represents the IMF sector structure.

participants, owing to their very low energy density. Because we are dealing with omnidirectional fluxes in our data, our conclusions apply to a sizeable volume of the outer magnetosphere: as a result of drift shell splitting, the electrons detected at one local time spread over a band of L-shells at other local times (Fig. 4), and a local measurement acquires global properties.

Data on the sector structure of the interplanetary field were obtained from the reports of Svaalgard (1976). Information on the state of the interplanetary medium was obtained from the NSSDC tape. The solar wind velocities for the recent past were obtained from Jack Gosling at Los Alamos. The best temporal resolution of data used for the present purposes was typically one hour for the

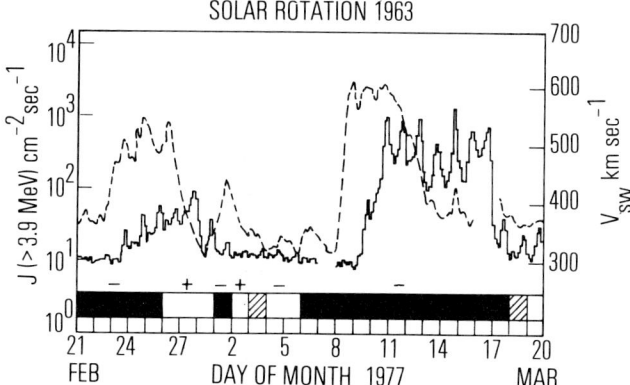

Fig. 6. Flux of energetic electrons (solid curve with logarithmic scale at left) as a function of time for solar rotation 1963. Three-hour averages of electron flux are plotted. Solar wind velocity (dashed curve with linear scale at right) is superimposed on these data. The pattern at the bottom of the figure represents the IMF sector structure.

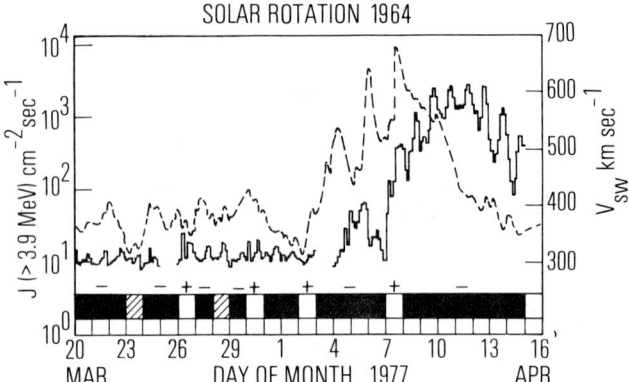

Fig. 7. Flux of energetic electrons (solid curve with logarithmic scale at left) as a function of time for solar rotation 1964. Three-hour averages of electron flux are plotted. Solar wind velocity (dashed curve with linear scale at right) is superimposed on these data. The pattern at the bottom of the figure represents the IMF sector structure.

solar wind data and one day for the data on the sector structure of the interplanetary magnetic field.

## Approach

In this study we are interested in the long-term behavior of the energetic electrons. Accordingly, we have chosen as the basic unit of data for most of our analyses the daily average of the electron flux and the corresponding daily averages of the parameters of the solar wind. The daily average was chosen in order to average out local time effects which are very important for energetic electrons at the synchronous orbit (Paulikas et al., 1968). A day is also the natural timescale on which to view the longer term dynamics of energetic electrons in the outer zone. Growth and decay of the energetic electron populations seems to occur on the time scale of many hours or days at least in part because diffusive effects appear to delay arrival of energetic electrons at the synchronous orbit from the inner magnetosphere where the peak of the energetic electron flux is usually found. For the present purpose, we have not eliminated adiabatic perturbations of electron flux from our data, although such a correction is possible. Such limited examination as we have made of the magnitude of adiabatic effects on the electron fluxes relative to non-adiatatic effects suggests that non-adiabatic effects are at least an order of magnitude more important in the dynamics of relativistic electrons at synchronous altitude at most times and totally dominate the dynamics during periods of electron flux increase.

The integrated effect of our data analysis approach is to view the magnetosphere as a black box, with the solar wind parameters as the input and the fluxes of energetic electrons at the synchronous altitude as the output. We seek information on the transfer function which will tell us the dependence of the generation of energetic electrons by magnetospheric processes on the parameters of the solar wind. The transfer function we seek thus incorporates implicitly within such details as the source functions of plasma in the magnetosphere, substorm processes, as well as subsequent transport and acceleration processes

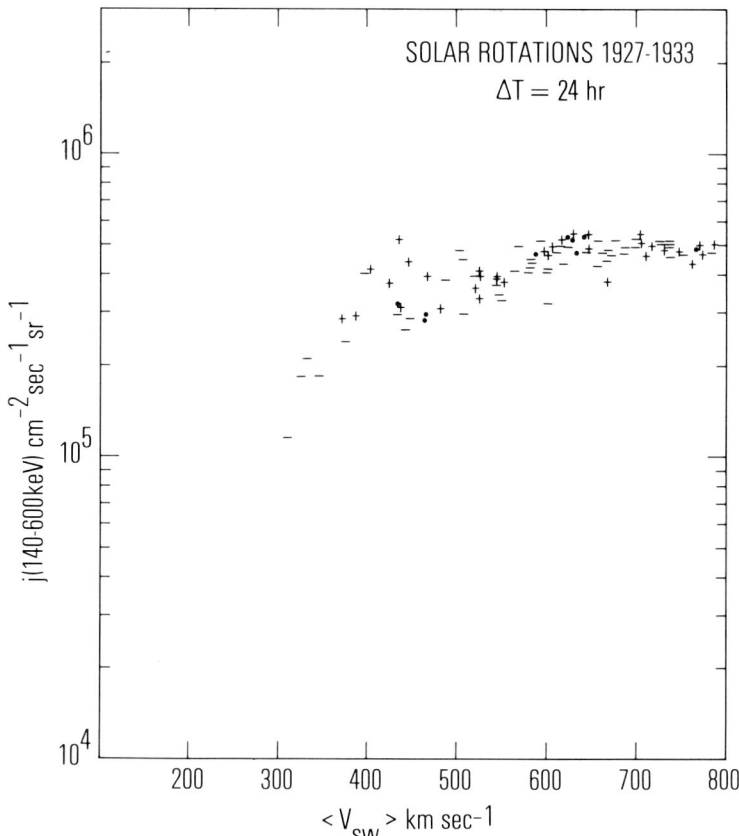

Fig. 8. Correlation of daily averages of 140-600 kev electron fluxes with daily averages of the solar wind velocity. The delay time between solar wind and energetic particle measurements is 12 hours. All data available for solar rotations 1927-1933 (July-December 1974) are included in this plot. The plus and minus signs indicate the sign of $E_y = -(V \times B)_y$ in geocentric solar-magnetospheric coordinates; data points for which the $E_y$ sign cannot be determined are simple points.

which, taken together, generate the high energy tail of the spectra illustrated in Fig. 3.

In this paper we have chosen the ATS-6 data for detailed study of the solar wind-energetic electron correlation. The ATS-1 and ATS-5 data base are used to extend our data base so that with only simple normalizations we are able to reconstruct the behavior of the energetic electron population for a large fraction (1967-1978) of a complete solar cycle and thus infer additional details about the behavior of energetic electrons in the outer zone over this time span.

## Effect of Solar Wind Parameters on Energetic Electrons

### Prototype Observations and Interpretations

Figures 5, 6 and 7 present data on energetic electrons and the solar wind velocity for each of three solar rotations. Figure 5 illustrates the case, common

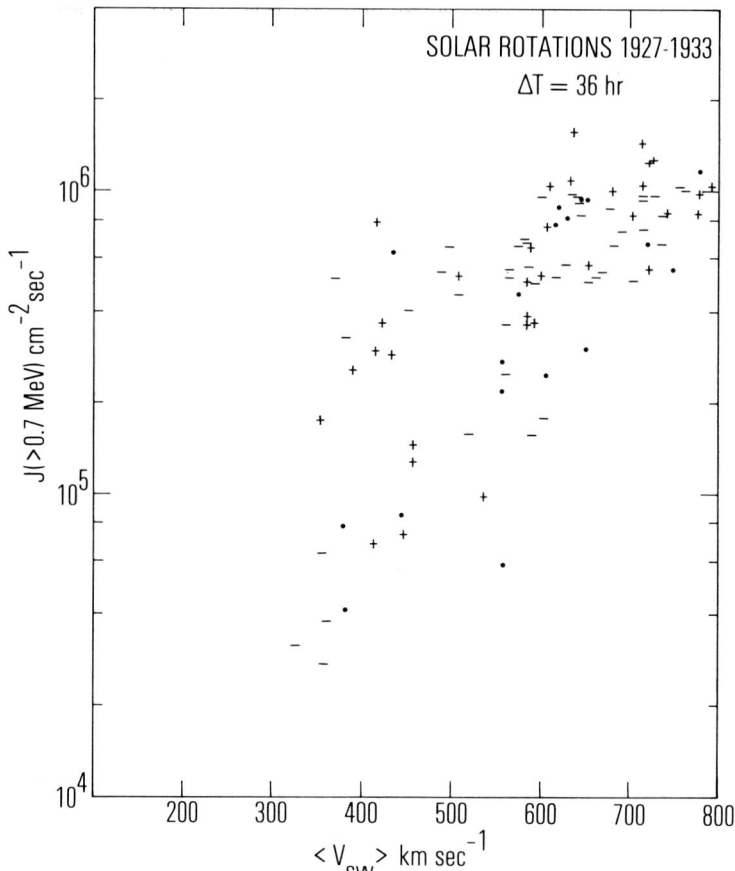

Fig. 9. Correlation of daily averages of greater than 0.7 MeV electron fluxes with daily averages of the solar wind velocity. The delay time between solar wind and energetic particle measurements is 24 hours. All data available for solar rotations 1927-1933 (July-December 1974) are included in this plot. The plus and minus signs indicate the sign of $E_y = -(V \times B)_y$ in geocentric solar-magentospheric coordinates; data points for which the $E_y$ sign cannot be determined are simple points.

in 1974 and 1975, when the sun was emitting two streams of plasma with unusually high velocity (Bame et al., 1976, Gosling et al., 1976, 1977, Feldman et al., 1978) and the subsequent response of the energetic electron population in the outer magnetosphere to these solar wind streams. Figures 6 and 7 show an opposite extreme, commonly noted in 1976 and 1977: the solar wind streams here are much weaker, reaching maxima of barely 600 km/sec, in contrast to the 800 km/sec peaks evident in Fig. 5.

In either event the response of magnetospheric energetic electrons is evident. Starting about a day or two after the solar wind stream first reaches the earth, the fluxes build up in correlation with the increasing velocity of the solar wind. The delay between arrival of a high speed stream and buildup of energetic fluxes appears to depend on the energy of the electrons under observation, with the delay of the response of the electrons increasing with electron energy. Subsequent to the passage of the peak in the solar wind velocity, electron fluxes in

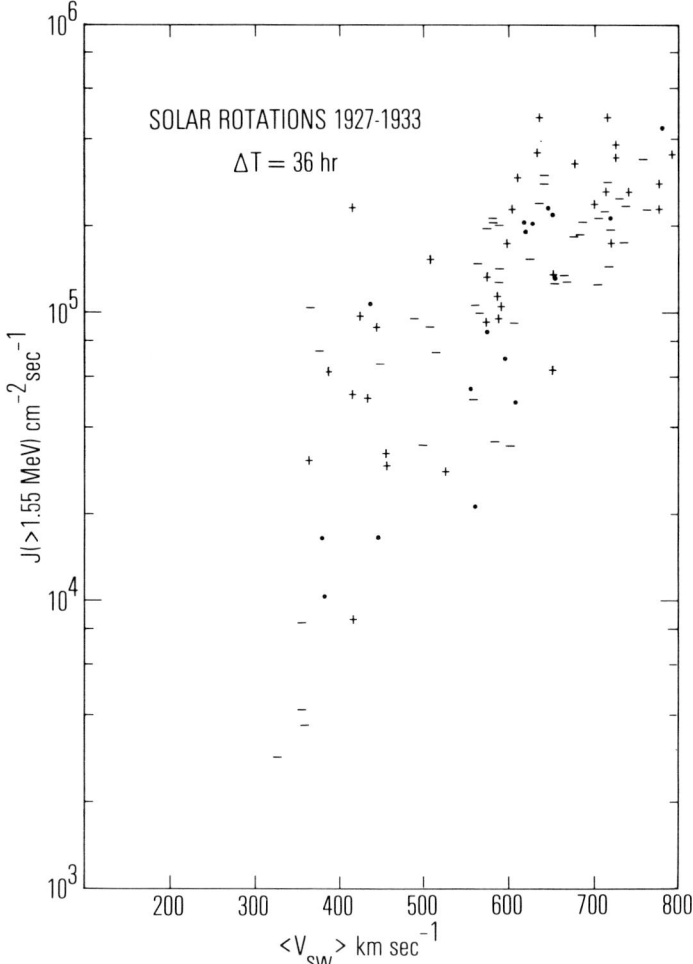

Fig. 10. Correlation of daily averages of greater than 1.55 MeV electron fluxes with daily averages of the solar wind velocity. The delay time between solar wind and energetic particle measurements is 36 hours. All data available for solar rotations 1927-1933 (July-December 1974) are included in this plot. The plus and minus signs indicate the sign of $E_y = -(V \times B)_y$ in geocentric solar-magnetospheric coordinates; data points for which the $E_y$ sign cannot be determined are simple points.

general do not track the decay in the velocity of the stream.

Similar comparisons have been made between energetic electron fluxes and other parameters which describe the solar wind, for example, solar wind energy density, solar wind pressure, the components of the interplanetary field, and the components of $-V \times B$. From these comparisons it soon became clear that the approximately exponential relationship between electron flux and solar wind velocity ($J \alpha \exp V_{sw}$) evident from inspection of Figures 5, 6 and 7 was the most pronounced signature of solar wind variability in the dynamics of energetic electrons.

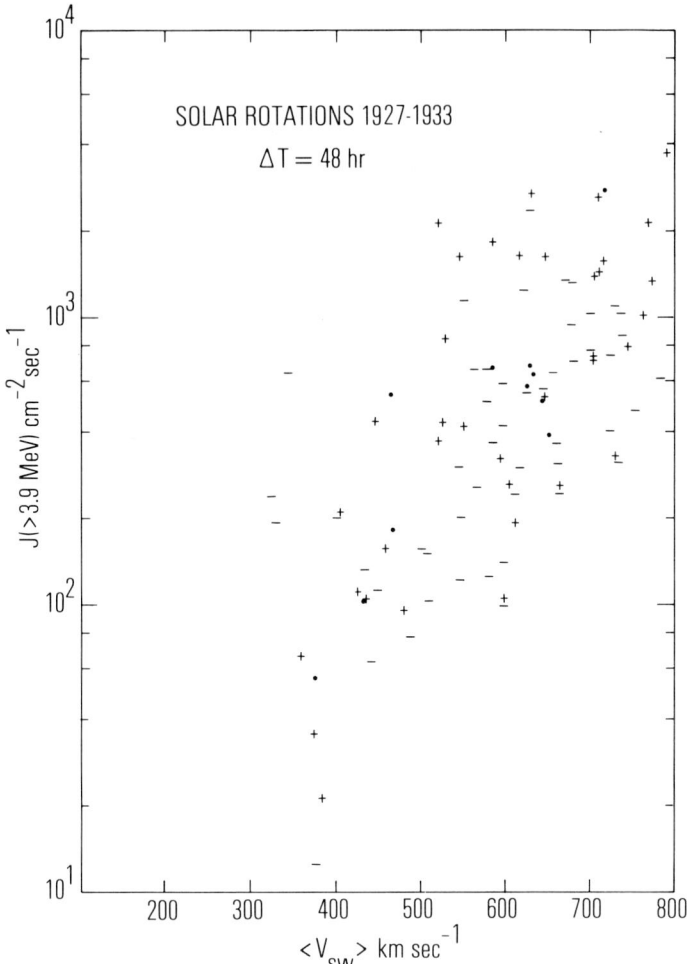

Fig. 11. Correlation of daily averages of greater than 3.9 MeV electron fluxes with daily averages of the solar wind velocity. The delay time between solar wind and energetic particle measurements is 48 hours. All data available for solar rotations 1927-1933 (July-December 1974) are included in this plot. The plus and minus signs indicate the sign of $E_y = -(V \times B)_y$ in geocentric solar-magnetospheric coordinates; data points for which the $E_y$ sign cannot be determined are simple points.

Examination of more than three years of ATS-6 data serves only to confirm the picture described above. The flux of energetic electrons at the synchronous altitude is strongly correlated with the presence of high speed streams in the solar wind. A well-ordered stream pattern gives rise to a regular sequence of electron flux growth and decay, a disordered flow pattern (i.e. a highly variable solar wind velocity) gives rise to a more chaotic behavior in the energetic electron population. The correlation between solar wind velocity and energetic population is most striking at the time of solar velocity increase. Once past the velocity peak, the behavior of the energetic electrons appears to be a combination of

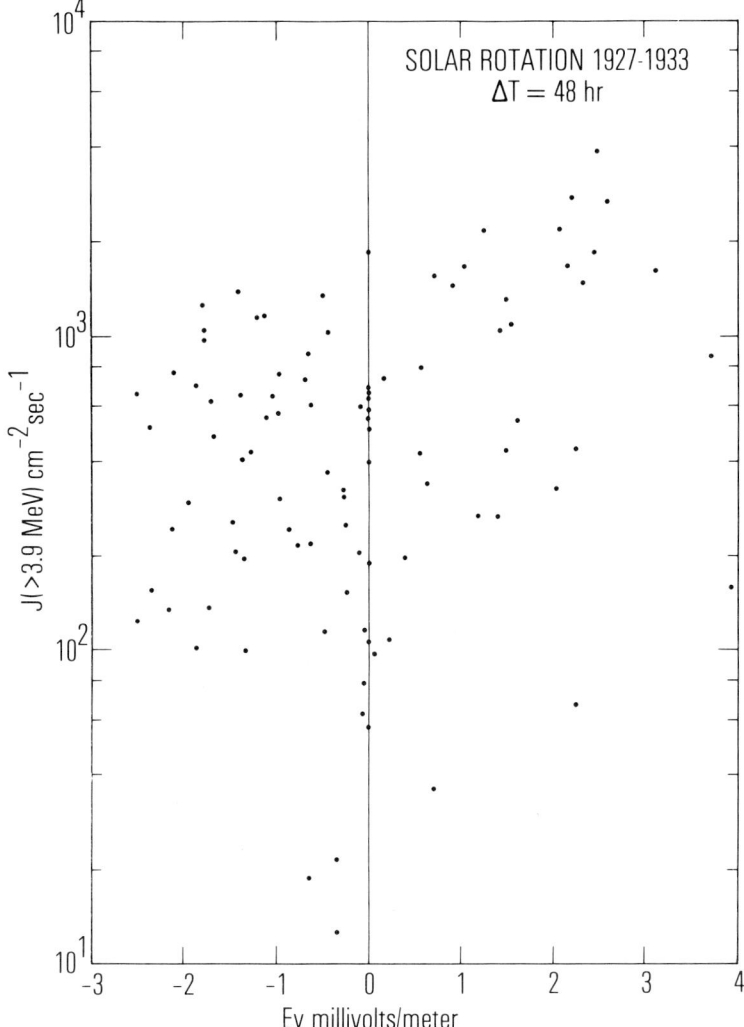

Fig. 12. Correlation of daily averages of greater than 3.9 MeV electron fluxes with $E_y = -(V \times B)_y$ in GSM coordinates. Data on this plot are identical to that presented in Figure 11.

continued lower level injection (or diffusion) of energetic electrons into the region of space accessible to our observations, ultimately followed by rapid decay of the electron flux.

Quantitative Comparisons

In order to arrive at a quantitative measure of the effect of solar wind velocity on the population of energetic electrons, we have made detailed correlations of the rise of electron fluxes with the solar wind velocities. As briefly noted above, we observe that a finite time delay appears to be required for energetic electrons to respond to changes in solar wind velocity. These

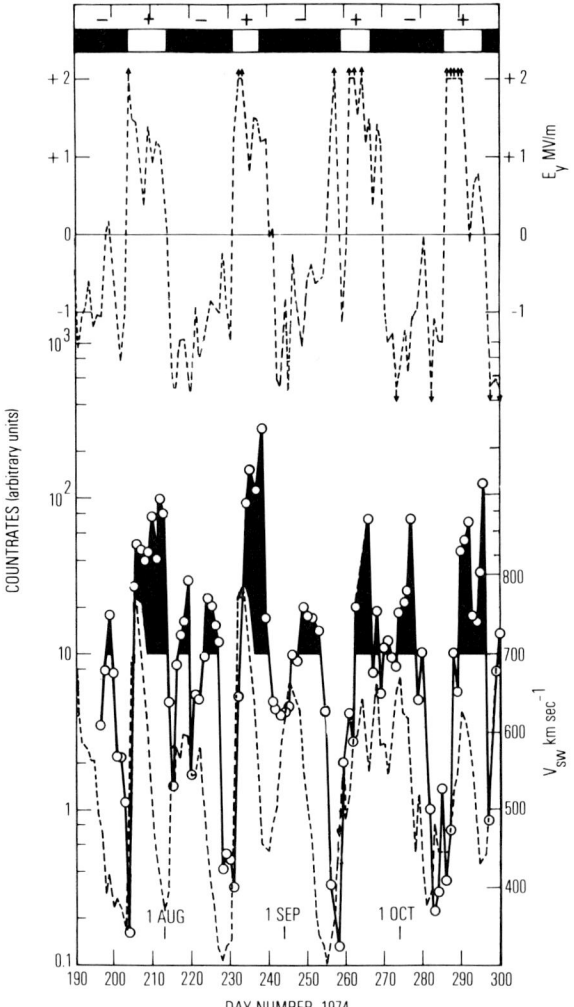

Fig. 13. Daily averages of energetic electron countrates observed in the late summer and fall of 1974 by ATS-6 are plotted as a function of Day Number, 1974. Also plotted, at the top of the figure, are the polarities of the interplanetary magnetic field as inferred by Svalgaard (1976), and the strength of the dawn-dusk electric field $E_y = -(V \times B)_y$. Local time for all particle data is local noon; the sector boundary transitions are assumed to occur at 0000 UT for the days indicated. Solar wind velocity is shown at the bottom of the figure, (dashed curve, bottom) referred to the right-hand scale. For emphasis we have shaded those portions of the curves where $E > 3.9$ MeV countrates exceed 10/sec.

differences in time required by the various electron energy channels to respond to changes in the solar wind can be taken to be a measure of the time the magnetosphere requires to generate energetic electrons and to transport these particles to the synchronous orbit. Thus, very approximately we find that about a day is needed to generate 140-600 keV electrons, while two days are required to generate > 3.9 MeV electrons. Such time delays have been incorporated in our

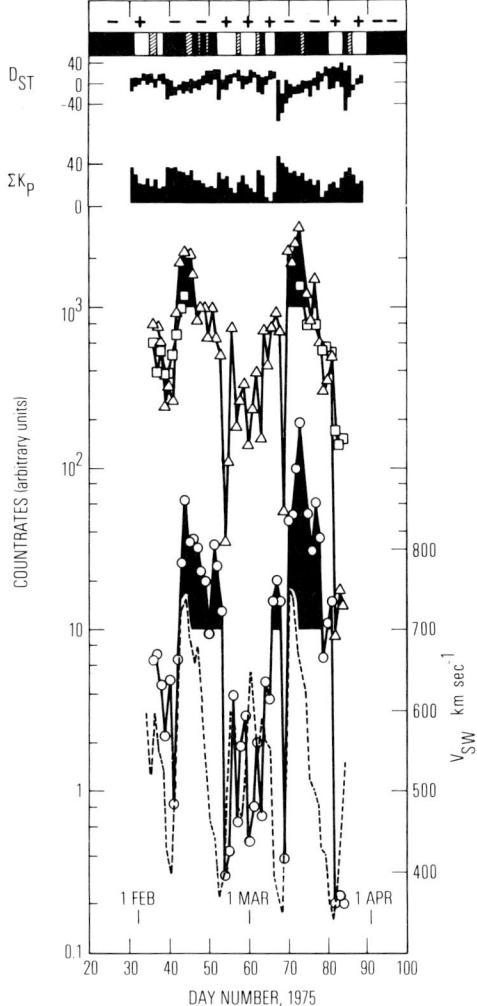

Fig. 14. Daily averages of energetic electron countrates observed in the spring of 1975 by ATS-6 and ATS-1. Circles are ATS-6 data for $E_e > 3.9$ MeV, triangles are ATS-6 data for $E_e > 1.55$ MeV and squares are ATS-1 data for $E_e > 1.9$ MeV. Shading for countrates above $10^3$ and 10 has been added for emphasis. Also plotted, at the top of the figure, are the polarities of the interplanetary field as inferred by Svalgaard (1976) and the velocity of the solar wind, (dashed curve, bottom), referred to the right hand scale. The daily sum of $K_p$ and the range of $D_{st}$ for each day is also shown at the top. The IMF sector structure during this period exhibited some days of mixed polarity, these days are indicated by cross-hatching.

correlation analysis and the results are presented in Figures 8 through 11. Similar plots have been made with other choices of time delays. We find that the delays used in Figures 8-11 minimize the scatter of the data points in that for both shorter or longer delays the correlation of electron fluxes with solar wind velocity is less clear cut.

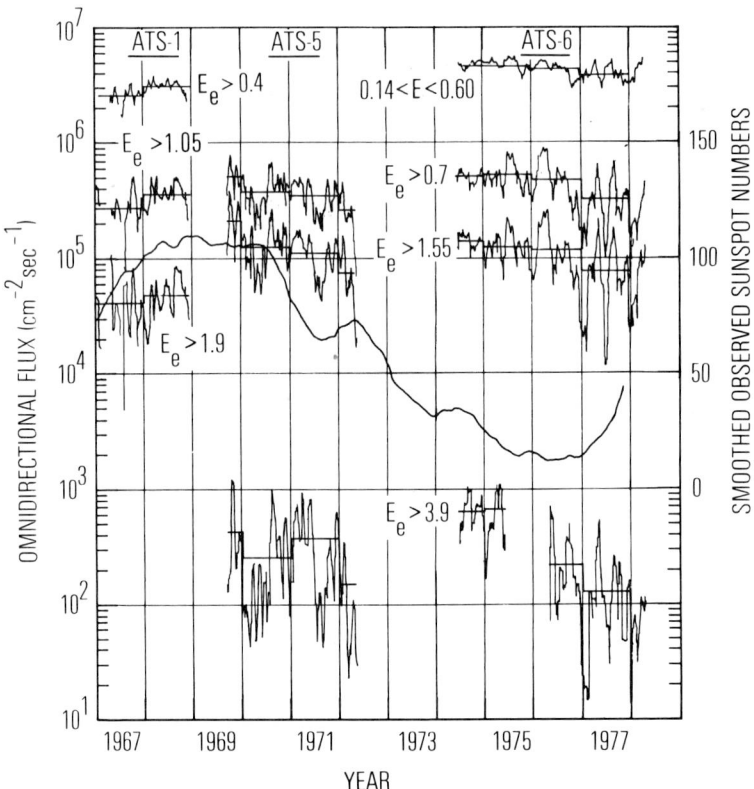

Fig. 15. ATS-1, and ATS-5, ATS-6 energetic electron fluxes (running 27 day averages) as a function of time. ATS-5 data was normalized to ATS-6 data in mid 1974. The energy thresholds for the ATS-1, ATS-5 and ATS-6 channels are shown on the figure. The flux averages for each year are also indicated (solid horizontal lines). Superimposed on this graph is the Zurich monthly sunspot number, referred to the linear scale on the right. Gap in the ATS-6 $E_e > 3.9$ MeV data in 1975 and 1976 is caused by our rejection of suspect data.

We have also looked for a correlation of the energetic electrons with $E_y = -(V \times B)_y$, following the lead of other workers. One such correlation plot is shown in Figure 12. As can be seen, the dependence of the flux of > 3.9 Mev electrons on $E_y$ is not nearly as strong as the dependence of the electron flux on the solar wind velocity. To be sure, the sign of $-(V \times B)_y$ viewed in geocentric solar-magnetospheric coordinates exerts a modulating influence even on the time scale of one day and appears to modulate the effectiveness of the solar wind in determining the maximum flux level that energetic electrons may reach. Examination of Figure 9-12 reveals that the maximum fluxes occur preferrentially when $E_y > 0$, (i.e. $B_z$ is pointing south), in agreement with the accepted view of interplanetary magnetic field - magnetosphere interaction. It could be argued that daily averages of $E_y$ suppress short period excursions of the interplanetary field and thus do not correctly represent the effects of short duration southward $B_z$. Examination of our IMF data base down to a temporal resolution of one hour shows that such a bias is not present, although examination of IMF data at higher time resolution has not been carried out.

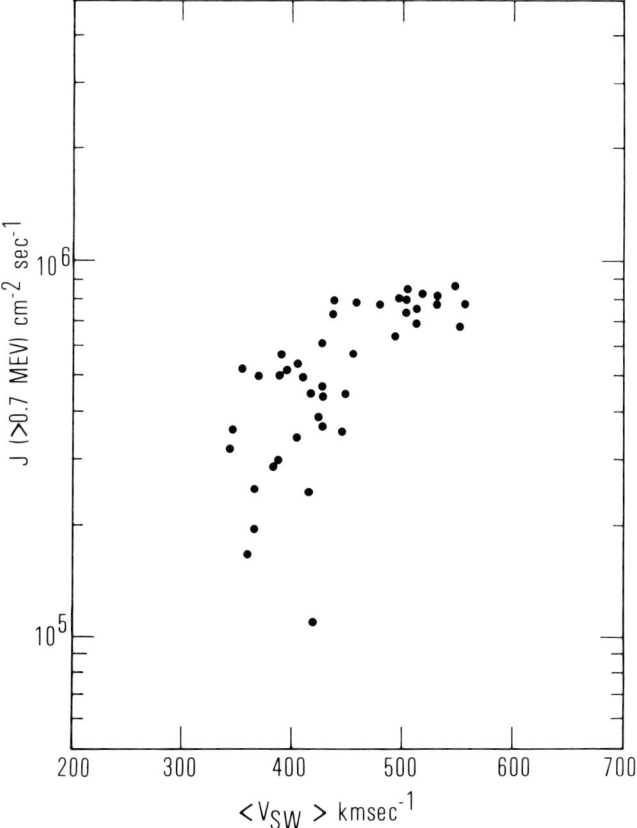

Fig. 16. 27-day average of energetic electons (E >0.7 MeV) as a function of corresponding averages of the solar wind velocity. All ATS-6 data for solar rotations 1927-1969 (June 1974 - August 1977) are included in this figure.

The picture which emerges is thus one where the velocity of the solar wind is the most important parameter in organizing the flux levels of energetic electrons in the outer magnetosphere, with the sign of the $E_y = -(V \times B)_y$ component modulating the efficiency of the solar wind velocity as a "generator" of energetic electrons.

Our earlier findings (Paulikas and Blake, 1976) can now be re-examined in this light. Figures 13 and 14 are reproduced from our earlier paper, but with the addition of the solar wind velocity. It is clear that the electron flux peaks coincide with the peaks in the velocity of the high speed solar wind imbedded within each sector. The arguments relating to seasonal effects (and thus an implied semi-annual variation) are still relevant, modified by our new appreciation of the explicit - and exponential - dependence of the energetic electron fluxes on solar wind velocity. Strong positive $E_y$, corresponding to southward directed $B_z$ appears to add a final factor of about two in energetic electron flux at the synchronous orbit.

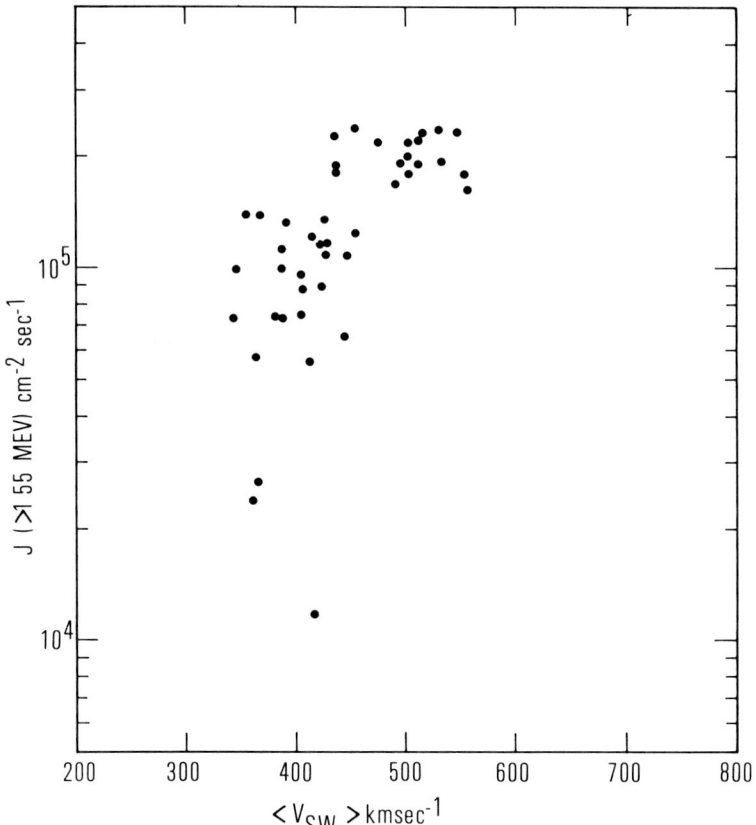

Fig. 17. 27-day averages of energetic electrons (E > 1.55 MeV) as a function of corresponding averages of the solar wind velocity. All ATS-6 data for solar rotations 1927-1969 (June 1974 - August 1977) are included in this figure.

### Long-Term Variability of Energetic Electron Fluxes

The time history of energetic electrons at the synchronous orbit since 1967 is presented in Fig. 15. Here we are presenting data obtained by ATS-1, ATS-5 and ATS-6, with the ATS-5 data having been normalized to the ATS-6 data so as to form a homogeneous data set.

The overall impression from Fig. 15 is that during the 1967-1978 time period, the energetic radiation is remarkably stable if sufficiently long time averages of data are examined. There are three types of variability which deserve attention:

1) The most marked variability, particularly noticeable at the higher electron energies and exaggerated by the fact that we are presenting running 27-day averages of the data, is associated with the 27-day solar rotation period and the effect, already described above, that discrete high-speed solar wind streams are efficient generators of energetic electrons.

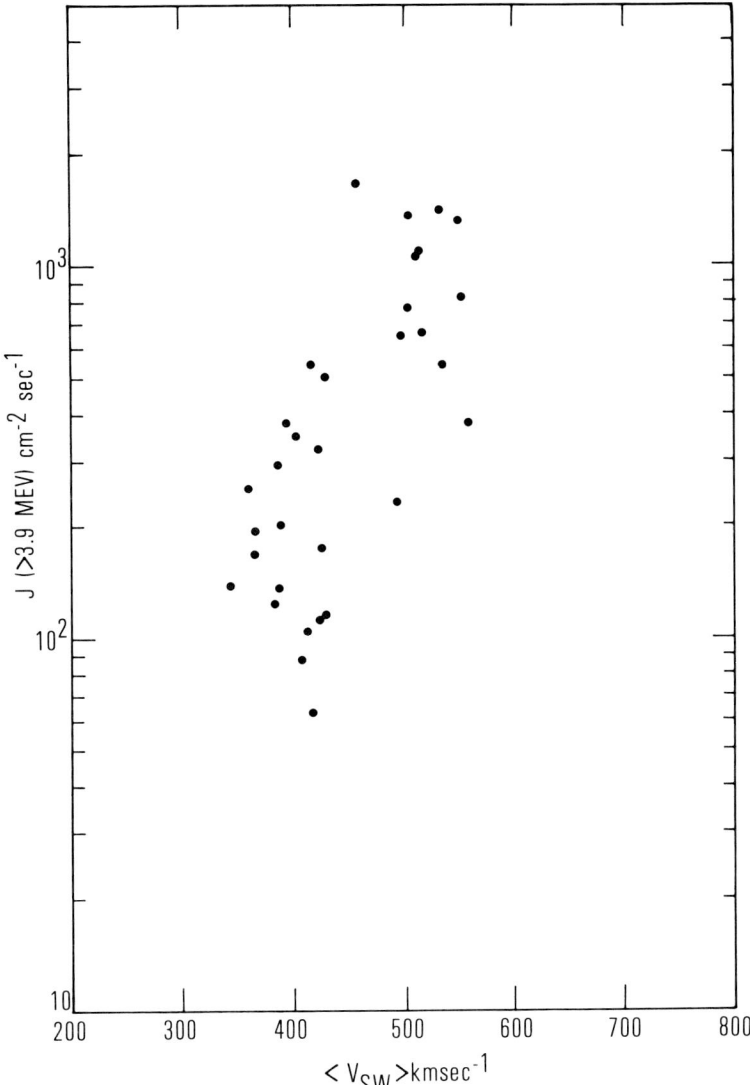

Fig. 18. 27-day averages of energetic electrons (E > 3.9 MeV) as a function of corresponding averages of the solar wind velocity. All ATS-6 data from June 1975 through August 1977 are included in this figure, except for a period from mid 1975 through early 1976.

Figures 16-18 present a correlation of 27-day averages of electron fluxes with corresponding averages of the solar wind velocity.
2) The general decrease in electron fluxes which began in 1976 appears to be associated with a decline in the average velocity of the solar wind. When we calculate semi-annual averages of the observed electron fluxes and compare these against averages of the solar wind (Figs. 19-21), we find, on this time scale a clear positive correlation between solar wind velocity and energetic electron fluxes.

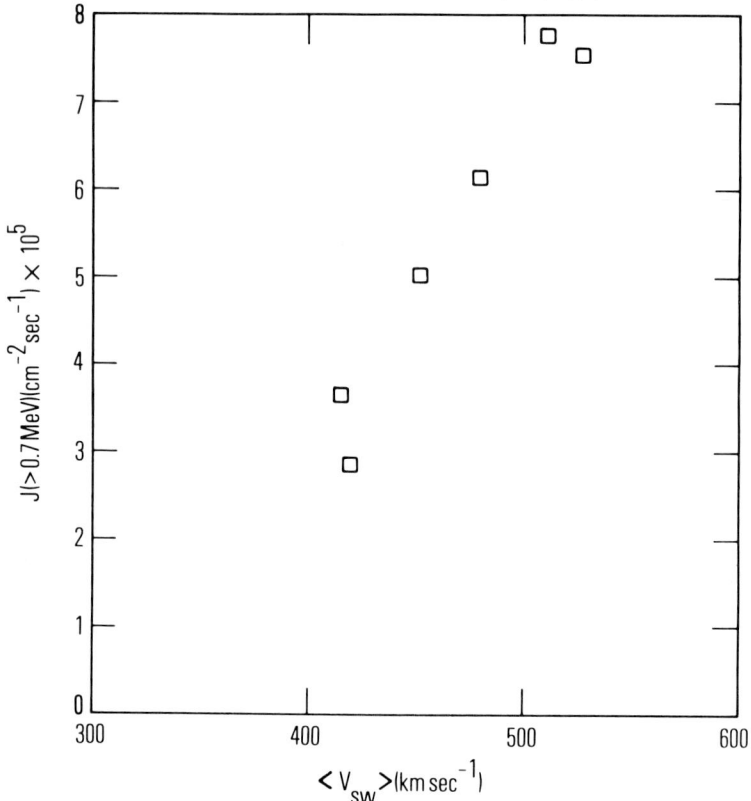

Fig. 19. Plot of the semi-annual average of the $E > 0.7$ MeV electron flux as a function of the semi-annual average of the solar wind velocity. ATS-6 data from mid 1974 through 1977 are included in this plot.

3) There is some, admittedly qualitative, evidence of a semi-annual effect. This effect is particularly noticeable in 1977 when the observations are not confused by spacecraft movement (which occurred in 1975 and 1976) or systematic changes in the solar wind velocity as occurred in 1976. A semi-annual variability in energetic electrons should be a natural consequence of the semi-annual variation in the strength of the interaction between the interplanetary magnetic field and the magnetosphere as described by Russell and McPherron.

Thus we see, that, when viewed on the time scale of a day or a solar rotation (27-days) or even on the time scale of half a year, we find that the correlation of energetic electron fluxes with solar wind velocity persists (Figs. 16 through 21). Similar results have been obtained for energetic protons in the outer magnetosphere by Baker et al. (1978). There is some evidence that the "initial state" of the magnetosphere also exerts a significant influence on the evolution of the population of energetic electrons under the impact of changing interplanetary conditions. As mentioned earlier $-V \times B$ effects can be seen in our data, but only as a modulating influence.

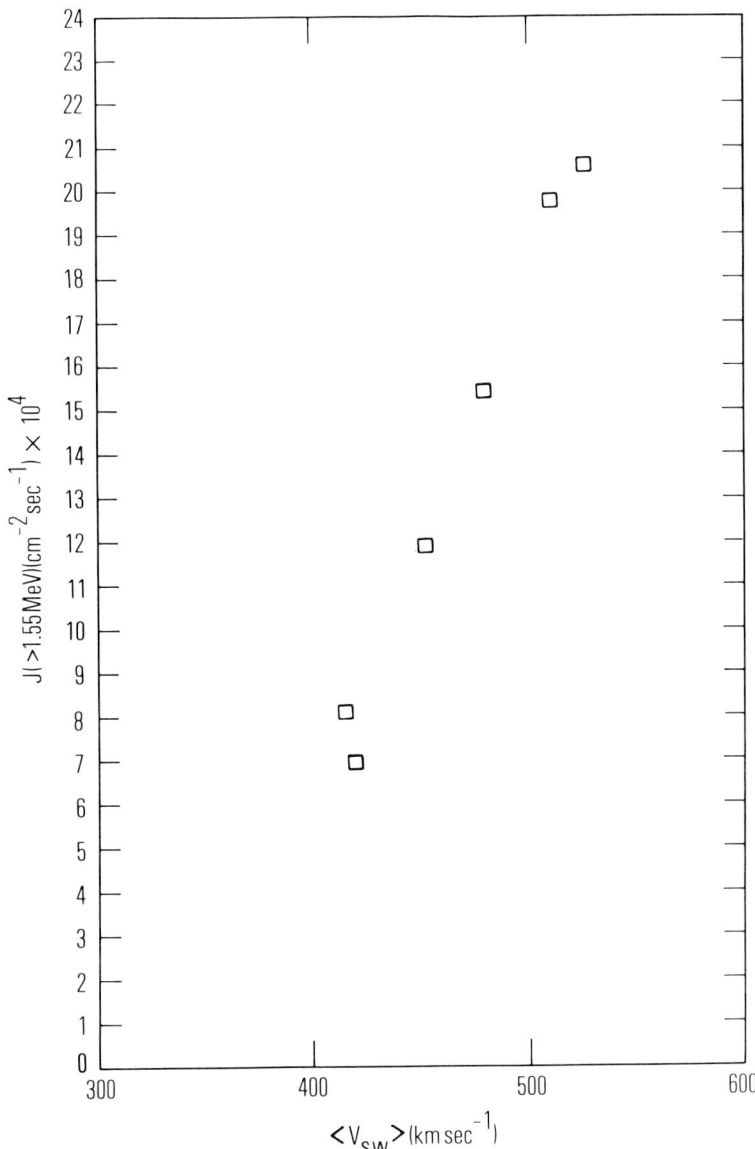

Fig. 20. Plot of the semi-annual average of the $E > 1.55$ MeV electron flux as a function of the semi-annual average of the solar wind velocity. ATS-6 data from mid 1974 through 1977 are included in this plot.

The data and results presented in this paper may still be modified by elimination of adiabatic effects from our data base and by a detailed analysis of the generation and transport process which affect energetic electrons. We have not explored the dependence of electric and magnetic fluctuations in the magnetosphere on solar wind parameters and how such fluctuations would affect the generation and transport of energetic electrons although it is already known

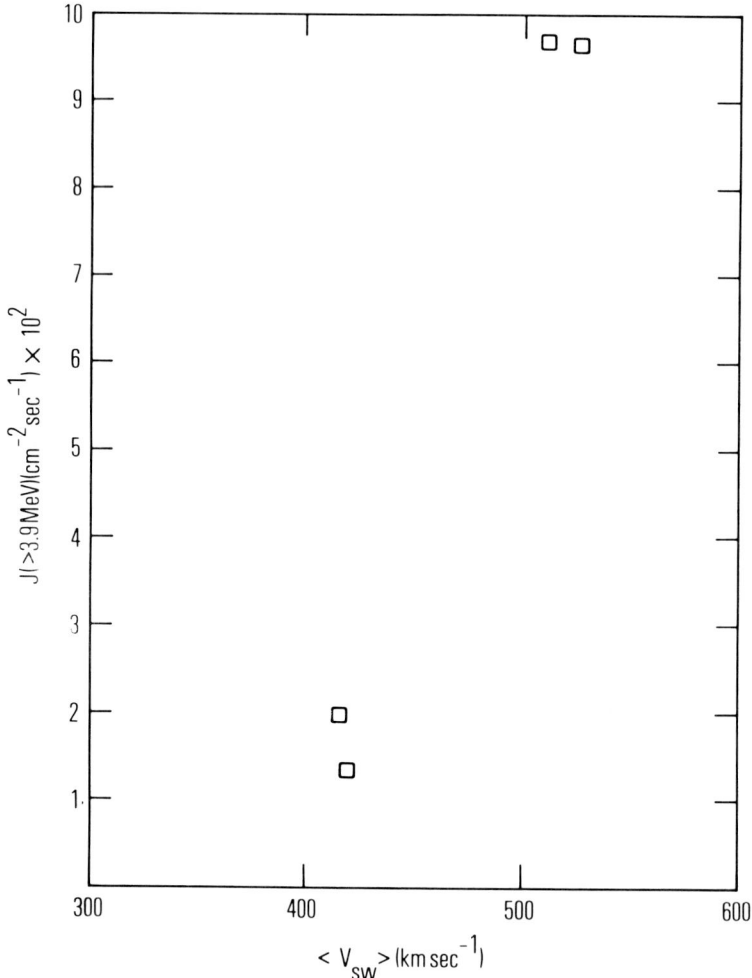

Fig. 21. Plot of the semi-annual average of the E > 3.9 MeV electron flux as a function of the semi-annual average of the solar wind velocity. ATS-6 data from mid 1974 through 1977 are included in this plot, excepting the semi-annual average for the last half of 1975 and the first half of 1976.

that some magnetic indices of interest scale as $B_z V^2$ (Svalgaard, 1977). Extension of our correlation to shorter timescales (< 1 hour) needs to be carried out, although we may be hampered in this respect by the very disparate timescales between the causes (substorms) of magnetospheric particle enhancements which occur in minutes or hours and the ultimate appearance of energetic electrons which occurs in days. However, it seems clear the the present results already offer some hope of both short-term and long-term prediction of the energetic electron radiation in the outer magnetosphere if a sufficiently accurate prediction of the parameters of the solar wind are available. We thus must look at one output of the variable sun - the solar wind - to understand the very variable magnetosphere.

Acknowledgements. We wish to thank Henry Hilton, Marie Wray, Gloria Holt and Adrienne Moon for their contributions to reduction and analysis of data spanning several spacecraft and more than a decade of time. This work was supported by NASA Contract NAS5-23788.

## References

Akasofu, S. -I., The roles of the north-south component of the interplanetary magnetic field on large-scale auroral dynamics observed by the DMSP satellite, Planet. Space Sci., 23, 1349, 1975.

Arnoldy, R. L., Signatures in the interplanetary medium for substorms, J. Geophys. Res., 26, 5189, 1971.

Baker, D. N., R. D. Belian, P. R. Higbie and E. W. Hones, Jr., High energy magnetospheric protons and their dependence on geomagnetic and interplanetary conditions, J. Geophys. Res., (to be published), 1978.

Bame, S. J., J. R. Asbridge, W. C. Feldman, and J. T. Gosling, Solar cycle evolution of high-speed solar wind streams, Astrophys. J., 207, 977, 1976.

Burch, J. L., Effects of interplanetary magnetic sector structure on auroral zone and polar cap magnetic activity, J. Geophys. Res., 78, 1047, 1973.

Burton, R. K., R. L. McPherron, and C. T. Russell, The terrestrial magnetosphere: A half-wave rectifier of the interplanetary electric field, Science, 189, 717, 1975.

Caan, M. N., R. L. McPherron, and C. T. Russell, Characteristics of the association between the interplanetary magnetic field and substorms, J. Geophys. Res., 82, 4837, 1977.

Crooker, N. U., J. Feynman, and J. T. Gosling, On the high correlation between long-term averages of solar wind speed and geomagnetic activity, J. Geophys. Res., 82, 1933, 1977.

Dessler, A. J., and T. W. Hill, Comment on "On the high correlation between long-term averages of solar wind speed and geomagnetic activity," by N. U. Crooker, J. Feynman and J. T. Gosling, J. Geophys. Res., 82, 5644, 1977.

Feldman, W. C., J. R. Asbridge, S. J. Bame, and J. T. Gosling, Long term variations of selected solar wind properties: IMP 6, 7 and 8 results, J. Geophys. Res., (to be published), 1978.

Fritz, T. A., J. P. Corrigan, and the ATS-6 Experimenters, Significant initial results from the environmental measurements experiment on ATS-6, NASA Technical Paper 1101, December 1977.

Garrett, H. B., A. J. Dessler and T. W. Hill, Influence of solar wind variability on geomagnetic activity, J. Geophys. Res., 79, 4603, 1974.

Gosling, J. T., J. R. Asbridge, and S. J. Bame, An unusual aspect of solar wind speed variations during solar cycle 20, J. Geophys. Res., 82, 3311, 1977.

Gosling, J. T., J. R. Asbridge, S. J. Bame, and W. C. Feldman, Solar wind speed variations: 1962-1974, J. Geophys. Res., 81 (28), 5061, 1976.

Kivelson, M. G. and C. T. Russell, Dependence of the polar cusp on the North-South component of the interplanetary magnetic field, J. Geophys. Res., 78, 3761, 1973.

Paulikas, G. A., and J. B. Blake, Modulation of trapped energetic electrons at 6.6 $R_e$ by the direction of the interplanetary magnetic field, Geophys. Res. Lett., 3, 227, 1976.

Paulikas, G. A., J. B. Blake and S. S. Imamoto, ATS-6 energetic particle radiation measurements at synchronous altitude, IEEE Trans. Aerospace and Electronic Systems, AES-11, 1138, 1975.

Paulikas, G. A., J. B. Blake, S. C. Freden and S. S. Imamoto, Observations of energetic electrons at synchronous altitude, 1. General features and diurnal variations, J. Geophys. Res., 73, 4915, 1968.

Russell, C. T., The solar wind and magnetospheric dynamics, in Correlated Interplanetary and Magnetospheric Observations, edited by D. E. Page, p. 3, D. Reidel Publishing Company, Dordrecht-Holland, 1974.

Russell, C. T. and R. L. McPherron, Semiannual variation of geomagnetic acitivity, J. Geophys. Res., 78, 92, 1973.

Russell, C. T., R. L. McPherron, and R. K. Burton, On the cause of geomagnetic storms, J. Geophys. Res., 79, 1105, 1974.

Singley, G. W. and J. I. Vette, A model environment for outer zone electrons, NASA Report NSSDC-72-13, December, 1972.

Svalgaard, L., An atlas of interplanetary sector structures 1947-1975, Stanford University Institute for Plasma Research, Report 648, February 1976, also subsequent private communications.

Svalgaard, L., Geomagnetic activity: Dependence on solar wind parameters, Coronal Holes and High Speed Wind Streams, ed. by J. B. Zirker, Colo. Asso. Univ. Press, Boulder, 1977.

PITCH ANGLE DISTRIBUTIONS OF > 30 keV ELECTRONS AT GEOSTATIONARY ALTITUDES

P. R. Higbie, D. N. Baker, E. W. Hones, Jr., and R. D. Belian

University of California, Los Alamos Scientific Laboratory
Los Alamos, New Mexico 87545

*Abstract*. The satellites 1976-059A and 1977-007A each carry energetic particle detectors which measure fluxes of electrons in the 30-300 keV energy range. Five separate sensors mounted at 30°, 60°, 90°, 120°, and 150° to the spacecraft spin axis provide two hundred samples of the three dimensional distribution function for every ten second spacecraft rotation. Spherical harmonic functions up to the fourth order have been fit to the observed pitch angle distributions. The second and fourth order coefficients obtained for these fits were averaged for each hour of local time. The probability distributions for the averaged harmonic coefficients were calculated and are presented as a function of local time. Possible relationships of these distributions to interplanetary conditions are discussed. Using the present analysis techniques, the intensity of electrons at the noon meridian is derived as a function of pitch angle and radial distance and is given by

$$j(\alpha, r) = 2.03 \times 10^8 (0.49 \sin^{4.78}\alpha + 0.51 \sin^{0.27}\alpha) e^{-r/1.60} \text{ el/cm}^2 \text{ sec sr.}$$

## Introduction

The theory that describes the motion of particles with different energies and pitch angles in a distorted magnetosphere is well known (e.g. Roederer, 1970). The particle distributions observed near the earth are the products of a series of stochastic processes. Variations in interplanetary conditions, i.e., the interplanetary field and the solar wind velocity and density, can cause compressions or rarefactions of the earth's magnetosphere inducing redistribution of particle fluxes (radial diffusion). Substorm processes inject particles into the midnight region thus providing a fresh source of particles, while wave-particle interactions scatter the particles causing them to be lost to the earth's atmosphere. If the detailed pitch angle distributions are known, then it may be possible to separate the changes, during geomagnetically disturbed periods, in the pitch angle distributions due to particles moving adiabatically, albeit in a more distorted magnetosphere, from those due to nonadiabatic processes such as interactions with the magnetopause or injections or scattering of particles near midnight.

Olson and Pfitzer (1977) have generalized McIlwain's (B,L) coordinate system to include the effects of the geomagnetic tail. They defined an average L which represents the average shell on which the observed particle move. They assumed an isotropic particle flux in calculating this parameter. This method could be improved by using observed pitch angle distributions.

West et al. (1973) made a survey of electron pitch angle distributions throughout the near ($< 20$ $R_E$) earth magnetosphere using data from Ogo 5. They found a normal distribution (i.e., maximum flux for 90° pitch angle particles) in the noon magnetosphere. Butterfly distributions (two maxima: between 0°-90° and 90°-180°) were found on the dusk side of the magnetosphere. The distributions near midnight were found to be isotropic when the field became taillike. Towards dawn the distribution was often normal or isotropic with only a weak tendency for butterfly distributions to exist. They explained the evolution of the postnoon butterfly distributions at ∼ 9 $R_E$ as being due to loss of 90° pitch angle particles at the magnetopause combined with drift shell splitting.

Kaye et al. (1978) made a statistical analysis of seven days of data from ATS 6. They defined an anisotropy parameter $A = J(90°)/J(40°) - 1$, which is negative for butterfly or "cigar" distributions and positive for normal or "pancake" distributions. They found that cigar distributions dominated in the dusk to postmidnight region and that the pancake distributions were most prevalent everywhere else. They also found that positive increases in A which occurred during injection events could be accounted for very well by the effects of an adiabatic compression. Similarly Lyons (1970) found that $> 200$ keV ions developed a cigar distribution when the local magnetic field decreased after a storm sudden commencement.

Baker et al. (1978) found that the existence of cigar anisotropies at L ∼ 6.6 in the local time interval of ∼ 20 to ∼ 03 hours preceded magnetospheric substorms. At the times of the injections, associated with these substorms, the distributions very quickly changed from cigarlike to pancakelike. The cigar phase was held to be a substorm precursor since this sequence was found in some 97 cases, whereas out of 17 cases for which there was no cigar phase only two were accompanied by substorm activity observed at earth and for these two cases the activity was weak.

In this paper we give a statistical analysis of pitch angle distributions as a function of local time and Kp, using simple parameters to characterize the distribution.

## Instrumentation and Analysis Techniques

The Charged Particle Analyzer (CPA) which is carried by the spacecraft 1976-059A and 1977-007A has been previously described in some detail (Higbie et al., 1978). The results given in this paper are based on data taken with the low energy detector (Lo-E) portion of the CPA. The Lo-E detector consists of five separate elements with viewing angles of 0°, ±30°, and ±60° with respect to the spacecraft equatorial plane. Each element has a 700 μ thick silicon semiconductor detector (0.5 $cm^2$ in area) which is collimated to a half angle of 2.6°. Several baffles are positioned along the barrel of the collimators to reduce the intensity of particles that enter outside the collimation angle and scatter off the walls of the collimator.

The sensors are shielded by enough material to stop 2 MeV electrons. The ±30° as well as the ±60° collimators are crossed to reduce weight and volume. Cross scattering between the two pairs of sensors is less than 0.1%. The geometrical factor for each element is $4 \times 10^{-3}$ cm$^2$ sr.

The spacecraft have a nominal spin period of ten seconds with their spin vectors pointed toward the earth. Both spacecraft are in geostationary orbit and, for the time periods covered in this paper, 1976-059 was at ~ 70°W and 1977-007 was at ~ 134°W. The corresponding geomagnetic latitudes are ~ 11° and ~ 5° respectively.

The five sensors are selected in sequence; while a given sensor is selected the counting rates for each of six energy windows are sampled for successive 8-msec periods. At the end of the sensor selection sequence, two 8-msec frames are skipped. Hence, thirty 8-msec samples are accumulated (five sensors times six levels each) every 256 msec. In one ten second rotation, therefore, two hundred samples (five elevation angles and forty azimuthal angles) are accumulated for each energy window. The energy windows all have a common upper energy cutoff of 300 keV and the lower thresholds are 30, 44, 65 95, 139 and 204 keV.

There are no magnetometers on these spacecraft, so the particle pitch angle distributions are calculated in a self consistent manner. The results of this calculation give the putative direction of the local magnetic field as well as a convenient parameterization of the shape of the electron pitch angle distributions, which can then be used to study the variations of the pitch angle distributions as a function of such variables as local time. Details of this calculation can be found in Higbie and Moomey [1977]. Briefly, spherical harmonic functions up to the fourth order are fit, using a linearized least squares method, to the two hundred samples taken for the lowest energy window during one spacecraft rotation, i.e. the observed counting rate at the elevation angles $\theta, \phi$ are

$$CR(\theta_i, \phi_i) = A_{oo} + \sum_m A_{1m} Y_{1m}(\theta_i, \phi_i) +$$

$$\sum_m A_{2m} Y_{2m}(\theta_i, \phi_i) + \sum_m A_{4m} Y_{4m}(\theta_i, \phi_i) \ .$$

where the Y's are spherical harmonic functions and the A's are calculated from the least squares fit. The A's are then used to calculate the quantities $C_n$, $\theta_n$, $\phi_n$ so the counting rate can be re-expressed by

$$CR(\theta, \phi) = G \sum C_n P_n(\cos\psi_n)$$

where the $P_n$'s are Legendre polynomials and

$$\cos\psi_n = \cos\theta \cos\theta_n + \sin\theta \sin\theta_n \cos(\phi - \phi_n).$$

The direction $(\theta_2, \phi_2)$ or $(\theta_4, \phi_4)$ is taken to be the direction of the symmetry axis of the distribution and is identified with the direction of the magnetic field. The $C_n$'s have been normalized by the factor G so that $C_o = 1$.

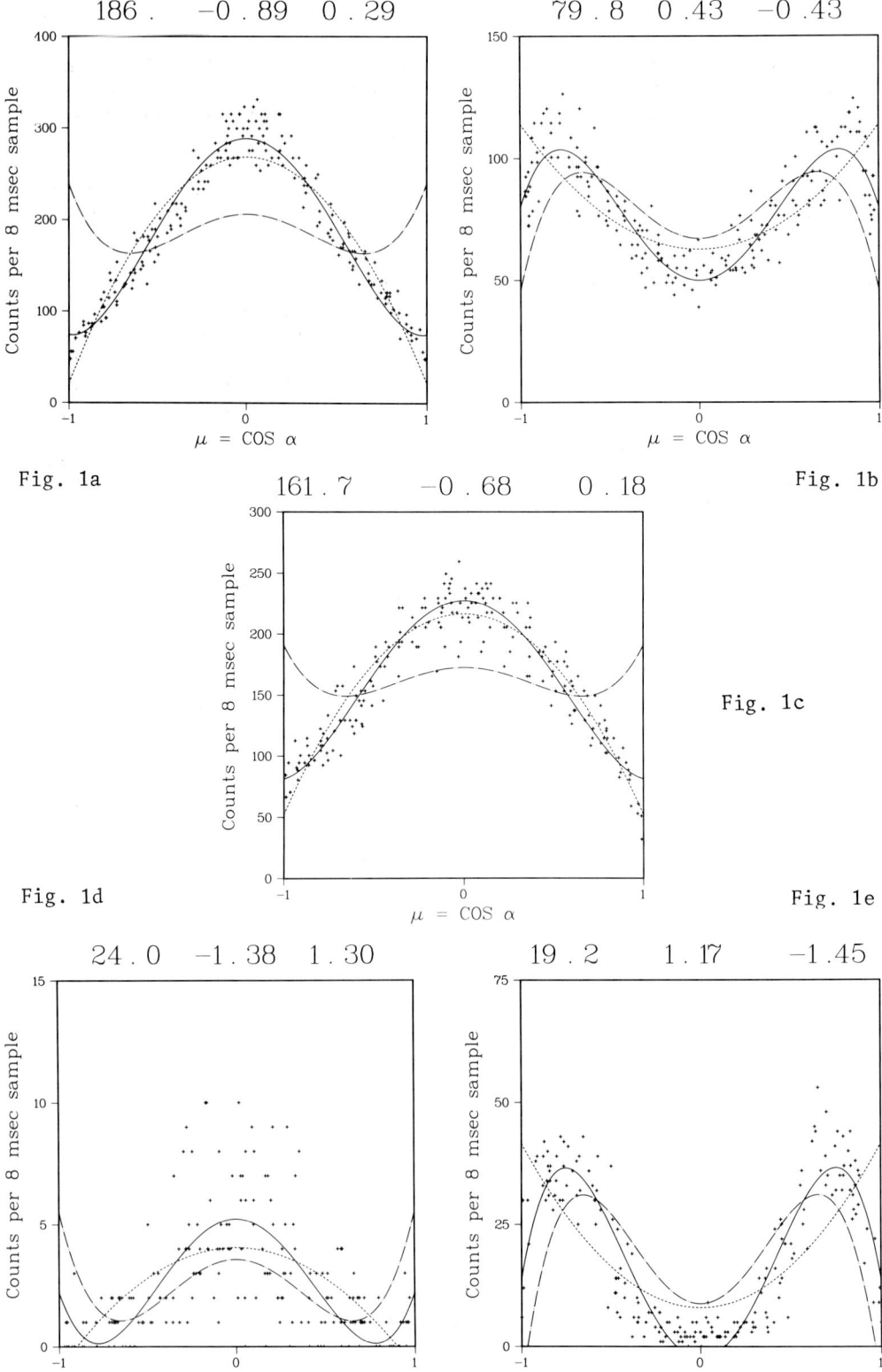

Fig. 1a  Fig. 1b  Fig. 1c  Fig. 1d  Fig. 1e

## Observations

Several sample distributions are illustrated in Figure 1. Although spherical harmonic functions are not "eigenfunctions" for whatever equation (e.g. the diffusion equation) describes the production of the pitch angle distribution, it can be seen that the second and fourth order harmonics fit the distributions reasonably well in Figures 1a, 1b and 1c. In Figure 1d, the true distribution is probably more sharply peaked than the fit suggests. This is partly due to the poor statistics, especially near $\mu = \pm 1$. In Figure 1e the fit becomes unphysical near $\mu = 0$ where it becomes negative.

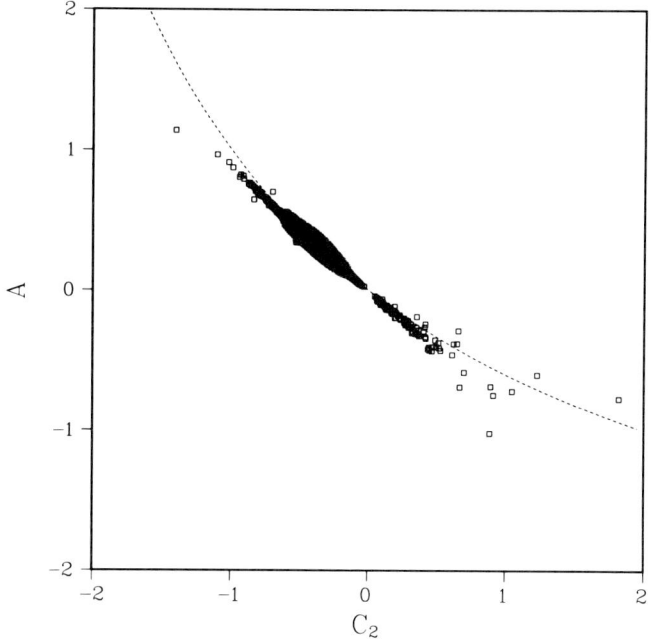

Fig. 2 The quantity $A = j(90°)/j(45°) - 1$ was calculated for a representative sample using the $C_2$ and $C_4$ parameters described in the text. For comparison, the theoretical line for a pure $C_2$ distribution is also shown.

---

Fig. 1 Each graph shows 200 samples of the 30-300 keV channel for electrons observed in a single spacecraft rotation. The dotted line represents the zeroth plus second order spherical harmonic fit to the data. The dashed line represents the zeroth and fourth order fits. The solid line is the zeroth, second and fourth order fits combined. The numbers at the top of the figures are G, $C_2$ and $C_4$ respectively. (a) and (c) are pancake distributions, (b) is a cigar distribution. (d) and (e) illustrates the limits of the fitting procedure (see text).

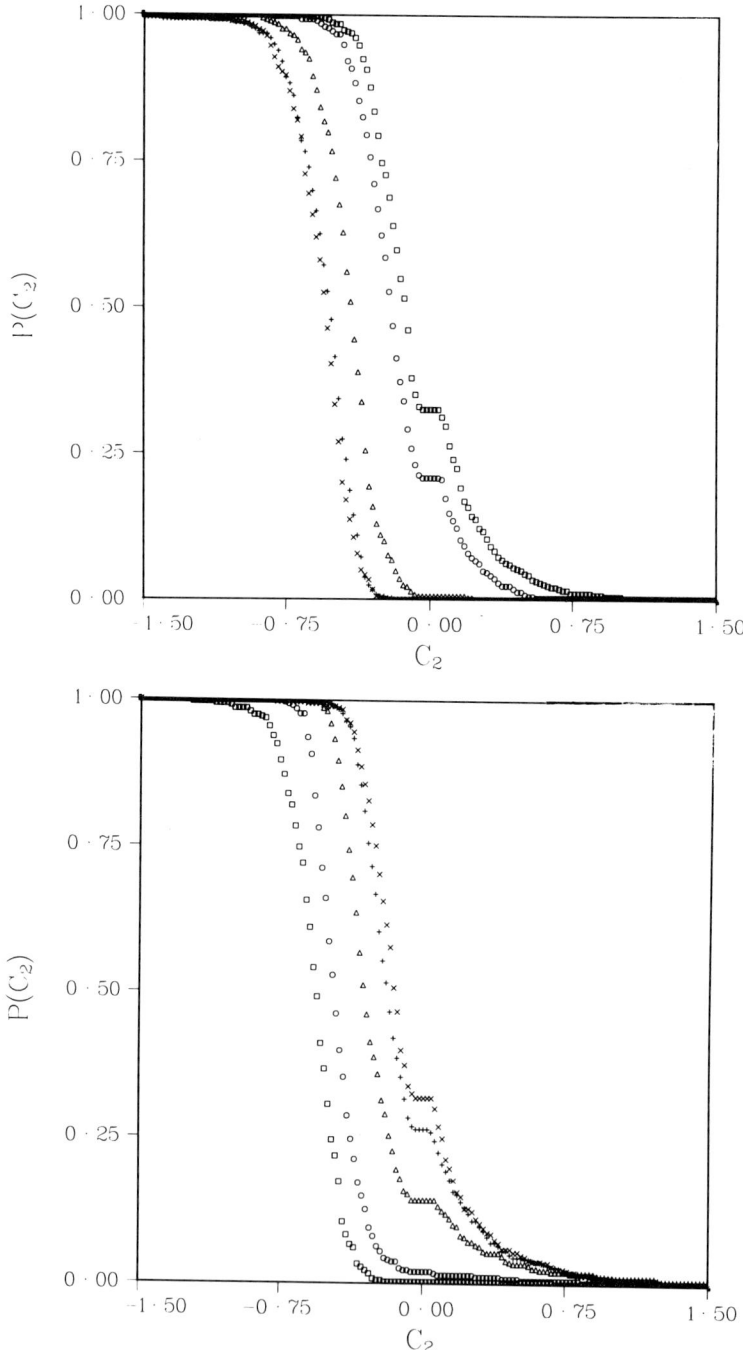

Fig. 3 Cummulative distributions are shown for various values of local time. The steps at $C_2 = 0$ are discussed in the text: (3a): LT 0-1 (□); 3-4 (O); 6-7 (Δ); 9-10 (+); 11-12 (X). (3b): LT 12-13 (□); 15-16 (O); 18-19 (Δ); 21-22 (+); 23-24 (X).

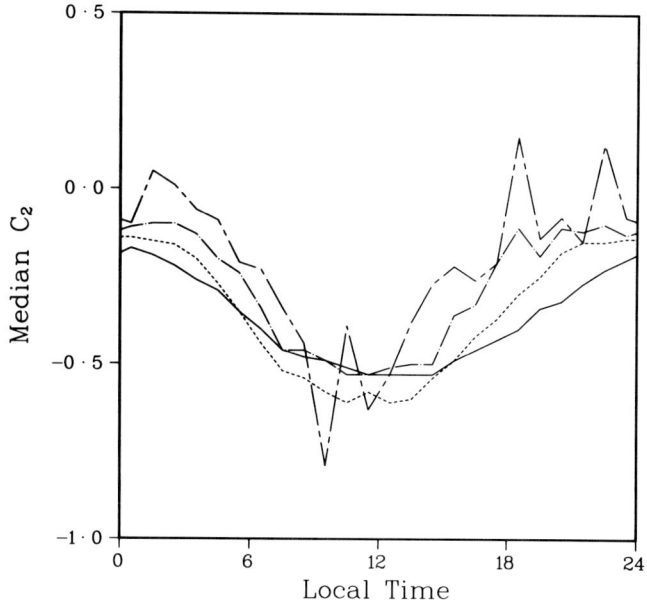

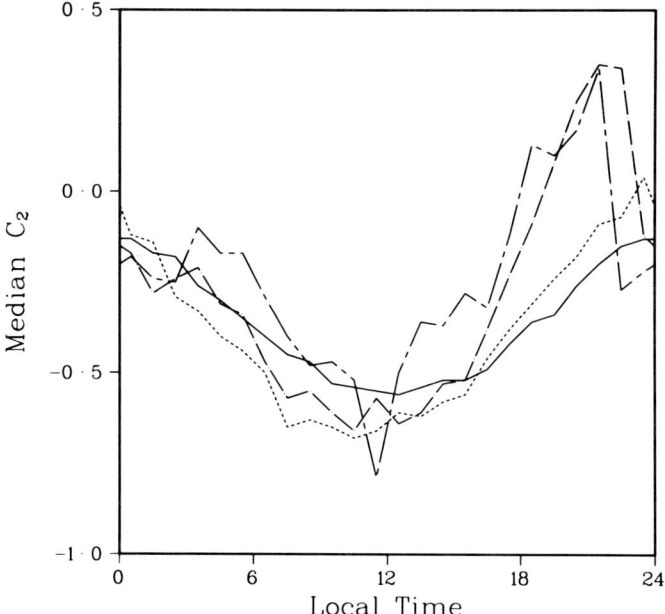

Fig. 4 Median values for $C_2$ for various Kp ranges vs local time. Kp 0 to 1+ (solid); 2- to 3+ (dotted); 4- to 5+ (dash-dot); 6- to 9+ (solid-dash). (4a) illustrates data from spacecraft 1976-059 ($\lambda_m$ = 11°) and (4b) illustrates data from spacecraft 1977-007 ($\lambda_m$ = 5°).

The quantities $C_0$, $C_2$ and $C_4$ were calculated for each ten second rotation of the spacecraft and then averaged for one hour intervals in local time. Specifically, the absolute values of $C_2$ and $C_4$ were averaged and the most frequently occurring sign for each was assigned to the corresponding average.

A commonly used anisotropy parameter is the quantity $A = J(90°)/J(45°) - 1$. $45°$ lies between a zero for the fourth order spherical harmonic at $30.56°$ and a relative maximum at $49.11°$. In Figure 2, the dashed line represents the A anisotropy parameter calculated assuming $C_4 \equiv 0$. The points are calculated from both $C_2$ and $C_4$ for data from 1976-059. The small scatter due to $C_4$ indicates that this particular parameter (A) is nearly equivalent to $C_2$. Most of the discussion will be confined to the $C_2$ parameter because the contribution from $C_4$ is relatively small. One year of data (April 1977 to April 1978) was analyzed for both 1977-007 and 1976-059 using one-hour averages obtained in the manner described above. In Figure 3 are plotted the cumulative distributions for $C_2$ parameterized by local time for all Kp conditions. It can be readily seen that positive, or cigarlike, anisotropies have the greatest probablity of occurrence in the 18-24-6 hour local time sector and that cigar distribution can be observed approximatley 25% of the time. The step at $C_2 = 0$ can be explained as follows. $C_2 = 0$ corresponds to an isotropic distribution, but the CPA usually records the loss cone of the distribution. Therefore, $C_2$ is never exactly zero and the flat step in the integral probability curve reflects this fact. This effect could be washed out by averaging the signed values of $C_2$ rather than averaging the magnitudes and using the majority sign as we have done.

The local time dependence of the medians of these distributions is shown in Figure 4. As the Kp index increases, the anisotropy in the interval 18-24-6 hours becomes less pancakelike and more cigarlike. However, this effect is relatively more pronounced on the dusk side of the magnetosphere where this tendency begins as early as noon for large values of Kp. At noon or slightly prenoon the distribution becomes even more pancake-shaped for moderate increases in Kp. This is especially obvious for 1977-007 which is at a lower L shell than 1976-059. There are only a few samples for Kp $\geq$ 6- at each local time which may produce some large statistical fluctuations for that curve.

Similar curves are shown in Figure 5 for the $C_4$ parameter. The systematics of these sets of curves is less clear, since the major effect of the $C_4$ parameter is to accentuate the pancakelike distribution described by $C_2$ (for $C_4 > 0$) and to fit cigar shapes (for $C_4 < 0$). However, there is an asymmetry in these curves with negative $C_4$ on the dusk side values shifting toward local noon for periods of increased geomagnetic activity; this corresponds to development of cigar distributions. A pancakelike development with increasing Kp is seen prenoon.

The quiet-time behavior of $C_2$ as observed by the two spacecraft is represented in Figure 6. The various percentiles represent the variation in the $C_2$ parameter (one sigma would correspond to percentile lines at 16 and 84%). The tendency for cigar distributions to form in the midnight sector is seen in the 90-percentile curve. This is not surprising since the Kp parameter is calculated on a three-hour time base and weak substorm activity occuring during this period may not

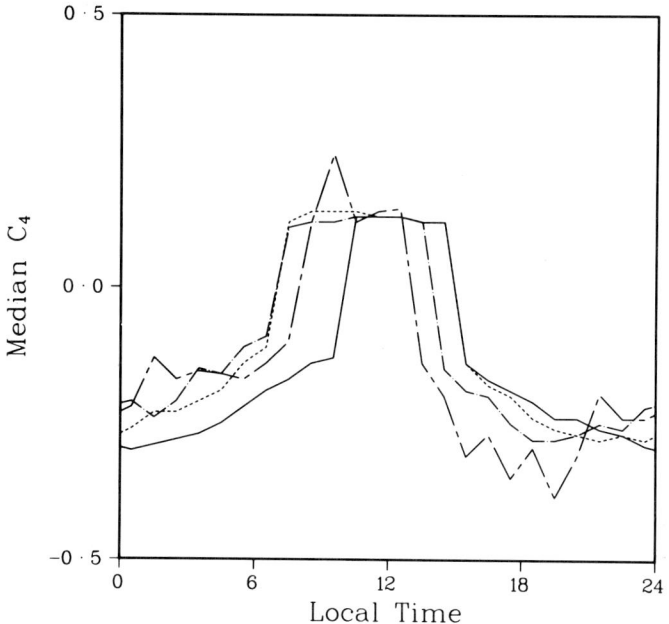

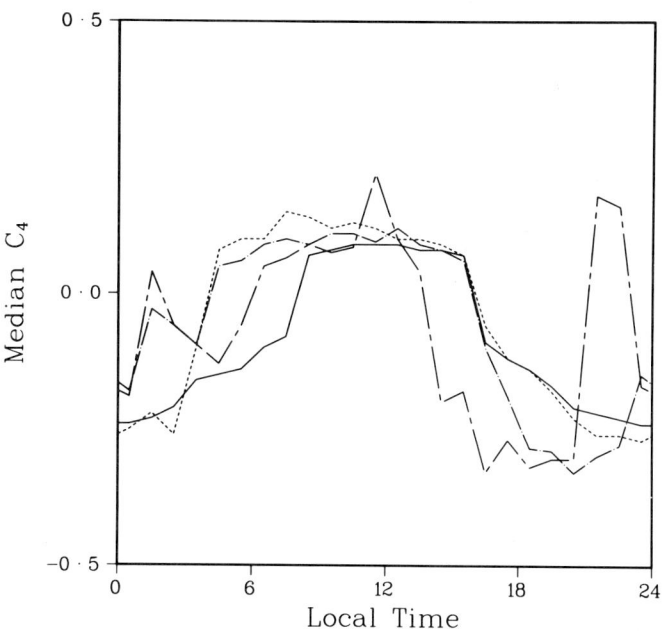

Fig. 5  Median values for $C_4$ are given for various Kp ranges vs local time.  (5a) spacecraft 1976-059, (5b) spacecraft 1977-007.  Kp ranges are coded as in Fig. 4.

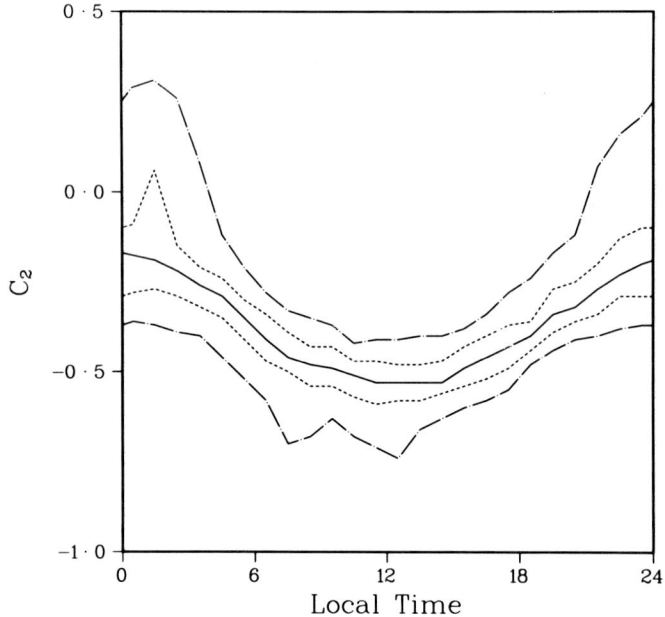

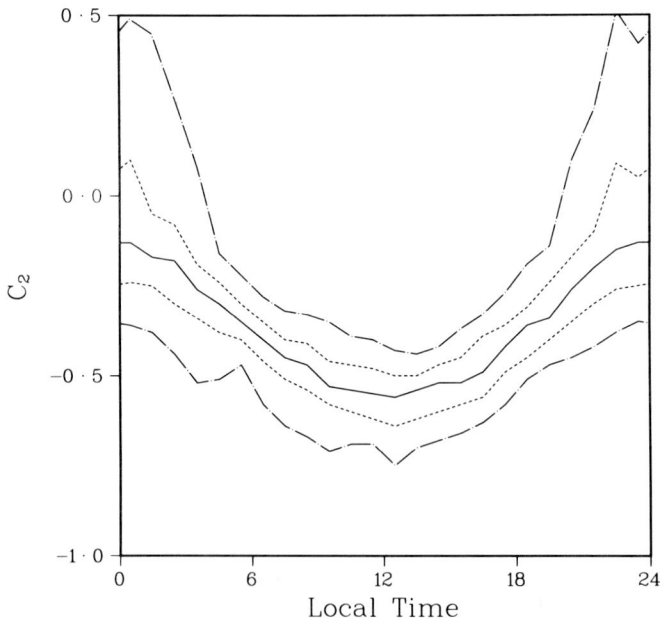

Fig. 6  Quiet time (Kp ≤ 1+) behavior of the $C_2$ parameter is shown by the median (solid line), 30 and 70 percentiles (dotted line), and 10-90 percentiles (dot-dashed line). (6a) is 1976-059 (6b) is 1977-007.

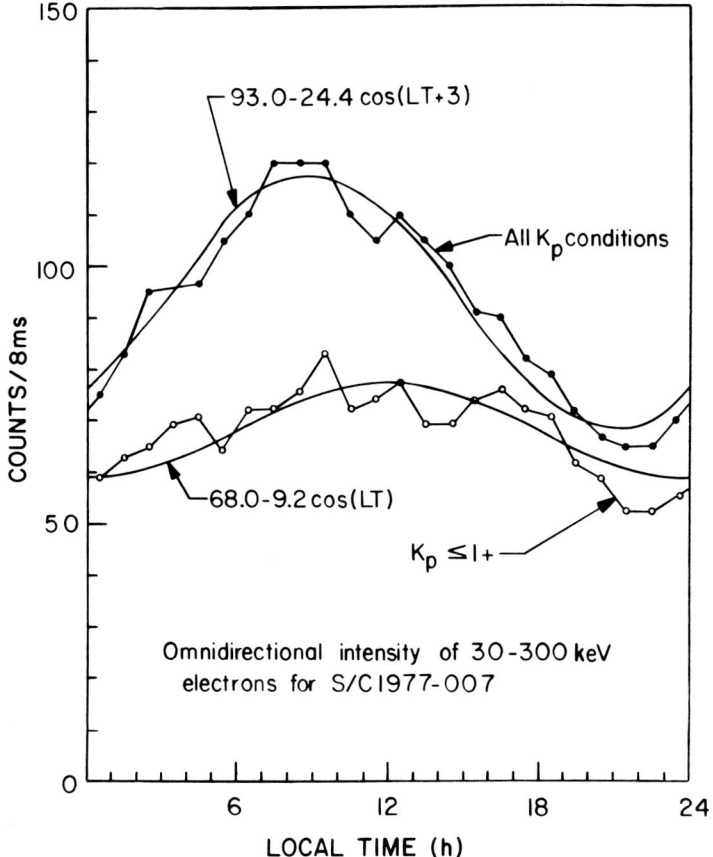

Fig. 7 Omnidirectional intensity (unnormalized for geometric factor) for 30-300 keV electrons are given for 1977-007.

affect the parameter greatly. From Figure 6 one can see that distinct cigar shapes ($C_2 > 0$) form less than 50% of the time for quiet conditions. Still there is a local time dependence such that the distributions are always less pancakelike at midnight than at noon.

Figure 7 illustrates the omnidirectionally averaged flux as a function of local time for 1977-007 for quiet times ($Kp \leq 1+$) and for all Kp conditions combined. There is a decided offset towards the prenoon sector when all Kp conditions are included. This may be a result of the interplanetary electric field which is expected to be large during geomagnetically active times, or it may be due to losses either at the magnetopause or to the atmosphere that are enhanced during disturbed times, or it may merely be the consequence of adding together "snapshots" of the magnetosphere. Injections in the midnight sector and effects of shell splitting on particles near noon at the time of magnetic perturbations may combine in such a way that the average picture gives an impression of a prenoon peak in intensity.

We will now consider only the quiet time behavior of the 30-300 keV electrons for 1977-007. Table 1 gives the results of a calculation

TABLE 1  INFERRED INTENSITIES AT THE NOON MERIDIAN

| $LT_o$ | $C_2$ | $C_4$ | G | $\alpha_o$ | $G(1+C_2P_2(\alpha_o)+C_4P_4(\alpha_o))$ | $\alpha_f$ | $R_f$ |
|---|---|---|---|---|---|---|---|
| 00 hr | -.13 | -.24 | 59 cts/8 msec | 10° | 40 cts/8 msec | 11.8° | 6.36 $R_E$ |
|  |  |  |  | 30 | 54 | 35.8 | 6.41 |
|  |  |  |  | 60 | 64 | 70.0 | 6.89 |
|  |  |  |  | 90 | 57 | 90.0 | 7.52 |
| 03 | -.22 | -.19 | 62 | 10 | 39 | 11.6 | 6.38 |
|  |  |  |  | 30 | 53 | 35.0 | 6.43 |
|  |  |  |  | 60 | 67 | 68.9 | 6.84 |
|  |  |  |  | 90 | 64 | 90.0 | 7.39 |
| 06 | -.38 | -.12 | 68 | 10 | 37 | 11.1 | 6.44 |
|  |  |  |  | 30 | 52 | 33.4 | 6.47 |
|  |  |  |  | 60 | 74 | 66.4 | 6.72 |
|  |  |  |  | 90 | 78 | 90.0 | 7.11 |
| 09 | -.50 | .09 | 75 | 10 | 44 | 10.4 | 6.54 |
|  |  |  |  | 30 | 51 | 31.3 | 6.54 |
|  |  |  |  | 60 | 77 | 62.6 | 6.62 |
|  |  |  |  | 90 | 96 | 90.0 | 6.78 |
| 12 | -.56 | .09 | 77 | 10 | 42 | 10.0 | 6.60 |
|  |  |  |  | 30 | 43 | 30.0 | 6.60 |
|  |  |  |  | 60 | 81 | 60.0 | 6.60 |
|  |  |  |  | 90 | 101 | 90.0 | 6.60 |

$P_2(\alpha) = 1/2(3\cos^2\alpha - 1)$

$P_4(\alpha) = 1/8(35\cos^4\alpha - 30\cos^2\alpha + 3)$

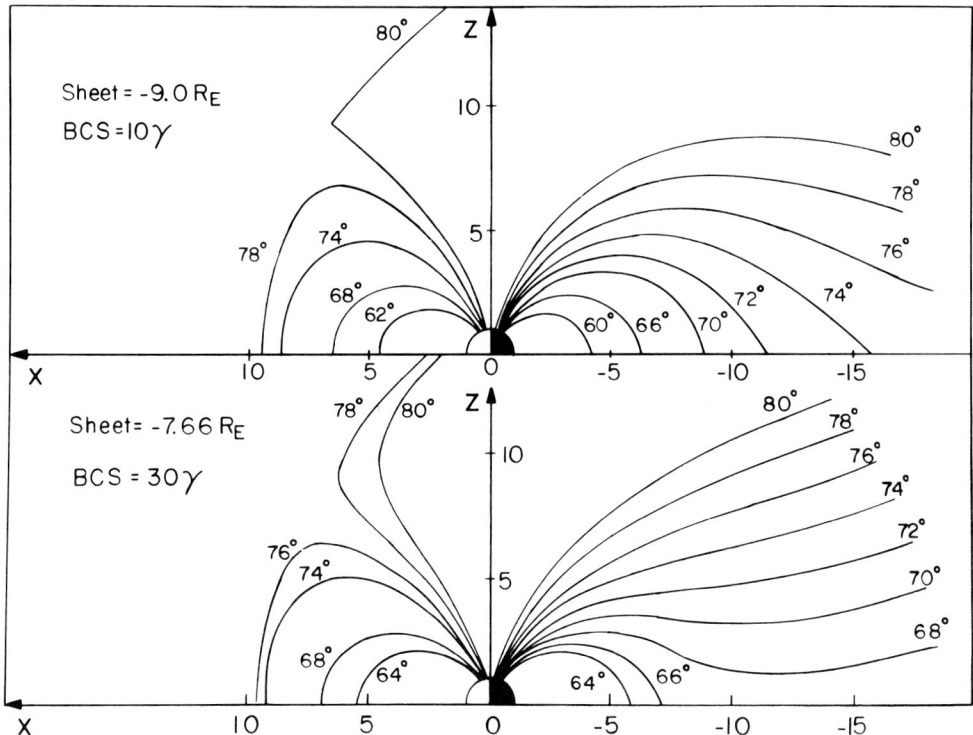

Fig. 8  Magnetic field line configurations for drift calculations described in the text. Numbers near curves represent the field line latitude at the intersection with the earth's surface.

for electrons drifting from various local times to the noon meridian. Column 1 gives the local time at which particles start. Columns 2, 3, and 4 give the median second and fourth spherical harmonic amplitudes and the average omnidirectional intensity respectively for the initial local time. The fifth column gives the initial pitch angle and the sixth column gives the intensity calculated for this angle. The last two columns give the pitch angle and radial distance to which the particles drift at noon. The noon pitch angles and radial distance were calculated using a computer program, DRIFT, developed by Roederer and Hones at the Los Alamos Scientific Laboratory. The field model used in this particular version of the program consists of an image dipole to which is added a Mead-Williams tail field. A copy of this program is available from one of the authors (EWH). The field geometry is illustrated in Figure 8. The upper panel illustrates the field lines lying in the noon-midnight meridian plane for a tail field strength of 10 γ and a radial distance to the current sheet of 9 $R_E$. The lower panel illustrates a distorted field generated by changing the tail field strength in the model to 30 γ and decreasing the current sheet distance to 7.66 $R_E$.

The data from this table are plotted in Figure 9 as a function of radial distance at the noon meridian. Note that the data given in

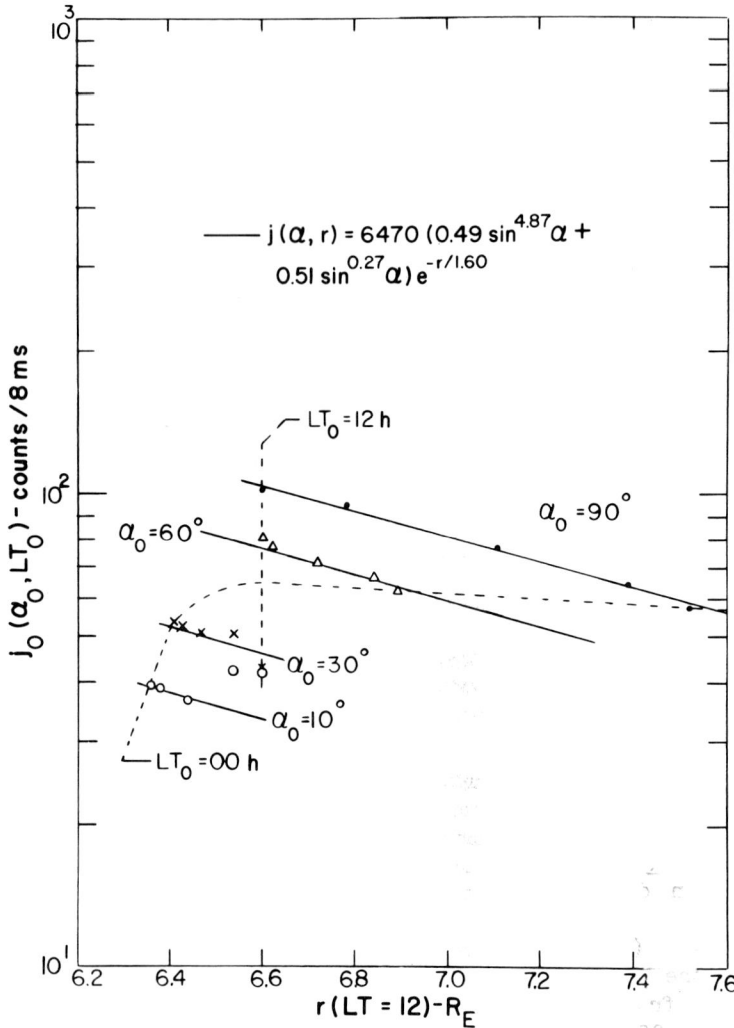

Fig. 9  Radial intensity (unnormalized for geometric factor) profile of electrons at noon meridian as deduced from the unnormalized omnidirectional intensity and pitch angle anisotropies at different local time using drift calculations.

Table 1 are sufficient to produce this kind of plot for any field model using a program similar to DRIFT. Neglecting the effects due to changes in the final pitch angle (these are all less than 10°) we find that the intensity for 30-300 keV electrons at noon can be represented by the equation:

$$j(\alpha,r) = 2.03 \times 10^8 (0.49 \sin^{4.78} \alpha + 0.51 \sin^{0.27} \alpha)e^{-r/1.60} \text{ el/cm}^2 \text{ sec sr}.$$

## Discussion

We have described an average or static picture of the low energy electron population at synchronous orbit. Of course the more inter-

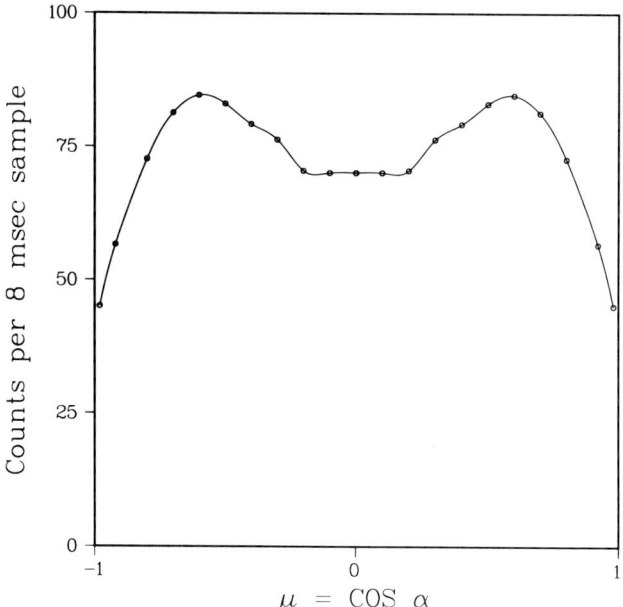

Fig. 10 Pitch angle distribution calculated for midnight using the radial electron intensity profile given in Figure 9 and the distorted field illustrated in the lower panel of Figure 8.

esting features are those involving dynamic processes; however the present picture may be adequate to develop or test various static models. We have presented results for a simple model for magnetically quiet times (Kp $\leq$ 1+). Some of the systematic behavior of this electron population during magnetically active times has also been presented.

Pfitzer et al. (1969) used data from Ogo 3 to determine the radial gradient of the flux for 50-120 keV energy electrons at noon local time and data from ATS-1 to determine the pitch angle distributions between 60° to 90° at the same local time. They used data from two periods near the dates February 7 and 15, 1967. We find their February 7 case agrees quite well with our quiet time results, although that was a geomagnetically disturbed day. However the Ogo radial profile was actually taken at 2100 UT on February 6 which was a period preceded by a six hour period with $K_p$ = 1. Furthermore, as our Figures 4 and 5 indicate, the pitch angle distributions show the least variability at noon.

Stevens et al. (1970) examined proton distributions in the energy range 0.060 to 3.3 MeV. Their pitch angle distributions are much more peaked ($\sim \sin^7 \alpha$) and their radial gradients are also steeper than we have found for electrons of comparable energies.

Kaye et al. (1978) examined electron and proton anisotropies at geostationary orbit for seven days of data in 1975. Five of these seven days are listed as disturbed days (Lincoln, 1975a,b). From their Figure 1, the probability of observing a cigar anisotropy is 0.4 (two cases out of five total). Our corresponding probability from Figure 3 is 0.31. Thus, their result for 32-51 keV electrons is in reasonable agreement with ours, but the probability for observing cigar anisotropies appears to be larger for more energetic electrons and for protons in the energy range 27-120 keV. This difference agrees with the results of Pfitzer et al. and Stevens et al. who find steeper radial gradients for these particles than for 30 keV electrons.

We have the following speculative view of energetic electron processes at the geostationary orbit. A more extensive exposition can be found in Baker et al. (1978). Electrons are injected into the region near midnight and, as they drift, form a distribution which maintains a pancakelike shape at all points around the geostationary orbit when the geomagnetic activity is low. If the interplanetary field turns southward the preexisting population responds on a time scale corresponding to the drift time of a particle with a given energy. For a 30 keV electron this would be on the order of hours, a 300 keV electron would respond on a time scale of tens of minutes. The response in the nightside magnetosphere is to form cigar shaped distributions because of the drift shell splitting process. The effect is more pronounced on the dusk side of the magnetosphere than on the dawn side since the effects may be masked by particle injections or nonadiabatic scattering in the tail near the midnight meridian. This response is illustrated in Figure 10 which shows the pitch angle distribution that results when the distribution shown in Figure 9 is allowed to drift to the midnight meridian through the distorted field illustrated in Figure 10. Thus we believe the shape of the pitch angle distribution to be a sensitive indicator of the build up of stress in the magnetotail due to a southward IMF.

Acknowledgments. We thank J. G. Roederer for his effort in developing the computer program used to calculate the electron drift paths. This work was performed under the auspices of the U.S. Department of Energy.

## REFERENCES

Baker, D. N., P. R. Higbie, E. W. Hones, Jr., and R. D. Belian, High-resolution energetic particle measurements at 6.6 $R_E$, 3, low energy electron anisotropies and short-term substorm predictions, J. Geophys. Res. (in press).

Higbie, P. R. and W. R. Moomey, Pitch angle measurements from satellites using particle telescopes with multiple view directions, Nucl. Instrum. Methods, 146, 439, 1977.

Higbie, P. R., R. D. Belian, and D. N. Baker, High-resolution energetic particle measurements at 6.6 $R_E$, 1, electron micropulsations, J. Geophys. Res. (in press).

Kaye, S. M., C. S. Lin, G. K. Parks, and J. R. Winckler, Adiabatic modulation of equatorial pitch angle anisotropy, J. Geophys. Res., 83, 2675, 1978.

Lincoln, J. V., Geomagnetic and solar data, J. Geophys. Res., 80, 1975, 2892(a) and 3284(b).

Lyons, L. R., Adiabatic evolution of trapped particle pitch angle distributions during a storm main phase, J. Geophys. Res., 82, 2428, 1977.

Olson, W. P. and K. A. Pfitzer, Magnetospheric magnetic field modeling, Report for Contract F44620-75-C-0033, McDonnell Douglas Astronautics, 1977.

Pfitzer, K. A., T. W. Lezniak, and J. R. Winckler, Experimental verification of drift-shell splitting in the distorted magnetosphere, J. Geophys. Res., 74, 1969, 4687.

Roederer, J. G., Dynamics of Geomagnetically Trapped Radiation, Springer-Verlag, New York, 1970.

Stevens, J. R., E. F. Martina, and R. S. White, Proton energy distributions from 0.060 to 3.3 MeV at 6.6 earth radii, J. Geophys. Res., 75, 1970, 5373.

West, H., Jr., R. Buck, and J. Walton, Electron pitch angle distributions throughout the magnetosphere as observed on Ogo 5, J. Geophys. Res., 78, 1064, 1973.

SOLAR ENERGETIC PARTICLES: FROM THE CORONA TO THE MAGNETOTAIL

E. C. Roelof

The Johns Hopkins University, Applied Physics Laboratory
Laurel, Maryland 20810

Abstract. Some new results concerning coronal and interplanetary propagation of energetic particles are discussed in the context of previous and current observations and interpretations. Evidence is presented that the bulk of protons $\sim 1$ MeV can be preferentially injected onto interplanetary field lines at coronal longitudes and latitudes well-removed ($\sim 60°$) from the site of the associated active region. For these protons, interplanetary propagation parallel and transverse to the interplanetary field evidently involves little scattering near 1 AU, so that parallel propagation is approximately "scatter free" and transverse propagation is primarily that due to the $\underline{E} \times \underline{B}$ drift. Upstream of the earth's bow shock, both intensities and anisotropies of $\sim 1$ MeV protons and electrons are significantly modified by the magnetosphere, and these effects may provide a probe of the structure of the bow shock and magnetosheath in regions not readily accessible to spacecraft.

## Introduction

My purpose in this paper is to offer some insights (gained from recent work by our group at JHU/APL and collaborators at other institutions) into the transport of solar energetic particles from the sun to the magnetosphere. As such, I will not be addressing directly the topic of solar particle entry into the magnetosphere. Rather I would like to describe the propagation environment of $\sim 1$ MeV protons, ions and electrons in the corona, in the interplanetary medium (both inside and outside 1 AU), and finally in the immediate vicinity of the earth's bow shock. In other words, I shall stress the nature of the "input function" for low-energy particles entering the magnetosphere, and I will emphasize particularly the characteristics of the particle flux anisotropy, since this may be a significant parameter in the entry process.

The non-homogeneity of coronal magnetic structure has been obvious from the earliest eclipse photographs. Although the first quantitative treatment of energetic particle propagation across the corona by Reid [1964] used homogeneous transport parameters, as did most subsequent attempts to model coronal particle propagation, observations of long-lived impulsive low energy particle events during the decline of Solar Cycle 19 and the rise of Solar Cycle 20 displayed interplanetary particle event histories which implied heterogeneous coronal structure [Fan et al., 1968; Lin et al., 1968; McCracken and Rao, 1970; Keath et al., 1971; Krimigis et al., 1971]. It is the lower energy particle observations (protons ≤ 10 MeV, electrons ≤ 1 MeV) in which coronal structure becomes most apparent because the measurable intensity at low energies is much longer lasting than the more impulsive high energy events.

One technique for revealing the influence of non-homogeneous coronal structure upon the injection of solar energetic particles is the "mapping" or ordering of particle flux by the estimated high coronal "connection longitude" (and latitude), of the large-scale interplanetary magnetic field (IMF) line through the spacecraft. Although in principle this mapping could be accomplsihed by integrating the magnetohydrodynamic (MHD) equations of motion for the expansion of the interplanetary plasma [Burlaga and Barouch, 1976; Pizzo, 1978], a very simple approximation was suggested by Snyder and Neugebauer [1966] and developed by Sakurai [1971] and Matsuda and Sakurai [1972]. This was the constant radial velocity approximation for the solar wind motion which Nolte and Roelof [1973a, 1973b] showed could be utilized to estimate the high coronal connection of the IMF (since the effects of plasma acceleration and corotation inside the MHD critical points tend to cancel each other). Using IMP 7/8 and Pioneer 10/11 solar wind velocities, Mitchell et al. [1979a] have shown that the constant radial velocity approximation can be applied consistently (outside of stream-stream interactions) from ~ 5 AU (Pioneer) to 1 AU (IMP), and that both velocity profiles are consistent with the coronal structures which they map back to. The high coronal connection longitudes and latitudes then label the IMF lines with their "natural" coordinates in the inner heliosphere, i.e., their high coronal foot points.

The applicability of IMF coronal connection coordinates to energetic particle analysis follows from the relative lack of particle scattering transverse to the IMF in the inner heliosphere. The strong and persistent field-aligned anisotropies in the rising phase of low-energy solar particle events imply that scattering is quite weak, and consequently the bulk velocity of the particles transverse to the IMF should be given to good approximation by the $\underline{E} \times \underline{B}$ drift. Observational verification of these statements on transverse and parallel particle propagation is presented in the following section, but in the remainder of this section, we shall examine the obvious consequences of these statements.

To say that particle streaming transverse to the IMF is primarily due to the $\underline{E} \times \underline{B}$ drift is equivalent to saying that particles stay on the same (moving) field line. Thus in the anisotropic rise phase of a solar particle event, the connecting longitude and latitude of the IMF, estimated from the solar wind speed measurement at a given time, also give the high coronal injection location of the energetic particles (neglecting any rearrangement of the high coronal field during the solar wind transit to the spacecraft). After the maximum of the event, when back-scatter from IMF structures beyond 1 AU have diminished the field-aligned anisotropy until it is comparable to (or less than) the transverse $\underline{E} \times \underline{B}$ drift anisotropy, the mapped particle fluxes represent the combined effects, on a given IMF line, of residual coronal injection as well as the return of the particles from beyond the spacecraft.

There is another type of energetic particle event in which the

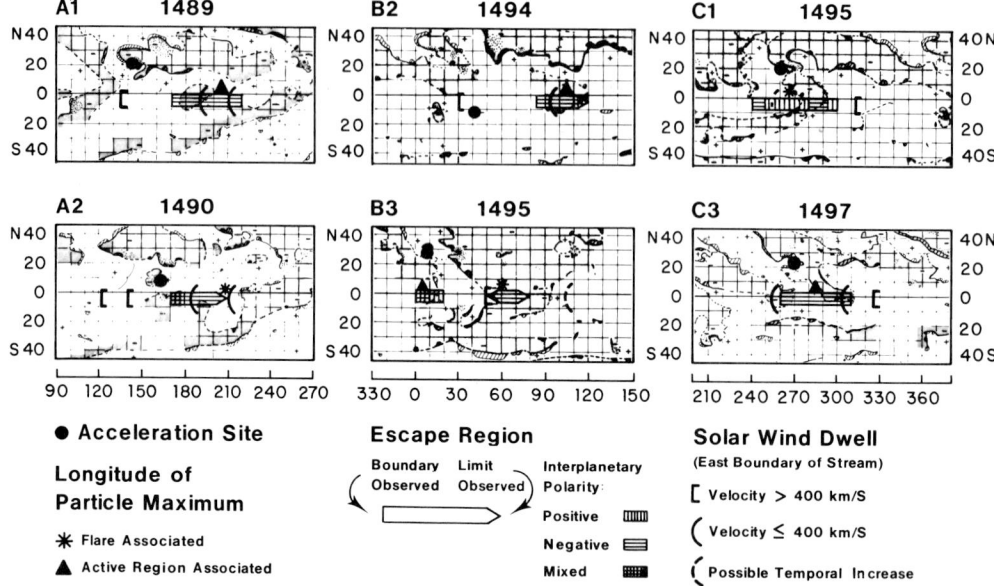

Fig. 1. Summary of "preferred escape" regions for 0.3-0.5 MeV protons observed by Mariner 4 in 1965. Panels consist of segments of chromospheric Hα synoptic charts [McIntosh and Nolte, 1975]. Shaded regions are those of well-defined negative magnetic polarity in the low corona. Heavy dots mark active regions most likely associated with particle increases. Particle escape regions are marked by bars at the heliographic latitude of Mariner 4 with IMF polarity indicated as in the legend. Eastern boundaries of coronal source regions for solar wind streams are indicated for high speed, low speed, and evolving streams. From Roelof et al. [1979].

mapped particle fluxes do not represent coronal injection by itself. Currently there is considerable interest in the possibility of acceleration of low energy particles in the interplanetary medium. Acceleration at solar wind stream interaction regions which develop strongly beyond ~ 2 AU [Barnes and Simpson, 1976] appears to share many characteristics with "shock spike" events well documented at 1 AU over the past decade [e.g., Sarris and Van Allen, 1974; Sarris et al., 1976; Pesses et al., 1978]. Acceleration of a more global nature has been postulated as an explanation for the "corotating" particle events associated with recurrent solar wind streams during the decline and minimum of the past two Solar Cycles [Bryant, et al., 1965; McDonald et al., 1976]. It is not yet clear whether localized (shock spike) acceleration is sufficient to explain the recurrent events, and whether the primary "source" of the particles therein is a long-lived outflow of solar energetic particles or in situ acceleration of solar wind plasma. However, in any of the above cases, coronal structure is the determining factor of the spatial distribution of any recurrent particle event, by means of IMF and solar wind stream structure and/or energetic particle emission. In such particle events, the "mapped" fluxes may not necessarily be pro-

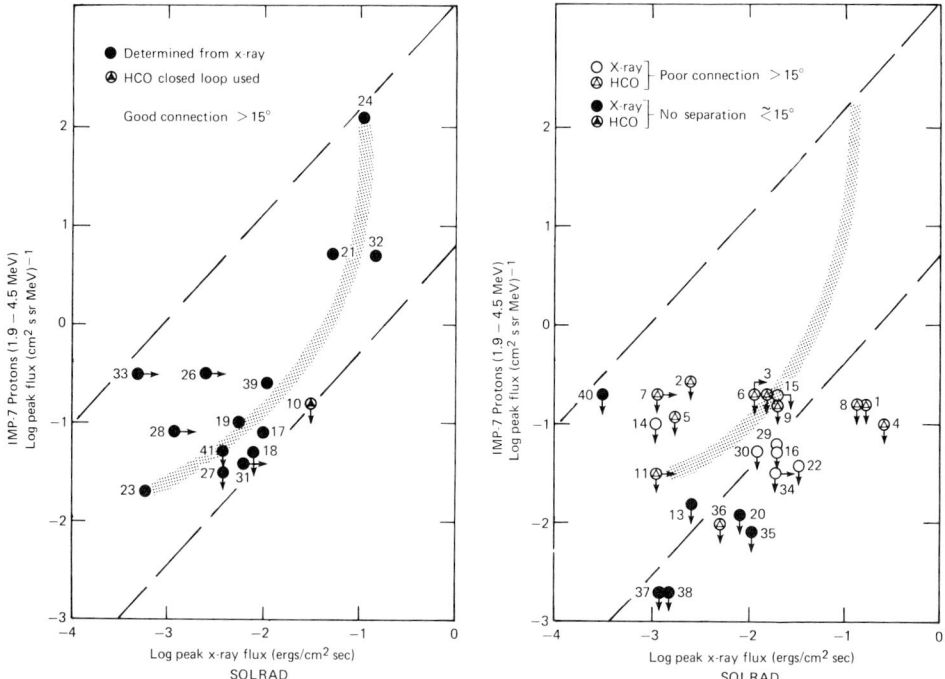

Fig. 2. Evidence for the role of x-ray bright loops in coronal particle transport. Peak proton flux from the JHU/APL detector on IMP 7 versus associated peak 1-8 Å SOLRAD flux plotted (a), on left, for the flares in which complexes of bright loops in the AS&E Skylab soft x-ray images extended from the flare site to the estimated low coronal foot-points of the large-scale IMF lines through IMP 7, and (b), on right, for the flares in which either the bright loop complex was not continuous, or in which the IMF connection fell with 15° arc-distance of the flare site. Upper limits on measured proton and/or x-ray fluxes are indicated by points with arrows. From Mitchell et al. [1979b].

portional to a coronal injection profile, but the mapping does accomplish the more general objective of ordering the particle fluxes in their "natural" coordinates of high coronal longitude and latitude.

From the above discussion, we see that detailed interpretation of mapped particle fluxes depends upon auxiliary information, e.g., flux anisotropies, solar wind structure, etc. In the examples we turn to now, the interpretation is usually straightforward, since they deal mainly with the rise of solar particle events.

## Coronal Transport and Interplanetary Injection

The Mariner 4 mission to Mars in 1965 offered the only opportunity to study low energy solar particle emission throughout the minimum between Solar Cycles 19 and 20. From the mapping of solar wind velocity, IMF polarity and 0.3-0.5 MeV proton fluxes over 9 solar rotations (Carrington 1489-1497), Nolte and Roelof [1977] demonstrated

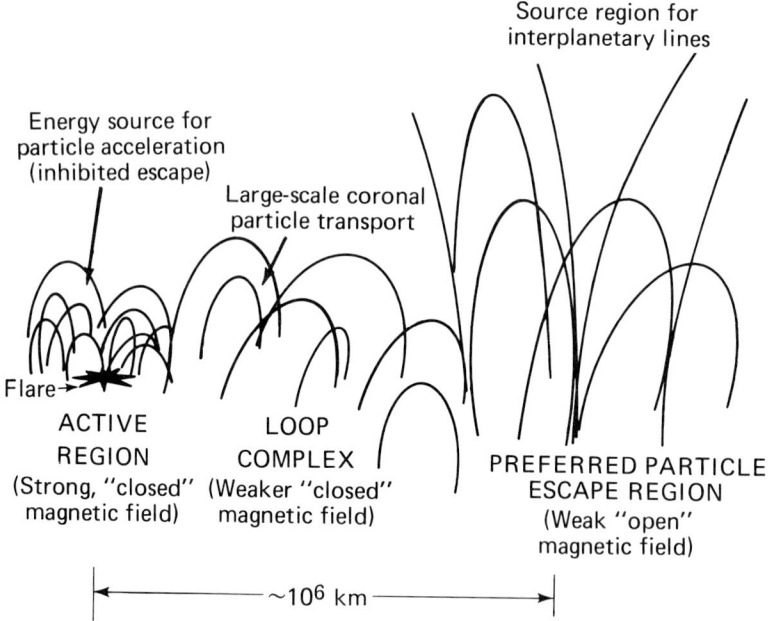

Fig. 3. Sketch of coronal magnetic field structure which would be consistent with characteristics of coronal proton propagation and escape presented in Figures 1 and 2.

statistically that the emission longitudes for high speed solar wind streams were not, in general, the same as those for energetic protons. In addition, the energetic protons did not appear to escape the corona at random locations. During the proton events the IMF polarity was negative 65% of the time, whereas the polarity over the entire period of 9 solar rotations was positive 60% of the time.

In an event-by-event study of the same Mariner 4 events, Roelof et al. [1979] found that the emission longitudes, or "preferred escape regions" for the energetic protons could be centered more than 60° west of the flare site. Moreover, the recurrent series of proton events noted by Krimigis [1969] was reflected in recurrent locations of preferred escape regions on successive rotations, even though the most probable acceleration sites for the events were different on the successive rotations. The graphical summary from this study is given as Figure 1. The charts are excerpts from the Hα Synoptic Charts prepared by McIntosh and Nolte [1975] for the indicated Carrington rotations. These charts utilize the continuity of Hα absorption features (filaments, filament channels, etc.) to outline the boundaries of large-scale magnetic polarity regions in the low corona. Inferred negative polarity regions are shaded. Stipple indicates Hα emission plages usually associated with active regions. The horizontal bar indicates the extent in heliolongitude of the particle escape region and is centered on the heliolatitude of the spacecraft. The bar has a vertical end where a spatial boundary to the particle population was identified by the mapping technique, and a pointed end indicating a limit to the westward extent (because the flare rise began when the connection lay within the preferred escape region), or a limit to the eastward extent (because the fluxes decayed below the detector background before an eastern boundary was detected). Polarity of the IMF is indicated within the bar according to the legend, and the eastern edges of solar wind stream emission regions are indicated by a bracket (peak > 400 km s$^{-1}$) or parenthesis (peak ≤ 400 km s$^{-1}$) at the "dwell" in heliolongitude which characterizes the decay of the stream [Roelof and Krimigis, 1973], and marks the high coronal boundary of the stream [Nolte et al., 1977].

In Figure 1, pairs of events from the three recurrent series of 1964-1965 are arrayed above each other. A filled dot marks the likely acceleration site of the 0.3-0.5 MeV protons. The heliolongitude of the mapped maximum in the particle fluxes is indicated by an asterisk if the increase could be time-associated with a flare, or by a filled triangle if there was no flare-association but rather a near-by active region which had produced flares prior to the particle event. The low level of activity during solar minimum reduced the ambiguity of the associations. While events C1 and C3 were recurrent in both acceleration site and preferred escape region, the preferred escape regions for events A1 and A2 were nearly identical even though the likely acceleration site had shifted 20° westward and 15° southward. In events B2 and B3, both acceleration sites and preferred escape regions had shifted ~ 30° eastward.

The dominance of negative IMF polarity in the preferred escape regions is evident, and the mapped polarities are in general agreement with equatorial Hα polarities (which are only weakly defined during solar minimum). The general pattern appears to be that the preferred escape regions are well-removed from the acceleration site (both in longitude and in latitude), but are well-connected to the IMF. Moreover, the recurrence trends during this solar minimum may be more due to the recurrence of the coronal structures delimiting the remote preferred escape regions than to recurrences in the longitudes of the acceleration sites. The role of interplanetary acceleration in these events would not seem to be significant, because each of the three recurrence pairs were made up of one impulsive (flare-associated) and one non-flare-associated event, both of which usually shared the same preferred escape region.

The interesting question then is what coronal structure determines the preferred escape regions? The information available on coronal structure in 1965 is quite limited compared to, say, the Skylab observations in 1973. At that time, the complement of more sensitive particle detectors on the IMP 7 spacecraft considerably reduced the ambiguity of the association between particle increases and the candidate flares. Consequently, Mitchell et al. [1979] undertook an analysis of the intensities of flare-associated ~ 1 MeV proton events in terms of the intervening coronal structure between the flare site and the IMF connection longitude at latitude.

Coronal propagation was studied for all flares of Hα importance ≥ 1 and/or for which a time-associated particle event was detected at any of the IMP 7, Pioneer 10 or Pioneer 11 spacecraft. The proton flare associations were established with the use of SOLRAD 1-8 Å x-ray fluxes, type II and IV radio emission, as well as > 0.22 MeV electron measurements from the JHU/APL detector on IMP 7. The coronal structure was identified from an atlas [Hanson and Roelof, 1979] of soft x-ray emission loops visible in the American Science and Engineering 2-32 Å and 44-54 Å grazing-incidence telescope images from Skylab during Carrington Rotations 1601-1610 (May 1973-February 1974), augmented by the potential-field calculations of open and closed field lines by Levine et al. [1977]. The nature of the coronal magnetic structure inferred from the soft x-ray images and potential field calculations was qualitatively categorized as "good" or "poor" connection, based on the degree of continuity of the loop structures between the flare site and the estimated IMF connection longitude of the spacecraft.

A suggestive pattern emerges when we compare in Figures 2a and 2b the relationship between peak 1.7-4.5 MeV proton flux versus peak 1-8 Å SOLRAD x-ray flux for the two categories of coronal magnetic structure ("good" or "poor" connection via a complex of loops over $> 15°$). Data points distinguish between structures inferred from the soft x-ray images or the potential-field (HCO) calculations. Points with downward-pointing arrows indicate the ambient proton flux if no proton increase was observed (and hence an upper limit to the flux) and arrows to the right are used on data points for which data gaps including the 1-8 Å x-ray flux peak allowed only a lower limit to be estimated. Comparison of the trends in Figures 2a and 2b reveal a quantitative difference between the two qualitative classifications of coronal structure. Of the 16 events in the "good" connection category (Figure 2a) only 4 do not produce proton fluxes above the ambient level at IMP 7. Of the remaining 25 flares where the coronal loop continuity is "poor" (Figure 2b), there is only a single moderate proton increase above the ambient level. Half of the upper proton flux limits in Figure 2b fall below the dashed line of slope unity, whereas all of the peak fluxes in Figure 2a lie above this line. Moreover the three largest x-ray peaks for the cases of "good" connection correspond to the three largest proton peaks in Figure 2a (as indicated by the stippled curve through all points determined by finite fluxes in both protons and x-rays); however the proton fluxes for the three largest x-ray peaks when the coronal connection was "poor" (Figure 2b) are upper limits which fall 2 orders of magnitude below the proton fluxes observed (for the same range of peak x-ray fluxes) when the coronal connection was "good".

In drawing conclusions from this result, one should remember that the 1-8 Å x-ray peak flux is considered a more appropriate indicator of electron acceleration, while type II and IV radio emission serves better for protons. However, if only the data points in Figures 2a and 2b are retained for which type II and/or IV emission was reported, there remain 8 points each in Figures 2a and 2b, but the pattern is unchanged, e.g., the three largest x-ray events in both Figures 2a and 2b were associated with radio emission. Also, Mitchell et al. [1979] draw attention to a peculiar sub-class of 6 events in which there was effectively no separation (< 15° arc length) between the IMF connection and the flare site (Figure 2b). None of these 6 events had measureable proton fluxes despite their proximity to the flare site; however, all were associated with low to moderate peak x-ray fluxes and none had reported type II or IV radio emission. It may be just

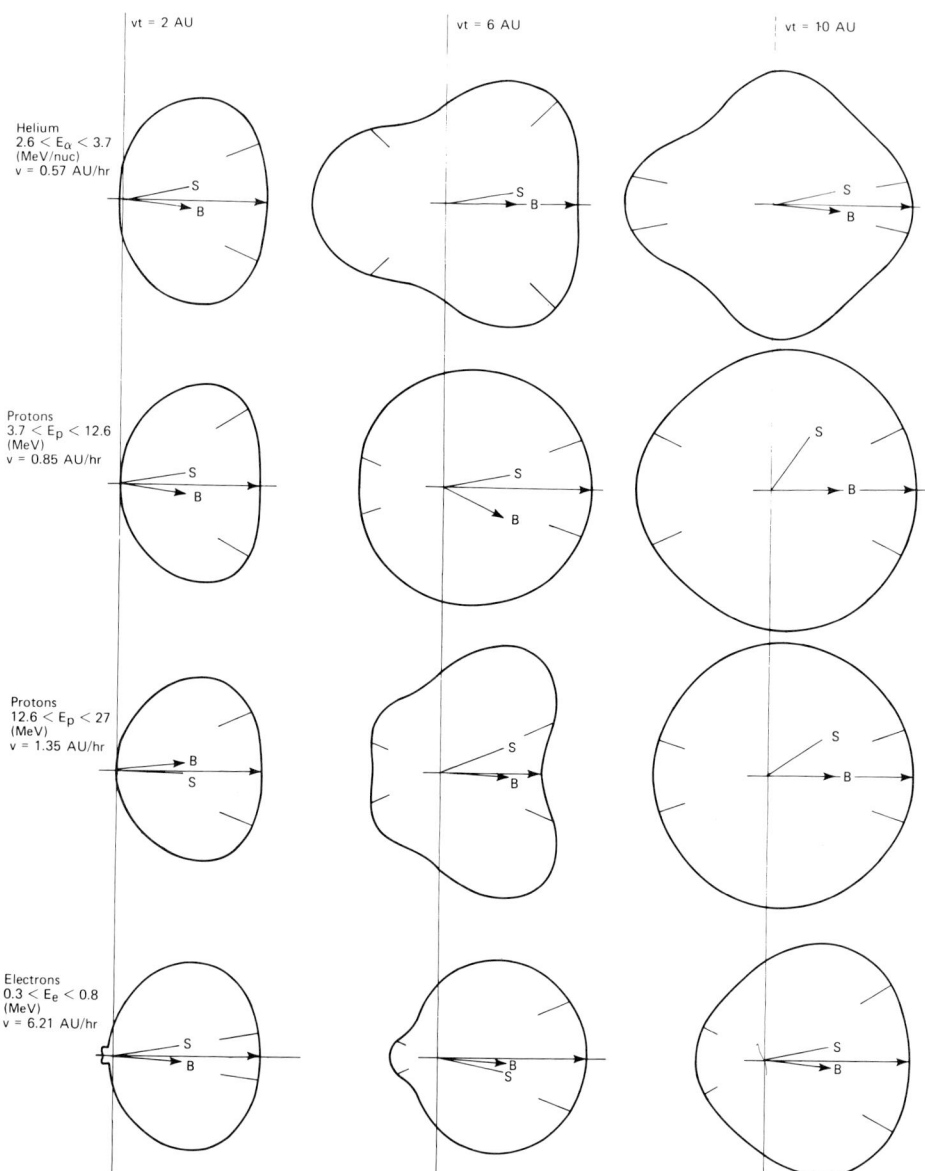

Fig. 4. Polar plots of particle flux intensity during the flare event of March 28, 1976 measured by the University of Kiel detectors on Helios 2 ~ 0.5 AU from the sun. Plots were selected at times corresponding to fixed values of particle distance travelled (velocity × time after flare) in order to demonstrate the relative lack of velocity and species dependence of interplanetary propagation. Prepared by Dr. G. Green.

an improbable coincidence that the 6 flares out of 41 for which IMP 7 was connected nearest to the flare site did not produce measurable protons. However, of the 15 events with "good" connection, about half of the observed proton events did not have reported type II and/ or IV radio emission and most of the flares with low to moderate x-ray emission were associated with proton events. By comparison, 5 of the 6 "no connection" upper limits in Figure 2b fall below all the corresponding "good connection" observed proton events in Figure 2a. It is the opinion of Mitchell et al. [1979b] that this limited evidence is balanced in favor of an actual effect in which there is relative inhibition of interplanetary particle injection in the immediate vicinity of the flare site. They summarize their deductions in the sketch shown in Figure 3. It appears obvious that the path of a particle from its acceleration site through the corona into the interplanetary medium depends upon the proportion of closed and open magnetic field lines in the corona. Complexes of closed loops are required to transport particles over large distances, but we will not observe those particles unless they gain access to open field lines. The efficient coronal transport of particles via loops may require a non-static field structure involving magnetic reconnection [Newkirk and Wentzel, 1978], while the non-homogeneous escape of particles from the corona would be a function of the distribution of more open structures, extreme cases of which are coronal holes [Noci, 1973; Krieger et al., 1973; Hundhausen, 1977; Levine, 1977]. The very reason for remote injection may be that the closed loops required for efficient energization of the particles can actually inhibit their escape near the flare site, but at the same time the loop complexes connect to regions of more predominantly open field lines where the bulk of the particles eventually seen in the interplanetary medium are injected.

The interpretation of remote injection in terms of the interplay of acceleration and coronal transport requires a careful evaluation of a wider body of observations, including the dependence on longitude of spectral index and composition [e.g., McCracken et al., 1971; Van Hollebeke et al., 1975; Perron et al., 1978]. However, one should bear in mind that the characteristics deduced from flux behavior near the peaks of events is influenced both by coronal and interplanetary propagation, whereas information contained in the onset and anisotropic rise phase is predominantly coronal [Reinhard and Wibberenz, 1974; Gold et al., 1977; Roelof and Krimigis, 1977].

If we try to examine the results of the studies of coronal propagation of ∼ 1 MeV protons by Roelof et al. [1979] and Mitchell et al. [1979], we can find a common pattern, even though the techniques differed and the periods analyzed were 8 years apart in Solar Cycle 20. Interplanetary injection of protons well away from the flare site is commonplace, and for some events, remote injection is actually the preferred mode in the corona. Thus the "anomalous distribution in longitude" established by Keath et al. [1971] for several flare proton events at energies > 7.5 MeV during the maximum of Solar Cycle 20 was also manifested in much smaller events (at energies a decade lower), both in the rise and the decay of that cycle. Although the evidence is far from definitive, there seems to be a possibility

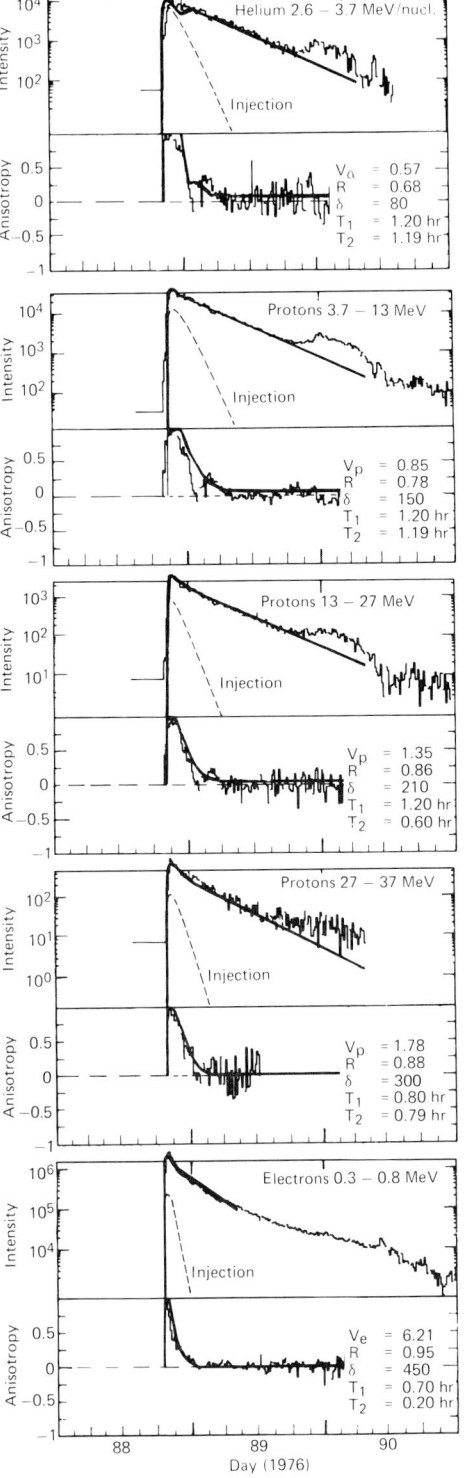

Fig. 5. Theoretical fits of "scatter-free" propagation to particle flux intensities and anisotropies for the event described in Figure 4. Dashed curves are the interplanetary injection history at the corona. A partially reflecting "boundary" was located at 1.4 AU, as inferred from solar wind velocities measured at Helios 2 and IMP 7/8.

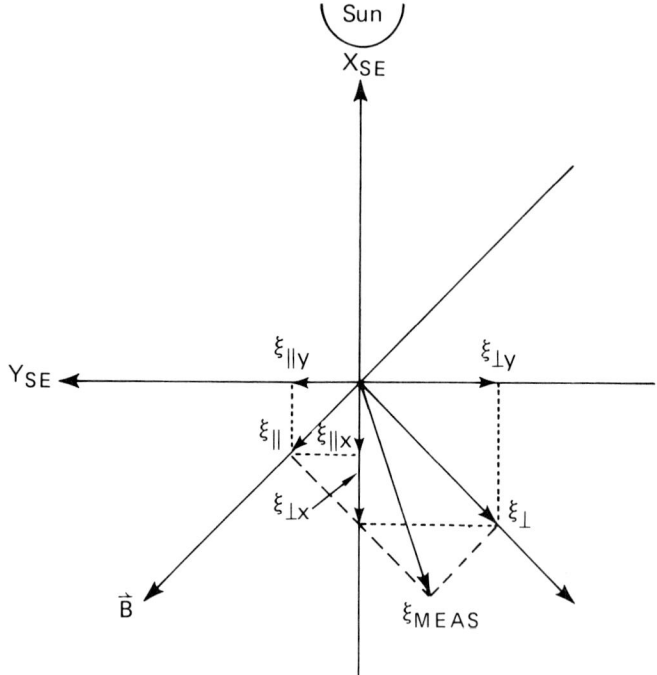

Fig. 6. Vector resolution of particle anisotropy ($\xi$) into components parallel ($\xi_\parallel$) and transverse ($\xi_\perp$) to the IMF. Further resolution into solar ecliptic sub-components $\xi_\parallel \to (\xi_{\parallel x}, \xi_{\parallel y})$, $\xi_\perp \to (\xi_{\perp x}, \xi_{\perp y})$ is appropriate for statistical analysis of parallel and transverse interplanetary propagation. From Zwickl and Roelof [1979].

that "anomalous" distributions may have been more the rule than the exception in Solar Cycle 20.

## Interplanetary Particle Propagation

Definitive deductions concerning coronal propagation of energetic particles cannot be made without a clear understanding of their propagation in the interplanetary medium. For that matter, we need the same understanding to test our concepts of the global structure of the outer heliosphere. Interestingly, the growth of our knowledge about particle propagation has been stimulated by the question of the inter-relationship of particle transport parallel to and transverse to the IMF.

Ahluwalia and Dessler [1962], in their discussion of the diurnal variation of galactic cosmic rays, first pointed out that the existence of the polarization field $\underline{E} = -(\underline{V} \times \underline{B})/c$ implied that electric field drift with velocity $\underline{V}_\perp = \underline{V} - \hat{\underline{B}}\ \hat{\underline{B}} \cdot \underline{V}$ should play a role in the transverse transport of energetic charged particles. Here $\underline{V}$ is the solar wind velocity, $\underline{B}$ the interplanetary magnetic field, and $\hat{\underline{B}} = \underline{B}/|\underline{B}|$. Subsequent theoretical discussion, beginning with Parker [1964] and Axford [1965], emphasized the role of diffusive transport at cosmic ray energies. However at much lower energies (protons ≲ 1 MeV), the often-observed "collimation" of persistent flux anisotropies along the IMF during the rising phase of solar particle events seemed inconsistent with an appreciable amount of scattering, either parallel or transverse at least near 1 AU [McCracken and Ness, 1966; Krimigis et al., 1971; Palmer et al., 1975; Zwickl et al., 1978]. Interest in the transverse transport of ∼ 1 MeV protons has been revived by studies of the anisotropies in non-impulsive events [Marshall and Stone, 1977, 1978; Zwickl and Roelof, 1979] which lack the strong collimation of solar flare injections and are related more to those in the decay phase of solar events as well as in the "corotating" events observable late in the solar cycle.

As an example of collimation in an impulsive solar event, Figure 4 shows an anisotropy analysis by Dr. Günter Green of the University of Kiel from a study we are doing of an event observed March 28, 1976 on the Helios 2 spacecraft about 0.5 AU from the sun. Plotted as polar diagrams as a function of arrival direction are the flux distributions in the ecliptic plane for alpha particles ∼ 3 MeV/nucleon, protons 3.7-12.6 MeV and 12.6-27 MeV, and relativistic electrons 0.3-0.8 MeV. The anisotropies are symmetrized by a 3-dimensional minimum-variance technique, and the directions to the sun (S) and along the IMF ($\underline{B}$) are indicated on the plots relative to the axis of minimum asymmetry. The unlabeled ticks on the contours delimit the cone of directions unmeasurable because of the elevation of $\underline{B}$ out of the ecliptic. Anisotropy diagrams are constructed for each species at three times during the event corresponding to distances travelled (velocity x time from flare) of vt = 2, 6 and 10 AU. The evolution of the angular distributions is remarkably similar among the different species despite a range of velocities of more than a decade. For distances travelled of 2 AU, all the particles are moving outward with no measurable backscatter from beyond 0.5 AU. However for vt = 6 AU, backscattered particles are present, but the anisotropy is still ∼ 2:1 outward. By vt = 10 AU (20 times the spacecraft's distance from the sun), the symmetrized anisotropy is considerably reduced and is becoming comparable to the anisotropy component transverse to the IMF.

The ordering of intensity (but not anisotropy) histories in terms of vt had been noticed in the flare particle event of 28 September 1961 observed on Explorer 12 by Bryant et al. [1962] and ascribed to diffusion with a rigidity independent mean-free-path. However, the prevalence of strongly anisotropic flare events led me to derive an equation appropriate to collimation [Roelof, 1969] in which the tendency for the particle to reduce its pitch angle in a diverging mean field in order to conserve the first adiabatic invariant (magnetic moment) competes with weak pitch-angle scattering due to irregularities in the total field. Numerical solutions to this equation have been developed by Earl [1976], but since the scattering must be quite weak (in order to produce the strong collimation), the more extreme approximation of "scatter-free" propagation [Nolte and Roelof, 1975; Roelof, 1979], in which the magnetic moment is strictly conserved, allows one to extract the essential details of the injection process without extensive calculations and with considerable flexibility in the mathematical representation of the magnetic field.

Fits of the scatter-free calculations to the intensity and aniso-

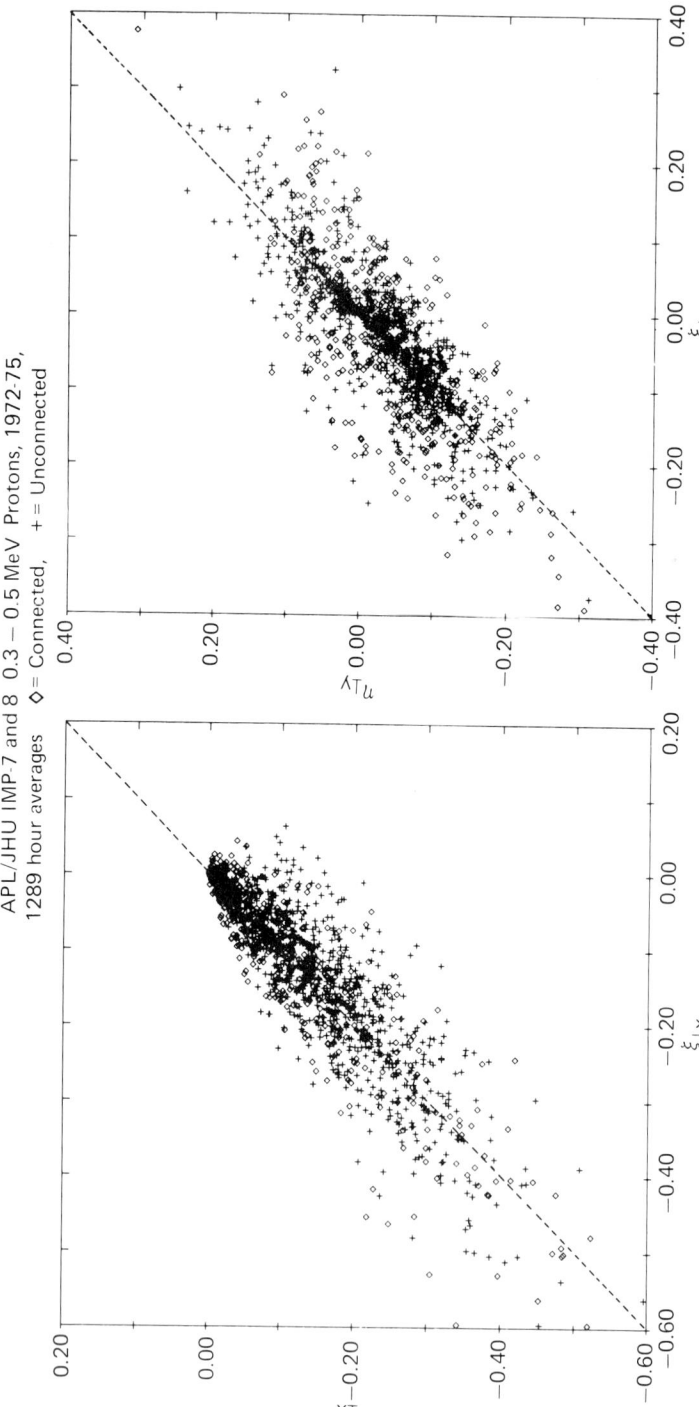

Fig. 7. Evidence that transverse interplanetary transport of low energy protons is predominantly due to $\underline{E} \times \underline{B}$ drift on a one-hour time scale. Scatter plots of observed transverse anisotropy components $\xi_{\perp x}$ and $\xi_{\perp y}$ (see Figure 6) versus theoretically predicted $\underline{E} \times \underline{B}$ drift anisotropy components $\eta_{\perp x}$ and $\eta_{\perp y}$ calculated from particle spectrum, solar wind velocity and IMF measured simultaneously. From Zwickl and Roelof [1979].

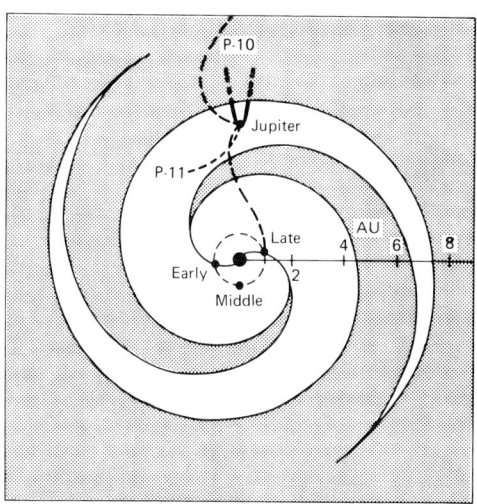

Fig. 8. Idealization of the large scale structure of the interplanetary medium in the heliographic equatorial plane during the period of very stable solar wind structure 1974-1976. The white areas represent the "cavities" formed by solar wind streams while the shading represents the stream interaction regions [Smith and Wolfe, 1977] which thicken with radial distance and eventually merge. Gold and Roelof [1979] have suggested that approximately scatter-free transport within the relatively undisturbed IMF of the cavities is consistent with the observed characteristics of Jovian electron propagation, and hence should describe the global propagation environment for low energy particles.

tropy amplitude histories of the March 28, 1976 are shown in Figure 5. All species represented in Figure 4 are given, as well as 27-37 MeV protons. The interplanetary injection history is modeled by a function of the form $\exp(-t/T_1)-\exp(-t/T_2)$, and the parameters R and $\delta$ are the effective "reflection coefficient" and "compression factor" of a reflective "boundary" at 1.4 AU. Comparison of Helios 2 and IMP 7/8 solar wind velocities for this period are consistent with a solar wind stream-stream interaction at that radius on March 29 which would connect via the IMF to Helios 2. Such a solar wind structure would produce a local compression of the IMF which would act as a rather effective "reflecting barrier" for all species of particles.

Turning now to the analysis of transverse propagation, it is necessary to chose the representation of the anisotropy vector with some care. If one wishes to study transverse propagation in detail on a time scale $\sim$ 1 hour, the motion of the field vector from hour to hour (often to positions well away from its nominal spiral direction), requires that the measured anisotropy vector ($\underline{\xi}$) be separated into orthognal components parallel ($\underline{\xi}_\parallel$) and transverse ($\underline{\xi}_\perp$) to the IMF for each measurement period. Then each component can be resolved separately into its projection along the axes of a "fixed" coordinate system, e.g., solar ecliptic coordinates. This resolution of $\underline{\xi}_\parallel \rightarrow$

($\xi_{\parallel x}$, $\xi_{\parallel y}$) and $\underline{\xi}_\perp \rightarrow$ ($\xi_{\perp x}$, $\xi_{\perp y}$) is diagrammed in Figure 6. The utility of this representation is that the parallel and transverse anisotropies from many hours of observation can be combined linearly for statistical analysis.

Zwickl and Roelof [1979] have applied this technique in their study of anisotropies of 0.3-0.5 MeV protons measured by the JHU/APL detectors on IMP 7/8. Using rather stringent solution criteria they selected $\sim$ 1300 statistically significant hourly measurements over the years 1972-1976 which minimized the probability of impulsive solar injection or influence of the magnetosphere. We shall return in the next section to the question of magnetospheric influence, but suffice it to say here that a number of self-consistent tests were applied to assure the best selection of anisotropies of purely interplanetary origin. They compared the observed transverse anisotropy $\underline{\xi}_\perp$ with that expected if the only means of transverse transport were the $\underline{E} \times \underline{B}$ drift. In that case, the anisotropy would be simply $\underline{n}_\perp = (\gamma + 1)(2/v) \underline{V}_\perp$, where $\gamma$ is the spectral slope of the particle intensity (differential in energy), v is the particle velocity and $\underline{V}_\perp$ is the component of the solar wind velocity transverse to the IMF. All quantities making up $\underline{n}_\perp$ were measurable on IMP 7/8 on an hourly average basis.

Their comparisons of the $X_{SE}$ and $Y_{SE}$ components of $\underline{\xi}_\perp$ and $\underline{n}_\perp$ are shown in Figure 7. One appreciates the degree of agreement when one realizes that the scattered points visible make up only $\sim$ 10% of all the points plotted; the remaining thousand are packed too densely along the lines $\xi_{\perp x} = \eta_{\perp x}$ and $\xi_{\perp y} = \eta_{\perp y}$ to be resolved. From this result, Zwickl and Roelof [1979] conclude that the $\underline{E} \times \underline{B}$ drift is the predominant mode of transverse interplanetary transport of $\sim$ 1 MeV protons, on the time scale of one hour and consequently these particles are constrained to move along the "same" field line in the inner heliosphere. Furthermore, this result implies that, to the accuracy that we can estimate the high coronal connection longitude and latitudes of field lines at 1 AU, we can say that the low energy particles on those field lines were injected into the interplanetary medium at those coronal locations.

To conclude this section, I would like to draw some inferences on the global characteristics of interplanetary propagation of low energy particles in the heliosphere. If particles propagate approximately scatter-free along the "same" field lines, then their time histories and transverse transport are determined by boundary conditions in the corona and the outer heliosphere. Following this line of reasoning, Gold and Roelof [1979] have suggested that the distribution of relativistic electrons from Jupiter observed from 0.3 AU to 7 AU is explicable in terms of the existence of "cavities" formed by the intersection of the reverse shock from one long-lived high speed solar wind stream with the forward shock of the following stream. Within these cavities, the particle propagation is approximately scatter-free, with reflection and transverse propagation taking place within the "walls" of the cavities formed by the stream-stream interactions which become strongly developed beyond $\sim$ 2 AU [Smith and Wolfe, 1977]. This cavity structure is sketched in the heliographic equatorial plane in Figure 8 for the case corresponding to 1974-1976

E. C. ROELOF

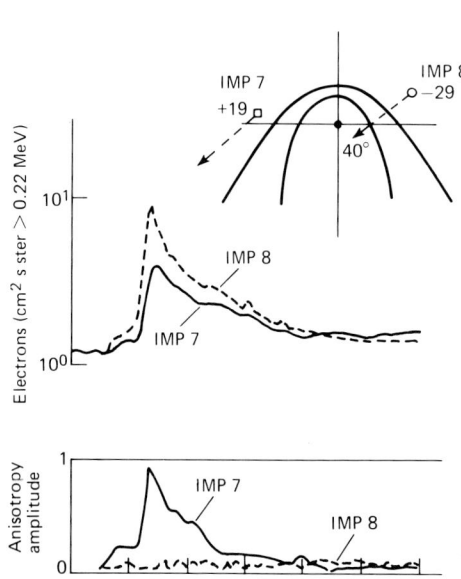

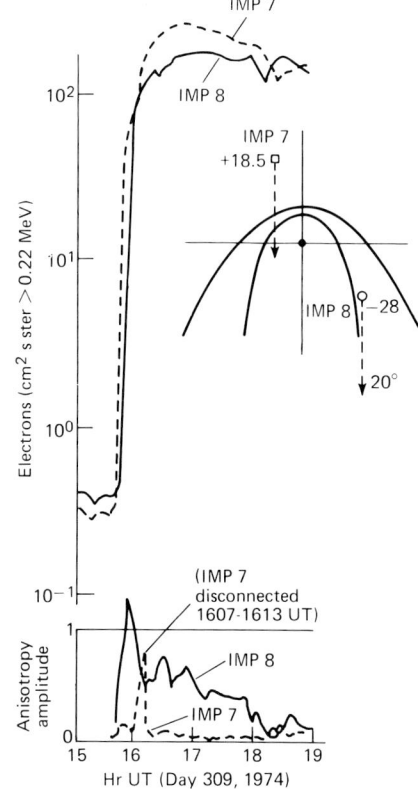

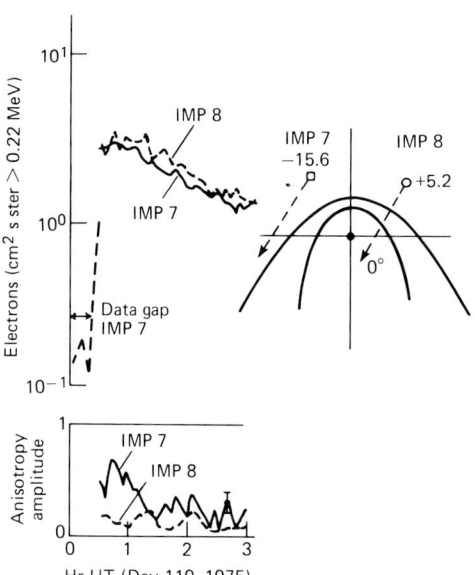

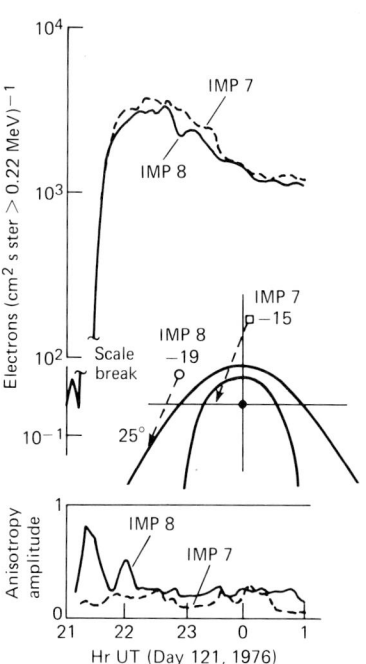

when there were two comparable high speed streams on opposite sides of the sun. The three-dimensional structure may be inferred from the banner-like configuration of the shock surfaces [Siscoe, 1976] and the shapes of the coronal hole boundaries which often mark the edges of the solar wind source regions [Nolte et al., 1977]. If this picture for propagation proves correct for Jovian electrons, it also should apply to all other species of energetic particles in the inner heliosphere, so that interplanetary propagation transpires within a region much more complex than the simple concentric "shell" used in theoretical calculations up until a few years ago.

## Interaction of Interplanetary Energetic Particles With the Magnetosphere

When studying $\leqslant 1$ MeV interplanetary particles upstream of the bow shock, a primary concern is that the earth is a prolific source of these particles. The emission of magnetospheric particles is a subject in itself, both for the earth and Jupiter, but one which I shall only mention in passing. One technique for reducing the "contamination" of near-earth interplanetary particle measurements by magnetospheric bursts, developed by C.-K. Ng of the University of Malaya and myself, is to estimate whether or not the IMF line through the spacecraft intersects the bow shock [Ng and Roelof, 1977; Roelof and Krimigis, 1977]. If it does not, the presence of magnetospheric particles is unlikely. We found that it was sufficient to make a simple linear extrapolation of the field direction from the spacecraft to the nominal bow shock surface measured by Fairfield [1971], which we considered to be a figure of revolution about the aberrated $X_{SE}$ axis.

However, if interplanetary particles are present, some interesting effects are measurable when the IMF intersects the bow shock and magnetospheric burst activity is low (e.g., when solar wind velocity and/or geomagnetic activity are low). Bostrom, et al. [1973] noted the anisotropy of > 200 keV solar protons was affected by the bow shock, and Palmer and Higbie [1978], Higbie and Palmer [1977], and Palmer and Craig [1978] have shown that Vela 5 and 6 observations of distortions of 0.5 MeV proton pitch angle distributions are consistent with non-random entry of the particles across the bow shock into the magnetosheath.

Relativistic electrons are strongly reflected from the magnetosphere. In Figure 9, taken from a forthcoming article by Ng and myself, the intensity and anisotropy histories observed in identical > 0.22 MeV electron channels from the JHU/APL detectors on IMP 7 and 8 are com-

Fig. 9. Effect of the magnetosphere on interplanetary propagation of relativistic electrons measured by intercalibrated JHU/APL detectors on IMP 7/8. Intensities and anisotropies are compared for four flare events in which the IMF from one IMP spacecraft intersected the nominal bow shock [Fairfield, 1971] while at the other IMP the IMF was unobstructed. Obstructed (dashed) intensities are higher and anisotropies are nullified, compared to those which are unobstructed (solid).

pared in four solar electron events in which the IMF connected one spacecraft to the bow shock but not the other. The spacecraft positions and IMF directions are sketched for each event. The intensity and anisotropy at the connected spacecraft is always drawn dashed, while the true interplanetary intensity and anisotropy are drawn with solid lines. The effect is dramatic. When connected to the bow shock, the reflection from the magnetosphere nullifies the strong interplanetary anisotropy and concomitantly increases the intensity by as much as a factor of two.

The reflection effect can explain the heretofore puzzling observations of solar electron events with impulsive time histories but negligible anisotropies even during the rise of the event [Allum, et al., 1971]. On the other hand, the reflection of solar particles offers the interesting possibility of probing the structure of the bow shock and magnetosheath, since the location of the IMF interaction during solar particle events is often in regions inaccessible to orbiting spacecraft.

## Conclusion

I have devoted the largest fraction of this paper to coronal propagation, the next largest to interplanetary propagation, the smallest to near-earth effects, and I still did not enter the magnetosphere proper. However I hope this paper will be useful to magnetospheric physicists not only in outlining some recent developments in solar and interplanetary particle physics, but also in stimulating some thoughts concerning analogs of magnetospheric processes which may be occurring in the corona and interplanetary space.

*Acknowledgements.* I am indebted to W. P. Olson, without whose encouragement this paper would not have appeared. This research was supported in part by the National Science Foundation (Atmospheric Sciences Branch Grant ATM-76-23816), by NASA (Grant NSG-7055), and by the Air Force Geophysics Laboratory via Task I of Contract N00024-78-C05384 between the Department of the Navy and The Johns Hopkins University.

## References

Ahluwalia, A. S. and A. J. Dessler, Diurnal variation of cosmic radiation intensity produced by a solar wind, Planet. Space Sci., 9, 195, 1962.

Allum, F. R., R. A. R. Palmeira, V. R. Rao, K. G. McCracken, J. R. Harries and I. Palmer, The degree of anisotropy of cosmic ray electrons of solar origin, Solar Phys., 17, 241, 1971.

Axford, W. I., The modulation of galactic cosmic rays in the interplanetary medium, Planet. Space Sci., 13, 115, 1965.

Barnes, C. W., and J. A. Simpson, Evidence for interplanetary acceleration of nucleons in corotating interaction regions, Astrophys. J., 210, 191, 1976.

Bostrom, C. O., J. A. Armstrong and E. C. Roelof, (Abstract), Anisotropy of low energy solar particles, EOS Trans. AGU, 54, 409, 1973.

Bryant, D. A., T. L. Cline, U. D. Desai and F. B. McDonald, Explorer

12 observations of solar cosmic ray and energetic storm particle after the solar flare of September 28, 1961, J. Geophys. Res., 67, 4983, 1962.

Bryant, D. A., T. L. Cline, U. D. Desai and F. B. McDonald, The continual acceleration of solar protons in the MeV range, Phys. Rev. Lett., 14, 481, 1965.

Burlaga, L. F. and E. Barouch, Interplanetary stream magnetism-Kinematic effects, Astrophys. J., 203, 257, 1976.

Earl, James A., The effect of adiabatic focusing upon charged-particle propagation in random magnetic fields, Astrophys. J., 205, 900, 1976.

Fairfield, D. H., Average and unusual locations of the earth's magnetopause and bowshock, J. Geophys. Res., 76, 6700, 1971.

Fan, C. Y., M. Pick, R. Pyle, J. A. Simpson and D. R. Smith, Protons associated with centers of solar activity and their propagation in the interplanetary magnetic field regions corotating with the sun, J. Geophys. Res., 73, 1555, 1968.

Gold, R. E. and E. C. Roelof, Jovian electron propagation via solar wind stream interaction regions, J. Geophys. Res., 84, submitted, 1979.

Gold, R. E., E. P. Keath, E. C. Roelof and R. Reinhard, Coronal structure of the April 10, 1969 solar flare particle event, Proc. 15th Int. Cosmic Ray Conf. (Plovdiv, Bulgaria), 5, 125, 1977.

Hanson, J. M. and E. C. Roelof, Synoptic charts of large-scale coronal x-ray structure during Skylab, April 1973-February 1974, Preprint, JHU/APL 78-05, Submitted to UAG Special Reports, WDC-A, 1979.

Higbie, P. R. and I. D. Palmer, The solar proton event of April 16, 1970, 2, transformation of pitch angle distribution across the earth's bow shock, J. Geophys. Res., 82, 2665, 1977.

Hundhausen, A. J., An interplanetary view of coronal holes, in Coronal Holes and High Speed Wind Streams, ed. J. B. Zirker, Colorado Associated University Press (Boulder), 225, 1977.

Keath, E. P., R. P. Bukata, K. G. McCracken and V. R. Rao, The anomalous distribution in heliocentric longitude of solar injected cosmic radiation, Solar Phys., 18, 503, 1971.

Krieger, A. S., A. F. Timothy and E. C. Roelof, A coronal hole and its identification as the source of a high velocity solar wind stream, Solar Phys., 23, 123, 1973.

Krimigis, S. M., Observations of low energy solar protons with Mariners 4 and 5, Trudi Mezdunaradnogo Seminara, in Proceedings of the Ioffe Physico-Technical Institute, (ed.) G. E. Kocharov, Acad. Sci., USSR, Leningrad, 43, 1969.

Krimigis, S. M., E. C. Roelof, T. P. Armstrong and J. A. Van Allen, Low energy ($\geqslant 0.3$ MeV) solar particle observations at widely separated points (> 0.1 AU) during 1967, J. Geophys. Res., 76, 5921, 1971.

Levine, R. H., Large-scale solar magnetic fields and coronal holes, in Coronal Holes and High Speed Wind Streams, ed. J. B. Zirker, Colorado Associated University Press (Boulder), 103, 1977.

Levine, R. H., M. D. Altschuler and J. W. Harvey, Solar sources of the interplanetary magnetic field and solar wind, J. Geophys. Res., 82, 1061, 1977.

Lin, R. P., S. W. Kahler and E. C. Roelof, Solar flare injection and propagation of low energy protons and electrons in the event 7-9 July, 1966, Solar Phys., 4, 338, 1968.

McCracken, K. G. and N. F. Ness, The collimation of cosmic rays by the interplanetary magnetic field, J. Geophys. Res., 71, 3315, 1966.

McCracken, K. G. and U. R. Rao, Solar cosmic ray phenomena, Space Sci. Rev., 11, 155, 1970.

McCracken, K. G., U. R. Rao, R. P. Bukata and E. P. Keath, The decay phase of solar flare events, Solar Phys., 18, 100, 1971.

McDonald, F. B., B. J. Teegarden, J. H. Trainor and T. T. Rosenvinge, The interplanetary acceleration of energetic nucleons, Astrophys. J., 203, L149, 1976.

McIntosh, P. S. and J. T. Nolte, H-alpha synoptic charts of solar activity during the first year of Solar Cycle 20, Rep. UAG-41, World Data Center A, NOAA, (Boulder, Colo.), 1975.

Marshall, F. E. and E. C. Stone, Persistent sunward flow of $\sim$ 1.6 MeV protons at 1 AU, Geophys. Res. Lett., 4, 57, 1977.

Marshall, F. E. and E. C. Stone, Characteristics of sunward flowing proton and alpha particle fluxes of moderate intensity, J. Geophys. Res., 83, 3289, 1978.

Matsuda, T. and T. Sakurai, Dynamics of the azimuthally dependent solar wind, Cosmic Electrodyn., 3, 97, 1972.

Mitchell, D. G., E. C. Roelof and J. H. Wolfe, Latitude dependence of solar wind velocity observed $\geqslant$ 1 AU, J. Geophys. Res., 84, submitted, 1979a.

Mitchell, D. G., J. M. Hanson, E. C. Roelof and R. H. Levine, Coronal bright x-ray loop structure and the propagation of solar flare protons, in preparation, 1979b.

Newkirk, Jr., G. and D. G. Wentzel, Rigidity-independent propagation of cosmic rays in the solar corona, J. Geophys. Res., 83, 2009, 1978.

Ng, C.-K. and E. C. Roelof, Reflection of $\sim$ 200 keV solar electrons and protons from the magnetosphere (Abstract), EOS Trans. AGU, 58, 1205, 1977.

Noci, G., Energy budget in coronal holes, Solar Phys., 28, 403, 1973.

Nolte, J. T. and E. C. Roelof, Large-scale structure of the interplanetary medium. I: High coronal source longitude of the quiet-time solar wind, Solar Phys., 33, 241, 1973a.

Nolte, J. T. and E. C. Roelof, Large-scale structure of the interplanetary medium. II: Evolving magnetic configurations deduced from multispacecraft observations, Solar Phys., 33, 483, 1973b.

Nolte, J. T. and E. C. Roelof, Mathematical formulation of scatter-free propagation of solar cosmic rays, Proc. 14th Int. Cosmic Ray Conf., (Munich), 5, 1722, 1975.

Nolte, J. T. and E. C. Roelof, Solar wind energetic particles, and coronal magnetic structure: The first year of solar cycle 20, J. Geophys. Res., 82, 2175, 1977.

Nolte, J. T., A. S. Krieger, E. C. Roelof and R. E. Gold, High coronal structure of high velocity solar wind streams, Solar Phys., 51, 459, 1977.

Palmer, I. D. and G. S. Craig, Distortion of interplanetary pitch distributions by earth's bow shock, J. Geophys. Res., 83, 5633, 1978.

Palmer, I. D. and P. R. Higbie, Magnetosheath distortion of pitch angle distributions of solar protons, J. Geophys. Res., 83, 30, 1978.

Palmer, I. D., R. A. R. Palmiera and F. R. Allum, Monte Carlo model of the highly anisotropic solar proton event of 20 April 1971, Solar Phys., 40, 449, 1975.

Parker, E. N., Theory of streaming of cosmic rays and the diurnal variation, Planet. Space Sci., 12, 735, 1964.

Perron, C., V. Domingo, R. Reinhard, and K.-P. Wenzel, Rigidity-independent coronal propagation and escape of solar protons and α particles, J. Geophys. Res., 83, 2017, 1978.

Pesses, M. E., J. A. Van Allen and C. K. Goertz, Energetic protons associated with interplanetary active regions, 1 to 5 AU, J. Geophys. Res., 83, 553, 1978.

Pizzo, V., A three-dimensional model of corotating streams in the solar wind, 1, theoretical foundations, J. Geophys. Res., 83, 5563, 1978.

Reid, G. C., A diffusive model for the initial phase of a solar proton event, J. Geophys. Res., 69, 2659, 1964.

Reinhard, R. and G. Wibberenz, Propagation of flare protons in the solar atmosphere, Solar Phys., 36, 473, 1974.

Roelof, E. C., Propagation of solar cosmic rays in the interplanetary magnetic field, Lectures in High Energy Astrophysics, Ed. H. Ogelman and J. R. Wayland, NASA SP-199, 111, 1969.

Roelof, E. C., The scatter-free approximation to energetic particle transport, Rev. Geophys. Space Phys., 17, in preparation, 1979.

Roelof, E. C. and S. M. Krimigis, Analysis and synthesis of coronal and interplanetary energetic particle, plasma and magnetic field observations over three solar rotations, J. Geophys. Res., 78, 5375, 1973.

Roelof, E. C. and S. M. Krimigis, Solar energetic particles below 10 MeV, Study of Travelling Interplanetary Phenomena/1977, ed. M. A. Shea, D. F. Smart and S. T. Wu, D. Reidel, (Dordrecht), 343, 1977.

Roelof, E. C., S. M. Krimigis, J. T. Nolte and J. M. Davis, Energetic solar particle events in 1965: Relationship to coronal magnetic structure, J. Geophys. Res., 84, submitted, 1979.

Sakurai, T., Quasi-radial hyperactivity approximation of the azimuthally dependent solar wind, Cosmic Electrodyn. 1, 460, 1971.

Sarris, E. T. and J. A. Van Allen, Effects of interplanetary shock waves on energetic charged particles, J. Geophys. Res., 79, 4157, 1974.

Sarris, E. T., S. M. Krimigis and T. P. Armstrong, Observations of a high-energy ion shock spike in interplanetary space, Geophys. Res. Lett., 3, 1976.

Siscoe, G. L., Three-dimensional aspects of interplanetary shock waves, J. Geophys. Res., 81, 6235, 1976.

Smith, E. J. and J. H. Wolfe, Pioneer 10, 11 observations of evolving solar wind streams and shocks beyond 1 AU, Study of Travelling Interplanetary Phenomena 1977, ed. M. A. Shea, D. F. Smart and S. T. Wu, D. Reidel (Dordrecht), 227, 1977.

Snyder, C. W. and M. Neugebauer, The relation of Mariner-2 plasma data to solar phenomena, in Solar Wind (ed. R. J. Mackin and M. Neugebauer), Pergamon Press, 25, 1966.

Van Hollebeke, M. A. I., L. S. MaSung and F. B. McDonald, The variation of solar proton energy spectra and size distribution with heliolongitude, Solar Phys., 41, 189, 1975.

Zwickl, R. D. and E. C. Roelof, Interplanetary propagation of < 1 MeV protons in non-impulsive energetic particle events, J. Geophys. Res., 84, submitted, 1979.

Zwickl, R. D., E. C. Roelof, R. E. Gold, S. M. Krimigis and T. P. Armstrong, Z-rich solar particle event characteristics 1972 - 1976, Astrophys. J., 225, 281, 1978.

# THE EFFECT OF MAGNETIC FIELD MODELS ON COSMIC RAY CUTOFF CALCULATIONS

K. A. Pfitzer

McDonnell Douglas Astronautics Company
5301 Bolsa Avenue
Huntington Beach, CA 92647

*Abstract.* The cosmic ray cutoffs were a long standing problem in cosmic ray physics and many diffusion and similar theories were developed to predict the access of cosmic rays to the observed latitudes. The Olson-Pfitzer, 1974, distributed current model of the magnetospheric magnetic field (which included the field depression of the quiet time ring current) brought the calculated cutoffs into much closer agreement with observation without the use of diffusion. However, calculated dawn dusk cutoffs and access to synchronous orbit were still not predicted correctly. These discrepancies were again frequently attributed to unknown processes (such as diffusion). The new tilt dependent distributed current model (which removes some of the high field regions present in the zero tilt model) brings the dawn dusk cutoff into agreement with observation and correctly predicts the access of cosmic rays to synchronous orbit. The model also predicts a relatively small tilt dependence in agreement with observation.

## Introduction

Magnetospheric magnetic field models have played an important role in the study of the entry of 1 to 100 MeV cosmic ray particles into the magnetosphere. Initially measured cutoffs over the plar cap agreed only partially with the early Stormer calculated values using the earth's main field. As the state of the art in magnetospheric magnetic field modeling improved by including the externally generated fields, the agreement between theory and data improved. However, when data were compared to entry calculations using the Mead-Williams (1965) and Olson's (1970) external field models, a persistant 4° to 6° difference between model and theory remained. Many experimentalists and theorists believed the magnetic field models to be sufficiently accurate and thus began to suggest anomalous diffusion mechanisms to explain the entry of cosmic rays to the observed latitudes (Lanzerotti, 1968; McDiarmid, et al., 1971; Morfill, 1973). These older models also had difficulty explaining the highly structured polar cap fluxes during the early phases of a solar cosmic ray event, again suggesting the presence of anomalous mechanisms.

The work of Sugiura (1973) showed that the Olson and Mead-Williams magnetic field models did not fit the magnetic field data very well in the inner magnetosphere. Olson and Pfitzer (1974), using Sugiura's work, developed a magnetic field model which for the first time included the effects of the diamagnetic plasma in the inner magnetosphere (the quiet time ring). The inclusion of this ring current depressed the field in the inner magnetosphere sufficiently to bring calculated and observed cosmic ray cutoffs into much closer agreement (Pfitzer, 1972; Masley, et al., 1973). There remained, however, some discrepancies; the calculated cutoffs in the dawn dusk meridian plane were still too high by 2° to 3°, and access to synchronous orbit could not be correctly predicted (Masley, 1975).

TABLE 1. Cutoff Determinations Using Various Models

|  | Gall, et al. 1968 | Olson 1970 Model | Olson-Pfitzer 1974 Model | Present Tilt Dependent Model | | |
| --- | --- | --- | --- | --- | --- | --- |
| Tilt | 0 | 0 | 0 | 0 | +35 | -35 |
| Noon | 72 | 74 | 69 | 70 | 68 | 68 |
| Midnight | 68 | 69 | 66 | 66 | 65 | 66 |
| Dawn | -- | -- | 70 | 67 | 67 | 67 |
| Dusk | -- | -- | 71 | 67 | 67 | 67 |

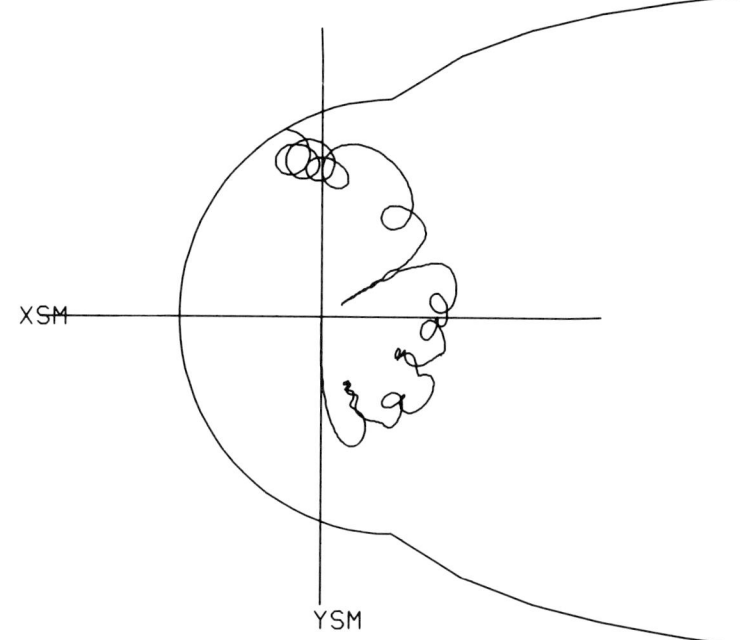

Fig. 1. Trajectory of a 5 MeV proton entering the daylit dawn side of the magnetosphere and entering the polar cap on the dusk meridian.

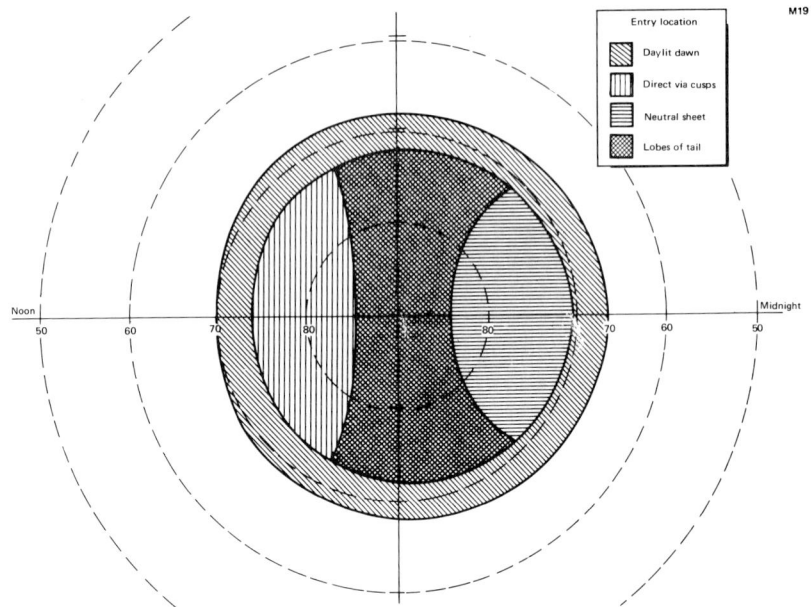

Fig. 2. Polar cap map showing the regions through which protons enter the magnetosphere before being observed at the polar cap.

Walker (1976) summarized the accuracy of the Olson-Pfitzer 1974 model. He showed that the model gave field values that were too large near the magnetopause, especially in the dawn-dusk region. Furthermore, the field was too large in the near tail region. These inaccuracies in the 1974 Olson-Pfitzer model appeared to be the probable cause for the remaining discrepancies between the observed and calculated cutoff values.

An improved version of the Olson-Pfitzer model is now available which includes the effects of the tilt of the earth's dipole axis and which has removed most of the problems found in the earlier model. In this paper we show that when this new accurate magnetic field model is used, the calculated and observed cutoff values agree with the experimental error without the need for invoking anomolous diffusion mechanisms. This tilt dependent model also permits a study of cutoffs versus the tilt of the dipole axis.

## Cutoff Calculations

The cutoffs are calculated by reverse integrating the trajectory of a charged particle. A proton with a negative charge is started at the satellite observation point with its velocity vector within a few degrees of the magnetic field vector (within the acceptance cone of the detector to which the calculation is compared) and an energy consistent with the observed energy. The trajectory of the particle is determined by a step-by-step integration of the Lorentz force equation using a fourth order Gill's method (Gill, 1951) Runge Kutta integration routine developed by Taylor (1967). The trajectory is followed

245

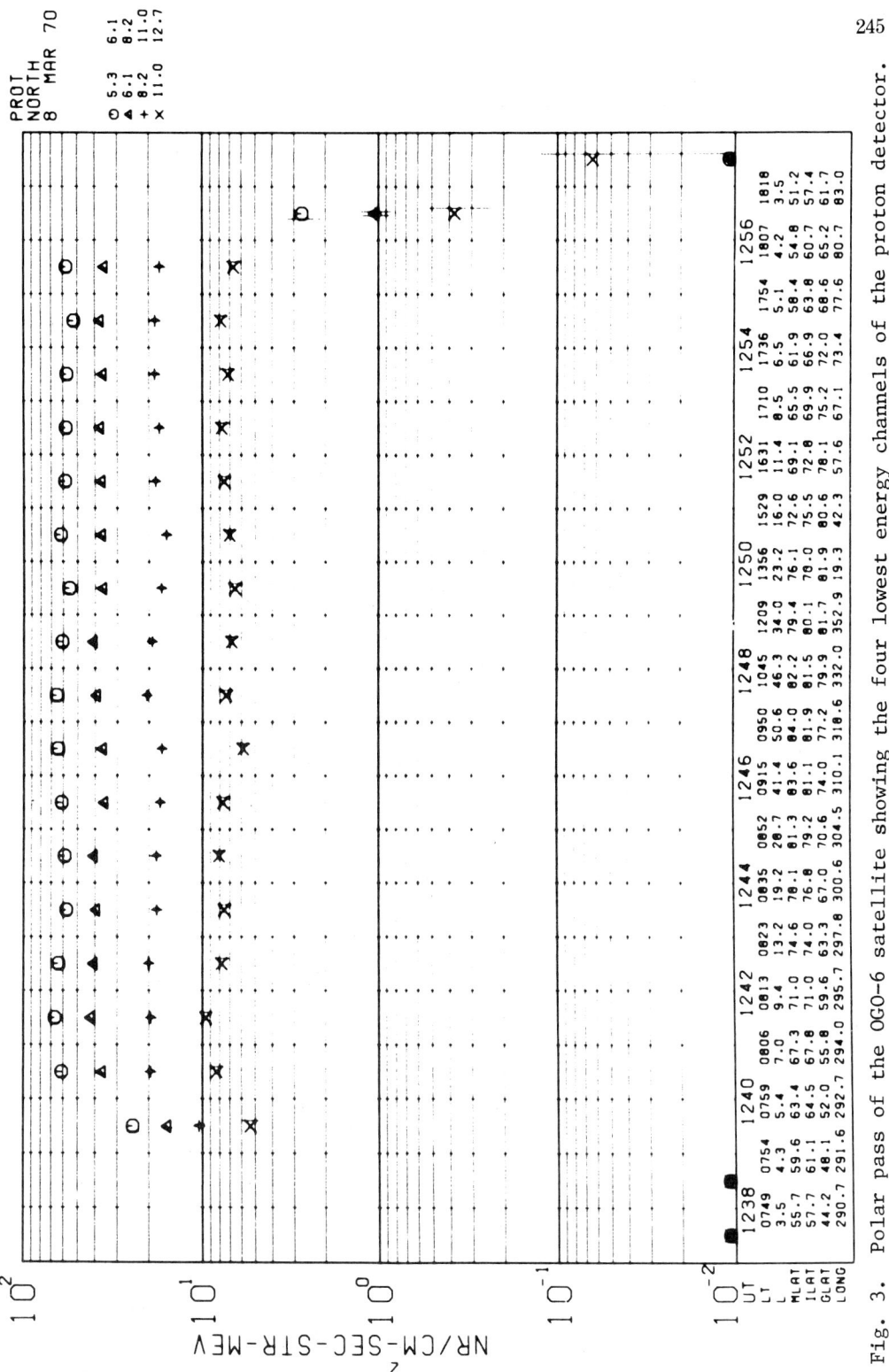

Fig. 3. Polar pass of the OGO-6 satellite showing the four lowest energy channels of the proton detector.

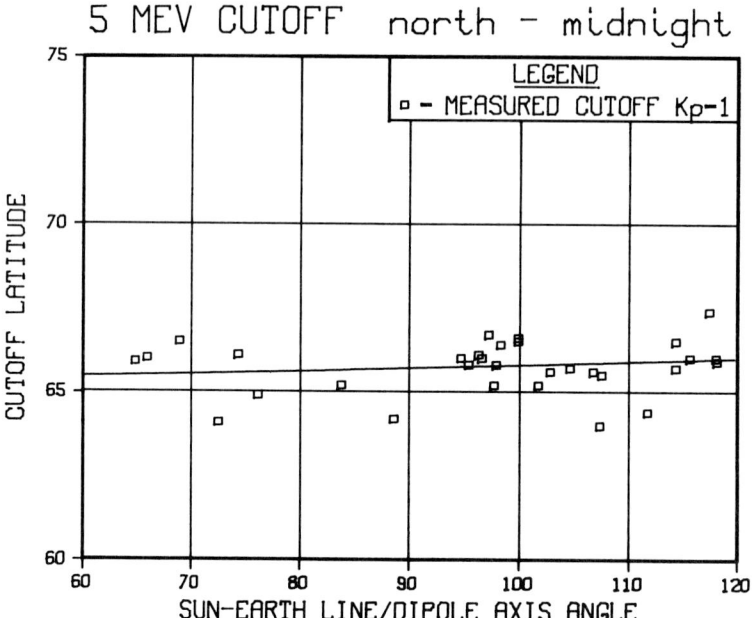

Fig. 4. Calculated and measured cutoffs. Calculated curve is the least squares fit in $K_p$ and tilt evaluated at $K_p=1$. ($\mu=90-\zeta$, where $\zeta$ is the sun-earth line/dipole axis).

until the particle either exits the magnetosphere, impacts the solid earth or exceeds the path length restrictions (is trapped in the earth's magnetic field). If the trajectory passes out of the magnetosphere, it is said to have originated from interplanetary space and that particle is said to have access to the specified latitude and pitch angle over the polar cap. By backward integrating a large number of trajectories for various latitudes, pitch angles, local times and dipole tilts, the minimum access latitude for a particle of a given energy and local time was determined. This minimum access latitude is the theoretically calculated cutoff value. Table 1 shows the results of these calculations for various older magnetic field models and the results of the present tilt dependent model calculations. Since most of the earlier calculations were performed with models whose field values were too large in some region of the magnetosphere (Walker, 1975), the cutoffs determined with the older magnetic field models have higher cutoff latitudes than the present calculations.

The trajectory calculations show that particles within a few degrees (4° for 5 MeV protons) of the cutoff latitude enter the magnetosphere via the daylight dawn side of the magnetopause. The particles closest to the cutoff latitude enter nearest the subsolar point; whereas, particles at higher latitudes, but still within the 4° band, enter more tailward along the dawn flanks. Figure 1 gives an example of the trajectory of a 5 MeV proton entering the dawn flanks and traveling to the polar cap.

These particles which enter via the dawn side of the magnetosphere

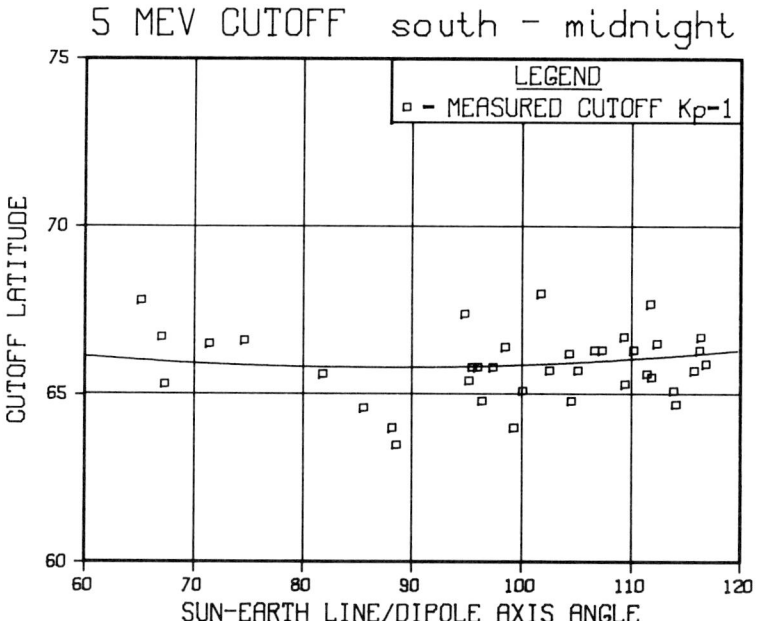

Fig. 5. Calculated and measured cutoffs. Calculated curve is the least squares fit in $K_p$ and tilt evaluated at $K_p=1$. ($\mu=90-\zeta$, where $\zeta$ is the sun-earth line/dipole axis).

populate a low latitude ring (4° wide for 5 MeV protons) around the polar cap. This ring has often been referred to as the low latitude polar cap. Figure 2 is a map of the polar cap showing the regions through which particles enter the magnetosphere before being observed over the polar cap. We note four separate regions: (1) the ring at low latitudes which has the dawn side as the entry location and whose entry direction is from the sunward direction; (2) the direction penetration region where particles enter the polar cap directly via the weak magnetic field regions of the magnetospheric cusps with an entry direction perpendicular to the ecliptic plane; (3) the lobe tail entry region where particles are observed to come up the lobes of the tail from the antisolar direction; and (4) the neutral sheet region where particles are observed to enter the magnetosphere via the distant flanks of the tail near the neutral sheet and whose entry is also from the antisolar direction. The regions shown in Figure 2 have been drawn using a limited number of trajectories and thus the boundaries may not be exact.

We note that the above specified polar cap regions map to unique entry regions on the magnetosphere with different entry vector requirements. Thus, during the early anisotropic phases of solar cosmic ray events, considerable structure is possible over the polar cap. The observed structure closely matches the calculated regions locations; however, a detailed verification of the polar cap structure is beyond the scope of this paper.

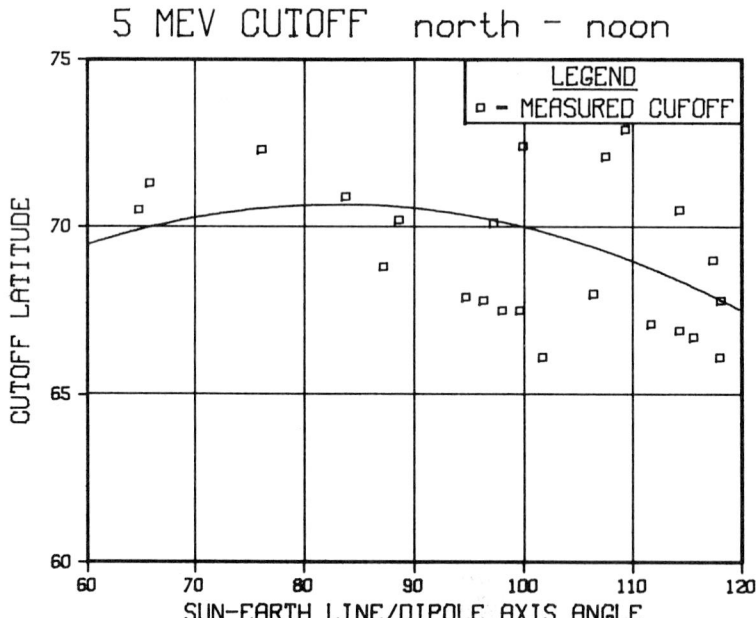

Fig. 6. Calculated and measured cutoffs. Calculated curve is the least squares fit in $K_p$ and tilt evaluated at $K_p=1$. ($\mu=90-\zeta$, where $\zeta$ is the sun-earth line/dipole axis).

The Data

The cutoff data used here are from the OGO-6 charged particle experiment which measures protons from 5.3 MeV to 78 MeV in nine ranges (A. J. Masley, principal investigator). Data from this experiment are available from the National Space Sciences Data Center. The instrumentation and calibration is described in detail by Masley (1975). All of the data were plotted across the polar cap during solar proton events. When the satellite enters the polar cap during a solar proton event, there is an abrupt (as large as 4 to 5 orders of magnitude) increase in the detector count rates (see Figure 3). The experimental cutoff latitude is defined as the latitude where the count rate reached 1/2 of the local maximum. Each entry and exit from the polar cap thus had assigned to it a cutoff latitude. A total of over 600 cutoff values were determined. These cutoff values covered a broad range of geomagnetic conditions and crossing local times.

In order to compare the measured cutoffs with the calculations, it is necessary to determine the cutoffs at magnetically quiet times (the models are valid only at quiet times) and at the various tilts of the earth's dipole axis.

In a first attempt to organize the data with geomagnetic activity, the value of $K_p$, the three hour planetary geomagnetic index, was associated with each measurement. Admittedly, the $K_p$ index is not a complete measure of the state of the magnetospheric magnetic field;

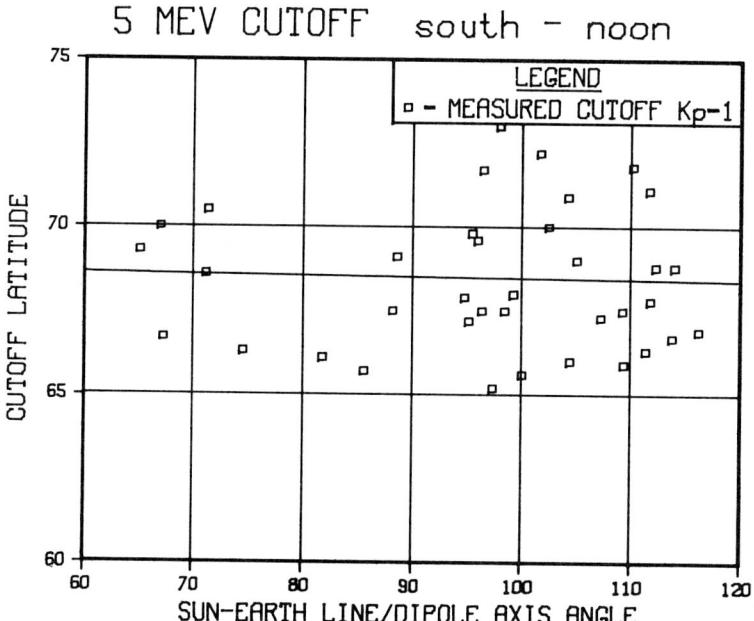

Fig. 7. Calculated and measured cutoffs. Calculated curve is the least squares fit in $K_p$ and tilt evaluated at $K_p=1$. ($\mu=90-\zeta$, where $\zeta$ is the sun-earth line/dipole axis).

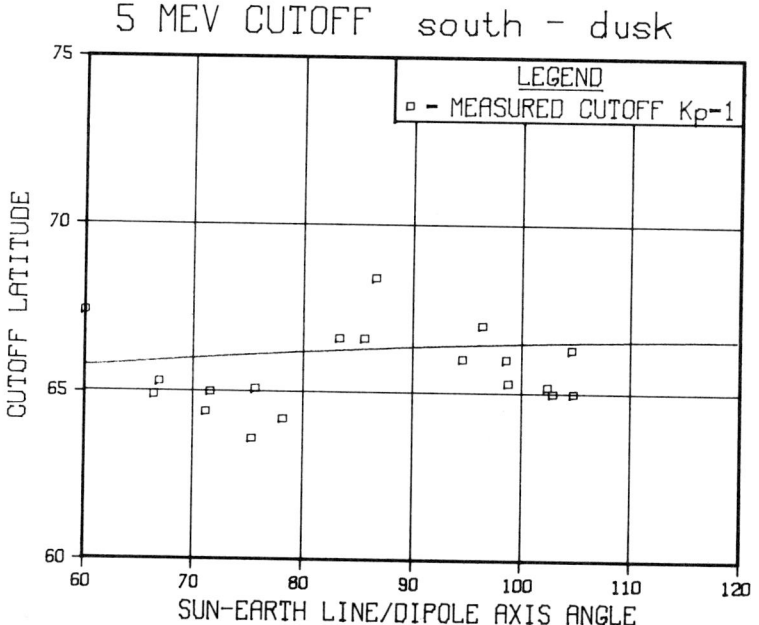

Fig. 8. Calculated and measured cutoffs. Calculated curve is the least squares fit in $K_p$ and tilt evaluated at $K_p=1$. ($\mu=90-\zeta$, where $\zeta$ is the sun-earth line/dipole axis).

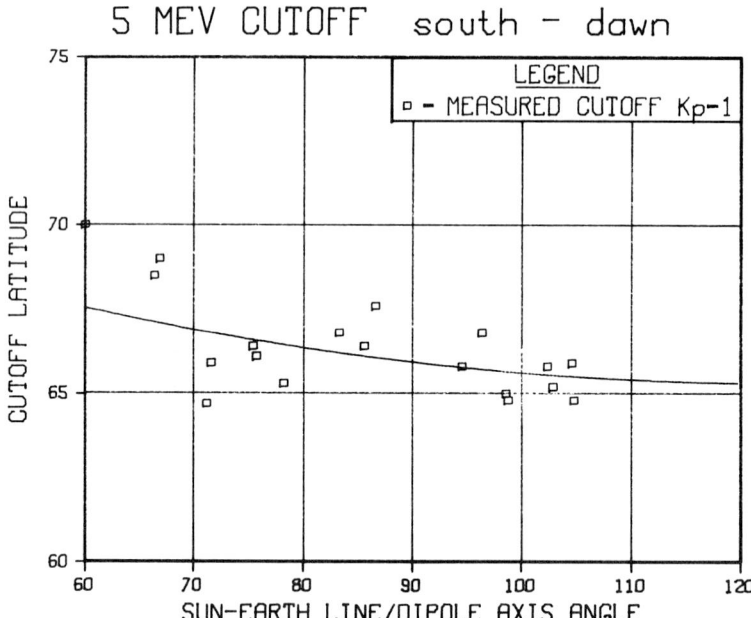

Fig. 9. Calculated and measured cutoffs. Calculated curve is the least squares fit in $K_p$ and tilt evaluated at $K_p=1$. ($\mu=90-\zeta$, where $\zeta$ is the sun-earth line/dipole axis).

however, the index is readily available and does improve the data organization. Also associated with each measurement is the angle, $\mu$, the complement of the angle between the north pointing dipole axis and the sun earth line (plus is toward the sun). The data were further separated into eight bins as to whether the cutoff determination was in the noon, midnight, dawn, or dusk, meridian and whether the data was for the northern or southern polar cap.

For each of these eight locations, the cutoff latitude, $\Lambda_c$, was least squares fitted by a polynomial in $K_p$ and $\mu$. The function

$$\Lambda_c(\mu, K_p) = a_1 + a_2 K_p + a_3 \mu + a_4 \mu K_p + a_5 K_p^2$$

was found to give the best fit to the data consistent with the scatter. In Figures 4 and 5, we show the observational data plotted for $K_p=1+1$, and the function $\Lambda_c$ plotted for $K_p=1$ [i.e., $\Lambda_c(\mu,1)$]. Figures 4 and 5, which are for $K_p=1+1$ in the northern and southern midnight polar caps show an excellent fit with minimal scatter and very little $\mu$ dependence. Figures 6 and 7 which are the cutoffs at noon show a large scatter, indicating that Equation (1) does not organize the data well. This should not be a surprising result since $K_p$ is primarily an index driven by the magnetotail and does not pretend to represent the noon magnetosphere. A better index for noon cutoffs might have been the standoff distance or $D_{st}$. Similar results were obtained at dusk and dawn (Figures 8 and 9).

TABLE 2. Calculated (New Tilt Dependent Model) and Observed (Best Fit for Tilt=0 and $K_p=0$) Cutoffs

|  | Best Fit Observation Tilt=0, $K_p=0$ | Calculation To Nearest Degree Tilt=0, $K_p=0$ |
|---|---|---|
| Noon | 70.4 ± 0.6 | 70 |
| Midnight | 66.4 ± 0.4 | 66 |
| Dawn | 67.5 ± 0.9 | 67 |
| Dusk | 67.5 ± 0.6 | 67 |

In Table 2 we summarize the calculated cutoff using the present tilt dependent model and the least squares value of the observation at $K_p=0$ and tilt equal zero ($\mu=0$). Note that in all cases observation and calculation agree within experimental error. The data also shows that the tilt dependence, if any, is small.

The magnetic activity is not adequately represented by the $K_p$ dependence and hence the small tilt dependences predicted by the calculation cannot be precisely verified by the data. The least amount of scatter is obtained in the midnight meridian where $K_p$ has some physical meaning. It is expected that if the proper organizational parameters were used (one which defines the state of the magnetospheric magnetic field) the predicted tilt dependence at noon is large enough to be easily detected by the OGO-6 data set.

## Summary

In this paper we have shown that by using an accurate model of the earth's magnetic field, the calculated and observed cosmic ray cutoffs agree within experimental error. Much of the early confusion in determining cutoffs and cosmic ray access can be attributed to models of insufficient accuracy. We have also shown that the predicted tilt dependence of the cutoffs is small and that this is consistent with the data. In future studies using organization parameters more representative of the magnetospheric magnetic field than the $K_p$ index used in this study, it should be possible to verify or disprove the small tilt dependence of the cutoffs predicted by the model. We have also briefly outlined the entry regions over the polar cap which helps to explain the high degree of structure in the polar cap during the early phases of solar proton events.

Acknowledgements. This work was supported by the Office of Naval Research under contract N00014-78-C-0215 and also by the Air Force Office of Scientific Research under contract F44620-75-C-0033.

## References

Gill, S., A process for the step-by-step integration of differential equations in an automatic digital computing machine, Proc. Cambridge Phil. Soc., 47, 96, 1951.

Lanzerotti, L. J., Penetration of solar protons and alphas to the geomagnetic equator, Phys. Rev. Lett., 21, 929, 1968.

Masley, A. J., W. P. Olson, and K. A. Pfitzer, Charged particle access calculations, Proc. Int. Conf. Cosmic Rays, 13th, 2, 1973.

Masley, A. J., Charged particle entry into the earth's magnetosphere and propagation to the polar region, Ph.D. Thesis, University of Melbourne, Australia, January 1975.

McDearmid, I. B., J. R. Burrows, and M. D. Wilson, Structure observed in solar particle latitude profiles and its dependence on rigidity, J. Geophys Res., 76, 227, 1971.

Morfil, G., Nonadiabatic particle motion in the magnetosphere, J. Geophys. Res., 78, 558, 1973.

Olson, W. P., A scalar potential representation of the tilted magnetic fields, paper WD-1332, McDonnell Douglas Astronautics Company, Huntington Beach, California, 1970.

Olson, W. P., and K. A. Pfitzer, A quantitative model of the magnetospheric magnetic field, J. Geophys. Res., 79, 3739, 1974.

Pfitzer, K. A., Comparison of the theoretically calculated high energy trapping boundary with experimentally determined values, EOS, Trans. AGU, 53, 1090, 1972.

Sugiura, M., and D. J. Poros, A magnetospheric field model incorporating OGO-3 and 5 magnetic field operations, Planet. Space Sci., 21, 763, 1973.

Taylor, H. E., Latitude local-time dependence of low energy cosmic-ray cutoffs in a realistic geomagnetic field, J. Geophys. Res., 72, 4467, 1967.

Walker, R. J., An evaluation of recent quantitative magnetospheric magnetic field models, Ref. Geophys. and Sp. Res., 14, 411-429, 1976.

Williams, D. J., and G. D. Mead, Night side magnetosphere configuration as obtained from trapped electrons at 1100 kilometers, J. Geophys. Res., 70, 3017, 1965.

OVERVIEW - ELECTRIC FIELDS

D. P. Cauffman

NASA/Headquarters
Washington, DC  20546

## Introduction

The electric fields session was designed to review progress in observations, theory, and modeling of magnetospheric electric fields, and to expose important new results.  It did not attempt to address difficulties in experiment techniques or to expose all new observations; for example, important progress in measurements of ionospheric electric fields by incoherent backscatter radar techniques and by ion drift measurements from the AE satellite were mentioned but not presented. This report will attempt to comment on the state and prospects of electric field research, with particular emphasis on highlights of this meeting and relevance to quantitative modeling of magnetospheric processes.

## Experiments

Experimental research in ionospheric electric fields now is mature. Ion drift and double probe techniques for satellite measurements and use of incoherent backscatter radars for ground-based measurements are now reliable, if not yet continuously available.  Notable successes in this area have been the discovery of the ionospheric convection pattern at high latitudes (Cauffman and Gurnett, 1971) and the relationship of asymmetries in the pattern to interplanetary magnetic field parameters (Heppner, 1972).

In the outer magnetosphere the double probe electric field measurement techniques that are presently available do not give accurate measurements at usefully high sensitivities under all conditions. However, in general, it is possible to distinguish when the measurements are degraded, and an important variety of good data sets are available. As discussed in this session by T. Aggson using IMP data, transient electric fields of as much as one-half volt per meter are found in the magnetosphere.  No completely satisfactory models for these exist yet.  The convection electric field in the outer magnetosphere can also be inferred from particle distribution function measurements. This technique was covered in another session and is not discussed here.

New results from ISEE presented by T. Aggson and GEOS presented by Pedersen and Grard represent a marked improvement and are just beginning to have an impact.  Within the next year comparisons should be available to evaluate the tradeoffs between two alternative experimental approaches tested on the ISEE-1 satellite; specifically, long cylindrical sensors

(Heppner et al, 1978) and shorter-baseline spherical ones employing guard rings and active biasing (Mozer, et al, 1978). Finally, there is a new technique using particle beams (Melzner et al, 1978) that has operated successfully on GEOS-2, which also carried double probes (Knott, 1975). Preliminary indications are that the two techniques agree (A. Pedersen, private communication, 1978). Although it has not yet been fully evaluated, there is some reason to believe that the new technique may offer a desirable alternative in some situations since the constraints are very different from those on the double probe technique. (The double probe technique is discussed by Fahleson, 1967).

Tables 1, 2, and 3 summarize electric field data sources: past, current, and planned. The data sets vary greatly in quality and availability. It is wise to enlist the aid of the principal investigators for validation. Particularly important data sets for future use include those from the EXOS-B, SCATHA, and Dynamics Explorer satellites. The Origins of Plasma in the Earth's Neighborhood (OPEN) mission will be the first opportunity for obtaining simultaneous high-quality data sets in various key regions of the magnetosphere.

## Theories and Models

Theories of electric fields in the magnetosphere are most developed for electric fields perpendicular to the geomagnetic field in the inner magnetosphere. Viable mechanisms have been proposed for solar wind/magnetosphere interactions and for ionosphere/magnetosphere interactions, but the former are limited to simple geometries and the latter are not yet capable of predicting the characteristics of electric fields parallel to the geomagnetic field. Most of the discussion of electric field theoretical work in this session was in the context of models.

Modeling is the art that casts theory in a form in which useful, if not realistic, boundary conditions can be imposed so that calculated predictions can be compared with experimental data. In this session T. Hill reviewed the status of models of the magnetospheric electric field. He reported that in the inner magnetosphere, calculation of the perpendicular electric field is presently limited less by the approximations involved in the governing equations than by the accuracy of the inner and outer boundary conditions. At both the ionosphere/magnetosphere and the solar wind/magnetosphere boundaries, inadequate theory limits quantitative modeling, although empirical models of the ionosphere are being constructed that may provide an adequate boundary condition for the perpendicular electric field in the magnetosphere. Quantitative models of the magnetotail are limited by lack of understanding of the dynamics of the tail. In summary, although recent years have seen initial efforts to apply modeling to magnetospheric electric fields, this use must be characterized as being in its infancy.

Several important new contributions to electric field modeling were presented in this session. H. Volland provided a framework for modeling global ionospheric electric fields. Hays and Roble modeled the transmission of ionospheric potentials to lower altitudes, an important accomplishment that gives a new perspective on space plasma physics and that this author predicts will form a cornerstone in future efforts to construct models of solar-terrestrial coupling. Mroz, et al., utilized the new Olson-Pfitzer magnetic field model formulation to compute the

TABLE 1. Past DC Electric Field Measurements

| Spacecraft Orbit (Km x Km, deg) | Launch (Mo/Da/Yr) | Stabili-zation [1] | Investigator Institution | Type [2] | Components | Axes | Base-line (m) | Sensi-tivity (mV/m) |
|---|---|---|---|---|---|---|---|---|
| Injun 5 (Expl 40) 2528 x 677 81° | 8/8/68 | mag. | Gurnett U. Iowa | SDP | 2 | 1 | 2.85 | 10 |
| OGO-6 1600 x 400 82° | 6/5/69 | 3AI | Aggson GSFC | CDP | 1 | 1 | 18.4 | 1.0 |
| IMP-6(I) (Expl 43) 31$R_E$ x 9865 37.7° | 3/13/71 | spin | Aggson GSFC | CDP | 3 | 3 | 75 35 5 | 0.1 |
| S$^3$A (Expl 45) 5.2$R_E$ x 200 3° | 11/15/71 | spin | Maynard GSFC | SDP | 1 | 1 | 5 | 0.2 |
| IMP-8(J) (Expl 50) 46$R_E$ x 22$R_E$ 28.7° | 10/26/73 | spin | Aggson GSFC | CDP | 2 | 1 | 106 | 0.1 |
| Hawkeye 1 (Expl 52) 20$R_E$ x 469 89.8° | 6/3/74 | spin | Gurnett U. Iowa | CDP | 2 | 1 | 43 | 0.1 |

[1] mag. - magnetically oriented
3AI - three axis inertial

[2] SDP - spherical double probes
CDP - cylindrical double probes

TABLE 2. Current DC Electric Field Measurements

| Spacecraft Orbit | Launch (Mo/Da/Yr) | Stabili-zation | Investigator Institution | Type[2] | Components | Axes | Baseline (m) | Sensitivity (mV/m) | Comments |
|---|---|---|---|---|---|---|---|---|---|
| S3-2 (S73-6) 900 x 230 97° | 12/3/75 | spin | Smiddy AFGL | SDP | 3 | 3 | 28 28 11 | 0.2 | |
| S3-3 (S74-2) 7856 x 246 97.5° | 7/8/76 | 3AI | Mozer UC Berkeley | SDP | 3 | 3 | 40 40 6 | 1.0 | |
| GEOS-1 $7.1R_E$ x 2000 26° | 4/20/77 | spin | Pedersen ESTEC | SDP | 2 | 1 | 42 | 0.3 | conductive s/c |
| GEOS-1 $7.1R_E$ x 2000 26° | 4/20/77 | spin | Melzner MPI Garching | Beam | 2 | n.a. | (1000) | 0.05 | 0.4-2.3 KV electrons |
| ISEE-1 $23R_E$ x 281 29° | 10/22/77 | spin | Mozer UC Berkeley | SDP | 2 | 1 | 73.5 | 0.1 | conductive s/c |
| ISEE-1 $23R_E$ x 281 29° | 10/22/77 | spin | Heppner GSFC | CDP | 2 | 1 | 179 | 0.1 | conductive s/c |
| GEOS-2 $6.6R_E$ x $6.6R_E$ 1° | 7/14/78 | spin | Pedersen ESTEC | ADP | 2 | 1 | 42 | 0.1 | conductive s/c |
| GEOS-2 $6.6R_E$ x $6.6R_E$ 1° | 7/14/78 | spin | Melzner MPI Garching | Beam | 2 | n.a. | (1000) | 0.05 | 0.4-2.3 KV electrons |
| EXOS-B $4.8R_E$ x 230 31° | 9/16/78 | spin | Obayashi U. Tokyo | CDP | 2 | 2 | 108 | 0.1 | conductive s/c |

TABLE 3. Planned DC Electric Field Measurements

| Spacecraft Orbit | Launch (Mo/Yr) | Stabili-zation | Investigator Institution | Type[2] | Compo-nents | Axes | Base-line (m) | Sensi-tivity (mV/m) | Comments |
|---|---|---|---|---|---|---|---|---|---|
| SCATHA (P78-2) $6.8R_E \times 4.4R_E$ 2.5° | 1/79 | spin | Aggson GSFC | CDP | 2 | 1 | 100 | 0.1 | conductive s/c |
| OFT-4 PDP 420 x 420 57° | 10/80 | 3AI | Shawhan U. Iowa | SDP | 1 | 1 | 4 | 2.0 | |
| Spacelab 2 PDP 420 x 420 57° | 4/81 | spin | Shawhan U. Iowa | SDP | 2 | 1 | 4 | 2.0 | |
| SM D/L 1100 x 290 2.9° | 3/81 | spin | Maynard GSFC | CDP | 3 | 3 | 35 33 6 | 0.05 | |
| DE-A $4.8 R_E \times 350$ 90° | 2/81 | spin | Shawhan U. Iowa | CDP | 3 | 2 | 172 6 | 20.0 | |
| DE-B 1200 x 275 90° | 2/81 | 3AI | Maynard GSFC | CDP | 3 | 3 | 20 20 20 | 0.1 | |
| OPEN-B (GTL) $240R_E \times 10R_E$ | 1/85 | | To be selected | | | | | | AO 6-79 |
| OPEN-C (PPL) $15R_E$ or $4R_E \times 300$ 90° | 7/85 | | To be selected | | | | | | AO 6-79 |
| OPEN-D (EML) $12R_E \times 2R_E$ 0° | 1/86 | | To be selected | | | | | | AO 6-79 |

electric field induced by temporal variation of the magnetic vector potential.

Several papers that were actually presented in other sessions should be mentioned for their significance for electric field modeling. E. Whipple introduced a general formulation involving the kinetic approach that greatly simplifies understanding the characteristics of particle populations throughout the magnetosphere. P. Smith presented a movie, showing simulated particle trajectories, that serves a similar purpose and demonstrates particle trapping, ring current formation and decay. J. Kisabeth modeled Birkeland currents using distributed currents instead of sheets, a major step toward realism. Leboeuf, et al., described a time-dependent two dimensional MHD simulation of solar wind interaction with a magnetic dipole that shows an unsteady magnetopause and that predicts turbulence in the magnetosheath.

## Modeling Philosophy

Several thoughts on what might best be termed the philosophy of modeling, that were stimulated by the discussion at the conference and from conversations that resulted from it, may be of interest to the reader.

### Explanatory Models

It is useful to distinguish between models that "explain" nature and those that "represent" nature. The goal of the physicist studying a physical system is to "explain" it by creating mathematical expressions to describe its characteristics and behavior--ideally, expressions that embody general, fundamental physical principles with a plausible, rigorously traceable set of assumptions leading to the equations employed, and that contain parameters whose values can be predicted once boundary conditions involving an equal number of unknowns are established. Creators of these _explanatory_ _models_ tend to push difficulties into the specification of the boundary conditions, and to produce explanations of simple special cases that may not be physical. However, they are desirable because they are instructive, and because generalization is straightforward (in principle)--one relaxes an assumption and calculates the consequences. Thus, complexity grows, built on a firm foundation.

In space plasma physics there is hope that fluid and kinetic modeling approaches may both reach the stage where realistic simulations based on explanatory models may be performed, involving three dimensional codes containing all the relevant physics run on appropriately dense grids with appropriately small time steps. These will assuredly be useful in studying particular features of the magnetosphere. Modeling the total magnetosphere using these approaches may be resource (i.e., computer time) limited.

### Representative Models

When confronted by a complex system such as the magnetosphere, it is not always possible to proceed directly to a description that satisfies the definition of an explanatory model given above. A useful intermediate step is the _representative_ _model_ that uses a different approach

to predicting the behavior or characteristics of a system. To build a
representative model, one characterizes each element of the system by a
plausible set of equations, further defines relationships between the
system elements, chooses dependent and independent variables, and uses
actual data, typical values, or average properties to give values to
the remaining parameters. The solutions that result, if in agreement
with observations, are said to "represent" the physical situation but
not, in general, to "explain" it, due to the following limitations
inherent in the method: (1) since the physics included was not derived
from first principles, it may be difficult to establish that all of the
right physics is embodied in the equations, and (2) since the number of
parameters can greatly exceed the number of unknowns, solutions will not,
in general, be unique. Nevertheless, the model may be of significant
use in the following applications: (1) to perform "sensitivity" studies
to identify key or ignorable parameters; or (2) to determine the consequences of various assumptions. For example, it may become obvious from
a representative model how to make assumptions that simplify the model,
or how model parameters are related to changes in boundary conditions.
The latter can guide experiments; the former aids in making the proper
assumptions for a rigorous theoretical framework. Representative models
are therefore useful as a way to make theory confront observations at
an earlier stage than otherwise possible when dealing with complex
systems. Used with care and discipline, a representative model should
evolve toward an explanatory model.

## Summary

Modeling of magnetospheric electric fields, while in its infancy,
is developing rapidly on many fronts employing a variety of approaches.
In the very near term the use of sophisticated graphics and extensions
of new concepts discussed in this session are predicted to result in
models mature enough to characterize realistically ionospheric and
auroral zone electric fields. In a few years simultaneous high-quality
data sets will be needed to test these predictions.

The general topic of magnetospheric electric fields is becoming of
central importance in understanding space plasmas. The conference
brought out clearly that lack of understanding of time varying current
systems is now limiting progress in magnetic field modeling. Understanding the origins and behavior of the electric fields that drive
these systems will be a major step forward. At present the electric
field is probably the most poorly measured or understood fundamental
parameter of the magnetosphere.

Acknowledgements. The author is indebted to R. Wolf co-chairman
of the electric fields session, for his work in organizing it, to
R. Wolf and W. Olson for reviewing this manuscript, and to the investigators listed in Tables 1, 2, and 3 for providing most of the information shown there. Discussions with R. Wolf concerning modeling
philosophy were particularly helpful. As the representative of one of
the sponsoring agencies, the author would also like to thank W. Olson
and his co-workers, including the AGU staff, for their work in organizing
the conference and this document. To the speakers and writers, however,
must go the ultimate credit for the quality and usefulness of the product.

This work was prepared as part of the author's official duties as an employee of the U. S. Government and, in accordance with 17 USC 105, is not available for copyright protection in the United States.

## References

Cauffman, D. P., and D. C. Gurnett, Double probe measurements of convection electric fields with the Injun 5 satellite, J. Geophys. Res. 76, 6014-6027, 1971

Fahleson, U., Theory of electric field measurements conducted in the magnetosphere with electric probes, Space Sci. Rev. 7, 238-262, 1967.

Heppner, J. P., Polar cap electric field distributions related to the interplanetary magnetic field direction, J. Geophys. Res. 77, 4877-4887, 1972.

Heppner, J. P., E. A. Bielecki, T. L. Aggson, and N. C. Maynard, Instrumentation for DC and low-frequency electric field measurements on ISEE-A, IEEE Trans. on Geoscience Electronics GS-16, 253-257, 1978.

Knott, K., ESA Scientific and Technical Review 1, 173-196, 1975.

Melzner, F., G. Metzner, and D. Antrack, The GEOS electron beam experiment S329, Space Sci. Instrumentation, in press, 1978.

Mozer, F. S., R. B. Torbert, U. V. Fahleson, C. G. Falthammar, A. Gonfalone, and A. Pedersen, Measurements of quasi-static and low-frequency electric fields with spherical double probes on the ISEE-1 spacecraft, IEEE Trans. on Geoscience Electronics GS-16, 258-261, 1978.

# SEMIEMPIRICAL MODELS OF MAGNETOSPHERIC ELECTRIC FIELDS

Hans Volland *

Radioastronomical Institute, University of Bonn
5300 Bonn, W.-Germany

Abstract. Electric fields are generated by the interaction between solar wind and the magnetosphere. These fields drive ionospheric and field-aligned electric currents. Three major types of global scale fields of magnetospheric origin have been identified: the convection field, the corotation field, and the polar cap field. The discovery of these fields as well as our present understanding of their spatial distribution is based on several direct and indirect observations, e.g., in situ measurements at ionospheric heights of electric fields and field-aligned electric currents, drift measurements, ground based magnetic observations, and the measurement of the structure of the plasmapause. Based on that knowledge, it is possible to construct simple semiempirical analytic models of electric potentials from which the electric field and current configurations can be derived uniquely if the electric conductivity of the dynamo layer is known. In this paper, simulations of electric field and current configurations of the three types of fields mentioned above are reviewed.

## Introduction

Large-scale electric fields and currents play a decisive role in the interaction between the solar wind and the magnetosphere, in particular in the transfer of energy and particles from the solar wind to the magnetosphere, and in the transport and acceleration of such particles within the magnetosphere itself. It is possible to discriminate between three major types of fields:

a) a convection field directed from dawn to dusk (Axford and Hines, 1961), and related with field-aligned electric currents within the auroral ovals (Zmuda and Armstrong, 1974). That field is apparently tied to the north- south component of the interplanetary magnetic field (Dungey, 1961),

b) a corotation field generated by unipolar induction within the corotating magnetospheric plasma (Alfvén, 1950). This field depends on the frame of reference and disappears for observers moving with the earth,

---

* This work has been performed while the author was at the Space Physics Research Laboratory, Department of Atmospheric and Oceanic Sciences, The University of Michigan, Ann Arbor, Michigan, 48019.

c) a polar cap field (Heppner, 1972) which is associated with ionospheric Hall currents encircling the poles (Svalgaard, 1968) and which is connected with the east-west component of the interplanetary field (IMF) (Friis- Christensen and al.,1972).

The physical processes involved in the generation of the polar cap field and of the convection field are still not well understood . Moreover, a self-consistent theory of plasma convection within the magnetosphere does not yet exist. Therefore, in order to describe magnetospheric electric fields, it is often convenient to eliminate a detailed consideration of magnetospheric plasma convection and to start with simplified models of electric field or current configurations (Vasylinas,1972). This is possible for stationary large-scale fields with periods larger than several hours. In this case, the electric field can be derived from a quasi-static potential $\Phi$ .

Electric potential fields of that kind have been developed by McIlwain (1974) from observed particle fluxes at synchronous orbit during and after substorms, and by Heppner (1977) from in situ electric field measurements on board of OGO-6. Numerical stationary models of ionospheric electric fields and currents based on observed field-aligned currents during quiet and disturbed conditions are due to Yasuhara et al. (1975). Yasuhara and Akasofu (1977), Nisbet et al.(1978), Nopper and Carovillano (1978), and Kamide and Matsushita (1978). Models inferred from ground based and satellite magnetic observations have been constructed by Hughes and Rostoker (1978). Analytic models of the magnetospheric convection field have been developed by Kawasaki and Fukushima (1974), Stern (1975), Gurevich et al. (1976), and Volland (1978). The polar cap field was modelled by Leontyev and Lyatsky (1974) and Volland (1975). The electric corotation field within a magnetic field of the Mead-type has been calculated by Schulz (1970).

In this paper, I shall review stationary models of the three types mentioned above which are based on simple analytic electric potential configurations. The electric field structure, the electric currents within the dynamo region, the field-aligned electric currents at the top of the dynamo layer, and the magnetic effects of the currents at the ground which can be derived from such potential fields, are compared with observations.

Time dependent solutions simulating substorm conditions or small-scale fields including inductive fields or parallel fields are not treated in this paper. Here, the reader is refered to Murphy et al. (1974), Heikkila and Pellinen (1977), Wolf (1975), and Harel et al. (1979) and to the review articles of Stern (1977), Fälthammar (1977), and Rostoker (1978).

## Basic theoretical concept

With the assumption that the geomagnetic field lines are electric equipotential lines outside the dynamo region, it is possible to map the electric potential $\Phi$ from the ionosphere into the magnetosphere and vice versa. From such potential fields, the electric currents as well as the field-aligned currents at the top of the dynamo region can be derived provided the total electric current is considered as source free, and the electric conductivity tensor $\underline{\sigma}$ of the dynamo region is known. The three-dimensional electric current configuration determines then the magnetic field observed at the ground (e.g. Richmond, 1974; Kisabeth, 1979).

Expressed in mathematical terms, I start from the two Maxwell-equations and Ohm's law, neglecting the time derivatives of the magnetic induction $\underline{B}$ and of the electric field $\underline{E}$:

$$\nabla \times \underline{E} = 0 \qquad (1)$$

$$\underline{j} = \underline{\sigma} \cdot (\underline{E} + \underline{U} \times \underline{B}_o) \qquad (2)$$

$$\nabla \times \underline{B} = \mu_o \underline{j} \qquad (3)$$

so that a chain of cause and events

$$\underline{E} \rightarrow \underline{j} \rightarrow \underline{B} \qquad (4)$$

exists. $\underline{j}$ is the electric current density, $\underline{U}$ the thermospheric wind, $\underline{B}_o$ the main geomagnetic field, and $\underline{B}$ the magnetic field due to the current ($B \ll B_o$). From (1) follows the potential field

$$\underline{E} = -\nabla \Phi \qquad (5)$$

The dynamo layer is approximated by a thin sheet at radial distance $r_o$. Above the dynamo layer the geomagnetic field lines as good electric conductors imply the orthogonality between $\underline{E}$ and $\underline{B}_o$:

$$\underline{E} \cdot \underline{B}_o = 0 \qquad (r > r_o) \qquad (6)$$

This allows mapping of $\underline{E}$ within magnetosphere and ionosphere. From the condition of a source free current, $\nabla \cdot \underline{j} = 0$, follows the field-aligned electric current at the top of the dynamo layer:

$$j_{||} \Big|_{r=r_o} = -\nabla \cdot \underline{I}_h / \sin\chi \qquad (7)$$

with

$$\underline{I}_h = \underline{\Sigma} \cdot (\underline{E} + \underline{U} \times \underline{B}_o) \qquad (8)$$

the horizontal height integrated dynamo current at altitude $r_o$. The electric conductivity tensor (already simplified) is

$$\underline{\Sigma} = \begin{pmatrix} \Sigma_p & \Sigma_h \\ -\Sigma_h & \Sigma_p \end{pmatrix} \qquad (9)$$

with height integrated Pedersen ($\Sigma_p$) and Hall ($\Sigma_h$) conductivity. $\chi$ is the angle of inclination of the geomagnetic field $B_o$. As boundary conditions, I assume

$$\Phi, I_\theta \quad \text{continuous} \tag{10}$$

where $I_\theta$ is the meridional component of $\underline{I}_h$ within the dynamo layer. Energy is transfered to the neutral thermosphere by these electric dynamo fields and currents according to

$$q = \underline{I}_h \cdot \underline{E} \tag{11}$$

(height integrated Joule heating), while the electric currents act like a momentum force to drive thermospheric winds:

$$\underline{m} = \underline{I}_h \times \underline{B}_o \tag{12}$$

(height integrated Ampere force).

As simplifications applied in the further approach, I neglect the Lorentz field $\underline{U} \times \underline{B}_o$ in (8). That Lorentz field can be subsequently determined from the impact of Joule heating and Ampere force on thermosphere dynamics (Volland, 1978a) and can be shown to be a secondary effect at least within the auroral ovals (Banks, 1977). Furthermore, I treat the geomagnetic field as inclined dipole field so that the solutions are valid only within the inner magnetosphere up to distances of about $r < 10\ r_o$. I assume the auroral zones to be of circular shape and centered at the magnetic poles. Finally, I assume constant electric conductivity elements $\Sigma_p$ and $\Sigma_h$, respectively, however with different values within the auroral zones and outside the auroral zones. These assumptions are made in order to obtain simple and tractable analytic solutions. They all can be relaxed, however, in more sophisticated numerical calculation; e.g., in section 5, I shall consider temporally and spatially dependent conductivities in order to determine the electric currents associated with the polar cap field.

I start in my treatment with the electric field which is considered as the external driving mechanism. Another approach is to reverse the chain in (4) to $\underline{j} \rightarrow \underline{E}$. That approach has been used by Yasuhara and Akasofu (1977), Nisbet et al.(1978), Kamide and Matsushita (1978), and by Nopper and Carovillano (1978). These authors start with the observed field-aligned electric currents of Iijima and Potemra (1976) and calculate $\underline{I}_h$ and $\underline{E}$ from equations (5) to (10). Although correct in principle, that approach has one disadvantage: the measurement of the field-aligned currents is by an indirect method with substantial error bars (only the magnetic fields due to these currents can be observed), whereas the electric fields are observed directly with much greater accuracy (Heppner, 1972; Cauffman and Gurnett, 1972).

## The convection field

The close relationship between the convection field and the north-south component of the IMF makes it plausible that the magnetosphere is partly open at least within the polar cap regions. Figure 1, adopted from Stern (1973), shows schematically a cross section of the geomagnetic dipole field immersed in the solar wind flow during away

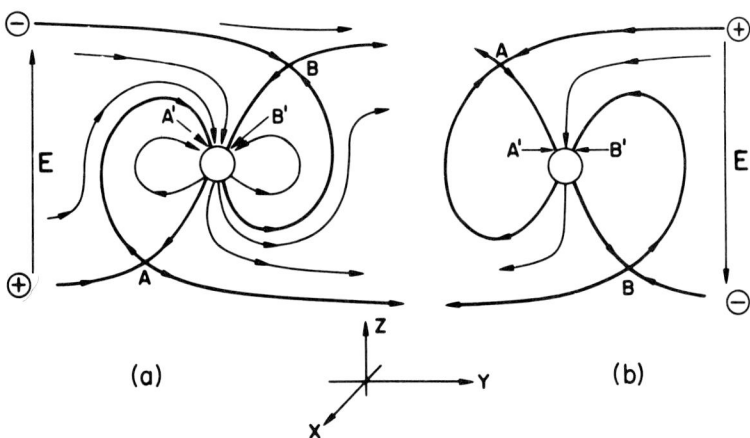

Fig. 1. Schematic view of the field of a dipole immersed in a constant field orthogonal to its axis with the solar wind moving into the plane of the drawing. (a) External field parallel to y axis. (b) External field antiparallel to y axis. (Adopted from Stern, 1973.)

(left) and during toward(right) polarity of the IMF as viewed from the sun. In both cases, the electric field $\underline{E} = -\underline{V} \times \underline{B}_i$ ($\underline{V}$: solar wind velocity; $\underline{B}_i$: IMF) of the solar wind is directed from dawn to dusk within the polar cap regions. This electric field is produced by a slight charge separation within the plasma so that it can cancel the Lorentz field $\underline{V} \times \underline{B}_i$. However, the open field lines which are good electric conductors are electrically connected with each other via the ionospheric dynamo layer. Therefore, electric currents flow along the open field lines and within the dynamo layer on both hemispheres in order to neutralize the polarization charge. These currents are directed into the ionosphere on the dawn side and out of the ionosphere on the dusk side. However, the solar wind plasma has a large dielectric constant, and the charge is immediately restored by a polarization current. Stern (1973) has estimated that the solar wind which passes through that magnetospheric window does not loose more than about 2 % of its kinetic energy to move the thermosphere via the Ampere force and to heat the thermosphere by Joule heating, and the primary electric field is reduced by not more than about 1 %.

The MHD-generator "solar wind–magnetosphere" behaves therefore like a current generator because it contains an internal resistance which is large compared with the load resistence "ionosphere", and the feedback from the ionosphere to the primary electric field is probably small. However, secondary electric polarization fields can be generated within the magnetosphere due to strong inhomogenities of the ionospheric electric conductivity within the auroral zones or near the food points of the plasmapause at ionospheric heights.

Sweeping of the field lines from the day side to the night side and reconnection there, together with mass conservation, causes a sunward flow of magnetospheric plasma within the area of closed field lines (Dungey, 1963). This flow is associated with an electric field within the inner magnetosphere which is also mainly directed from dawn to dusk. Therefore, at the border between open

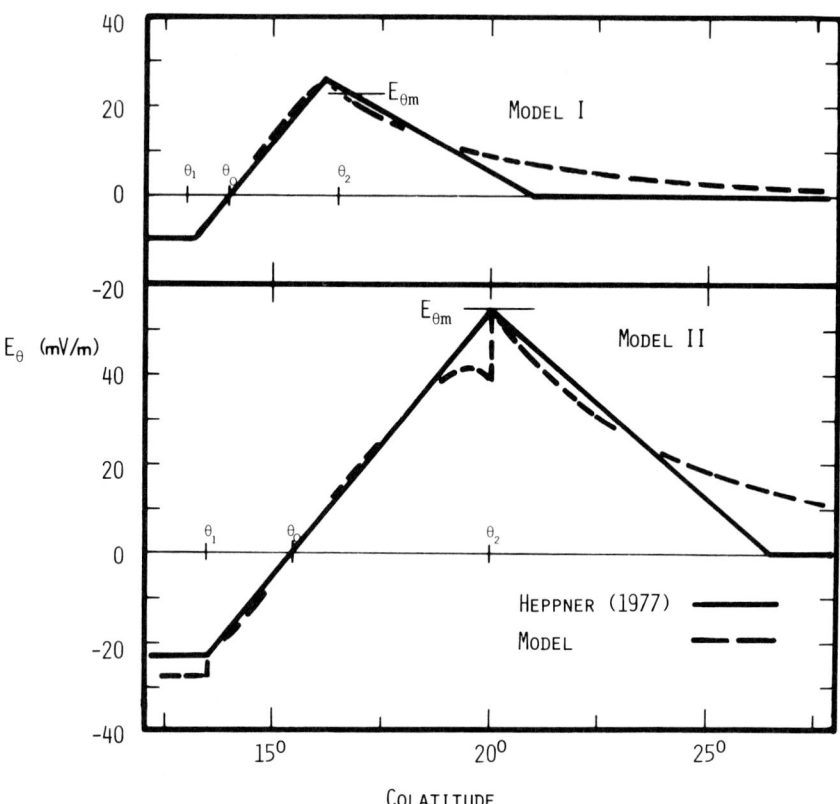

Fig. 2. Meridional electric field strength at ionospheric heights during dawn ($\tau_m$ = 1800 LT) versus co-latitude. The solid lines are taken from Heppner (1977) and show two typical configurations which are the averages from dawn and dusk during quiet and moderately disturbed conditions (model C(o) and model A from Heppner). The dashed lines give model simulations. The numbers $\theta_1$, $\theta_0$, $\theta_2$, indicate the range of the transition from negative to positive values of $E_\theta$, and the location of the reversal.

and closed field lines a reversal of the electric field must occur.

The sunward flow of the plasma is responsible for charge separation on the dusk and on the dawn side of the partial ring current which generates an additional secondary electric field. That field is directed from dusk to dawn and weakens the primary field at lower latitudes (Schield et al., 1969; Wolf, 1975). Charge neutralization is provided by two pairs of field-aligned electric currents which flow in opposite direction to, and equatorward of, the primary field-aligned currents.

There exist reasons in favour of the argument that another mechanism may act additionally to generate the convection field. It is loosely called viscosity-like interaction and may cause charge se-

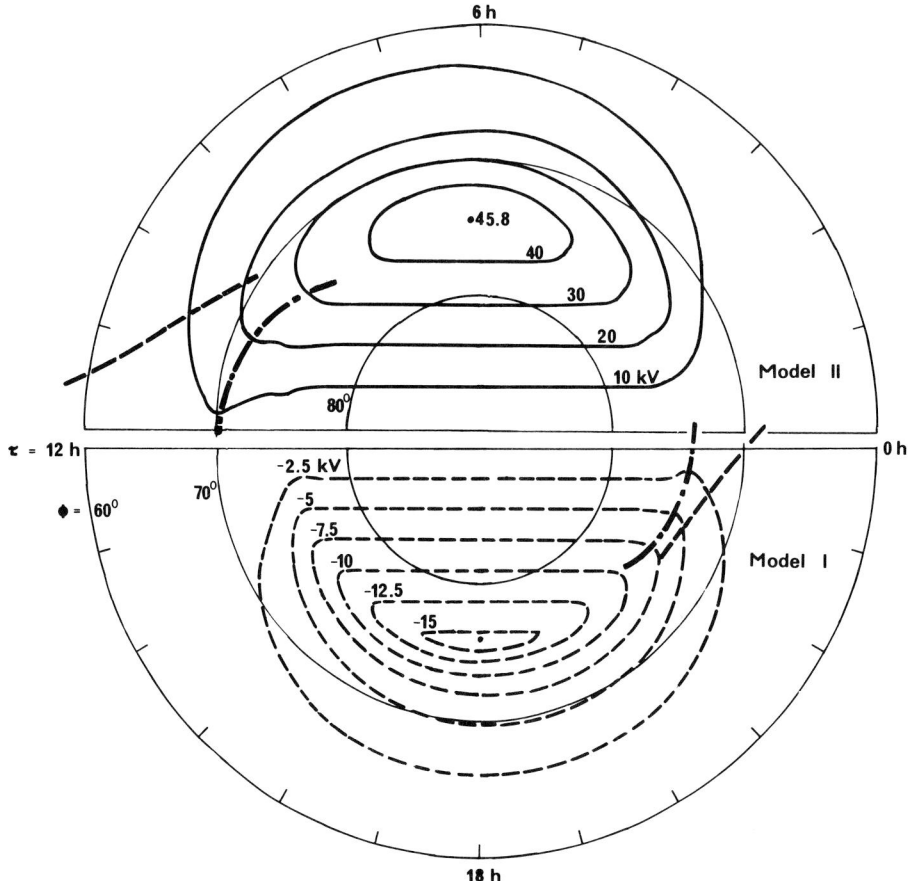

Fig. 3. Equipotential lines of the electric potential at ionospheric heights on the northern hemisphere versus local time and latitude determined from the two models of Figure 2. Upper half of the figure gives the configuration of model II on the morning side ; lower half gives the configuration of model I on the evening side. The thick dash-dot lines indicate the locations of zero meridional field strength. The thick dashed lines are the locations of the Harang discontinuity taken from Maynard (1974).

paration within the boundary of the magnetosphere (Axford, 1969; Coleman, 1971). That mechanism generates essentially the same field configuration as the MHD generator effect.

The field configuration outlined above is just the field which Heppner (1972, 1977), Cauffman and Gurnett (1972), and Heelis et al. (1976a) observe at higher latitudes and at ionospheric heights. Figure 2 shows two typical cases of Heppner's (1977) observations of the meridional component $E_\theta$ at dawn versus invariant colatitude on the northern hemisphere for quiet conditions (Kp ≃ 0) and for moderately disturbed conditions (Kp ≃ 2 - 5). This field remains nearly constant over the polar cap, reverses within the auroral ovals, and decays rapidly beyond the lower latitude border of the auroral ovals. With increasing geomagnetic activity, the amplitude increases, and the transition zone widens and shifts to lower latitudes. But the general form remains unaltered.

Sometimes, intense electric fields are observed near the ionospheric projection of the plasmapause in pre-midnight regions (Heelis et al.,1976b; Smiddy et al.,1977; Maynard, 1978). Probably these intense fields are coupled with sharp decreases in the mid-latitude trough F2 layer electron density (Köhnlein and Raitt, 1977). Calculations by Banks and Yasuhara (1979) suggest that the field-aligned electric currents in those regions remain small. Therefore the enhanced electric field can be considered as a secondary field which is generated in order to compensate the decreased electric conductivity so that the meridional ionospheric current remains continuous (Boström, 1964).

Applying equations (5) to (10), the field configurations of Figure 2 can be simulated by the following set of analytic formulae for the electric potential (Volland, 1978b):

$$\Phi_c = -iA_j \left\{\frac{L}{L_j}\right\}^{p_j/2} \exp(i\tau) \tag{13}$$

for $\theta \leq \theta_1$ ($j=1$) ; $\theta \geq \theta_2$ ($j=2$)

and

$$\Phi_c = -i\left\{B_j + \left|1-L/L_j\right|\left\{1-\frac{2}{2+n_j}\left|\frac{1-L/L_j}{1-L_o/L_j}\right|^{n_j/2}\right\}C_j\right\}\exp(i\tau) \tag{14}$$

for $\theta_1 \leq \theta \leq \theta_o$ ($j=1$) ; $\theta_o \leq \theta \leq \theta_2$ ($j=2$).

Here, it is

$$L = \frac{r}{r_o \sin^2\theta} \quad ; \quad L_j = \frac{1}{\sin^2\theta_j} \quad (\theta_j = \theta_1, \theta_o, \theta_2)$$

$$n_1 = 4 \quad ; \quad n_2 = 1.5\,(1+i) \quad ; \quad p_1 = -1 \quad ; \quad p_2 = 4$$

$$B_1 = A_1 \quad ; \quad C_1 = \bar{p}_1 A_1/2 \quad ; \quad B_2 = 2C_2/\bar{p}_2 = A_2$$

$$C_2 = -\frac{2/\bar{p}_1 + (L_o/L_1-1)n_1/(2+n_1)}{2/\bar{p}_2 + (L_o/L_2-1)n_2/(2+n_2)} C_1$$

$$\bar{p}_1 = \frac{\Sigma_p p_1 + i(\Sigma_h - \bar{\Sigma}_h)/\cos\theta_1}{\Sigma_p} \quad ; \quad \bar{p}_2 = \frac{\Sigma_p p_2 + i(\Sigma_h - \bar{\Sigma}_h)/\cos\theta_2}{\Sigma_p}$$

$\Sigma_p$, $\Sigma_h$ and $\bar{\Sigma}_p$, $\bar{\Sigma}_h$ are Pedersen and Hall conductivity within and outside the auroral zones, respectively. r is the radial distance, $r_o$ the distance of the dynamo layer, $\theta$ the geomagnetic colatitude, $\tau$ the magnetic local time, L the shell parameter. $\phi(L)$ is parallel to the L-shell and therefore fulfills condition (6). The auroral zone lies within the boundaries $(\theta_1, \theta_2)$. $\theta_o$ is the colatitude of zero meridional field strength (see Figure 2). Since $\phi$ is complex, only the real part

$$\text{Real } (\phi) = |\phi| \cos (\tau - \tau_m) \tag{15}$$

has physical meaning ($|\phi|$ is the magnitude of $\phi$, $\tau_m$ is the local time of maximum). Furthermore, the potential is symmetric with respect to the equator, so that only the northern hemisphere shall be considered in the following.

In order to simulate the observations of Heppner (1977), I did two model calculations. In model I, valid during quiet conditions, the conductivities were assumed to be constant everywhere with the values $\bar{\Sigma}_p = \Sigma_p = 15$ mhos; $\bar{\Sigma}_h = \Sigma_h = 10$ mhos, and $A_1 = 14.8$ kV. In model II, valid during moderately disturbed conditions, I added a region of enhanced conductivities at the location of the auroral zone between $\theta_1$ and $\theta_2$ with values $\Sigma_p = 22$ mhos; $\Sigma_h = 24$ mhos, leaving $\bar{\Sigma}_p$ and $\bar{\Sigma}_h$ as in model I, and $A_1 = 44.1$ kV.

Note that $A_1$ is the only remaining free parameter in (13) and (14) which together with the number $p_1 = -1$ in (13) determines the primary driving field within the polar caps. Its structure corresponds to an asymptotically uniform field across the polar cap regions of the outer magnetosphere (Hill and Rassbach, 1975; Schulz, 1976). The choice of $p_2 = 4$ was necessary to simulate the observed configuration of the plasmapause and of the $S_q^p$ current (Volland, 1975).

The dashed curves in Figure 2 derived from the model calculations show the meridional electric field strength during dawn ($\tau_m = 18$ LMT). We notice discontinuities of $E_\theta$ at the borders of the auroral zone at $\theta_1$ and $\theta_2$ in model II. These discontinuities result from the discontinuity of the conductivities and indicate the existence of a secondary electric polarization field. Such field is generated to maintain the continuity of the meridional electric current (Boström, 1964). It weakens the total field within the auroral zones and strengthens it within lower latitudes. It is responsible for the increase of the effective conductivity (Cowling conductivity) within the auroral zones and thus for the strong polar electro jets (Fukushima, 1971). In model I, it was assumed that $\Sigma_p = \bar{\Sigma}_p$. Therefore, no discontinuity in $E_\theta$ occurs.

Figure 3 shows the equipotential lines of the electric potential of the two models plotted versus local magnetic time within the northern hemispheric ionosphere. Since the model electric field is harmonic with a period of one day, the evening pattern is a repetition of the morning pattern, apart from a sign reversal. Therefore, only the evening pattern for model I, and only the morning pattern of model II is shown.

According to Fukushima (1971), the magnetic effects on the ground due to overhead currents represent mainly Hall currents because the magnetic effect of the Pedersen current within the ionosphere is compensated to a large extent by the magnetic effect of the field-aligned current. Therefore, Figure 3 also represents in a first approximation the equiva-

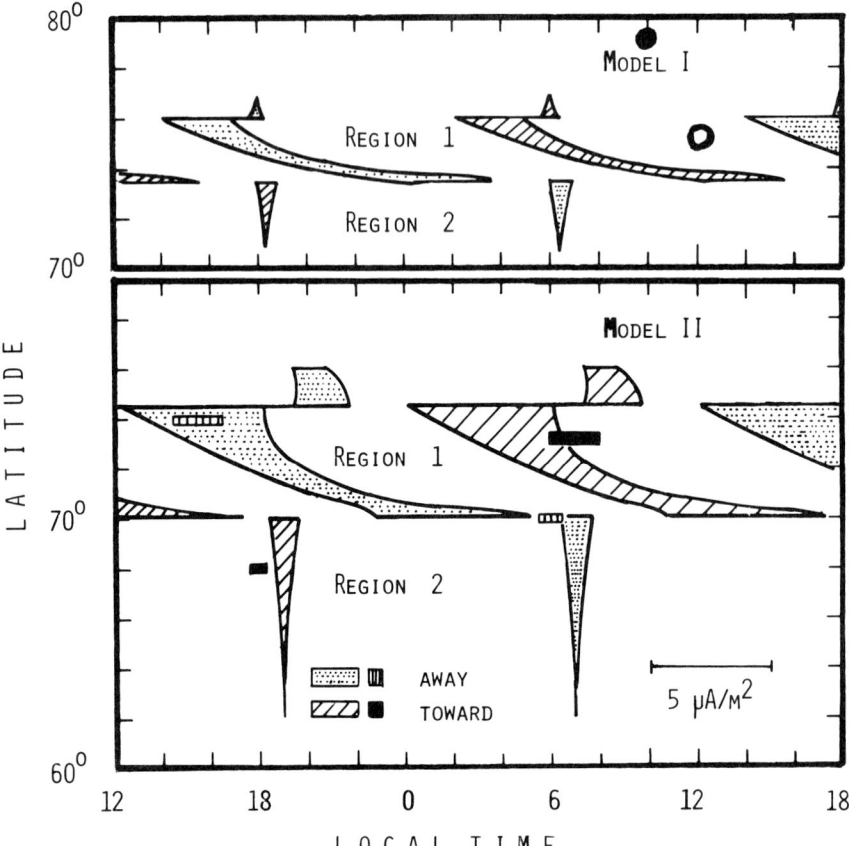

Fig. 4. Model field-aligned electric current density versus local time and latitude on the northern hemisphere. The hatched regions indicate flow into the ionosphere; the dotted regions indicate flow out of the ionosphere. The center of the horizontal extension of the areas gives the local time of maximum flow. The horizontal extension is a measure of the magnitude of the maximum flow (see scale on the right). The rectangles are due to observations of Iijima and Potemra (1976;Figure 3) and give maximum amplitudes of flow as functions of local time and latitude within their regions 1 and 2. The open and solid circles in the upper figure indicate maximum out and in flow due to the polar cap field during toward polarity.

lent current system of the Hall current. Such system can be derived from a stream function

$$D_c = -\bar{\Sigma}_h \Phi_c \qquad (16)$$

Between two streamlines in Figure 3 flows a current of 25 kA and 100 kA, respectively. The stream configuration resembles the $S_q^p$ current system during quiet conditions, and the DP2 current system during disturbed conditions (Nishida and Kokubun, 1971).

The thick dash-dotted lines in Figure 3 indicate the theoretical location of zero meridional field strength within the auroral zones. Along these lines, one expects a reversal of the zonal plasma flow, sometimes called the Harang discontinuity. The observed Harang discontinuity during quiet and during disturbed conditions is indicated by the thick dashed lines, taken from Maynard (1974). The discrepancy in latitude between theoretical and observed curves is due to the neglect of the eccentric location of the auroral ovals with respect to the magnetic poles in the models. A further discrepancy not yet explained comes about from the comparison of the potential differences between dawn and dusk of 30 kV and 92 kV, respectively, in the models (or 23 kV and 76 kV in Heppner's (1977) analysis), and the corresponding Kp values stated by Heppner (1977) to be Kp = 0 and 2 -5, respectively. Using a relationship between $\phi_c$ and Kp given by Kivelson (1976; her eq.11), one finds, instead, the numbers Kp = 3.1(2.0) and 6.5 (6.2) from the potential differences in Figure 3. (Numbers in parentheses are from Heppner's data).

The field-aligned currents which can be derived from eq.(7) are shown in Figure 4 versus local magnetic time with respect to geomagnetic colatitude. Hatched areas indicate currents flowing into the ionosphere, dotted areas indicate currents flowing out of the ionosphere. The horizontal length of the hatched or dotted areas is a measure of the maximum magnitude of the field-aligned current density. The center of the horizontal length gives the time of maximum flow. We notice in Figure 4 two separate regions which are nearly in antiphase: one pair of currents in region 1 flows into the ionosphere on the dawn side and out of the ionosphere on the dusk side at auroral zone latitudes. It corresponds to the discharging current system indicated in Figure 1. The second pair flows into the opposite direction beyond the low latitude border of the auroral zone (region 2 in Figure 4). This pair of currents may be identified as that current system which closes the partial ring current (Schield et al.,1969). Both current systems have been observed (Iijima and Potemra, 1976) and are in basic agreement with the calculations.

Strictly speaking, that part of the electric field which is associated with the field-aligned currents in region 2 outside the auroral zones is a secondary field generated by charge separation within the Alfvén layer. It thus acts like a driving field within the area of closed field lines. One can define the primary convection field as that part of the field which has zero field-aligned currents beyond the low latitude border of the auroral zone. In my model, this field configuration is determined by the parameter $p_2 = 1$ (Volland, 1975).

Figure 5 shows the excess longitudinal current

$$\Delta I_\lambda = - (\Sigma_h - \bar{\Sigma}_h)E_\theta + (\Sigma_p - \bar{\Sigma}_p)E_\lambda \qquad (17)$$

due to the enhanced electric conductivity within the auroral zone for model II. This excess current is a measure of the polar electrojet. The locations of zero zonal flow are indicated by the dashed lines. They are not identical with the locations of the Harang discontinuity in Figure 3 because of the Pedersen term in (17). The total current flowing within the auroral zone is about 300 kA, and the time of maximum eastward flow is at about 1600 LMT and near the low latitude border of the auroral zone. That current consists mainly of the Hall component. Only

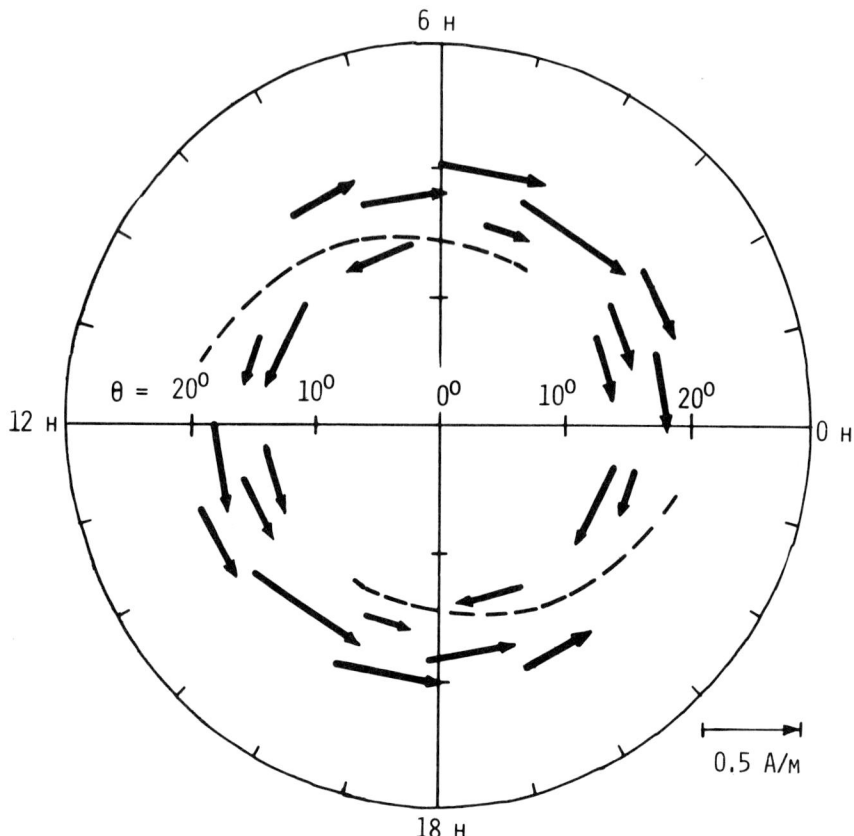

Fig. 5. Model II: zonal excess current within the auroral zone due to the enhanced electric conductivities (polar electrojets). The dashed lines separate the eastward flowing from the westward flowing electrojet.

near the location of the field reversal of $E_\theta$ does the Pedersen current become dominant.

Because of the large Hall component, the influence of the polar electrojet on Joule heating is relatively minor. In Figure 6, the height integrated Joule heating averaged over magnetic local time,

$$\bar{q} = \frac{1}{2} \text{Real} \, (\underline{I}_h \cdot \underline{E}^*) , \qquad (18)$$

is plotted versus magnetic colatitude for models I and II. ($\underline{E}^*$ is the conjugate complex of $\underline{E}$.) The thin horizontal lines in Figure 6 indicate the location of the auroral zone. The polar cap region, the auroral zone region, and the lower latitude region contribute nearly equally to Joule heating. The Joule heat input in Figure 6, averaged over the sphere, is 0.12 mW/m$^2$ and 1.2 mW/m$^2$, respectively. Therefore, even during quiet conditions the contribution of Joule heating is a significant fraction of the solar XUV heat input above 120 km altitude, and it exceeds XUV heating during disturbed conditions.

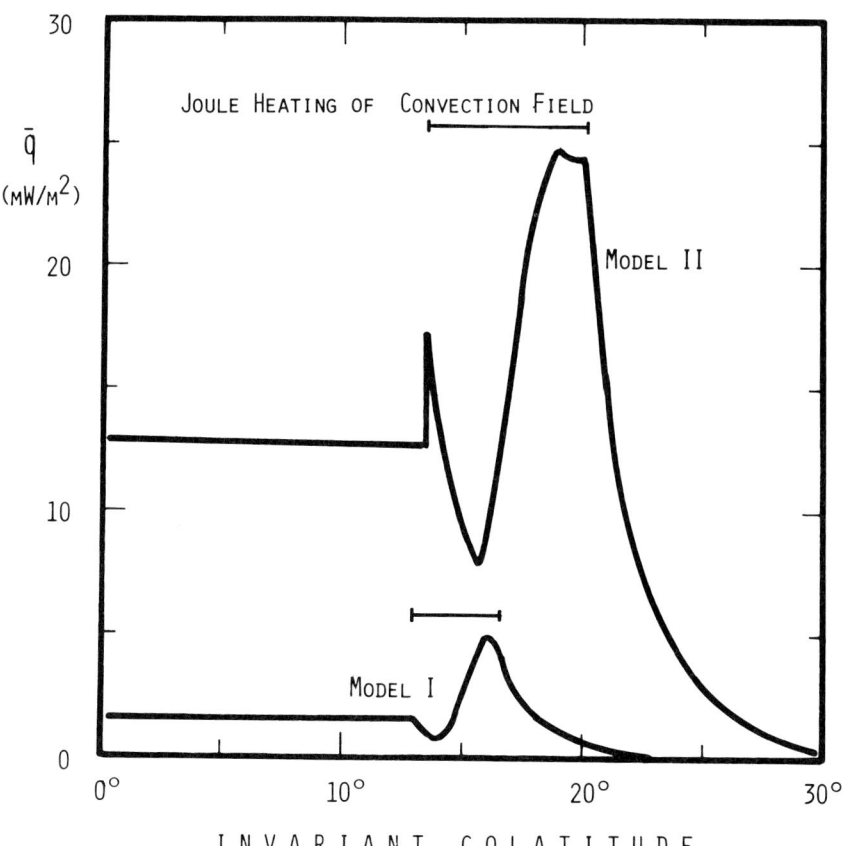

Fig. 6. Joule heating of the convection field: Local time averaged term $\bar{q}$ versus invariant colatitude determined from models I and II in Figure 2. Horizontal thin lines indicate range of auroral zone.

Maximum Joule heating at the low latitude border of the auroral zone is due to the maximum electric field strength there, rather than due to the enhanced electric conductivity. If one neglects in model II the enhanced region of electric conductivity within the auroral zone, the difference in $\bar{q}$, as compared to $\bar{q}$ in Figure 6, is only minor. Likewise, the field-aligned electric currents are rather insensitive to an increase of the electric conductivity. The reason for that is the following: an enhanced Pedersen conductivity within the auroral zone is responsible for a secondary meridional electric field which is opposite to the primary field. Thus, the enhancement of the conductivity is compensated by the reduction of the electric field which leaves the meridional Pedersen current unaltered. It is this component of the Pedersen current which contributes mainly to Joule heating. On the other hand, the same secondary electric field gives rise to a zonal Hall current which contributes significantly to the polar electrojet without feedback to either Joule heating or field-aligned currents (Boström, 1964; Fukushima, 1971).

Joule heating as well as the Ampere force drive thermospheric winds and influence thermospheric composition. These effects are considered by Mayr et al.(1978), Straus and Schulz (1976), and Volland (1978a).

### The corotation field

It is well known that an electric field $\underline{E}$ and the geomagnetic field $\underline{B}_o$ measured in the frame of the ionosphere transform into

$$\underline{E} = \underline{E}' + \underline{V} \times \underline{B}'_o$$
$$\underline{B}_o = \underline{B}'_o \tag{19}$$

where $\underline{E}'$ and $\underline{B}'_o$ are the fields measured in a frame moving with nonrelativistic velocity $\underline{V}$. The plasma of the inner magnetosphere apparently moves nearly rigidly with the earth. Its zonal velocity is therefore

$$V_\lambda = \Omega r \sin\theta \tag{20}$$

where $\Omega = 7.29 \times 10^{-5} \, s^{-1}$ is the angular frequency of the earth's rotation. If no source exists for $\underline{E}$ in the frame of the ionosphere ($\underline{E} = 0$), it is

$$\underline{E}' = - \underline{V} \times \underline{B}_o \tag{21}$$

in a nonrotating frame fixed to the magnetosphere (Alfvén, 1950).

If for convenience, $\underline{B}_o$ is considered as a coaxial field, $\underline{E}'$ can be derived from the potential

$$\Phi_r = - \Phi_{ro}/L \tag{22}$$

with $\Phi_{ro} = \Omega r_o^2 B_{oo} = 90$ kV, and $B_{oo} = 3 \times 10^{-5}$ T the geomagnetic field at the equator. This potential is symmetric with respect to the equator and decreases with increasing L.

An ionized particle within the magnetosphere possessing thermal energy moves along equipotential shells of the combined corotation field and the convection field. Since $\Phi_r$ decreases with L, while $\Phi_c$ increases for $L < L_2$, one expects an inner region where the equipotentials are closed and an outer region where they are open. The last closed equipotential is believed to be the plasmapause during steady state conditions (Nishida, 1966; Kivelson, 1976). That last closed equipotential shell can be found from the equation (Volland, 1975)

$$2L^{1+p/2} \sin(\tau-\delta) - pL_m^{1+p/2} + (p+2)L\, L_m^{p/2} = 0 \tag{23}$$

with

$$\left. \begin{array}{l} \delta = - \arg(A_2) \\ \\ p = p_2 \end{array} \right\} \text{ from (13)}$$

and

$$2/(p+2)$$

$$L_m = \left( \frac{2 \Phi_{ro} L_2^{p/2}}{pA_2} \right)$$

$L_m$ being the maximum elongation of the plasmapause within the equatorial plane of the magnetosphere, and $\delta$ being the phase of the maximum elongation with respect to 1800 LT. The observed configuration of the plasmapause (e.g., Carpenter and Park, 1973) fits well to the parameter $p = p_2 = 4$ in (23). This is one of the reasons for choosing this number in (13).

## The polar cap field

According to Heppner's (1972) observations, during towards polarity of the IMF, the electric field in the polar cap regions weakens on the dawn side in the northern hemisphere and on the dusk side in the southern hemisphere. It increases on the dusk side in the north and on the dawn side in the south. During away polarity, that effect reverses. It can be described by a zonally independent meridional field superimposed on the convection field. This field is directed to the equator during toward polarity and directed to the poles during away polarity. It owes its existence to the antisymmetric structure of the magnetic field configuration indicated in Figure 1. Figure 1 also shows that it is strictly speaking the east-west component of the IMF which is responsible for this field (Friis-Christensen et el.,1972).

A meridional field gives raise to a zonal Hall current

$$I_\lambda = - \bar{\Sigma}_h E_\theta \qquad (24)$$

which in turn generates magnetic field variations. At the poles and at the ground, these variations are directed downward during toward polarity and directed upward during away polarity. It was just this magnetic effect observed at polar stations which led Svalgaard (1968) and Mansurow (1969) to the discovery of the polar cap field. The Hall current is essentially a partial ring current encircling the poles near $10^\circ$ colatitude (Svalgaard, 1973). The ring current has its maximum during local noon and during summer.

In order to simulate this field, the following analytic expressions for the potential can be used (Volland, 1975):

$$\Phi_p = \Phi_{po} \begin{cases} \dfrac{1}{\sqrt{L_3 L_4}} - \dfrac{1}{L} & \theta \lesseqgtr \theta_3 \\[1em] \dfrac{(1-\sqrt{L_4/L})^2}{\sqrt{L_3 L_4}(1-\sqrt{L_4/L_3})} & \text{for} \quad \theta_3 \lesseqgtr \theta \lesseqgtr \theta_4 \\[1em] 0 & \theta_4 \lesseqgtr \theta \lesseqgtr 90^\circ \end{cases} \qquad (25)$$

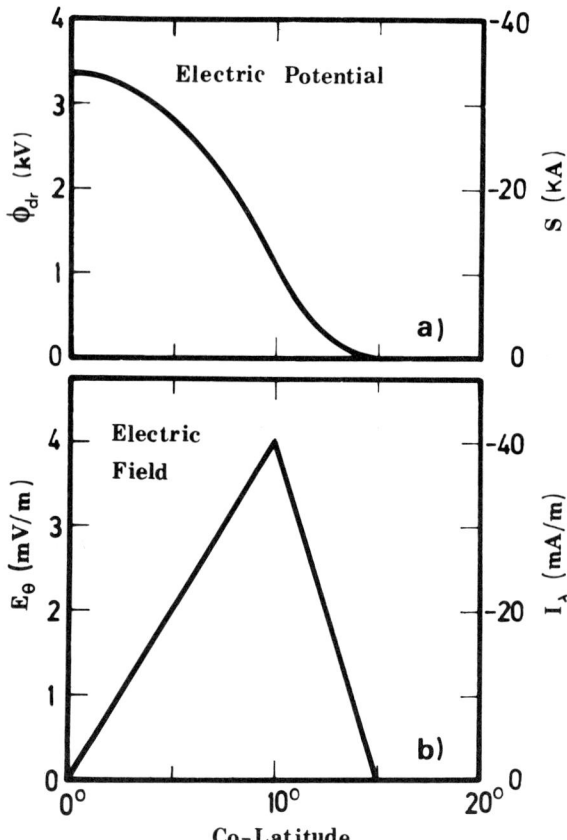

Fig. 7. (a) Electric potential, and (b) meridional component of the electric field strength at ionospheric heights generated by differential superrotation of the northern hemisphere polar cap magnetospheric plasma during toward-sector polarity of the interplanetary magnetic field. The right abscissas indicate stream function and zonal electric Hall current. During away polarity, all signs must be reversed.

and
$$\Phi_p(180°-\theta) = -\Phi_p(\theta) \tag{26}$$

It is $|\Phi_{po}| \simeq 75$ kV

and $L_j = 1/\sin^2\theta_j$ ; $L_3 \simeq 10°$ ; $L_4 \simeq 15° \simeq L_1$

On the northern hemisphere, $\Phi_{po}$ is positive for toward polarity and negative for away polarity. On the southern hemisphere, the sign reverses. Comparison between (22) and (25) shows that during toward polarity, the northern polar cap plasma superrotates with respect to the earth, the southern polar cap plasma rotates in retrograde direction. During away polarity, the direction of the rotation changes. If $\Phi_{po} = -90$ kV, the plasma within the northern polar cap would be fixed in a nonrotating frame of reference.

Figure 7 shows $\Phi_p$ and $E_\theta$ derived from (25) versus colatitude during toward polarity on the northern hemisphere.

In order to simulate the electric currents driven by the polar cap field, one has to take into account the spatial and temporal variation of the electric conductivities within the polar cap regions. I use the following formulae for $\Sigma_p$ and $\Sigma_h$ which may approximately be valid during equinox conditions:

$$\Sigma_p = \overline{\Sigma}_p \sin\theta (1-\cos\tau)/\sin\theta_4$$

$$\Sigma_h = \overline{\Sigma}_h \sin\theta (1-\cos\tau)/\sin\theta_4 \qquad \text{for } \theta \leq \theta_4 \qquad (27)$$

where $\overline{\Sigma}_p = 15$ mhos, $\overline{\Sigma}_h = 10$ mhos, the undisturbed values used in model I, and $\tau$ the local magnetic time. The conductivities are therefore zero at the pole and at midnight. They maximize at noon. Thus the Hall current configuration is in general agreement with Svalgaard's (1973) findings.

The formulae for the field-aligned currents derived from (7), (8), (25), and (27) are now some lengthly expressions which can be condensed to the following approximate form

$$j_\| \approx \frac{2\Phi_{po} \overline{\Sigma}_p}{r_o^2 \sin\theta_4} \begin{cases} -3\sin\theta(1-\cos\tau - \overline{\Sigma}_h \sin\theta/\overline{\Sigma}_p) & \theta < \theta_3 \\ & \text{for} \\ \dfrac{\sin\theta_3(\sin\theta - \sin\theta_3)}{(\sin\theta_4 - \sin\theta_3)^2}(1-\cos\tau) & \theta_3 \leq \theta \leq \theta_4 \end{cases} \qquad (28)$$

with maximum currents of the order of 0.2 $\mu A/m^2$ at $\theta_3$ at about 1100 LMT, and at $\theta_4$ at noon. At midnight, the current is nearly zero. During toward polarity and on the northern hemisphere, the current at $\theta_4$ is directed out of the ionosphere (see Figure 4). On the southern hemisphere, the direction reverses.

Field-aligned electric currents at the polar border of the auroral zones at noon which depend on the polarity of the IMF have in fact been observed (Potemra and Saflekos, 1978). Their resemblance to the theoretical prediction in Figure 4 is not unique because the measurements are still preliminary.

## Conclusion

Following an idea of Vasyliunas (1972) that a detailed consideration of the magnetospheric plasma convection can be eliminated in the modeling of global scale fields and currents within the dynamo layer and within the inner magnetosphere, I have simulated three major magnetospheric fields: the convection field, the corotation field, and the polar cap field. By introducing a simple semiempirical analytical function for the electric potential of each field type, electric field and current configurations can be derived uniquely if the electric conductivities of the dynamo layer are known. The potentials are chosen such that their derived fields and currents fit optimally to the available various direct and indirect observations like in situ measurements at ionospheric heights of electric fields and field-aligned electric currents, drift measurements, observations of the plasmapause, and ground based magnetic observations.

This method is limited to the description of stationary global fields and currents with time scales of the order of one day or larger. It fails to simulate short periodic and/or local phenomena such as substorms.

## Literature

Alfvén, H., Cosmic electrodynamics, Clarendon Press, Oxford, 1950.

Axford, W.I., Magnetospheric convection, Rev. Geophys. Space Phys., 7, 421, 1969.

Axford, W.I., and C.O. Hines, A unifying theory of high-latitude geophysical phenomena and geomagnetic storms, Can.J.Phys., 39, 1433, 1961.

Banks, P.M., Observations of Joule and particle heating in the auroral zone, J.Atm.Terr.Phys., 39, 179, 1977.

Banks, P.M., and F. Yasuhara, Electric fields and conductivity in the nighttime E-region: a new magnetosphere-ionosphere-atmosphere coupling effect, Geophys. Res.Lett. 6, 1979 (in print).

Boström, R., A model of the auroral electrojet, J.Geophys.Res., 69, 4883, 1964.

Carpenter, D.L., and C.G. Park, On what ionospheric workers should know about the plasmapause-plasmasphere, Rev. Geophys. Space Phys., 11, 133, 1973.

Cauffman, D. P., and D. A. Gurnett, Satellite measurements of high latitude convection fields, Space Sci. Rev., 13, 369, 1972.

Coleman, P. J., A model of the geomagnetic cavity, Radio Sci, 6, 321, 1971.

Dungey, J. W., Interplanetary magnetic field and the auroral zones, Phys.Rev.Lett., 6, 47, 1961.

Dungey, J.W., The structure of the exosphere or adventures in velocity space, in *Geophysics-The Earth's Environment*, pp. 6o3, Gordon and Breach, New York, 1963.

Fälthammar, C.-G., Problems related to macroscopic electric fields in the magnetosphere, Rev. Geophys. Space Phys. 15, 457, 1977.

Friis-Christensen, E., K. Lassen, J. Wilhjelm, J.M. Wilcox, W. Gonzalez, and D.S. Colburn, Critical component of the interplanetary magnetic fields responsible for large geomagnetic effects in the polar cap, J.Geophys. Res., 77, 3371, 1972.

Fukushima, N., Electric current systems for polar substorms and their magnetic effect below and above the ionosphere, Radio Sci., 6, 269, 1971.

Gurevich, A.V., A.L. Krylov, and E.E. Tsedilina, Electric fields in the earth's magnetosphere and ionosphere, Space Sci.Rev. 19, 56, 1976.

Harel, M., R.A. Wolf, P.H. Reiff, and M. Smiddy, Computer modeling of events in the inner magnetosphere, this volume, 1979

Heelis, R. A., W.B. Hanson, and J.L. Burch, Ion convection velocity reversals in the dayside cleft, J.Geophys.Res., 81, 3803, 1976a.

Heelis, R. A., R.A. Spiro, W.B. Hanson, and J.L. Burch, Magnetosphere-ionosphere coupling in the mid-latitude trough, Trans.Am.Geophys. Union, 57, 990, 1976b.

Heikkila, W. J., and R. J. Pellinen, Localized induced electric field within the magnetotail, J. Geophys. Res. 82, 1610, 1977.

Heppner, J. P., Polar-cap electric distributions related to the interplanetary magnetic field direction, J. Geophys. Res., 77, 4877, 1972.

Heppner, J. P., Empirical models of high-latitude electric fields, J. Geophys. Res., 82, 1115, 1977.

Hill, T. W., and M. E. Rassbach, Interplanetary magnetic field direction and the configuration of the dayside magnetosphere, J. Geophys. Res., 80, 1, 1975.

Hughes, T. J., and G. Rostoker, A model three-dimensional current system for steady state conditions in the magnetosphere, J. Geophys. Res. 83, 1978 (in preparation).

Iijima, T., and T. A. Potemra, The amplitude distribution of field-aligned currents at northern high latitudes observed by Triad, J. Geophys. Res., 81, 2165, 1976.

Kamide, Y., and S. Matsushita, Simulation studies of ionospheric electric fields and currents in relation to field-aligned currents: 1. Quiet period, J. Geophys. Res., submitted 1978.

Kawasaki, K., and N. Fukushima, Equivalent current pattern of ionosphere and field-aligned currents generated by geomagnetic $S_q^P$-field with closed equatorward auroral boundary, Report Ion. Space Res. Japan, 28, 187, 1974.

Kisabeth, J.L., On calculating magnetic and vector potential fields due to large-scale magnetospheric current systems and induced currents in an infinitely conducting earth, this volume, 1979.

Kivelson, M. G., Magnetospheric electric field and their variation with geomagnetic activity, Rev. Geophys. Space Phys., 14, 189, 1976.

Köhnlein, W., and W.J. Raitt, Position of the mid-latitude trough in the topside ionosphere as deduced from ESRO 4 observations, Planet. Space Sci., 25, 600, 1977.

Leontyev, S. V., and W. B. Lyatsky, Electric field and currents connected with y-component of interplanetary magnetic field, Planet. Space Sci., 22, 811, 1974.

Mansurov, S.M., New evidence of a relationship between magnetic fields in space and on earth, Geomagn. Aeron., 9, 622, 1969.

Maynard, N. C., Electric field measurements across the Harang discontinuity, J. Geophys. Res., 79, 4620, 1974.

Maynard, N. C., On large scale poleward-directed electric fields at sub-auroral latitudes, Geophys. Res. Lett., 5, 617, 1978.

Mayr, H. G., and I. Harris, Some characteristics of electric field momentum coupling with the neutral atmosphere, J. Geophys. Res., 83, 3327, 1978.

Mayr, H.G., I. Harris, and N.W. Spencer, Some properties of upper atmosphere dynamics, Rev. Geophys. Space Phys., 16, 1978.

McIlwain, C.E., Substorm injection boundaries, in Magnetospheric Physics, ed. B.M. McCormac, D. Reidel, Dordrecht-Holland, p.143, 1974.

Murphy, C.M., C. S. Wang, and J.S. Kim, Inductive electric field of a field-aligned current system, J. Geophy. Res., 79, 2901, 1974.

Nisbet, J.S., M. J. Miller, and L.A. Carpenter, Currents and electric fields in the ionosphere due to field-aligned auroral currents, J. Geophys. Res., 83, 2647, 1978.

Nishida, A., Formation of the plasmapause or magnetospheric plasma knee by the combined action of magnetospheric convection and plasma escape from the tail, J. Geophys. Re., 71, 5669, 1966.

Nishida, A., and S. Kokubun, New polar magnetic disturbances, $S_q^p$, SP, DPC, and DP2, Rev. Geophys. Space Phys.,9, 417, 1971.

Nopper, R.W., and R.L Corovillano, Polar-equatorial coupling during magnetically active periods, Geophys. Res. Lett., 5, 699, 1978.

Potemra, T.A., and N.A. Saflekos, Pattern of Birkeland currents, ionospheric currents and polar cap convection and their dependence upon the interplanetary magnetic field, JMU/APL-Report 78/10, The John Hopkins University, Applied Physics Laboratory, Laurel, Md., 1978.

Richmond, A.D., The computation of magnetic effects of field-aligned magnetospheric currents, J. Atm. Terr. Phys., 36, 245, 1974.

Rostoker, G., Electric and magnetic fields in the ionosphere and magnetosphere, J. Geomag. Geoelectr., 30, 67, 1978.

Schield , M.A., J.W. Freeman, and A.D. Dessler, A source for field-aligned currents at auroral latitudes, J. Geophys. Res., 74, 247, 1969.

Schulz, M., Plasma convection in the tail of a rotating planetary magnetosphere, Rep. SAMSO-TR-76-221, Space Sci. Lab., Aerosp. Corp., El Segundo, Calif., 1976.

Schulz, M., Compressible corotation of a model magnetosphere, J. Geophys. Res., 75, 6329, 1970.

Smiddy, M., M.C. Kelley, W. Burke, F. Rich, R. Sagalyn, B. Shuman, R. Hays, and S. Lai, Intense poleward-directed electric fields near the ionospheric projection of the plasmapause, Geophys. Res. Lett. 4, 543, 1977.

Stern, D.P., The motion of a proton in the equatorial magnetosphere, J. Geophys. Res., 80, 595, 1975.

Stern, D.P., Large scale electric fields in the Earth's magnetosphere, Rev. Geophys. Space Phys., 15, 156, 1977.

Straus, J.M., and M. Schulz, Magnetospheric convection and upper atmospheric dynamics, J. Geophys. Res., 81, 5822, 1976.

Svalgaard, L., Sector structure of the interplanetary magnetic field and daily variation of the geomagnetic field at high latitudes, Geophys.Pap. R.-6, Danish Meteorol. Inst., Copenhagen, 1968.

Svalgaard, L., Polar cap magnetic variations and their relationship with the interplanetary magnetic sector structure, J. Geophys. Res., 78, 2064, 1973.

Vasyliunas, V.M., The interrelationship of magnetospheric processes, in Earth's Magnetospheric Processes, ed. by B.M. McCormac, p. 29 R.Reidel, Dordrecht, Netherlands, 1972.

Volland, H., Models of the global electric fields within the magnetosphere, Ann. Geophys., 31, 159, 1975.

Volland, H., Magnetospheric electric fields and currents and their influence on large scale thermodynamic circulation and composition, Space Sci. Rev., submitted 1978 a.

Volland, H., A model of the magnetospheric electric convection field, J. Geophys. Res., 83, 2695, 1978 b.

Wolf, R.A., Ionosphere-magnetospheric coupling, Space Sci. Rev., 17, 537, 1975.

Yasuhara, F., and S.-I. Akasofu, Field-aligned and ionospheric electric fields, J. Geophys. Res., 82, 1279, 1977.

Yasuhara, F., Y. Kamide, and S.-I. Akasofu, Field-aligned and ionospheric currents, Planet. Space Sci., 23, 1355, 1975.

Zmuda, A.J., and J.C. Armstrong, The diurnal flow pattern of field-aligned currents, J. Geophys. Res., 79, 4611, 1974.

QUASISTATIC ELECTRIC FIELD MEASUREMENTS ON THE GEOS-1 AND GEOS-2 SATELLITES

Arne Pedersen and Réjean Grard

Space Science Department of ESA, ESTEC, Noordwijk, The Netherlands.

Abstract. Quasi-static electric fields are measured on GEOS-1 and GEOS-2 with a 42 m long spinning double probe antenna. Most of the measurements presented in this paper have been collected with GEOS-1 in the morning, noon and afternoon local time sectors. In the front side of the magnetosphere the electric field is, on average oriented from dawn to dusk, and the associated plasma drift is sunward; a reversal of the electric field direction is often observed in the local time morning hours. Impulsive fields observed in the early morning can be interpreted as plasma injection events from the magnetotail. Fluctuations and pulsations with periods of one to several minutes and amplitudes of several mV/m are often detected in the frontside magnetosphere hours and even days after disturbed periods.

## Introduction

A double probe antenna for measuring quasistatic electric fields forms part of the scientific payload of GEOS-1, operational from May 1977 to June 1978, as well as GEOS-2 launched in July 1978. GEOS-1 did not reach the planned geostationary orbit but was injected into a 12 hour elliptical orbit with apogee near 7 $R_E$, and local time of apogee near 18 h at the start of the mission. GEOS-2 was successfully injected into a geostationary orbit and is at present positioned at $37^o$ E longitude. Both satellites carry identical antennas made of two spherical sensors separated by a distance of 42 m. A description of the technique used and a presentation of the initial results are contained in Pedersen et al. (1978).

Electric fields are measured in a plane perpendicular to the spin axis of the satellite. Because of asymmetries in the probe current-voltage characteristics due to the influence of the supporting booms on GEOS-1, the electric field in the sunward direction can be determined with an accuracy of only 1 mV/m, whereas the component perpendicular to this direction (approximately from dawn to dusk) can be evaluated with an error less than 0.5 mV/m. These accuracies refer to the spin averaged absolute levels of the field components; relative time variations of the electric field of a few tenths of mV/m can be detected as long as the density and temperature of the surrounding plasma remain unchanged. The numbers quoted above pertain to conditions where the plasma density is relatively low, of the order of a few per $cm^3$, but considerably more accurate measurements are possible in denser media. Typical electron densities at 6-7 $R_E$ are in the range 5-50 $cm^{-3}$, as reported by Decreau et al. (1978).

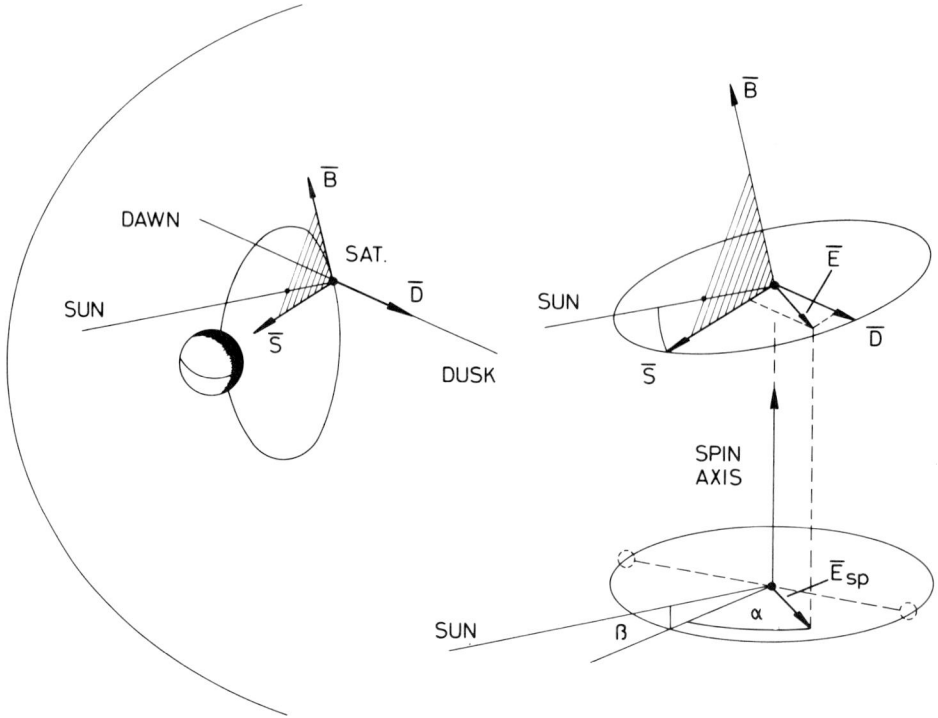

Fig. 1. Data is presented in the SDB coordinate system. The B-axis is along the magnetic field $\bar{B}$; the S-axis is in the plane defined by $\bar{B}$ and the satellite-sun direction, and the D-axis completes the set. The electric field $\bar{E}$ is assumed to be nearly perpendicular to $\bar{B}$. The measurements yield the projection of the electric field in the spin plane, $\bar{E}_{sp}$. The orientation of this component is defined by an angle $\alpha$ measured with respect to the direction of the antenna at its closest approach to the satellite-sun line, which forms an angle $\beta$ with the spin plane.

GEOS-2 has modified electric field sensors with additional voltage controlled guards to diminish the effect of the asymmetry mentioned previously. This arrangement has definitely improved the quality of the measurements, but the associated accuracy will only be assessed after an evaluation of a number of data sets taken under different plasma conditions. It has also been possible with GEOS-2 to collect data during a series of eclipses which had a maximum duration of 72 minutes; in absence of photoemission the probe symmetry was improved further as seen from initial data samples.

GEOS-2 also included another successful electric field experiment based on the injection of accelerated electrons in a direction perpendicular to the local magnetic field and the detection of the displacement of the returning beam due to plasma drifts (Melzner et al., 1978). Early comparisons of limited data sequences have already demonstrated the agreement of the two techniques.

Further support for the quality of the spherical double probe is

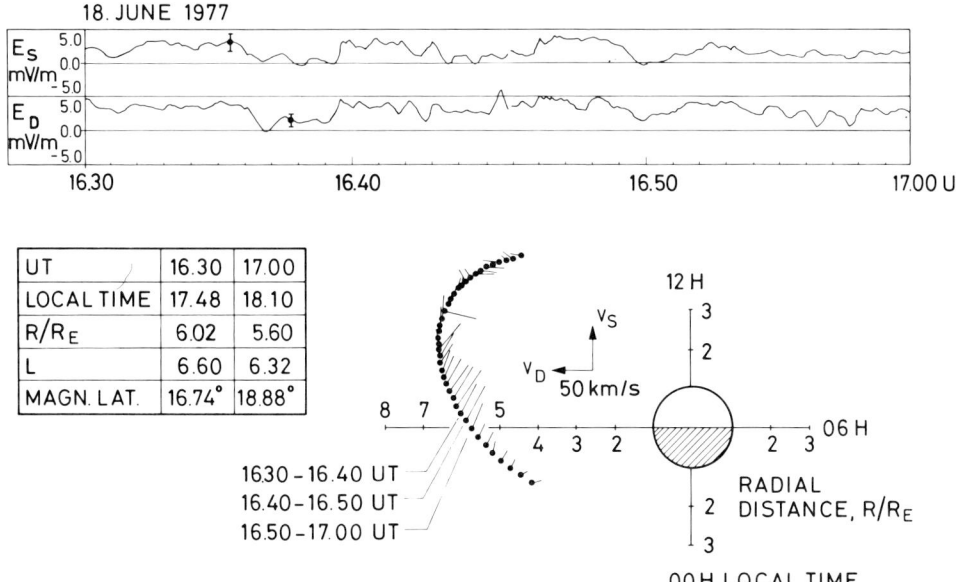

Fig. 2. Measurements of the quasi-static electric field in the afternoon sector near L = 6.5. $E_S$ and $E_D$ are sunward and dawn to dusk fields respectively. The magnetic field was very quiet and varied from 150 γ to 200 γ over the time shown.

given in Mozer et al. (1978). A similar antenna with an effective length of 72 m is presently operational on the ISEE-1 satellite; measurements in the solar wind are in excellent agreement with electric fields independently derived from the measured values of the relative plasma drift and ambient magnetic field.

This paper is a survey of the GEOS-1 data which have been analysed so far. Complete data sets, including magnetic field information only exist for periods of typically 8 hours per day at times when GEOS-1 apogee was in the afternoon, noon and morning sectors. The measurements made with GEOS-1 on the nightside and the performance of the GEOS-2 instrument have presently only be evaluated from spot checks.

## Coordinates for Data Presentation

It is assumed in this paper that the conductivity parallel to the Earth's magnetic field $\bar{B}$ is very large and the component of the electric field in this direction is negligible. The data is presented in the so-called SDB coordinate system illustrated in Fig. 1. The B-axis is parallel to the instantaneous dc magnetic field; the S-axis is in the plane defined by $\bar{B}$ and the satellite-sun direction, and the D-axis, which completes this orthogonal reference system, is in the general dawn to dusk direction. The assumption about the perpendicularity of the electric and magnetic field vectors, together with the definition of the coordinate system, have the consequence that the component of $\bar{E}$ along the B-axis is zero. The validity of this assump-

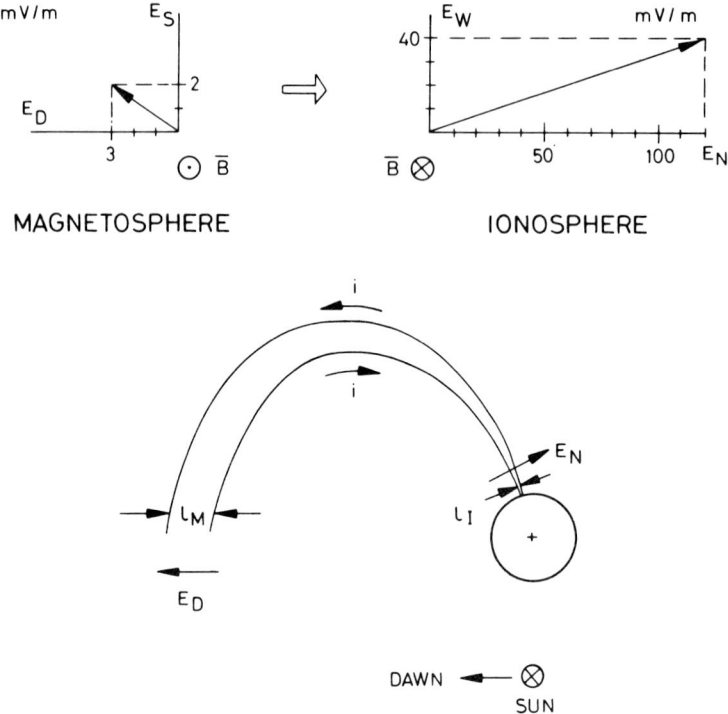

Fig. 3. Projection into the conjugated ionosphere of the electric field observed in the magnetosphere between 1700 and 1800 LT, on 18 June 1977.

tion will be assessed later. The presentation of the data can therefore be limited to the display of the components $E_S$ and $E_D$ along the S- and D-axes, respectively.

This reference system is consequently allowed to rotate because one axis follows the $\bar{B}$ vector, but is fixed in inertial coordinates. The electric field due to the satellite motion through the Earth magnetic field is of the order 0.17 mV/m near apogee for GEOS-1, and is approximately equal to 0.3 mV/m in the geostationary orbit for GEOS-2.

### The Afternoon Sector

Large dawn to dusk electric fields have been observed on several days near 18 h local time for L-values between 5 and 7.5. An example of such a large field is shown in the top pannel of Fig. 2 which displays data of 18 June 1977. A characteristic feature is the presence of the fluctuations of several mV/m on time scales from a fraction of a minute to several minutes; the average electric field is in this case clearly dawn to dusk and sunward. The error bars represent the accuracies of the instanteneous values; the relative variations are significant to within a few times the line thickness as the plasma conditions did not change during the interval shown. The environment could indeed be monitored by measuring the floating

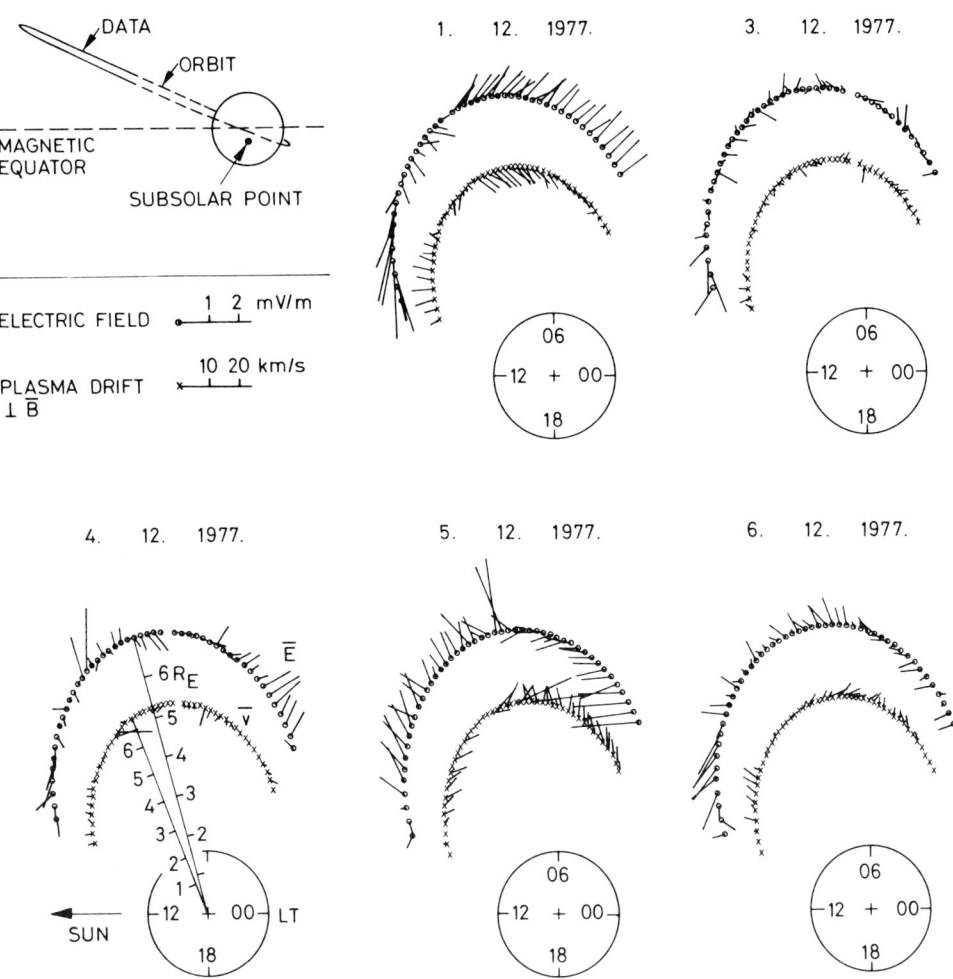

Fig. 4. Ten minute averages of electric fields and plasma drifts for a sequence of days in December 1977. Electric field and drift vectors are represented in the S-D plane; the origins of these vectors are aligned along the spacecraft orbit, which is shown as a plot of radial distance vs local time. The scale for radial distance differs for electric field and drift velocity in order to improve legibility.

potential of the satellite with respect to one electric field probe (Decreau et al., 1978). More detail about the accuracy of the measurements can be found in Pedersen et al. (1978).

The bottom diagram is a vectorial representation of the plasma drift, $\bar{v} = \bar{E} \times \bar{B}/B^2$, averaged over 10 minute periods; this vector is also perpendicular to $\bar{B}$ and is described by its two components $v_S$ and $v_D$. The dots indicate both the origins of the vectors and the locations (radial distance vs local time) of the measurements. It is seen that the plasma drift is predominantly sunward at around 1800 LT.

After analysing a limited number of days in June and July 1977 when

the apogee of GEOS-1 was near 1800 LT, it appears that there are
typical days or sequences of a few days when large dawn to dusk
electric fields are observed; in between there are days with very small
electric fields often below the threshold of the measurements. During
June-July 1977 the Earth's magnetic field was very quiet and no
striking changes in magnetospheric conditions could be associated with
the occurence of large electric fields. So far no correlations have
been made between measurements of solar wind parameters and measurements in the afternoon sector on GEOS-1.

Electric fields observed in the ionosphere have in the past been
mapped to the magnetospheric equator in an attempt to obtain information about the general magnetospheric convection pattern. Due to
lack of knowledge about electric fields parallel to $\vec{B}$ this can only be
done under the assumption that each magnetic field line is equipotential, i.e. the conductivity along $\vec{B}$ is infinitely large. The
result of such mapping (e.g. Mozer and Lucht, 1974) is in a broad
sense in good agreement with commonly used convection models (Axford
and Hines, 1961; McIlwain, 1972).

Fig. 3 is an illustration of the inverse operation namely a
projection of the average field observed near 1800 LT, on 18 June 1977
down to the ionosphere. This results in an electrif field direction in
good agreement with average observations made near the conjugate point
on rockets and with Barium releases during relatively quiet conditions
(Fahleson et al, 1971). However, the derived magnitude of the ionospheric electric field exceeds any value observed during quiet or
slightly disturbed times by at least a factor of 6. One may attempt to
reconcile the magnetospheric and ionospheric observations using the
following model. Consider that the magnetospheric dawn to dusk electric field $E_D$ extends over a given distance $l_M$ in the vicinity of the
geomagnetic equator, as illustrated in Fig. 3; the equivalent generator which can support the same field across the same distance must
develop a voltage $\Delta V_M = E_D l_M$. Assume also that this voltage drives,
as indicated, a current loop that closes at a lower altitude in the
ionosphere; the direction of this current flow is chosen on the basis
of observations of field aligned currents in the vicinity of 1800 LT
(Iijima and Potemra, 1978).

The voltage can then be written

$$\Delta V_M \approx \Delta V_\parallel + \Delta V_I$$

where $\Delta V_\parallel$ is the sum of voltage drops along the magnetic field lines,
and $\Delta V = E_N l_I$ is the voltage drop along the path of length $l_I$ that
closes the loop in the ionopshere.

With reference to Fig. 2 we find that $E_D$ is of the order 3 mV/m over
at least a distance $l_M = 1\ R_E$ which leads to $\Delta V_M = 19$ kV, and corresponds to a distance $l = 160$ km in the ionosphere ($l_M/l_I \approx 40$).
Steady northward electric fields of the order 20 mV/m have been observed on several rocket flights over south-north distance of 100-
200 km. Taking $l_I$ to be 160 km and a typical value $E_I = 20$ mV/m, we
find $\Delta V_I = 3.2$ kV.

The model which is the basis for the above equation then predicts:

$$\Delta V_\parallel = <E_\parallel> \cdot .1 = 15.8 \text{ kV}$$

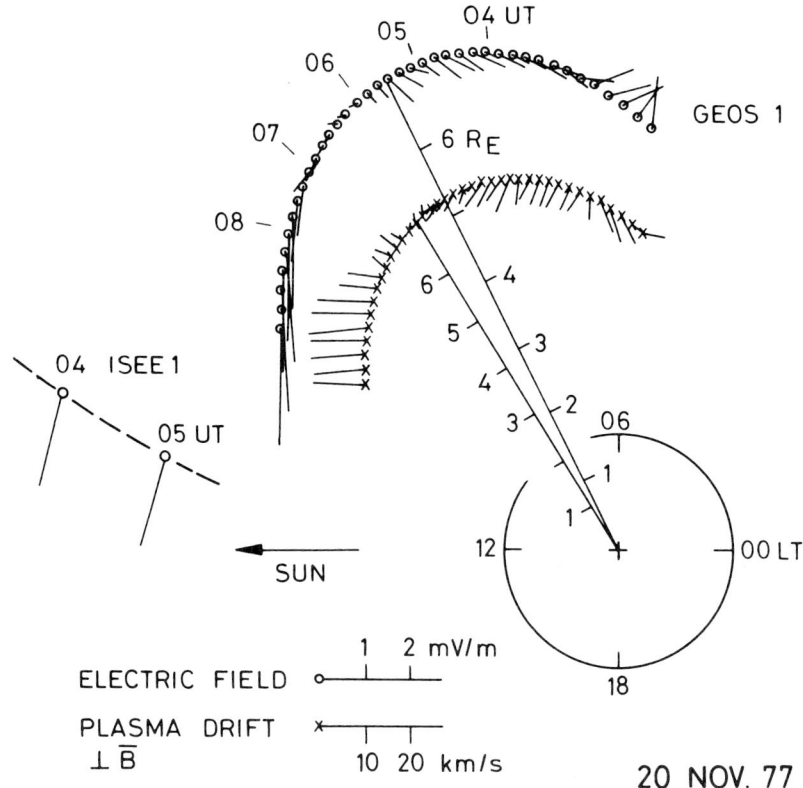

Fig. 5. Comparison between GEOS-1 and ISEE-1 electric field data; see Fig. 4 for complementary information.

where $\langle E_\parallel \rangle$ is the average parallel electric field along the total length, $l_\parallel$, of parallel field lines involved in the current loop of Fig. 3. If we take the field lines concerned to be at L = 6 and L = 7 respectively, then $l_\parallel \approx 130,000$ km, which results in $\langle E_\parallel \rangle = 0.12$ mV/m. Much larger parallel electric fields have indeed been measured in the auroral ionosphere (Mozer, 1976) and it follows that the average over the remaining magnetospheric part of the field lines are even less than 0.12 mV/m. Although this model is simple, this exercise indicates that energy is fed from the magnetosphere into the ionosphere, and it gives an order of magnitude of the voltage drop along the magnetic field lines which connect the auroral regions to the magnetic equator.

## The Frontside Magnetosphere

Fig. 4 displays 10 minute averages of electric field and plasma drift for several days in December 1977. The top left panel is a projection of the orbit in a plane perpendicular to the magnetic equator, which shows that the magnetic latitude near apogee was of the order 20-25°. This sequence of measurements gives a feeling for

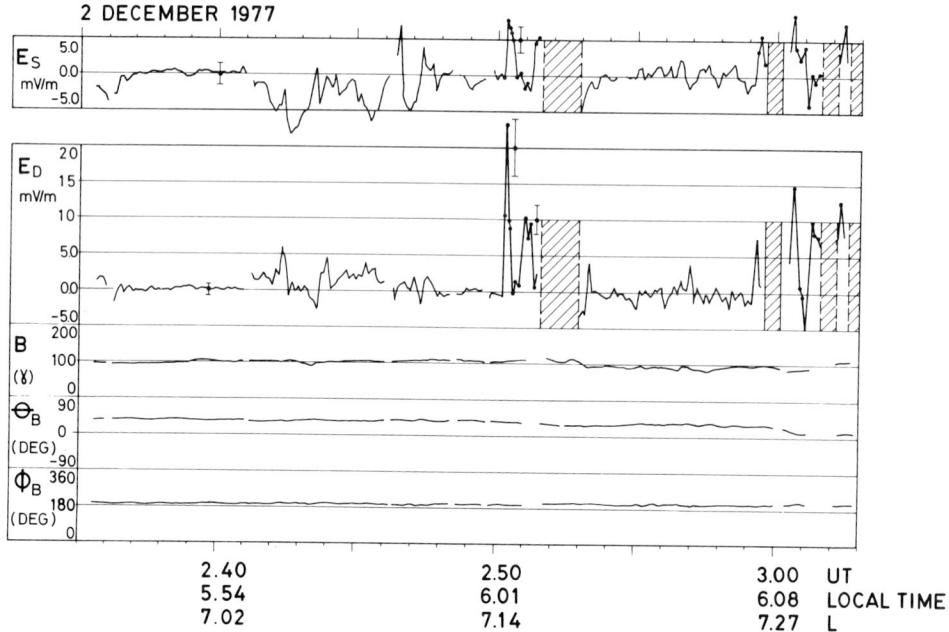

Fig. 6. The quasi-static electric field during a substorm. Large dawn to dusk components of the electric field were measured near 0240 UT and 0300 UT. Near these times the electric field was perturbed and often varied within a spin period (6 s) to the extent that $\bar{E}$ could not be measured (cross hatching). B, $\theta_B$ and $\phi_B$ are magnetic field magnitude, angle above VD plane (VDH coordinates), and angle of VD component relative to V.

the day to day variability of the electric field and it also indicates the existence of a dawn to dusk electric field or sunward plasma flow on the frontside magnetosphere for L = 6 to 7.

On 20 November 1977 the nearly identical electric field experiments on GEOS-1 and ISEE-1 yielded data from orbits which came close in space and time as shown in Fig. 5. Electric fields averaged over 10 minute periods are displayed along the GEOS-1 orbit; associated plasma drifts are also plotted (the orbit described by the origin of the drift vectors is drawn on a different scale for legibility). The ISEE-1 data have been reported by Mozer et al. (1979).

It is seen that both spacecraft measure electric fields oriented in the general dawn to dusk direction with magnitudes of the order of 2 mV/m. This comparison not only demonstrates the agreement between the two sets of data, but also establishes the fact that substantial dawn to dusk electric fields, and associated sunward plasma drifts prevail over large regions in the frontside magnetosphere for periods of several hours.

### The Morning Sector

As mentioned in the previous section, with reference to Fig. 4, the electric field is generally oriented from dawn to dusk around

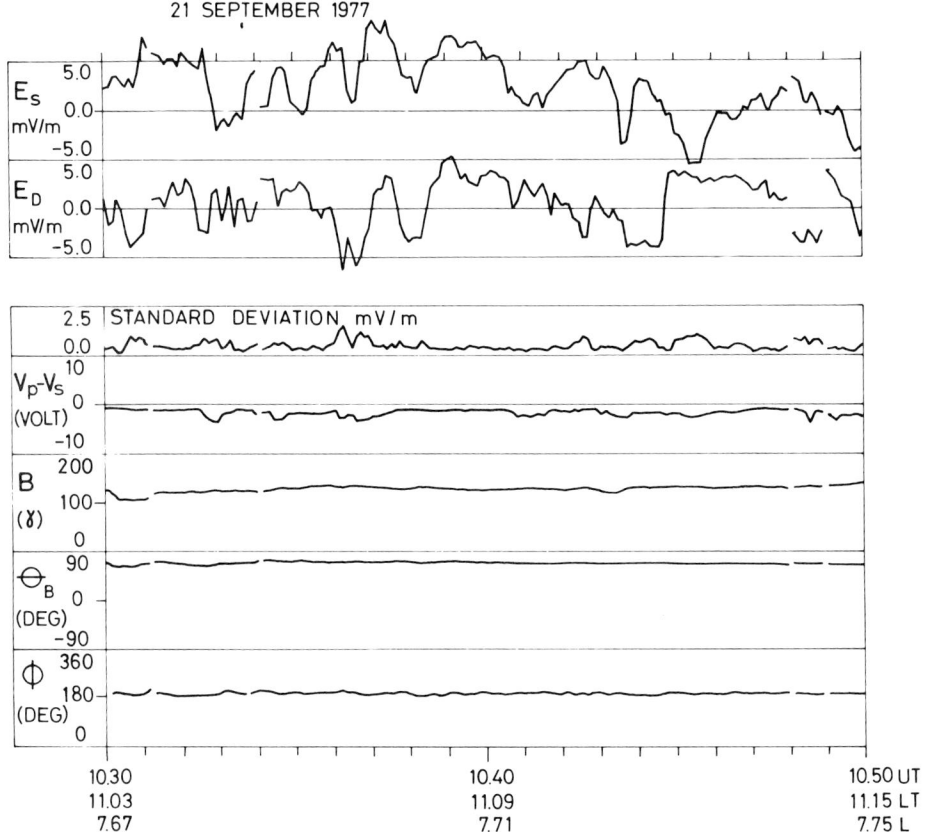

Fig. 7. Fluctuating electric field observed in the vicinity of the magnetopause near noon LT. The magnetopause moved out and crossed the GEOS-1 position at 1020 UT. $V_p-V_s$ is probe-satellite potential difference; the probe is biased with a current source to bring its potential ($V_p$) near to that of the plasma. The standard deviation gives the accuracy of the least square fit used to derive the electric field once per spin.

noon. However, transition to a dusk-dawn electric field, or anti-sunward plasma flow, is often occurring toward the morning hours. The local time for the field turn-over varies from day to day and is typically 0900.

This feature is important because it indicates that the sunward plasma flow in the central regions of the magnetosphere turns into an anti-sunward direction along the flanks of the magnetosphere as shown in the early model of Axford and Hines (1961).

Near local noon, the electric field is dawn to dusk, i.e. the plasma flow is sunward. In the case of the ISEE-1 data from 20 November 1977 the plasma was flowing sunward all the way up to the magnetopause (Mozer et al., 1979).

The data collected on 20 November 1977 (Fig. 5) show a dawn to dusk electric field over most of the GEOS-1 orbit with the exception

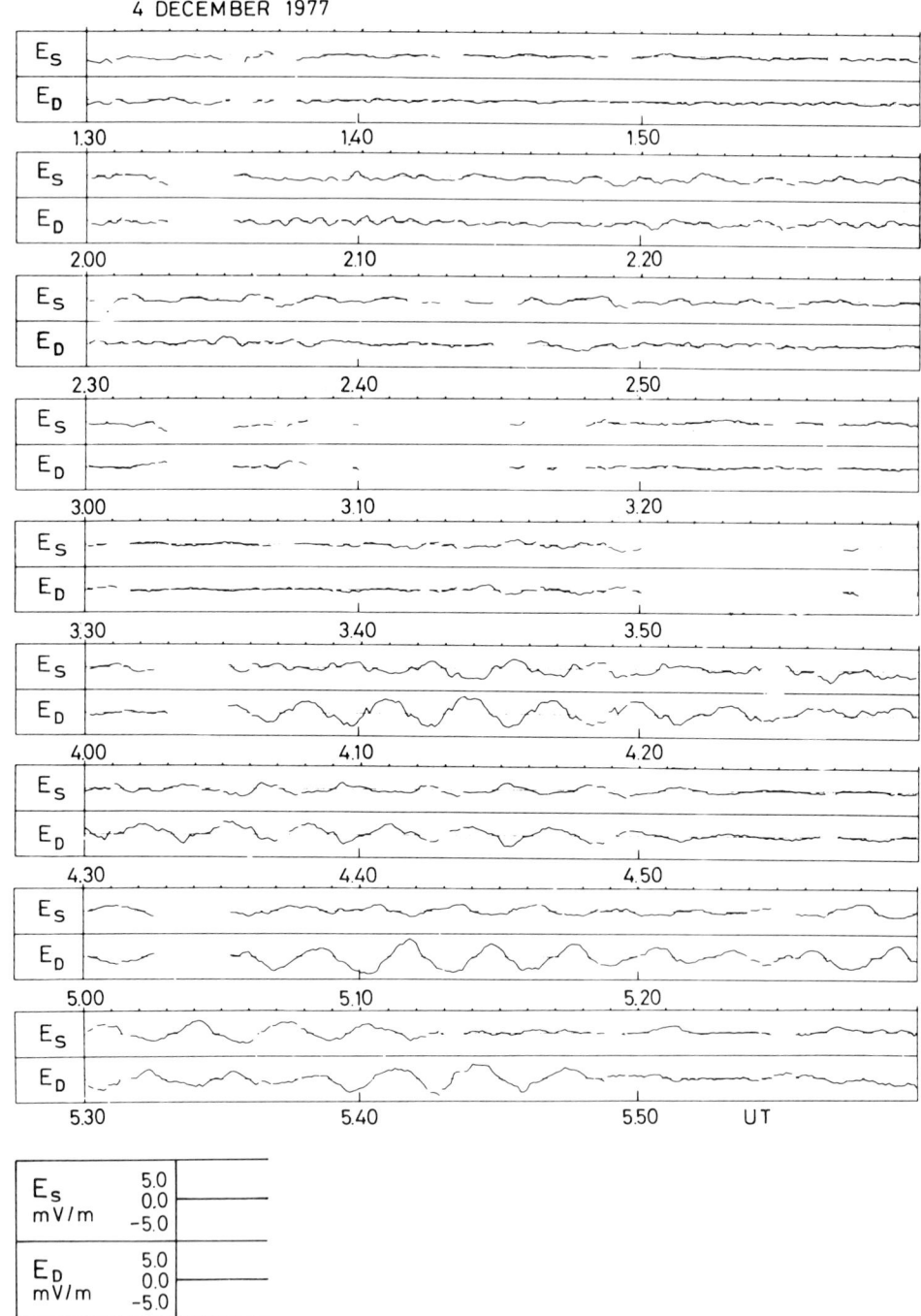

Fig. 8. The $E_S$ and $E_D$ components of the electric field for nearly one full pass of GEOS-2 on 4 December 1977. PC4-PC5 wave events are observed after about 0400 UT.

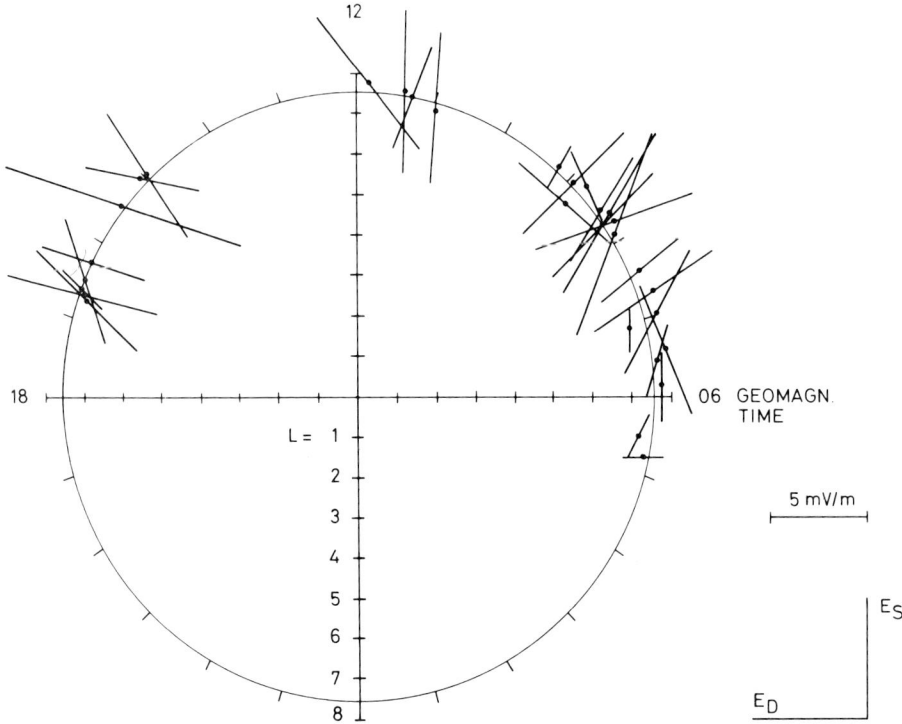

Fig. 9. The orientation of the PC5 wave electric fields for a number of events on GEOS-1 shown in a geomagnetic time versus L-value diagram.

of a short interval between 0600 and 0630 LT where the electric field has an opposite direction. The solar wind magnetic field has a southward component on that day and magnetic field line reconnection is also observed near the front magnetopause by the electric and magnetic field experiments on ISEE-1 and 2 (Mozer et al., 1979). A large region with sunward flow is expected as a consequence of an effective reconnection process near the front magnetopause.

It is likely that the dominant dusk-dawn electric fields, i.e. antisunward flows seen in the morning sector for some days (Fig. 4) result from an absence of, or less effective, reconnection process; the plasma then returns near the magnetosphere flanks rather than to flow to the magnetopause.

## The Night Sector during Quiet Conditions

Very little data has been systematically analysed due to the short time elapsed since GEOS-1 has entered the magneto-tail and GEOS-2 has been launched. After evaluation of several data quicklooks, one can nevertheless make the following comments:
a) Electric field pulsations with amplitudes of the order 1 mV/m and periods of 1-5 minutes are frequently observed. These pulsations are

generally linearly polarised along the azimuthal direction.
b) On many days, the dc electric field is oriented from dawn to dusk and has a magnitude of approximately 1mV/m.
c) On several other days, on the contrary, the field has a very small amplitude, less than 0.5 mV/m and its orientation changes from day to day. This observation made with GEOS-1 seems to be confirmed by GEOS-2 data, especially during eclipses when the accuracy of the measurement is of the order 0.1 mV/m.

It should be mentioned however, that the evaluation of the data requires a careful interpretation when the ambient plasma is hot and tenuous, a situation which is frequently encountered in that part of the magnetosphere.

## Substorms

Electric fields larger than 20 mV/m have been observed during substorms. These unexpectedly large fields saturate the main data channel but can however be telemetered via an auxiliary channel with less sensitivity and time resolution (500 mV/m full range instead of 10; 11 samples per second instead of 32). Fig. 6 shows an example of such large electric fields observed on 2 December 1977. Measurements transmitted through the auxiliary channel are represented by lines marked with dots; results are not plotted when the field is so unstable that its magnitude and orientation cannot be evaluated with confidence. These large electric fields have typical duration of 20-30 s and are oriented from dawn to dusk, which may be associated with impulsive injections of plasma from the magnetotail; the drift velocity associated with these phenomena is typically 200 km/s. It is significant that in Fig. 6 there are onsets near 0250 and 0300 UT with subsequent impulses and strong fluctuations. Thus each typical event seems to have a duration of several minutes and consists of a series of discrete enhanced plasma flows. These events are detailed in Pedersen et al. (1978) and are similar to observations made on the Imp satellite by Aggson and Heppner (1977).

Multiplying the 30 s duration of an impulse by a derived drift speed of 200 km/s, one can infer that GEOS-1 was passed by a plasma structure with a dimension of the order of 1 $R_E$ along the direction of the flow. It is difficult, however, to say more about the configuration of this plasma cloud or stream.

## Electric Field Fluctuations and Pulsations

Electric field fluctuations with time scales of 0.5-5 minutes seem to have larger amplitudes, and to occur more often at local times around noon. In many cases these variations display a very clear structure.

Fluctuations increase in amplitude after major magnetospheric disturbances and at times when the magnetopause moves close to the satellite position, situations which occurred on the 29 July and 21 September 1977 when GEOS-1 actually crossed the magnetopause. Fig. 7 shows an example of such time-varying electric fields close to the inner edge of the magnetopause; the magnetopause crossings will be the subject of separate publications.

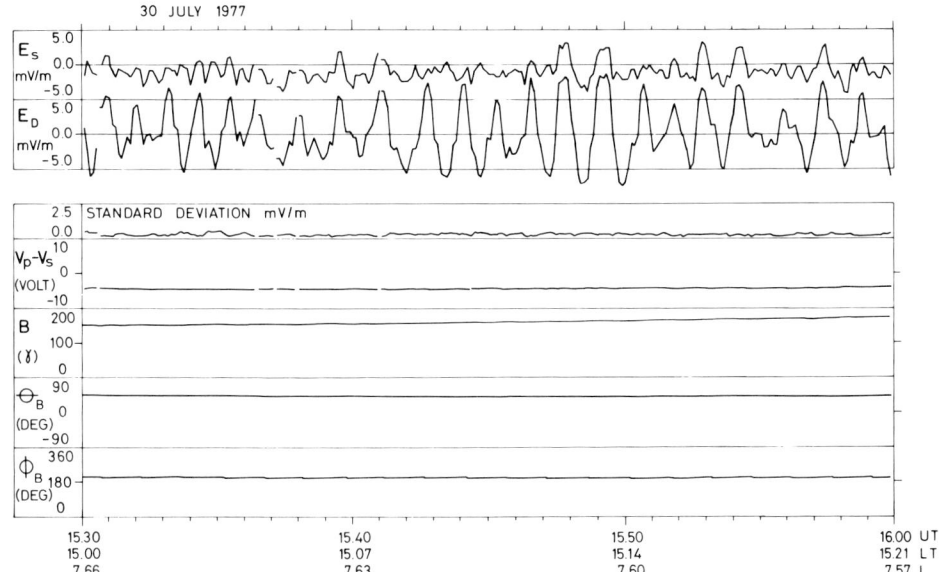

Fig. 10. Electric field pulsations with a period of approximately 1.5 minutes and a maximum amplitude of 7 mV/m observed on 30 July 1977, following a major magnetospheric disturbance on the previous day.

Waves in the PC4 and PC5 ranges, i.e. with periods in the ranges 45-150 s and 150-600 s respectively, have previously been detected with magnetic sensors on the ground and in space, sometimes simultaneously (Kokubun et al., 1976, Lanzerotti et al., 1976, Cummings et al., 1978). The electric component of these waves, however, was observed for the first time on GEOS-1 (Pedersen et al, 1978). Both PC4 and PC5 pulsations were observed on 4 December 1977, as illustrated in Fig. 8. Data are missing when their quality is poor, when a diagnostic programme is performed with the electric sensor a few minutes after the beginning of every hour, and when an active experiment using the antenna for wave transmission is operated. A PC4 event with a period of approximately 60 s and an amplitude of 1 mV/m is seen at around 0210 UT. At 0340 there is a gradual rise of a PC5 pulsations peaking at 0410 UT with a period of 180 s and an amplitude of 3 mV/m. There is a recurrence at intervals of 30 to 60 minutes. The average dc electric field is small since the two components of the pulsations are nearly symmetrical around the horizontal axes.

At the time of this review, no complementary magnetic field data is available for these PC4-PC5 events, and a full treatment including the calculation of the Poynting vector has not been carried out yet.

A total of 30 events have been identified so far in the front side of the magnetosphere and Fig. 9 shows a summary of these observations. The location of the spacecraft, L value vs geomagnetic time, is indicated by a dot; the length of the bar is proportional to the peak to peak amplitude of the pulsation. The electric field is linearly polarized within the measurement uncertainty, and its orientation is in most cases radial with a few exceptions in the morning hours. PC4 and PC5 events are not simultaneous with major substorms, but seem to appear several hours later and also on the following days. One example is given in Fig. 10, where a pulsation with an amplitude of 7 mV/m and a period of 90 s is seen on the day following the storm of 29 July 1977.

Dramatic changes in the particle population are associated with these events. The fluxes of $H^+$ and $O^+$ ions have been observed on GEOS-1 to vary with the same period but with a phase shift of $90°$ (D. Young, private communication). Comparison between GEOS-1 data and ionosphere drift measurements in the auroral zone also resulted in the identification of the same PC5 events (R. Greenwald, private communication).

Simultaneous satellite observations of the magnetic components of PC4-PC5 waves on several geostationary satellites have demonstrated that the phase velocity is away from the front of the magnetosphere, (Hughes et al., 1978). This conclusion is also reached by Olson and Rostoker (1978) on the basis of ground observations supporting the theory that Kelvin Helmholtz instabilities are generated near the front magnetopause. Certain frequency components are expected to come into resonance conditions and give rise to the nearly monochromatic PC4-PC5 events when the waves travel away from the noon sector.

## Summary

The processing of GEOS-1 data has just entered a routine phase and GEOS-2 was launched a few months ago. Nevertheless it is possible to point out several new features which have resulted from the first data analysis. The main characteristic of the electric fields observed so far is their frequent variability. The amplitude of the fluctuations is generally large with respect to the mean value and it is necessary to carry out a careful averaging in order to obtain a picture of magnetospheric convection.

The typical features of the dc electric field in a quiet or slightly disturbed magnetosphere at L values between 6.5 and 7.5 are listed below for various local time sectors:
(i) Afternoon: A dawn to dusk electric field of 1 to 3 mV/m is frequently observed for periods of up to several hours. The strength of this field varies from day to day, and it is sometimes below the threshold of measurements ( 0.5 mV/m).
(ii) Frontside: Dawn to dusk fields of the order of 1-2 mV/m seem to be always present around noon; they are associated with sunward flows and have been detected all the way up to the magnetopause.
(iii) Morning: The extent of the dawn to dusk fields towards the morning hours changes from one day to the next. An inversion of the field direction occurs between 0700 and 1100 LT. Future studies using combination of data from GEOS-1, GEOS-2 and ISEE-1 will aim to clarify if the inversion to dusk-dawn fields and associated anti-sunward flows dominate when conditions for field line reconnection near the front magnetopause are less favourable.
(iv) Night: Few data have been analysed in the night sector but irregular dawn-dusk electric fields with amplitudes of the order of 1 mV/m seem to prevail on certain days whereas at other times the electric field is considerably smaller.

Some features of the electric field have a time scale of the order of 1 minute:

(i) Substorm events: Very large dawn-dusk electric fields of 20 mV/m have been observed in the morning and night sectors. These events consist of a series of pulses, typically 30 s long, distributed over a period of several minutes. These events are probably associated with plasma injection from the magnetotail.

(ii) Fluctuations and pulsations: The electric field is turbulent near the magnetopause where fluctuations of 5-10 mV/m may have durations of one to several minutes. Pulsations in the PC4-PC5 range with amplitudes of several mV/m have been observed outside L = 7.0 over the whole front magnetosphere from 0600 to 1800 LT. The wave electric fields appear to be close to linearly polarized and its direction is radial, except in a few cases in the morning where it is azimuthal. Observations of a similar periodicity in GEOS ion data and conjugated ground radar drift measurements are promising subjects for future correlated studies.

Acknowledgements. The authors are indebted to Mr. P. Jones and Mr. C. Dutton who have developed the computer software used for processing the electric field data.

## References

Aggson, T.L. and J.P. Heppner, Observations of large transient magnetospheric electric fields, J. Geophys. Res., 82, 5155-5165, 1977.

Axford, W.I. and C.O. Hines, A unifying theory of high latitude geophysical phenomena and geomagnetic storms, Can. J. Phys., 39, 1433-1464, 1961.

Cummings, W.D., S.E. DeForest and R.L. McPherron, Measurements of the Poynting vector of standing hydromagnetic waves at geosynchronous orbits, J. Geophys. Res., 83, 697-706, 1978.

Decreau, P.M.E., J. Etcheto, K. Knott, A. Pedersen, G.L. Wrenn and D.T. Young, Multi-experiment determination of plasma density and temperature, Space Sci. Rev., 22, 633-645, 1978.

Fahleson, U., C-G. Fälthammar, A. Pedersen, K. Knott, G. Brommundt, G. Schumann, G. Haerendel and E. Rieger, Simultaneous electric field measurements in the auroral ionosphere by using three independent techniques, Radio Science, 6, 233-245, 1971.

Fahleson, U., C-G. Fälthammar and A. Pedersen, Electric field and plasma measurements in the autoral ionosphere prior to a magnetic substorm, Space Research, 471-475, D. Reidel Publ. Co., Dordrecht, The Netherlands, 1975.

Hughes, W.J., R.L. McPherron and J.N. Barfield, Geomagnetic pulsations observed simultaneously on three geostationary satellites, J. Geophys. Res., 83, 1109-1116, 1978.

Iijima, T. and T.A. Potemra, Large-scale characteristics of field-aligned currents associated with substorms, J.Geophys. Res., 83, 599-615, 1978.

Kokubun, S., R.L. McPherron and C.T. Russell, OGO5 observations of PC5 waves, Ground-Magnetosphere correlations, J. Geophys. Res., 81, 5141-5149, 1976.

Lanzerotti, L.J., H. Fukunishi and C.G. Maclennan, Observations of magnetohydrodynamic waves on the ground and on a satellite, J. Geophys. Res., 81, 4537-4545, 1976.

McIlwain, C.E., Plasma convection in the vicinity of the geosynchronous orbit, in Earth's Magnetospheric Processes, edited by B. McCormac, 268-279, D. Reidel Publ. Co., Dordrecht, The Netherlands, 1972.

Melzner, F.S., G. Metzner and D. Antrack, The GEOS electron beam experiment S-329, Space Sci. Instru., 4, 45-56, 1978.

Mozer, F.S. and P. Lucht, The average auroral zone electric field, J. Geophys. Res., 79, 1001-1006, 1974.

Mozer, F.S., Magnetospheric dc electric fields; present knowledge and outstanding problems to be solved during the IMS, in The scientific satellite programme during the IMS, 101-131, D. Reidel Publ. Co., The Netherlands, 1976.

Mozer, F.S., R.B. Torbert, U. Fahleson, C-G. Fälthammar, A. Gonfalone and A. Pedersen, Measurements of quasi-static and low frequency fields with sperical double probes on the ISEE-1 spacecraft, IEEE Trans. on Geoscience Electr., 16, 258-261, 1978.

Mozer, F.S., R.B. Torbert, U. Fahleson, C-G. Fälthammar, A. Gonfalone and A. Pedersen, Electric field measurements in the solar wind, bow shock, magnetosheath, magnetopause and magnetosphere, Proceedings of ESLAB Symposium, Innsbruck, 1978. To be published in Space Science Reviews, 1979.

Olson, J.V. and G. Rostoker, Longitudinal phase variations of PC4-5 micropulsations, J. Geophys. Res., 83, 2481-2488, 1978.

Pedersen, A., R. Grard, K. Knott, D. Jones, A. Gonfalone and U. Fahleson, Measurements of quasi-static electric fields between 3 and 7 earth radii on GEOS-1, Space Sci. Rev., 22, 333-346, 1978.

# GENERATION OF THE MAGNETOSPHERIC ELECTRIC FIELD

T. W. Hill

Space Physics and Astronomy Department, Rice University
Houston, Texas 77001

Abstract. The boundary conditions on the magnetospheric electric field are determined by the solar-wind/magnetosphere interaction, the dynamics of the magnetospheric tail, and the atmosphere/magnetosphere interaction. This paper surveys various mechanisms that are suspected of playing a role in the establishment of these boundary conditions.

The hypothesis that merging of terrestrial and interplanetary magnetic fields controls the potential drop across the polar cap has received considerable observational support. In particular, the control exerted by the interplanetary magnetic-field direction over polar cap convection patterns appears to favor the merging hypothesis vis-à-vis alternative mechanisms involving quasi-viscous momentum transfer at the magnetopause, although the latter undoubtedly plays some role. The merging process has been studied theoretically in considerable detail, but the idealized geometry of the models and their assumption of time-independence have until now precluded their direct application to the magnetopause.

Magnetic merging may also play a crucial role in the dynamics of the magnetospheric tail. Here the assumption of time-independence is particularly embarassing; induction electric fields are known to be essential in the tail dynamics, especially during substorms when they exceed the potential field in magnitude. Such induction fields have not yet been successfully incorporated in the merging theory.

The electric coupling between the atmosphere and magnetosphere, insofar as it results from ionospheric conduction currents and associated Birkeland (magnetic-field-aligned) currents, is understood relatively well, even quantitatively. There is, however, no consensus as to how this coupling is modified by parallel (magnetic-field-aligned) electric fields, nor as to how such fields are generated in the collisionless magnetospheric plasma.

To the extent that these interactions can be simulated by prescribed boundary conditions, the self-consistent distributions of plasmas and electric fields can be calculated quantitatively and realistically within the sub-auroral ionosphere and magnetosphere.

## Introduction

The potential electric field ($\underline{E} = -\nabla\phi$) in the magnetosphere satisfies two boundary conditions, the outer boundary being the magnetosphere/solar-wind interface (magnetopause) and the inner boundary being the magnetosphere/atmosphere interface (ionosphere).

The distribution of the imposed potential between the two boundaries affects, and is affected by, the configuration and motion of plasma in the magnetosphere.

The mutually consistent distributions of magnetospheric plasmas and electric fields within eight or nine Earth radii distance, in response to a given potential distribution at the boundaries, are understood in considerable detail. Quantitative numerical models have been developed that are capable of simulating the response of the inner magnetosphere to an externally imposed electric potential distribution. I will touch only briefly on the conceptual framework of these models; they are discussed in detail elsewhere in this volume.

I will concentrate here on the processes by which the potential distribution is established at the boundaries of this inner region. Here our understanding is in a relatively primitive state, especially with regard to the day side solar-wind/magnetosphere interaction and the dynamics of the magnetospheric tail. Viable mechanisms have been proposed for the solar-wind/magnetosphere interaction, and some of them have been treated quantitatively but within simplified geometries. There exists little consensus, however, as to which mechanism is dominant, nor any quantitative models that deal with the three-dimensional geometry of the magnetopause. Likewise the dynamics of the magnetospheric tail is not understood to the point of constructing quantitative comprehensive models. In the tail, especially during substorms, the non-potential (i.e., solenoidal) electric field may become dominant, thus complicating the problem.

I will also describe briefly the mechanisms by which the atmosphere influences the magnetospheric electric field, and the extent to which these mechanisms have been, or may soon be, incorporated into the quantitative simulation work mentioned above.

The reader is referred to a recent review paper by Stern (1977) for interesting historical perspectives on the subject, for a thorough bibliography, and for a more detailed discussion of some of the concepts described below.

## Convection in the Inner Magnetosphere

A simplified view of the magnetosphere is shown in Figure 1 (a -- equatorial cross section; b -- noon midnight meridian cross section). The dashed contours represent the boundaries of what I have called the inner magnetosphere. The outer boundary is a magnetic flux shell that lies sufficiently inside the magnetopause that the solar-wind interaction, including the dynamics of the magnetospheric tail, may be summarized, insofar as its effect on the interior region is concerned, by the specification of a prescribed, albeit time-dependent, electric potential distribution on the boundary. The inner boundary is taken to be the ionosphere, and its effect on the magnetospheric electric field may be summarized conveniently, if not completely, by the specification of a prescribed, albeit time-dependent, two-dimensional (height-integrated) conductivity tensor. These two boundaries contain a vast region of space within which the motion of magnetospheric plasma, and the associated electric field, can be calculated quantitatively and realistically if the boundary conditions are specified. The boundary conditions can, in fact, be specified more or

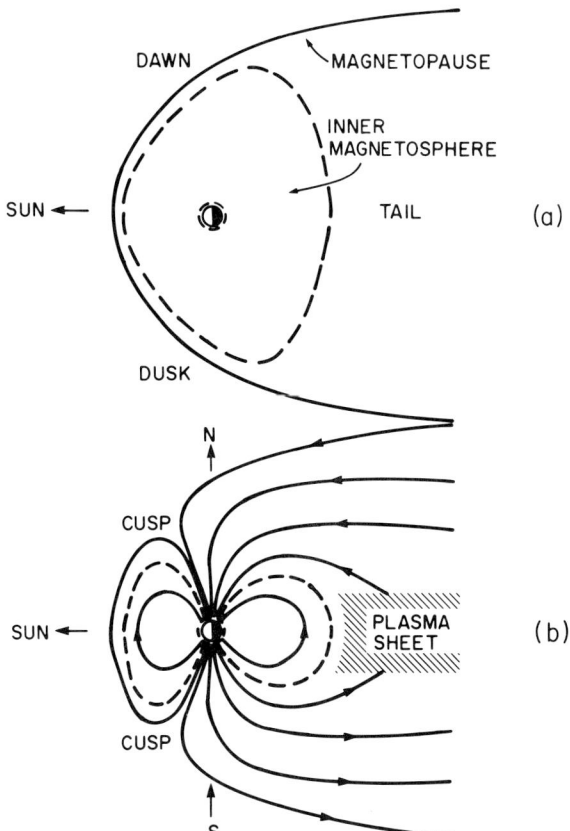

Fig. 1. Simplified cross sections of the magnetosphere in (a) the equatorial plane and (b) the noon-midnight meridian plane. The dashed lines represent the boundaries of the "inner magnetosphere" within which the self-consistent calculation of magnetospheric electric fields has reached a relatively high level of sophistication. The noon meridian section illustrates the closed magnetosphere topology in which magnetic field lines do not connect the earth with interplanetary space.

less adequately by combined spacecraft and ground-based observations, and the resultant magnetospheric convection patterns can then be calculated in sufficiently quantatitative detail to be compared meaningfully with other simultaneous data. The historical development of these calculations is described in the papers by Vasyliunas (1970, 1972) and by Wolf (1970, 1974, 1975), Jaggi and Wolf (1973) and Harel and Wolf (1976).

Conceptually, we can view the logic of these models as follows. We are given the potential distribution on the outer boundary, the conductivity distribution on the inner boundary, and the assumption that the magnetic field lines (as given by an empirical field model) are perfect conductors (i.e., equipotentials). We can then calculate the electric field distribution that would exist <u>in vacuo</u>, i.e., if the magnetosphere were devoid of plasma and the ionospheric conductivity were zero. This "first iteration" field would immediately reveal two inconsistencies: (1) the ionosphere would conduct both Pedersen and Hall currents that would, in general, not close within the ionosphere, so that they would polarize and thus modify the imposed electric field, except to the extent that they can be closed in the magnetosphere through the intermediary of Birkeland (magnetic-field-aligned) currents; (2) the magnetospheric plasma itself would polarize because of the charge-dependent magnetic gradient and curvature drifts, and would thus modify the imposed electric field except to the extent that the polarization current can be closed in the ionosphere through Birkeland currents. Given an initial distribution of magnetospheric plasma, and the equations of motion that satisfies, we can then iterate so as to minimize these inconsistencies and thus find the mutually consistent distributions of plasma and potential in the inner magnetosphere.

If we then allow the boundary conditions to change on a time scale that is slower than the Alfvén-wave travel time between ionosphere and magnetosphere (about a minute) but faster than the longitudinal drift period (several hours), we can calculate a series of consecutive self-consistent but non-steady-state configurations that represent the time-dependent motions in the inner magnetosphere during, for example, a magnetospheric substorm. The outer boundary conditions in this case may include a prescribed induction electric field due to rapid changes in the magnetospheric tail magnetic field in addition to the externally-imposed electrostatic potential distribution.

This kind of self-consistent calculation (much easier said than done) has now been applied to the simulation of a particular magnetospheric substorm event, with some success at reproducing the observed behavior of the inner magnetosphere. These results are described elsewhere in this volume (Harel et al., 1979).

The feasibility of such quantitative models relies in large part on two facts: (1) the behavior of plasma in the inner magnetosphere can be described by simple guiding-center equations of motion, and (2) empirical models exist that describe quantitatively the magnetic field configuration in this region of space.

The global numerical models have not yet included (1) the dynamo effect of upper atmospheric winds, (2) the loss of magnetospheric plasma through precipitation and charge exchange, or (3) the effects of parallel (magnetic-field-aligned) electric fields. The inclusion of the first two effects is mainly a computational problem; in the case of parallel electric fields, however, we are not in a theoretical position to predict confidently even their occurrence, much less their magnitude and configuration.

In spite of these limitations, the relative success of numerical simulation studies of the inner magnetosphere stands in sharp contrast to our meager understanding of the physical processes that establish the boundary conditions, especially at the outer boundary.

### Solar-Wind/Magnetosphere Interaction

Quantitative global models of the solar-wind/magnetosphere interaction do not yet exist, partly because the global geometry is complicated, but more importantly because the basic physical

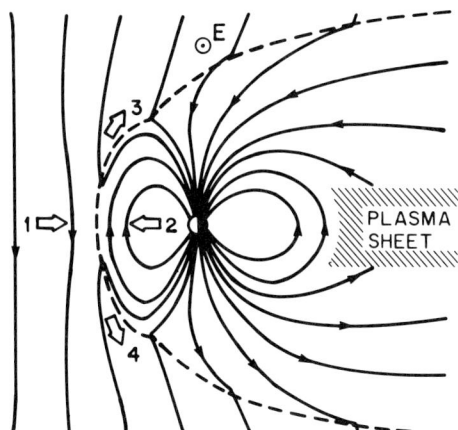

Fig. 2. A noon-midnight meridian section illustrating the open magnetosphere topology in which the polar-cap magnetic field lines connect the earth with interplanetary space (cf. Figure 1b). Open arrows indicate the plasma motion associated with magnetic merging at the day side magnetopause.

mechanisms of this interaction are not adequately understood. Proposed mechanisms for this interaction can be classified according to whether or not they require the existence of magnetic field lines connecting Earth with interplanetary space, that is, whether they utilize the open (Figure 2) or closed (Figure 1b) model of the magnetosphere.

The basic problem is to explain the transfer of solar-wind momentum that drives the antisunward convection of plasma across the polar regions of the Earth. The qualitative pattern of the flow, as first deduced by Axford and Hines (1961) from ground-based observations, is depicted in Figure 3. The region of antisunward flow in the ionosphere is generally called the polar cap (the region within the dashed boundary in Firgure 3a). In the closed model, the polar-cap boundary maps along "closed" field lines to a viscous or quasi-viscous boundary layer inside the equatorial magnetopause (Figure 3b). In the open model, the polar cap maps along "open" field lines into interplanetary space (Figure 2). The solar-wind momentum transfer establishes the electrostatic potential drop across the polar cap, and the return (sunward) flow is controlled by the low latitude fringing field which is determined in large part by the distribution of ionospheric conductivity.

We know from correlative observations (see, for example, Heppner, 1972; Burch, 1974; Fairfield, 1977, and references therein) that the potential distribution is controlled to a large extent by the direction of the interplanetary magnetic field, in particular by its two components perpendicular to the solar-wind flow direction. These correlations seem to indicate that a direct interconnection of the geomagnetic and interplanetary magnetic fields is responsible for establishing the potential distribution in the magnetosphere by mapping the solar-wind motional electric field down along open field lines (Figure 2). This mode of momentum transfer (Dungey, 1961), in turn, requires the phenomenon of magnetic "merging" whereby

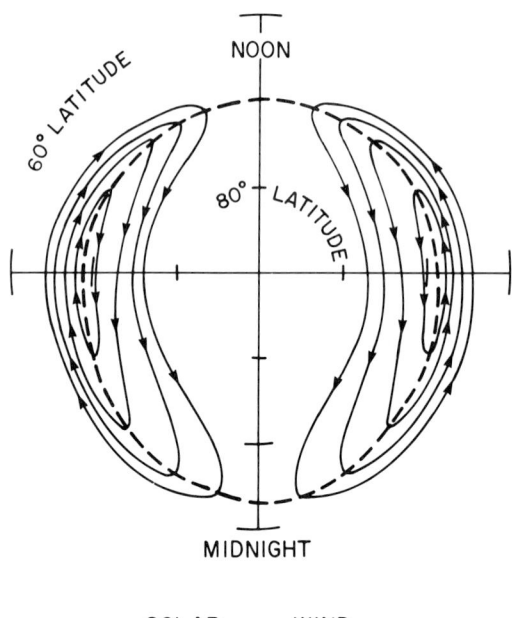

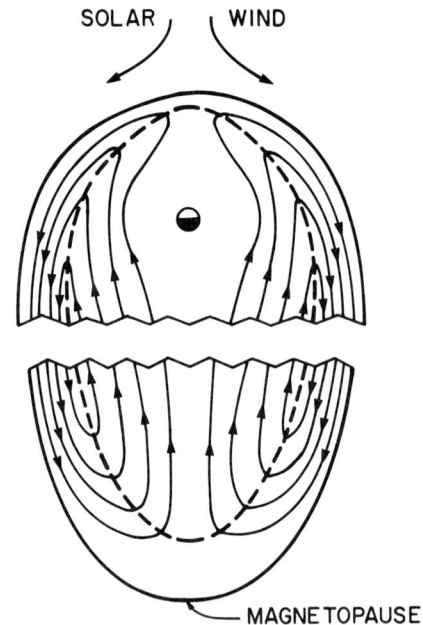

Fig. 3. High-latitude plasma convection pattern as viewed (a) on the earth's surface (polar projection) and (b) in the equatorial plane of the magnetosphere. Antisunward convection occurs in the polar cap [within the dashed boundary in (a)]; this polar cap maps magnetically to a quasi-viscous boundary layer in the equatorial plane within the closed magnetosphere model [outside the dashed boundary in (b)]. In the open model, the polar cap maps magnetically into interplanetary space.

interplanetary and geomagnetic field lines become interconnected at the day side magnetopause (Figure 4). Given this interconnection, solar-wind momentum can be transmitted directly to the polar cap by the magnetic stress developed along these open field lines.

In a closed magnetosphere model, momentum must be transferred across the magnetopause through a quasi-viscous process. Ordinary (collisional) viscosity is inadequate by many orders of magnitude. The apparent viscosity may, in general, be produced either by the absorption of hydromagnetic wave momentum or by the anomalous diffusion of solar-wind plasma across the magnetopause ("anomalous" in the sense that the mean free path due to wave-particle interactions is much less than that due to Coulomb collisions). The observed interplanetary-magnetic-field control of polar cap convection patterns has been a stumbling block to the development of closed-model momentum transfer models.

Magnetic merging and anomalous diffusion are both difficult to treat theoretically, largely because both processes violate the frozen-in-flux principle (Alfvén and Fälthammar, 1963) whereby the plasma populations of two different magnetic flux tubes cannot intermix. This frozen-in-flux condition can be written

$$\underline{E} + \underline{v} \times \underline{B} = 0$$

where $\underline{E}$ and $\underline{v}$ are the electric field and bulk plasma velocity (measured with respect to a common frame of reference), and $\underline{B}$ is the magnetic field. Wave momentum transfer, on the other hand, may or may not require a violation of frozen-in-flux, depending on the nature of the waves involved.

## Magnetic Merging at the Magnetopause

Magnetic merging is the process whereby magnetic flux tubes flow between two topologically distinct sectors of the magnetic field, e.g., from regions 1 and 2 into regions 3 and 4 in Figures 2 and 4. [A "flux tube" is defined here as a collection of charged particles whose guiding centers share a common bundle of field lines -- the motion of a flux tube is then defined in terms of the motion of a collection of particles and does not rely on the ambiguous (and somewhat disreputable) concept of field-line motion.] The fact that such motion implies violation of Equation (1) in a steady state can be appreciated by referring to the flow and field configuration of Figure 4. Note first that Equation (1) implies $E_{\parallel} \equiv \underline{E} \cdot \underline{B}/|\underline{B}| = 0$. Then Faraday's law would imply that, for a steady state, the component of $\underline{E}$ perpendicular to the plane of the figure is constant. On the other hand, the same component of $\underline{v} \times \underline{B}$ cannot be a constant because both components of $\underline{B}$ in the plane of the figure vanish at the "neutral line" (which extends perpendicular to the figure through the point labelled "N"), but do not vanish elsewhere. This conclusion still holds if we add a uniform component of $\underline{B}$ perpendicular to the plane of the figure to allow for merging between magnetic fields that are not strictly antiparallel, which is the general case at the day side magnetopause. In this general case the "neutral line" is no longer "neutral" ($\underline{B} \neq 0$ there), but the same line still separates magnetic-field sectors that are topologically distinct. (Tradition has somehow identified this line as feminine, so that it is customarily called, not a separator, but a separatrix.) Thus Equation (1) cannot hold throughout the flow field of Figure 4 in a steady state, and we have two basic alternatives: either (1) the component of $\underline{E}$ along the separatrix vanishes, in which case Equation (1) holds there but must be violated in the external flow field, or (2) the component of $\underline{E}$ along the separatrix is nonzero, such that Equation (1) holds in the external flow field but must be violated near the separatrix. The first option can be shown to imply that $E_{\parallel}$ in the external flow is, not merely nonzero, but of a sufficient magnitude to completely decouple the plasma motions inside and outside the magnetosphere. We would thus sacrifice the primary argument for adopting the open magnetosphere topology in the first place, namely, the idea that magnetospheric convection is driven by the transfer of solar-wind momentum along interconnected field lines. Thus the second option is widely espoused, that an open magnetosphere requires, for all practical purposes, the occurrence of magnetic merging at the magnetopause (see, however, the dissenting view of Heikkila, 1975).

Existing theoretical work on the merging process can, in many cases, be described as quantitative, but only as applied to highly idealized geometries. The bulk of this work has utilized the magnetohydrodynamic formalism as applied to the two-dimensional geometry of Figure 4; this work has been synthesized in the comprehensive review paper by Vasyliunas (1975). Parallel work from a collisionless ("single-particle") approach has been presented by Alfvén (1968), Dessler (1968, 1971), Speiser (1970), Eastwood (1972), Hill (1973a, 1975), and Francfort and Pellat (1976).

The theory of magnetic merging remains a highly controversial subject. Nevertheless, the various theories seem to have converged on the following fundamental points: (1) the outflow (regions 3 and 4 in Figure 4) occurs at a speed comparable to the Alfvén speed as measured in the inflow region (regions 1 and 2 in Figure 4); this in turn implies that (2) the inflow speed (the "merging speed") is smaller than the local Alfvén speed by a ratio comparable to the tangent of the angle at which the field lines intersect the magnetopause; (3) the Alfvén speed is self-adjusting such that the merging speed is determined more by the dynamics of the macroscopic flow field than by the dynamics of the "diffusion region" i.e., the small region surrounding the separatrix wherein corrections to Equation (1) become important; but that (4) the allowable range of merging speeds depends on the relative orientation of the two opposing fields, such that merging is most efficient when the two fields are antiparallel and becomes less efficient as the fields become more nearly parallel.

The above results can be applied qualitatively to the day side magnetopause, resulting in what I consider to be the two decisive victories for the open magnetosphere model: (1) the magnitude of the observed polar-cap potential drop (~ 50 kV) can be produced with reasonable merging speeds -- something like 10% of the interplanetary magnetic flux that would intersect the magnetospheric cross section is required to merge with the geomagnetic field; and (2) the open model provides a natural explanation for the observed control of polar-cap convection by the transverse components of the solar-wind magnetic field, both the north-south component which controls the magnitude of the potential drop, and the east-west component which controls the dawn-dusk asymmetries of the convection pattern (see the review by Fairfield, 1977, and references therein).

In spite of these successes, there is one apparent setback that has been pointed out by Heikkila (1975). The merging process is inherently dissipative in that $\tilde{j} \cdot \tilde{E} > 0$ where $\tilde{j}$ and $\tilde{E}$ are the electric current and electric field at the magnetopause. The required magnitudes of $\tilde{j}$ and $\tilde{E}$ imply a readily observable rate of particle acceleration and/or heating in the merging process. In the simple Dungey picture (Figure 2), this energization would be manifested in an accelerated outflow of plasma away from the equator near the dayside magnetopause, and this outflow is not observed by spacecraft that have seemingly looked in the right place at the right time with the right instrumentation (Haerendel et al. 1978). Although the temporal and angular resolution of the spacecraft measurements may be considered marginal for the detection of this outflow, a more natural explanation of this apparent discrepancy is probably that merging occurs near the cusps (marked in Figure 1b) rather than at the nose of the magnetosphere as depicted in Figure 2. This explanation was proposed by Haerendel et al. (1978) and its consequences for the magnetospheric convection pattern have been explored by Crooker (1979) Crooker has shown that, if merging occurs at the cusps rather than at the nose of the magnetosphere then a large fraction of the subsolar region of the magnetopause would not be magnetically connected to the merging region, for typical interplanetary magnetic-field orientations, and thus would not be expected to exhibit the accelerated outflow that would be expected in the case of a purely southward interplanetary field as shown in Figure 2. Thus the cusp merging hypothesis accounts for the fact that the merging-accelerated plasma is apparently observed at low altitudes along cusp field lines (Reiff et al., 1977; Hill and Reiff, 1977), but not over a wide area of the subsolar magnetopause (Haerendel et al., 1978).

The basic merging hypothesis thus appears to be capable of surviving the challenge put forth by Heikkila (1975) on the basis of the disagreement between predicted and observed outflows on the day side magnetopause. This question is not really settled observationally, however, and it is clear that the resolution of the conflict will, at the very least, bring significant revisions in our ideas regarding the location of merging and the resultant flow geometry.

Crooker (1979) has also shown that the hypothesis of merging at the cusps would provide a comprehensive (although still qualitative) explanation of the observed dawn-dusk asymmetries of the polar-cap convection pattern and their dependence on the dawn-dusk component of the interplanetary magnetic field (Figure 5).

There remains, however, no realistic quantitative model of the merging process as applied to the solar-wind/magnetosphere interaction. There are three main obstacles to the development of such a model: (1) the present theories all apply to highly idealized geometries that cannot be applied unambiguously to the complicated three-dimensional flow and field geometry at the magnetopause; (2)

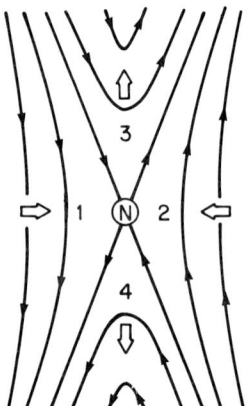

Fig. 4. A detail of the merging region shown at the dayside magnetopause in Figure 2. Open arrows denote plasma flow and "N" marks the neutral line or separatrix. The same figure, rotated 90°, would illustrate the geometry of magnetic merging in the magnetospheric tail.

even if such a model were available, we lack a quantitative empirical model against which the theory might be tested; and (3) the magnetopause merging process may well be inherently time-dependent, or even turbulent (see below), whereas the available theories deal primarily with a steady state laminar flow configuration.

Diffusive Momentum Transfer

As is illustrated in Figure 6, the convection potential drop that is established in a diffusive boundary layer is proportional to the thickness of that layer, and hence to the distance that particles can diffuse across the magnetic field while they flow downstream along the magnetopause. A quantitative treatment by Eviatar and Wolf (1968) has shown that an ion-cyclotron instability due to the interpenetration of solar-wind and magnetospheric plasmas may produce cross-field diffusion at a rate comparable to the Bohm limit (Bohm et al., 1949), which corresponds to a random walk of particles by a significant fraction of a gyroradius per gyroperiod. Evidence of such rapid diffusion has been tentatively identified along cusp field lines by Reiff et al. (1977). However, diffusion at even this extreme rate apparently fails to account for the magnitude of the observed convection potential drop by perhaps an order of magnitude (Hill and Wolf, 1977).

The Bohm limit applies to microscopic diffusion of individual particles within a flux tube. This upper limit can, in principle, be circumvented if one allows, not just particles, but entire flux tubes (with transverse dimensions of many gyroradii) to diffuse by means of a stochastic convection process (i.e., a flux-tube interchange process -- see Gold, 1959). In this kind of process Equation (1) may continue to hold locally within the diffusing flux tubes but must be violated at some point along the flux tubes (see below).

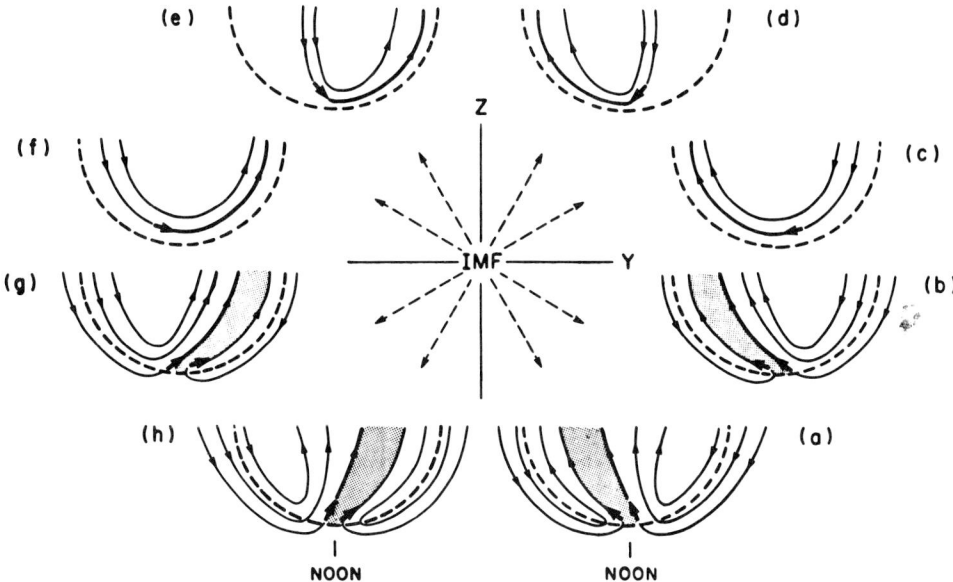

Fig. 5. Predicted convection patterns in the nothern-hemisphere day side polar cap, for different orientations of the interplanetary magnetic field (IMF) component transverse to the solar wind (Crooker, 1979). The Y axis is dusk and the Z axis is north in the IMF direction indicator; the convection patterns are all viewed from above the north pole with the sun toward the bottom of the figure. In each case the dashed contour represents the polar-cap boundary, within which the field lines are open to interplanetary space. The predictions are based on a model in which merging occurs at the cusps but not at the nose of the magnetosphere. Note that the antisunward convection within the polar cap is enhanced overall when the IMF has a strong southward (-Z) component, and is enhanced on the (dawn/dusk) side when the IMF has a (+/-) Y component (the dawn-dusk asymmetry being reversed in the southern hemisphere). Both of these correlations are confirmed by observations as cited in the text.

A model of this type has been proposed by Lemaire and Roth (1978), who envision a process by which solar-wind flux tubes containing higher than average plasma momentum flux penetrate the magnetopause and become trapped within to become closed geomagnetic flux tubes loaded with solar-wind plasma. A point that was not recognized explicitly by LeMaire and Roth is that their process requires magnetic merging to occur at two points along the flux tube in order to dissociate the intruding flux tube from its interplanetary roots. This point was explicitly recognized by Haerendel et al. (1978), who proposed a similar process to occur especially in the neighborhood of the cusps as the result of turbulent convection there. In the view of Haerendel et al., merging occurs at the cusp as a by-product (albeit a necessary by-product) of turbulent eddy flow of plasma in the cusp indentation of the magnetopause.

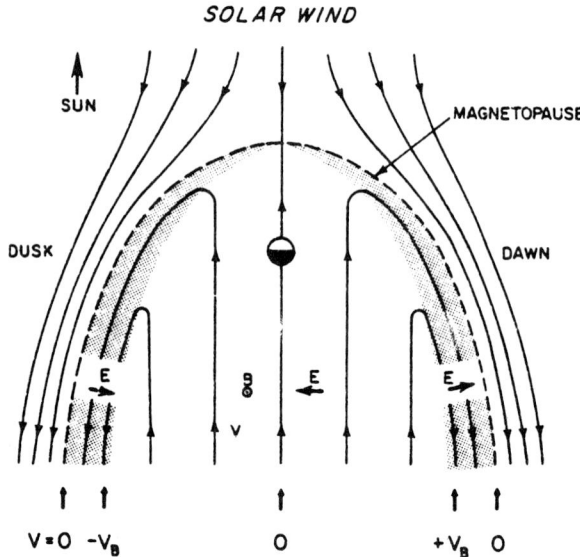

Fig. 6. A detail of the front half of Figure 3b, showing the quasi-viscous boundary layer (shaded) just inside the magnetopause. The thickness of this layer determines the magnitude $V_B$ of the potential drop across it, and the total polar cap potential drop is just $2V_B$ for the symetric case illustrated here.

In a variation of the idea of viscous momentum transfer, Johnson (1978) has proposed that relatively stagnant magnetosheath plasma enters onto closed field lines near the cusp, presumably through a turbulent convection process of the type envisioned by Haerendel et al. Johnson would derive the momentum of the boundary-layer flow from the internal energy of this initially stagnant but hot plasma as it expands adiabatically downstream along the flanks of the tail, rather than by direct transfer of magnetosheath momentum across the magnetopause locally at the flanks.

Very little theoretical work has been devoted to quasi-viscous momentum transfer through the stochastic convection process mentioned above, compared, for example, to the extensive theoretical literature on magnetic merging. The observed control of the interplanetary magnetic field over polar-cap convection patterns had, after all, apparently established magnetic merging as the dominant transfer process. However, theoretical interest in quasi-viscous momentum transfer processes has been revived by recent observations (Eastman et al., 1976; Crooker, 1977) which have confirmed that the low-latitude boundary-layer flow, discovered by Freeman et al. (1969) under unusually disturbed conditions, is in fact a persistent feature of the magnetosphere under more typical conditions as well. There is, as yet, no conclusive evidence that this boundary-layer flow occurs on closed geomagnetic field lines (Reiff et al., 1978), but it does resemble very closely the viscous boundary layer postulated by Axford and Hines (1961).

It may turn out that the correct model is something of a hybrid, wherein stochastic flux-tube convection describes the entry of solar-wind plasma into the magnetosphere (and the concomitant momentum transfer), but the direction of the interplanetary magnetic field determines (through the merging process) how this momentum is transmitted to the ionosphere to drive the polar-cap convection.

Wave Momentum Transfer

The diffusion processes described above violate the frozen-in-flux condition (1), either throughout the boundary layer (particle diffusion) or at least at the perimeter of the boundary layer (flux-tube diffusion). Hydromagnetic waves can, in principle, transfer momentum across the boundary without violating (1). For example, the velocity shear at the magnetopause may drive a Kelvin-Helmholtz instability (Dessler and Fejer, 1963, and references therein) which would transport solar-wind momentum (but not particles) across the magnetopause. This idea has been largely abandoned in recent work because magnetopause observations have generally not been interpreted as supporting the idea of Kelvin-Helmholtz instability, and because the recent boundary-layer observations cited above have indicated that the momentum transfer is probably accomplished through the same process that injects solar-wind particles across the magnetopause. The Kelvin-Helmholtz instability has not, however, been definitively ruled out as a potentially important momentum transfer mechanism.

Prospects for Quantitative Modeling

We are clearly not in a position even to formulate, much less solve, a realistic quantitative model of the solar-wind/magnetosphere interaction. Nevertheless, work has begun on the development of the numerical software that will be required to perform such an analysis. Initial experiments with a two-dimensional magnetohydrodynamic code are described elsewhere in this volume (Lebouef et al., 1979). The implicit assumption in this work is that our understanding of the physics of the solar-wind/magnetosphere interaction will be sufficiently improved in a few years to make intelligent use of the numerical machinery once it has been developed.

## The Magnetospheric Tail

Whatever the mechanism, we know with some certainty that the solar wind typically imposes a potential drop of the order of 50 kV across the polar cap. This is the rate at which geomagnetic flux is typically being added to the magnetospheric tail (1 volt = 1 weber/sec). At this rate, the entire magnetic flux of the Earth would be carried away by the solar wind in about a day and a half, were it not for the return (sunward) half of the convection cycle depicted in Figure 2. The Earth is, of course, too good a conductor to let that happen, but the interesting conclusion from this is that the magnetosphere must reverse the antisolar convection (and the associated Poynting flux) on a time scale much less than a day, lest the geomagnetic field suffer severe (factor-of-two) distortions (which are not observed). It is remarkable, if not significant, that the magnetospheric tail reverses the convection on a time scale that is just marginally sufficient to prevent such gross distortions of the geomagnetic field. The time scale on which this occurs is about an hour, and the process by which it occurs is called a magnetospheric substorm.

A review of the substorm dynamics of the magnetospheric tail is beyond the scope of this paper. Satellite observations have provided a fairly complete phenomenological description of the process, and have been reviewed by Russell and McPherron (1973) (see also McPherron et al., 1973). We know that the antisunward convection transports magnetic flux and plasma into the magnetospheric tail where the associated field and plasma energy are stored. The tail consists of two oppositely-directed "lobes" of magnetic flux (Figure 1b), each containing about 1/10 of the geomagnetic dipole flux, the two lobes being separated by the electric current flowing in the plasma sheet (Hill, 1974, and references therein). The convection energy is stored in the magnetic field of the tail, which contains about 5/6 of the total energy of the tail (the other 1/6 being primarily particle energy in the plasma sheet).

When the polar-cap convection is enhanced, we know that the tail magnetic flux and magnetic energy increase for a period of the order of an hour before establishing the necessary enhanced return flow -- this period of tail growth is called the substorm growth phase. The return flow is established impulsively in a process called (not very appropriately, from this point of view) the substorm expansion phase. During this phase, lasting only a few minutes, the tail injects some of its excess energy into the inner magnetosphere, and presumably ejects a like amount downstream into interplanetary space. This energy is injected partly in the form of magnetic flux returning to the inner magnetosphere (the distended tail magnetic field lines are said to collapse toward a more dipole-like configuration), and partly in the form of accelerated plasma-sheet particles that are injected into the ring current.

It is clear that the expansion phase, by which the tail establishes the return convective flow, is a violent process involving a rapid and large-scale reconfiguration (or partial collapse) of the tail magnetic field. It is less clear whether or not this reconfiguration is accomplished through magnetic merging at a neutral line within the tail (as in Figure 4, rotated by 90°). Such merging has been proposed by a number of theories (Atkinson, 1966; Axford, 1967; Dessler, 1968; Coroniti and Kennel, 1972; Hill, 1973b; Hill and Reiff, 1978) and appears to be consistent with a number of satellite observations (Nishida and Hones, 1974, and references therein), but the question of observational support for this idea is controversial (cf., Lui et al. 1977; Nishida and Russell, 1978).

In any case, the rapid field reconfiguration cannot be described in terms of a steady-state merging model that assumes $\nabla \times \underline{E} = 0$ (as in most of the merging literature). The rotational electric field associated with the magnetic-field reconfiguration (through Faraday's law) has a total electromotive force that must exceed the steady polar-cap convection potential drop by a factor of the order of the ratio of growth-phase to expansion-phase time constants, a factor of at least five. We have no self consistent (even qualitative) model that incorporates such large inductive electric fields, nor do we have any theoretical consensus even as to why the return convective flow is established impulsively rather than continuously, although a number of ideas have been advanced in the papers cited above.

## The Atmosphere/Magnetosphere Interaction

Apart from the violent substorm behavior of the magnetospheric tail, the magnetosphere has two ways to modify or redistribute the electric potential imposed upon it by the solar wind: (1) the magnetospheric plasma can polarize, giving rise to the "shielding" effect whereby the convection electric field tends to be excluded from low latitudes, and (2) the ionospheric plasma can polarize, giving rise to the "line-tying" effect whereby the magnetospheric convection is coupled to the motion of the un-ionized upper atmosphere through ion-neutral molecule collisions.

The mechanisms of polarization are, of course, different in the two cases. The ionosphere polarizes through the divergence of (ohmic) Pedersen currents. These ionospheric currents, being dissipative in character, tend always to reduce the convection electric field as measured in the frame of reference of the neutral atmosphere, or, equivalently, to reduce the relative velocity between the ionized and un-ionized constituents of the upper atmosphere. The magnetosphere, on the other hand, polarizes through the divergence of magnetic gradient- and curvature-drift currents, which process tends to reduce the convection electric field inside of the drift shell on which such divergence occurs, and to increase it correspondingly outside.

The shielding effect, first elucidated by Schield et al. (1969), has been treated quantitatively in the papers cited above in the section on inner magnetospheric convection. The line-tying effect was described by Axford and Hines (1961) and has been discussed subsequently by several authors (see Hill et al., 1975 and references therein) insofar as it tends to resist the primary driving force of magnetospheric convection. In addition to resisting magnetospheric convection generally, the line-tying effect has (at least) the following important consequences: (1) it imposes the corotational flow that dominates magnetospheric convection at low latitudes; (2) it drives an antisunward flow of the neutral upper atmosphere across the polar cap, the inertia of which makes it even more difficult to reverse the convective flow; (3) it tends to counteract the shielding effect, to the extent that ionospheric polarization is able to cancel the polarization of magnetospheric plasma; (4) it partially imposes on the magnetosphere the motions associated with thermally-driven upper-atmospheric winds, including the winds that arise as by-products of magnetospheric convection (through electron precipitation or Joule heating) as well as winds driven by non-magnetospheric causes; and (5) it causes the ionosphere to reflect incoming Alfvén waves (Mallinckrodt and Carlson, 1978), which may affect the reconfiguration of the magnetospheric tail during substorms (Hill and Reiff, 1978).

The existence of parallel (magnetic-field-aligned) electric fields at low altitudes in the auroral zone, inferred earlier from their effects on auroral particle beams (Evans, 1974), has now been confirmed by direct measurements in situ (e.g., Mozer et al., 1977). Proposed mechanisms for maintaining such parallel fields within an essentially collisionless plasma are numerous and will not be reviewed here (see Block and Fälthammar, 1976; Fälthammar, 1977). Generally these mechanisms require either a large Birkeland current density or otherwise highly anisotropic particle beams, and both of these conditions are typically satisfied near the edge of the polar cap and along auroral-zone field lines.

It is not clear what net effect these parallel electric fields should have on the convection electric field in the magnetosphere. There are two opposing effects: (1) the parallel fields accelerate auroral particles and hence produce localized conductivity enhancements in the ionosphere, the effect of which is to strengthen the coupling between the ionosphere and the atmosphere through line-tying; (2) the parallel fields also produce a magnetic-field-aligned potential drop, which tends to decouple the convection electric field in the ionosphere from that in the magnetosphere. It is customary to assume tacitly that the latter effect dominates such that the net effect of a parallel electric field is to decouple the magnetospheric convection from the atmosphere, but this has never been explicitly demonstrated to my knowledge. The first effect has been partly incorporated into the quantitative numerical simulation models described above, but the second effect has not.

## Conclusion

I will not close with the customary list of outstanding problems for future research -- the list can be found scattered throughout the foregoing text. Instead I will close with the opinion that the magnetospheric electric field is both the most understood and the least understood aspect of magnetospheric physics. The calculation of inner magnetospheric electric fields has reached a level of sophistication at which the accuracy of the results is limited by the accuracy with which boundary conditions can be specified, rather than by the accuracy of the approximations involved in the governing equations (except possibly the $E_{\parallel} = 0$ approximation). On the other hand, the location of the boundary on which boundary conditions are specified is not merely a matter of convenience but rather a matter of necessity. There is a great deal of fundamental physics to be understood before we will be ready to extend the dashed line in Figure 1.

**Acknowledgments.** I wish to thank R. A. Wolf, A. J. Dessler, D. J. Southwood, and P. H. Reiff for their valuable comments. This work was supported in part by the National Science Foundation (Grant ATM77-12619) and by the National Aeronautics and Space Administration (Grant NGR44-006-137).

## References

Alfvén, H., Some properties of magnetospheric neutral surfaces, J. Geophys. Res., 73, 4379, 1968.
Alfvén. H., and C.-G. Fälthammar. Cosmical Electrodynamics, Oxford University Press, 1963.

Atkinson, G., A theory of polar substorms, J. Geophys. Res., 71, 5157, 1966.
Axford, W. I., Magnetic storm effects associated with the tail of the magnetosphere, Space Sci. Rev., 7, 149, 1967.
Axford, W. I., and C. O. Hines, A unifying theory of high-latitude geophysical phenomena and geomagnetic storms, Can. J. Phys., 39, 1433, 1961.
Block, L. P., and C.-G. Fälthammar, Mechanisms that may support magnetic-field-aligned electric fields in the magnetosphere, Ann. Geophys., 32 161, 1976.
Bohm, D., E.H.S. Burhop, and H.S.W. Massey, The use of probes for plasma exploration in strong magnetic fields, in Characteristics of Electrical Discharges in Magnetic Fields, ed. by A. Guthrie and R. K. Wakerling, pp. 13-76, McGraw-Hill, New York, 1949.
Burch, J. L., Observations of interactions between interplanetary and geomagnetic fields, Rev. Geophys. Space Phys., 12, 363, 1974.
Coroniti, F. V., and C. F. Kennel, Changes in magnetospheric configuration during the substorm growth phase, J. Geophys. Res., 77, 3361, 1972.
Crooker, N. U., Explorer 33 entry layer observations, J. Geophys. Res., 82, 515, 1977.
Crooker, N. U., Dayside merging and cusp geometry, J. Geophys. Res., (in press), 1979.
Dessler, A. J., Magnetic merging in the magnetospheric tail, J. Geophys.Res., 73, 209, 1968.
Dessler, A. J., Vacuum Merging: A possible source of the magnetospheric cross-tail electric field, J. Geophys. Res., 76, 3174, 1971.
Dessler, A. J., and J. A. Fejer, Interpretation of Kp index and M-region geomagnetic storms, Plan. Space Sci., 11, 505, 1963.
Dungey, J. W., Interplanetary magnetic field and the auroral zones, Phys. Rev. Letters, 6, 47, 1961.
Eastman, T. E., E. W. Hones, Jr., S. J. Bame, and J. R. Asbridge, The magnetospheric boundary layer: site of plasma, momentum, and energy transfer from the magnetosheath into the magnetosphere, Geophys. Res. Lett., 3, 685, 1976.
Eastwood, J. W., Consistency of fields and particle motion in the 'Speiser' model of the current sheet, Planet. Space Sci., 20, 1555, 1972.
Evans, D. S., Precipitating electron fluxes formed by a magnetic field-aligned potential difference, J. Geophys. Res., 79, 2853, 1974.
Eviatar, A. and R. A. Wolf, Transfer processes at the magnetopause, J. Geophys. Res., 73, 5561, 1968.
Fairfield, D. H., Electric and magnetic fields in the high-latitude magnetosphere, Rev. Geophys. Space Phys., 15, 285, 1977.
Fälthammar, C.-G., Problems related to macroscopic electric fields in the magnetosphere, Rev. Geophys. Space Phys., 15, 457, 1977.
Francfort, P. and R. Pellat, Magnetic merging in collisionless plasmas, Geophys. Res. Lett., 3, 433, 1976.
Freeman, J. W., Jr., C. S. Warren. and J. J. Macguire, Plasma flow directions at the magnetopause on January 13 and 14, 1967, J. Geophys. Res., 73, 5719, 1968.
Gold, T., Motions in the magnetosphere of the Earth. J. Geophys. Res., 64, 1219, 1959.
Haerendel, G., G. Paschmann, N. Sckopke, H. Rosenbauer, and P. C.

Hedgecock, The frontside boundary layer of the magnetosphere and the problem of reconnection, J. Geophys. Res., 83, 3195, 1978.

Harel, M., and R. A. Wolf, Convection, in Physics of Solar Planetary Environments, Vol. II, (ed. D. J. Williams), p. 617, Amer. Geophys. Un., Washington, D. C., 1976.

Harel, M., R. A. Wolf, P. H. Reiff, and M. Smiddy, Computer modeling of events in the inner magnetosphere, (this volume), 1979.

Heikkila, W. J., Is there an electrostatic field tangential to the dayside magnetopause and neutral line?, Geophys. Res. Lett., 2, 154, 1975.

Heppner, J. P., Polar cap electric field distributions related to the interplanetary magnetic field direction, J. Geophys. Res., 77, 4877, 1972.

Hill, T. W., A three-dimensional model of magnetic merging at the magnetopause, Radio Sci., 8, 915, 1973a.

Hill, T. W., A mechanism for the growth phase of magnetospheric substorms, Planet. Space Sci., 21, 1307, 1973b.

Hill, T. W., Origin of the plasma sheet, Rev. Geophys. Space Phys., 12, 379, 1974.

Hill, T. W., Magnetic merging in a collisionless plasma, J. Geophys. Res., 80, 4689, 1975.

Hill, T. W., and P H. Reiff, Evidence of magnetospheric cusp proton acceleration by magnetic merging at the dayside magnetopause, J. Geophys. Res., 82, 3623, 1977.

Hill, T. W., and P. H. Reiff, Plasma sheet dynamics and magnetospheric substorms, J. Geophys. Res., (submitted), 1978.

Hill, T. W., and R. A. Wolf, Solar-wind interactions, in The Upper Atmosphere and Magnetosphere, (ed. F. S. Johnson), p. 25, Nat. Acad. Sci., Washington, 1977.

Hill, T. W., A. J. Dessler, and R. A. Wolf, Mercury and Mars: The role of ionospheric conductivity in the acceleration of magnetospheric particles, Geophys. Res. Lett., 3, 429, 1976.

Jaggi, R. K., and R. A. Wolf, self-consistent calculation of the motion of a sheet of ions in the magnetosphere, J. Geophys. Res., 78, 2852, 1973.

Johnson, F. S., The driving force for magnetospheric convection, Rev. Geophys. Space Phys., 16, 161, 1978.

Leboeuf, J. N., T. Tajima, C. F. Kennel, and J. M. Dawson, Global simulation of the time dependent magnetosphere, (this volume), 1979.

Lemaire, J., and M. Roth, Penetration of solar wind plasma elements into the magnetosphere, J. Atmos. Terr. Phys., 40, 331, 1978.

Lui, A.T.Y., C.-I Meng, and S.-I. Akasofu, Search for the magnetic neutral line in the near-earth plasma sheet, 2. systematic study of IMP6 magnetic field observations, J. Geophys. Res., 82, 1547, 1977.

Mallinckrodt, A. J., and C. W. Carlson, Relations between transverse electric fields and field-aligned currents, J. Geophys. Res., 83, 1426, 1978.

McPherron, R. L., C. T. Russel, and M. P. Aubry, Satellite studies of magnetospheric substorms on August 15, 1968; 9. Phenomenological model for substorms, J. Geophys. Res., 78, 3131, 1973.

Mozer, F. S., C W. Carlson, M. K. Hudson, R. B. Torbert, B. Parady, J. Yatteau, and M. C. Kelly, Observations of paired electrostatic shocks in the polar magnetosphere, Phys. Rev. Lett., 38, 292, 1977.

Nishida, A., and E. W. Hones, Jr., Association of plasma sheet thinning with neutral line formation in the magnetotail, J. Geophys. Res., 79, 535, 1974.

Nishida, A., and C. T. Russell, On the expected signatures of reconnection in the magnetotail, J. Geophys. Res., 83, 3890, 1978.

Reiff, P. H., T. W. Hill, and J. L. Burch, Solar-wind plasma injection at the dayside magnetospheric cusp, J. Geophys. Res., 82, 479, 1977.

Reiff, P. H., T. W. Hill, and J. L. Burch, Reply, J. Geophys. Res., 83, 229, 1978.

Russell, C. T., and R. L. McPherron, The magnetotail and substorms, Space Sci. Rev., 15, 205, 1973.

Schield, M. A., J. W. Freeman, and A. J. Dessler, A source for field-aligned currents at auroral latitudes, J. Geophys. Res., 74, 247, 1969.

Speiser, T. W., Conductivity without collisions or noise, Planet. Space Sci., 18, 613, 1970.

Stern, D. P., Large-scale electric fields in the Earth's magnetosphere, Rev. Geophys. Space Phys., 15, 156, 1977.

Vasyliunas, V. M., Mathematical models of magnetospheric convection and its coupling to the ionosphere, in Particles and Fields in the Magnetosphere, (ed. B. M. McCormac), p. 60, Reidel, Dordrecht, 1970.

Vasyliunas, V. M., The interrelationship of magnetospheric processes, in Earth's Magnetospheric Processes, (ed. B. M. McCormac), p. 29, D. Reidel, Dordrecht, 1972.

Vasyliunas, V. M., Theoretical models of magnetic field-line merging, 1, Rev. Geophys. Space Phys., 13, 303, 1975.

Wolf, R. A., Effects of ionospheric conductivity on convective flow of plasma in the magnetosphere, J. Geophys. Res., 75, 4677, 1970.

Wolf, R. A., Calculations of magnetospheric electric fields, in Magnetospheric Physics, (ed. B. M. McCormac), p. 167, Reidel, Dordrecht, 1974.

Wolf, R. A., Ionosphere-magnetosphere coupling, Space Sci. Rev., 17, 537, 1975.

# INDUCED MAGNETOSPHERIC ELECTRIC FIELDS

G. J. Mroz, W. P. Olson, and K. A. Pfitzer

McDonnell Douglas Astronautics Company, 5301 Bolsa Avenue, Huntington Beach, California 92647

Abstract. Both the magnetic field and the vector magnetic potential, $\vec{A}$, have been determined by integrating over the major magnetospheric current systems. The temporal variation associated with the daily precession of the geomagnetic dipole axis is contained in the representation of $\vec{A}$. The induced electric field given by $\partial \vec{A}/\partial t$ is significant throughout most of the magnetosphere ($\sim 1$ mVm$^{-1}$). The reaction of the magnetospheric plasma to this induced field and the resulting total electric field are determined. The motion of low energy charged particles in such a nonconservative field is considered.

## Introduction

The understanding of the magnetospheric electric field, $\vec{E}$, remains incomplete. Although new data on the electric field are becoming available, there is currently no quantitative description of the various sources of the magnetospheric electric field. It is known that $\vec{E}$ can have at most three (3) sources, charge separation, time varying magnetic fields, and a component from the relativistic transformation of magnetic fields between translating coordinate systems. Of these, the quantitative description of the time variations on the magnetospheric magnetic field, $\vec{B}$, is at present most complete. Several quantitative models of $\vec{B}$ are available. Recently models of the quiet time "tilted" field from the magnetopause, ring, and tail currents has been developed (Olson and Pfitzer, 1977.) These models implicitly contain a time dependence since the "tilt" angle changes as the earth rotates about its axis.

The Olson-Pfitzer (1977) model of the magnetospheric field considers all the major magnetospheric current systems as sources of the total external field. The procedure establishes models of the current system; the Biot-Savart law is applied to this current system to yield the magnetic field, thus

$$\vec{B}(\vec{r}) = \frac{\mu_0}{4\pi} \int \vec{J}(\vec{r}') \times \frac{(\vec{r}-\vec{r}')}{|\vec{r}-\vec{r}'|^3} dV'.$$

Because of the complexity of this current system, the integral is performed numerically. Similarly, and with very little additional cost in computer time, the vector potential for this field can also be obtained,

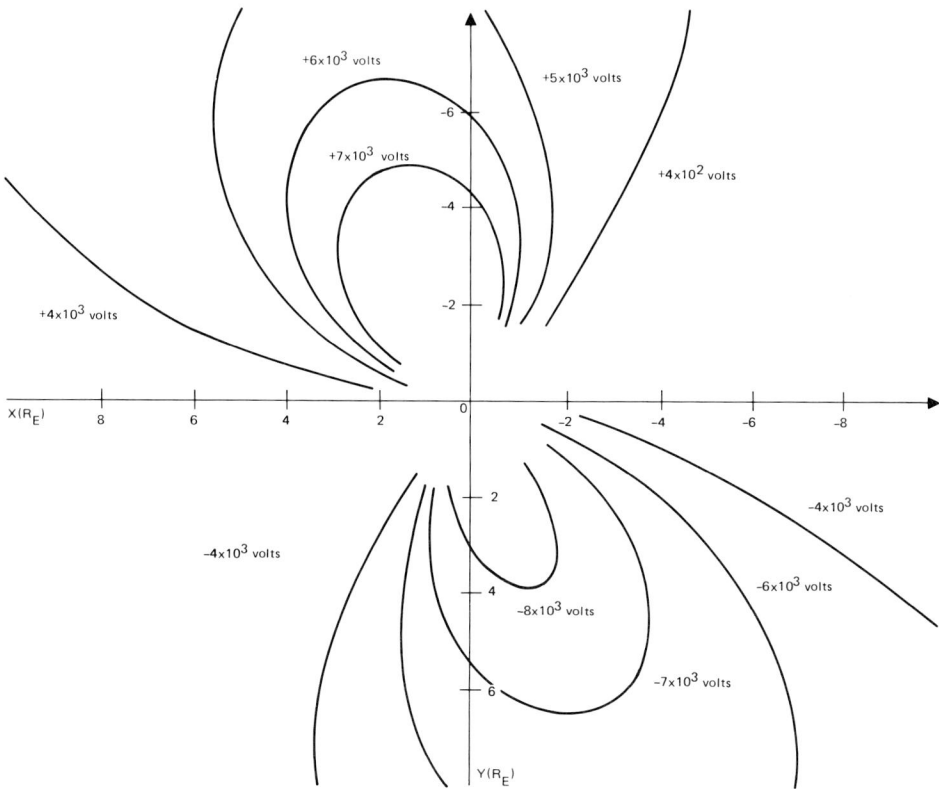

Fig. 1. Equipotential contours in the equatorial plane assuming an equipotential surface at R=1 Re LAT=0° deg., day=172, UT=0.

$$\overline{A}\ (\overline{r}) = \frac{\mu_0}{4\pi} \int \frac{\overline{J}\ (\overline{r}')}{|\overline{r}-\overline{r}'|}\ dV'.$$

Since the magnetospheric current system model includes a dependence on the daily precession of the earth's dipole moment, this particular time dependence of the magnetic field and the associated vector potential are properly modeled. The induced component of the magnetospheric electric field resulting from this dependence of the magnetic field can be obtained from the vector potential

$$\overline{E}_I = -\frac{\partial \overline{A}}{\partial t}.$$

Because of the presence of charged particles (low density plasma) throughout the magnetosphere (this medium is the source of the current system), local electric fields resulting from this medium must also be considered.

Indeed, the total electric field at a given location will not only include the contribution from the local time dependence of the

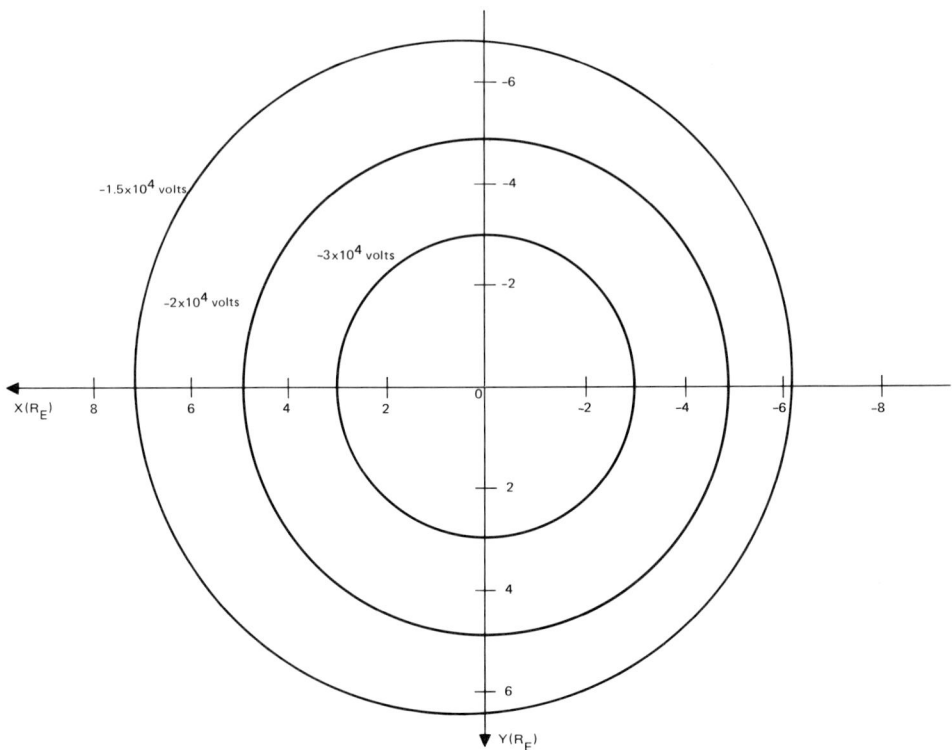

Fig. 2. Equipotential contours in the equatorial plane assuming a rotating, uniformly magnetized sphere potential at boundary, lat=0°, day=172, UT=0.

magnetic field, but also contributions from other regions of the magnetosphere transmitted to this location by the medium, to the extent that it is conductive. Thus, the total electric field,

$$\overline{E}_T = - \nabla\phi - \frac{\partial \overline{A}}{\partial t},$$

at a given point in the magnetosphere includes contributions from the induced field from other regions of the magnetosphere, linked to the given location by the conductivity of the medium, as well as any contributions from local charge separation in the medium. All this is contained in the scalar potential component, $\overline{E}_S = -\nabla\phi$ of the total field.

## Modeling the Electric Field

A procedure developed by Hones and Bergeson (1965) was applied here to evaluate the scalar potential component of the electric field. In this procedure, it is assumed that in the region of interest, charged particle densities are high enough to assure sufficient conductivities along the magnetic field lines such that $\overline{E}_T \cdot \hat{B} = 0$.

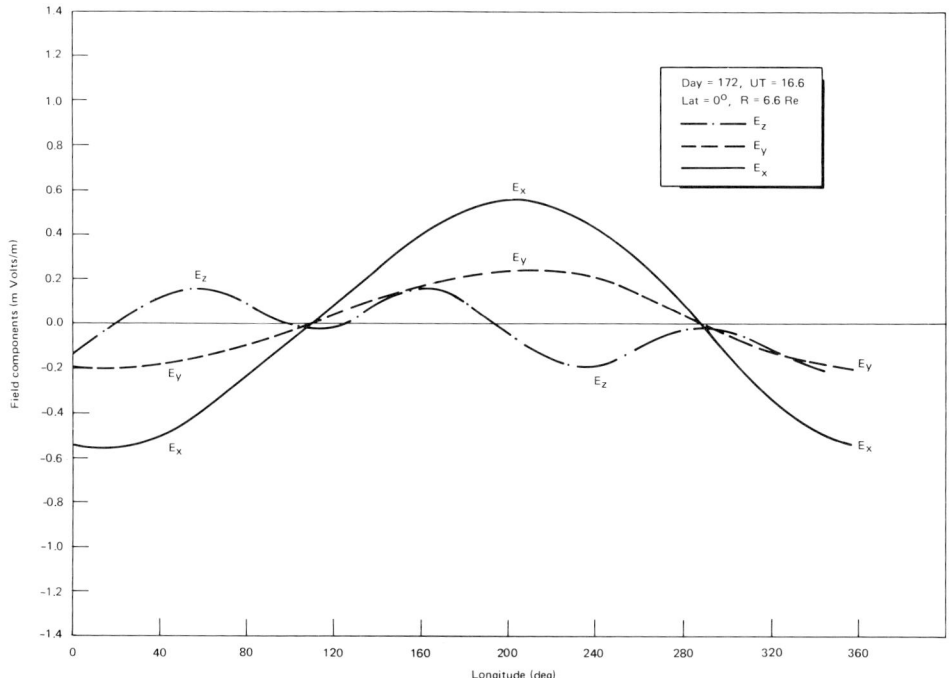

Fig. 3. Total electric field experienced by a charged particle rotating with the earth at synchronous orbit.

Then

$$\frac{d\phi}{ds} = \frac{\partial A_{\shortparallel}}{\partial t},$$

where $\frac{d\phi}{ds}$ is the scalar potential gradient along the field line, and $\frac{\partial A_{\shortparallel}}{\partial t}$ is the component of $\overline{E}_I$ parallel to the magnetic field line. This equation can now be integrated along the field line to give the scalar potential at a given point in the magnetosphere. Thus,

$$\phi(r) = \phi_B - \frac{\partial}{\partial t} \int_{FL} \overline{A} \cdot \overline{ds},$$

where the line integral is along the field line, and $\phi_B$ is the potential at some appropriate boundary.

Options for the selection of the appropriate boundary condition for $\phi$ include the following:

1. Experimentally determined.
2. Constant over an appropriate surface.
3. Uniformly magnetized rotating sphere (Hones and Bergeson, 1965),

$$V = \overline{U} \cdot \overline{A},$$

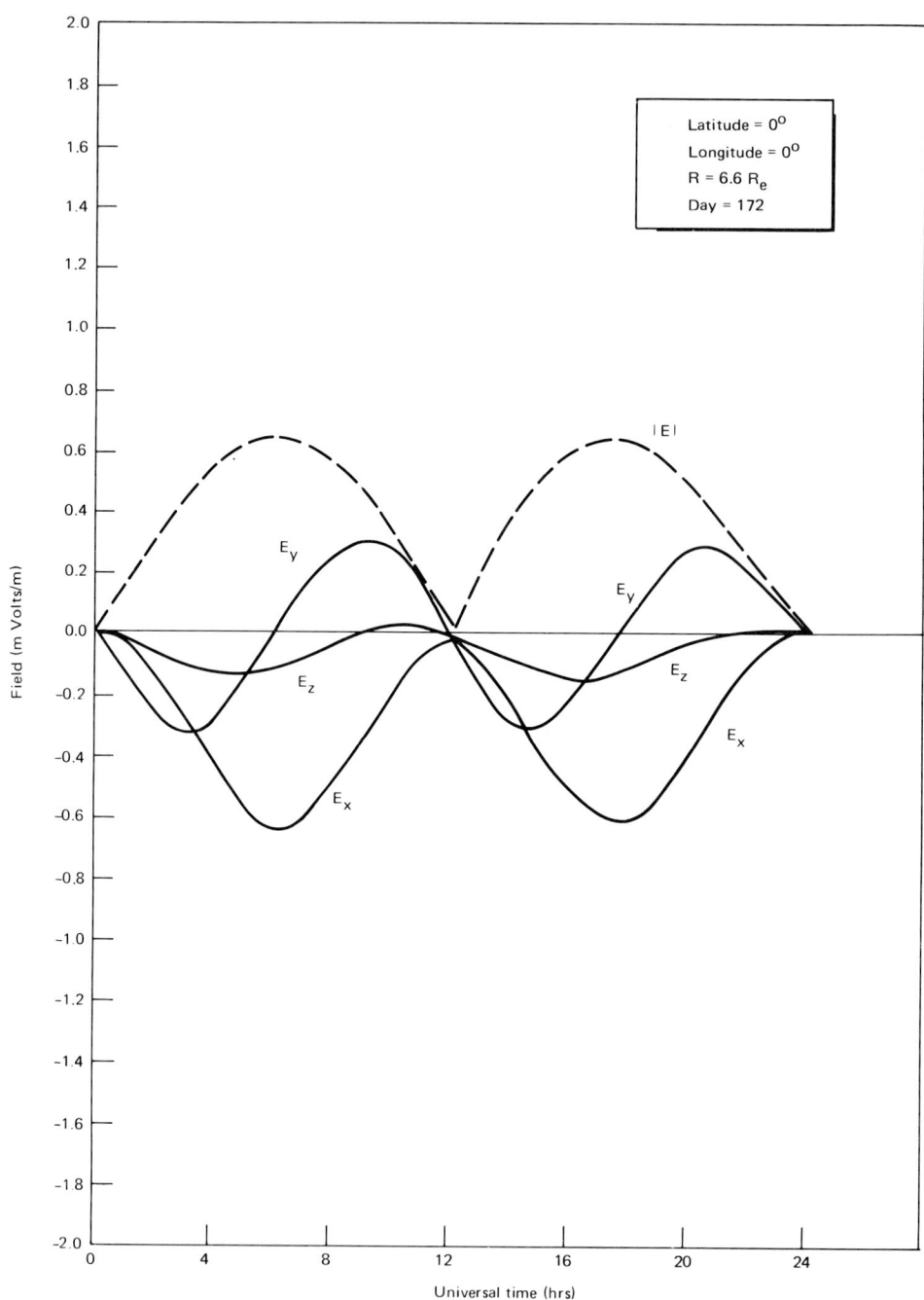

Fig. 4. Total electric field experienced by a charged particle rotating with earth at synchronous orbit.

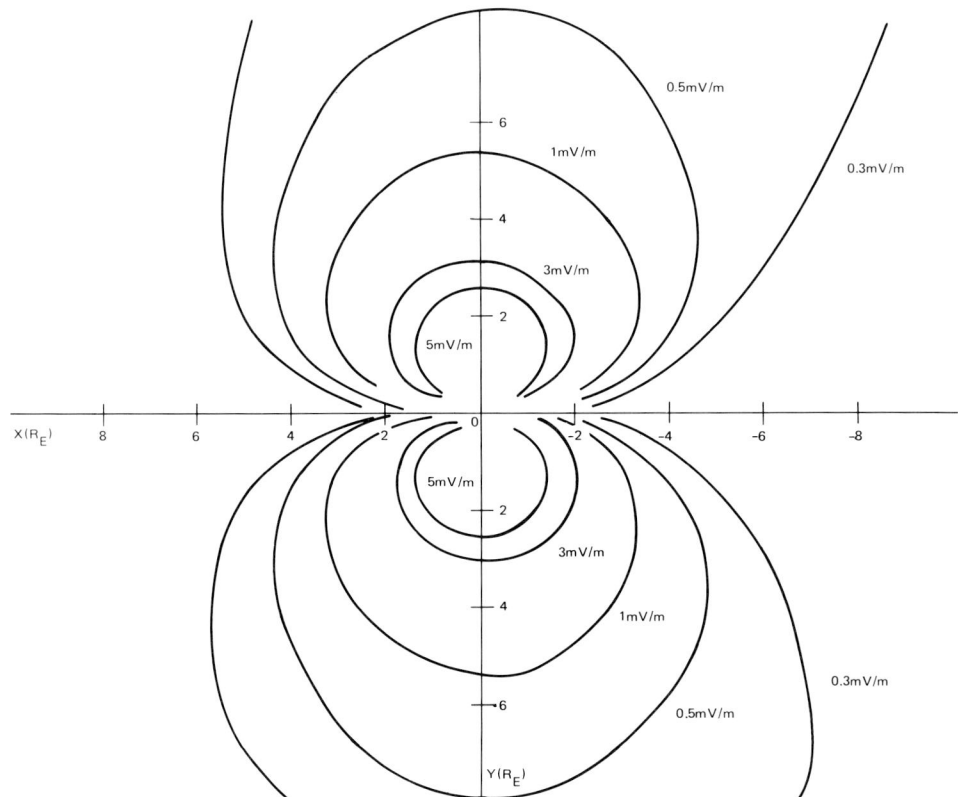

Fig. 5. Constant $E_{total}$ contours in a fram rotating with the earth. UT=0, day=172, equatorial plane, X is through Greenwich equipotential boundary.

where

$V$ = the potential at point $\bar{r}$,
$\bar{U} = \bar{w} \times \bar{r}$,
$\bar{A} = \bar{\mu} \times \bar{r}/R_s^3$,

$\bar{w}$ = angular velocity
$\bar{\mu}$ = magnetic moment
$R_s$ = radius of sphere

## 4. Self-consistent determination

In the present work, options 2 and 3 were employed to evaluate the sensitivity of the electric field model to both the choice of the boundary condition and the non-local contributions to the induced electric field. Some results of this investigation are illustrated in Figures 1 and 2. In both figures, equipotential contours are plotted in the earth's equatorial plane. The positive x direction in

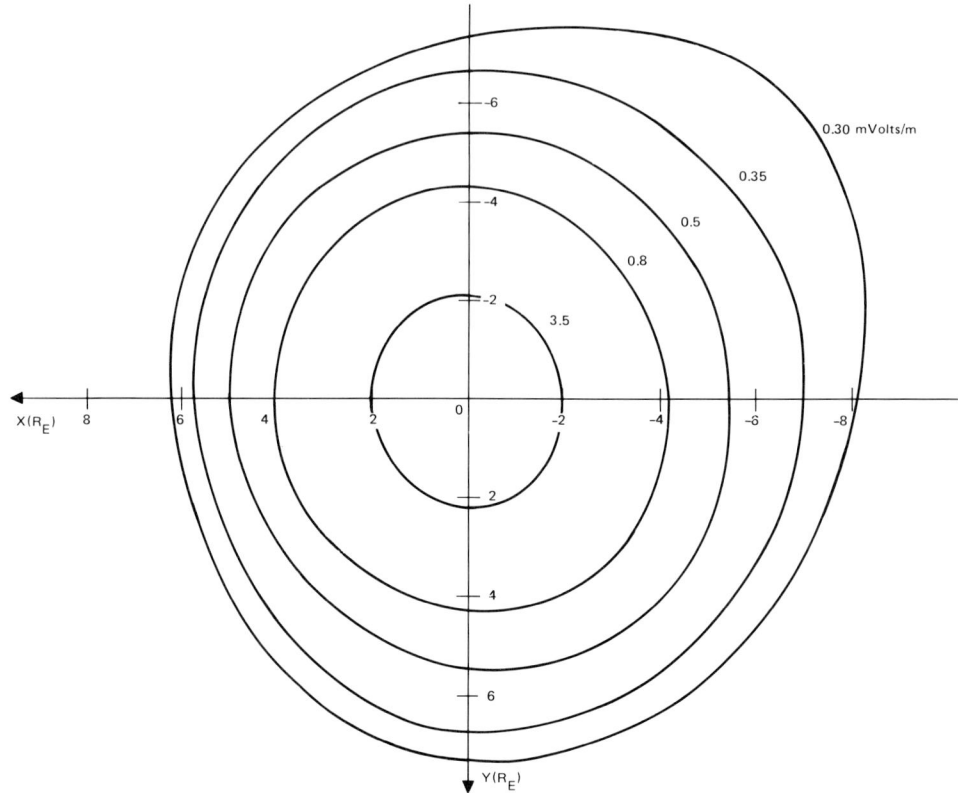

Fig. 6. Constant $E_{total}$ contours in an inertial frame, UT=0 hrs, day=172, equatorial plane, X is through Greenwich, rotating magnetized sphere.

both plots is through the Greenwich meridian. In Figure 1, a uniform (constant) potential surface was used at R =1 Re, the induced electric field (i.e., $\partial A_{||}/\partial t$) is due to both the dipole field and the diurnal variations in the current systems. (The coordinate system is inertial; i.e., the geographic coordinate system, as described, coincides with the inertial system at the specified time, day 172, UT=0 hrs.) This figure clearly illustrates the mapping of the induced electric field, as it occurs along the magnetic field line, to the equatorial plane; without this field, the equatorial plane would be an equipotential surface.

Figure 2 illustrates the same contours with a rotating uniformally magnetized sphere as the boundary condition. The sensitivity to boundary conditions is clearly evident. The assymetry in the equipotential contour in Figure 2 is due to the induced electric field resulting from time variation of the current system.

Comparison of these figures suggests that if the ionospheric electric field is variable, the total electric field in the magnetosphere will also vary substantially with time. Thus, through this mechanism, the ionosphere may exert some influence on the dynamics

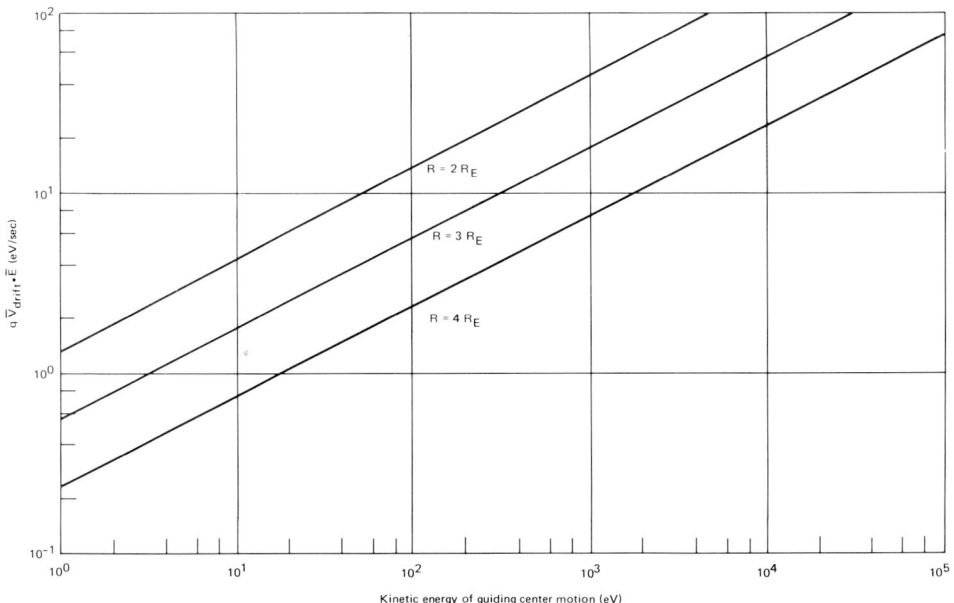

Fig. 7. Time rate of change of proton kinetic energy due to change in guiding center motion contribution (UT-16.6 hrs, day = 172, equatorial plane, longitude = 0°

of magnetospheric fields and low energy plasma.

Also, Figure 1 indicates the change in $\phi$ between the ionosphere and the equatorial plane when the ionospheric value of $\phi$ is taken to be zero. Thus, magnetic field lines are not equipotentials even when only the "small" induced fields like the wobbling dipole are present.

The various components of the total electric field which would be experienced by a charged particle rotating with the earth in geosynchronous orbit are illustrated in Figures 3 and 4. The uniformally magnetized, rotating sphere boundary condition was used for these calculations. Figure 3 displays the field as a function of longitude; Figure 4, the electric field as a function of time. Contours of constant $|\bar{E}_{Total}|$ are shown in Figures 5 and 6 for the two boundary conditions discussed above. (Figure 5, the equipotential boundary, and Figure 6, the rotating magnetized sphere.)

## Charged Particle Energization

To indicate how a field such as the one illustrated in Figure 3 and 4 would contribute to charged particle acceleration, the time rate of change of a proton's kinetic energy was calculated using the formula from Northrop (1963)

$$\frac{d \langle w \rangle}{dt} = q \bar{E} \cdot \bar{V} + M \frac{\partial B}{\partial t},$$

where $\langle w \rangle$ is the kinetic energy averaged over a gyration period, M is

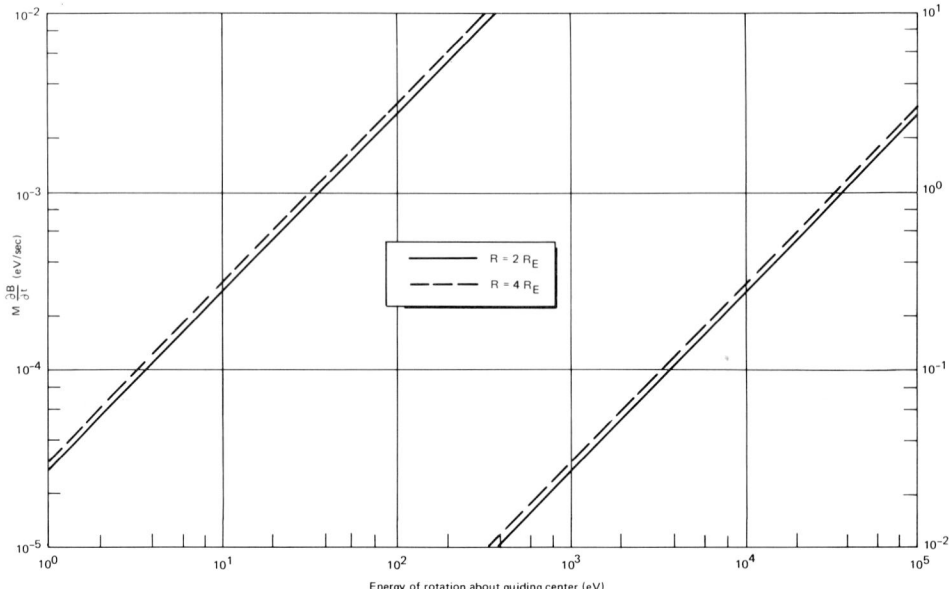

Fig. 8. Time rate of change of proton kinetic energy due to changes in rotation about the guiding center (UT=16.6, day=172, equatorial plane, longitude=0°).

the magnetic moment of the proton, q its charge, and $\overline{V}$ the velocity of its guiding center. The results of this calculation are shown in Figures 7 and 8. Figure 7 shows the energy gained by the guiding center motion as a function of the guiding center kinetic energy, and similarly, Figure 8 shows the energy change in the motion about the guiding center.

## Conclusions

An approach to modeling the electric field in the magnetosphere has been considered and contributions from the diurnal variation in the major magnetic current systems have been evaluated. The presence of charged particles in the magnetosphere complicates the problem and requires the evaluation of non-local contributions to the electric field at a point.

The electric fields resulting from the diurnal variations in the magnetospheric magnetic fields have been shown to be of the order of 0.1 m Volt/m and could contribute significantly to charged particle energization. The presence of other, much larger electric fields is the magnetosphere is recognized. However, their duration is usually much shorter than that of the fields discussed in this paper, and the cumulative effect of these small fields may be important. Some of the considerations of the boundary value problem, and the associated non-local contributions to the electric field at the point, discussed in this paper, will be just as important in considering the electric fields from other sources.

Acknowledgements: This work was supported by Office of Naval Research under contract N00014-78-C-0215, by the Air Force Office of Scientific Research under contract F44620-75-C-0033, and by the Defense Nuclear Agency under contract DNA001-77-C-0240.

## References

Olson, W. P., and K. A. Pfitzer, Magnetic field modeling, McDonnell Douglas Astronautics Company Annual Scientific Report to AFOSR, January 1977.

Hones, Jr., E. W., and J. E. Bergeson, Electric field generated by a rotating magnetized sphere, J. Geophys. Res., 70, 4951, 1965.

Northrop, T. G., The adiabatic motion of charged particles, John Wiley and Sons (Interscience), New York, 1963.

COUPLING OF MAGNETOSPHERIC ELECTRICAL EFFECTS INTO THE GLOBAL ATMOSPHERIC ELECTRICAL CIRCUIT

P. B. Hays and R. G. Roble

Space Physics Research Lab, University of Michigan, Ann Arbor, Michigan 48105 and National Center for Atmospheric Research*, Boulder, Colorado, 80307

Abstract. A quasi-static model of global atmospheric electricity is used to examine the electrical coupling between magnetospheric processes in the upper atmosphere and atmospheric electricity in the lower atmosphere.

## Introduction

Within recent years it has been suggested that a solar-terrestrial coupling occurs through variations in atmospheric electricity (Markson, 1971; 1978; Herman and Goldberg, 1978; Park, 1976; and Sartor, 1965). Measurements of ground currents and potential gradients have shown atmospheric electrical responses to solar flares (Cobb, 1967; Reiter, 1969; 1971), geomagnetic and auroral activity (Cobb, 1967; Reiter, 1972; Lobodin and Paramonov, 1972), solar sector boundary crossings (Park, 1977; Reiter, 1977), and solar cycle variations (Mühleisen, 1977; Israel, 1971). These solar- and upper atmosphere-induced variations are superimposed upon complex electrical variations associated with meteorological and anthropogenic processes in the lower atmosphere (Israel, 1971).

To study the electrical processes in the lower atmosphere and the coupling between solar- and upper atmosphere-induced variations upon the global electrical circuit, we have constructed a quasi-static model of global atmospheric electricity (Hays and Roble, 1978). In this paper, we review the essential features of the model and discuss the results obtained thus far on the effects of magnetospheric convection and substorms on the global atmospheric electrical circuit.

## Numerical Model

The numerical model of global atmospheric electricity that is used to examine the electrical coupling between the earth's upper and lower atmosphere has been described in detail by Hays and Roble (1978) and Roble and Hays (1978). A schematic diagram of the global quasi-static model is shown in Fig. 1. It is assumed that thunderstorms act as dipole generators, each with a positive center at the top of

---

*The National Center for Atmospheric Research is sponsored by the National Science Foundation.

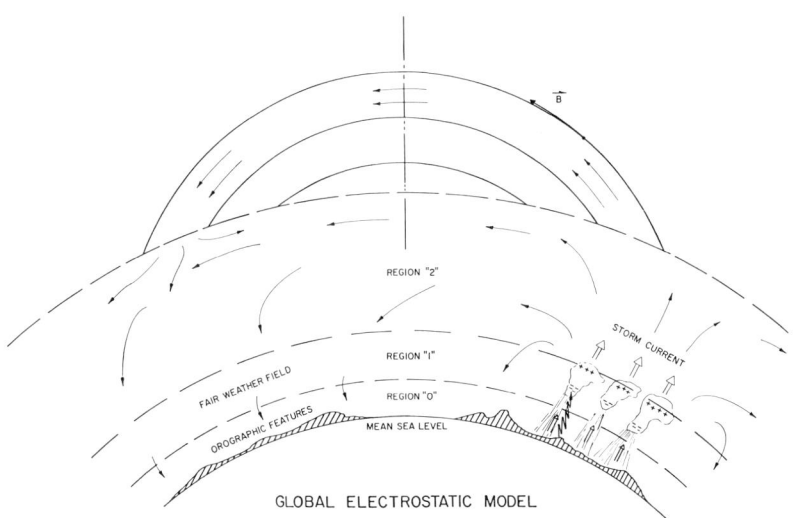

Fig. 1. Schematic diagram of the global model of atmospheric electricity.

the cloud and a negative center a few kilometers lower than the positive center. In fair-weather regions far away from the storm centers, the distribution of the electrostatic potential above the earth is determined by the current return from the sources to the earth's surface. The geometry of the model is based upon an atmosphere broken into four coupled regions, as illustrated in Fig. 1. Region 0 represents the lower tropospheric boundary region which has a variable conductivity in the horizontal and vertical; it also includes the earth's orography. Region 1 represents the upper troposphere below the negative current source region within the thunderstorm and region 2 represents the stratosphere and mesosphere above the positive current source region of the thunderstorm. For regions 0, 1, and 2 it is assumed that the electrical conductivity is isotropic. Region 3 represents the ionosphere and magnetosphere above the dynamo region where the conductivity is anisotropic and where magnetically conjugate regions are connected along geomagnetic field lines through the magnetosphere. The mathematical details of the model, the boundary conditions, and the matching conditions between various regions are described by Hays and Roble (1978).

The height and latitudinal distribution of electrical conductivity used for the model calculations are shown in Figs. 2a and 2b, respectively. The vertical profiles are compared to the measurements of 1) Morita et al. (1971), 2) Sagalyn (1960), 3) Mozer and Serlin (1969), 4) Cole and Pierce (1965), and 5) Tran and Polk (1972). We assume that the electrical conductivity in the upper atmosphere is maintained by cosmic ray activity which varies by about a factor of two between the equator and pole (Israel, 1971). Since the conductivity is maintained primarily by small mobile ions, it varies as the square root of the cosmic ray ion production rate. Also shown in Fig. 3 are assumed variations of electrical conductivity for a sudden increase in cosmic ray ionization resulting from a solar flare and a subsequent Forbush decrease (Forbush, 1966). The results of solar flare induced

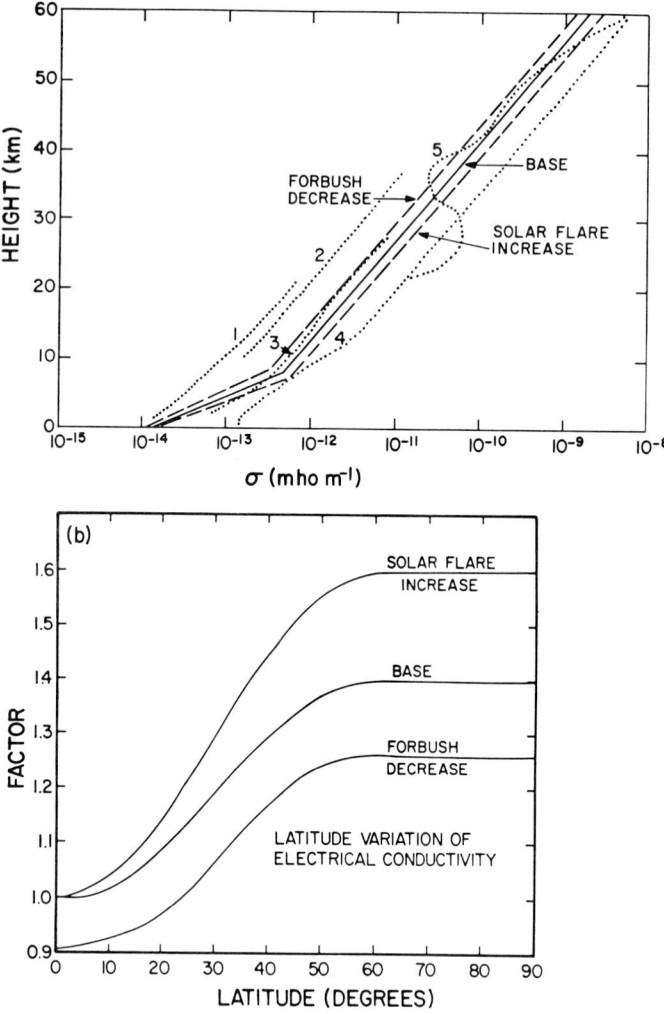

Fig. 2 Model electrical conductivity variations with (a) altitude and (b) latitude. The assumed variations for a solar flare increase and subsequent Forbush decrease are also shown.

perturbations have been discussed by Hays and Roble (1978).

On a global scale there is no net current source within the model and the thunderstorms act as pumps, recirculating current from the lower to the upper atmosphere. We arbitrarily consider 2000 individual thunderstorms occurring at 1900 UT and randomly distribute the point thunderstorms in latitude and longitude in accordance with the hourly probability of thunderstorm occurrence as defined by Crichlow et al. (1971). Northern hemisphere summer months are assumed; therefore at the 1900 UT maximum in ionospheric potential, thunderstorm activity occurs over Central America, Florida and the Rocky Mountains, in addition to the Amazon basin and parts of Central Africa. It is assumed that at 1900 UT, 600, 100, 1000, 280, and 20 thunderstorms

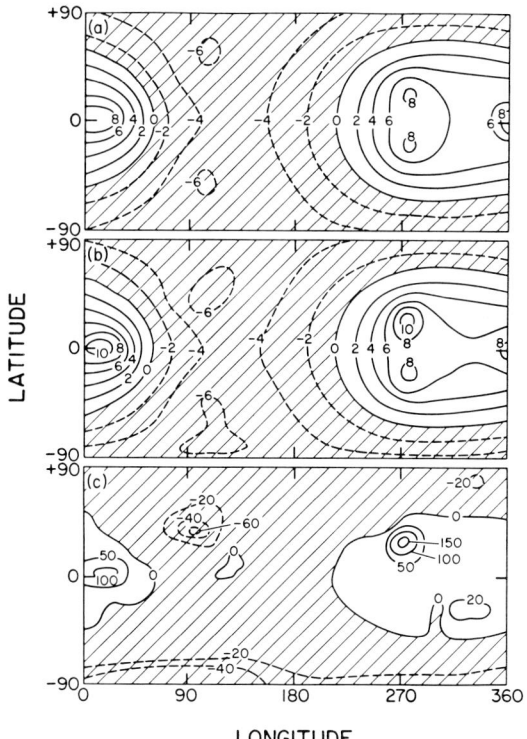

Fig. 3. Contours of calculated potential difference ($\Phi - \Phi_\infty$) along various conductivity surfaces at approximately (a) 105 km, (b) 50 km, and (c) 25 km. All values are in volts when multiplied by $10^3$.

are randomly distributed in latitude and longitude in the vicinity of Africa, Asia, Central America, Argentina, and the Alps, respectively. The storm centers are also randomly distributed in the vertical.

## Model Solution for the Lower Atmosphere

The calculated electric potential along the $\sigma = 7.3 \times 10^{-12}$ mho m$^{-1}$ constant conductivity surface, which occurs at approximately 25 km at the equator and slopes downward to 23 km in both polar regions, is shown in Fig. 3c. The calculated fair-weather ionospheric potential is $\Phi_\infty = 291,000$ V and the quantity ($\Phi - \Phi_\infty$) along the constant conductivity surface is plotted. Over the thunderstorm regions of Central America, Africa, and Argentina, the value of ($\Phi - \Phi_\infty$) is positive and electric current flows upward from the thunderstorm region to ionospheric heights. The maximum value of ($\Phi - \Phi_\infty$) is 16,000 V whereas minimum values occur over the mountainous fair-weather region, with -5700 V over both Tibet and Antarctica. The calculated value of ($\Phi - \Phi_\infty$) along the $\sigma = 4.74 \times 10^{-10}$ mho m$^{-1}$ surface near 50 km in the equatorial region is shown in Fig. 3b. The potential distribution has spread considerably from the concentrated regions over the thunderstorms at 25 km to a more uniform distribu-

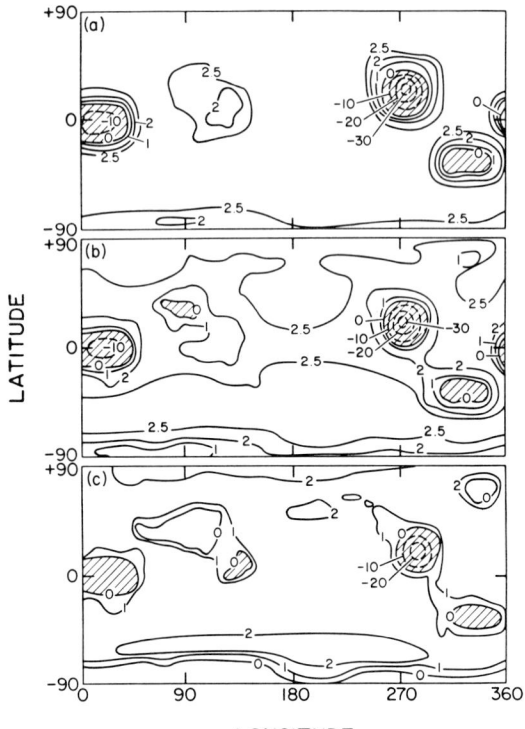

Fig. 4. Contours of calculated potential Φ along various height surfaces at approximately (a) 8 km, (b) 4 km, and (c) 2 km. All values are in volts when multiplied by $10^5$.

tion at 50 km. The calculated value of $(\Phi - \Phi_\infty)$ along the $\sigma_m = 4.54 \times 10^{-6}$ mho m$^{-1}$ conductivity surface, which is the base of the magnetosphere in region 3 of the model at roughly 105 km, is shown in Fig. 3a. $\Phi_\infty$ is the global mean average along this surface. The maximum value of $(\Phi - \Phi_\infty)$, which occurs over the thunderstorms in Africa, is 975 V and the minimum value is -600 V over the Central Pacific region. For these model calculations it is assumed that the geomagnetic and geographic poles are coincident. At the geomagnetic equator the field lines are horizontal and the vertically directed electric current is restricted from spreading laterally. The maximum potential develops in this region. In Central America, the thunderstorm region is displaced from the geomagnetic equator and therefore currents flow from the storm region along the geomagnetic field line into the conjugate hemisphere. The potential perturbation then attenuates as it penetrates to lower altitudes in the conjugate hemisphere.

The calculated potential Φ along the constant conductivity surface $\sigma_1 = 4.3 \times 10^{-13}$ mho m$^{-1}$, which is approximately 8 km in the equatorial regions, is shown in Fig. 4a. Under each thunderstorm region the potential is strongly negative. The minimum value of several million volts occurs under the Central America thunderstorm region. In the fair-weather region away from the thunderstorms the calculated poten-

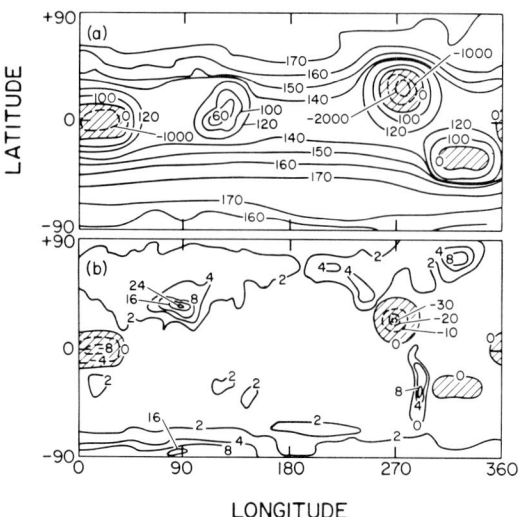

Fig. 5. Contours of calculated (a) ground potential gradient V m$^{-1}$ along the earth's orographic surface and (b) ground current, in A m$^{-2}$ when multiplied by $10^{-11}$.

tial is positive with respect to the ground, indicating a return current to ground. The calculated potentials at 4 km and 2 km respectively are shown in Fig. 4b and 4c. They are similar to the potential calculated on the 8 km surface but at these lower altitudes the distortion effects of the earth's orographic features become evident. The potential becomes zero when a surface intersects a high mountain that protrudes above that altitude, and the potential becomes greatly distorted when the surface is near an orographic feature such an Antarctica, Greenland, or the Himalayas. On the 2 km surface the outline of continental features begin to appear.

The calculated potential gradient and the ground current along the

Table 1. Comparison of Mühleisen (1977) estimate of global circuit parameters with model calculations at 1900 UT.

| | Mühleisen (1977) | Model |
|---|---|---|
| Number of thunderstorms | 1500–1800 | 2000 (specified) |
| Mean value of current intensity over one thunderstorm cell | 0.5–1.0 A | 0.51 A |
| Global current | 800–1800 A | 1065 A |
| Ionospheric potential | 180–400 kV (range) 278 kV (mean) | 291 kV |
| Columnar resistance | $1.3 \times 10^{17}$ $\Omega m^2$ | $1.39 \times 10^{17}$ $\Omega m^2$ |
| Total resistance | 230 $\Omega$ | 273 $\Omega$ |

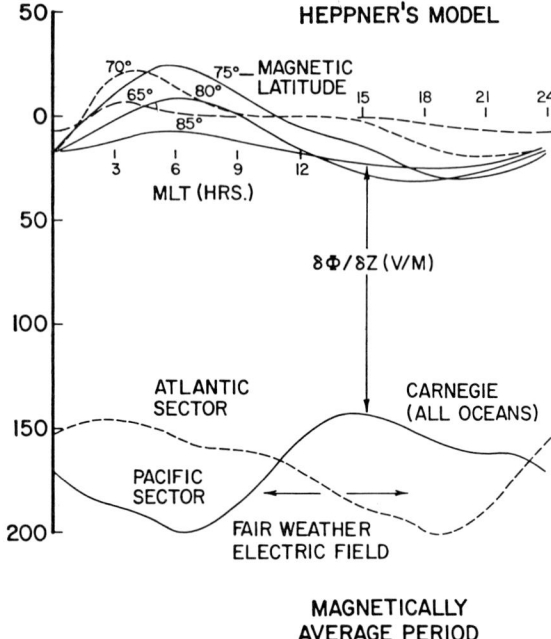

Fig. 6. Calculated diurnal variation of ground electric potential gradient (V m$^{-1}$) as a function of magnetic local time. The upper curves are calculated diurnal variations at various magnetic latitudes of the effect of the ionospheric potential distribution, using Heppner's model for a magnetically average period, upon the ground potential gradient. The lower curves are the diurnal (UT) variation of the ground potential gradient measured during the Carneige expedition in magnetic local time for stations in the Atlantic and Pacific sectors. The total potential gradient variation is determined by the difference between the upper and lower curves as indicated by the vector $\delta\Phi/\delta z$.

earth's orographic surface are contoured in Fig. 5a and 5b, respectively. The highest positive electric potential gradient and electric current flow occur in Antarctica and other high mountainous regions. These occur because of the high conductivity at mountaintop level and the effect of orography in distorting the electric potential and in amplifying the electric field. The current into the ground in the fair-weather regions is balanced by a current from the ground into the thunderstorm regions, leaving no net current flow on a global scale. The electric field at sea level is 134 V m$^{-1}$ at the equator and 173 V m$^{-1}$ in the polar regions, which compares well with the measurements of 120 V m$^{-1}$ over the equator and 155 V m$^{-1}$ at 60° latitude that were made by the Carnegie expedition (Israel, 1971). Table 1 compares the parameters of the model with the estimate of global circuit parameters made by Mühleisen (1977). There are considerable uncertainties in these global estimates, yet the comparison of parameters is encouraging.

## Coupling of Magnetospheric Effects into the Global Circuit

At high geomagnetic latitudes, there exists a plasma circulation pattern that results from the solar wind interaction with the earth's geomagnetic field (Axford and Hines, 1961; Axford, 1969). This interaction establishes a dawn to dusk electric potential drop of about 50-250 kilovolts across the polar cap. Electric fields over the polar cap have been measured by electrostatic probes on satellites (Heppner, 1972) and on balloons (Mozer and Serlin, 1969), drift measurements of whistler ducts (Carpenter, et al., 1972), drifts of barium clouds (Haerendel and Lüst, 1970; Meriwether, 1973), drifts of ionospheric ions (Heelis, et al., 1976) and incoherent scatter radar (Banks, et al., 1973) From these measurements a basic pattern of electric potential distribution has emerged as a persistent feature of the polar ionosphere. This electric potential pattern was modeled empirically by Heppner (1977) for both geomagnetic quiet and disturbed periods. During geomagnetic quiet periods the dawn to dusk potential drop may increase to 150-250 kV. The downward mapping of this ionospheric potential pattern has been considered by Park (1976). His calculations show that the large horizontal scale ionospheric potential perturbations map effectively downward in the direction of decreasing conductivity to the earth's surface. Because of the high electrical conductivity at the earth's surface the horizontal electric field turns vertical producing perturbations in the vertical potential gradient at the earth's surface. The magnitude of the perturbation of potential gradient at the earth's surface at high latitudes is about ± 20% of the fair weather potential gradient established by worldwide thunderstorms.

For our calculations, we use the empirical model of Heppner (1977) to specify the ionospheric potential distribution in the high latitude ionosphere. This potential distribution is the upper boundary condition which is coupled into the global model of atmospheric electricity as described by Roble and Hays (1978). Since the problem is linear the solutions obtained for the downward mapping of ionospheric potential and the charging of the lower atmosphere by thunderstorms described in the previous section can be superimposed. Our calculations also consider the effects of the Earth's orography and the tilted geomagnetic and geographic poles.

The calculations of the ground electric potential gradient variation using Heppner's magnetically averaged period model is shown in Fig. 6 for various magnetic latitudes. The magnetospheric potential pattern remains sun-aligned in geomagnetic coordinates. Therefore, as the Earth rotates around its geographic pole the ionospheric potential maps directly downward and the potential perturbation pattern moves over the earth's surface in a complex but systematic fashion during the day. The calculated ground potential gradient variation around various high magnetic latitude circles are plotted as functions of magnetic local time in Fig. 6. For early magnetic local times the ionospheric potential perturbations to the earth's potential gradient are positive. For later magnetic local times the perturbations are negative. The Carnegie expedition has shown that at low and middle latitudes there is a universal time variation of

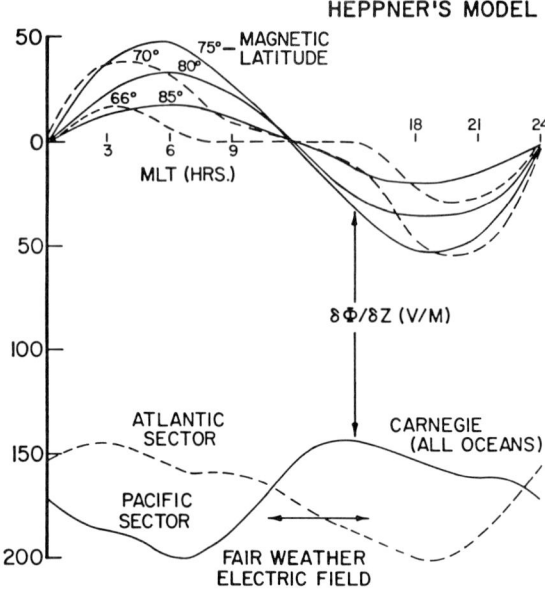

Fig. 7. Same caption as for Fig. 6 except using Heppner's ionospheric potential distribution model for a magnetically disturbed period.

potential gradient at the earth's surface due to the diurnal variation of thunderstorm frequency (Israel, 1971). The maximum electric potential gradient occurs at 1900 UT and the minimum occurs near 0400 UT. This potential pattern is plotted in magnetic local time for the Atlantic and Pacific sectors in Fig. 6 where it is seen that different diurnal variations in magnetic local time are obtained for the two sectors. The total ground potential gradient variation for a station in the Pacific sector is determined by the difference of the variation of the upper curve with the lower solid curve in Fig. 6 as indicated by the line with arrows giving $\delta\phi/\delta z$. The same variation for an Atlantic sector station is obtained from the difference of the upper curve with the lower dashed curve in Fig. 6. The results show that for an Atlantic sector station the diurnal variation of potential gradient is small whereas the diurnal variation is much larger for a Pacific sector station. Thus, measurements made at high latitudes should be interpreted in terms of variations of ionospheric potential in addition to thunderstorm frequency variations. Kasemir (1972) has reported that the diurnal UT variation of potential gradient at both Thule, Greenland and South Pole stations are about 30% less than the global low latitude UT variation attributed to variations in thunderstorm frequency.

## Magnetospheric Substorms

A similar calculation of the effect of ionospheric potential variations on the ground potential gradient for magnetically disturbed periods is shown in Fig. 7. The dawn to dusk ionospheric potential drop across the polar cap for the disturbed model is 140 kV. The results are similar to those discussed in the previous section; only the amplitude is greater indicating a greater upper atmospheric influence during magnetically disturbed periods. In our calculations we consider the effects of the earth's orography and the tilted geographic and geomagnetic poles. Acting alone, the positive and negative potential perturbations at high latitudes to the ionospheric potential at the upper boundary cause a downward and upward current respectively to flow from the ionosphere to the ground. The model requires that the divergence of the current must be equal to zero, therefore, any imbalance in the downward and upward current system must be compensated by the global circuit which causes a change in the ionospheric potential. Such an imbalance may occur when the positive/negative current is aligned over Antarctica, where the surface conductivity on the high mountain plateau is large, and when the negative/positive current is aligned over a cloud-covered ocean, where the surface conductivity is low. For certain large geomagnetic substorms with the upper boundary potential pattern properly aligned over the continent and ocean the current imbalance may require as much as a 5-10% change in the ionospheric potential maintained primarily by thunderstorms.

## References

Axford, W. I. and C. O. Hines, A unifying theory of high latitude geophysical phenomena and geomagnetic storms, Can. J. Phys. 39, 1433, 1961.

Axford, W. I., Magnetospheric convection, Rev. Geophys. Space Phys. 7, 421, 1961.

Banks, P. M., J. R. Doupnik and S.-I. Akasofu, Electric field observations by incoherent scatter radar in the aurora zone, J. Geophys. Res. 78, 6607, 1973.

Cauffman, D. P. and D. A. Gurnett, Satellite measurements of high latitude convection electric fields, Space Sci. Rev. 13, 369, 1972.

Carpenter, D. L., K. Stone, J. C. Siren and T. L. Crystal, Magnetospheric electric fields deduced from drifting whistler paths, J. Geophys. Res. 77, 2819, 1972.

Cole, R. K., Jr. and E. T. Pierce, Electrification in the earth's atmosphere for altitude between 0 and 100 kilometers, J. Geophys. Res. 70, 2735-2749, 1965.

Cobb, W. E., Evidence of a solar influence on the atmospheric electric elements at Mauna Loa Observatory, Mon. Weather Rev. 95, 905, 1967.

Chricklow, W. Q., R. C. Davis, R. T. Davis and M. W. Clark, Hourly probability of world-wide thunderstorm occurance, Office of telecommunication/ITS research report 12, Boulder, Colorado, April 1971.

Forbush, S. E., Time variations of cosmic rays, Handbuch der Physik, Geophysik III, J. Bartels, ed. Springer-Verlag, New York, 1966.

Haerendel G. and R. Lüst, Electric fields in the ionosphere and magnetosphere, in Particles and Fields in the Magnetosphere, ed. B. M. McCormac, D. Reidel, Dordrecht, Netherlands, 1970.

Hays, P. B. and R. G. Roble, A quasi-static model of global atmospheric electricity I. The lower atmosphere, J. Geophys. Res, in press, 1978.

Heelis, R. A., W. B. Hanson, J. L. Burch, Ion convection velocity reversals in the dayside cleft, J. Geophys. Res. 81, 3803-3809, 1976.

Heppner, J. P., Electric field variations during substorms--OGO-6 measurements, Planet. Space Sci. 20, 1475, 1972.

Heppner, J. P., Empirical models of high-latitude electric fields, J. Geophys. Res. 82, 1115, 1977.

Herman, J. R. and R. A. Goldberg, Initiation of non-tropical thunderstorms by solar activity, J. Atmos. Terr. Phys. 40, 121-134, 1978.

Israel, H., Atmospheric Electricity, Vol. 1 and Vol. 2, translated from German, Israel Program for Scientific Translations, Jerusalem, 1973.

Kasemir, H. W., Atmospheric electric measurements in the Arctic and Antarctic, Pure Appl. Geophys. 100, 70, 1972.

Lobodin, T. V. and N. A. Paramonov, Variation of atmospheric electric field during aurorae, Pure. Appl. Geophys. 100, 167, 1972.

Markson, R., Considerations regarding solar and lunar modulation of geophysical parameters, Atmospheric electricity and thunderstorm, Pure Appl. Geophys. 84, 161, 1971.

Markson, R., Solar modulation of atmospheric electrification and possible implications for the sun-weather relationship, Nature 273, 103, 1978.

Meriwether, J. W., J. P. Heppner, J. D. Stolarik and E. M. Wescott, Neutral winds above 200 km at high latitudes, J. Geophys. Res. 78, 6643, 1973.

Morita, Y., H. Ishirawa and M. Kanada, The vertical profiles of small ion density and electric conductivity in the atmosphere in 19 km, J. Geophys. Res. 76, 3431, 1971.

Mozer, F. S. and R. Serlin, Magnetospheric electric field measurements with balloons, J. Geophys. Res. 74, 4739, 1969.

Mühleisen, R., The global circuit and its parameters, in Electrical Processes in Atmospheres, Sternkopff Verlag, Ed. H. Dolezalek and R. Reiter, Darmstadt, 467, 1977.

Olson, D. E., The evidence for auroral effects on atmospheric electricity, Pure Appl. Geophys. 84, 118-138, 1971.

Park, C. G., Solar magnetic sector effects on the vertical atmospheric electric field at Vo stak, Antarctica, Geophys. Res. Lettrs. 3, 475-478, 1976a.

Park, C. G., Downward mapping of high-latitude electric fields to the ground, J. Geophys. Res. 81, 168, 1976b.

Reiter, R., Solar flares and their impact on potential gradient and air-earth current characteristics at high mountain stations, Pure Appl. Geophys. 72, 259, 1969.

Reiter, R., Further evidence for impact of solar flares or potential gradient and air-earth current characteristics at high mountain stations, Pure Appl. Geophys. 86, 142, 1971.

Reiter, R., Case study concerning the impact of solar activity upon potential gradient and air-earth current in the lower troposphere, Pure Appl. Geophys. 94, 218, 1972.

Reiter, R., Atmospheric electricity activities of the institute for atmospheric environmental research, Electrical Processes in Atmospheres, Steinkopff Verlag, ed. H. Dolezalek and R. Reiter, Darnstadt, 759, 1977.

Roble, R. G. and P. B. Hays, A quasi-static model of global atmospheric electricity II. Electrical coupling between the upper and lower atmosphere, J. Geophys. Res., submitted, 1978.

Sagalyn, R. C., Atmospheric electricity, U.S. Air Force Handbook of Geophysics, McMillan, 1960.

Sartor, D., Induction charging thunderstorm mechanism, Problems of Atmospheric and Space Electricity, S. C. Coroniti, ed. Elsevier, Pub. Co., Amsterdam 307, 1965.

Tran, A. and C. Polk, Lowest (40 km-70 km) ionospheric conductivity profiles and Schumann (ELF) resonances, Report to Air Force Cambridge Labs, on contract F-19628-70-C-0090, University of Rhode Island, Dept. of Electrical Engineering, August, 1972.

OVERVIEW - LOW ENERGY PARTICLES

E. C. Whipple, Jr. and S. E. DeForest

University of California, San Diego
La Jolla, California 92093

   In the section on low energy particles a total of eight papers were presented. Two of these dealt directly with low energy particles per se: the paper by Young on the ion composition in the magnetosphere, and the paper by Garrett on a model of the low energy particle environment. The other papers dealt with magnetospheric boundary layer regions and other structures influenced predominantly. This division of papers illustrates the need that exists for a comprehensive survey of low energy particle data in the magnetosphere. A great amount of data has been accumulated over the last several years. However, quantitative descriptive models simply do not exist except for the one described by Garrett. A systematic compilation of existing data organized by region and by activity level would be of great value.
   There is a question as to the definition of what is meant by "low energy particles." Experimentalists tend to use their own instrumentation as a reference, so that "high energy" refers to the range above their experiment, and "low energy" to the range below. However, at the meeting there seemed to be a concensus that a good definition of low energy particles is the range from 0 keV to 100 keV. This is roughly the range over which the electric field drift is important for particles in the magnetosphere.
   An important subdivision of low energy particles is the class of what might be called "very low energy" or thermal partcles, those with energies below about 100 eV. Historically, this category of particles has been extremely difficult to describe quantitatively because of spacecraft charging effects and also because of contamination of spacecraft produced photoelectrons and secondary electrons. These problems are more fairly well understood qualitatively, but it has still been difficult to obtain accurate quantitative self-consistent thermal electron and ion concentrations and temperatures. An important step in this direction is the realization that electrostatically clean spacecraft like GEOS and ISEE are required for good very low energy particle measurements.
   The review of ion composition measurements by Young brought out the variability of the ion composition as well as the importance of the heavier ions at certain times and in certain regions. One's first reaction is that the magnetosphere is even more complex than has been realized; however, the heavy ions are actually significant clues to the source regions of the magnetospheric plasma and should shed light on the plasma injection and loss mechanisms. There is a need for ion composition measurements in the energy range between 20 keV and 100 keV where there is presently no information. What about negative ions?

If the ionospheric source for ions extends as low as 100 km, then there may be negative ions injected into the magnetosphere at times. No information exists at present.

Garrett's review of the particle environment emphasized the need for and the lack of quantitative models of the low energy plasma. He stressed the fact that information on velocity distributions is essential since these tend in general not to be Maxwellian.

Heikkila discussed the energy budget in the current systems as the magnetospheric plasma flows through various regions. His claim that the necessary dissipation of $10^{11}$ to $10^{12}$ watts across the magnetopause has not been observed is so far unanswered. He emphasized the need for modeling the bow shock and magnetosheath regions as well as the magnetopause and magnetosphere interior.

Two papers dealt with the relationships between geomagnetic indices and magnetospheric quantities. Slavin and Holzer described a useful correlation between the transfer of magnetic flux on the dayside of the magnetosphere, the AL index and magnetic field enhancements in the magnetotail. Maezawa pointed out that care should be used in using the AU, the Am, and the AL indices for modeling work since each of these has a different functional dependence on solar wind parameters.

Papers by Lemaire, Eastman and Hones, and Voigt and Fuchs dealt with boundary layer phenomena. Lemaire suggested that the boundary layer could be the region where blobs of plasma in the solar wind lose their excess momentum. The braking distance for such a blob is on the order of the thickness of the boundary layer inside the magnetospause. Eastman and Hones reviewed the signatures that are to be expected across a boundary layer for various degrees of particle access, and for different transport mechanisms.

Voigt and Fuchs emphasized that the "merging" process should occur primarily at higher latitudes in the cusp region rather than in the equatorial regions.

ION COMPOSITION MEASUREMENTS IN MAGNETOSPHERIC MODELING

D.T. Young

Physikalisches Institut, University of Bern
3012 Bern, Switzerland

Abstract. In recent years several high altitude satellites have carried energetic ion mass spectrometers measuring magnetospheric ion composition up to $\sim 20$ keV. This energy range often includes most of the particles responsible for magnetospheric dynamics in the regions surveyed. A key result is that during magnetic disturbances the hot magnetospheric plasma contains, in addition to $H^+$, typically $\sim 30\%$ of heavier ions made up predominantly of $O^+$ with smaller amounts of $He^{++}$ and $He^+$. In this paper mass spectrometer instrumentation and the attendant problems of design and measurement are discussed. Several recent results are presented emphasizing near-equatorial composition measurements of the plasmasphere, magnetic field-aligned ion beams and aspects of the dynamics of trapped ion fluxes. Included is a brief discussion of current active modeling experiments involving artificially injected ions as tracers.

## Introduction

Recently it has been established that ions heavier than protons make up roughly 1% $\sim$ 10% (by number) of quiescent magnetospheric particle populations from a few eV (Young et al., 1977) through the keV range (Geiss et al., 1978) up to several MeV (Fritz, 1976). Furthermore, during magnetic disturbances the heavy component apparently increases to 10% $\sim$ 50% or more of the total over much of this energy range (Shelley et al., 1976; Geiss et al., 1978; Fritz and Wilken, 1976). The traditional role assigned to heavier ions in the magnetosphere and aurora was based on their potential use as passive tracers of particle origins and processes (cf. Axford, 1970; Cornwall, 1972; Blake, 1973), a possibility which still exists. However, given the mounting evidence that heavy ions are at times a significant fraction of the total, it is necessary to consider as well their role as active determinants of magnetospheric dynamics.

Measurement of hot plasma composition, from a few eV to 16 keV, in the high altitude equatorial magnetosphere began with the launches of GEOS-1 (20 Apirl 1977), ISEE-A (22 October 1977) and GEOS-2 (14 July 1978). Only a small fraction of this data has thus far been analyzed and it seems premature to discuss "models" of magnetospheric composition - although this is clearly a goal for the near future. The inverse problem - namely the effect of composition on models of magnetospheric proces-

TABLE 1. Summary of instruments performing ion composition measurements relevant to magnetospheric studies. Values of mass resolution and $H^+$ sensitivity (at minimum energy) are in some cases rough approximations based only on published spectra and may be greater than stated here.

| Space-craft | Launch Date (Yr.) | Orbit incl. (°) | Orbit perigee × apogee | Plasma population covered[a] | Analyzer type[b] | Energy range (keV) | Mass range (AMU) | Mass resolution (M/ΔM) | $H^+$ sensitivity (cm$^{-3}$) | Pitch angle[c] (°) | Reference |
|---|---|---|---|---|---|---|---|---|---|---|---|
| OGO-1 | 1964 | 31 | 280×149000 | P | RF | ≲.01 | 1–45 | 9 | 1 | ram | Tylor et al., 1965 |
| OGO-5 | 1968 | 31 | 300×145000 | P | EMC | ≲.01 | 1,4,16 | 1 | $10^{-1}$ | ram | Harris and Sharp, 1969 |
| OV1-18 | 1969 | 99 | 550 | M | EMW | 1–9 | 1–16 | 1 | $10^{-2}$ | <50 | Sharp et al., 1974 |
| ISIS-2 | 1971 | 88 | 1400 | I | EMC | ≲.01 | 1–64 | 16 | 1 | ram | Hoffman et al., 1974 |
| 1971-089A | 1971 | 93 | 800 | M* | EMW | 0.7–12 | 1–32 | 2 | $10^{-2}$ | <50 | Shelley et al., 1972 |
| S3-3 | 1976 | 98 | 350×8000 | M* | EMW | 0.5–16 | 1–≳32 | 2 | $10^{-2}$ | 0–180 | Shelley et al., 1976 |
| GEOS-1 | 1977 | 26 | 2000×38000 | P,M | EMC | 0–16 | 1–>140 | 4 | $10^{-3}$ | ∼90[d] | Balsiger et al., 1976 |
| ISEE-A | 1977 | 26 | 500×131000 | P,M,S,T | EMC | 0–16 | 1–>140 | 4 | $10^{-3}$ | ∼90 | Shelley et al., 1978 |
| GEOS-2 | 1978 | 0 | 35700 | P,M | EMC | 0–16 | 1–>140 | 4 | $10^{-3}$ | ∼90[d] | Balsiger et al., 1976 |
| SCATHA | 1979 | 8 | 28000×43000 (nominal) | P, M | EMC | 0–0.1 | 1,2,4,16 | 8 | $10^{-1}$ | 0–180 | Reasoner[e] |
| | | | | | EMC | 0.1–32 | 1–>140 | 4 | $10^{-3}$ | 0–180 | NSSDC77-03[f] (Johnson) |

a) Plasma population: I–ionosphere, P–plasmasphere, M–magnetosphere, M*–precipitated, S–solar wind, T–magnetotail
b) Analyzer type: RF–Bennett type, EMC–curved electric and magnetic, EMW–electric with Wien filter
c) Pitch angle: local pitch angle coverage; ram–relies on satellite velocity ram effect.
d) During special periods GEOS can cover 0–180 pitch angles.
e) D.L. Reasoner, Private communication (1978).
f) From Vette and Vostreys (1977).

ses - is even more difficult to assess. However, certain aspects of the latter problem have been studied theoretically and are reviewed by Cornwall and Schulz (1978).

This review is therefore more topical than comprehensive. It concentrates on 3 areas involving measurement of natural plasmas (the plasmasphere, field-aligned ion beams and trapped ion fluxes) and a fourth, the measurement of artificially injected ions. It is likely that results in these four areas could have considerable impact on the modeling of ion acceleration, injection and loss processes in the magnetosphere.

Reviews by Shelley (1978), Prangé (1978) and Johnson et al. (1976) have summarized composition of the hot magnetospheric plasma. Most of the data then available for review were obtained from measurements at relatively low altitudes ($\leq$ 8000 km) within the energy range 0.5 - 16 keV (see Table 1). In contrast to this data set, many (but not all) models of plasma convection, diffusion and loss have dealt primarily with particles mirroring near the equatorial plane. Thus, in addition to being timely, the GEOS results reviewed here serve to complement the lower altitude satellite results and extend measurements to energies < 500 eV. An important set of observations, from the ISEE-A mass spectrometer, are not yet available for inclusion. Ion composition above 0.1 MeV has been reviewed recently (Fritz, 1976; Cornwall and Schultz, 1978; Spjeldvik, 1978) and will not be discussed.

## Instrumentation

Much of the newly acquired information on hot plasma composition comes as the result of improved instrumentation. For the first time a single spectrometer can encompass a wide range of ion mass extending over an energy range from $\sim$ 1 eV to beyond 10 keV (Balsiger et al., 1976; Shelley et al., 1976). In general there exist several distinct types of mass spectrometers, namely conventional types (limited to ion energies of a few eV), energetic mass spectrometers (operating below $\sim$ 20 keV) and solid state detectors (generally useful for this purpose only above $\sim$ 100 keV). There is at present a serious gap in energy coverage between 20 and 100 keV which may conceiveably be filled by future instruments (e.g. Gloeckler, 1977).

Table 1 gives a summary of mass spectrometers which have been used to measure composition of the magnetospheric plasma. Not included are low altitude ionospheric instruments, solid state detectors and those devoted exclusively to solar wind studies. Since the technology behind energetic mass spectrometers is basically more complex than that of the ubiquitous electrostatic analyzer (ESA) and was developed later, it is not clear that composition instruments have yet reached a comparable level of sophistication. The highest mass resolution in use at present is $\sim$ 4 in the GEOS-type spectrometers carried by GEOS and ISEE. This is not sufficient, for example, to resolve less abundant ionospheric ions (e.g. $^{14}N$ from $^{16}O$) or minor solar wind species (the silicon and iron groups) which may be present in the magnetosphere.

One perhaps obvious constraint on composition instruments are the multiple trade-offs among the three parameters of sensitivity, time resolution and mass resolution. Because ions travel more slowly than electrons, available fluxes are correspondingly lower. In addition, certain ion species (e.g. $He^{++}$ or $O^{++}$) have concentrations $10^{-2}$ to $10^{-4}$ that of $H^+$. Thus, a very high detector sensitivity is required, particularly for high altitude observations. This can be obtained by increasing either the sensitive area of the detector, often with a corresponding loss in mass resolution, or the measurement time which in turn leads to a loss in time resolution. By way of their basically simpler optical systems, ESA's usually have greater sensitivity for a given weight. This allows, for example, multi-ESA instruments with relatively fast sampling rates to determine the ion distribution function in 3 dimensions within a reasonable length of time.

Until recently ion mass and charge were not, for the most part, measureable so it was reasonable to interpret data from ESA's in terms of protons. Energetic ion mass spectrometers have shown that whereas the "proton assumption" is not entirely wrong, seldom is it the whole story. It is well known that an ESA underestimates ion number density (density refers to number density throughout this paper) when a significant number of heavy ions are present, because the ion mass and hence velocity is unknown. A more subtle effect occurs with ESA's using channel electron multipliers as detectors (nearly all of them do) due to their lower efficiency for heavy ions at energies below 5 keV (Burrous et al., 1967). Details depend on ion mass and energy and whether pre-multiplier acceleration is used but, for example, 1 keV $O^+$ ions have $\sim 30\%$ lower probability of being registered by a channel electron multiplier than the same energy protons. The overall effect for an ESA sampling a 1 keV plasma with equal number of $O^+$ and $H^+$ ions would be to underestimate the total ion density by 40%. This is not intended to imply that all results with electrostatic energy analyzers are incorrect, but only to point out that under certain conditions composition measurements may be necessary to obtain the correct density. In fact, we clearly need complementary measurements: electrostatic energy analyzers giving rapid 3-dimensional scans of the entire plasma distribution and mass analyzers giving the same data for several ion species but necessarily at a reduced rate.

Composition measurements suitable for quantitative modeling focus attention on the abundances of different ion species and particularly on the absolute and relative accuracy of these measurements. Absolute calibration of a mass spectrometer is in principle the same as for any particle detector: a relation (the "geometric factor") is established between a known incident particle flux and the measured counting rate. With the GEOS mass spectrometers for example, we estimate the absolute accuracy of the flux calibration to be $\pm 50\%$ and the relative accuracy to be $\pm 20\%$. Absolute calibration must be extrapolated to the very lowest energies. This has been checked for the GEOS spectrometer at energies $\sim 1$ eV using measurements from 4 other GEOS-1 experiments (Decreau et al., 1978). At an energy of 16 keV the measured fluxes agree to within $\pm 50\%$ of those determined by the lowest energy channel (24 - 32 keV) of the MPAE Lindau experiment S-321 on GEOS-1 (A. Korth, private communication). Furthermore, the GEOS mass spectrometer is tested regularly in flight to check detector discrimination thresholds and the voltages determining mass peak locations. The latter have been found to be stable to better than 0.3% during 1 year of GEOS-1 operations (Geiss et al., 1978).

Much larger errors than those discussed above may occur when data, in the form of detector counting rates as a function of location in mass-energy-direction space, are converted into particle distribution functions whose moments give the parameters (e.g. total density, composition, temperature, etc.) of interest. This is a problem of interpretation which exists for all particle data, but is particularly critical for mass spectrometers if we want to discuss, as is often the case for example, the "percentage" of plasma of solar wind origin (i.e. the $^4He^{++}/H^+$ ratio) in the magnetosphere. For example, at very low energies (< 100 eV) measurements must be transformed to the frame of reference of the unperturbed plasma i.e. the effects of satellite potential and velocity must be removed. This is particularly important for mass spectrometer measurements because both ion charge and mass enter the equations describing the distribution of ion flux reaching the satellite (cf. Whipple et al., 1974). An additional problem arises when the ion pitch angle distributions are different for different ion species (cf. Figure 3 in section 3.2) in which case it may not be sensible to discuss abundances based on simple integrals of the locally observed fluxes. Finally, a similar problem occurs in comparing abundances of distributions with markedly different temperatures: at any one time the magnetospheric plasma from 1 eV to > 50 keV can be characterized by several Maxwellian distributions (Garrett, 1977). Should composition then be discussed on the basis of Maxwellian or other model populations, or by simple lumping of energy ranges (e.g. 0-10 eV or 1-20 keV) or should an integral be formed over all available data? For early GEOS data the second method has been used, with a lower energy cutoff of 1 keV (cf. Figure 9) because the ion flux tends to have a strong minimum at this energy (Balsiger et al., 1978). With more sophisticated routines now being developed the first method should become feasible. Obviously for quantitative modeling purposes as well the first method is preferable.

Recent Measurements of Magnetospheric Composition

<u>The plasmasphere</u>

Unlike the keV plasma, high altitude thermal plasma has been surveyed extensively by classical mass spectrometers on OGO-1 and 5 (Table 1). Average ion composition was reported to be $H^+ : He^+ : O^+ \sim 1000 : 10 : 1$ (Chappell et al., 1970), similar to the earlier OGO-1 data (Taylor et al., 1965). Although the OGO results, particularly those from OGO-5, contributed greatly to the understanding of plasmasphere dynamics, little was published on the composition data per se although the above result is widely quoted in the literature.

Using the GOES-1 spectrometer, Young et al., (1977) reported significant amounts of $He^{++}$ and $O^{++}$ ($D^+$ may be present as well, see Geiss et al., 1978) in addition to $H^+$, $He^+$ and $O^+$. Ionic abundances were found to

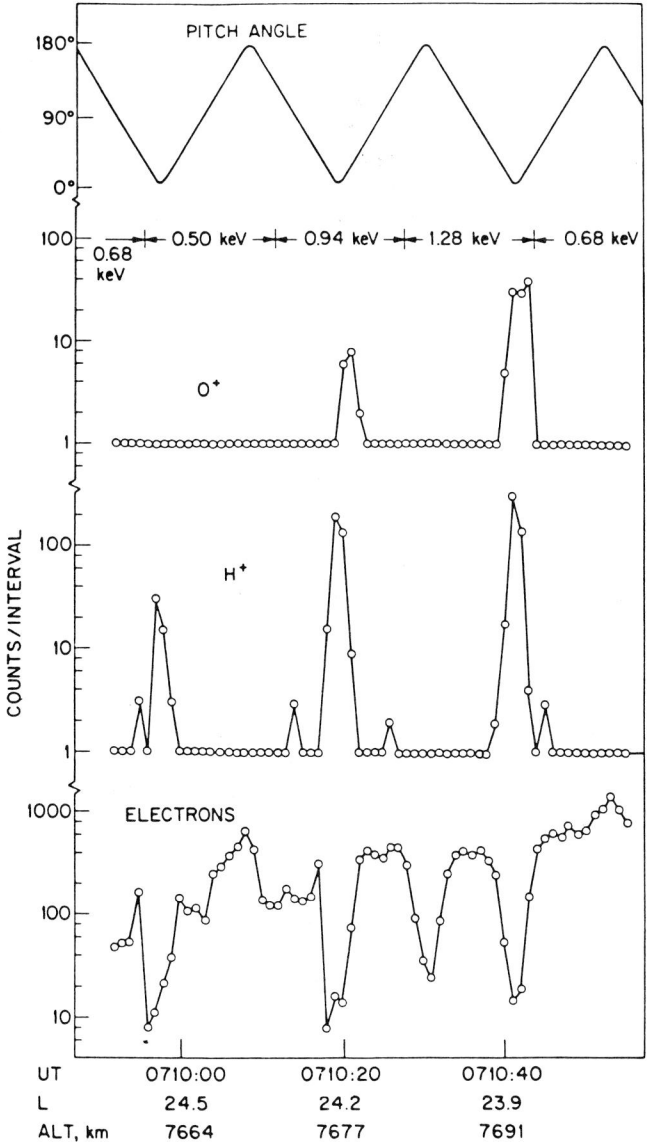

Fig. 1. Data from satellite S3-3 orbit 67 on July 17, 1976. The upper panel shows the pitch angle of the measured particles. The two center panels show data from the mass spectrometer at the indicated energies, and the lower panel shows electron fluxes in the energy range from 0.37 to 1.28 keV. The relative temporal precision of the plots is about one second. (From Shelley et al., 1976.)

differ significantly from the reported OGO result with typical composition of $H^+ : He^+ : O^+ : O^{++} : He^{++} \simeq 1000 : 100 : 10 : 3 : 1$. The discrepancy by a factor $\sim 10$ between OGO and GEOS measurements of $He^+$ and $O^+$ is not yet explained. Contributing factors may be the lack of angular and energy measurements with the OGO spectrometers such that part of the heavy ion distributions, which are narrow relative to $H^+$, were missed - although an order of magnitude effect seems unlikely. In addition GEOS tended to charge positively by $\gtrsim 1$ Volt in the outer plasmasphere and beyond (Decreau et al., 1978) whereas OGO was expected to be charged negatively (Harris and Sharp, 1969; Whipple et al., 1974). This would bias GEOS data slightly toward heavier ions which, for given velocity, more easily overcome the satellite potential barrier.

Although $He^{++}$ and $O^{++}$ are present in small amounts in the topside ionosphere equatorward of the trough region (Hoffman et al., 1974) it was surprising to find relatively large amounts (occasionally $O^{++}/O^+ \sim 1$) in the high altitude plasmasphere. In addition to in situ production by UV and particle impact ionization, coupled with the long lifetime of the doubly charged state against charge exchange, thermal diffusion may be significant in transporting doubly charged ions to high altitudes (Geiss et al., 1978). This should be particularly effective for enriching $O^{++}$ relative to $O^+$ because thermal diffusion goes roughly as $Q^2$ and the parent $O^+$ ion has relatively small scale height. Diffusion calculations by Geiss et al. indicate that $O^{++}/O^+$ may reach values $\sim 1$ within several days, in agreement with the GEOS-1 observations.

Magnetic field-aligned ion beams

Ion beams accelerated away from the ionosphere along the magnetic field direction were first observed by the S3-3 satellite mass spectrometer (Shelley et al., 1976). The example shown in Figure 1 is typical of the composition in these beams, which are observed in the 3000 - 8000 km altitude range on > 60% of all S3-3 orbits (Ghielmetti et al., 1978). The $H^+$ fluxes are generally $\sim 5 \times 10^7$ ions $(cm^2 sec\ ster\ keV)^{-1}$ with $O^+$ somewhat less intense and $He^+$ contributing < 1%. The majority of upward ion flows are field-aligned, but conical distributions (showing a local minimum in flux intensity along $\underset{\sim}{B}$ with a maximum at smaller pitch angles) are observed as well. Ghielmetti and co-workers studied 370 cases of upward directed ion flows above a flux threshold of $\sim 2 \times 10^6$ ions $(cm^2 sec\ ster\ keV)^{-1}$. They concluded that the main acceleration region was probably located at an altitude of $\sim 1\ R_E$ and that the spatial distribution of occurrence frequency agreed reasonably well with statistical location of the auroral oval (beams were not observed below invariant latitudes of 60°). The distribution of upward ion flows was strongly peaked on the dusk side of the magnetosphere, which is also suggestive of an auroral origin for the acceleration. The S3-3 satellite also carried ESA's covering energies down to 90 eV with which narrow, field-aligned upward moving ion beams were observed (Minzera and Fennel, 1977). At lower altitudes (400 - 600 km) a rocket experiment carrying ESA's measured intense upward moving ion fluxes at 90 - 500 eV ($\sim 10^8$ ions $(cm^2 sec\ ster\ keV)^{-1}$ with conical pitch angle distributions (Whalen et al., 1978). Apparently the ISIS satellites have observed upward accelerated ion beams as well (Ungstrup et al., 1977). In a study of earlier data from ESRO 1A, Hultqvist and Borg (1978) reported earthward accelerated

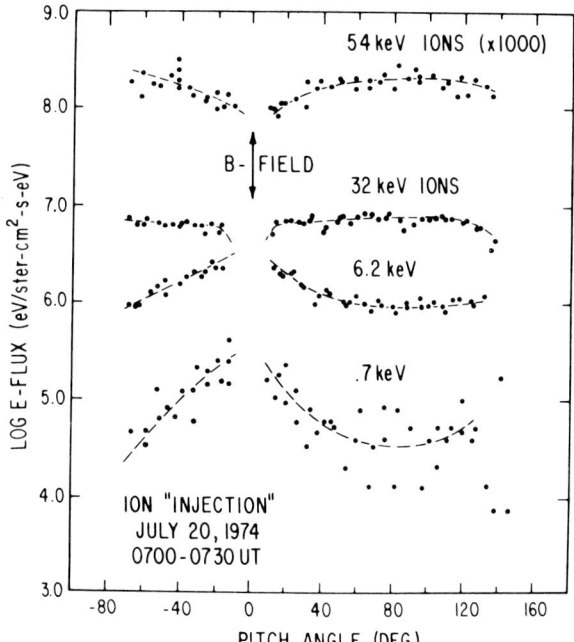

Fig. 2a. Ion pitch angle distributions from ATS-6 UCSD electrostatic analyzer following a substorm injection of plasma at 0630 UT. Negative pitch angle corresponds to look directions on the earthward side of field line. Note that the full width at 10% of maximum is $\sim \pm 50°$ for the 0.7 keV ions. (From Mauk and McIlwain, 1975.)

beams of 1 and 6 keV ions at altitudes of 1000 km in association with "inverted V" events (these in turn are thought to be the source of discrete auroral arcs). Since none of the several experiments were simultaneous it is not clear whether all are observing aspects of the same phenomena, although Hultqvist and Borg suggest that upward and downward ion beams could exist simultaneously at different altitudes on the same field line.

The conical or flat pitch angle distributions seen by S3-3 and ISIS satellites and the rocket experiment have been interpreted as being consistent with perpendicular heating of ions to $\sim$ 100 eV at lower (500 - 2000 km) altitudes, possibly by the electrostatic ion cyclotron instability (Ungstrup et al., 1977; Wahlen et al., 1978) or by turbulence in electrostatic shocks (Torbert and Mozer, 1978). Ungstrup et al. (1977) have pointed out that because the $O^+$ cyclotron wave is unstable to a smaller current (Kindel and Kennel, 1971) the instability may work its way down the field line to the $H^+ - O^+$ transition region (at $\sim$ 1000 km) at which point $O^+$ would be selectively heated and accelerated upwards. Further acceleration of the ions to kilovolt energies by parallel electric fields could then explain the narrower field-aligned beams seen at S3-3. It seems generally agreed that the evidence nearly all points to-

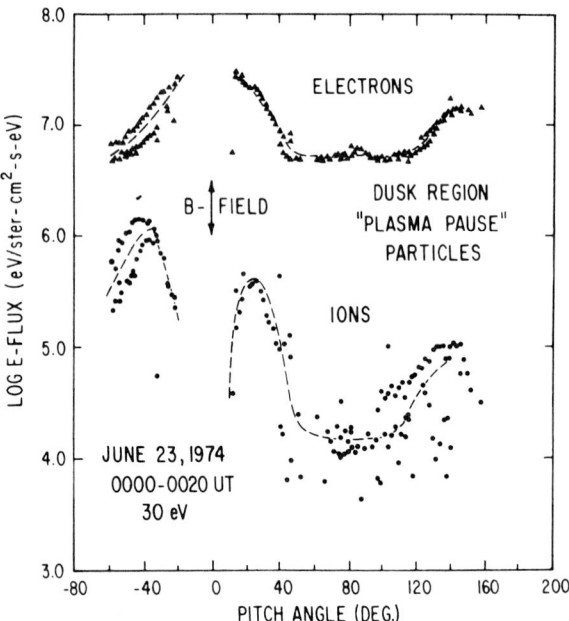

Fig. 2b. Pitch angle distribution for field-aligned "source cone" ions of 30 eV from ATS-6 UCSD instrument. This event apparently is not associated with substorm injection. Note that full width of ion peak is $\sim \pm 50°$, also the asymmetry in the peak and the well developed, very wide "loss cone". (From Mauk and McIlwain, 1975.)

ward the upward ion beams being yet another manifestation of the classical auroral acceleration process.

Turning to higher altitude, near-equatorial data, Mauk and McIlwain (1975) reported ATS-6 observations of "source cone" distributions of low to medium energy (.01 $\sim$ 3 keV) ions (Figure 2a and 2b). Although no extensive study of these data has appeared, the ion distributions were fairly broad ($\sim$ 80° full width at 10% of maximum) with $\sim$ 20° loss cone-like structures (the atmospheric loss cone is $\lesssim$ 5°). Mauk and McIlwain remarked that the source cone distributions were persistent features associated with substorm injection events and with the dusk bulge region of the plasmasphere. McIlwain (1976) reported observations of bouncing ion clusters whose bounce period led him to suggest that they might be $O^+$ ions accelerated away from the auroral ionosphere. Borg et al. (1978) have summarized initial results of 0.3 - 20 keV ion and electron pitch angle measurements with a set of ESA's on GEOS-1. They conclude that field-aligned ions are present at GEOS-1 although the ions have much broader pitch angle distributions and occur less frequently than would be expected from the S3-3 results.

Normally the GEOS-1 mass spectrometer views local pitch angles in the range 70° $\lesssim \alpha \lesssim$ 110° as a consequence of the orbit inclination of 26.5° and orientation of the satellite spin axis parallel to that of the earth's. Under certain circumstances, namely during special attitude

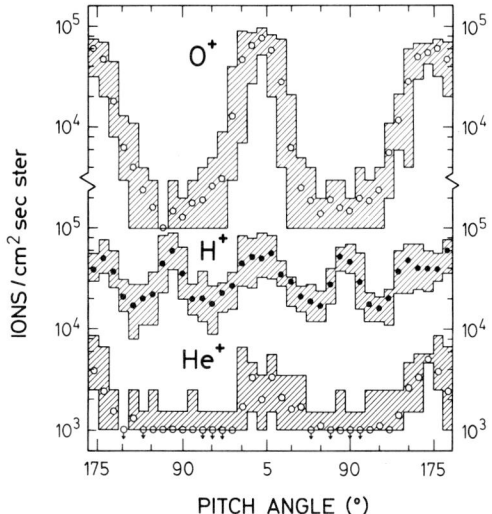

Fig. 3. Pitch angle distribution and composition of ∿ 50 eV ions observed with GEOS-1 mass spectrometer. Shaded regions represent the extreme values over 8 satellite revolutions (∿ 48 sec). Symbols show average values, divisions on the abscissa correspond to pitch angles of 30, 60, 120 and 150 degrees as viewed by detector. Full width of $O^+$ peak at 10% is ∿ ±50°. (From Geiss et al., 1978.)

manoeuvres or magnetic disturbances, the spectrometer scans nearly the full range of pitch angles for extended (≳ 1 hour) periods. The distribution shown in Figure 3 (Geiss et al., 1978) was obtained during a moderately active period when the satellite spin axis was titled over. In this case the field-aligned beams had mean energies ∿ 50 eV although instances of energies ≳ 1 keV were also observed. Number density ratios were $O^+$ : $H^+$ : $He^+$ : $O^{++}$ ∿ 20 : 4 : 1 : 1 in the field-aligned component. Note that the $H^+$ distribution contains a trapped component and that the beam width at 10% intensity is nearly the same as that reported by Mauk and McIlwain (1975). During magnetically disturbed periods, particularly the early main phase of storms, a variety of field-aligned phenomena have been observed over an energy range from 10's of eV to 16 keV. An example of one type is the "conical" distribution shown in Figure 4. If the maximum at 120° pitch angle is mapped down the field line to its mirror point, then it originated about 3500 km below the satellite, i.e. still some 4 $R_E$ above the ionosphere. The field-aligned structures observed at GEOS tend to vary rapidly with time, typically changing from field-aligned flows as in Figures 3 and 4, to flows at large angles to the magnetic field, and back again on a time scale of minutes to tens of

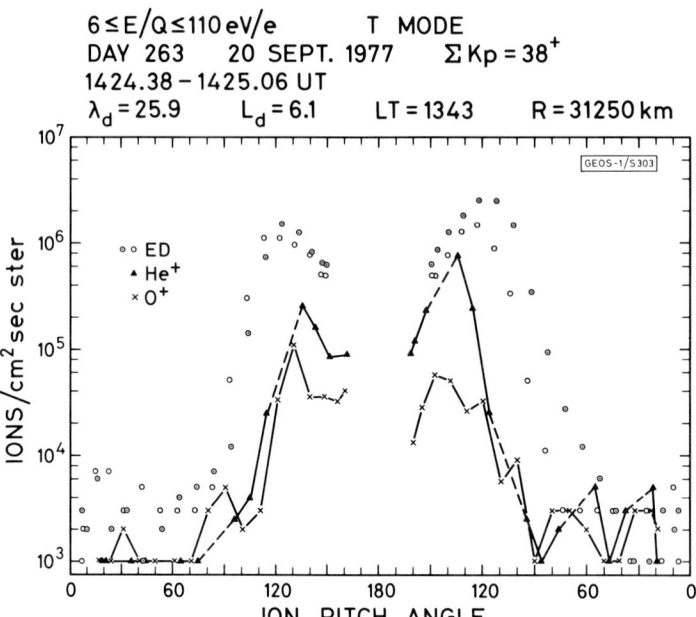

Fig. 4. Pitch angle distribution and composition of 35 eV ions observed at GEOS-1. Ions are moving away from the earth, the H$^+$ ions with a velocity of $\sim$ 40 km/sec (for a mean energy of 35 eV peaked at 60° to the field line). Output of energy detector (ED) corresponds approximately to H$^+$ flux. (Not published previously.)

minutes. One possible explanation is that the variations are due to irregular motion of field-aligned structures past the satellite at high velocity. This would account for the apparent combination of field-aligned and perpendicular (to $\underset{\sim}{B}$) velocity components which are often seen.

Composition of the field-aligned ions at GEOS is variable and no systematic study has yet been made. It appears, however, that O$^+$ is generally present and is occasionally a large component of the beams. If composition is taken as a signature of the source region, then many beams originate near $\sim$ 1000 km, below which O$^+$ becomes the dominant species. Following this line of thought, the conical beam shown in Figure 4 is of some interest although we caution that so far it is a sample of one. Note that in Figure 4 He$^+$ rather than O$^+$ is the second most abundant ion, the reverse of the more normal trend in beam composition (cf. Figures 1 and 3). This signature is probably more characteristic of composition at very high altitudes (several thousand km) than the topside ionosphere and, taken together with the shape of the pitch angle distribution, suggests the possibility of very high altitude acceleration of ions perpendicular to $\underset{\sim}{B}$. Since ion cyclotron waves are good canditates for the acceleration mechanism, a search should be made for the appropriate wave signals at GEOS-1 during these events.

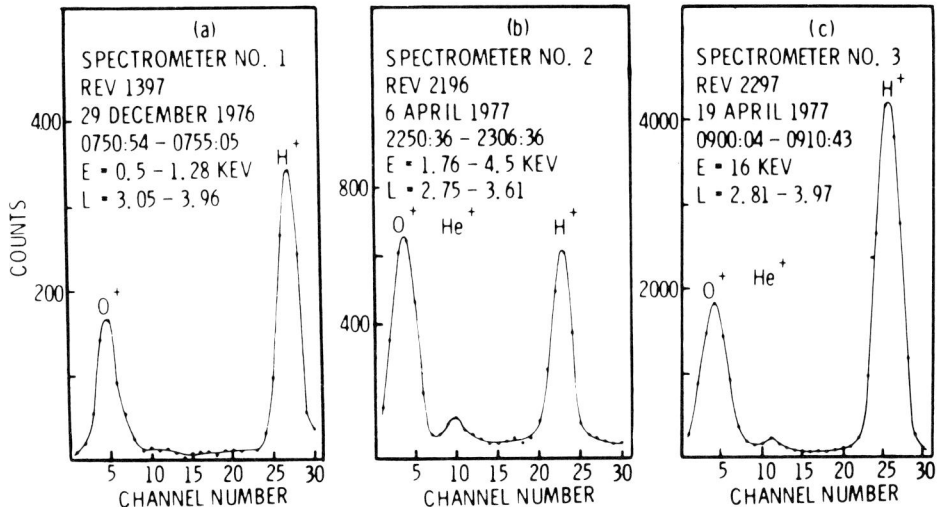

Fig. 5a,b,c. Ring current mass spectra from S3-3 integrated over various energy channels. Data are from the main phase of storms on 29 December 1976, 6 April 1977, and 19 April 1977 taken at magnetic local times of 0200, 2100 and 2400. Maximum equatorial pitch angles scanned were 15° - 21°, 20° - 41° and 15° - 31° respectively. (Johnson et al., 1977.)

Composition of trapped ions

Trapped energetic (keV) ions of ionospheric origin were first observed in the inner (L = 2.6 - 4.0) ring current by the S3-3 mass spectrometer (Johnson et al., 1977). Near apogee the S3-3 spectrometer can observe ions with equatorial pitch angles between 50° and 15° at L values of 2.6 to 4.0 respectively. Using data obtained during 3 magnetic storms (Figures 5a, b, c and 6a, b, c) Johnson et al. found an $O^+/H^+$ density ratio of $\approx 2$ in the range 0.5 - 16 keV. During the recovery phase of the 29 December 1976 storm, $O^+$ and $He^+$ fluxes (the latter were relatively weak during the main phase of all 3 storms) became the dominant ions relative to $H^+$. This one case seems to be in qualitative agreement with charge exchange lifetimes at these energies (Tinsley, 1976).

GEOS-1 has returned a fairly comprehensive set of observations of trapped ion composition below 16 keV. It surveyed within 30° of the magnetic equator for L = 3 to 8 at all local times. An example of GEOS-1 data plotted in a 3-D format is shown in Figure 7. These rather typical spectra illustrate a general feature of magnetospheric storm-time composition: in addition to ionospheric $O^+$ there is also a significant amount of solar wind $He^{++}$. Indeed, both $He^+$ and $He^{++}$ ions seem to be a much more common feature in GEOS than in S3-3 data. The most probable reason is a difference in instrument sensitivities between the GEOS and S3-3 spectrometers (Ghielmetti et al. gave a threshold of $\approx 2 \times 10^6$ ions $(cm^2 sec\ ster\ keV)^{-1}$ compared with the GEOS sensitivity of $\approx 2 \times 10^3$ ions $(cm^2 sec\ ster\ keV)^{-1}$). A second factor may be differences in $O^+$ and $He^{++}$ pitch angle distributions.

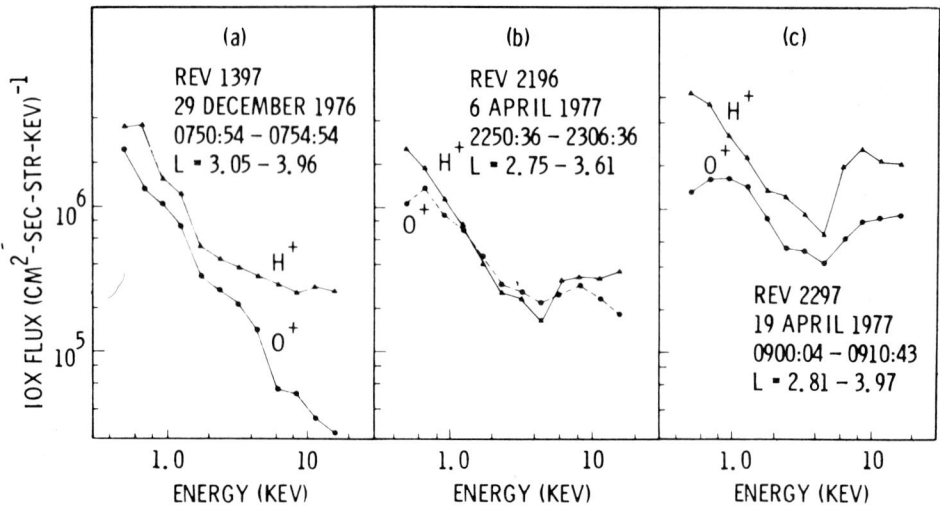

Fig. 6a,b,c. Energy spectra corresponding to Fig. 5a,b,c.

Geiss et al. (1978) studied 11 magnetic storms with GEOS-1 and concluded that on the average the solar wind and ionosphere contribute about equally to plasma injected into the equatorial magnetosphere during storms, although in the 2 hour samples from individual storms either source may dominate (Table 2).

A more sophisticated analysis of average composition would include the actual ion drift paths and estimates of ion injection and loss rates along the paths, but an interesting aspect of the average composition may be seen by first summarizing Table 2 as follows:

$$H^+ : O^+ : He^+ : He^{++} = \begin{cases} 100 : 6 : 2.7 : 0.7 & \text{(quiet)} \\ 100 : 22 : 1.3 : 1.8 & \text{(storm)} \end{cases}$$

We see that during storms $O^+$ and $He^{++}$ increase about the same relative to $H^+$ whereas $He^+$ decreases. This might be expected from the emerging two-source (ionosphere and solar wind) picture for the origin of hot storm-injected plasma. What is striking about the average composition, however, is the <u>increasing</u> percentage of $H^+$ in going from magnetically disturbed to quiet conditions as well as the <u>decreasing</u> $He^{++}/H^+$ ratio. If charge exchange with neutral hydrogen is the main loss mechanism, then $H^+$ and $He^{++}$ should both decrease relative to $O^+$ by about a factor of 3 per day (based on an average energy per charge of 7 keV at L = 6.6 and data of <u>Tinsley</u>, 1976). (Similar arguments about ring current decay have led <u>Lyons</u> (1977), <u>Tinsley</u> and others to conclude that the recovery phase ring current should be mostly $He^+$, see discussion below.)

Bearing in mind that data in Table 2 are averaged over selected periods and may not be entirely representative of "average" storm and quiet composition, it is nonetheless useful to consider an illustrative ex-

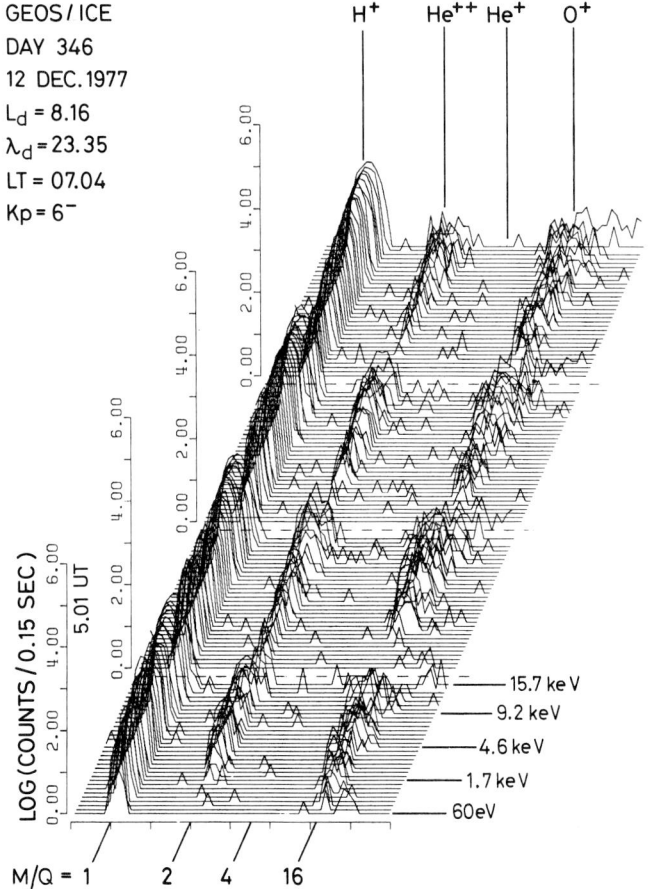

Fig. 7. Three-dimensional L Mode mass-energy spectra from GEOS-1 spectrometer. Single 6-minute scans of the 25 eV – 16 keV energy range are separated by dashed lines. Zero and 1 count per sample (0.15 sec) are plotted as zero on the (logarithmic) ordinate. Abscissa gives mass per charge in units of 8 mass steps. Values of $L_d$, $\lambda_d$ and LT are referenced to UT at start of plot. All GEOS (1 and 2) mass spectrometer data is currently available in this format. (From Geiss et al., 1978).

ample of the effect of charge exchange on the 4 main ion species. Suppose that we take the storm time composition of Table 2 and allow it to decay, without further injections, for 5 days by charge exchange alone. Then we find:

$$H^+ : O^+ : He^+ : He^{++} = 100 : 22 : 1.3 : 1.8 \quad \text{(storm)}$$

$$\underset{5 \text{ days}}{\Big\downarrow} \text{charge exchange at } L = 6.6$$

$$= 0.13 : 6.6 : 2.0 : 0.06 \quad \text{(decayed)}$$

TABLE 2. Average characterisitcs of ion composition in the region Ld = 6.9 - 8.2 and energy per charge range 1 - 16 keV/e. Averages based on six 2-hour periods, average total energy is weighted by number density. Kp average was 1+ for 48 hours preceeding each of the six 2-hour samples for magnetic quiet. (From Geiss et al., 1978).

| | | Number density ($10^{-2}$ cm$^{-3}$) | | | | Average total energy (keV) | | | | Contribution to excess energy density (percent) | |
|---|---|---|---|---|---|---|---|---|---|---|---|
| | | $H^+$ | $He^{++}$ | $He^+$ | $O^+$ | $H^+$ | $He^{++}$ | $He^+$ | $O^+$ | Solar plasma | Terrestrial plasma |
| Storm day-side | Average | 57 | 1.1 | 0.8 | 15 | 6.0 | 11.8 | 5.1 | 4.7 | 45 | 55 |
| | Range | (20-120) | (0.2-3.6) | (0.4-1.2) | (2.1-30) | (5.0-8.5) | (10.2-18.8) | (3.9-7.3) | (4.1-6.1) | | |
| Storm night-side | Average | 59 | 1.0 | 0.7 | 11 | 6.2 | 14.5 | 4.2 | 4.3 | 40 | 60 |
| | Range | (25-135) | (0.6-1.5) | (0.3-1.8) | (3.0-22) | (3.6-9.6) | (10.2-20.2) | (2.8-5.0) | (3.9-4.9) | | |
| Magnetic quiet | Average | 15 | 0.1 | 0.4 | 0.9 | 6.8 | 13.9 | 5.8 | 5.1 | -- | -- |
| | Range | (8.2-25) | (0.02-0.2) | (0.2-0.6) | (0.1-1.6) | (5.4-8.6) | (9.6-19.6) | (3.8-8.6) | (3.8-7.2) | | |

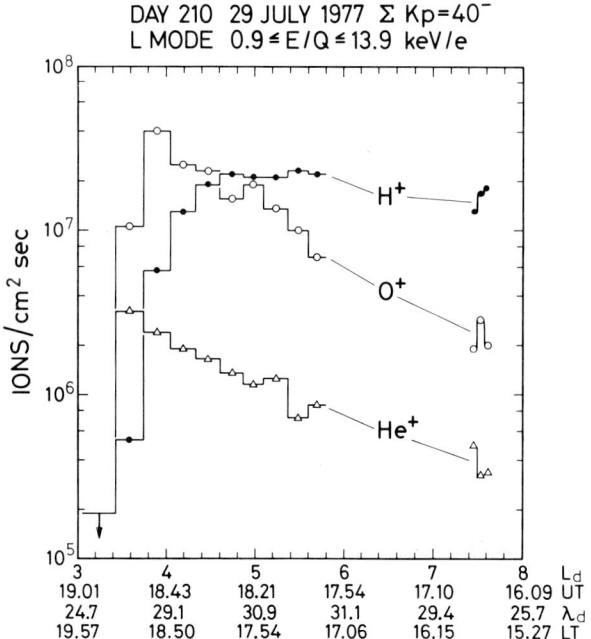

Fig. 8. Radial profile of ion composition from GEOS-1 during 29 July 1977 magnetic storm. Each point is an average over 25 energy channels between 0.9 and 13.9 keV. (From Geiss et al., 1978.)

Under this assumption, the amounts of $H^+$ and $He^{++}$ decrease dramatically as expected, and $He^+$ increases because of the decay of $He^{++}$ to $He^+$. (For an average ion energy of 7 keV/e, only 48% of the $He^{++}$ seen by the GEOS spectrometer contributes to $He^+$ detected due to the spectrometer's upper detection limit of 16 keV for $M/Q \leq 4$ and the doubling of ion energy per charge with the $He^{++} \rightarrow He^+$ conversion.)

Much of the GEOS data is at odds with a simple decay picture: injections clearly occur which keep the total flux within a factor of 5 of storm levels (Table 2) and, more to the point, $H^+$ is always observed to be the dominant ion during quiet periods with little injection (cf. Figure 9). Maintenance of the "quiet" flux and composition levels would seem to require primarily a solar wind origin for injected plasma. However, for the $\lesssim$ 20 keV ions discussed here we have $\tau_{H^+} \sim \tau_{He^{++}} \ll \tau_{He^+}$ (Tinsley, 1976). Continuous addition of a few percent $He^{++}$ will therefore lead to an equilibrium $He^+/H^+$ ratio much larger than observed. Selective depletion of $He^+$ or acceleration to > 16 keV (e.g. by wave particle interactions) must then be invoked to keep the $He^+/H^+$ ratio from becoming more than the few percent observed.

Relatively pure $H^+$ may be injected directly from terrestrial sources, although as we have seen there is a strong tendency for $O^+$ and $He^+$ to be injected as well (cf. Figures 1, 3 and 4). Two other (indirect) terrestrial sources should be mentioned in this regard: Hill (1974) discussed the polar wind as a possible source for the plasma sheet but found it negligible compared to solar wind. In any case, the polar wind can be expected to be rich in He$^+$ (with a large He$^+$/O$^+$ ratio) again contrary to the "quiet" observations of Table 2. Recently Freeman et al. (1977) suggested that cold plasma stripped off the plasmasphere and convected out to the dayside magnetopause might form a "circulation" pattern through acceleration in the dayside magnetosheath and, via the plasma mantle, be carried back into the plasmasheet and thence to the inner magnetosphere (having gained several keV around this loop). However, GEOS-1 data suggests that He$^+$/H$^+$ ~ 10% in the plasmasphere (Young et al., 1977; Decreau et al., 1978) so that any circulation of the type proposed by Freeman et al. must contribute very little to the total ion flux.

Presently there is no simple answer to the H$^+$ dilemma as posed by observations presented here. We emphasize, however, that a more complete study (presently being undertaken by Balsiger et al., 1978) is needed to establish this as a general result. Resolution of this problem may well require modeling of particle transport and loss processes on a scale of days to explain fully the observed "quiet" composition.

Figure 8 gives a radial profile of ion composition through the ring current region in the dusk local time sector about 12 hours after a storm began (maximum Dst was - 100 nT at 6 - 7 hours UT on 29 July). One limitation of the GEOS-1 mass spectrometer within the inner ring current is of course the relatively low upper energy limit of 16 keV. In the range $L_d$ = 3.4 to 4.6, O$^+$ is the dominant ion with a mean energy of 3.0 keV and density of 2.0 cm$^{-3}$. Similar data sets from the ring current region are not always so simple in appearance, and in particular the large O$^+$ fluxes at $L \simeq 4$ are not common. Generally in terms of density H$^+ \gtrsim$ O$^+$ $\gg$ He$^+$. Furthermore, at lower altitudes there is a characteristic "soft" component in the ion spectra with mean energies of 1 keV or less, similar to (but softer than) the low-energy portion of Figures 6a, b,c.

Significant in both S3-3 and GEOS-1 ring current measurements is the fact that He$^+$ has never been observed to be the dominant ion and has a low relative abundance (< 10% of total density) within the L values and energy ranges of the instruments (Sharp et al., 1977; Geiss et al., 1978). This seems to contradict the conclusions of Tinsley (1976), Lyons and Evans (1976) and Lyons (1977) that in the recovery phase of storms the ring current should be predominantly He$^{++}$ because it has by far the longest lifetime against charge exchange. Several difficulties arise in attempting to compare results from S3-3, GEOS-1 and Explorer 45 data sets, however, including the question of field line mapping in the presence of a strong ring current (Lyons, 1977). Furthermore, the GEOS-1 and S3-3 spectrometers were not designed for studies of the inner ring current in terms of shielding against penetrating background radiation, upper energy range and time resolution. Lastly, studies of storm-time ring current must insure that the composition is not altered by unobserved injections of new plasma.

The geostationary orbit is an important observation post, covering all local times once per day in a region of the outer magnetosphere nominally connected to the auroral zones. For this reason we present preliminary composition data from the first 4 days of GEOS-2 operations (Fig-

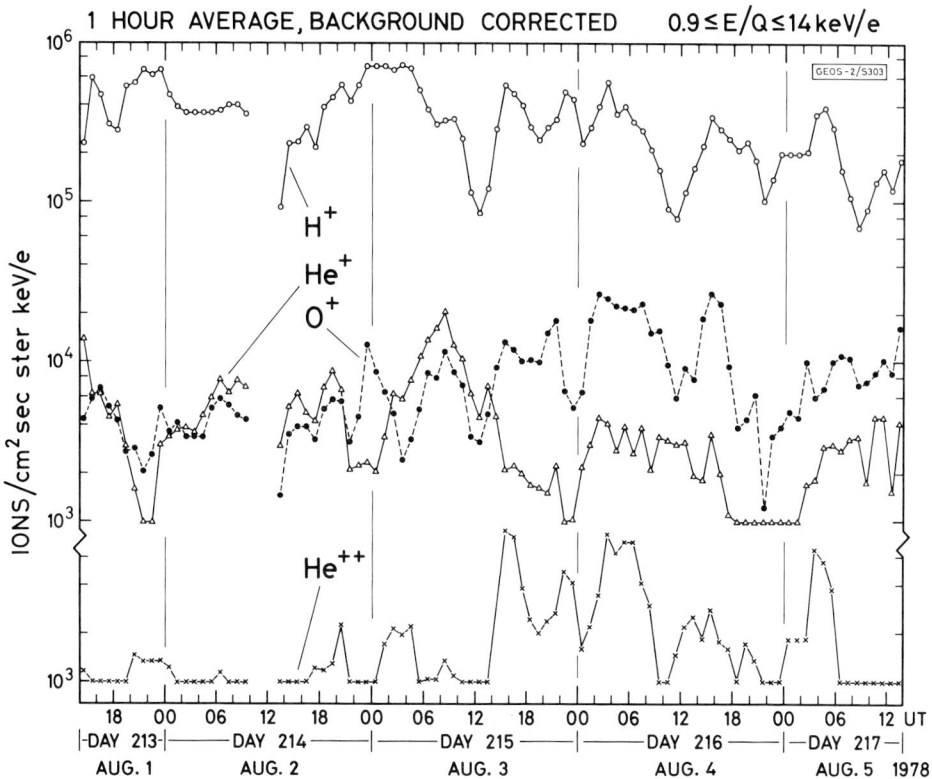

Fig. 9. One-hour averaged data from the GEOS-2 mass spectrometer following initial switch-on at 1400 UT on 1 August 1978. Averages are over 0.9 - 14 keV and are corrected for background. Local time is given approximately by UT + 40 minutes. A flux of $10^3$ (cm$^2$sec ster keV/e)$^{-1}$ corresponds to an average background limit of 1 count/sec. (Not published previously.)

ure 9) when magnetic activity was low to moderately disturbed. Note that $H^+$ is the dominant ion during this period. There are several clear flux enhancements in $H^+$, $O^+$ and $He^+$, notably at 1300 UT on Day 215, 0000 UT and 1200 UT on Day 216, and ca. 0200 UT on Day 217. Further study is necessary to explore species dependent dispersions in arrival time within each of the enhancements (ion drift velocity depends only on energy per charge). However, correlation (or lack of it) between different species may in itself be significant. For example, during the period before ca. 1200 UT on Day 215 magnetic activity is low (Kp $\leq$ 3o), $He^{++}$ fluxes are weak or absent and there is a consistent tracking of $O^+$ and $He^+$ fluxes. After ca. 1200 UT on Day 215 Kp goes above 3o, rises steadily to 6- on Day 216 and remains significantly higher than before. During this disturbed period $He^{++}$ fluxes are much stronger, suggesting injections, whereas the $O^+$ and $He^+$ ions no longer track - $O^+$ is generally more intense and $He^+$ is weaker than before. A similar trend is observed by

GEOS-1 during storms (H. Balsiger, private communication) but its interpretation is not clear.

Auroral zone sounding rockets have detected precipitating $He^{++}$ (Wahlen et al., 1971) and $^3He$ (Bühler et al., 1976) in solar wind-like fractions. A more unexpected observation of precipitating ions was made by Shelley et al. (1972) who found large fluxes of energetic $O^+$ ions at low altitudes (Table 1) during magnetic storms. Shelley and co-workers summarized a total of 11 magnetic storms covering all local times (Shelley et al., 1974). Using 3 orbits of the 1971-098A satellite per storm, they found significant ($\gtrsim 2 \times 10^6$ ions $(cm^2 sec\ ster\ keV)^{-1}$) $O^+$ fluxes on every pass. The spatial distributions of $O^+$ and $H^+$ fluxes are interesting in several respects: a) $O^+$ ion fluxes tend to occur equatorward of $H^+$ ions by $\sim 5°$ of invariant latitude and both extend over 30° to 40° (cf. Figure 8 from GEOS data); b) the average $O^+$ energy tends to increase (i.e. the spectrum hardens) with increasing latitude contrary to what might be expected from convection or radial diffusion; c) fluxes are a factor of $\sim 10$ more intense at local midnight relative to noon and d) $O^+$ fluxes may dominate $H^+$ over a fairly wide range of latitudes. GEOS-1 data has not yet been analyzed in sufficient detail to present complementary equatorial results. However we note that points (a) and (b) are, roughly speaking, also true for GEOS data whereas (d) apparently is not. Sharp et al. (1976) noted that the intensity of precipitating $O^+$ ions correlates well with the auroral AE index and Dst (and $H^+$ somewhat less so), implying that $O^+$ may be injected rapidly into the equatorial magnetosphere in response to increased magnetic activity. This could be expected in view of the apparent auroral nature of upward ion beams. Furthermore it is in keeping with GEOS observations of rapid injection of plasma (solar and terrestrial) in response to magnetic activity.

## Artificial Ion Injection Experiments

Chemical release experiments, particularly those involving barium, are a well known technique for probing the ionosphere and inner magnetosphere (cf. Michel, 1973; Wescott et al., 1975). One variant of this method is to explode a barium shaped charge at high ($\geq 500$ km) altitude, thus creating an ion jet parallel to the magnetic field with velocity greater than the escape velocity (11 km/sec). Thus far optical means have been used to track the jets, but a potentially more sensitive method is in situ detection of the ions with a satellite borne mass spectrometer (Young, 1976). In this way ion acceleration, pitch angle scattering and drifts could be measured. This method will be tried during the next year with separate releases of $Ba^+$ and $Li^+$ (the latter is not a shaped charge release) at the conjugate "foot" of the GEOS-2 magnetic field line. The method's main limitation is that large scale convective drifts may carry the ions away from the satellite field line during their rather long ($\geq 1$ hour in the absence of acceleration) transit from ionosphere to equatorial magnetosphere. In view of the frequently observed acceleration of heavy ionospheric ions away from the earth during magnetically active periods and their broad pitch angle distributions, artificially injected beams (if they can be detected near the equatorial plane) have considerable potential as a probe of these processes.

Chemical releases involve the injection of large numbers ($\sim 10^{23}$) of ions at one time. A less intense but better controlled and repeatable injection experiment will be attempted using a xenon ion gun (Masek and Cohen, 1978) on the SCATHA satellite due to be launched early in 1979. The collimated $Xe^+$ beam intended for control of satellite potential, can deliver a maximum of 2 mA at 2 keV (equivalent to $10^{16}$ ions released/ sec). The gyroradius of 2 keV $Xe^+$ ions at L = 6.6 is $\sim$ 700 km so the beam winds around inside a magnetic flux tube of $\sim$ 3000 km diameter although it fills only a limited region of velocity space.

If the $Xe^+$ beam disperses a $r^{-2}$ then it would be "visible" to the GEOS spectrometer only at distances $\leq$ 100 km. However, the magnetic field, under ideal conditions, acts as a mass spectrometer and will focus the beam after 1/2 gyration (i.e. first order focus at 180°) or a full gyration (perfect focus at 360°). Thus, rather than being uniformly dispersed inside a 3000 km diameter flux tube, the ions tend to be concentrated in an outer shell and an inner core, centered on the satellite. Using data from Masek and Cohen (1978) the shell flux is estimated to be at the limit of detection for GEOS, whereas the core flux after one gyroperiod is $\sim$ 100 times the detection limit. Focussing at 360° also works in the presence of a uniform perpendicular electric field. Although these rough calculations suggest that a beam transmission experiment is borderline over distances > 1 $Xe^+$ gyroradius, the two satellites can in principle be manoeuvred to pass within a few hundred kilometers of one another (both have thruster systems and large fuel reserves for maneouvering).

An important effect which aids in detection of $Xe^+$ or other heavy ions and is useful for studying electric fields is the large ion drift velocity relative to thermal velocity. For example, at geostationary orbit (B $\sim$ 100 nT) an electric field of 1 mV/m gives an $\underline{E} \times \underline{B}$ drift of 10 km/sec, corresponding to 19% of the speed and 3.5% of the energy of a 2 keV $Xe^+$ ion. The resulting anisotropy in $Xe^+$ arrival direction and energy displacement should enable electric fields to be measured as an integrated effect over 100 $\sim$ 1000 km of beam propagation.

## Concluding Remarks

In the introduction and instrumentation sections we emphasized that measurement of the composition of kilovolt magnetospheric plasma is still in its early stages. The primary data sets, namely from S3-3, GEOS-1 and 2 and ISEE-A, continue to be built, but are far from being digested in a systematic way. It is already clear, however, that heavy ions often constitute a non-negligible fraction of the total, especially during disturbances, and that this warrants the attention of the modeling community even at this early stage. Furthermore, modeling of the low energy (< 100 keV) magnetospheric particle environment is also just beginning (see review by Garrett, this volume) and now is perhaps the best time to incorporate the salient features of ion composition into quantitative models.

Acknowledgements. I wish to thank Prof. J. Geiss and Drs. H. Balsiger, C.R. Chappell, L.R. Lyons and E. Whipple for helpful discussions. Dr. L. Weber and Mrs. R. Bänninger provided valuable assistance in data analysis. GEOS magnetometer data was provided in advance of publication by Prof. F. Mariani. This work was supported under grant 2.886.77 of the Swiss National Science Foundation.

## References

Axford, W.I., On the origin of radiation belt and auroral primary ions, in Particles and Fields in the Magnetosphere, edited by B.M. McCormac, p. 46, D. Reidel, Dordrecht, Holland, 1970.

Balsiger, H., P. Eberhardt, J. Geiss, A. Ghielmetti, H.-P. Walker, D.T. Young, H. Loidl, and H. Rosenbauer, A satellite-borne ion mass spectrometer for the energy range 0 to 16 keV, Space Sci. Instrum., 2, 499, 1976.

Balsiger, H., P. Eberhardt, J. Geiss, and D.T. Young, Storm time injection of 0.9 - 16 keV/e solar and terrestrial ions into the high altitude magnetosphere, To be submittted to J. Geophys. Res., 1978.

Blake, J.B., Experimental test to determine the origin of geomagnetically trapped radiation, J. Geophys. Res., 78, 5822, 1973.

Borg, H., L.-A. Holmgren, B. Hultqvist, F. Cambou, H. Reme, A. Bahnsen, and G. Kremser, The keV plasma experiment on GEOS-1, KGI preprint 78 : 303, July, 1978; to be published in Space Sci. Rev.

Bühler F., W.I. Axford, H.J.A. Chivers, and K. Marti, Helium isotopes in an aurora, J. Geophys. Res., 81, 111, 1976.

Burrous, C.B., A.J. Lieber, and V.T. Zaviantseff, Detection efficiency of a continuous channel electron multiplier for positive ions, Rev. Sci. Instrum., 38, 1477, 1967.

Chappell, C.R., K.K. Harris and G.W. Sharp, A study of the influence of magnetic activity on the location of the plasmapause as measured by OGO 5, J. Geophys. Res., 75, 1970.

Cornwall, J.M., Radial diffusion of ionized helium and protons: A probe for magnetospheric dynamics, J. Geophys. Res., 77, 1756, 1972.

Cornwall, J.M., and M. Schulz, Physics of heavy ions in the magnetosphere, Aerospace Corp. Rept. SSL-78 (3960-05)-1, 1978, to be published in Solar System Plasma Physics, edited by C.F. Kennel and L.J. Lanzerotti, 1978.

Decreau, P.M.E., J. Etcheto, K. Knott, A. Pedersen, G.L. Wrenn, and D.T. Young, Multi-experiment determination of plasma density and temperature, (preprint) presented at 13th ESLAB Symposium, Innsbruck, Austria, June 1978, to be published in Space Sci. Rev.

Freeman, J.W., H.K. Hills, T.W. Hill, and P.H. Reiff, Heavy ion circulation in the earth's magnetosphere, Geophys. Res. Lett., 4, 195, 1977.

Fritz, T.A., Ion Composition, Physics of Solar Planetary Environments, edited by D.J. Williams, p. 716, Am. Geophys. Union, Washington, D. D.C., 1976.

Fritz, T.A., and B. Wilken, Substorm generated fluxes of heavy ions at the geostationary orbit, Magnetospheric Particles and Fields, edited by B.M. McCormac, p. 171, D. Reidel, Dordrecht, Holland, 1976.

Garrett, H.B., Modeling of the gesynchronous orbit plasma enviroment - Part 1, Air Force Geophysics Laboratories Rept. AFGL-TR-77-0288, 1977.

Geiss, J., H. Balsiger, P. Eberhardt, H.-P. Walker, L. Weber, D.T. Young, and H. Rosenbauer, Dynamics of magnetospheric ion composition as observed by the GEOS mass spectrometer, (preprint) paper presented at 13th ESLAB Symposium, Innsbruck, Austria, June 1978, to be published in Space Sci. Rev.

Ghielmetti, A.G., R.G. Johnson, R.D. Sharp, and E.G. Shelley, The latitudinal, diurnal, and altitudinal distributions of upward flowing energetic ions of ionospheric origin, Geophys. Res. Lett., 5, 59, 1978.

Gloeckler, G., A versatile detector system to measure the charge states, mass compositions and energy spectra of interplanetary and magnetospheric ions, (abstract), 15th International Cosmic Ray Conference, Conference Papers, Vol. 9, Abst. T51, 207, Bulg. Acad. Sci., Plovdiv, August 1977.

Harris, K.K., and G.W. Sharp, OGO V ion spectrometer, Trans. IEEE Geosci., GE-7 (2), 93, 1969.

Hill, T.W., Origin of the plasma sheet, Rev. Geophys. and Space Phys., 12, 379, 1974.

Hoffman, J.H., W.H. Dodson, C.R. Lippincott, and H.D. Hammack, Initial ion composition results from the ISIS 2 satellite, J. Geophys. Res., 79, 4246, 1974.

Hultqvist, B., and H. Borg, Observations of energetic ions in inverted V events, Planet. Space Sci., 26, 673, 1978.

Johnson, R.G., R.D. Sharp, and E.G. Shelley, Composition of hot plasmas in the magnetosphere, Physics of the Hot Plasma in the Magnetosphere, edited by B. Hultqvist and L. Stenflo, p. 45, Plenum Press, New York, 1976.

Johnson, R.G., R.D. Sharp, and E.G. Shelley, Observations of ions of ionospheric origin in the storm-time ring current, Geophys. Res. Lett., 4, 403, 1977.

Kindel, J.M., and C.F. Kennel, Topside current instabilities, J. Geophys. Res., 76, 3055, 1971.

Lyons, L.R., An alternative analysis of low- and high-altitude observations of ring current ions during a storm recovery phase, J. Geophys. Res., 82, 2367, 1977.

Lyons, L.R., and D.S. Evans, The inconsistency between proton charge exchange and the observed ring current decay, J. Geophys. Res., 81, 6197, 1976.

Masek, T.D., and H.A. Cohen, Satellite positive-ion-beam system, J. Spacecraft and Rockets, 15, 27, 1978.

Mauk, B.H., and C.E. McIlwain, ATS-6 UCSD auroral particles experiment, IEEE Trans. on Aerospace and Electronic Systems, 11, 1125, 1975.

McIlwain, C.E., bouncing clusters of ions at seven earth radii (abstract), EOS, Trans. Am. Geophys. Union, 57, 307, 1976.

Michel, K.W., Fluorescent ion jets for studying the ionosphere and magnetosphere, Acta Astronautica, 1, 37, 1974.

Mizera, P.F., and J.F. Fennel, Signatures of electric fields from high and low altitude particle distributions, Geophys. Res. Lett., 4, 311, 1977.

Prangé, R., Energetic (keV) ions of ionospheric origin in the magnetosphere, a review, to appear in Annales de Géophysique, 34, 1978.

Sharp, R.D., R.G. Johnson, E.G. Shelley, and K.K. Harris, Energetic $O^+$ ions in the magnetosphere, J. Geophys. Res., 79, 1844, 1974.

Sharp, R.D., R.G. Johnson, and E.G. Shelley, The morphology of energetic $O^+$ ions during two magnetic storms: temporal variations, J. J. Geophys. Res., 81, 3283, 1976.

Sharp, R.D., E.G. Shelley, and R.G. Jonson, A search for helium ions in the recovery phase of a magnetic storm, J. Geophys. Res., 82, 2361, 1977.

Shelley, E.G., Heavy ions in the magnetosphere, (preprint), invited review presented at International Symposium on Solar-Terrestrial Physics, Innsbruck, Austria, June, 1978.

Shelley E.G., R.G. Johnson, and R.D. Sharp, Satellite observations of energetic heavy ions during a geomagnetic storm, J. Geophys. Res., 77, 6104, 1972.

Shelley, E.G., R.G. Johnson, and R.D. Sharp, Morphology of energetic $O^+$ in the magnetosphere, Magnetospheric Physics, edited by B.M. McCormac, p. 135, D. Reidel, Dordrecht, Holland, 1974.

Shelley, E.G., R.D. Sharp, and R.G. Johnson, Satellite observations of an ionospheric acceleration mechanism, Geophys. Res. Lett., 3, 654, 1976.

Shelley, E.G., R.D. Sharp, R.G. Johnson, J. Geiss, P. Eberhardt, H. Balsiger, G. Haerendel, and H. Rosenbauer, Plasma composition experiment on ISEE-A, IEEE Trans. on Geosci. Elec., GE-16, 266, 1978.

Spjeldvik, W.N., Expected charge states of energetic ions in the magnetosphere, (NOAA/SEL Preprint 417), invited review presented at International Symposium on Solar-Terrestrial Physics, Innsbruck, Austria, June 1978.

Taylor, H.A., Jr., H.C. Brinton, and C.R. Smith, Positive ion composition in the magnetosphere obtained from the OGO-A satellite, J. Geophys. Res., 70, 5769, 1965.

Tinsley, B.A., Evidence that the recovery phase ring current consists of helium ions, J. Geophys. Res., 81, 6193, 1976.

Torbert, R.B., and F.S. Mozer, Electrostatic shocks as the source of discrete auroral arcs, Geophys. Res. Lett., 5, 135, 1978.

Ungstrup, E., D.M. Klumpar, and W.J. Heikkila, Heating of ions to superthermal energies in the topside ionosphere by electrostatic ion-cyclotron waves, Univ. Texas at Dallas, preprint, 1977.

Vette, J.I., and R.W. Vostreys, Report on active and planned spacecraft and experiments, Nat. Space Science Data Center Publ., NSSDC/WDC-A-RS 77-03, 181-183, September 1977.

Wescott, E.M., H.C. Stenbaek-Nielsen, T.N. Davis, W.B. Murcray, H.M. Peek, and P.J. Bottoms, The L = 6.6 Oosik barium plasma injection experiment and magnetic storm of March 7, 1972, J. Geophys. Res., 80, 951, 1975.

Whalen, B.A., J.R. Miller, and I.B. McDiarmid, Evidence for a solar wind origin of auroral ions from low-energy ion measurements, J. Geophys. Res., 76, 2406, 1971.

Whalen, B.A., W. Bernstein, and P.W. Daly, Low altitude acceleration of ionospheric ions, Geophys. Res. Lett., 5, 55, 1978.

Whipple, E.C., J.M. Warnock, and R.H. Winckler, Effect of satellite potential on direct ion density measurements through the plasmapause, J. Geophys. Res., 79, 179, 1974.

Young, D.T., Proposed $Ba^+$ ion-tracer conjugacy and related studies using the GEOS satellite, in The Scientific Satellite Program During the International Magnetospheric Study, edited by K. Knott and B. Battrick, p. 397, D. Reidel, Dordrecht, Holland, 1976.

Young, D.T., J. Geiss, H. Balsiger, P. Eberhardt, A. Ghielmetti, and H. Rosenbauer, Discovery of $He^{2+}$ and $O^{2+}$ ions of terrestrial origin in the outer magnetosphere, Geophys. Res. Lett., 4, 561, 1977.

QUANTITATIVE MODELS OF THE 0 TO 100 KEV MID-MAGNETOSPHERIC PARTICLE ENVIRONMENT

H. B. Garrett

Air Force Geophysics Laboratory
Hanscom AFB, MA 01731

Abstract. Quantitative modeling of the near-earth plasma environment in the 0-100 keV energy range has made significant strides in the last few years in response to the needs of spacecraft users and to the substantial improvements in the modeling of the earth's electric and magnetic fields. Concurrent with this has been an extensive growth in the observational data base necessary to evaluate these models. In this paper, current models of the plasmasphere and near-geosynchronous environment are reviewed and compared with data. The review indicates that although significant problems are still not resolved, quantitative low-energy plasma models are at the stage magnetic field models were a decade ago and that substantial gains in the next few years are to be expected.

## Introduction

In this review we will delineate four categories of models based primarily on the degree of empirical and theoretical input. We will limit the scope of the review to those plasmas controlled primarily by ionospheric or auroral dynamics. That is, we will consider only the low energy (0-100 keV) charged particle population in the plasmasphere and the near-earth plasma sheet regions. We will also ignore variations in ionic composition (see Young, 1978, for a detailed treatment of variations).

## Types of Quantitative Models

### Definitions

Although it is difficult to distinguish between models, a useful classification can be devised based on the level of theoretical or empirical input to the model. For example, the simplest model is a statistical compendium or histogram of various parameters as a function of space and time. Such a model has little theoretical input, being based on actual measurements. Consideration of basic physical principles makes possible the derivation of simple analytic expressions capable of simulating changes in the environment - the second type of model. Third is a static field model - a model which employs theory to predict the trajectories of particles in static electric and mag-

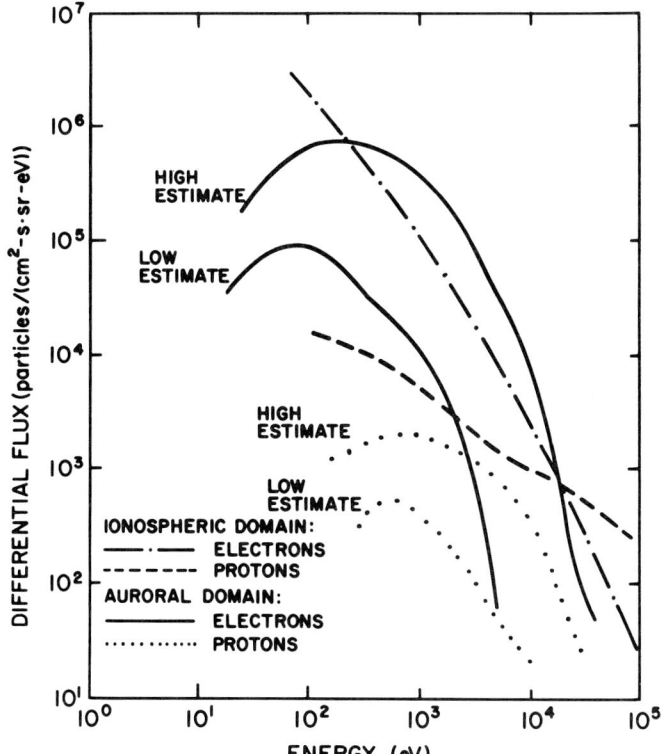

Fig. 1. Plot of the composite electron and proton spectra in the inner magnetosphere and plasma sheet (adapted from Chan et al., 1977).

netic field models. Finally, the most complete model from a theoretical standpoint is a full, 3-dimensional, time-dependent model capable of taking into account particle injection events.

Statistical Models

Statistical models, as defined here, are compendiums or histograms of various plasma parameters based on actual data. The basic example of this type of model are the composite or average distribution functions generated by Chan et al. (1977) for various magnetospheric and solar wind regions. The electron and ion spectra for the inner magnetosphere and plasma sheet are shown in Figure 1. Such descriptions are of great value in many applications (e.g., dosage calculations).

In order to adequately model the magnetosphere, a prohibitive number of distribution functions are needed as a function of time and spatial coordinates. Further, the distribution function itself requires a detailed description. DeForest and McIlwain (1971) have simplified the problem for geosynchronous orbit by describing the plasma in terms of the first 4 moments of the distribution function.

$$\langle N_i \rangle = \text{number density for species i (number/cm}^3\text{)} = n_i \qquad (1)$$

$\langle NF_i \rangle$ = number flux for species i (number/cm²sec-sr)  (2)

$$= \frac{n_i}{2\pi} \left(\frac{2kT_i}{\pi m_i}\right)^{1/2}$$

$\langle P_i \rangle$ = pressure for species i (dynes/cm²) = $n_i k T_i$  (3)

$\langle EF_i \rangle$ = energy flux for species i (ergs/cm²sec-sr)  (4)

$$= \frac{m_i n_i}{2} \left(\frac{2kT_i}{\pi m_i}\right)^{3/2}$$

The results on the right are for a Maxwell-Boltzmann distribution:

$$f_i(V) = n_i \left(\frac{m_i}{2\pi kT_i}\right)^{3/2} e^{-mV_i^2/2kT_i}$$

where: $n_i$ = number density of species i

$m_i$ = mass of species i

$T_i$ = temperature of species i

$V_i$ = velocity of species i

k = Boltzmann constant

f = distribution function in sec³/km⁶

The description of the plasma in terms of these quantities is quite useful, as not only are they physically meaningful in their own right, but they can be used to derive a Maxwellian or 2-Maxwellian distribution of the environment (see Garrett and DeForest, 1978). Rough estimates by DeForest and McIlwain (1971) of the local time variations of the 4 moments are tabulated in Table 1 for data from the geosynchronous ATS-5 satellite in 1970. The great spatial and temporal variations of the geosynchronous plasma are clearly demonstrated in this table, particularly the 2 orders of magnitude change in the electron moments.

Vasyliunas (1968), Stevens et al. (1977), Su and Konradi (1977), and Garrett et al. (1978) have all carried out statistical studies similar to those of DeForest and McIlwain (1971). Vasyliunas (1968) extensively analyzed the low energy electrons (~ 40 eV to ~ 2 keV) in the evening sector using OGO 1 and OGO 3. A small part of his results for the statistical distribution of various electron parameters within the plasma sheet are plotted in Figure 2. Garrett et al. (1978) have constructed histograms of the electron and ion temperatures, current, and spacecraft potential for 10 minute averages of ATS-5 (1969-1970) and ATS-6 (1974 and 1976) data as functions of local time and $K_p$. Some of the results are

TABLE 1. 50-ev to 50,000-ev Spectral Integrals

| | Electrons | | | | | | Protons | | | | | |
|---|---|---|---|---|---|---|---|---|---|---|---|---|
| | 0000 LT | 0300 LT | 0600 LT | 1200 LT | 1800 LT | 2100 LT | 0000 LT | 0300 LT | 0600 LT | 1200 LT | 1800 LT | 2100 LT |
| **Number Density, particles/cm³** | | | | | | | | | | | | |
| Minimum | 0.07 | 0.22 | 0.17 | 0.06 | 0.02 | 0.04 | 0.7 | 0.6 | 0.7 | 0.3 | 0.5 | 0.6 |
| Maximum | 8.3 | 4.8 | 2.9 | 1.2 | 1.9 | 4.4 | 3.8 | 3.5 | 1.9 | 2.2 | 2.2 | 2.4 |
| Typical | 2.0 | 2.0 | 1.2 | 0.4 | 0.10 | 0.4 | 1.2 | 1.2 | 1.2 | 0.9 | 0.8 | 1.1 |
| **Energy flux, erg/cm² sec ster** | | | | | | | | | | | | |
| Minimum | 0.21 | 0.38 | 0.42 | 0.26 | 0.04 | 0.10 | 0.16 | 0.13 | 0.05 | 0.13 | 0.14 | 0.14 |
| Maximum | 9.4 | 15.2 | 14.6 | 2.3 | 1.01 | 7.2 | 0.61 | 0.47 | 0.47 | 0.76 | 0.63 | 0.85 |
| Typical | 3.0 | 3.0 | 1.5 | 1.0 | 0.40 | 0.5 | 0.3 | 0.3 | 0.3 | 0.22 | 0.30 | 0.30 |
| **Number flux, 10⁶ particles/cm² sec ster** | | | | | | | | | | | | |
| Minimum | 15 | 37 | 32 | 15 | 2 | ? | 6 | 4 | 4 | 2 | 4 | 5 |
| Maximum | 1510 | 1020 | 832 | 122 | 162 | 864 | 25 | 17 | 16 | 15 | 23 | 25 |
| Typical | 300 | 300 | 200 | 70 | 30 | 60 | 12 | 10 | 8 | 7 | 8 | 10 |
| **Pressure, 10⁻¹⁰ dynes/cm²** | | | | | | | | | | | | |
| Minimum | 2.7 | 6 | 6 | 4 | 0.4 | 1 | 66 | 51 | 25 | 31 | 54 | 56 |
| Maximum | 190 | 266 | 173 | 25 | 14 | 128 | 235 | 196 | 169 | 242 | 255 | 327 |
| Typical | 50 | 60 | 30 | 12 | 7 | 8 | 140 | 120 | 90 | 80 | 90 | 120 |
| **Ratio of total particle pressure to the magnetic field pressure** | | | | | | | | | | | | |
| Minimum | 0.13 | 0.13 | 0.08 | 0.04 | 0.09 | 0.09 | | | | | | |
| Maximum | 2.7 | 1.07 | 0.52 | 0.21 | 0.62 | 1.06 | | | | | | |
| Typical | 0.4 | 0.4 | 0.2 | 0.12 | 0.2 | 0.25 | | | | | | |

Table 1. Minimum, maximum, and typical values of the 4 moments of the electron and proton distribution functions as a function of local time. The data are from the geosynchronous ATS-5 satellite for 1970 (DeForest and McIlwain, 1971).

plotted in Figure 3. Note the difference between T(AVG) and T(RMS). These "temperatures" are defined by:

$$T(AVG) = \frac{\langle P \rangle}{\langle N \rangle} \tag{5}$$

$$T(RMS) = \frac{\langle EF \rangle}{2 \langle NF \rangle} \tag{6}$$

Only if the plasma is distributed according to a Maxwell-Boltzmann distribution will these quantities actually be temperatures such that T(AVG) = T(RMS) - a condition that seldom holds in the magnetosphere as demonstrated in Figure 3.

The preceding examples have been constrained to the geosynchronous and near-geosynchronous orbit. Although the plasmasphere, plasma sheet, magnetopause, and solar wind have been observed to penetrate to geosynchronous orbit, we cannot obtain a complete picture of the near-earth magnetosphere if we limit our data to the geosynchronous region. In particular, the models have not been extended to other latitudes and radial distances. A useful, though somewhat objective way of extending the models to other spatial positions is to extrapolate the results from an eccentric, inclined satellite in the manner of the intensity plots of high energy particles. A particularly good example is given in Figure 4 for data from a main phase decrease (Frank, 1967). It is a meridian plot in the generalized $R - \lambda_m$ coordinate system (see Roederer, 1970, for an explanation of various coordinate systems) of low energy ions (200 eV $\leq$ E $\leq$ 50 keV). An important observation derived from such

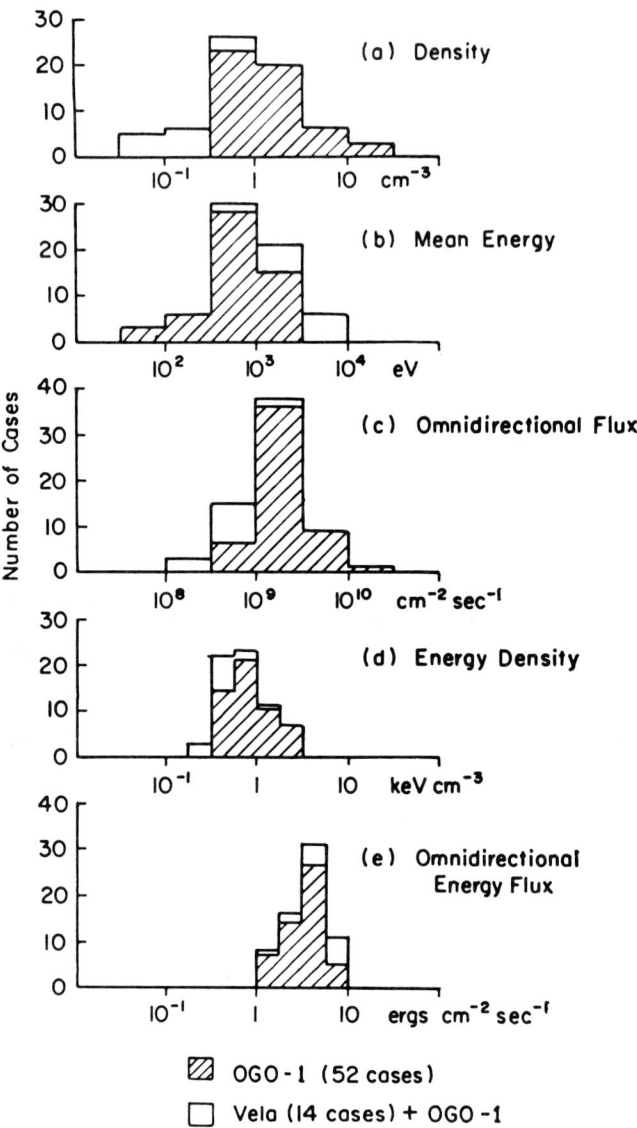

Fig. 2. Observed distributions of various electron parameters within the plasma sheet from OGO-1 and VELA (Vasyliunas, 1968).

studies is the pronounced movement and fluidity of the inner boundary with geomagnetic activity.

The other region of concern in studying the low energy, near-earth environment, is the plasmasphere. (See reviews by Chappell, 1972, and Carpenter and Park, 1973). As the plasmasphere is linked to the ionosphere and contained on closed field lines, defining the distribution function at one point allows its determination (in principle) along the entire flux tube. Carpenter (1966), using ground-based whistler studies, delineated the approximate shape and density of the plasmaspheric

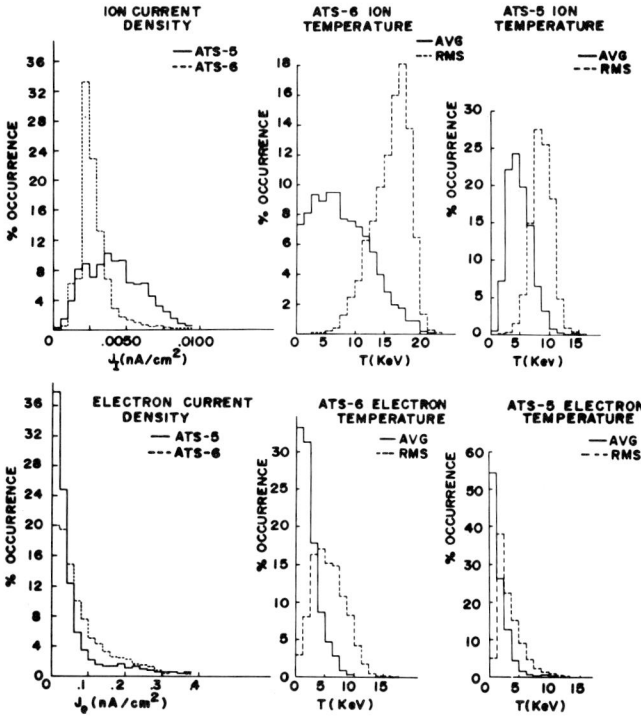

Fig. 3. Occurrence frequencies of ATS-5 (1969 and 1970) and ATS-6 (1974 and 1976) electron and ion current densities, T(RMS), and T(AVG) according to Garrett et al. (1978).

electron population (Figure 5a and 5b). Chappell et al. (1970) have made extensive in situ measurements of the positive ion populations with OGO-5. Figure 6 illustrates these results for an inbound pass on the dayside for $H^+$, $He^+$, and $O^+$ showing the sharp drop-offs in various constituents with increasing L.

The plasmasphere, because of its elongation near dusk, often crosses the geosynchronous orbit. Lennartsson and Reasoner (1978) have employed ATS-6 data to construct a statistical model of these crossings. They have tabulated the number of encounters of the ATS-6 satellite with ion populations in the energy range 1-30 eV and densities between $1 \leq n \leq 10$ $cm^{-3}$ ('weak warm' plasma events) and $n > 10$ $cm^{-3}$ ('intense warm' plasma events). Their results are given in Figures 7a and 7b as a function of local time and percent occurrence.

## Analytic Simulations

A major difficulty in the preceding statistical models is the inability to model the effects of complex time variations while maintaining the simultaneous behavior of the necessary descriptive parameters. For the geosynchronous orbit, DeForest and Wilson (1976) have solved this problem by the expediency of providing computer tapes of the detailed spectra for several typical plasma events. This way of modeling,

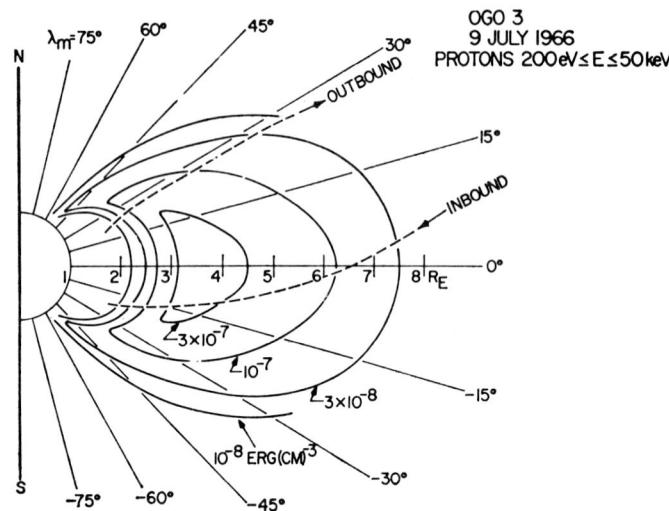

Fig. 4. Contour plot of constant proton (200 eV ≤ E ≤ 50 keV) energy density in the R-$\lambda_m$ coordinate system on July 9, 1966 (Frank, 1967).

though precise and explicitly including time variations, lacks compactness and does not directly provide many of the required parameters. A solution to this problem of the trade-offs between accuracy and massive amounts of data has been the introduction of analytic equations capable of modeling specific parameter variations - our second type of model.

For geosynchronous orbit, Mauk and McIlwain (1974) have generated an analytic expression for the injection boundary, $L_b$ (in $R_e$) - the boundary

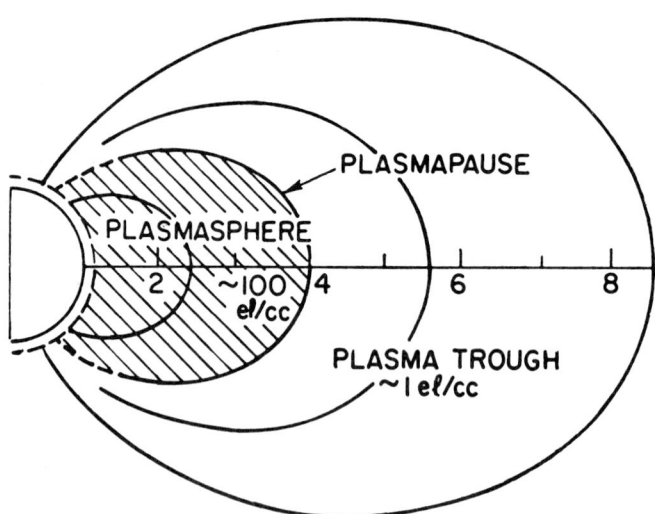

Fig. 5. Idealized meridional cross-section of the magnetosphere near 1400 LT showing the location of the plasmasphere and plasma trough (Carpenter, 1966).

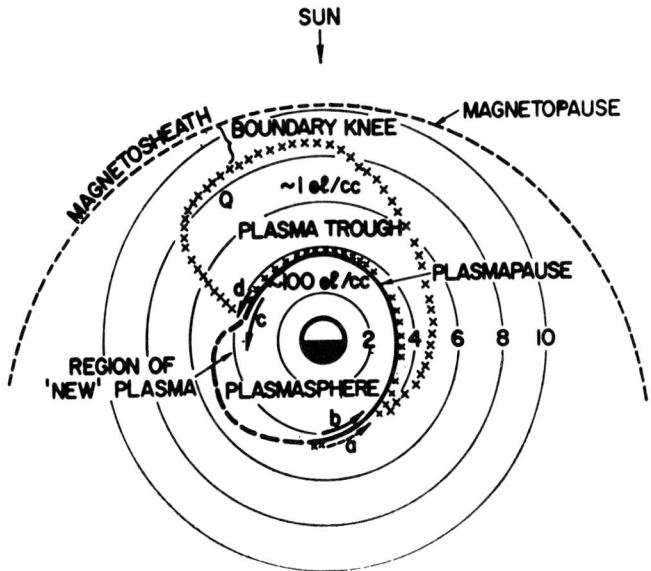

Fig. 5b. Equatorial cross-section of the magnetosphere illustrating the approximate locations of the plasmasphere and plasma trough in LT and $R_e$.

along which "hot" plasma appears at and near geosynchronous orbit at the beginning of a substorm - in terms of local time LT and $K_p$ for the evening-midnight quadrant (Figure 8):

$$L_b = (122-10\ K_p) / (LT - 7.3) \tag{7}$$

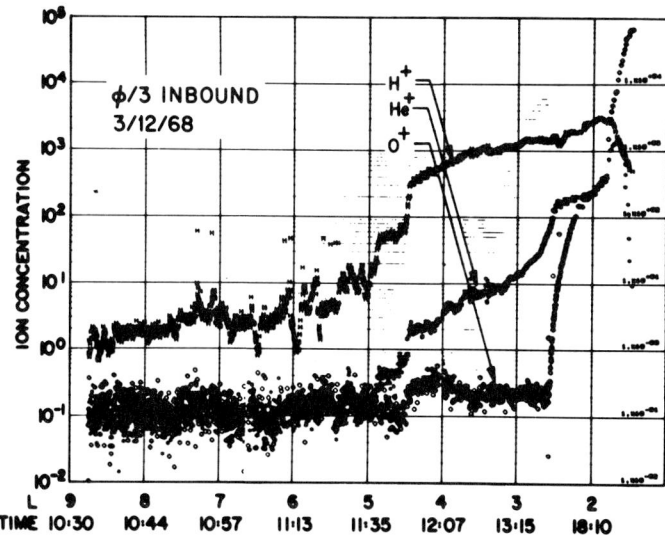

Fig. 6. Data from an inbound pass of OGO-5 on the dayside magnetosphere showing the concentrations of $H^+$, $He^+$, and $O^+$ as a function of L-value and LT (Chappell et al., 1970).

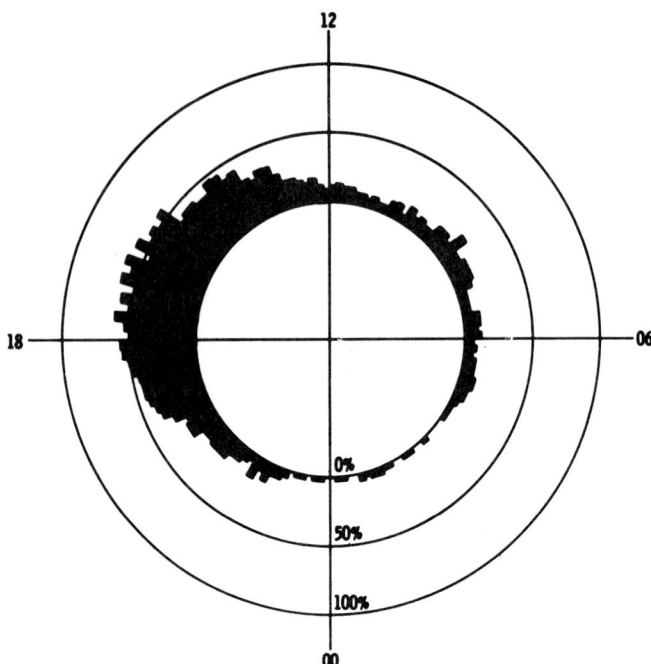

Fig. 7a. The distribution of weak warm plasma events ($1 \leq n \leq 10$ cm$^{-3}$, KT = 1-30 eV) as a function of local time (Lennartsson and Reasoner, 1978).

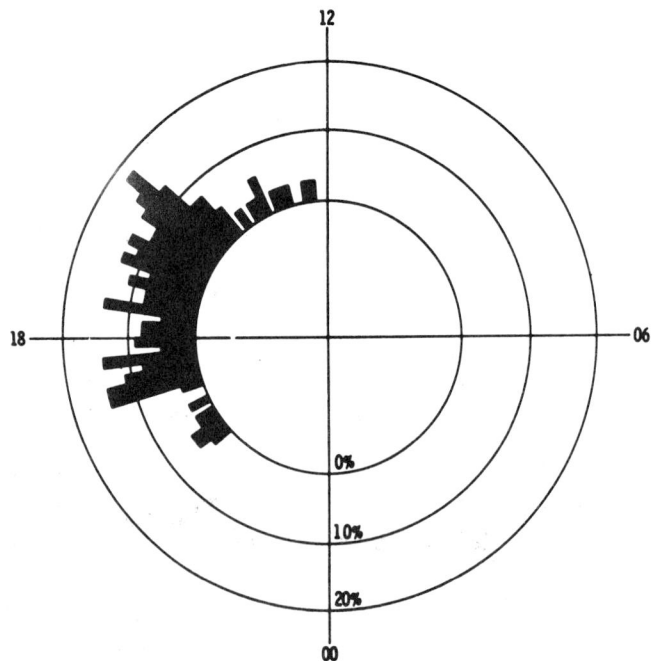

Fig. 7b. The distribution of intense warm plasma events (n > 10 cm$^{-3}$, KT = 1-30 eV) as a function of local time (Lennartsson and Reasoner, 1989).

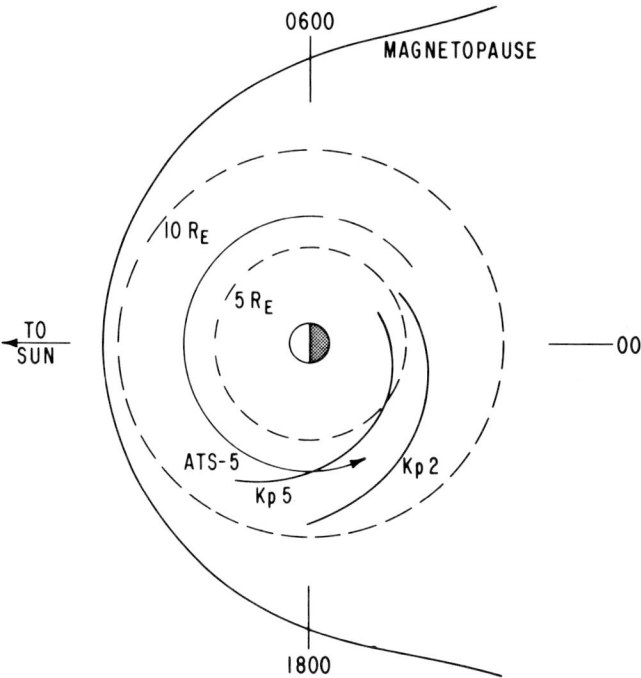

Fig. 8. The low energy plasma injection boundary for a $K_p$ of $5_o$ and $2_o$ as derived from Equation 7 (Mauk and McIlwain, 1974).

This formula is derived from ATS-5 data and based in part on a formula for the plasmapause position in the midnight-dawn sector from Carpenter and Park (1973) (see also Freeman, 1974; Kivelson and Southwood, 1975b; Kivelson, 1976; Lemaire, 1976):

$$L_{pp} = 5.7 - 0.47 \, K_p \tag{8}$$

(as $K_p$ is taken to be the maximum value of $K_p$ in the preceding 12 hours in the Carpenter-Park formula and the instantaneous value for the Mauk-McIlwain formula, these equations should be treated as rough approximations). As will be shown later, such formulas can be used as the basis of much more sophisticated models of the magnetosphere.

Garrett and DeForest (1978) (see also Su and Konradi, 1977) have introduced a very compact way of modeling the geosynchronous plasma environment. They fit the first 4 moments of the distribution function for 10 selected days of ATS-5 data with an equation linear in $A_p$ (daily average $a_p$) and varying diurnally and semidurnally in local time, LT:

$$M_i \, (A_p, LT) = (a_o + a_1 \, A_p) \, (b_o + b_1 \cos (\tfrac{2\pi}{24} \, (LT + t_1)) \tag{9}$$
$$+ \, b_2 \cos (\tfrac{4\pi}{24} \, (LT + t_2)))$$

where: $M_i$ = moment i

$a_o, a_1, b_o, b_1,$

$b_2$, $t_1$, $t_2$ = fitted parameters

The days were selected to cover a wide range of geomagnetic activity and so that a plasma injection was initiated when the ATS-5 satellite was near local midnight. The use of the 4 moments allows the calculation of a variety of parameters including the determination of a "2 Maxwellian" omnidirectional distribution function (Garrett, 1977; Garrett and DeForest, 1978):

$$f_2(V) = \left(\frac{m_i}{2\pi K}\right)^{3/2} \left(\frac{n_{1i}}{T_{1i}^{3/2}} e^{-m_i V^2/2kT_{1i}} + \frac{n_{2i}}{T_{2i}^{3/2}} e^{-m_i V^2/2kT_{2i}}\right) \qquad (10)$$

where: $n_{1i}$, $n_{2i}$ = 2 population densities (n/cm$^3$)

for constituent i

$T_{1i}$, $T_{2i}$ = 2 population temperatures

for constituent i

Figure 9 (Garrett and DeForest, 1978) illustrates the variations in the 4 moments as described by Equation 9 for 2 levels of $A_p$ (15 and 207) as a function of local time. The model has been demonstrated to adequately simulate the geosynchronous environment for low to moderately high levels of geomagnetic activity. At the highest levels it is biased towards the particular active days studied. The simulation has permitted limited predictions of the geosynchronous parameters (Garrett, 1977) though it should be remembered that the simulation is biased towards injections when the satellite was near local midnight.

Static Models

Statistical models and simulations can go only so far in modeling the magnetosphere. Unfortunately, they are currently confined primarily to geosynchronous orbit and the plasmasphere. It is possible, however, to extend geosynchronous results to other latitudes and radial distances by use of static models of the magnetosphere. Static models (e.g. models which assume fixed magnetic and electric fields and then determine the 3-dimensional drifts of the particles) can be said to have originated with Alfvén (1939, reprinted 1970). Modern examples are those of Kavanagh et al. (1968), Roederer and Hones (1970), Wolf (1970), Chen (1970), McIlwain (1972), Williams et al. (1974), Konradi et al. (1975), Stern (1975), and Walker and Kivelson (1975) for the magnetosphere and Chappell et al. (1970) for the plasmasphere (note: these are time-independent models). As most of these models were developed for the purpose of understanding specific satellite data, they allow a comparison with observations.

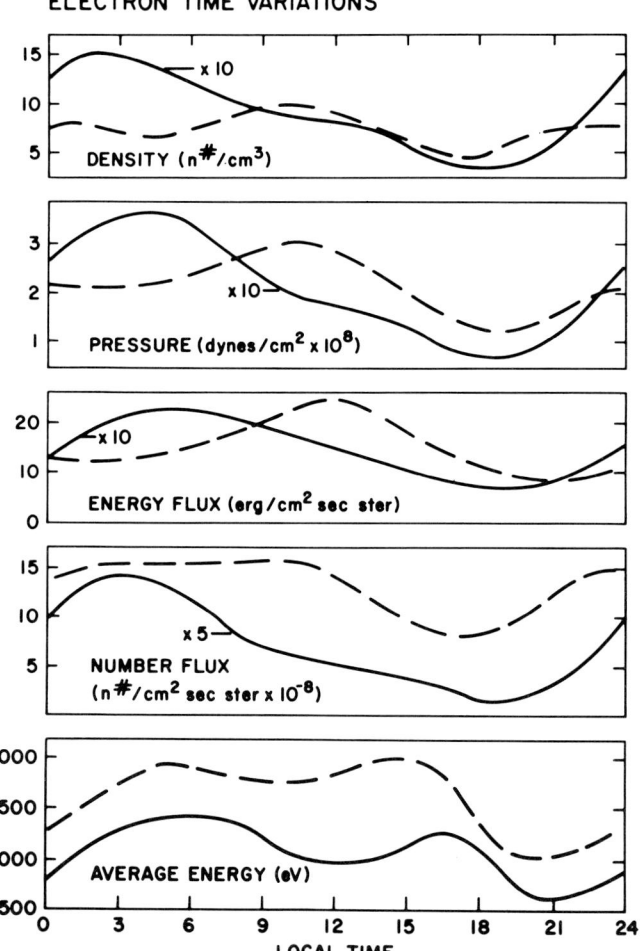

Fig. 9a. Local time variations in the 4 moments and mean energy of the electrons as predicted by the Garrett and DeForest (1978) model for $A_p = 15$ (solid lines) and $A_p = 207$ (dashed lines).

As a typical example of this class of models, McIlwain (1972) has carefully analyzed ATS-5 observations and generated a "best fit" to the particle observations by varying the electric field model. His predictions of electron and ion trajectories typical for this type of model are given in Figures 10a and b. McIlwain has by this method succeeded in "dissecting" ATS-5 spectrograms, tracing the particles back to their origins.

McIlwain (1972) notes that his model, assuming constant fields, cannot account for particle trapping. Varying fields are necessary to bring in fresh fluxes of particles. Konradi et al. (1975) have combined Mauk and McIlwain's (1974) analytic equation (Equation 7) with a static field model to generate quantitative models of the magnetosphere that can account for the arrival of new plasma. Using the magnetic

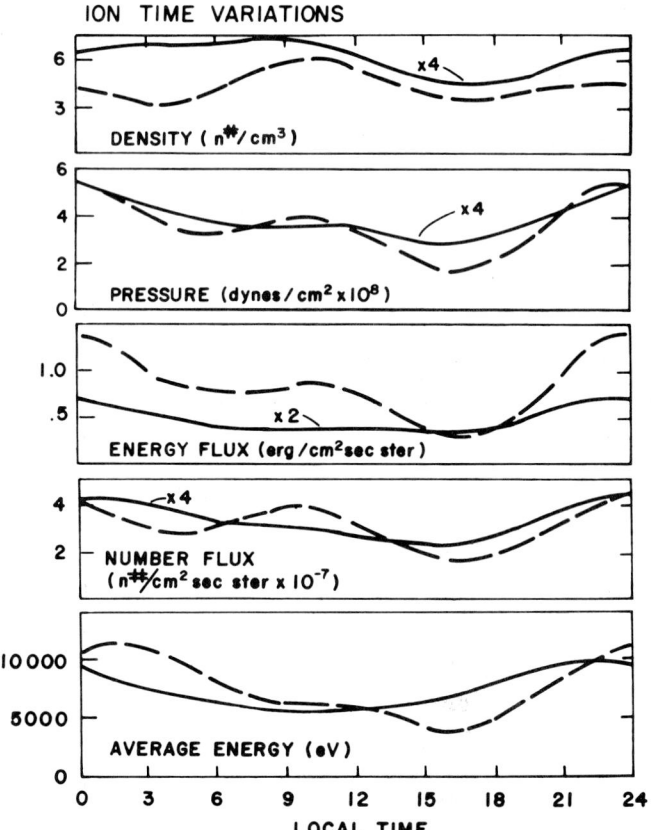

Fig. 9b. Same as 14a for ions.

field model of Roederer (1970) and the $E^3$ electric field model of McIlwain (1972), Konradi et al. (1975) assumed particles originated on or anti-earthward of the Mauk and McIlwain boundary. Particles were either traced backwards in time to their origin or forward from the boundary to the desired satellite position. The results, compared with data from the Explorer 45 satellite (apogee 5.19 $R_e$, perigee altitude 268 km), are presented in Figures 11a and 11b. The results are in good qualitative agreement (note, however, the lack of drift echoes in the electron observations) and tend to support their contention that particles are injected along a limiting boundary. Their results are supported by Williams et al. (1974) who find that the creation of a plasma cloud, followed by convection, best fits Explorer 45 observations

Walker and Kivelson (1975) carry out somewhat similar calculations as Konradi et al. (1975) for electrons. They, however, assume an initia uniform particle distribution in the region from which the particles are convected and then suddenly increase a uniform cross-magnetosphere electric field. Results of OGO 3 and ATS 1 for variations in the particles in universal time as the initial distribution is altered at various locations along the orbit are given in Figure 12. Their work supports the contention of Kivelson and Southwood (1975a) that localized

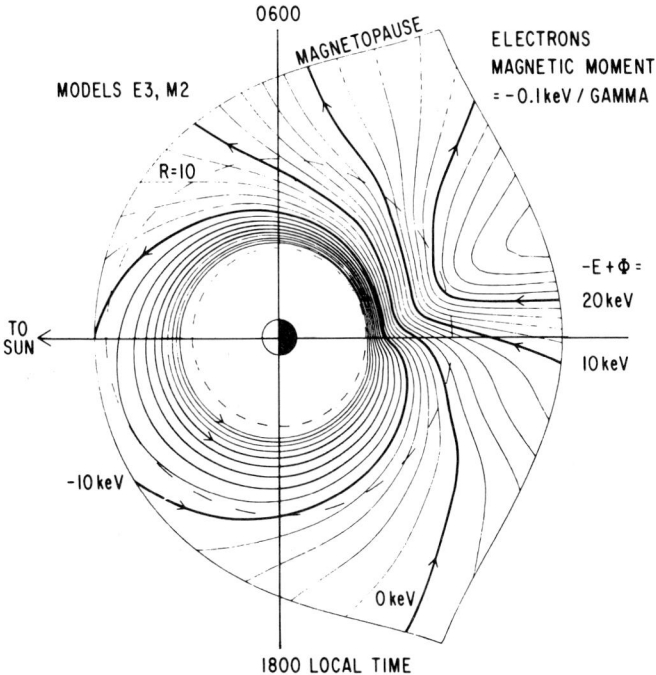

Fig. 10a. Trajectories for electrons having $\mu = -0.1$ keV/$\gamma$ in the McIlwain model (McIlwain, 1972).

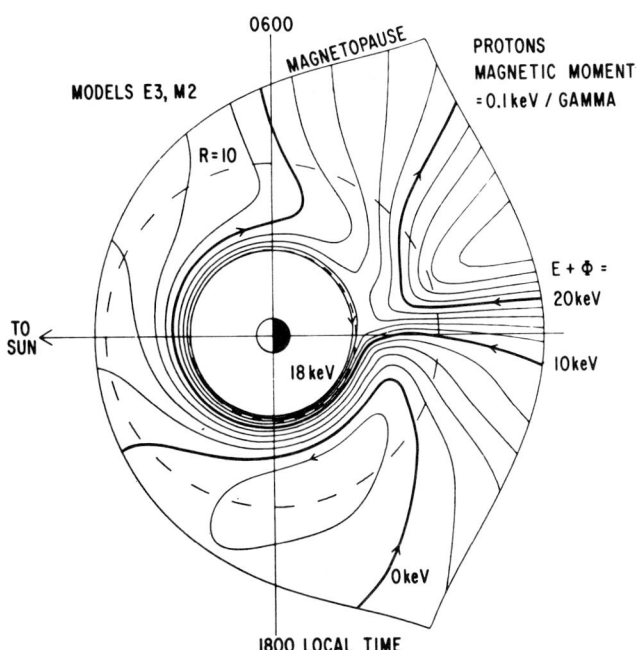

Fig. 10b. Same as 16a for protons having $\mu = 0.1$ keV/$\gamma$

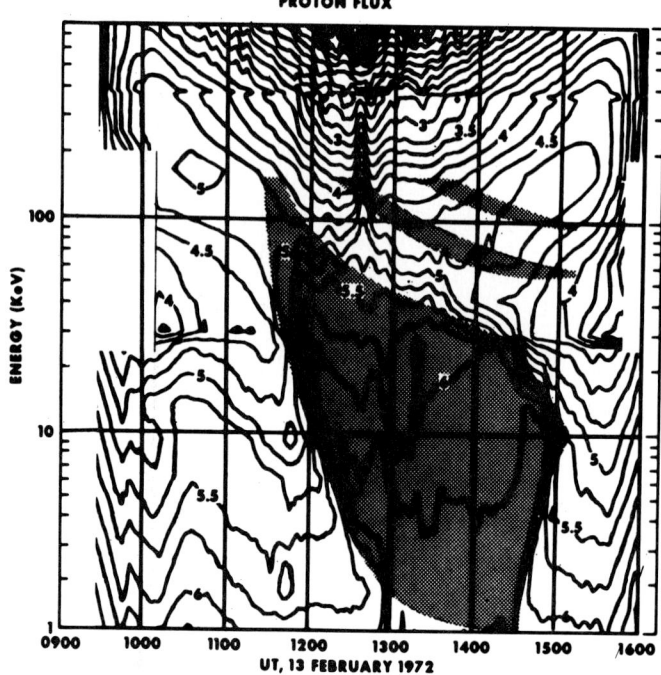

Fig. 11a. Spectrogram of 81° pitch angle protons on February 13, 1972. The shaded areas are the dispersion plot and two echoes for protons injected instantaneously along the injection boundary at 1116 UT and then drifting according to the assumed static model (Konradi et al., 1975).

particle injection is not necessary to account for observations (see also Smith et al., 1978).

Chappell et al. (1970) and, to a lesser extent, Wolf (1970) and Chen (1970) generate models of the plasmasphere particle trajectories. Chappell et al. (1970) (based on work by Brice, 1967, and Nishida, 1966) have produced qualitative plots of the motion of flux tubes in the plasmasphere (qualitative as the effects of ionospheric conductivity have not been included). Sample results are given in Figure 13 where the distance between the dots gives the drift distance of a flux tube in one hour.

A powerful technique for modeling the behavior of the low energy plasma in static fields has been presented recently by Whipple (1978). The scheme is based on the observation that the trajectories of particles are straight lines in a coordinate system consisting of the electric potential (U), magnetic field intensity (B), and the modified longitudinal invariant (K) (Roederer, 1970). The equations of motion become:

$$\frac{dB}{dt} = W \qquad (12)$$

$$\frac{dU}{dt} = -(\frac{\mu}{e})W \qquad (13)$$

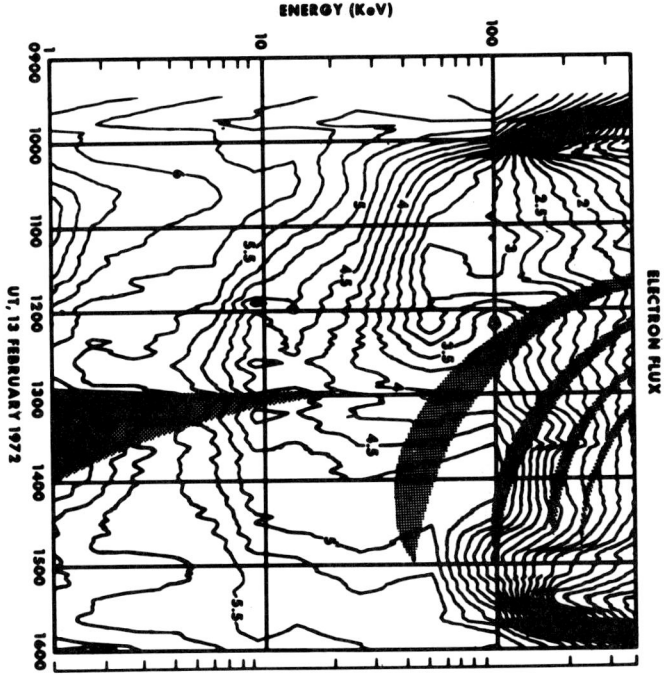

Fig. 11b. Same as Figure 11a for electrons.

$$\frac{dK}{dt} = 0 \qquad (14)$$

where: W is a generalized velocity function:

$$W = [(\vec{\nabla} U \times \vec{\nabla} B) \cdot \vec{n}_k] \, / \, (\vec{B} \cdot \vec{n}_k)$$

$\vec{n}_k$ = unit vector perpendicular to the local constant-K surface

$\mu = \frac{1}{2} m V^2 \, / \, B$

e = particle charge

The boundary on which W vanishes divides the magnetosphere into two regions and forms a natural boundary for dealing with particle orbits. A sample result is given in Figure 14 where several different particle trajectories are given. One through 5 are open trajectories which connect to the boundaries (magnetopause or tail boundary), 6 through 9 are closed, 10 through 12 are in the drift loss cone. The two boundary lines represent the 2 approximate dawn-to-earth and earth-to-dusk lines for W = 0 - a trajectory ending on one of these lines enters the other half of the magnetosphere.

Whipple's method makes possible very rapid calculation of particle trajectories once the W = 0 boundaries are defined. Given the distribution function along one boundary, it is straightforward to determine particle distributions at other locations. Although still

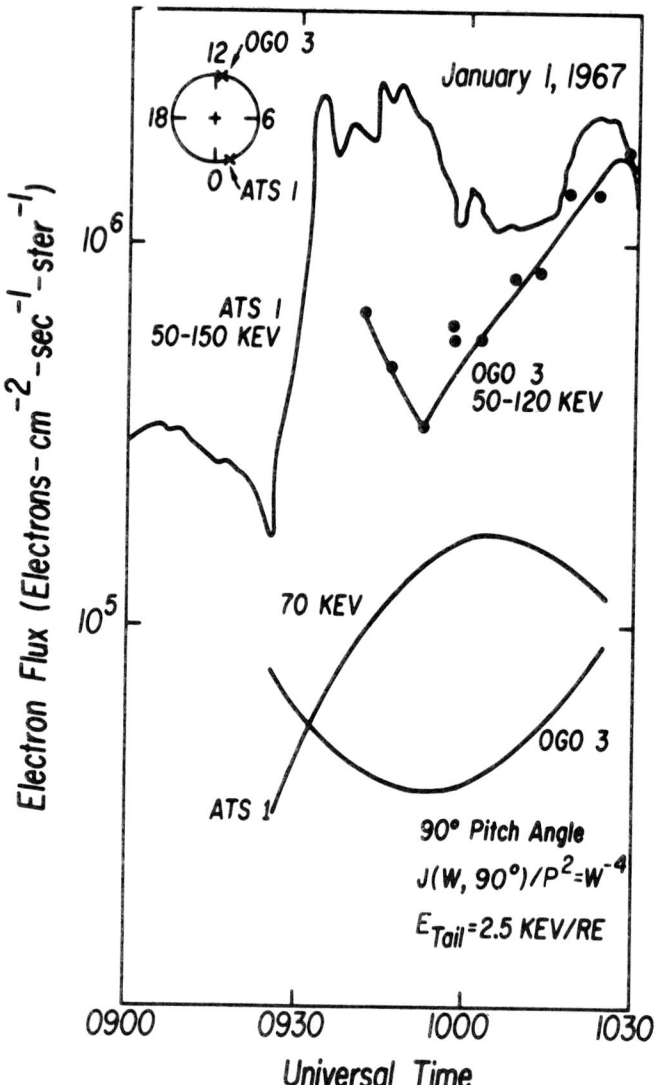

Fig. 12. ATS-1 and OGO-3 integral electron observations are compared to calculated flux changes for 70 keV electrons of 90° pitch angle at the position of the two satellites (Walker and Kivelson, 1975).

in a formative stage, this approach promises a real theoretical simplication by translating complex energy-dependent trajectories into straight lines by putting the complexity into a one time calculation of appropriate boundaries and coordinate transformations.

Time-Dependent Models

All static models suffer from one simple difficulty - the magnetosphere is not static. It is easy to show (Williams et al., 1974) that

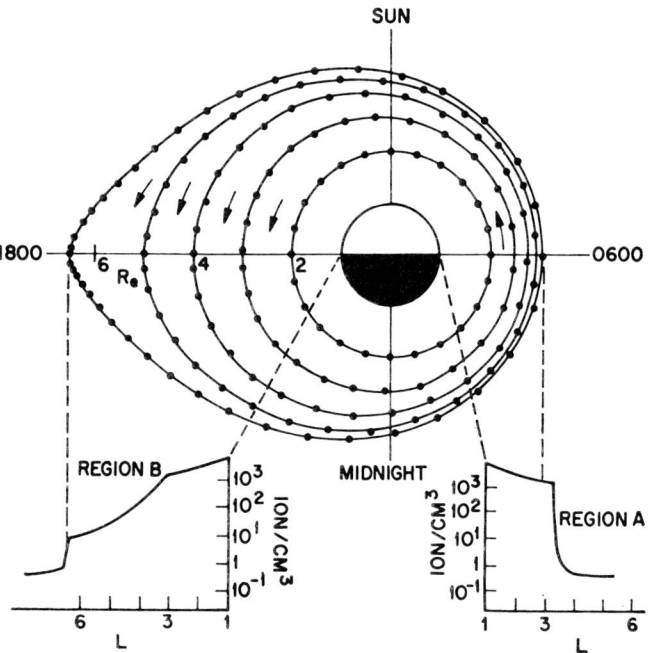

Fig. 13. The motion of flux tubes in the plasmasphere as calculated by Chappell et al. (1970). The distance between dots gives the approximate drift distance of the flux tubes in one hour.

particles (particularly electrons) circulating in static models will eventually be lost by pitch angle diffusion or through various boundaries. The injection event, marked by rapidly changing electric and magnetic fields, is the assumed source of new particles, though as

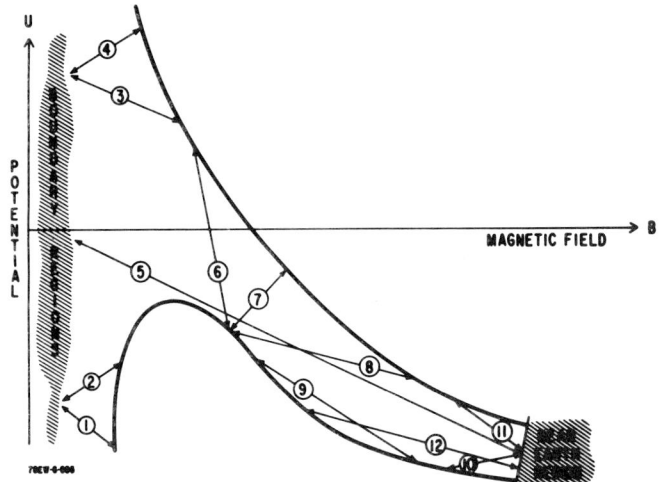

Fig. 14. Particle trajectories in U-B space (Whipple, 1978). See text for explanation of trajectories and boundaries.

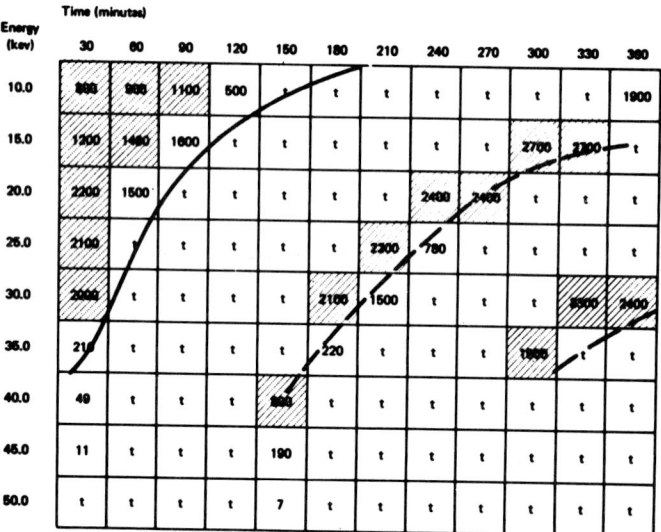

Fig. 15a. Proton flux values (particles/cm$^2$-sec-ster-eV) as a function of time following injection (in this case, when the satellite was at midnight) and energy (Roederer and Hones, 1974). This presentation is essentially identical to a spectrogram.

discussed, it is still questioned whether the particles are from a uniform source or appear suddenly in situ. Thus, to properly model the magnetosphere, time-dependency is necessary.

Roederer and Hones (1974) have developed an effective time-dependent model of the magnetosphere by an adaptation of their 1970 static model (Roederer and Hones, 1970). They have, however, added a time-varying uniform dawn-dusk field and a localized, azimuthal field. An analysis of field variations, adjusted to yield numerical results comparable with the ATS-5 data of DeForest and McIlwain, indicated that a rise time of 10 minutes followed by a slow decay (2 hours) was necessary to account for ATS-5 observations. Only 90° pitch angles and energies above 10 keV were considered.

The Roederer-Hones model quantitatively reproduces the ATS-5 data. Flux values (particles/cm$^2$-sec-ster-eV) for protons and electrons observed by a satellite passing through midnight at the time of onset are presented in Figures 15a and b; "t" indicates particles that were trapped before the onset. Non-hatched values represent particles initiating in the plasma sheet, while hatched regions correspond to particles injected through an acceleration channel between 0° and 10°E solar magnetospheric longitude. Roederer and Hones' results indicate that no electrons are captured into stably trapped orbits - an observation in agreement with DeForest and McIlwain's observations (see comments under the Konradi et al., 1975 model).

Although the Roederer-Hones model is very successful in fitting the ATS-5 data, it is not intended to be self-consistent. That is, the fields the particles themselves generate and the background fields must agree. The Rice University group, under R. A. Wolf, has carried out a

| Energy (Kev) \ Time (minutes) | 30 | 60 | 90 | 120 | 150 | 180 | 210 | 240 | 270 | 300 | 330 | 360 |
|---|---|---|---|---|---|---|---|---|---|---|---|---|
| 10.0 | 2500 | 5500 | 4830 | 3000 | 2200 | 1700 | 1300 | 1100 | 600 | 400 | 200 | 300 |
| 15.0 | 410 | 1400 | 1100 | 760 | 560 | 450 | 380 | 270 | t | t | t | t |
| 20.0 |  | 500 | 280 | t | t | t | t | t | t | t | t | t |
| 25.0 |  | 180 |  | t | t | t | t | t | t | t | t | t |
| 30.0 |  |  | t | t | t | t | t | t | t |  | 310 |  |
| 35.0 |  |  |  |  |  |  |  |  |  |  | t | t |
| 40.0 |  |  |  |  |  |  |  |  |  |  |  |  |
| 45.0 |  |  |  |  |  |  |  |  |  |  |  |  |
| 50.0 |  |  |  |  |  |  |  |  |  |  |  |  |

Fig. 15b. Same as 15a, only for electrons.

systematic development program aimed at ultimately developing a self-consistent, 3-dimensional model of magnetospheric storm processes. Chen (1970), Wolf (1970), Chen and Wolf (1972), Jaggi and Wolf (1973), and Wolf (1974) are all papers in this series. More recently are Southwood (1977) and Harel et al. (1978).

Figure 16 is a schematic of the Rice program. As originally outlined by Vasyliunas (1970), they start with a model of the earth's electric and magnetic field and calculate the plasma distribution and pressure from which the pressure gradients, and hence, the currents perpendicular to the magnetic field are calculated. From the divergence of these currents, they obtain the field aligned currents flowing between the magnetosphere and the ionosphere. Assuming closed current loops, the electric fields in the ionosphere are calculated. The ionospheric electric fields are then mapped back into the magnetosphere. The process is iterated at each stage until the initial and final fields agree. A storm has been modeled by taking actual substorm electric field measurements from the S3-3 satellite. (See paper by Wolf and Harel at this conference). Although the model is still being developed, results to date are quite promising. It is significant that this model makes a serious attempt to include internally-consistent fields and plasma flows.

Early results for the plasmasphere, Chen and Wolf (1972), are presented in Figure 17. This figure illustrates the distortion of the plasmasphere boundary following a factor-of-two decrease in the convection field. Such detached regions have actually been observed (see Chappell, 1974).

Problem Areas

Although only briefly covered in the preceding discussions, several problem areas still exist in the quantitative modeling of the magnetospheric plasmas. Although most are traceable to the complexity of the

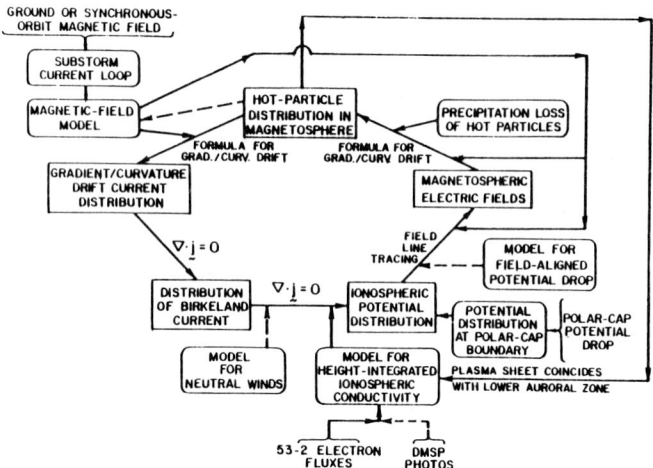

Fig. 16. Schematic representation of the Rice University magnetospheric modeling program (based on Vasyliunas, 1970).

system itself, several of the problems are related to the needs of the users and to problems that could be resolved by the community at large. Several of these areas are discussed with the aim of identifying group actions that could ameliorate them.

Probably the single overriding user problem at present is the lack of adequate statistical tables of the parameters necessary to define the environment along various orbits. The data are available for such an analysis and the parameters have been defined. Specifically, the 4 moments of the distribution function and the derived "temperature" (T(RMS) is currently favored) are the most requested user parameters in the absence of the detailed distribution function. These data are required at all spatial locations and during all conditions, though obviously much spatial and temporal averaging is required. The inability of the techniques used in constructing the highly successful radiation models to model the low energy population has contributed to this problem. Fortunately, the geosynchronous environment is currently being studied in this manner and should serve as a guide for future work. Considering the potential diversity of future near-earth missions, a variety of such models, perhaps specifically tailored to given orbits, are required.

Considering the complexity in time and space of the near-earth plasma environment, the accurate study of the environment demands multiple satellite studies. In particular, the whole concept of trajectory-tracing must be validated for if it is possible, then the number of satellites necessary to monitor the near-earth environment (a necessity until accurate geomagnetic prediction programs are available) would be greatly reduced. Taking into account the cost of satellite monitors, this alone could justify the entire magnetospheric modeling effort.

Large spatial and temporal variations in the magnetosphere are still only tentatively understood. The injection process, the most important assumption in plasma modeling, is at best a subject of controversy as demonstrated by the models presented in this article. As yet (hopefully this conference will resolve the issue) no standardization of the

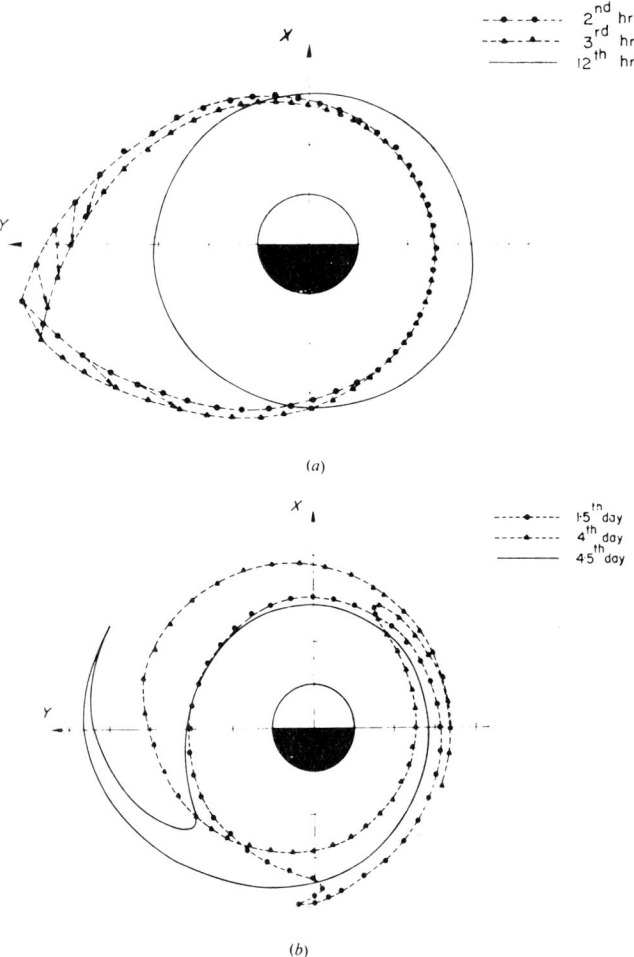

Fig. 17. The distortion of the plasmapause at various times following a factor of two decrease in the convection electric field at time t = 0.

model fields, particularly the electric fields, has been attempted between the various models. Complicating the issue is that at various times the perturbations of the geomagnetic field are so large as to cause the solar wind to appear at geosynchronous orbit. Plasma dropouts, the movement of a satellite into the high latitude tail, are relatively common, short duration ( ~ 10 minute) events which can also significantly skew statistical studies.

Various side issues also cloud the basic problem of low-energy particle modeling (many might argue that such side issues are actually the main issues). Field-aligned or parallel electric fields and particle fluxes are still comparative unknowns. Parallel particle fluxes may, for example, significantly affect spacecraft charge buildup (DeForest, 1977). Particle diffusion is another problem that has only been briefly touched in the models (see Roederer, 1970). This problem has been particularly acute for the low-energy electron population. Recently, the growing volume of information on the particle compositions in the near-earth plasma environment has raised entirely new issues. It now appears that composition must be an important consideration.

Finally, a variety of orbit-energy configurations exist for which no data are available. Considering the possible drought in the 1980's of near-earth probes as the shuttle is developed, it behooves us to carefully review the current state of the existing data base and, in the light of this information, more carefully plan future missions.

## Conclusions

The current state of quantitative low-energy plasma models is much like that of magnetic field models a decade ago or that of electric field models five years ago in that no standard models yet exist. There are still almost as many models as researchers. There does, however, seem to be a convergence on certain accepted magnetic and electric field models (Roederer's magnetic field or those of Olson and Pfitzer and the electric field of McIlwain). The concept of an injection appears to be gaining acceptance and, finally, the necessary parameters are being defined. New theoretical directions (particularly those of Wolf and Whipple) which are closely coupled with quantitative outputs promise even better results than have already been attained.

In the future the input and output of the various models should be more closely tailored to potential users. All the models developed to date are valuable to large groups outside the original designer. The fact that they are not in wide use is directly attributable to the lack of a common mode of data input and output between the designers and the potential users. As an example, several of the particle-tracing programs could be "black-boxed" to allow the computation of the particle distributions likely to be encountered along various orbits, for which only limited data is available, by extrapolating existing data or a general model.

In conclusion, the quantitative modeling of the low-energy plasma environment is in a very healthy state. Models, in general, can reproduce observations. A major problem is standardization but as in the case of magnetic and electric field models, this will come with time and experience. Finally, user needs must be carefully considered in this age of more competitive funding.

## References

Alfvén, H., A theory of magnetic storms and of the aurorae, Proceedings of the Royal Swedish Academy of Sciences, 1939, Reprinted in EOS, Trans. Am. Geophys. Union, 51, 180-194, 1970.

Brice, N. M., Bulk motion of the magnetosphere, J. Geophys. Res., 72, 5193, 1967.

Carpenter, D. L., Whistler studies of the plasmapause in the magnetosphere. 1. Temporal variations in the position of the knee and some evidence on plasma motions near the knee, J. Geophys. Res., 71, 693-709, 1966.

Carpenter, D. L., and C. G. Park, On what ionospheric workers should know about the plasmapause-plasmasphere, Rev. Geophys., 81, p. 133-154, 1973.

Chan, K. W., D. M. Sawyer, and J. I. Vette, A model of the near-earth plasma environment and application to the ISEE-A and -B orbit, NSSDC/WDC-A-R&D 77-01, July 1977.

Chappell, C. R., K. K. Harris, and C. W. Sharp, The morphology of the bulge of the plasmasphere, J. Geophys. Res., 75, 3848, 1970.

Chappell, C. R., Recent satellite measurements of the morphology and dynamics of the plasmasphere, Rev. Geophys., 10, 951-979, 1972.

Chappell, C. R., Detached plasma regions in the magnetosphere, J. Geophys. Res., 79, 1861-1870, 1974.

Chen, A. J., Penetration of low-energy protons deep into the magnetosphere, J. Geophys. Res., 75, 2458-2467, 1970.

Chen, A. J., and R. A. Wolf, Effects on the plasmasphere of a time-varying convection electric field, Planet. Space Sci., 20, 483-509, 1972.

DeForest, S. E., Specification of the geosynchronous plasma environment, AFGL-TR-77-0031, 1977.

DeForest, S. E., and C. E. McIlwain, Plasma clouds in the magnetosphere, J. Geophys. Res., 76, 3587-3611, 1971.

DeForest, S. E., and A. R. Wilson, A preliminary specification of the geosynchronous plasma environment, DNA 3951T, 1976.

Frank, L. A., On the extraterrestrial ring current during geomagnetic storms, J. Geophys. Res., 72, 3753-3767, 1967.

Freeman, J. W., $K_p$ dependence of the plasma sheet boundary, J. Geophys. Res., 79, 4315, 1974.

Garrett, H. B., Modeling of the geosynchronous orbit plasma environment-Part I, AFGL-TR-77-0288, 1977.

Garrett, H. B., and S. E. DeForest, An analytical simulation of the geosynchronous plasma environment, submitted to Planet. Space Sci., 1978.

Garrett, H. B., G. Mullen, E. Ziemba, and S. E. DeForest, Modeling of the geosynchronous orbit plasma environment - Part II, to appear as AFGL In-House Report, 1978.

Harel, M., R. A. Wolf, and P. H. Reiff, Preliminary report of the first computer run simulating the substorm-type event of 19 September 1976, Appendix A, Annual NSF report for grants ATM 74-21185 and ATM 74-21185A01, 1978.

Jaggi, R. K., and R. A. Wolf, Self-consistent calculation of the motion of a sheet of ions in the magnetosphere, J. Geophys. Res., 78, 2852-2866, 1973.

Kavanagh, L. D., Jr., J. W. Freeman, Jr., and A. J. Chen, Plasma flow in the magnetosphere, J. Geophys. Res., 73, 5511, 1968.

Kivelson, M. G., and D. J. Southwood, Local-time variations of particle flux produced by an electrostatic field in the magnetosphere, J. Geophys. Res., 80, 56, 1975a.

Kivelson, M. G., and D. J. Southwood, Approximations for the study of drift boundaries in the magnetosphere, J. Geophys. Res., 80, 3528, 1975b.

Kivelson, M. G., Magnetospheric electric fields and their variations with geomagnetic activity, Rev. Geophys., 14, 189-197, 1976.

Konradi, A., C. L. Semar and T. A. Fritz, Substorm-injected protons and electrons and the injection boundary model, J. Geophys. Res., 80, 543-552, 1975.

Lemaire, J., Steady state plasmapause positions deduced from McIlwain's electric field models, J. Atmos. Terr. Phys., 38, 1041-1046, 1976.

Lennartsson, W., and D. L. Reasoner, Low-energy plasma observations at synchronous orbit, J. Geophys. Res., 83, 2145-2156, 1978.

Mauk, B. H. and C. E. McIlwain, Correlation of $K_p$ with the substorm plasma sheet boundary, J. Geophys. Res., 79, 3193-3196, 1974.

McIlwain, C. E., Plasma convection in the vicinity of the geosynchronous orbit, in Earth's Magnetospheric Processes, edited by B. M. McCormac, p. 268, D. Reidel, Dordrecht, Netherlands, 1972.

Nishida, A., Formation of plasmapause, or magnetospheric plasma knee by the combined action of magnetospheric convection and plasma escape from the tail, J. Geophys. Res., 71, 5669, 1966.

Roederer, J. G., Dynamics of geomagnetically trapped radiation in Physics and Chemistry in Space, Vol. 2, edited by J. G. Roederer and J. Zahringer, Springer, New York, 1970.

Roederer, J. G., and E. W. Hones, Jr., Electric field in the magnetosphere as deduced from asymmetries in the trapped particle flux, J. Geophys. Res., 75, 3923-3926, 1970.

Roederer, J. G., and E. W. Hones, Jr., Motion of magnetospheric particle clouds in a time-dependent electric field model, J. Geophys. Res., 79, 1432-1438, 1974.

Smith, P. H., N. K. Beutra, and R. A. Hoffman, Motions of charged particles in the magnetosphere under the influence of a time-varying large scale convection electric field, Proceedings of the Quantitative Modeling of Magnetospheric Processes Conference, edited by W. A. Olson, 1978.

Southwood, D. J., The role of hot plasma in magnetospheric convection, J. Geophys. Res., 35, 5512-5520, 1977.

Stern, D. P., The motion of a proton in the equatorial magnetosphere, J. Geophys. Res., 80, 595-599, 1975.

Stevens, N. J., R. R. Lovell, and C. K. Purvis, Provisional specification for satellite time in a geomagnetic substorm environment, Proc. of the Spacecraft Charging Conference, AFGL-TR-77-0051/NASA-TMX-73537, 1977.

Su, S.-Y. and A. Konradi, Description of the plasma environment at geosynchronous altitude, NASA Johnson Technical Note, P-10, 1977.

Vasyliunas, V. M., A survey of low-energy electrons in the evening sector of the magnetosphere with OGO 1 and OGO 3, J. Geophys. Res., 73, 2839-2884, 1968.

Vasyliunas, V. M., Mathematical models of the magnetospheric convection and its coupling to the ionosphere, in Earth's Particles and Fields in the Magnetosphere, edited by B. M. McCormac, D. Reidel, Dordrecht, Holland, 60, 1970.

Walker, R. J., and M. G. Kivelson, Energization of electrons at synchronous orbit by substorm-associated cross-magnetospheric electric fields, J. Geophys. Res., 80, 2071-2082, 1975.

Whipple, E. C., (U, B, K) Coordinates: A natural system for studying magnetospheric convection, submitted to J. Geophys. Res., 1978.

Williams, D. J., J. N. Barfield, and T. A. Fritz, Initial Explorer 45 sub-storm observations and electric field considerations, J. Geophys. Res., 79, 554, 1974.

Wolf, R. A., Effects of ionospheric conductivity on convective flow of plasma in the magnetosphere, J. Geophys. Res., 75, 4677-4698, 1970.

Wolf, R. A., Calculation of magnetospheric electric fields, in Magnetospheric Physics, edited by B. M. McCormac, D. Reidel Publishing Co., Dordrecht, Holland, 167-177, 1974.

Young, D. T., The role of ion composition measurements in magnetospheric modeling, Proceedings of the Quantitative Modeling of Magnetospheric Processes Conference, edited by W. P. Olson, 1978.

ENERGY BUDGET FOR SOLAR WIND-MAGNETOSPHERIC INTERACTIONS

Walter J. Heikkila

Center for Space Sciences, University of Texas at Dallas
Richardson, Texas 75080

Abstract. The principal electromagnetic force on a collisionless plasma is the magnetic force $\underline{J} \times \underline{B}$ due to currents within the plasma (in addition to inertial forces). Such current systems must be closed, because we can neglect the displacement current. Also, a current in the presence of an electric field is required to energize or de-energize the plasma, according to the value of $\underline{E} \cdot \underline{J}$. Using these two principles, energy relationships for the bow shock, magnetosheath, magnetopause, boundary layer, and plasma sheet are discussed. In the bow shock it is proposed that there are two currents, back-to-back, one of which provides the random energization needed so that the entropy can increase in the shock, while a polarization current causes the plasma ions to lose energy. Until a more quantitative model of these two currents is developed it is difficult to say how much electromagnetic and plasma energy is involved. Currents in the magnetosheath constitute a generator-motor pair; although the total amount of power may be high, the average energy gain or loss per particle associated with $\underline{E} \cdot \underline{J}$ may be small because of the large number of particles. Magnetopause observations from several spacecraft do not provide any evidence of plasma energization except for the appearance of quite energetic particles (>20 keV) at too low a rate to be consistent with reconnection. On the contrary, most of the plasma particles apparently lose energy and momentum in going through the magnetopause into the low latitude boundary layer. Within the boundary layer there must be westward current which opposes the magnetospheric electric field; it may be here that the electromagnetic power for auroral phenomena on the night side is created. The electromagnetic power dissipated in the plasma sheet is about $10^{11}$ watts, but varies widely dependent on solar wind conditions. In general, gradient and curvature drifts of plasma particles are important in providing a mechanism for the currents; this means that the plasma flow is not consistent with frozen-in magnetic field.

## Introduction

This paper attempts to present some fundamental considerations about plasma flow and energy, resulting from the fact that solar wind flows around, and into, the magnetosphere. It is not a review paper, and some of the points may be controversial. Qualitative comments, relating to basic physics on a global scale, are advanced; this physics could be made quantitative when a specific event, or satellite orbit, is considered.

## Plasma Flow

Throughout most of the magnetosphere the plasma is essentially collisionless; in such a plasma individual particles are subject to only the electromagnetic force, the Lorentz force $\underline{F}$ on a particle with charge q and velocity $\underline{v}$,

$$\underline{F} = q(\underline{E} + \underline{v} \times \underline{B}) \tag{1}$$

$\underline{E}$ and $\underline{B}$ are the electric and magnetic fields, in SI units. Atomic, nuclear, and gravitational forces are assumed to be negligible in this paper. When summed over all particles, the net effect is to give a body force on the plasma

$$\underline{F}^P = \Sigma_\sigma n^\sigma \underline{F}^\sigma = \rho \underline{E} + \underline{J} \times \underline{B} \tag{2}$$

(see Schmidt, 1966, p. 64). Here $\rho$ and $\underline{J}$ are the charge and current densities, and $\sigma$ specifies summation over all particles, which can be of several kinds (at least two). Since the charge densities in a plasma are quite small we can neglect $\rho\underline{E}$ (except in thin regions of charge separation as in double layers) and so the magnetic force

$$\underline{F}^M = \underline{J} \times \underline{B} \tag{3}$$

is the principal body force on the plasma.

There are also inertial forces which affect the flow pattern. One can often treat the plasma as a fluid with a velocity $\underline{V}$; these inertial forces can be considered as a pressure force with tensor component $P_{ik}$, so that the equation of motion of the fluid is

$$\Sigma_\sigma m^\sigma n^\sigma \frac{dV_k}{dt} + \frac{\partial P_{ik}}{\partial x_i} = F_k^P \tag{4}$$

(see Schmidt, 1966, p. 67). Individual particles can go across the magnetic field only by gradient and curvature drifts (in addition to the electromagnetic drift velocity $\underline{V}^E = \underline{E} \times \underline{B}/B^2$); these drifts produce a resultant current $\underline{J}$. Thus, the pressure force is mainly involved in changes of the fluid velocity along field lines; for example, the plasma pressure in regions of high magnetic field strength serves to accelerate the flow toward regions of weaker field, because of changes in pitch angle, associated with conservation of the first adiabatic invariant.

Consequently, the magnetic force $\underline{J} \times \underline{B}$ is primarily involved in determining the flow pattern across the magnetic field. If this force is directed normal to the velocity vector the flow will be deflected; for a positive value of $\underline{V} \cdot \underline{F}$ the flow will speed up, while for a negative value the speed will diminish. The current which provides this force is described by the continuity relation

$$\text{div } \underline{J} = -\frac{\partial \rho}{\partial t} \tag{5}$$

However, by using Poisson's equation, we note that $\frac{\partial \rho}{\partial t} = \text{div } \underline{J}_D$, where $\underline{J}_D$ is the displacement current, which is negligible: "...the neglect of the displacement current separates the theory of cosmic electrodynamics from that of radio astronomy." (Dungey, 1958, p. 8) Consequently, div $\underline{J}$ $\simeq$ 0, and all current systems in the plasma are closed (very nearly). This is indicated in Figure 1 (the indicated energy relationships are explained in the next section).

By noting that the magnetic field strength increases behind the bow shock it is obvious that currents must flow in the bow shock. Also, since the plasma flow is diverted around the magnetospheric obstacle we note that there must be currents in the magnetosheath itself, to give the necessary force $\underline{F}^M = \underline{J} \times \underline{B}$. It is a very important task of future modeling to find out what is the complete path of all currents in the bow shock and magnetosheath; closure of currents within the magnetosphere proper may also need further explanation.

## Plasma Energy

We can use another principle, that of conservation of energy, to find out more about the physics of the magnetosphere.

Energy in a plasma can appear in two forms, kinetic and electromagnetic. When a charged particle is moved through a displacement $\delta \underline{r}$ in a time $\delta t$, with $\delta \underline{r} = \underline{v}\delta t$, the rate of work done by electrical forces (summed over all particles) is

$$\varepsilon = \Sigma_\sigma \underline{F}^\sigma \cdot \underline{v}^\sigma = \underline{E} \cdot \Sigma_\sigma q^\sigma \underline{v}^\sigma = \underline{E} \cdot \underline{J}. \qquad (6)$$

This simple relation completely describes whether a plasma is gaining or losing energy (random or ordered) in the presence of an electromagnetic field. This agrees with Ohm's law for a linear medium $\underline{J} = \underline{\sigma} \cdot \underline{E}$ (where $\sigma$ is the [tensor] conductivity), but it is more fundamental. A negative value for this quantity, $\underline{E} \cdot \underline{J} < 0$, corresponds to a generator of electromagnetic energy, while a positive value, $\underline{E} \cdot \underline{J} > 0$, corresponds to a load on the electromagnetic system (see Figure 1). Poynting's theorem describes the sources (or sinks) of this energy. The beauty of the circuit description is that cause and effect are clearly separated.

## Solar Wind

We must try to model the magnetosphere in three dimensions; only then can we be sure how various current systems can be closed. In the solar wind the orientation of the interplanetary magnetic field (IMF) relative to the solar wind velocity vector is arbitrary. When we are concerned primarily with the bow shock and the magnetosheath this arbitrary direction can be constructed from two fields, one with no component along the solar wind velocity vector (so that $\underline{V}^{SW} \cdot \underline{B}^{SW} = 0$), and the other with only a component along the velocity vector ($\underline{V}^{SW} \times \underline{B}^{SW} = 0$). These two directions correspond to perpendicular ($\theta_{Bn} = 90°$) and parallel ($\theta_{Bn} = 0°$) shocks, respectively, at the subsolar point (see e.g. the reviews by Greenstadt, 1976, and Formisano, 1977).

When we are concerned with interactions with the magnetosphere proper, then the north-south relationships become important, i.e., the value of $B_z$ in the solar wind may be significant.

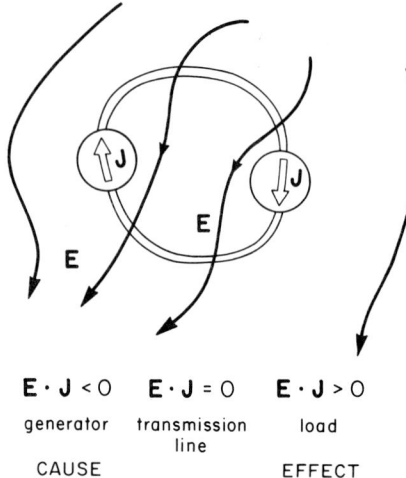

Fig. 1. Magnetospheric currents must be closed, since we can neglect the displacement current. In the presence of an electric field, regions in which the current opposes the electric field ($\underline{E} \cdot \underline{J} < 0$) behave as generators (the cause), while regions in which the current goes with the electric field ($\underline{E} \cdot \underline{J} > 0$) act as loads (the effect).

In the first case there is an interplanetary electric field in the earth's frame of reference; by Maxwell's equations we know that the tangential component is continuous across any surface, such as the bow shock. In the latter case, there is no electric field in the solar wind; the normal component of the magnetic field is continuous.

## Bow Shock

In the interplanetary medium the solar wind velocity exceeds both the Alfvén and magnetosonic velocities. Accordingly, there must be a bow shock; after the shock, the plasma velocity becomes sub-sonic. This can be accomplished only by an irreversible process, with an increase in entropy (Courant and Friedrichs, 1948, p. 121).

(a) Perpendicular shock. In this case we have a tangential component of the electric field at the subsolar point, $\underline{E}_t = -\underline{V}^{SW} \times \underline{B}^{SW}$, which is continuous across the shock. Since the plasma speed is essentially $|\underline{V}| = |\underline{E} \times \underline{B}/B^2| = E/B$, this requires that $|\underline{B}|$ must increase behind the shock in order to have a reduced velocity. Hence currents must flow in the bow shock, according to Ampere's law. A possible direction of this current $\underline{J}_\perp^{BS}$ can be determined from the requirement that there must be thermal dissipation in the plasma, in order to have an increase in entropy (see Formisano, 1977, p. 66). Thus it appears that $\underline{E}^{BS} \cdot \underline{J}_\perp^{BS} > 0$, and $\underline{J}_\perp^{BS}$ flows in the $\underline{B} \times \nabla B$ direction as a generalized gradient-drift current. For the case of a southward interplanetary magnetic field, $\underline{J}_\perp^{BS}$ flows eastward, in the same sense as the magnetopause current $\underline{J}^{MP}$. The associated energization of plasma particles has been observed quite regularly.

On the other hand, the post-shock energy spectrum consists of two parts, with most of the plasma ions losing energy (Montgomery et al., 1970). The latter can be accomplished through a polarization current, $J_2^{BS}$, with $\underline{E} \cdot \underline{J}_2^{BS} < 0$: a polarization current must exist because of the increasing strength of the magnetic field (Schmidt, 1966, p. 160). Thus it appears that there are in fact two currents within the shock, one after the other. In this way, the sunward force $\underline{J}_2^{BS} \times \underline{B}$ can be balanced by an opposing force due to $\underline{J}_1^{BS}$, to hold the shock stationary (but ram and random plasma pressure forces must also be taken into account). Also, with only one current there would be nothing to prevent the associated magnetic field appearing ahead of the shock in the solar wind, which is impossible by definition of a shock. In our model $\underline{J}_1^{BS}$ would have exactly the right magnetic field in front of the shock to cancel the contributions to the magnetic field of all downstream currents such as $\underline{J}_2^{BS}$, magnetosheath currents, magnetopause current, and so on.

An additional feature that may be important is an electric field normal to the shock, as has been visualized for the magnetopause (see the review by Willis, 1978). This may account for the electron heating, as observed, as a transfer of energy from the solar wind ions.

An important consideration is how each current is closed; is the first current $\underline{J}_1^{BS}$ the load for the second current $\underline{J}_2^{BS}$? Or do they close somewhere far downstream, through still other currents? Also, what determines how the current carriers for $J_1$ and $J_2$ are selected? Is this selection accomplished on the basis of mass, energy, momentum, pitch angle, or some other consideration?

Until we know the answer to these questions it is difficult to say what is the total amount of energy dissipation associated with $\underline{E} \cdot \underline{J}$ at the bow shock. However, we can try to estimate, at least in order of magnitude. Under average conditions the interplanetary electric field is about $1.4 \times 10^{-3}$ volts/m (Heikkila, 1975); this represents a total potential drop of about $5 \times 10^5$ volts across the bow shock (taking the width as 50 $R_e$, in the dawn-dusk meridian). If we estimate the bow shock current as being about $5 \times 10^{-3}$ amp/m, corresponding to a change in the magnetic field of 5 nT, then the total amount of electromagnetic dissipation is about $10^{12}$ watts.

Since the total number of particles taking part in the bow shock interaction is very large (more than $10^{29}$ per second) the average energization is small. However, only a small fraction of the particles undergo this energization, so that the energization per particle might amount to several hundred electron volts, perhaps much more for some particles. Some of the energized particles are observed as forerunners in the solar wind, but it seems likely that most of them appear downstream in the magnetosheath.

By coincidence, this power turns out to be about equal to the power that would be associated with reconnection on the dayside magnetopause; thus the comparison between the bow shock and the magnetopause in terms of energy relationships should be very illuminating.

(b) Parallel shock. Here, the tangential component of the electric field in the bow shock vanishes, $\underline{E} \cdot \underline{J} = 0$, and there is no energization of plasma particles; the normal component of the magnetic field is continuous. Since there is nevertheless the requirement that the shock must reduce the velocity to subsonic conditions, the only thing that the plasma can do is to scatter the particles in pitch angle. The velocity reduction can be achieved simply as a result of the average increase in pitch angle, without any decrease (or increase) in the energy of the particles.

However, observations suggest that a parallel shock may not be time-independent (Greenstadt, 1976; Formisano, 1977). Such a shock is very complex.

## Magnetosheath

We know that there must be currents within the magnetosheath, from the fact that there must be $\underline{J} \times \underline{B}$ forces to slow down and divert the flow (across the magnetic field lines), and then to accelerate the flow farther downstream. Only a very small amount of the energy that is associated with this flow enters the magnetosphere, so that we can neglect this interaction with the magnetosphere as a first approximation when considering the magnetosheath itself.

Let us for simplicity consider first the case of a southward IMF (similar considerations apply for any transverse $\underline{B}$ satisfying $\underline{V}^{SW} \cdot \underline{B}^{SW} = 0$). A westward current near the subsolar regions will produce a sunward force, needed to slow down the plasma. Since the electric field is eastward this current acts as a generator with $\underline{E}^{MS} \cdot \underline{J}_1^{MS} < 0$.

Farther downstream the plasma flow speeds up; since the magnetic field must have a southward component (by assumption) this requires a dawn-dusk electric current $\underline{J}_2^{MS}$. Here the plasma acts as a load on the electromagnetic system. Thus, the magnetosheath behaves as a generator-motor pair, as indicated in Figures 1 and 2.

A substantial amount of power is involved. Something like $10^{12} - 10^{13}$ watts (again for average conditions) goes into particle energization. However, a large number of particles is also involved, so that the average de-energization associated with $\underline{J}_1^{MS}$, and average energization associated with $\underline{J}_2^{MS}$, may be quite small, perhaps on the order of a few electron volts. Additional power is associated with pressure forces, related to changes in pitch angle, but without particle energization.

With an IMF directed along the solar wind velocity, the situation is again more complicated, as we noted in discussing the bow shock. The pitch angle distribution will be broader after the shock, in view of the reduced flow. In the magnetosheath the flow speed is further reduced: this can be effected by an increasing strength of the magnetic field, leading to increasing pitch angles of the plasma particles, by conservation of the first adiabatic invariant. It thus appears that the flow, if it follows magnetic field lines (as assumed, for example, by Spreiter and Alksne, 1969), might converge, rather than diverge as it should do by analogy with fluid flow. A plasma force directed across the magnetic field must be involved in diverting the flow, so that again there must be an electric current. By symmetry, we must have a current vortex, circulating clockwise as viewed from the sun, to produce a force pointing out from the axis of symmetry (Figure 3). This current will enhance the magnetic field strength near the axis and a consequent further reduction in the plasma velocity as noted earlier.

Again, the plasma speeds up downstream, both by observations and by analogy with ordinary fluid behavior. The acceleration of the flow

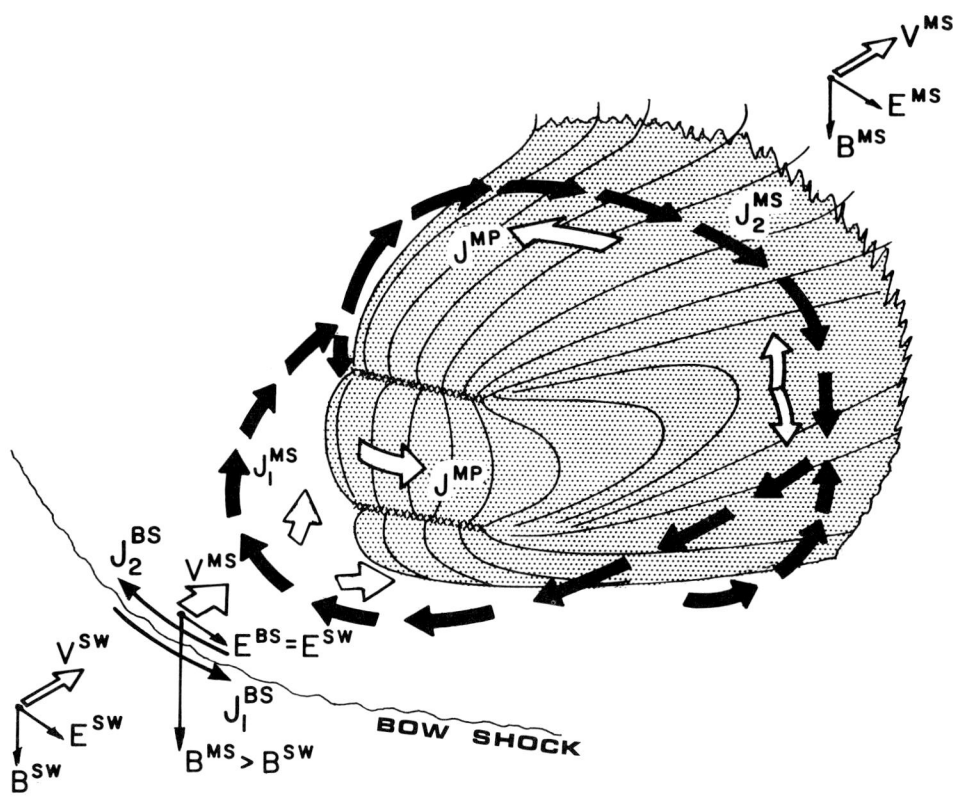

**SOUTHWARD IMF**

Fig. 2. With a southward interplanetary magnetic field, there are two currents in the bow shock, back-to-back, energizing or de-energizing particles. Within the magnetosheath the plasma is first slowed down in the sub-solar regions, and while farther back it is speeded up again; thus the magnetosheath current circuit behaves as a generator-motor pair.

can be accomplished by inertial forces, with decreasing pitch angles in an expanding magnetic field. The diversion to a flow velocity more nearly parallel to the anti-solar direction requires another current, $\underline{J}_2^{MS}$, which circulates in the opposite sense from $\underline{J}_1^{MS}$ (Figure 3).

At this moment it is impossible to say what energy is associated with these currents, because we know nothing about the electric field. It seems likely that a time independent solution may be impossible because of the difficulty of matching boundary conditions at the bow shock. However, it seems safe to say that although the total amount of power may again be quite large, nevertheless the average energization or de-energization of plasma particles may be small; in any case, very little of this energy gets into the magnetosphere itself.

Magnetic gradient and curvature drifts are the most likely explanation for the origin of the currents, in all cases that we have considered. This means that the plasma flow is not $\underline{V}^E = \underline{E} \times \underline{B}/B^2$, but is instead (with a guiding center approximation)

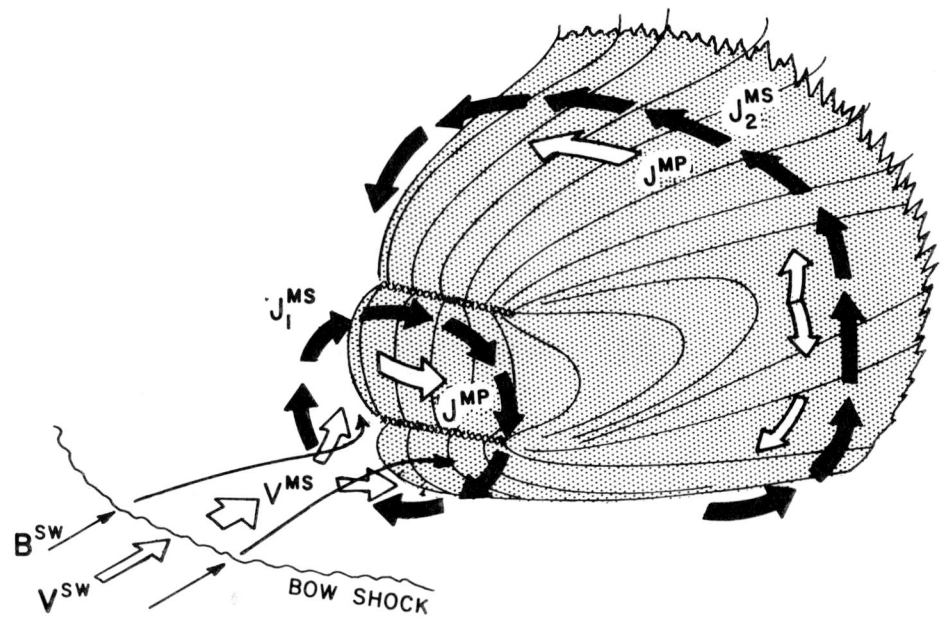

**RADIAL IMF
AWAY SECTOR**

Fig. 3. If the interplanetary magnetic field points radially out from the sun, in the direction of the solar wind velocity, plasma speed in the magnetosheath is controlled mostly by changes in pitch angle. Circulating currents are required to produce the outward and inward forces necessary for the observed fluid-like flow.

$$\underline{v}^D = \frac{\underline{E} \times \underline{B}}{B^2} + \frac{W \sin^2 \alpha \; (\underline{B} \times \nabla B)}{q B^3} + \frac{2W \cos^2 \alpha (\underline{B} \times \underline{R})}{q B^2 R^2} \quad (7)$$

Here W is the kinetic energy of the particles, $\alpha$ is the pitch angle, and $\underline{R}$ is the radius of curvature of the magnetic field lines, pointing to the center of curvature. The particles go in different directions, depending on the energy, pitch angle, and charge. In particular, the principle of frozen-in magnetic fields does not appear to hold. This can be seen most clearly in Fig. 3 where the plasma flow in the subsolar magnetosheath diverges, whereas the magnetic field converges, relative to the axis of symmetry (assuming of course that Figure 3 is a correct description of the flow).

## Magnetopause

Observations made with Heos-2 (Haerendel et al., 1978), Exp 33 (Crooker, 1977), Imp-6 (Eastman and Hones, 1978), and ISEE 1 and 2 (Paschmann et al., 1978) are very convincing that the plasma is generally not energized at the magnetopause (see Heikkila 1975, 1978, for a more thorough discussion). It is true that weak fluxes of particles at higher energies (> 20 keV) are present at all times, but the energy involved does not appear to be sufficient to salvage standard reconnection models of the magnetosphere. Furthermore, these energetic particles are not predicted by reconnection theories. If anything, there may be some loss of energy, and certainly of directed flow. Since the magnetopause surface current flows eastward, this observation is compatible with a small westward electric field on these crossings. Such a field could be produced by induction due to a moving current, by Lenz's law. If this is so, then we need to be cautious about accepting steady state models. Sonnerup (1977) has stated that "it is unlikely that they may be described, even approximately, by a sequence of steady-state configurations."

## Boundary Layer

Observation indicates that the boundary layer, just inside the magnetopause, is on closed magnetic field lines (Palmer and Hones, 1978). Plasma is observed to flow anti-sunward, which is consistent with an electric field with a dusk-dawn component. Such an electrostatic field can be produced by charges on the sides of the boundary layer, as shown in Figure 4 (see also Figure 3 of Heikkila, 1978). These same charges can produce the cross-tail field, which is observed to have a potential difference of about 50 kV, on the average (McCoy et al., 1975).

A force is required on the plasma in the boundary layer, directed inward from the magnetopause, so that the flow can everywhere be tangential to the curved magnetopause (Martyn, 1951). This requires a westward current on the front; the strength of this current is about $10^7$ amperes, based on $3 \times 10^{25}$ protons per second flowing at 50 km/s in a 100 nT magnetic field, assuming a radius of curvature of $10^8$ m of the magnetopause and boundary layer. This current must flow in opposition to the magnetospheric electric field, created by the charges on the insides of the boundary layer (Figure 4), and thus it constitutes a generator. It may be here that the electromagnetic energy is produced to power all auroral phenomena, as is indicated by the circuit diagram in Figure 4. This amounts to about $5 \times 10^{11}$ watts on the average, but is known to vary widely; the strength of the southward component of the IMF is observed to regulate this power by some mechanism, as yet uncertain.

Since the dayside magnetopause is an equipotential surface, it is likely that the magnetopause over the tail may also be an equipotential surface. Indeed, Heos-2 crossings do not indicate any sudden decrease in particle energies at the magnetopause over the magnetotail, but rather a slow decrease through the plasma mantle apparently due to velocity dispersion. If this were true, then the generator for auroral activity would not be provided by the surface current over the lobes of the magnetotail, as proposed by Akasofu (1977).

## Plasma Sheet

The extension of the magnetotail depends in part on the cross-tail current (the so-called neutral sheet current) in the plasma sheet. This current, which is included in magnetospheric models, is on the order of 30 ma/m. In the presence of the cross-tail electric field this

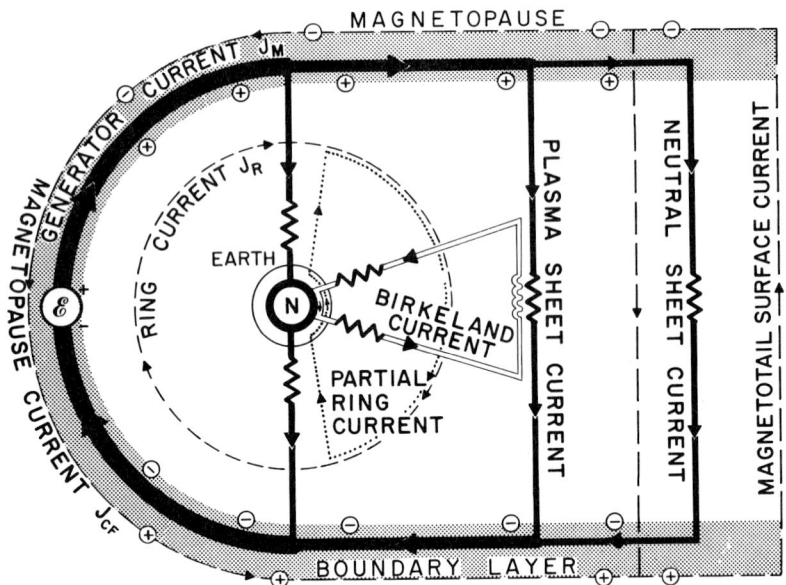

Fig. 4. In the boundary layer there must be a westward current, to produce the inward force necessary so that the flow can be everywhere tangential to the magnetopause. In the presence of the cross-tail electric field this region acts as a generator. It is here that the energy necessary to power all auroral processes is created.

represents a load amounting to $5 \times 10^9$ watts per earth radius; this load extends at least as far as the moon's orbit, for a total power of more than $10^{11}$ watts. All these figures can vary by at least one or two orders of magnitude, depending on solar wind conditions. The maximum input of solar wind particles to the plasma sheet that we could possibly consider is that contained in a cylinder parallel to the solar wind velocity across which there is the average value of the cross-tail potential difference (Heikkila, 1975); this is about $4 \times 10^{27}$ sec$^{-1}$, on the average. However, the phase space density of particles in the plasma sheet is about an order of magnitude less (Hones et al., 1971) so that it is likely that only about $4 \times 10^{26}$ sec$^{-1}$ actually enter (see the review by Hill, 1974). This is comparable to the number of particles which flow through the plasma mantle (Rosenbauer et al., 1975) and boundary layer (Eastman et al., 1978). An average energization of about 10 keV per particle is implied. This may be divided unequally between the electrons and positive ions, as is suggested by adiabatic theory, so that electrons would gain a few keV in energy, and protons the rest, perhaps 15 keV. If the first adiabatic invariant is conserved, the mirror points for the particles must come down; this is true even if the mechanism is the transverse betatron acceleration mechanism. For the longitudinal Fermi process this is obviously true. The particles that cause the auroras have these kinds of energies; thus, it appears likely that the main energization mechanisms must operate far out in the magnetotail (see Heikkila et al., 1979).

There may also be some additional acceleration (or deceleration) of particles in precipitation into the atmosphere by parallel electric fields (Falthammar, 1977).

An induced electric field must be involved in substorms (Heikkila and Pellinen, 1977). It seems likely that it would be associated with ion-tearing instability of the magnetotail (Schindler, 1974; Galeev and Zalenyi, 1977; Galeev, 1978). However, they considered the problem only in two dimensions, but three dimensional considerations may be of vital importance (Sonnerup, 1977). A very important requirement of future modeling is that of producing models of the very extended tail, just prior to the onset of a substorm, showing the high curvature in the neutral sheet. Ideally, such models should be differentiable, so that gradient and curvature drifts of particles could be calculated.

## Conclusions

Thus far the bow shock and magnetosheath currents have not been included in magnetospheric models. This is regrettable, since it is magnetosheath plasma that interacts with the magnetosphere, not solar wind plasma. By modeling these outer regions we may learn something about what kinds of currents are associated with either a closed or an open magnetosphere. A very powerful tool is the fact that all current systems must be closed; by looking at current circuits we can distinguish between cause and effect.

Acknowledgements. I wish to thank Prof. F. S. Johnson for helpful discussions. Tom Hill and Ray Walker made valuable comments as referees. This work was supported by National Science Foundation Grant No. ATM78-03603.

## References

Akasofu, S.-I., Physics of Magnetospheric Substorms, D. Reidel Pub. Co., Dordrecht-Holland, 1977.
Courant, R., and K. O. Freidricks, Supersonic Flow and Shock Waves, Interscience Publishers, Inc., New York, 1948.
Crooker, N. U., Explorer 33 entry layer observations, J. Geophys. Res. 82, 515-522, 1977.
Dungey, J. W., Cosmic Electrodynamics, Cambridge University Press, Cambridge, 1958.
Eastman, T. E., and E. W. Hones, Jr., Characteristics of the magnetospheric boundary layer and magnetopause layer as observed by IMP 6, submitted to J. Geophys. Res., 1978.
Falthammar, C.-G., Problems related to macroscopic electric fields in the magnetosphere, Rev. Geophysics and Space Physics 15, 457-466, 1977.
Formisano, V., The physics of the Earth's collisionless shock wave, Journal de Physique, 38, C6-65 - C6-88, 1977.
Galeev, A. A., Reconnection in the magnetotail, Academy of Sciences, USSR, Space Research Institute publication No. 260, 1978.
Galeev, A. A., and L. M. Zeleny, Magnetic reconnection in a space plasma, Academy of Sciences, USSR, Space Research Institute publication No. 249, 1977.
Greenstadt, E. W., Phenomenology of the Earth's bow shock system: A

summary description of experimental results, in B. M. McCormac (ed.) *Magnetospheric Particles and Fields*, 13-28, D. Reidel Publishing Co., Dordrecht-Holland, 1976.

Haerendel, G., G. Paschmann, N. Sckopke, and H. Rosenbauer, The frontside boundary layer of the magnetosphere and the problem of reconnection, J. Geophys. Res. 83, 3195-3216, 1978.

Heikkila, W. J., Is there an electrostatic field tangential to the dayside magnetopause and neutral line?, Geophys. Res. Lett. 2, 154, 1975.

Heikkila, W. J., Electric field topology near the dayside magnetopause, J. Geophys. Res., 83, 1071-1078, 1978.

Heikkila, W. J., and R. J. Pellinen, Localized induced electric field within the magnetotail, J. Geophys. Res. 82, 1610-1614, 1977.

Heikkila, W. J., R. J. Pellinen, C.-G. Falthammar and L. P. Block, Potential and inductive electric fields in the magnetosphere during auroras, submitted for publication in Planet. Space Sci., 1978.

Hill, T. W., Origin of the plasma sheet, Rev. Geophys. Space Phys. 12, 379, 1974.

Hones, E. W., Jr., J. R. Asbridge, S. J. Bame, and Sidney Singer, Energy spectra and angular distributions of particles in the plasma sheet and their comparison with rocket measurements over the auroral zone, J. Geophys. Res., 76, 63-87, 1971.

Martyn, D. F., The theory of magnetic storms and auroras, Nature, 167, 92, 1951.

McCoy, J. E., R. P. Lin, R. E. McGuire, L. M. Chase, and K. A. Anderson, Magnetotail electric fields observed from lunar orbit, J. Geophys. Res. 80, 3217-3224, 1975.

Montgomery, M. D., J. R. Asbridge, and S. J. Bame, Vela 4 plasma observations near the Earth's bow shock, J. Geophys. Res. 75, 1217-1231, 1970.

Palmer, I. D., and E. W. Hones, Jr., Characteristics of energetic electrons in the vicinity of the magnetospheric boundary layer at Vela orbit, J. Geophys. Res. 83, 2584-2596, 1978.

Paschmann, G., N. Sckopke, G. Haerendel, J. Papamastorakis, S. J. Bame, J. R. Asbridge, J. T. Gosling, E. W. Hones, Jr., and E. R. Tech, ISEE plasma observations near the subsolar magnetopause, submitted to Space Science Reviews, 1978.

Rosenbauer, H., H. Grunwaldt, M. D. Montgomery, G. Paschmann, and N. Sckopke, Heos 2 plasma observations in the distant polar magnetosphere: the plasma mantle, J. Geophys. Res. 80, 2723-2737, 1975.

Schindler, K., A theory of the substorm mechanism, J. Geophys. Res. 79, 2803, 1974.

Schmidt, G., *Physics of High Temperature Plasmas: an Introduction*, Academic Press, New York, 1966.

Spreiter, J. R., and A. Y. Alksne, Plasma flow around the magnetosphere, Rev. of Geophys. 7, 11-50, 1969.

Sonnerup, B. U., Magnetic field reconnection, to appear in *Solar System Plasma Physics — A Twentieth Anniversary Review*, C. F. Kennel, L. J. Lanzerotti and E. N. Parker, eds., North Holland Publishing Co., 1978.

Willis, D. M., The magnetopause: microstructure and interaction with magnetospheric plasma, Journ. Atmos. and Terres. Physics, 40, 301-322, 1978.

# THE MAGNETOPAUSE LAYER AND PLASMA BOUNDARY LAYER OF THE MAGNETOSPHERE

T. E. Eastman

EG&G, Inc., P. O. Box 809
Los Alamos, New Mexico 87544

E. W. Hones, Jr.

University of California, Los Alamos Scientific Laboratory
P. O. Box 1663, Los Alamos, New Mexico 87545

*Abstract*. Due to recent availability and analyses of high time resolution satellite data (including IMP 6 and the ISEE-1 and -2 satellite pair), the study of the magnetopause and boundary layer has entered a period of renewed activity. Plasma observations from the VELA satellites first established the presence of magnetosheath-like plasma with reduced density and flow velocity in a relatively thin ($\lesssim$ 1 $R_E$) layer bordering the plasma sheet at low latitudes and bordering the lobe environment at high latitudes. Recent analyses of HEOS 2, Explorer 33 and IMP 6 data have established the presence of this "plasma boundary layer" (PBL) over the entire sunward magnetosphere near the magnetopause. This review gives a brief summary of recent published results on two distinct regions of the PBL: that bordering open field line regions and that bordering closed field line regions. The magnetopause layer (i.e., current layer) can usually be identified by a change in magnetic field direction and cannot be uniquely identified by any other field or plasma parameters. Immediately earthward of this magnetopause current layer, a PBL of magnetosheath-like plasma is usually observed that has dominantly magnetosheath-like energy spectra and flow characteristics. Observed plasma boundary layer thicknesses are highly variable and are generally much larger than the magnetopause layer thicknesses even near the subsolar region. Several suggested source mechanisms for the plasma boundary layer are discussed and compared.

## Introduction

Quantitative magnetospheric magnetic field models include, either directly or indirectly, a magnetopause current system (see the review by Roederer, 1975). The magnetopause currents are part of the basic boundary conditions of the magnetosphere and they contribute significantly to the outer magnetospheric field. Due to recent availability and analyses of high time resolution satellite data (including IMP 6 and the ISEE-1 and -2 satellite pair), the study of the magnetopause and plasma boundary layer (PBL) has entered a period of renewed activity. The many labels applied to the PBL may be simply describing a common region maintained by various entry and transport processes. We

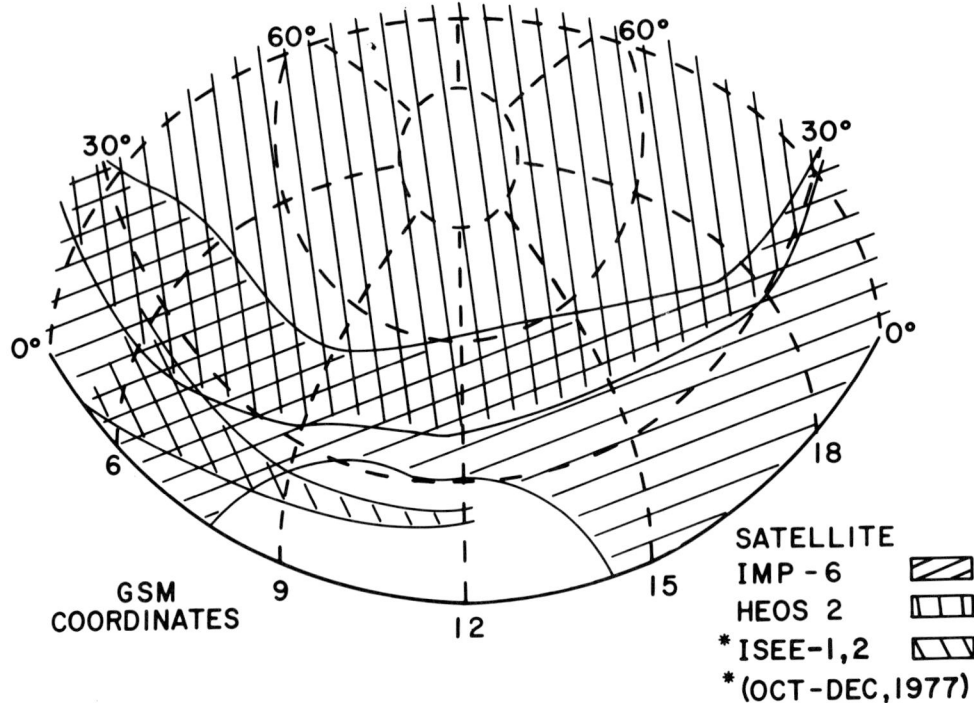

Fig. 1. Satellite coverage of the sunward magnetopause and plasma boundary layer (PBL) for IMP 6, HEOS 2 and the ISEE-1 and -2 satellite pair. ISEE coverage is given only for October-December, 1977.

use the term "plasma boundary layer" (PBL) for lack of a more physically meaningful label that can only be chosen after the magnetopause interaction processes have been clearly and uniquely identified.

Satellite coverage of the sunward magnetopause and PBL is shown in Figure 1 for four satellites that include both plasma and field measurements. HEOS 2 provides excellent coverage of the higher latitude PBL whereas IMP 6 gives complementary coverage at lower latitudes. The ISEE spacecraft have recently provided dual spacecraft sampling with a spacecraft separation distance of $\sim$ 250 km. This separation distance will be gradually increased to provide a further delineation of space and time variations near the magnetopause and other transition regions. For the locations of more tailward PBL observations, the reader is referred to Figure 1 in Sckopke and Paschmann (1978).

Plasma observations from the VELA satellites first established the continued presence of magnetosheath-like plasma with reduced density and flow velocity in a relatively thin ($\lesssim 1\ R_E$) layer bordering the plasma sheet at low latitudes and bordering the lobe environment at high latitudes [Hones et al., 1972; Akasofu et al., 1973]. A recent study of the tailward portion of the PBL based on IMP 7 data has been reported by Scarf et al. (1977). Recent analyses of HEOS 2 data [Rosenbauer et al., 1975; Paschmann et al., 1976; Haerendel et al., 1978], Explorer 33 data [Crooker 1977] and IMP 6 data [Eastman et al.,

1976; Eastman and Hones, 1978] have established the presence of this plasma boundary layer over the entire sunward magnetosphere near the magnetopause.

A confusing variety of labels have been applied to the PBL due to a wide variety of spacecraft trajectories, data sets and time resolution. Analyses of the IMP 6 [Eastman and Hones, 1978], VELA [Palmer and Hones, 1978] and HEOS 2 data [Paschmann et al., 1976] suggest an observational description in terms of a low latitude boundary layer (LLBL) bordering a closed field line region (the plasma sheet or outer ring current region) and a high latitude boundary layer (HLBL or plasma mantle) bordering a region of probably open field lines (the extended polar cap region or tail lobe environment). The distinction between these two regions of the PBL is based on the comparative profiles of observed plasma density and thermal energy in addition to energetic electron distributions. Both the density and thermal energy decrease with increasing distance inward from the magnetopause in the HLBL [Paschmann et al., 1976]. VELA magnetosheath-lobe environment crossings that we have checked also show this HLBL signature. The VELA energetic electron data indicate that this region is on open field lines (Palmer and Hones, 1978). In contrast, IMP 6 crossings of the LLBL and VELA magnetosheath-plasma sheet crossings show an increase in thermal energy, along with the density decrease, with increasing distance inward from the magnetopause. Energetic electron data from IMP 6 (Eastman and Hones, 1978) and VELA (Palmer and Hones, 1978) indicate that the LLBL is on closed field lines.

Magnetopause layer (i.e., the current layer), LLBL and HLBL are regions for which convenient empirical definitions can be specified (Eastman and Hones, 1978). It should be noted that the magnetopause layer and the PBL are related but (usually) distinct regions so that any magnetopause model is incomplete without a corresponding treatment of the PBL.

## Observations

General characteristics of the PBL can only be evaluated by comparing many crossing examples since boundary motions and spatial variations are often dominant (as in multiple magnetopause crossings). Our description of LLBL characteristics is based on over 100 IMP 6 crossings of the magnetosphere's sunward surface. HEOS 2 results are based on a similarly large set of magnetopause and PBL crossings (e.g., see Haerendel et al., 1978). Fortunately, the ISEE-1 and -2 satellite pair are presently providing high time resolution measurements that can often separate the effects of boundary motion. However, space and time variations are sometimes large enough in the vicinity of the magnetopause that even careful analysis of the ISEE data does not always lead to a clear separation of these variations (see Paschmann et al., 1978).

Some characteristics of the LLBL as observed by IMP 6 are illustrated in Figure 2. The overall density and velocity decrease from the magnetopause layer to the inner extent of the LLBL (going from right to left in the left side of the figure) is accompanied by an increase in thermal energy and continued magnetosheath-level low frequency magnetic field fluctuations, given by the standard deviation $SD_B$. The plasma β (the ratio of plasma to magnetic field energy density) usually drops to

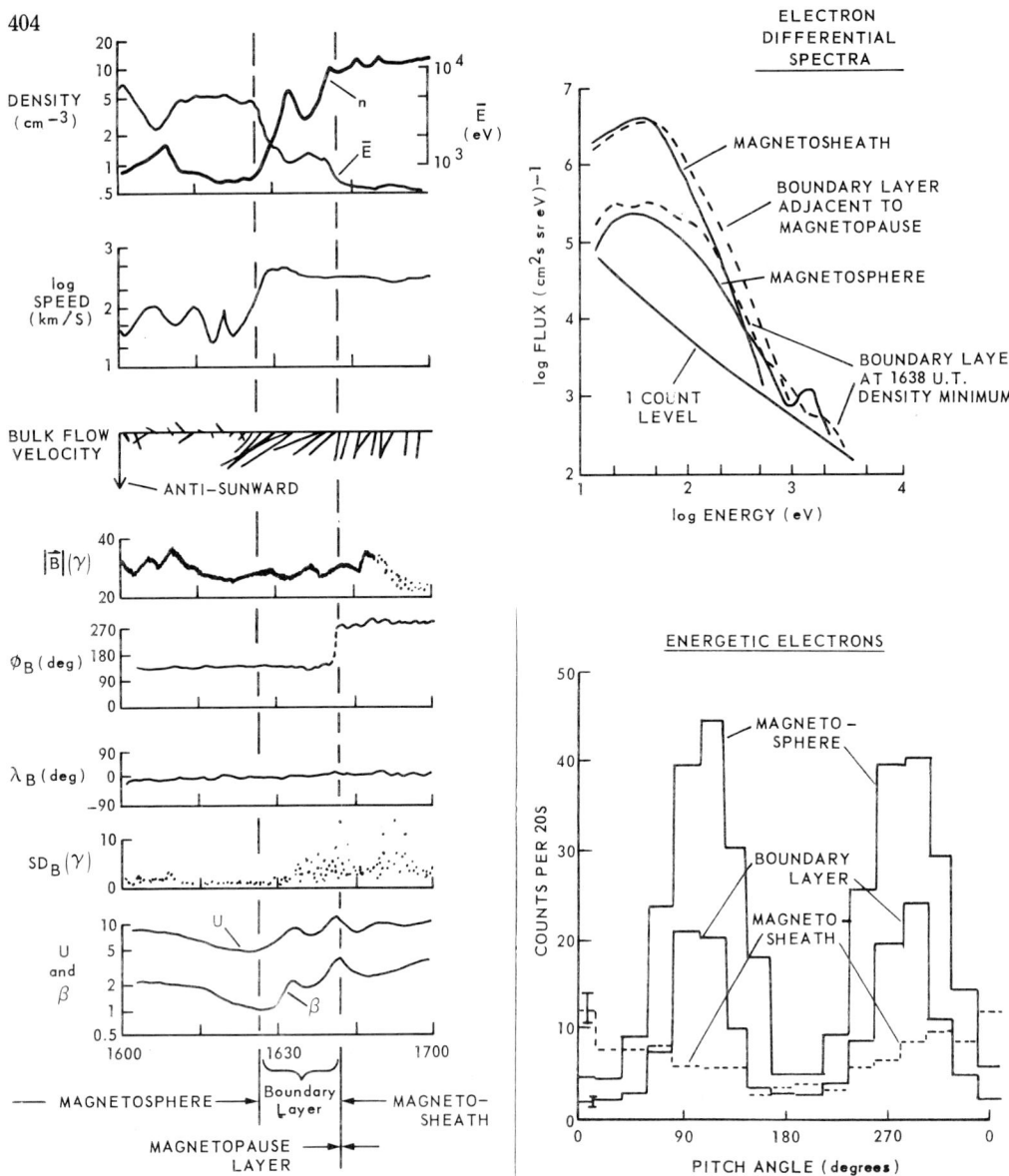

Fig. 2. Basic characteristics of the LLBL are illustrated by this March 4, 1972 IMP 6 crossing. Various plasma regions are identified at lower left based on both plasma ion and magnetic field signatures as shown in the eight plots on the left side of the figure. U and β denote the total energy density (keV/cm³) and plasma β, respectively. Energetic electron pitch angle distributions are shown at the lower right hand side of the figure. These pitch angle distributions indicate that the LLBL is on closed field lines. Electron differential spectra shown at upper right illustrate the similarity of LLBL energy spectra with spectra from the adjacent magnetosheath. Electron spectra at the 1638 UT density minimum, however, resemble spectra from the nearby magnetosphere. This crossing occurred at r =11 $R_E$, $\phi_{GSM} = 328°$ and $\lambda_{GSM} = 44°$.

$\lesssim 1$ in the inner portions of the LLBL consistent with the decay of field fluctuations. This is because, for low β, the magnetic field is dominant and is not readily perturbed by the plasma. Within the boundary layer the plasma flow directions are more variable and usually shift into a direction (for the ecliptic plane flow component) that is farther from the $X_{GSM}$-axis than the nearby magnetosheath flow direction. Some crossings show a significant cross-field velocity component; for example, the angle between the plasma flow and the field direction for the March 4, 1972 crossing shown in Figure 2 is $\gtrsim 20°$. Many cases of higher cross-field flow components have been sampled by IMP 6 in the LLBL as well as by HEOS 2 (Haerendel et al., 1978). The electron differential spectra shown at the upper right side of Figure 2 show the commonly observed similarity of electron spectra sampled on each side of the magnetopause layer. Although periods of local density minima in the LLBL can often have magnetospheric-like spectra, as shown in this example for a period near 1638 UT, electron spectra within the LLBL are often virtually indistinguishable from those of the nearby magnetosheath. This spectral similarity of plasma on opposite sides of the magnetopause layer with a gradation towards magnetospheric spectra farther into the LLBL has been noted in data from both IMP 6 and HEOS 2 (e.g., the October 10, 1973 crossing, Haerendel et al., 1978). Energetic electron pitch angle distributions are often pancake-shaped in the LLBL and adjacent magnetosphere (suggesting a closed field line region). A streaming distribution as shown is often observed in the adjacent magnetosheath.

IMP 6 plasma, energetic electron and magnetic field observations provide several results about the structure of the low latitude boundary layer and magnetopause layer in the region covered by IMP 6 as shown in Figure 1 (see Eastman and Hones, 1978):

1. In all IMP 6 crossings, some magnetosheath-like plasma is observed earthward of the magnetopause layer. The spectral intensities of LLBL electrons close to the magnetopause layer are often virtually indistinguishable from those of the adjacent magnetosheath electrons. Ion spectral intensities in the LLBL and local magnetosheath are also very similar. Many crossings show a significant magnetospheric contribution to the ion and electron spectra, especially farther earthward from the magnetopause layer and near relative density minima within the LLBL.
2. High temporal resolution (three-second average) data reveal that in 24 out of 40 IMP 6 magnetopause crossings, no distinct changes in density or electron spectra are observed at the magnetopause layer.
3. The LLBL thickness is highly variable and, generally, is much greater than the magnetopause layer thickness.
4. Nominal LLBL thickness values based on 90 IMP 6 boundary crossings show no statistically significant correlation with latitude, $K_p$, hourly averages of IMF $B_y$ or IMF $B_z$ or with the locally measured z-component of the magnetosheath magnetic field. Space and/or time variations of the LLBL thicknesses may conceal any real correlation.
5. Observed LLBL bulk plasma flow almost always has an anti-sunward component, even in the near noon region equatorward of the cusp, and often has a significant cross-field component.
6. Energetic electron (47 to 350 keV) pitch angle distributions indicate that the low latitude boundary layer is on closed field lines.

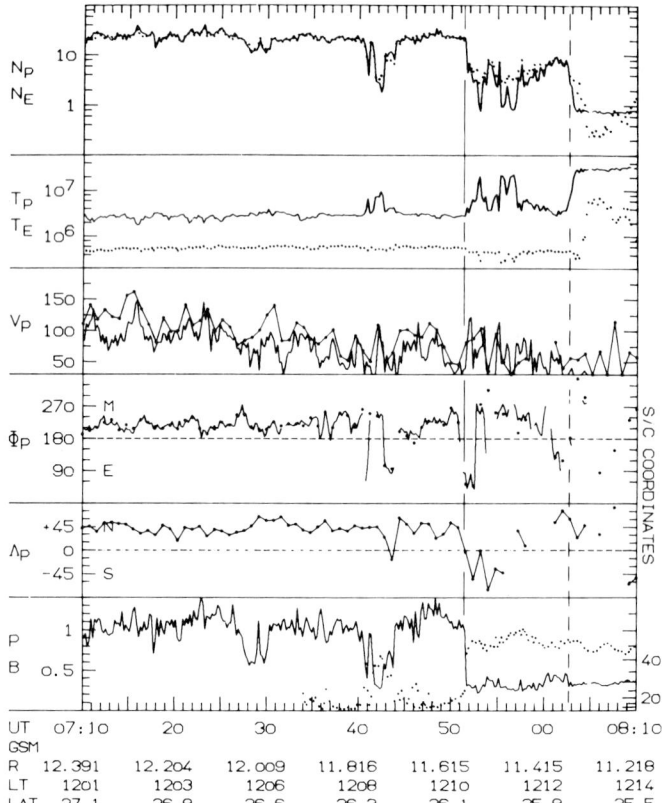

Fig. 3. Two- and three-dimensional plasma parameters and magnetic field pressure for the ISEE-1 crossing of November 3, 1977. Solar magnetospheric (GSM) coordinates are used for spacecraft positions with R (in R), local time (LT in hours) and latitude (in degrees). Solid lines denote protons and dotted lines denote electrons in the plots of density (N in cm$^{-3}$) and temperature (T in Kelvin). Bulk flow directions are given in terms of azimuth ($\Phi$) and elevation ($\Lambda$) in spacecraft coordinates (close to solar ecliptic coordinates) based on both the 2D (solid line) and 3D (dots) moment analysis. M, E, N or S denote morning, evening, northward or southward flow components, respectively. Total plasma pressure (solid line) and magnetic field pressure (dotted line) are given in units of $10^{-8}$ dynes/cm² (left scale). The dotted curve also gives the magnetic field strength (in gammas) by using the quadratic right hand scale. The magnetopause is marked by a solid vertical line and the inner surface of the plasma boundary layer is marked by a dashed line. Magnetosheath plasma is sampled prior to 0751 UT and magnetospheric plasma is sampled after 0803 UT. This figure is from Paschmann et al. (1978).

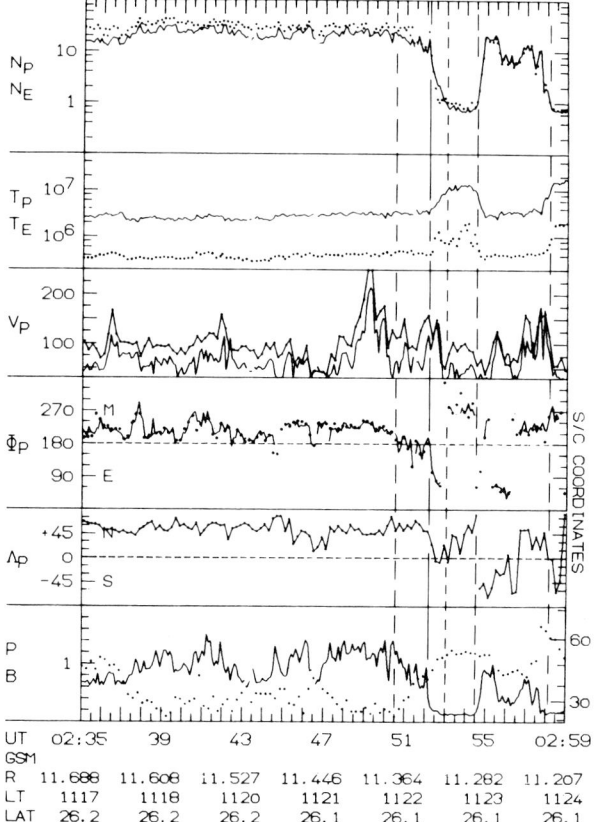

Fig. 4. Two- and three-dimensional plasma parameters and magnetic field pressure for the ISEE-2 crossing of November 8, 1977. This figure uses the same format as Figure 3 except that an isolated plasma boundary layer is marked off by vertical dashed lines at 0254:50 and 0258 UT. The magnetopause layer is centered on the vertical solid line at 0252 UT. This figure is from Paschmann et al. (1978).

These observations are generally consistent with the HEOS 2 observations. However, Haerendel et al., (1978) report that the apparent plasma flow in the LLBL opposed the external flow in approximately 25% of all crossings. Such reversed flows in the LLBL have not yet been noted in the higher time resolution IMP 6 data except for brief intervals during three crossings near the cusp region. Multiple crossings of the inner PBL surface could explain the occasional HEOS 2 observations of sunward component flow in the LLBL; however, forthcoming analysis of ISEE data should resolve this difference. Another difference is that HEOS 2 data commonly show a density plateau in the PBL with a sudden density change at the magnetopause layer and at the inner extent

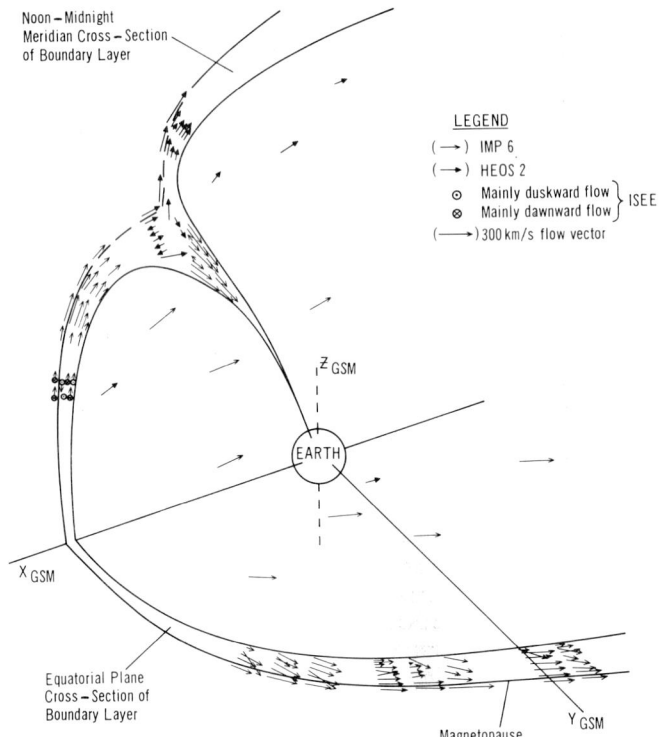

Fig. 5. Summary of plasma flow in the sunward plasma boundary layer as observed by IMP 6, HEOS 2, ISEE-1 and ISEE-2. Approximate projections of observed flow vectors are given on the noonmidnight meridian and equatorial plane cross-sections of the PBL based on satellite crossings with an earth centered angle of < 15° from the X-Z and X-Y planes, respectively. Flow vectors marked between the two cross-sections are observed PBL flow vectors projected onto the magnetopause surface. The magnetosheath bulk flow adjacent to the magnetopause was assumed to be parallel to the magnetopause surface for this plot. This drawing has been made to scale except that the PBL is enlarged ∿50%.


of the PBL. An ISEE-1 crossing on November 3, 1977 is shown in Figure 3 that clearly shows a density plateau signature. Such density plateau signatures are rarely observed by IMP 6; instead, the majority of IMP 6 crossings reveal no distinct changes in density or electron spectra at the magnetopause layer. The ISEE observations suggest a spatial change in PBL structure with increasing distance from the subsolar region when compared to the IMP 6 observations. This is further emphasized in the ISEE-2 crossing of November 8, 1977 shown in Figure 4. This crossing shows a PBL that is detached from the magnetopause layer, a situation that has not yet been clearly found in the IMP 6 data. The isolated PBL in this ISEE-2 crossing has magnetosheath-like spectra (Paschmann et al., 1978) although it is clearly located on magnetospheric field lines (see Russell and Elphic, 1978). Paschmann et al. (1978) describe a possible temporal model for these isolated PBL regions in which "plasma entry and/or transport along the boundary are 'switched' on and off." The flow parameters shown in Figure 4 during the detached PBL segment or plasma "intrusion" show a dawn-dusk flow reversal that suggests a vortex of plasma flow. Such possible vortices are evident in other segments of the ISEE PBL crossings near the noon meridian at 25° latitude. Paschmann et al. (1978) point out that the apparent PBL thickness may be produced by "temporally limited plasma entry" and/or the passage of plasma vortices.

A comparison of the ISEE and IMP 6 results shows that the PBL structure within $\sim 30°$ of the subsolar point is significantly different than its structure along the flanks farther from the subsolar region. The relationship between the available IMP 6, HEOS 2 and ISEE observations may be seen by comparing observed plasma flow directions from all four spacecraft for various crossings as shown in Figure 5. All flow vectors drawn in the PBL in the equatorial plane or noon-midnight meridian plane are based on projecting observed flow vectors based on crossings that occurred within 15° of the GSM X-Y or X-Z plane, respectively. Other flow vectors are estimated projections of observed flow vectors projected onto the magnetopause surface. Substantial changes in plasma flow directions are frequently observed in two regions: the subsolar region and the cusp regions. The remaining HLBL and LLBL regions consistently show anti-sunward flow that is closely field-aligned in the HLBL (Rosenbauer et al., 1975) and generally field-aligned in the LLBL (except near the ecliptic plane and in the subsolar region) although significant fluctuations in the observed LLBL plasma flow field are superimposed on this pattern.

## Discussion

Classic merging signatures as predicted by most steady-state reconnection models (e.g., Levy et al., 1964) are singularly absent in the HEOS 2 (Haerendel et al., 1978) and IMP 6 data (Eastman and Hones, 1978). However, Crooker (1978) has proposed a geometrically updated reconnection model that does not predict accelerated net plasma flows along the dayside magnetopause. The model envisioned by Crooker (1977) is bimodal in that it incorporates an open field line boundary region maintained by reconnection processes overlaying a PBL on closed field lines. Hones (1976) has presented a related picture that, in addition, incorporates the relationship of magnetotail processes. The closed PBL in Crooker's model would be directly exposed to the oncoming magnetosheath plasma only in regions where reconnection processes were not occurring. Over the magnetopause region sampled by IMP 6, the PBL would be normally exposed, whereas in the cusp region and the HLBL, boundary layers maintained by reconnection processes would dominate.

Reconnection could be significant, even in the subsolar region, if "patchy," forced reconnection can operate as proposed recently by Sonnerup (1978) and Schindler (1978). Patchy reconnection would not result in the unobserved large scale laminar reconnection pattern. Schindler has developed a "patchy" reconnection picture which differs in some respects from the Lemaire et al. (1978) model of impulsive penetration of solar wind irregularities into the magnetosphere. These impulsive penetration models could be used to explain the disordered flow pattern often observed by ISEE at 25° latitude near the noon meridian (Paschmann et al., 1978). Haerendel (1978) has described a picture of eddy convection (leading to localized, sporadic reconnection) to explain HEOS 2 particle and field observations in the outer polar cusp (Paschmann et al., 1976). This model could also be used to explain the highly variable flow pattern observed by ISEE (see Figures 3 and 4) in the subsolar region.

The impulsive penetration models and turbulence models emphasize the three-dimensional, non-steady-state observed character of the magnetopause and PBL. Mechanisms that provide for highly space-and-time dependent regions of magnetosheath plasma penetration into the PBL could explain many of the IMP 6 observations including the observation that the PBL thickness is highly variable and, generally, is much larger than the magnetopause layer thickness. However, although the boundary layer plasma flow has significant fluctuations compared to the nearby magnetosheath plasma flow, the flow field is still always antisunward and is generally well ordered. Further ISEE observations may show that the PBL along the flanks of the magnetosphere, when analyzed in an averaged rest frame of the moving plasma, has a turbulent flow pattern that reflects conditions at the time of magnetosheath plasma penetration of the magnetopause. If not, a diffusion process may yet provide a viable explanation if it can explain the thick PBL/thin current layer combination.

Roederer (1977) gave an excellent review in which he identified the number 2 problem of the International Magnetospheric Study as the identification of "the mechanisms for the entry of solar wind plasma into the magnetosphere." He also listed a closely related problem as problem number 1; namely, the "effect of the interplanetary magnetic field on topography, topology and stability of the magnetospheric boundary." It should be noted that the complexities involved in a study of the magnetopause and PBL, reflected in the trend towards time dependent, three-dimensional models, make it difficult to identify a unique mechanism for a given set of observations. However, the available observations place severe constraints on any viable self-consistent model for magnetopause currents. Such a model must include the effects of plasma and field on both sides of the boundary without assuming the frozen-field approximation and must incorporate the plasma boundary layer within its theoretical framework.

<u>Acknowledgments.</u> The authors thank Drs. K. Schindler, B. Sonnerup, N. Crooker and D. Fairfield for helpful comments and suggestions. They acknowledge, gratefully, the assistance of Dr. D. N. Baker in reviewing this paper. This work was performed under the auspices of the U.S. Department of Energy.

## References

Akasofu, S.-I., E. W. Hones, Jr., S. J. Bame, J. R. Asbridge, and A. T. Y. Lui, Magnetotail and boundary layer plasmas at geocentric distance of 18 $R_E$: Vela 5 and 6 observations, J. Geophys. Res. $78$, 7257, 1973.

Crooker, N. U., Explorer 33 entry layer observations, J. Geophys. Res. $82$, 515, 1977.

Crooker, N. U., The magnetospheric boundary layers: A geometrically explicit Model, J. Geophys. Res. $82$, 3629, 1977.

Crooker, N. U., Dayside merging and cusp geometry, submitted to J. Geophys. Res., 1978.

Eastman, T. E., E. W. Hones, Jr., S. J. Bame, and J. R. Asbridge, The magnetospheric boundary layer: Site of plasma, momentum and energy transfer from the magnetosheath into the magnetosphere," Geophys. Res. Lett. 3, 685, 1976.

Eastman, T. E. and E. W. Hones, Jr., Characteristics of the magnetospheric boundary layer and magnetopause layer as observed by IMP 6, submitted to J. Geophys. Res., 1978.

Haerendel, G., G. Paschmann, N. Sckopke, H. Rosenbauer, and P. C. Hedgecock, The frontside boundary layer of the magnetosphere and the problem of reconnection, J. Geophys. Res. 83, 3195, 1978.

Haerendel, G., Microscopic plasma processes related to reconnection, J. of Atm. Terr. Phys. 40, 343, 1978.

Hones, E. W., Jr., J. R. Asbridge, S. J. Bame, M. D. Montgomery, S. Singer, and S.-I. Akasofu, Measurements of magnetotail plasma flow made with VELA 4B, J. Geophys. Res. 77, 5503, 1972.

Hones, E. W., Jr., "The magnetotail: Its generation and dissipation," in Physics of Solar Planetary Environments, p. 558, ISSTP Proceedings, Boulder, CO, 1976.

Lemaire, J., M. J. Rycroft and M. Roth, Control of impulsive penetration of solar wind irregularities into the magnetosphere by the interplanetary magnetic field direction, Aeronomica Acta, No. 193, Belgian Space Aeronomy Institute, 1978.

Levy, R. H., H. E. Petschek and G. L. Siscoe, Aerodynamic aspects of the magnetospheric flow, AIAA J. 2, 2065, 1964.

Palmer, I. D., and E. W. Hones, Jr., Characteristics of energetic electrons in the vicinity of the magnetospheric boundary layer at Vela orbit, J. Geophys. Res. 83, 2584, 1978.

Paschmann, G., G. Haerendel, N. Sckopke, H. Rosenbauer, and P. D. Hedgecock, Plasma and magnetic field characteristics of the distant polar cusp near local noon: The entry layer, J. Geophys. Res. 81, 2883, 1976.

Paschmann, G., N. Sckopke, G. Haerendel, J. Papamastorakis, S. J. Bame, J. R. Asbridge, J. T. Gosling, E. W. Hones, Jr. and E. R. Tech, ISEE plasma observations near the subsolar magnetopause, submitted to Space Science Reviews, 1978.

Roederer, J. G., Proposed work on quantitative magnetospheric B-field models for the international magnetospheric study, University of Denver, June 1975.

Roederer, J. G., Global problems in magnetospheric plasma physics and prospects for their solution, Space Science Reviews 21, 23, 1977.

Rosenbauer, H., H. Grunwaldt, M. D. Montgomery, G. Paschmann, and N. Sckopke, HEOS 2 plasma observations in the distant polar magnetosphere: The plasma mantle, J. Geophys. Res. 80, 2723, 1975.

Russell, C. T. and R. C. Elphic, Initial ISEE magnetometer results: Magnetopause observations, submitted to Space Science Reviews, 1978.

Scarf, F. L., L. A. Frank and R. P. Lepping, Magnetosphere boundary observations along the IMP 7 orbit 1. Boundary locations and wave level variations, J. Geophys. Res. 82, 5171, 1977.

Schindler, K., On the role of irregularities in plasma entry into the magnetosphere, to be published, 1978.

Sckopke, N. and G. Paschmann, The plasma mantle: A survey of magnetotail boundary layer observations, J. of Atm. and Terr. Phys. 40, 261, 1978.

Sonnerup, B. U. O., Transport mechanisms at the magnetopause, invited paper presented at the AGU Chapman Conference on Magnetospheric Substorms and Related Plasma Processes, Los Alamos, NM, October, 1978.

THE MAGNETOSPHERIC BOUNDARY LAYER : A STOPPER REGION FOR A GUSTY SOLAR WIND

J. Lemaire

Institut d'Aéronomie Spatiale, 3 Avenue Circulaire,
B - 1180 Brussels, Belgium.

Abstract. Considering that the solar wind plasma density and its momentum density are inhomogeneous over characteristic distances smaller than the magnetopause diameter, filamentary irregularities or clouds of solar wind particles can dent the magnetopause surface and eventually penetrate impulsively into the magnetosphere. Gusty penetration of solar wind plasma irregularities depends on the excess of momentum density and on the orientation of the diamagnetic currents (carried by the plasma inhomogeneities) with respect to the Chapman-Ferraro currents at the magnetopause. The relatively thick Plasma Boundary Layer PBL observed in the frontside magnetopause region can be considered as a stopper region where most of the irregularities lose their excess kinetic energy by Joule dissipation of depolarization Birkeland currents flowing in and out of the polar cusp ionosphere. Expansion of the volume confining the engulfed magnetosheath-like plasma drives field aligned motion of the plasma front surfaces down into the lower cusp regions. The density inside the plasma elements decreases as a result of the expansion along interconnected magnetic field lines. The thickness and the density distribution in the Boundary Layer depends on the irregularity spectrum of the solar wind plasma interacting with the geomagnetic field at any instant of time.

## Introduction

There are many examples of relatively thick Boundary Layers formed just below or above surfaces separating two different kinds of fluids. A rather illustrative example is the Boundary Layer observed at sea surface where bubbles of air are mixed up with water when a gusty wind blows over the sea (S.A. Thorpe 1978, personal communication).

The concentration of air engulfed in this Boundary Layer decreases with depth. The thickness of this transition layer is highly variable and depends on the characteristics of the wind streaming above the surface. Different theories for the formation of this boundary layer have been advocated. Numerous experimental difficulties hinder the determination of the detailed structure for this Boundary Layer.

This reminds us of the difficulties encountered by magnetospheric physicists because of the coarse time resolution in experimental data in the study of the Boundary Layer immediately earthward of the Magnetopause surface. The theories advanced for the Magnetospheric Boundary Layer, also, are in a controversial phase of development.

Although there are some analogies between the Magnetospheric Boundary Layer described in the following section and the hydrological Boundary Layer, any closer comparison cannot be sought. The analogies should be considered only as illustrations, since the physics involved in both cases is different.

## Observations

Since detailed observations of the magnetospheric Boundary Layer are given in the paper by Eastman and Hones in this issue, it will suffice to list them briefly for later reference. More details can be found in Freeman et al. (1968), Akasofu et al. (1973), Rosenbauer et al. (1975), Paschmann et al. (1976, 1979), Eastman et al. (1976), Crooker (1977), Eastman and Hones (1978), Haerendel et al. (1978).

1) Magnetosheath-like plasma is present earthward of the magnetopause. The region where this relatively cold solar wind plasma has been observed is known as the Magnetospheric Boundary Layer. Other terms are also found in the literature referring to different parts of the magnetopause region where the observations were made. We will use like Eastman and Hones (1979), PBL as a generic term for Plasma Boundary Layer. It includes the High Latitude Boundary Layer (HLBL) and the Low Latitude Boundary Layer (LLBL) as introduced by Paschmann et al. (1976) and Haerendel et al. (1978).
2) The low latitude portion of the PBL (i.e. the LLBL) is located in a region of generally closed geomagnetic field lines.
3) The plasma density decreases progressively with depth earthward of the magnetopause.
4) The average energy of the ions in the LLBL increases progressively with depth from magnetosheath values to magnetospheric values.
5) The electron energy spectra in the LLBL are often virtually indistinguishable from those of the adjacent magnetosheath electrons.
6) The plasma flow velocity often has a significant component perpendicular to the local geomagnetic field direction. The plasma flows generally away from the subsolar point.
7) The value of the bulk flow velocity progressively decreases with depth.
8) Magnetosheath-level fluctuations in the magnetic field intensity and direction are usually present in the Boundary Layer. The standard deviation of these low frequency field fluctuations decay in the inner portions of the LLBL.
9) The thickness of the Boundary Layer is rather variable and ranges between ~ 100 km to several thousands of kilometers. There is a tendency for having thicker Boundary Layers at larger distances from the subsolar point. No correlation between the thickness or presence of a Boundary Layer and the Interplanetary Magnetic Field direction has so far been clearly identified.
10) Sometimes significant density enhancements are observed in the Boundary Layer with at least partly detached magnetosheath-like plasma regions. Fully detached intrusions of magnetosheath-like plasma have been observed by the ISEE satellites near the noon meridian at ~ 25° GSM Latitude.

It is this set of observations that theories are supposed to account for. An example from each of two general classes of theoretical models (steady state & non-steady state) are briefly reviewed in the next section.

## Steady state model

Eastman et al (1976) suggested that the magnetospheric Boundary Layer observed by IMP6 and VELA satellites along the flanks of the magnetosphere is formed by magnetosheath plasma which has diffused into the magnetosphere and which is streaming parallel to the magnetopause across closed magnetospheric field lines.

The model of Eastman et al. (1976) is based on laboratory studies illustrated in fig. 1 and reported by Baker and Hammel (1965). Indeed it has been shown experimentally that a plasma stream directed into a strong magnetic field region can easily move across the magnetic field lines, at least when the walls of the vacuum chamber are made of insulating material; in other words, when $\Sigma_p$, the transverse conductivity integrated along the magnetic field lines (Pedersen conductivity), has a sufficiently low value. On the contrary, when the walls are good conductors of electricity, Baker and Hammel (1965) show that the plasma stream does not penetrate significantly across the transverse magnetic field, but is deflected toward the $-\vec{v} \times \vec{B}$ direction, where $\vec{v}$ is the initial stream velocity and $\vec{B}$ the magnetic field intensity. This extreme case, where $\Sigma_p \cong \infty$, is abusively called the frozen-field approximation (or the ideal MHD-approximation). When $\Sigma_p$ is large, the polarization charges that induce electric fields ($\vec{E} = -\vec{v} \times \vec{B}$) at the location of the plasma stream, are rapidly neutralized by field-aligned currents and by the shorting effect of the walls. The electric polarization charges are carried continuously towards the surface of the plasma element by polarization drifts in opposite directions for the electrons and ions. Note that the resulting polarization current ($\vec{J}_p$) flows in a direction opposite to the locally induced and time dependent electric field ($\vec{E}$). As a consequence $\vec{J}_p \cdot \vec{E}$ is a negative quantity as it is the case for MHD generators (Heikkila, 1978).

When the Pedersen conductivity, however, is reduced to sufficiently low values (as in the ionosphere of the Earth), the plasma stream can penetrate a significant distance into the magnetic field region. Momentum is transferred to the walls and kinetic energy is dissipated by Joule heating in the resistive part of the circuit where field-aligned currents are closed by transverse currents.

Eastman et al. (1976) considered that the magnetosheath plasma is a similar stream of plasma moving across the geomagnetic field. The polarization electric field is perpendicular to the magnetopause. The field-aligned currents flow up and down in the high latitude ionosphere which is a load for the magnetospheric current system depicted in Fig. 3 of Eastman et al. (1976).

The initial penetration of plasma into the geomagnetic field to produce a relatively thick boundary layer is supposed to proceed by diffusive processes (e.g. Eviatar and Wolf, 1968). Note that the Boundary Layer model of Eastman et al. (1976), as well as related approaches by Coleman (1970) or Cole (1974) are steady state models. Since it implies Diffusive Penetration of solar wind particles across the magnetopause, we will call it "DP model" for future reference.

## Impulsive penetration model

At the symposium on the "Magnetopause Regions" (Amsterdam, September 1976), Lemaire and Roth proposed a non-steady state model to describe

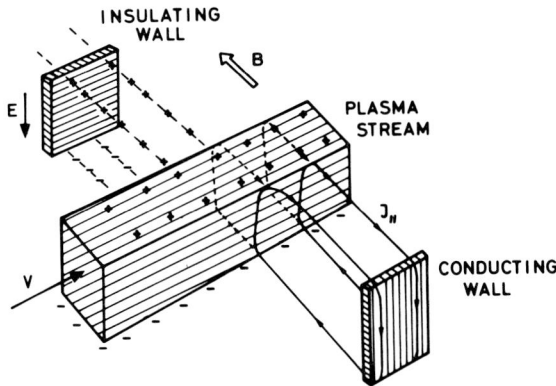

Fig. 1. Simplified model of a collisionless plasma crossing a transverse magnetic field showing (a) the charging up effect of an insulating wall and (b) the depolarizing effect of a conducting wall.

the interaction between the solar wind and the geomagnetic field. They have suggested that the solar wind is formed of small scale filaments or irregularities which are thrown into the magnetosphere because of their excess of momentum density.

The Impulsive Penetration (IP) model of Lemaire and Roth (1978) is also based on the physical principles illustrated in fig. 1. The main difference between this IP-model and the DP-model is that the latter could be active even in a steady state situation, while the former requires plasma inhomogeneities in the solar wind.

Figure 2 illustrates the penetration mechanism according to the IP model. A filamentary solar wind structure (whose equatorial cross section is represented by the hatched areas) is convected across the Bow Shock and across the Magnetopause because of its excess of momentum. The IMF is supposed to be southward to favor penetration in the magnetosphere (see Lemaire et al., 1978). The diamagnetic currents ($\vec{J}_d$) circling around the surface to maintain the total pressure balance, combine with the oppositely directed Chapman-Ferraro currents ($\vec{J}_{CF}$) to weaken the $\vec{J} \times \vec{B}$ force at the location where the filament impacts on the magnetopause. As a consequence, the plasma element is expanded and accelerated toward the inside of the magnetosphere.

Once engulfed in the region of closed magnetospheric field lines, the magnetosheath-like plasma element is slowed down by transferring its excess of momentum to the ionospheric plasma in the throat region; the excess of kinetic energy is dissipated by Joule heating in the dayside cusp ionosphere where Titheridge (1976) has found a large and well defined peak in ionospheric temperatures at 1000 km and 400 km altitude (Lemaire and Roth, 1978).

The plasma elements with the largest total momentum will be stopped deeper into the geomagnetic field than those with a smaller total momentum. The larger the total momentum the deeper the intruding plasma irregularity can penetrate into the magnetosphere before it is slowed down as described by Lemaire (1977), Lemaire and Roth (1978), Lemaire et al. (1978).

For instance a solar wind irregularity of 10000 km diameter with an original momentum excess of 5% penetrates through the magnetopause with an excess bulk flow velocity $V = 20$ km s$^{-1}$. If it carries a magnetic induction $B_i$ of 30 nT the induced convection electric field $\underline{E} = - \underline{V} \times \underline{B}_i^i = 0.6$ mV/m. If its density is 5 cm$^{-3}$, half of its kinetic energy will be dissipated in the ionosphere ($\Sigma_p = 0.2$ Siemens) before it has penetrated a distance of 2 Earth radii into the magnetosphere.

A plasma irregularity with a momentum density smaller than the average would have a smaller velocity than the magnetosheath plasma when it has traversed the Bow Shock. Therefore this plasma hole will not be able to reach the average magnetopause surface ; it will be convected around the magnetosphere at larger radial distances (i.e. in the middle of the magnetosheath layer). A sudden reduction of the average momentum density in the solar wind leads therefore to the formation of a new magnetopause at larger average radial distances from the Earth.

The dayside Boundary Layer is then considered as the stopper region of solar wind plasma irregularities of all sizes and of all momentum densities exceeding the average solar wind value.

## Discussion

Laboratory experiments by Bostik (1956) and Baker and Hammel (1965) support the idea that magnetized and non-magnetized plasma elements can actually move across large transverse magnetic fields when the conductivity is not too large somewhere along the magnetic field lines (see fig. 1). Theoretical studies by Dolique (1963) and Schmidt (1960, 1966) have confirmed this idea on which both the DP model and the Impulsive Penetration model are based.

Let us now compare the results predicted by both models with the observations reported above and made in the Boundary Layer of the dayside Magnetosphere.

1) Both models account for the presence of magnetosheath-like plasma earthward of the magnetopause.

In the DP model an unspecified diffusion process is responsible for continuous particle penetration, while in the IP model the magnetosheath plasma penetrates impulsively into the magnetosphere wherever and whenever the solar wind carries irregularities with a sufficient excess of momentum density.

The steady state DP model should work at any time even when there are no irregularities in the solar wind. Sometimes the Boundary Layer is not observed adjacent to the magnetopause or is so thin that it might have escaped detection. This raises the question of why the diffusion process in the DP model is inactive on some occasions ?

The absence of a well resolved PBL or the observation of a PBL detached from the magnetopause layer is easier to justify in the IP model by the occasional absence of small scale irregularities in the solar wind during some period of time preceeding the observation.

2) In the DP model as well as in the IP model, the PBL can be located in a region of generally closed geomagnetic field lines. However, for the DP model the magnetopause is a smooth surface permeable to particle diffusion. In the IP model the magnetopause is assumed to be an almost

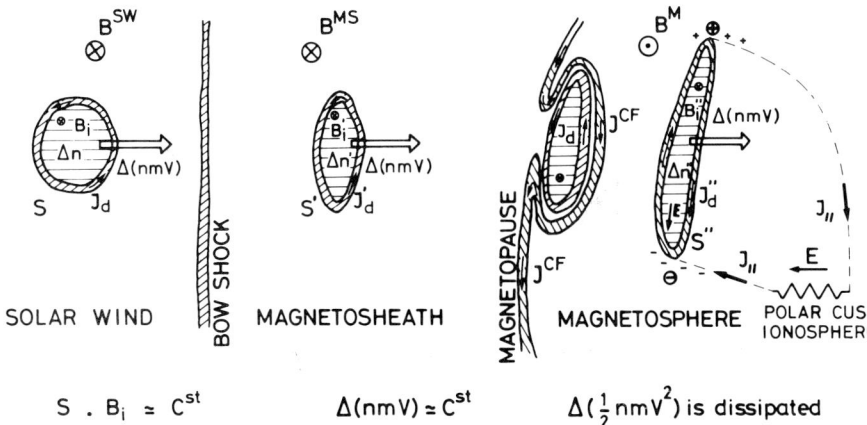

Fig. 2. Sequence of events representing the positions of a solar wind plasma irregularity penetrating through the Bow Shock and Magnetopause. S, S', S" are the successive cross sections of the intruding filament. $\vec{B}_i$, $\vec{B}_i'$, $\vec{B}_i"$ the magnetic inductions inside the element. $\vec{J}_d$, $\vec{J}_d'$, $\vec{J}_d"$ the sums of magnetisation, grad-B and curvature currents. $J_{CF}$ is the Chapman-Ferraro current density. $\vec{B}^{SW}$, $\vec{B}^{MS}$, $\vec{B}^M$ are the magnetic field intensities in the solar wind, magnetosheath and magnetosphere respectively; $\vec{E}$ is the polarisation electric field ($-\vec{V} \times \vec{B}_i$) induced in the magnetosphere by the plasma element moving with the velocity $\vec{V}$. As soon as the plasma element is engulfed in a region with finite integrated Pedersen conductivity, the excess kinetic energy of the intruding irregularity can be dissipated by Joule heating ; the excess of momentum is transfered to the ionospheric plasma in the throat region.

closed and bumpy surface with localized regions where plasma filaments hang out of the magnetosphere. In these regions magnetic flux interconnection occurs between the IMF and the geomagnetic field.
3) The plasma density decreases progressively with depth in both models. For the DP model this is a straightforward consequence of the particle diffusion mechanism. In the IP model it is due to the spreading of the plasma elements as they penetrate deeper into the magnetosphere. Indeed, when a filament is engulfed in the geomagnetic field, the magnetic field lines can cross the volume where the magnetosheath-like plasma is confined as illustrated in fig. 3 (Lemaire, 1977).

Charge separation electric fields prevent the electrons from escaping faster than the slower ions (Kan, 1975, Swift, 1975; Lemaire and Scherer, 1978). These charge separation electric fields have a parallel (field aligned) component which maintains the local and global neutrality of the plasma. Nevertheless, this $E_\parallel$-field cannot prevent the expansion of the plasma volume in directions parallel to the magnetic field. Such an expansion drives the motion of the edges of the confined plasma element towards the low altitude cusp regions where Carlson and Tobert (1977) have detected impulsive magnetosheath-like plasma precipitation. It is clear that the number density of the engulfed magnetosheath-like plasma decreases with time as a consequence of its field-aligned expansion.

Plasma filaments observed near the inner side of the magnetospheric

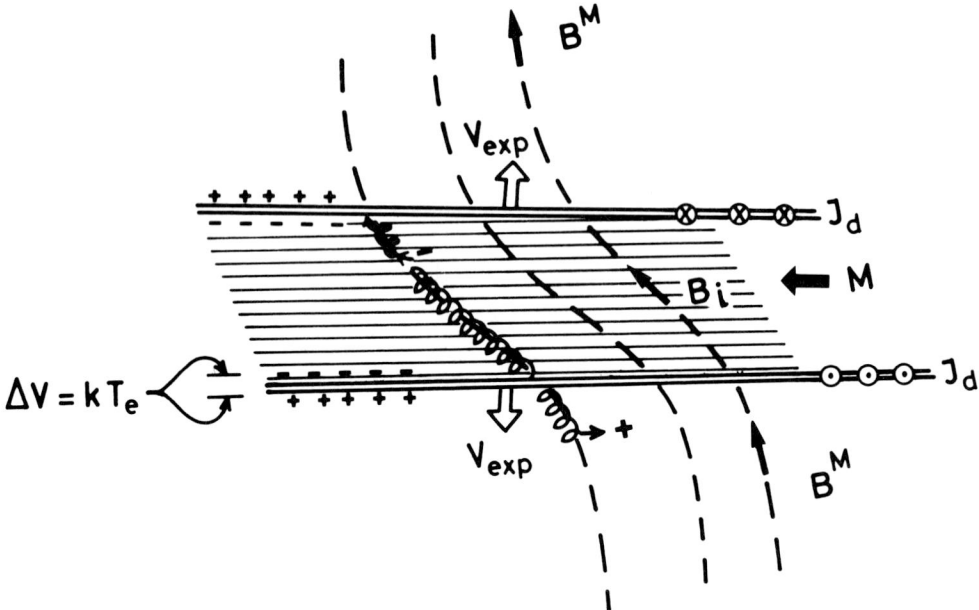

Fig. 3. Cylindrical plasma element engulfed in an external magnetic field ($\vec{B}^M$). The magnetisation ($\vec{M}$) produced by the surface currents ($\vec{J}_d$) is not parallel nor antiparallel to $\vec{B}^M$. The magnetic field induction ($\vec{B}_i$) inside the filament is not aligned with $\vec{B}^M$. The magnetic field lines (dashed lines) traverse the boundaries of the plasma element. The electrons are prevented to escape across this boundary by electrostatic potential barriers which preserve the global quasi-neutrality in the magnetosheath-like plasma element. The 100 eV - 10 keV ions can easily traverse this potential barriers and are precipitated into the polar cusp ionosphere where they have been observed with an energy dependent time dispersion by Carlson and Torbert (1978). The plasma pressure drives expansion of the volume element in directions parallel to the magnetic field lines. It is only when the plasma front surfaces have reached the low altitude cleft region that the precipitated magnetosheath electrons (confined behind the potential barrier) can be detected.

Boundary Layer are in a more advanced expansion phase than those which are closer to the magnetopause since the latter entered more recently. We think that this explains why the density in the Boundary Layer gradually decreases with depth.

There are cases that show density plateaus (usually in the subsolar region) where the density changes by steps in the Boundary Layer (Haerendel et al., 1978). The IP model can account for such observations as evidence for penetration of an extended irregularity (10000 km or more in diameter) behind the magnetopause. One can also view this as the formation of a new magnetopause at a larger average radial distance as described above.

4) The progressive increase of the average ion energy with depth (as observed in the LLBL) is associated with the decrease in the number

density of magnetosheath-like particle discussed above. Indeed when the density of the cold magnetosheath-like ions decreases compared to the hot magnetospheric particles, the average ion temperature necessarily increases. Note that magnetospheric ions can easily diffuse into and out of the engulfed filaments along interconnected magnetic field lines (see fig. 3) since their energy is much larger than the electrostatic potential (50 - 100 Volts) existing at the surface to keep the magnetosheath-like plasma globaly neutral. This explains why both magnetospheric and magnetosheath ions are detected at the same time and at the same places throughout the PBL.

5) Magnetosheath-like electrons are confined within the engulfed plasma element by the electrostatic potential barriers of 50 - 100 Volts, mentioned above. Therefore, the energy spectrum of the confined electrons remains virtually indistinguishable to the spectrum of the adjacent magnetosheath electrons.

To explain the electron energy spectra observed in the PBL with the DP model, the infered diffusion mechanism must scatter equally well electrons of any energy and of any pitch angle. Furthermore, the diffusion coefficient for all these electrons should be almost equal to the diffusion coefficient for the solar wind ions, indeed the electron Boundary Layer has nearly the same thickness as the ion Boundary Layer.

6) The observations of the plasma flow velocity in the Boundary Layer are also consistent with the theory of impulsive penetration. The bulk flow of an intruding filament can have components both perpendicular and parallel to the geomagnetic field direction. In the IP model the velocity vector is directed away from the subsolar point in agreement with the observations in the Low Latitude Boundary Layer. Note that the observed plasma bulk flow velocity is not necessarily parallel to the magnetopause as expected from the steady state DP model.

As a consequence of the field-aligned expansion mentioned above, the plasma flow velocity in the Boundary Layer can assume large values parallel and antiparallel to the local magnetic field direction.

7) As a consequence of the electrodynamic coupling between the intruding plasma elements and the polar cusp ionosphere, the excess of momentum carried by the plasma elements is transfered impulsively to the ionospheric plasma in the throat region. Repetitive action of this impulsive momentum transfer can push ionospheric plasma across the polar cap and permanently drive the well known two-cell convective flow pattern at high latitudes.

As a result of this momentum transfer, the cross-B velocity of the intruding plasma elements decreases with depth earthward of the magnetopause, in accordance with the Boundary Layer observations. The larger the integrated Pedersen conductivity, the faster the bulk flow velocity will decrease. When the transverse conductivity is infinite somewhere along the interconnected magnetic field lines, the plasma bulk flow speed decreases abruptly at the magnetopause (see, Willis, 1978).

Diffusion processes invoked in the DP model can also account for velocity shears in the Boundary Layer. In both models the predicted differential velocity structure is similar to that envisaged by Axford and Hines (1961) for a viscous-like interaction.

8) The magnetic induction ($\vec{B}$) measured inside the high $\beta$-plasma region can be quite different in intensity and direction from the external

geomagnetic field ($\vec{B}^M$) (see Lemaire, 1977). This feature is illustrated in fig. 3 where we have shown the distribution of magnetic field lines produced by a cylindrical current system whose axis of symmetry is tilted by a large angle with respect to the external magnetic field direction ($\vec{B}^M$). The diamagnetic field perturbation produced is usually large when the kinetic pressure (nkT) of the plasma is of the same order of magnitude as $B^2/2\mu_o$, the magnetic field pressure (i.e. when $\beta \gtrsim 1$). Since the plasma density and pressure decrease as a consequence of the field-aligned expansion,$\beta$ decreases and the pressure gradients driving the magnetisation currents smooth out gradually. The result is that the diamagnetic field perturbations produced by these surface currents also die out as a function of penetration depth. This corresponds to what is observed at the inner portions of the Boundary Layer where the standard deviation of low frequency magnetic fluctuations generally decay with distance from the magnetopause.

In the DP model irregular magnetic field variations should not be present or they must be interpreted as consequences of motions of the magnetopause current system. Forward and backward wavy motions of the average magnetopause as originally assumed, require large plasma bulk flow velocity components normal to the magnetopause, as well as quite large accelerations of the adjacent magnetosheath and magnetospheric plasmas. Even the recent ISEE observations probably can be interpreted in the frameworks of both alternatives.

9) The thickness of the Boundary Layer is not estimated in the DP model. In the IP model the frontside magnetospheric Boundary Layer is considered as the stopper region of all solar wind irregularities with excesses of momentum (Lemaire, 1977) and with appropriate magnetisation directions (Lemaire et al. 1978). For an excess density (and momentum density) of only 5%, a plasma irregularity of 10000 km diameter, breaking through the magnetopause with an initial speed of 20 km/sec will be stopped in about 30 minutes if the integrated Pedersen conductivity is 0.2 Siemens. During this slowing down period, the plasma center of mass has penetrated nearly 2 Earth Radii behind the average magnetopause (Lemaire, 1977). Smaller irregularities with smaller excesses of momentum are stopped in a shorter distance i.e. closer to the average position of the magnetopause. It appears therefore that the thickness of the dayside Boundary Layer is expected to be highly variable and will mainly depend on the spectral distribution of the momentum density irregularities in the solar wind.

According to Lemaire et al. (1978) the orientation of the IMF also controls access of solar wind irregularities to the interior of the magnetosphere. For instance when $B_z$ has a large southward component, penetration is greatly favored in the vicinity of the subsolar point. Furthermore, when By is positive (away IMF polarity) the post-noon quadrant of the northern hemisphere is favored as well as the pre-noon sector of the southern hemisphere. On the contrary when By < o (toward IMF polarity) the reverse is true : i.e. penetration is easier in the northern pre-noon and southern post-noon quadrants.

The lack of correlation between the IMF direction and the average thickness of the Boundary Layer may be due to the averaging over all quadrants and should be reexamined in the light of these conclusions. The presence of a Boundary Layer at certain latitudes and local times should in principle depend on the IMF direction. But the thickness is mainly dependent on the irregularity spectrum in the solar wind at any instant of time.

10) In the impulsive penetration model significant density enhancements are expected when a large scale filament has recently been injected into the Boundary Layer. It is more difficult to explain partly detached magnetosheath-like plasma-regions in the frame work of the steady-state DP model.

Conclusions

From the preceeding discussion it can be seen that both the DP-model and the IP-model for the Boundary Layer formation can account for a number of the observations. However, there are some features that only a non-steady state model (e.g. the Impulsive Penetration model) can explain satisfactorily.

It is probably not yet possible to determine definitively the relative importance of both mechanisms in transfering particles, momentum and energy from the solar wind to the magnetosphere. The relative importance of steady state or non-steady state merging processes is even more difficult to answer, since reliable description and definitions of these processes are still under debate.

Acknowledgments. The author thanks Drs. J. Eastman, E. Wipple, and M. Roth for helpful comments and suggestions.

References

Akasofu, S.-I., E.W. Hones, Jr., S.J. Bame, J.R. Asbridge, and A.T.Y. Lui, Magnetotail and boundary layer plasmas at a geocentric distance of ~18 $R_E$ : Vela 5 and 6 observations, J. Geophys. Res., 78, 7257-7274, 1973.

Axford, W.I., and C.O. Hines, A unifying theory of high-latitude geophysical phenomena and geomagnetic storms, Canad. J. Phys., 39, 1433-1464, 1961.

Baker, D.A., and J.E. Hammel, Experimental studies of the penetration of a plasma stream into a transverse magnetic field, Phys. Fluids, 8, 713-722, 1965.

Bostick, W.H., Experimental study of ionized matter projected across a magnetic field, Phys. Rev., 104, 292-299, 1956.

Carlson, C.W., and R.B. Torbert, Solar wind ion injections in the morning auroral oval, (preprint 1977).

Coleman, P., Jr., Tangential drag on the geomagnetic cavity, Cosmic Electrodyn., 1, 145-159, 1970.

Coles, K., Outline of a theory of solar wind interaction with the magnetosphere, Planet. Space Sci., 22, 1075-1088, 1974.

Crooker, N.U., Explorer 33 entry layer observations, J. Geophys. Res., 82, 515-522, 1977.

Crooker, N.U., The magnetospheric boundary layers : a geometrically explicit model, J. Geophys. Res., 82, 3629-3633, 1977.

Dolique, J.-M., Pénétration d'un faisceau neutralisé ions-électrons dans une barrière magnétique. Conditions portant sur la hauteur de la barrière, Compt. Rend. Acad. Sci. Paris, 256, 3984-3987, 1963.

Dolique, J.-M., Pénétration d'un faisceau neutralisé ions-électrons dans une barrière magnétique. Condition portant sur le gradient du champ magnétique, Compt. Rend. Acad. Sci. Paris, 256, 4170-4172, 1963.

Eastman, T.E., E.W. Hones,Jr., S.J. Bame, and J.R. Asbridge, The magnetospheric boundary layer : site of plasma, momentum and energy transfer from the magnetosheath into the magnetosphere, Geophys. Res. Letters, 3, 685-688, 1976.

Eastman, T.E. and E.W. Hones, Jr., Characteristics of the magnetospheric boundary layer and magnetopause layer as observed by IMP 6, submitted to J. Geophys. Res., 1978.

Eviatar, A., and R.A. Wolf, Transfer processes in the magnetopause, J. Geophys. Res., 73, 5561-5576, 1968.

Freeman, J.W., Jr., C.S. Warren, and J.J. Maguire, Plasma flow directions at the magnetopause on January 13 and 14, 1967, J. Geophys. Res., 73, 5719-5731, 1968.

Haerendel, G., G. Paschmann, N. Sckopke, H. Rosenbauer, and P.C. Hedgecock, The frontside boundary layer of the magnetosphere and the problem of reconnection, J. Geophys. Res., 83, 3195-3216, 1978.

Heikkila, W.J., Energy budget for the magnetosphere, (paper presented at AGU Conference on "Quantitative modelling of magnetospheric processes", La Jolla), 1978.

Kan, J.R., Energization of auroral electrons by electrostatic shock waves, J. Geophys. Res., 80, 2089-2095, 1975.

Lemaire, J., Impulsive penetration of filamentary plasma elements into the magnetospheres of the Earth and Jupiter, Planet. Space Sc., 25, 887-890, 1977

Lemaire, J., and M. Roth, Penetration of solar wind plasma elements into the magnetosphere, J. Atmos. Terr. Phys., 40, 331-335, 1978.

Lemaire, J., M.J. Rycroft, and M. Roth, Control of impulsive penetration of solar wind irregularities into the magnetosphere by the interplanetary magnetic field direction (to be published in Planet. Sp. Sci.), Aeronomica Acta A n° 193, 1978.

Lemaire, J., and M. Scherer, Field aligned distribution of plasma mantle and ionospheric plasmas, J. Atmos. Terr. Phys., 40, 337-342, 1978.

Paschmann, G., G. Haerendel, N. Sckopke, H. Rosenbauer, and P.C. Hedgecock, Plasma and magnetic field characteristics of the distant polar cusp near local noon : The entry layer, J. Geophys. Res., 81, 2883-2899, 1976.

Paschmann, G., N. Schopke, G. Haerendel, J. Papamastorakis, S.J. Bame, J.R. Asbridge, J.T. Gosling, E.W. Hones, Jr., and E.R. Tech, ISEE plasma observations near te subsolar magnetopause ; submitted to Space Science Reviews, 1979.

Rosenbauer, H., H. Grünwaldt, M.D. Montgomery, G. Paschmann, and N. Sckopke, HEOS 2 plasma observations in the distant polar magnetosphere : the plasma mantle, J. Geophys. Res., 80, 2723-2737, 1975.

Schmidt, G., Plasma motion across magnetic fields, Phys. Fluids, 3, 961-965, 1960.

Schmidt, G., Physics of High Temperature Plasmas, Academic Press, New York and London, 1966.

Swift, D.W., On the formation of auroral arcs and acceleration of auroral electrons, J. Geophys. Res., 80, 2096-2108, 1975.

Titheridge, J.E., Ionospheric heating beneath the magnetospheric cleft, J. Geophys. Res., 81, 3221-3226, 1976.

# EMPIRICAL RELATIONSHIPS BETWEEN INTERPLANETARY CONDITIONS, MAGNETOSPHERIC FLUX TRANSFER, AND THE AL INDEX

James A. Slavin and Robert E. Holzer

Department of Earth and Space Science and Institute of Geophysics and Planetary Physics, University of California, Los Angeles, California 90024

Abstract. Holzer and Slavin (1978) have found that the transfer of magnetic flux to and from the dayside magnetosphere as inferred from observed displacements of the magnetopause surface is correlated with both the magnitude of the auroral zone magnetic index AL and the incident flux of southward IMF. Empirical expressions specifying the rate at which magnetic flux is eroded in terms of interplanetary parameters and the rate of magnetic flux return as a function of AL have been developed. These relations are then used to predict magnetotail magnetic field enhancements from interplanetary and ground based data during an interval of substorm activity. The total magnetic flux in the tail is increased during intervals when the amount of flux transferred into its volume by dayside erosion exceeds the flux lost to the dayside by magnetospheric convection. Using Ogo-5 tail observations it is found for the sample events considered that these magnetic field enhancements can be described by empirical expressions for the magnetic flux transfer rates.

## Introduction

The magnetosphere relaxes into its minimum energy, or ground state, configuration when the Z component of the interplanetary magnetic field in GSM coordinates is northward for a period of several hours or more (Akasofu and Kamide, 1976; Holzer and Slavin, 1978a; and references therein). A southward IMF enhances the power input from the solar wind and alters the positions and intensities of the magnetospheric current systems (e.g. Coroniti and Kennel, 1972; Perreault, 1974). In seeking quantitative measures of the effects of interplanetary medium conditions on the magnetosphere it is more convenient to consider the changes in the magnetic fields than deal explicitly with the magnetospheric current systems. Thus, a southward flux of IMF field lines impinging on the forward magnetopause via dayside reconnection is expected to result in a decrease in the magnetospheric magnetic flux passing through the earth's magnetic equatorial plane and an increase in the magnetotail, or open, flux (e.g. Levy et al, 1964). However, this situation is not stable and magnetospheric processes act to transfer excess open flux back to a closed configuration.

Measurements of net flux transfer have been made by observing changes in the size of the polar cap (Burch, 1972; Akasofu and Kamide, 1976), the diameter of the tail and its magnetic fields (Caan et al, 1973; Maezawa, 1975), and the position of the forward magnetopause with a knowledge of solar wind dynamic pressure (Aubry et al, 1970; Holzer and Slavin, 1978a). Each of these methods has certain limitations. While the surface magnetic field is well known, the polar cap measurements are limited by polar orbiter periods, $\sim 100$ minutes, and the uncertainty in the location of the last closed field line. Observations by satellites in the tail provide a continuous measure of local tail flux density, but are hampered by lack of knowledge of the variations in tail radius. The boundary shape and magnetic fields of the forward magnetosphere are well defined, but satellites relatively infrequently encounter the magnetopause at least twice during an expansion or a contraction. In this paper empirical relationships between magnetic flux transfer as inferred from Ogo 5 magnetopause observations, the interplanetary medium, and the auroral AL index are presented (Holzer and Slavin, 1978a, b). These findings are then applied to 3 months of ground based and interplanetary observations on a time scale of hours to days and to magnetotail observations by Ogo 5 during a well documented interval of substorm activity on a shorter time scale of $\sim 10$'s of minutes as a preliminary study of their potential usefulness in modeling magnetospheric dynamics.

## Rates of Flux Erosion and Return

The rate at which magnetic flux is transferred to the tail, or eroded, is controlled predominantly by conditions in the interplanetary medium while the rate at which flux is returned from the tail is largely determined by internal magnetospheric processes. Imbalances between these two convective rates occur for a number of reasons both internal and external to the magnetosphere and result in a net transfer of flux,

$$\Delta\Phi = |\delta\Phi_e - \delta\Phi_r| \qquad (1)$$

where $\delta\Phi_e$ and $\delta\Phi_r$ are the total flux eroded and returned, respectively, in any given interval of time (Coroniti and Kennel, 1973; Holzer and Reid, 1975). When the amount of flux eroded exceeds the amount returned, a contraction of the dayside magnetosphere occurs. Similarly, expansions happen when the opposite situation is true. Under certain chance circumstances these flux transfer events are observed by satellites as multiple magnetopause encounters from which net flux transfers may be calculated (Aubry et al, 1970; Holzer and Slavin, 1978a). In doing so the solar wind dynamic pressure and/or the magnetic field intensity near the magnetopause is examined to separate the effects of compressions and rarefactions due to changing $P_{sw}$ from the contractions and expansions due to alterations in the magnetospheric current systems. With this method net transfers of about $2 \times 10^{15}$ to $2 \times 10^{16}$ Maxwells over periods of $\sim 20$ to $\sim 200$ minutes may be observed.

In 1968 and 1969 nine expansion events were observed by Ogo 5 (Holzer and Slavin, 1978a). In each case Ogo 5 was moving outbound and observed multiple crossings of the magnetopause over a distance greater than $0.5 R_e$ after any changes in solar wind dynamic pressure were taken into consideration. With a knowledge of the shape of the magnetopause and the magnetic flux density near this boundary, it was then possible to calculate the net additional magnetic flux consistent with the observed expansion of the forward magnetospheric cavity. Because the Z component of the IMF was predominantly northward during these events, the observed net flux transfers were assumed equal to the total flux returned. The flux returned was found to be correlated with the time integral of the AL index (Allen et al, 1968, 1969) over the temporal extent of the event

$$\delta \Phi_r = 1.1 \times 10^{12} \int AL \, dt \, (\gamma\text{-min}) + 4.0 \times 10^{15} \text{Mx} \qquad (2)$$

with a correlation coefficient of 93% (Holzer and Slavin, 1978b). This empirical relationship between the returned flux and the AL index suggests that the current intensity in the westward electrojet as measured by this auroral magnetic index is proportional to the rate of flux return from the tail. This finding is consistent with the various current disruption and reconnection models of tail relaxation initiated by Atkinson (1967). At this juncture both the "reconnective" (e.g. Nishida and Russell, 1978; and references therein) and "wave" (e.g. Chao et al, 1977; and references therein) models of substorms are too qualitative in nature for the results presented here to favor one or the other.

Ten contraction events were also observed by the Ogo 5 satellite during times when interplanetary observations were being made by Heos 1, Explorer 35 and/or Explorer 33. As both erosion and return of flux were significant during these events, the total flux eroded was set equal to the flux returned as estimated with the AL index and equation (2) plus the net transfer observed by the satellite. The total flux eroded was then compared with the applied flux of southward IMF field lines across the total width of the dayside magnetosphere,

$$\Phi_{ap} = \int \bar{B}_z V_{sw} W \, dt \qquad (3)$$

where $\bar{B_z}$ is the southward component of the IMF in GSM coordinates, $V_{sw}$ is the solar wind velocity, and W is the full width of the dayside magnetosphere which is assumed constant at $30R_e$. A significant correlation coefficient of 85% is found for the least square linear regression

$$\delta\Phi_e = 0.12\Phi_{ap} + 1.2 \times 10^{16} Mx \qquad (4)$$

Such a finding is consistent with dayside reconnection being the cause of the flux erosion (e.g. Levy et al, 1964; Burch, 1974). As some reconnection is expected theoretically for a weakly northward IMF (Sonnerup, 1974) it should be noted that equation (3) may be a slight underestimate of the impinging flux with the proper orientation for merging. While both the empirical relationships presented are based on a small set of data points, the correlations found are encouraging and will be further examined when more data become available from missions such as ISEE. Thus, equations (2) and (4) suggest that both $\Phi_{ap}$ and AL may be directly linked to changes in the configuration of the magnetosphere through physically reasonable empirical relationships.

## Long Term Events

During isolated intervals of geomagnetic activity it is expected that the total flux eroded and returned should be equal insofar as the periods considered begin and end with the magnetosphere in the ground state. The "isolated intervals" will be defined as periods beginning and ending with $B_z>0$ and $|AL|<20\gamma$. Accordingly, over the extent of these disturbed periods it would be expected that the time integral of the AL index will be proportional to the applied southward flux if the relationships in the preceding section are correct.

56 such intervals of activity in the first 3 months of 1969 were identified by Holzer and Slavin (1978b) and the returned flux as inferred from the AL index plotted in Figure 1 against the applied flux as determined by Heos 1 and Explorer 35. When both satellites were observing simultaneously, the average southward component of the IMF was used in computing the applied flux. The length of the intervals ranged from 4 to 126 hours with a mean of 26 hours. Six minute averages of $B_z$ in GSM coordinates were used in determining $\bar{B_z}$ for computing $\Phi_{ap}$. As shown a correlation coefficient of 94% was found along with an overall erosion efficiency of 18%. Based on both the flux transfer events observed by Ogo 5 and this statistical study, it is concluded that the average erosion efficiency, $\delta\Phi_e/\Phi_{ap}$, is $0.2\pm0.1$.

The empirical relationships presented specify the rates of erosion and return on at least a statistical basis. In

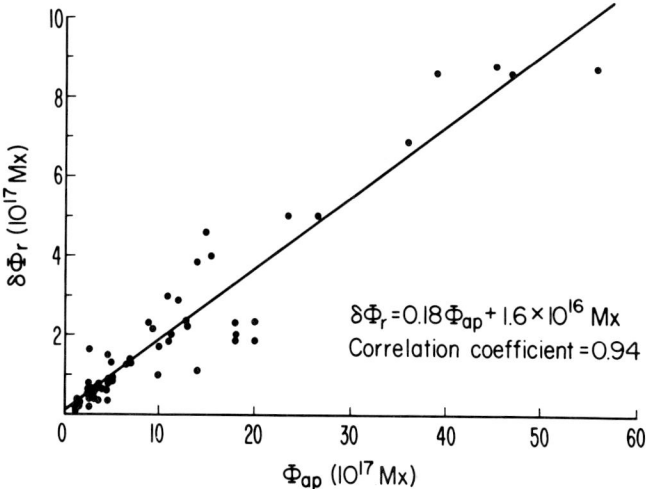

Fig. 1. The total magnetic flux returned during 56 complete intervals of geomagnetic activity as inferred from the AL index is plotted against the applied flux of southward IMF field lines.

individual cases and on time scales of less than an hour a number of problems arise with the use of such formulae. The positions of the westward electrojet and the ground observing stations are sometimes such that the electrojet current is either not observed or seriously underestimated (e.g. Akasofu and Lepping, 1977). As shown in Figure 2 the IMF orientation and magnitude observed by a single near-earth satellite is sometimes not representative of what the bulk of the magnetopause experiences. During 19 of the 56 complete intervals examined data were available from both Explorer 35 and Heos 1. However, the values of the applied flux observed by each frequently differed significantly and the correlation coefficient was a marginal 71%. It is interesting to speculate that the continued study of ground based magnetic indices may result in the ability to use the entire magnetosphere as a probe measuring the eastward, or merging, electric field in the solar wind. As such it would have the advantages of low sensitivity to small scale spatial inhomogeneities, linearity, and low cost.

### Short Term Events

As already pointed out, even if the relationships presented are physically correct, the way in which the IMF and the westward electrojet are observed may limit their use quantitatively on more than a statistical basis. However, due to the potential usefulness of having a continuous measure of the state of the magnetosphere with respect to magnetic flux transfer it is desirable to examine these em-

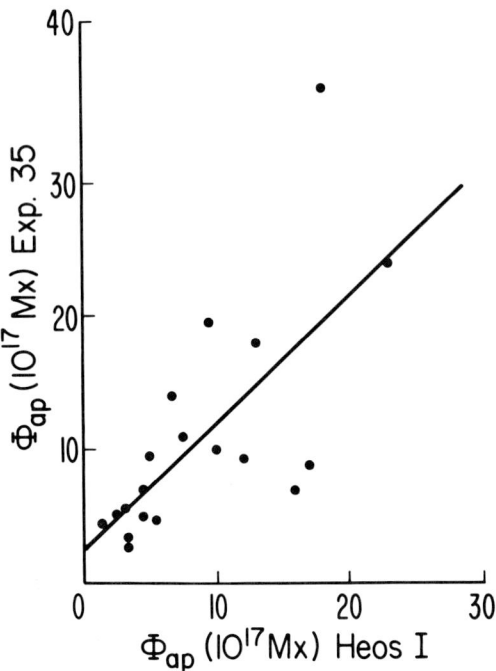

Fig. 2. During 19 complete intervals both Explorer 35 and Heos 1 were simultaneously observing the solar wind. For each case $\Phi_{ap}$ measured by Explorer 35 is plotted against $\Phi_{ap}$ observed by Heos 1.

pirical relations on a shorter time scale.

Preliminary to a more comprehensive study, an interval of substorm activity on August 9, 1968 when Ogo 5 was in the magnetotail and Explorer 34 in the interplanetary medium has been examined. The magnetic field observations by Ogo 5 shown in Figure 3 display classic tail substorm signatures (Russell et al, 1971; Russell and McPherron, 1973; Pytte and West, 1978). At the time of the observations Ogo 5 was above the expected position of the neutral sheet near the noon-midnight meridian 10-16$R_e$ down the tail on an inbound pass. On three separate occasions the tail field intensity increases while becoming less dipolar followed by an expansion of the plasma sheet over the satellite and a return to the more dipolar presubstorm configuration. In both Figures 3 and 5 solid vertical lines denote substorm onsets determined with ground based magnetic observations by Pytte and West (1978). The sudden decrease in $B_x$ following the onset(s) of each substorm is due to the diamagnetic effect of the plasma sheet engulfing the satellite.

Using the empirical relations presented and observations of the interplanetary medium and the AL index in 12 minute averages, the rates of flux erosion and return have been cal-

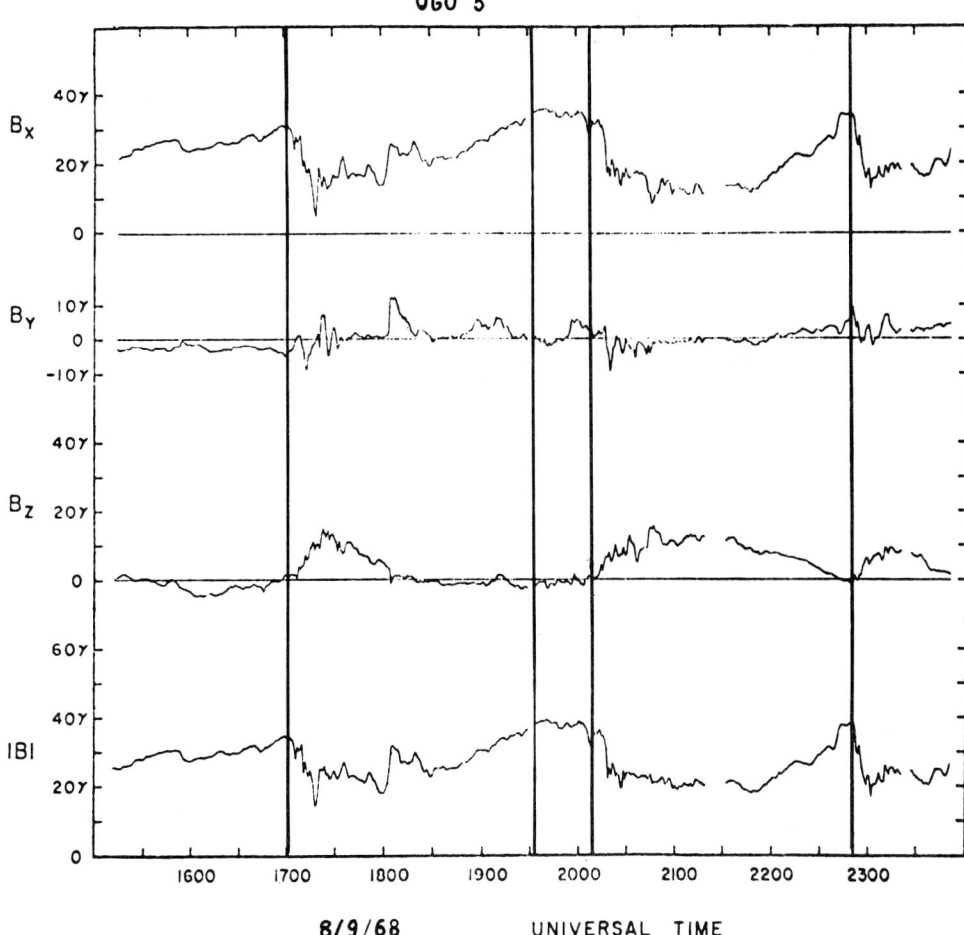

Fig. 3. Ogo 5 magnetic field observations in the tail in GSM coordinates during an interval of substorm activity.

culated and plotted for this period of time. The rate of erosion is then

$$\frac{d\Phi_e}{dt}(\frac{Mx}{sec}) = 0.2(1.9 \times 10^{10} cm) B_z^-(\gamma) V_{sw}(km/sec) \quad (5)$$

and the rate of return

$$\frac{d\Phi_r}{dt}(\frac{Mx}{sec}) = 1.8 \times 10^{10} AL(\gamma) \quad (6)$$

As shown in Figure 4 each of the three substorms is preceded by southward IMF and erosion. About 30-60 minutes after the southward turning of the IMF enhanced flux return as inferred from the AL index begins. The total magni-

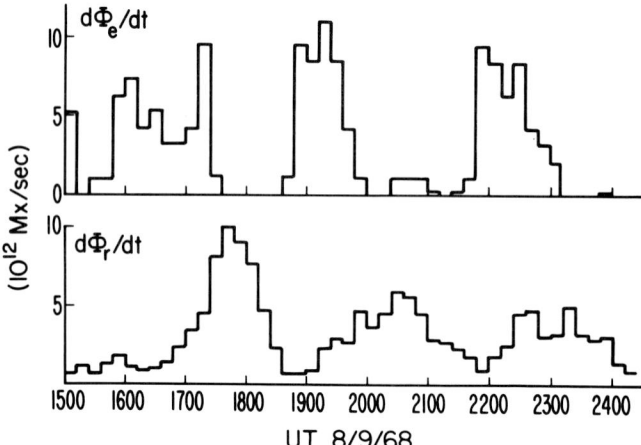

Fig. 4. Rates of flux erosion and return as a function of time inferred from $V_{sw}B_z^-$ observations and the AL index.

tudes of the inferred flux erosion and return for each substorm agree to within 10-25%.

By subtracting the flux returned from the amount eroded in Figure 4 it is possible to predict the net transfer to the tail. The predicted tail lobe flux as a function of time is set equal to the observed lobe flux at 1500 UT, calculated under the assumption of cylindrical tail geometry with a constant $20R_e$ radius, plus the net flux transfer inferred from Figure 4. As equations (5) and (6) describe transfer rates for the forward magnetosphere, a 24 minute shift toward earlier times in the bottom plot of Figure 5 has been made to allow for convection propagation time between the tail and the forward magnetosphere (e.g. Rostoker et al, 1972). In considering Figure 5 it must be noted that the observed tail flux, assumed to be $\frac{\pi}{2}B_xR_T$, seriously underestimates the actual lobe flux during the times that Ogo 5 is in the plasma sheet, approximately ∿1705 to ∿1800 and ∿2015 to ∿2230 as indicated by onboard particle instruments. As shown in Figure 5 the predicted flux as a function of time is in at least qualitative agreement with the in situ observations although the predicted amplitude is up to ∿30% smaller than what is observed. The agreement is in fact very good considering the simplicity of the relations used and the assumptions made. Better fits to the observations are possible if small adjustments are made in the lag times between return and erosion of flux, but doing so at this time would be arbitrary without a study of many such tail events. Hence, an assessment of the quantitative precision possible with empirical expressions of the type presented must await the conclusion of a more comprehensive study.

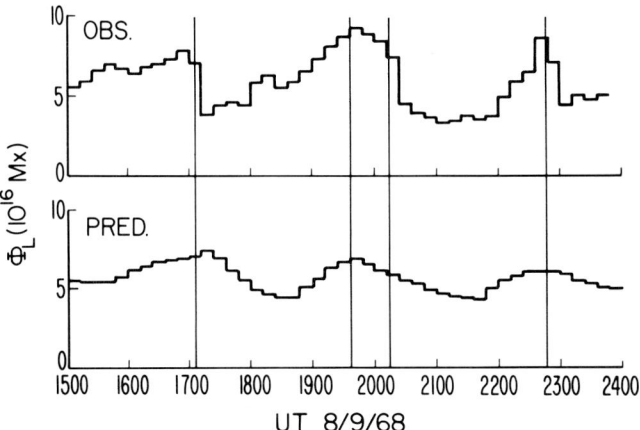

Fig. 5. The upper plot depicts the observed flux in the northern tail lobe for a cylindrical tail, $\Phi_L = \frac{\pi}{2} R_t^2 B_x$. In the lower frame the lobe flux predicted from the inferred transfer rates in Figure 4 is displayed as a function of time.

## Discussion

Many correlative studies of how geomagnetic activity relates to interplanetary parameters have been conducted over the years (e.g. Hirshberg and Colburn, 1969; Arnoldy, 1971; Garrett et al, 1974a; and references therein). The goal of such studies is usually to identify those solar wind conditions which correlate best with the occurrence and intensity of magnetic disturbances at the earth's surface and then to suggest physical processes which might explain the observations. However, the inference of magnetospheric processes from these correlations is severely limited by the statistical interdependence among the various solar wind parameters and lack of quantitative knowledge concerning the magnetospheric sources of ground observed "geomagnetic activity."

An alternate approach to this problem is to assume a physical model by which interplanetary conditions affect magnetospheric processes and then compare quantitative modeling predictions with observations. In particular, this method has been used by Perreault (1974), Burton et al (1975), and Akasofu (1978) in studying magnetospheric energy dissipation, the ring current and the AE index respectively. The work reported in this paper also adopts this method of investigating magnetospheric processes. Holzer and Slavin (1978a) made measurements of magnetic flux transfer in the magnetosphere. The Dungey (1961) concept of reconnection in magnetospheric dynamics requires $\delta\Phi_e$ to be determined predominantly by $\Phi_{ap}$. Holzer and Slavin (1978a,b) found $\delta\Phi_e$ and $\Phi_{ap}$ to be in fact correlated with the best linear regression given in equation (4). Similarly, the current disruption model of magnetotail relaxation predicts an enhancement of the westward electrojet as the tail flux decreases. A correlation between $\delta\Phi_r$ and $\int AL dt$, where the AL index introduced by Davis and Sugiura (1966) is designed to be a measure of the maximum westward electrojet current density, was sought and found. The empirical relationships found have then been used to predict both long and short term flux transfer which are compared with observations.

In developing even simple empirical models it is necessary to consider the ways in which both interplanetary conditions and westward electrojet intensity are inferred from observation. As discussed in a number of papers (e.g. Garrett, 1974b), the sparse network of auroral zone magnetic observatories can on occasion seriously underestimate electrojet current strength due principally to the latitudinal motion of the current accompanying changes in the configuration of the magnetosphere. In addition to being physically reasonable, the use of time integrals of the AL index minimizes this problem and provides a more uniform measure of electrojet intensity than using AL index values at a single given time or "peak" values. Similarly, as shown in Figure 2 and discussed in other studies (e.g. Pytte, 1978; Perrealut and Kamide, 1978) the values of $\Phi_{ap}$ measured by different earth orbiting satellites can differ significantly. It should also be noted that hourly, or even longer averages of the north-south IMF component are often used (e.g. Akasofu, 1978) in place of the average southward magnetic field in correlations and models involving dayside reconnection. The two values are equivalent only during intervals with no periods of northward IMF as is often not the case. Because the IMF changes north-south polarity on time scales of minutes to hours, the use of hourly averages of $B_z$ as opposed to $\overline{B_z}$, determined with $B_z$ averages over much less than an hour, can result in relations and correlations that are accurate when the IMF orientation is southward and steady, but fail when the field changes north-south direction more frequently. Ultimately, the successful understanding and modeling of magnetospheric processes will depend at least in part on the ability to quantitatively measure both the interplanetary environment of the magnetosphere and the intensity of the individual ionospheric-magnetospheric current systems.

As quantitative modeling of the magnetosphere and its processes progresses it becomes necessary to have a continuous measure of the state of the system with respect to magnetic flux transfer, or equivalently, energy. One example of this requirement is the work of Chanteur et al (1977) on high energy electron drift echoes observed at geostationary orbit. In that study empirical relations were used to infer magnetopause stand-off distance from the $K_p$ index and tail magnetic field intensity from the AE index for the purpose of studying the effect of changing magnetospheric configuration on particle drift shells. In fact, the understanding of many substorm related processes might be furthered by an index describing the state of the magnetosphere. Such an index must be based on both magnetospheric and interplanetary parameters. The empirical expressions presented suggest the possibility of creating such an index which could then be used formulating time dependent magnetospheric models.

Acknowledgments. The authors are pleased to acknowledge the use of interplanetary medium data sets prepared by the space physics group at the UCLA Institute of Geophysics and Planetary Physics. We are also indebted to the co-investigators Paul J. Coleman, Jr., Thomas A. Farley, and Darrell L. Judge for providing us with Ogo 5 fluxgate magnetometer data. The research reported in this paper was supported by National Science Foundation grant ATM 75-01431 and National Aeronautics and Space Administration grant NGR 05-007-276. Institute of Geophysics and Planetary Physics publication 1869.

## References

Akasofu, S.-I., Interplanetary energy flux associated with magnetosphere substorms, preprint, Geophysical Institute, University of Alaska, 1978.

Akasofu, S.-I., and Y. Kamide, Substorm energy, Planet Space Sci., 24, 223, 1976.

Akasofu, S.-I., and R.P. Lepping, Interplanetary magnetic field and magnetospheric substorms, Planet. Space Sci., 25, 895, 1977.

Allen, J.H., C.C. Abston, and L.D. Morris, Auroral Electrojet Magnetic Activity Indices, World Data Center Report UAG-29, Boulder, Colorado, 1968.

Allen, J.H., C.C. Abston, and L.D. Morris, Auroral Electrojet Activity Indices, World Data Center Report UAG-31, Boulder, Colorado, 1969.

Arnoldy, R.L., Signatures in the interplanetary medium for substorms, J. Geophys. Res., 76, 2189, 1971.

Atkinson, G., An approximate flow equation for geomagnetic flux tubes and its application to polar substorms, J. Geophys. Res., 72, 5373, 1967.

Aubry, M.P., C.T. Russell, and M.G. Kivelson, Inward motion of the magnetopause before a substorm, J. Geophys. Res., 75, 7018, 1970.

Burch, J.L., Rate of erosion of dayside magnetic flux based on a quantitative study of the dependence of polar cusp latitude on the IMF, Radio Sci., 8, 955, 1973.

Burch, J.L., Observations of interactions between interplanetary and geomagnetic fields, Rev. Geophys. Space Phys., 12, 363, 1974.

Burton, R.K., R.L. McPherron, and C.T. Russell, An empirical relationship between interplanetary conditions and Dst, J.Geophys.Res., 80, 4204, 1975.

Caan,M.N., R.L.McPherron, and C.T.Russell, Solar wind and substorm-related changes in the lobes of the geomagnetic tail, J.Geophys.Res., 78, 8087, 1973.

Chanteur,G., R.Gendrin, and S.Perraut, Experimental study of high-energy electron drift echoes observed on board ATS 5, J.Geophys.Res., 82, 523, 1977.

Chao,J.K., J.R.Kan, A.T.Y.Lui, and S.-I. Akasofu, A model for thinning of the plasma sheet, Planet.Space Sci., 25, 703, 1977.

Coroniti,F.V., and C.F.Kennel, Changes in magnetospheric configuration during the substorm growth phase, J.Geophys. Phys., 77, 3361, 1972.

Coroniti,F.V., and C.F.Kennel, Can the ionosphere regulate magnetospheric convection?, J.Geophsy.Res., 78, 2837, 1973.

Davis,T.N., and M.Sugiura, Auroral electrojet activity index AE and its universal time variations, J.Geophys. Res., 71, 785, 1966.

Dungey,J.W., Interplanetary magnetic field and the auroral zones, Phys.Rev.Letts., 6, 37, 1961.

Garrett,H.B., A.J.Dessler, and T.W.Hill, Influence of solar wind variability on geomagnetic activity, J.Geophys.Res., 79, 4603, 1974a.

Garrett,H.B., The role of fluctuations in the IMF in determining the magnitude of substorm activity, Planet. Space Sci., 22, 111, 1974b.

Hirshberg,J., and D.S.Colburn, Interplanetary field and geomagnetic variations - A unified view, Planet.Space Sci., 17, 1183, 1969.

Holzer,T.E., and G.C.Reid, The response of the dayside magnetosphere-ionosphere system to time-varying field line reconnection at the magnetopause, J.Geophys.Res., 80, 2031, 1975.

Holzer,R.E., and J.A.Slavin, Magnetic flux transfer associated with expansions and contractions of the dayside magnetosphere, J.Geophys.Res., 83, 3831, 1978a.

Holzer,R.E., and J.A.Slavin, A correlative study of magnetic flux transfer in the magnetosphere, In press, J. Geophys.Res., 1978b.

Levy,R.H., J.E.Petschek, and G.L.Siscoe, Aerodynamic aspects of the magnetospheric flow, AIAA J., 2, 2065, 1964.

Maezawa,K., Magnetotail boundary motion associated with geomagnetic substorms, J.Geophys.Res., 80, 3543, 1975.

Nishida,A., and C.T.Russell, On the expected signatures of reconnection in the magnetotail, J.Geophys.Res., 83, 3890, 1978.

Perreault,P.D., On the relationship between IMF and magnetospheric storms and substorms, Ph.D. thesis, University of Alaska, 1974.

Perreault,P.D., and Y.Kamide, Reply, J.Geophys.Res., 83, 2709, 1978.

Pytte,T., Comment on "A dusk-dawn asymmetry in the response of the magnetosphere to the IMF $B_z$ components" by P.D. Perreault and Y.Kamide, J.Geophys.Res., 83, 2707, 1978.

Pytte,T., and H.I.West, Ground-satellite correlations during presubstorm magnetic field configuration changes and plasma sheet thinning in the near-earth magnetotail, J.Geophys.Res., 83, 3791, 1978.

Rostoker,G., H.-L.Lam, and W.D.Humer, Response time of the magnetosphere to the interplanetary electric field, Can. J.Phys., 50, 544, 1972.

Russell,C.T., R.L.McPherron, and P.J.Coleman,Jr., Magnetic field variations in the near geomagnetic tail associated with weak substorm activity, J.Geophys.Res., 76, 1823, 1971.

Russell,C.T., and R.L.McPherron, The magnetotail and substorms, Space Sci.Rev., 15, 205, 1973.

Sonnerup,B.V.O., Magnetopause reconnection rate, J.Geophys.Res., 79, 1546, 1974.

STATISTICAL STUDY OF THE DEPENDENCE OF GEOMAGNETIC ACTIVITY ON SOLAR
WIND PARAMETERS

Kiyoshi Maezawa

Institute of Space and Aeronautical Science, University of Tokyo
Komaba, Meguro-ku, Tokyo 153, Japan

Abstract. Dependence of geomagnetic activity on various solar wind
parameters is examined statistically using three geomagnetic disturbance indices, (1) dayside auroral zone disturbance index, AU, (2)
nightside auroral zone disturbance index, AL, and (3) subauroral zone
disturbance index, am. It is found that the dependence on the solar
wind parameters is different for different geomagnetic indices, qualitatively as well as quantitatively. The main findings are: (1) AL and
am are almost proportional to $V^2$, where V is the solar wind velocity,
while AU is almost linear to V. (2) In spite of this difference,
dependence on the magnitude of the IMF, B, is almost the same for the
three indices. (3) The solar wind density has a strong effect on am
but has little effect on AL. The magnitude of the solar wind density
effect on am is little affected by the IMF latitude angle. (4) Effect
of the variability of the IMF is seen for all the indices but the
effect is the largest on am. These findings are discussed in terms of
a composite model of the solar wind-magnetosphere interaction mechanisms.

## Introduction

Geomagnetic activity observed on the ground represents a best
available tool for estimating quantitatively the amount of the solarwind energy that flows into the magnetosphere. This is because the geomagnetic activity has been constantly monitored for many years worldwide and has been represented in the form of several standard geomagnetic indices. Thus a number of authors have studied the dependence
of geomagnetic indices on the solar wind parameters to obtain a clue to
identify the solar-wind-magnetosphere interaction mechanism. It has
been established through these studies that geomagnetic activity is
under strong influence of the southward component of the IMF ( interplanetary magnetic field). This fact has been taken as evidence that
the main mechanism of the solar wind energy transfer to the magnetosphere is the magnetic field reconnection (Dungey, 1961).

There have been continuing efforts to correlate geomagnetic activity
with other solar wind/IMF parameters. For example, dependence of the
geomagnetic activity on the solar wind velocity, V, has been established
(Snyder et al., 1963; Murayama and Hakamada, 1975). The dependence on
the IMF variability, σ, has also been reported (Ballif et al., 1967;
Garrett et al., 1974). The functional form of the dependence on V and

the existence of the σ effect have been a matter of discussion (Hirshberg and Colburn, 1969; Garrett et al., 1974; Murayama and Hakamada, 1975; Burton et al., 1975). Although there has been little indication that the substorm activity as represented by the AE index is correlated with the solar wind density, Svalgaard (1977) recently showed in a conclusive way that the world-wide index am is influenced by the solar wind density.

There would be at least two reasons why the analyses made so far led to diverse results on the dependence of geomagnetic activity on the solar wind parameters. Firstly, different geomagnetic indices have been used by different authors. Since there may be more than one mechanism by which the solar wind energy is transferred to the magnetosphere, different magnetic indices may represent different members or aspects of several interaction mechanisms. Secondly, solar wind parameters are generally intercorrelated among themselves (Wilcox et al., 1967; Hirshberg and Colburn, 1969). This means that careful statistics using a large number of data would be required to separate out the effect of each solar wind parameter.

Taking into consideration the situation described above, we have performed a statistical analysis of the dependences of three different geomagnetic indices on several solar wind/IMF parameters using about 10 years' data of the solar wind observation (Maezawa, 1978). Conclusion has been reached that the effects of the solar wind parameters are indeed different for different geomagnetic indices.

In this paper, these results are reviewed, with emphasis on different functional dependences of the AU and AL indices on the solar wind velocity. On the basis of these results, geomagnetic indices are normalized for solar-wind velocity and IMF-magnitude dependences, and the effects of the solar wind density and IMF variance are examined in detail.

## Data Analyzed

To represent geomagnetic activity, we utilize three indices AU, AL, and am, of which the first two are measures of auroral zone geomagnetic activity (associated mainly with substorms) and the last one is the index constructed by Mayaud (1968) to improve the planetary disturbance index ap (Kp) taking account of the inhomogeneity of the station distribution. These geomagnetic indices represent geomagnetic disturbances registered at different regions on the surface of the earth. The AL and AU indices are mainly contributed by the westward and eastward electrojets, which flow in the nightside and dayside auroral zones, respectively. On the other hand, the am index is calculated from the subauroral zone magnetic records, where direct influence of the above electrojets is smaller, so that the index is more sensitive than AL and AU to other world-wide geomagnetic variations.

The solar wind and IMF data used are about 10 years' data set of three-hourly values calculated from original hourly data provided by WDC-A, Rockets and Satellites. Five solar wind/IMF variables, namely, the IMF magnitude B, the sine of the IMF latitude angle $\theta$, the IMF relative variability $\sigma/B$ where $\sigma$ is the IMF variance within three hours, the solar wind velocity V, and the solar wind density n are selected as key parameters. The three hour averaged value of $\sin \theta$ is defined as $\sin \theta \equiv B_z/B$, where $B_z$ is the three-hour averaged solar-magnetospheric z (north-south) component of the IMF. The IMF variance $\sigma$ is defined as

$\sigma^2 = \sigma_x^2 + \sigma_y^2 + \sigma_z^2$ in terms of the variances of the three orthogonal components.

In a preliminary stage of analysis, we noted that the solar wind proton temperature T also had a strong correlation with geomagnetic activity. However, when the partial correlation analysis was applied, the above correlation was found to be merely an apparent one arising from the correlation of T with other solar wind parameters. Hence we will not consider T as a key parameter.

### Dependence of Geomagnetic Indices on the IMF Magnitude and the Solar Wind Velocity

We will first try to separate out the dependence of geomagnetic indices on the IMF magnitude B and the solar wind velocity V. In order to eliminate the effects of possible intercorrelation among the solar wind parameters, the whole data set is divided into $4 \times 4 \times 4 \times 3 = 192$ subsets, namely into 4 bins of B, 4 bins of V, 4 bins of $\sin\theta$, and 3 bins of $\sigma/B$. (Solar wind density effect will be discussed later) The ranges of parameter values corresponding to these bins are listed in Table 1. These ranges are determined in such a way that each bin contains approximately the same number of data. Each bin will be designated by the abbreviation of the parameter and by the bin number as V1, ST4, S2, etc.

In Fig. 1, averaged values of AU (left panel) and AL (right panel) are plotted against B for various ranges of V and for a fixed range of $\theta (-0.3 < \sin\theta < 0)$. For the sake of convenience the sign of AL is reversed i.e. taken positive here. Different characters are used for plotting the points corresponding to different bins of V, as indicated in the upper left corner of each panel. Dashed lines and solid lines are for the S1 and S3 bins of the IMF relative variability, respectively. (The result for the S2 bin is not given here to avoid further complexity.) It is seen that both AU and AL show almost linear dependences on B. The dependence on the solar wind velocity can be recognized from the fact that the higher positions are occupied in the graph by the lines corresponding to higher solar wind velocities. Dashed lines are gener-

TABLE 1. Range of Parameters Corresponding to Each Bin

| SW/IMF Parameters | Abbreviation | Bins | | | |
|---|---|---|---|---|---|
| | | 1 | 2 | 3 | 4 |
| $B(\gamma)$ | B | $B < 4.5$ | $4.5 \leq B < 5.7$ | $5.7 \leq B < 7.5$ | $7.5 \leq B$ |
| V(km/s) | V | $V < 360$ | $360 \leq V < 420$ | $420 \leq V < 480$ | $480 \leq V$ |
| $\sin\theta$ | ST | $\sin\theta < -0.3$ | $-0.3 \leq \sin\theta < 0$ | $0 \leq \sin\theta < 0.3$ | $0.3 \leq \sin\theta$ |
| $\sigma/B$ | S | $\sigma/B < 0.4$ | $0.4 \leq \sigma/B < 0.6$ | $0.6 \leq \sigma/B$ | |

B: Magnitude of the IMF
V: Solar wind velocity
$\theta$: Latitude angle of the IMF
$\sigma$: Variance (within three hours) of the IMF

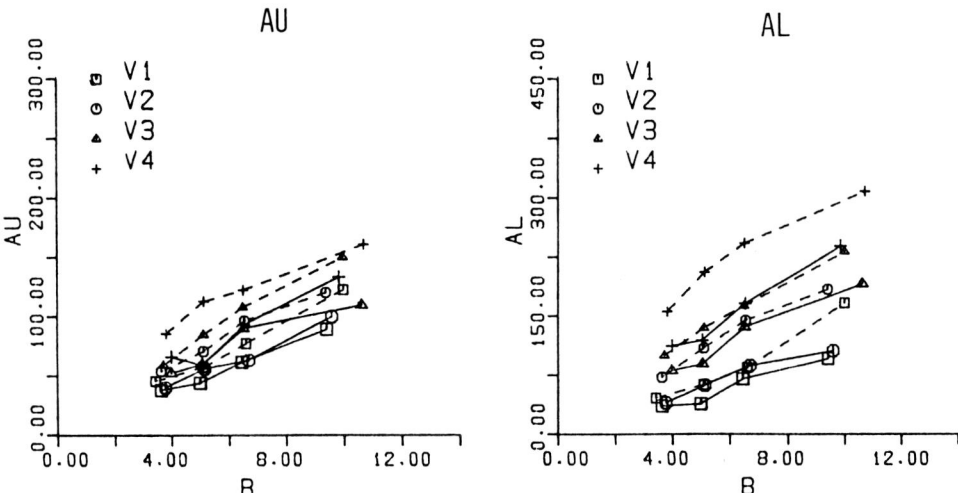

Fig. 1. Dependence of AL and AU indices on the IMF magnitude, B, for a fixed range of $\sin\theta$ ($-0.3 < \sin\theta < 0$). Different characters used for the plot represent different ranges of the solar wind velocity, V (see Table 1). Solid lines and dashed lines correspond to different ranges of the IMF variability $\sigma/B$ (see text).

ally situated at higher levels than the solid lines with the same symbol, indicating that the parameter $\sigma/B$ also affects the geomagnetic activity. The same characteristics have been observed for other ranges of $\sin\theta$ and for the am index as well (Maezawa, 1978). We may also note a slight tendency that for the uppermost line (dashed line for V4) the rate of increase of AL and AU declines with increasing B. This reflects the general tendency that AU and AL have what we may call the saturation effect, i.e., their rate of increase with any solar wind parameter is suppressed when geomagnetic activity as represented by these indices becomes very high (Maezawa, 1978). With the exception of this effect the basic relationship between each of these indices and B seems linear.

The dependence of AU and AL on V is explicitly shown in Fig. 2 using the same format as Fig. 1 except that the different lines in each panel now correspond to different ranges of B and $\sigma/B$. At first sight the dependence on V looks similar to the dependence on B examined earlier. However, a careful examination of the figure would show that the slope of the curves is much steeper for AL than for AU. As a matter of fact, if we extrapolate the curve linearly toward low velocity values, AL curves tend to cross the V-axis around the value $V \simeq 200$ km/s, while extrapolation of AU curves passes near the origin or even crosses the AU-axis. Alternatively, if we fit these curves by a power law AL (or AU) $\propto V^n$, it can be shown that $n \simeq 2$ for AL and $n \simeq 1$ for AU (Maezawa and Nishida, 1977, see the discussion below) The dependence of the am index on the solar wind velocity is almost identical to that of AL with $n \simeq 2$.

If AU and AL are both linear to B but have different dependences on V as discussed above, the ratio AL/AU would have a distinct character that it depends on V, but not on B. This expectation is confirmed in

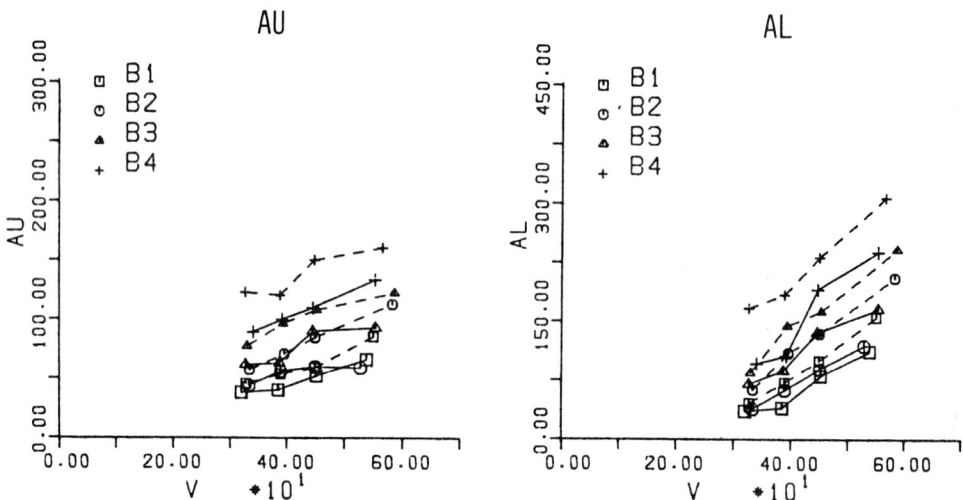

Fig. 2. Dependence of AL and AU indices on the solar wind velocity, V, for the range $-0.3 < \sin\theta < 0$. Different characters used for the plot correspond to different ranges of the IMF magnitude B.

Fig. 3, where the ratio AL/AU is plotted against B in the lower panels and against V in the upper panels. The result is presented for all the ranges of $\sin\theta$ (ST1 ~ ST4 from left to right). In the lower panels, each curve is almost parallel to the B-axis, i.e., the ratio AL/AU is not systematically dependent on B. On the other hand, in the upper panels, a linear dependence of the ratio AL/AU on V is evident. Overlapping of the lines in these panels confirm that the ratio is not dependent on B. Thus we may conclude that approximately AU $\propto$ BV and AL $\propto$ BV$^2$.

In order to further substantiate the above conclusion, and to evaluate the relative magnitude of the dependences on other solar wind parameters, we will utilize a standard statistical technique below. Generally if a statistical variable Y is a function of several statistical variables $X_i$ in the form

$$Y = \prod_i (X_i)^{P_i} \quad (1)$$

where $\prod$ denotes the multiplication of all the terms, the relationship can be reduced to a linear expression

$$\log Y = \sum_i P_i \log X_i \quad (2)$$

in terms of logarithm of variables. Thus the exponents $P_i$ can be evaluated in general by a linear regression analysis of log Y on log $X_i$. An advantage of this method is that by applying a <u>partial</u> regression analysis, the complicated effect of intercorrelation among $X_i$'s is reduced automatically. This method is used to obtain the best power law fit to the dependence of geomagnetic indices on the solar wind parameters.

In Table 2 we list the values of exponents obtained in this way for

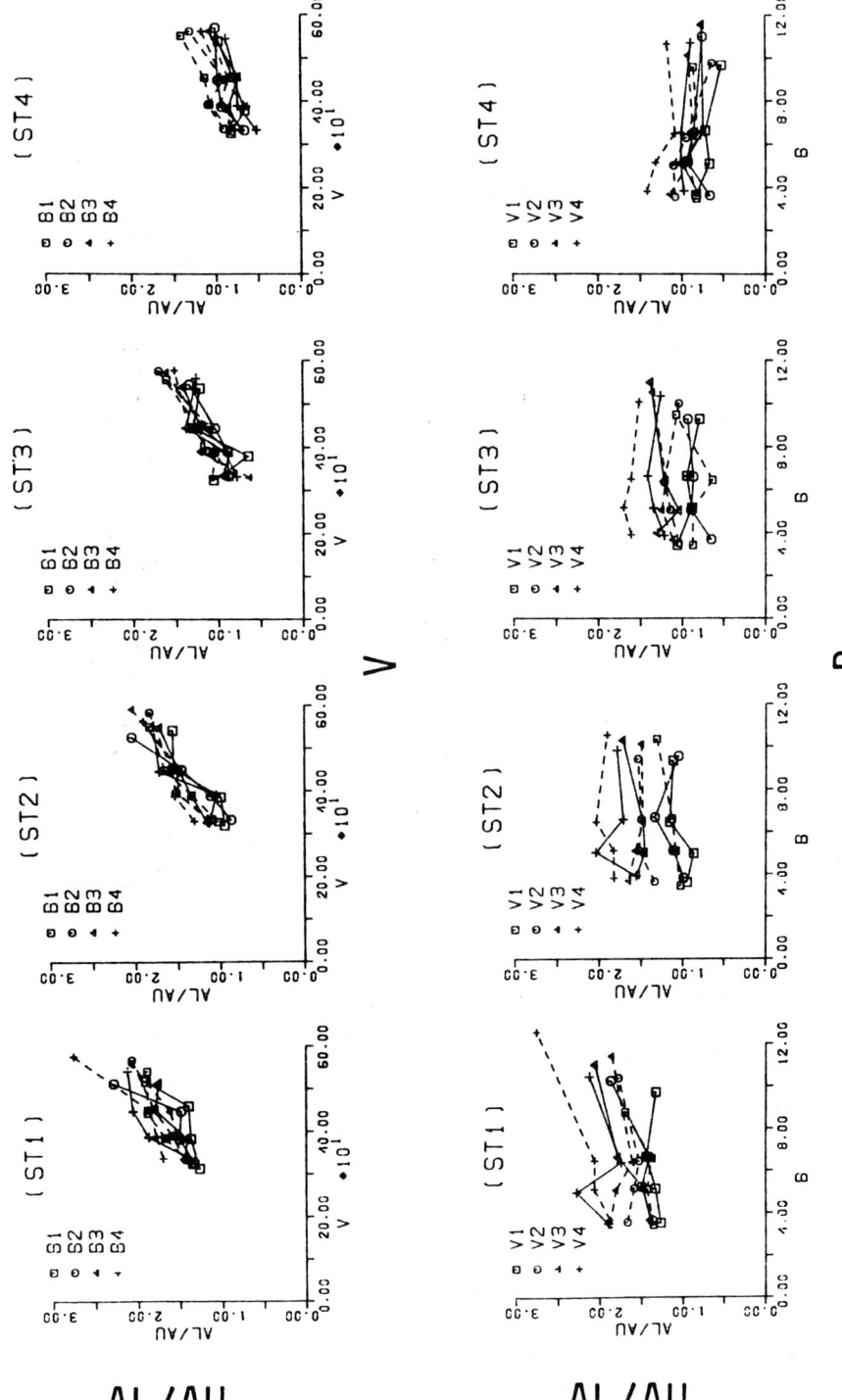

Fig. 3. Dependence of the ratio AL/AU on the IMF magnitude (lower panels) and on the solar wind velocity (upper panels). Four different ranges of $\sin\theta$ are shown from left to right.

TABLE 2.  Exponents for Power Law Approximation

|    | B | V | $-\sin\theta$ | $\sigma/B$ | n | T |
|----|---|---|---|---|---|---|
| AL | 0.85 ±0.06 | 2.08 ±0.12 | 0.54 ±0.03 | 0.18 ±0.05 | 0.07 ±0.03 | −0.05 ±0.04 |
| AU | 0.67 ±0.08 | 1.15 ±0.16 | 0.34 ±0.04 | 0.12 ±0.07 | 0.13 ±0.04 | 0.00 ±0.05 |
| am | 1.03 ±0.05 | 2.34 ±0.10 | 0.37 ±0.02 | 0.36 ±0.04 | 0.20 ±0.03 | −0.10 ±0.03 |

the three geomagnetic indices AL, AU, and am. Confidence intervals are indicated in the table for the 95 % confidence level. Since we no longer need to divide the data set into many subclasses the solar wind temperature is also included as a parameter. Only the data for which the IMF was directed southward ($B_z < -0.5\gamma$) have been analyzed because the assumed power law dependence form $(-\sin\theta)^p$ is applicalbe only when $\theta < 0$.

As seen in Table 2, the exponent obtained for the IMF magnitude is close to unity particularly for AL and am. Although the exponents for AL and AU are smaller than unity, this would probably be caused by the 'saturation effect' of AL and AU noted earlier in Fig. 1 and in Maezawa (1978). This effect has been found to be more pronounced for AU than for AL. On the other hand, no similar effect has been found for the am index.

The exponent values obtained for the dependence on V confirm the results obtained in the preceding analyses. The exponent for V is close to unity for the AU index, while for the other two indices it is almost equals two. For the ratio AL/AU we obtain the relation $AL/AU \propto B^{0.18}V^{0.93}$, which is in reasonable agreement with the conclusion drawn from Fig. 3.

Let us now turn to the exponents obtained for other variables. First, all the indices show a moderate dependence on $\sin\theta$ and the exponent for AL is the largest of the three indices, indicating that the dependence on the direction of the IMF is the strongest for the AL index. However, as discussed in the next section, the functional form of the $\sin\theta$ dependence is much affected by the superposed effects of solar wind density and IMF variance. Therefore we would not discuss the $\sin\theta$ dependence in more detail at this stage.

For the remaining solar wind parameters, the regression coefficients are smaller than those for B, V, and $\sin\theta$. It may be noted that for the am index the regression coefficients on n and $\sigma/B$ are relatively large when compared to those for AL and AU. This point will be the main subject of the next section. It will be shown that the effects of n and $\sigma/B$ on am become particularly important when the IMF is directed northward.

### The Solar Wind Density and the IMF Variance Effects

The effects of the solar wind density n and the IMF variability $\sigma$ have been a matter of discussion (Garrett et al., 1974; Svalgaard, 1977). In this section we will examine these effects in detail as a

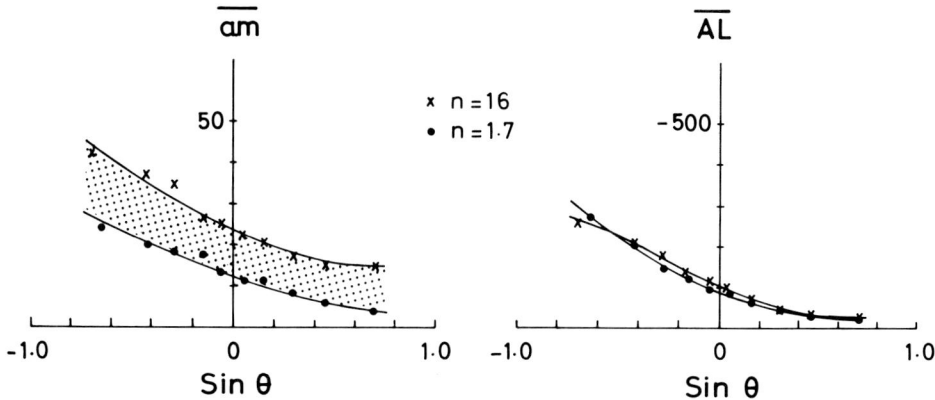

Fig. 4. The normalized indices $\overline{AL}$ and $\overline{am}$ are plotted against $\sin\theta$. Dots correspond to the case of low solar wind density ($n \simeq 1.7$). Crosses corresponds to the case of high solar wind density ($n \simeq 16$).

function of $\sin\theta$. Since these effects are found to be rather minor when compared to V and B dependences examined earlier, we will first normalize the index values for the V and B dependences using the results obtained earlier. Namely, the AL and am indices will be normalized by multiplication of a factor $(B_0 V_0^2 / BV^2)$, where $B_0$ and $V_0$ are constants ($B_0 = 6.0\gamma$, $V_0 = 430$ km/s) and B and V are observed values of the IMF magnitude and the solar wind velocity, respectively. (The AU index has been studied in a similar way by multiplication of a factor $(B_0 V_0 / BV)$. However, its dependences on n and $\sigma/B$ are found to be midway between AL and am, and it will not be shown in the following.) The n and the $\sigma/B$ effects on the above reduced indices will be studied separately, because n and $\sigma/B$ are found to be correlated little with each other (Correlation coefficient is 0.07.)

Fig. 4 shows the dependence of normalized indices $\overline{am}$ (left) and $\overline{AL}$ (right) on $\sin\theta$ for two representative ranges of n. The analysis has been made separately for ten bins of the solar wind density, but only the results for two extreme bins of n are plotted for simplicity. In each panel, crosses represent the case of a high solar wind density ($n = 16/cm^3$ on the average) and dots represent the case of a low solar wind density ($n = 1.7/cm^3$ on the average). Separation between these two curves, which is represented by shaded area in the case of $\overline{am}$, expresses the magnitude of the n dependence. Comparison of the two panels immediately shows that the dependence on n is by far the greater for $\overline{am}$ than for $\overline{AL}$. It is also noted that the extent of the n dependence of $\overline{am}$ does not decrease appreciably with increasing $\theta$. The n dependence represents an important contribution to the variability of $\overline{am}$ when the IMF is directed strongly northward, whereas the value of $\overline{am}$ itself decreases with increasing $\theta$ for a fixed value of n. Thus the mechanism of the solar wind density effect on $\overline{am}$ would be independent of the dayside magnetic field reconnection, the rate of which would be strongly dependent on $\sin\theta$.

In order to demonstrate the effect of n more clearly, $\overline{am}$ and $\overline{AL}$ are plotted against the solar wind density in Fig. 5 for the most northward range of the IMF investigated. ($\sin\theta = 0.65$ on the average.) A

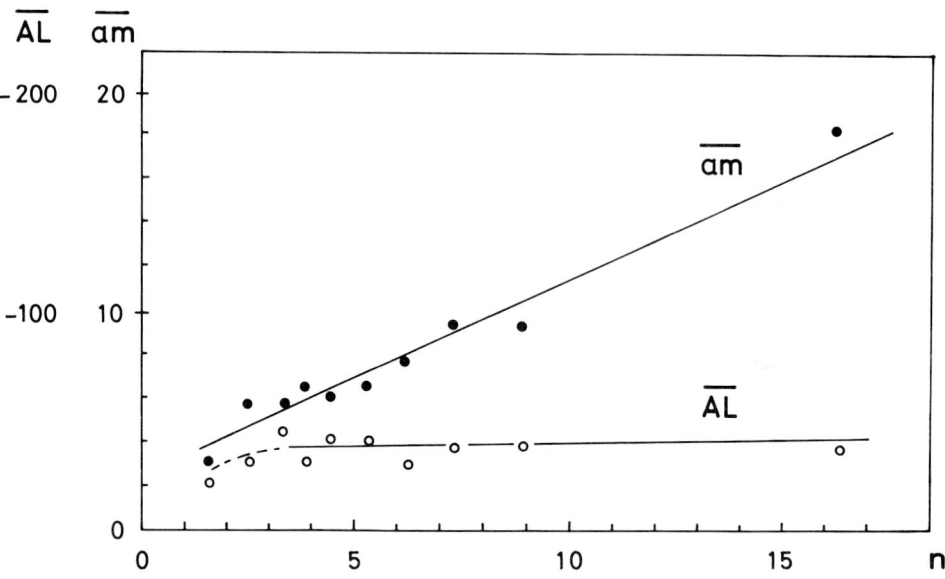

Fig. 5. $\overline{AL}$ and $\overline{am}$ are plotted against solar wind velocity for the most northward orientation of the IMF studied ($\sin\theta \sim 0.67$).

strong, almost linear dependence on n is noted for $\overline{am}$, but little dependence is noted for $\overline{AL}$. Whatever the mechanism of the n dependence of $\overline{am}$ would be, it can scarecely affect the nightside auroral zone activity represented by $\overline{AL}$.

Let us now examine the effect of the IMF relative variability. The normalized indices $\overline{am}$ and $\overline{AL}$ are plotted against $\sin\theta$ in the left and right panels, respectively, of Fig. 6 in a similar way as in Fig. 4. In each panel dots represent the cases where $\sigma/B$ is extremely low ($\sigma/B = 0.20$ on the average) and crosses represent the cases where $\sigma/B$ is extremely high ($\sigma/B = 0.82$ on the average). It is seen that the dependence of $\overline{am}$ on $\sigma/B$ is larger than that of $\overline{AL}$. The $\sigma/B$ dependent part represented by the shaded area is an important constituent of $\overline{am}$ variability when the IMF is directed northward just as the n dependent part was. On the other hand, these characteristics are not observed for $\overline{AL}$ plotted in the right panel; it is seen that the separation between two lines diminishes rapidly with increasing $\theta$ in the range $\theta > 0$.

Finally, we suggest that determination of the dependence of geomagnetic activity level on $\theta$, which is important in terms of the reconnection model, should be done carefully because its dependence is different depending on the value of n and $\sigma/B$. To see this, let us take the ratio between the magnitude of $\overline{am}$ at $\sin\theta = -0.6$ to that at $\sin\theta = 0$, as a measure of the steepness of the dependence of $\overline{am}$ on $\sin\theta$. This ratio generally increases with decreasing $\sigma/B$; it is about 1.7 for the case $\sigma/B \simeq 0.8$, 2.8 for the case $\sigma/B = 0.2$ and as high as 4.2 when the $\overline{am}$ value is linearly extrapolated to the case $\sigma/B = 0$. The similar ratio taken for $\overline{AL}$ is 2.2 for the case $\sigma/B = 0.8$, 3.4 for the case $\sigma/B = 0.2$, and 4.5 when $\overline{AL}$ is linearly extrapolated to the case $\sigma/B = 0$. Therefore the dependence of geomagnetic activity on $\sin\theta$ for the ideal case $\sigma/B=0$

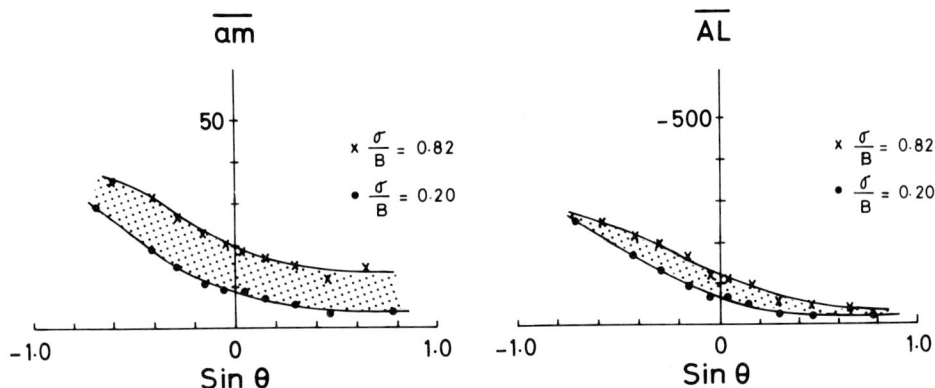

Fig. 6. The normalized indices $\overline{AL}$ and $\overline{am}$ are plotted against $\sin\theta$. Dots and crosses correspond to a low value (~ 0.2) and a high value (~ 0.82) of the IMF relative variability, $\sigma/B$, respectively.

would be much steeper than what we would expect from the analysis of the average case $\sigma/B = 0.5$.

## Summary and Discussion

The result of analysis presented in this paper may be summarized as follows:
(1) The dependence on the IMF magnitude B is almost linear for all the investigated indices AL, AU, and am, as long as the geomagnetic activity is not very high.
(2) Instead of the above, the dependence on the solar wind velocity V is different for the three indices. AU is roughly proportional to V, while AL and am are roughly proportional to the square of V.
(3) The dependence of am on the solar wind density differs qualitatively as well as quantitatively from that of AL. The effect of n is much larger on am than on AL. The magnitude of its effect on am does not decrease noticeably when the IMF is directed northward.
(4) The dependence of am on the IMF relative variability $\sigma/B$ is also different from that of AL. The magnitude of the dependence is larger for am than for AL.
(5) The functional form of the dependence of geomagnetic indices on $\sin\theta$ depends on the values of $\sigma/B$ and n. When the effects of $\sigma/B$ and n are removed, geomagnetic indices would show a much stronger dependence on $\sin\theta$ than has been derived.

Let us first discuss points (1) and (2). We note that almost all reconnection theories proposed so far predicted that the maximum reconnection rate is determined essentially by the Alfvén velocity $V_A$ of the medium (Petschek,1964; Yeh and Axford, 1970; Sonnerup, 1970). According to this idea the maximum reconnectable magnetic flux is proportional to $B \cdot V_A = B^2/\sqrt{4\pi\rho}$, which is not correlated in any direct way with the solar wind velocity. Thus the simple reconnection theories fail to account for the observed relationship.

On the other hand, some people have taken a simpler view that the reconnection speed is controlled by the solar wind velocity (Gonzalez and

Mozer, 1974; Maezawa, 1976; Svalgaard, 1977). In this view, the convection electric field resulting from the reconnection is proportional to B·V. This idea is fevolable for explaining the linear dependence of the AU index on V and B found above, if the convection electric field in the magnetosphere is directly mapped to the auroral zone ionosphere. Since the dayside auroral conductivity is largely controlled by sunlight which has no relation to the solar wind, the dayside auroral zone current intensity is expected to be determined by the convection electric field.

On the other hand, the relationship AL $\propto$ BV$^2$ cannot be explained only by the B times V dependence of the convection electric field assumed above. If we retain the above assumption for the convection electric field, we should require additionally that the night side auroral zone conductivity depends on the solar wind velocity through yet other mechanism(s). This would mean that the particle precipitation from the near-earth tail region would be controlled by the solar wind velocity.

It should be emphasized that even the mechanism of the formation of the plasma sheet has not been well understood. Thus the above view is so far based on no firm theoretical ground. However, it would be appropriate to point out that there is ample possibility that the particle population, their pitch-angle distribution or configuration of the near-earth plasma sheet is affected by the solar wind velocity. The Kelvin-Helmholtz instability occuring at the magnetopause would be a possible candidate for this mechanism, but much work would be needed before this view point can be advanced.

The large n dependence of the am index (item 3) cannot be explained by the reconnection model alone, either. Since this effect is not strongly related to the latitude angle of the IMF, this phenomenon should be explained by an interaction mechanism which is operative even when the IMF is directed northward. The only known mechanism by which the solar wind density variation affects the state of the magnetosphere is the overall compression and the expansion of the magnetosphere by the solar wind. The ground magnetic effects of this interaction is worldwide in nature and seen irrespective of the latitude angle of the IMF (Nishida and Maezawa, 1971). Further, the magnitude of this effect (of the order of 10γ) would be sufficient for explaining the dependence of am on n. Therefore we propose that the am index is contributed much by this mechanism. Part of large σ/B dependence of am (item 4) may also be explained by this mechanism since the pressure variability in the solar wind is correlated with the IMF variance (Garrett et al., 1974).

An important conclusion of our analysis is that more than one solar wind-magnetosphere interaction mechanism is responsible for geomagnetic activity and the effects of these mechanisms are reflected in different geomagnetic indices in different ways. In this connection, the AE index defined as AE = AU - AL would not be suitable for the analysis of the solar wind-magnetosphere interaction because its constituents, AU and AL, have different dependences on the solar wind velocity. It is also suggested that the effects of the IMF variance and the solar wind density should be carefully removed before one utilizes geomagnetic indices (particularly planetary indices such as am, ap, Kp etc.) to investigate the dependence of geomagnetic activity on the IMF latitude angle.

Acknowledgements. The solar wind IMF data used in this analysis have been provided by WDC-A Rockets and Satellites.

## References

Ballif, J. R., D. E. Jones, P. J. Coleman, Jr., L. Davis, Jr., and E. J. Smith, Transverse fluctuations in the interplanetary magnetic field: a requisite for geomagnetic variability, J. Geophys. Res., 72, 4357, 1967.

Burton, R. K., R. L. McPherron, and C. T. Russell, The terrestrial magnetosphere: A half-wave rectifier of the interplanetary electric field, Science, 189, 717, 1975.

Dungey, J. W., Interplanetary magnetic field and the auroral zone, Phys. Rev. Lett., 6, 47, 1961.

Garrett, H. B., A. J. Dessler, and T. W. Hill, Influence of solar wind variability on geomagnetic activity, J. Geophys. Res., 79, 4603, 1974.

Gonzalez, W. D., and F. S. Mozer, A quantitative model for the potential resulting from reconnection with an arbitrary interplanetary magnetic field, J. Geophys. Res., 79, 4186, 1974.

Hirshberg, J., and D. S. Colburn, Interplanetary field and geomagnetic variations - a unified view, Planet. Space Sci., 17, 1183, 1969.

Maezawa, K., Magnetospheric convection induced by the positive and negative Z components of the interplanetary magnetic field: quantitative analysis using polar cap magnetic records, J. Geophys. Res., 81, 2289, 1976.

Maezawa, K., Dependence of geomagnetic activity on solar wind parameters: a statistical approach, Solar Terrestrial Environmental Research in Japan, 2, 103, 1978.

Maezawa, K., and A. Nishida, Inferences of solar wind velocity from geomagnetic indices, Transactions American Geophysical Union, 58, 760, 1977.

Mayaud, P. N., Indices Kn, Ks, et Km, 1964-1967, Centres National de la Recherche Scientifique, Paris, 1968.

Murayama, T., and K. Hakamada, Effects of solar wind parameters on the development of magnetospheric substorms, Planet. Space Sci., 23, 75, 1975.

Nishida, A., and K. Maezawa, Two basic modes of interaction between the solar wind and the magnetosphere, J. Geophys. Res., 76, 2254, 1971.

Petschek, H. E., Magnetic field annihilation, in AAS-NASA Symposium on the Physics of Solar Flares, ed. by W. N. Hess, NASA SP-50, pp. 425-439, 1964.

Snyder, C. W., M. Neugebauer, and U. R. Rao, The solar wind velocity and its correlation with cosmic-ray variations and with solar and geomagnetic activity, J. Geophys. Res., 68, 6361, 1963.

Sonnerup, B. U. Ö., Magnetic-field re-connecxion in a highly conducting imcompressible fluid, J. Plasma Phys., 4, 161, 1970.

Svalgaard, L., Geomagnetic activity: dependence on solar wind parameters SUIPR Report No. 699, Stanford University, 1977.

Wilcox, J. M., K. H. Schatten, and N. F. Ness, Influence of interplanetary magnetic field and plasma on geomagnetic activity during quiet-sun conditions, J. Geophys. Res., 72, 19, 1967.

Yeh, T., and W. I. Axford, On the re-connexion of magnetic field lines in conducting fluids, J. Plasma Phys., 4, 207, 1970.

A MACROSCOPIC MODEL FOR FIELD LINE INTERCONNECTION BETWEEN THE MAGNETO-
SPHERE AND THE INTERPLANETARY SPACE

G.H. Voigt and K. Fuchs

Angewandte Geophysik, Technische Hochschule Darmstadt,
Alexanderstrasse 35, D-6100 Darmstadt, West Germany

Abstract. An 'open' magnetospheric $\underline{B}$-field model has been derived from the 'closed' Voigt-1972-model by assuming that the time-dependent merging process will result in a quasi-static picture of interconnected field lines.
The model describes the macroscopic influence of the IMF on the global magnetospheric field topology. Numerical calculations reproduce the latitudes of the poleward and equatorward boundaries of polar cusp electron precipitation as functions of the amount and orientation of the IMF. The results are in agreement with published measurements of these boundaries. Their latitudinal shifts depend on variations of the solar wind pressure, the dipole tilt angle, and the IMF conditions.
The amount of magnetic flux penetrating the magnetopause indicates clearly that the main region for field line interconnection with the magnetosheath is not the subsolar point at the frontside of the magnetopause, but rather the polar cusp region or the so-called entry layer.

## 1. Introduction

We know that the magnetosphere reacts on changes of the solar wind pressure as well as on changes of the amount and orientation of the interplanetary magnetic field (IMF).
Changes of the global field topology which are due to variations of the solar wind pressure can be commonly described with 'closed' magnetospheric models which do not need any sort of field line interconnection to the interplanetary space. A model of this type includes physical parameters which can reproduce magnetospheric situations under steady-state conditions. Successive variations of the model parameters lead to series of model states which are in agreement with observed long-time variations in the magnetosphere.
On the other hand, changes of the $\underline{B}$-field topology which are due to the influence of the IMF require a quantitative magnetospheric model which is capable of computing field line interconnection between magnetospheric and interplanetary field lines. In addition to the physical parameters mentioned above, an 'open' model of this type must include parameters which allow for changes of the boundary conditions on the magnetopause.
Unfortunately, we are far away from understanding the plasma processes which result in the situation of interconnected field lines. On the other hand, there is some indirect evidence for field line interconnection to the IMF.

The polar cusp shifts equatorward during periods of southward IMF and moves poleward during periods of northward IMF [Kivelson et al., 1973; Yasuhara et al., 1973]. A similar argumentation holds for the size and position of the nightside part of the auroral oval [Kamide et al., 1977]. Moreover, Aubry et al. [1970] were able to demonstrate that a reversal of the vertical component of the IMF from northward to southward results in an earthward motion of the dayside magnetopause during periods of constant pressure of the solar wind. In this case, the decrease of the stand-off distance associated with an equatorward shift of the polar cusp is not due to a compression of the whole magnetospheric cavity; it is rather due to field line erosion on the magnetopause [Burch, 1973; Kamide et al., 1976].

It is the purpose of this paper to describe a quantitative magnetospheric $\underline{B}$-field model which is capable of reproducing the observed displacement of the polar cusp in dependence on the IMF orientation. This 'open' model can be derived from a 'closed' model [Voigt, 1972] when the boundary condition, $\underline{B}_\perp = 0$ at the magnetopause, is replaced by the less restrictive condition that the normal component of the total field penetrates the boundary continuously.

The model includes two 'constants of interconnection' which enable us to describe static field line interconnection and to change the boundary conditions which results in a 'closed' as well as in an 'open' magnetosphere.

## 2. Method

For the construction of the 'open' magnetospheric field model, we ignore the characteristics of the time-dependent plasma process called 'field line reconnection' which refers to a process whereby magnetic energy is dissipated at a magnetic neutral surface or line. Instead we reduce ourselves to the term 'field line interconnection' by assuming that the reconnection process will result from time to time in a static picture of interconnected field lines.

Moreover, we do not consider the thermal plasma inside and outside the magnetosphere. Therefore, the model is not self-consistent with respect to the plasma and the magnetic field. Finally, we neglect the microstructure of the magnetopause in order to be able to describe the interconnection of field lines as a boundary value problem of the potential theory.

Let us assume that $\underline{B}_s$ represents the magnetic field of the earth's dipole or the field of currents distributed within the magnetosphere. $\underline{B}_s$ must be shielded against the interplanetary space by an additional Chapman-Ferraro field $\underline{B}_{cfs}$ to such an extent that only a certain proportion $C_d$ of the normal component $\hat{n} \cdot \underline{B}_s$ can penetrate the magnetopause. Here, $\hat{n}$ is the unit vector perpendicular to the magnetopause. Thus, we find the boundary condition

$$\hat{n} \cdot (\underline{B}_s + \underline{B}_{cfs}) = C_d \, \hat{n} \cdot \underline{B}_s \quad . \tag{1}$$

We call $C_d$ a 'constant of interconnection' and assume that this constant is the same for all $\underline{B}_s$. The Chapmann-Ferraro field $\underline{B}_{cfs}$ can be derived from a scalar potential $U_{cfs}$ by

$$\underline{B}_{cfs} = -\nabla U_{cfs} \quad . \tag{2}$$

Following (1) and (2), we find the boundary condition for $U_{cfs}$ at the magnetopause as follows:

$$\frac{\partial}{\partial n} U_{cfs} = (1. - C_d) \, \hat{\underline{n}} \cdot \underline{B}_s \quad . \tag{3}$$

Moreover, each $U_{cfs}$ must meet Laplace's equation. The same procedure is valid in the outer space by shielding the IMF against the interior of the magnetospheric cavity. We assume a homogeneous interplanetary field $\underline{B}_{imf}$ in the outer space shielded by the corresponding Chapman-Ferraro field $\underline{B}_{cfi}$ which is given by

$$\underline{B}_{cfi} = -\nabla U_{cfi} \quad . \tag{4}$$

We use an other 'constant of interconnection', $C_{imf}$, in order to take into account different plasma conditions on both sides of the magnetopause. In analogy to (3), we find at the magnetopause the boundary condition

$$\frac{\partial}{\partial n} U_{cfi} = (1. - C_{imf}) \, \hat{\underline{n}} \cdot \underline{B}_{imf} \quad . \tag{5}$$

A superimposition of all $\underline{B}_s$, $\underline{B}_{cfs}$, and $\underline{B}_{imf}$, $\underline{B}_{cfi}$ yields the final configuration $\underline{B}_{inner}$ within the magnetospheric cavity, and $\underline{B}_{outer}$ in the interplanetary space:

$$\underline{B}_{inner} = \sum_s \underline{B}_s + (1. - C_d) \cdot \sum_s \underline{B}^*_{cfs} + C_{imf} \cdot \underline{B}_{imf} \tag{6}$$

$$\underline{B}_{outer} = \underline{B}_{imf} + (1. - C_{imf}) \cdot \underline{B}^*_{cfi} + C_d \cdot \sum_s \underline{B}_s \quad . \tag{7}$$

In (6) and (7), the terms of the sum with the index s represent the field of the earth's dipole and the field of currents distributed inside the magnetospheric cavity. Moreover, $\underline{B}^*_{cfs}$ and $\underline{B}^*_{cfi}$ denote the Chapman-Ferraro fields of the 'closed' model [Voigt, 1972].

The 'closed' model is specified by $C_d = 0$ in equation (3) and $C_{imf} = 0$ in equation (5). In this case, the magnetopause is parallel to the field lines, and we have no interconnection to the IMF.

On the other hand, the values $C_d = C_{imf} = 1$ describe the extreme case of neglecting the effect of the solar wind plasma on the magnetosphere. In this case, we have a totally open magnetosphere, and the magnetopause does not exist [Voigt, 1978].

The numerical values of both constants of interconnection lie independently in the ranges

$$0 \leq C_d \leq 1 \tag{8}$$

$$0 \leq C_{imf} \leq 1 \tag{9}$$

For a detailed understanding of formulas (6) and (7) in the light of the potential theory, the reader may be referred to Voigt [1978].

Roederer pointed out in a recent review [Roederer, 1977], that the real magnetospheric $\underline{B}$-field topology in a steady-state configuration lies obviously somewhere between the extreme cases given by $c_d = c_{imf} = 0$ and $c_d = c_{imf} = 1$. The real situation depends on the orientation of the IMF and on the plasma processes on both sides of the magnetopause.

Our theory does not yield an explanation of the plasma processes involved. Therefore, we try to find the real $\underline{B}$-field topology by varying both constants of interconnection, $c_d$ and $c_{imf}$, until the model magnetosphere is in best correspondence with a measured $\underline{B}$-field topology.

This semi-empirical procedure will be outlined in Section 4; it yields numerical values for $c_d$ and $c_{imf}$ which are valid for averaged magnetospheric conditions.

## 3. Model Parameters

The magnetopause of the magnetospheric model must have a given geometry in order to allow for three-dimensional, analytical solutions of Neumann's boundary value problems specified by equations (3) and (5). Therefore, the magnetopause is represented as a half-sphere on the dayside connected to a semi-infinite circular cylinder on the nightside.

The earth's dipole is placed eccentrically with respect to the center of the sphere. This enables us to fit the curvature of the dayside magnetopause on spacecraft observations of boundary crossings as they have been reported by Fairfield [1971]. The magnetic configuration depends on the predefined shape and position of the magnetopause. On the other hand, Walker [1976] compared the geometries of the model magnetopause with the boundary derived from Fairfield's [1971] measurements and found a good correspondence on the dayside for magnetospheric quiet conditions.

The physical parameters of the present model are the same which we know from the closed model [Voigt, 1976]:

1. stand-off distance to the subsolar point,
2. radius of the tail at $X_{GSE} = -10$ Re ,
3. tail field intensity,
4. ring current $D_{st}$-index,
5. dipole tilt angle.

In this paper, we are interested in the influence of the IMF on the magnetospheric $\underline{B}$-field topology. Therefore, the model parameters noted above remain constant in the following calculations. Their numerical values refer to a quiet magnetosphere [Voigt, 1976].

## 4. Results

In order to demonstrate the effect of field line interconnection on the global magnetospheric field topology, we simplify the situation in the first step saying that both constants of interconnection are identical: $c_d = c_{imf} = c_0$.

Figures 1 to 3 show the field line topologies of the model in the noon-midnight meridian plane. The field lines are spaced every $2°$ in latitude, starting at 66°. Only field lines with one or two ends on the earth have been plotted; pure interplanetary field lines without connection to the dipole have been omitted.

The model parameters in all Figures refer to a quiet magnetosphere; they are given by stand-off distance = 11 Re, tail radius = 16 Re, tail field intensity = 20 nT, $D_{st}$-index = 0 nT, tilt angle = 0°, IMF intensity = 5 nT.

In Figure 1, the homogeneous IMF is southward directed. The closed magnetosphere is given by $C_o = 0$. In this case, the field line configuration is independent of the IMF strength and direction, since there is no interconnection between the magnetospheric and the interplanetary field. Therefore, the magnetopause is parallel to the field lines. The intersection latitude of the last closed field line is 80°.

The other two pictures in Figure 1 show the field line topology with interconnection to the IMF for $C_o = 0.1$ and $C_o = 0.2$. In these cases, the magnetosphere is open and high-latitude dipole connected field lines extend into interplanetary space. The kinks in the field lines are due to the thin Chapman-Ferraro current layer on the magnetopause. The intersection latitude of the last closed dayside field line decreases from 76° in case of $C_o = 0.1$ to 74° in case of $C_o = 0.2$.

With increasing constant of interconnection, the number of open field lines increases and the equatorward boundary of the polar cusps move equatorward. From the point of view of the field line patterns in Figure 1, we are coming to the conclusion that the main region for field line interconnection is not the subsolar point at the frontside of the magnetopause, but rather the polar cusp region.

The $C_o = 0.2$ configuration in Figure 1 indicates the formation of a X-type neutral line at the center of the plasma sheet. The existence of a neutral line of this type depends on the orientation of the IMF (see Figure 2) and could be regarded as the starting point for a time-dependent substorm process.

Figure 2 shows a comparison between the southward and the northward directed IMF. The 'open' $C_o = 0.2$ configuration is identical in Figures 1 and 2. A field line pattern which seems to be a 'closed' model occurs only when the IMF is strongly northward directed. The physical distinction between the closed model in Figure 1 and the model with northward IMF in Figure 2 is as follows: The closed magnetosphere in Figure 1 is specified by $C_o = 0$. Therefore, the normal component $B_\perp$ of the total field is equal to zero at the magnetopause, and no field line can penetrate the boundary from the inside to the outside and vice versa. In contrast, the magnetosphere with northward IMF in Figure 2 is specified by $C_o = 0.2$. Therefore, the normal component $B_\perp$ is not equal to zero at the magnetopause. Interplanetary field lines penetrate the boundary in this case and join closely to the magnetospheric field lines; they have not been plotted since they are not connected to the earth's dipole.

Figure 3 shows the effect of other IMF orientations marked by the arrows. In these cases, we no longer have symmetry of the B-field configuration in the northern and southern hemispheres in spite of the zero dipole tilt angle. We can expect differences in the shape of the polar cusps in both hemispheres when interconnection of field lines is observed.

Note that north-south asymmetries of the polar cusp field topology

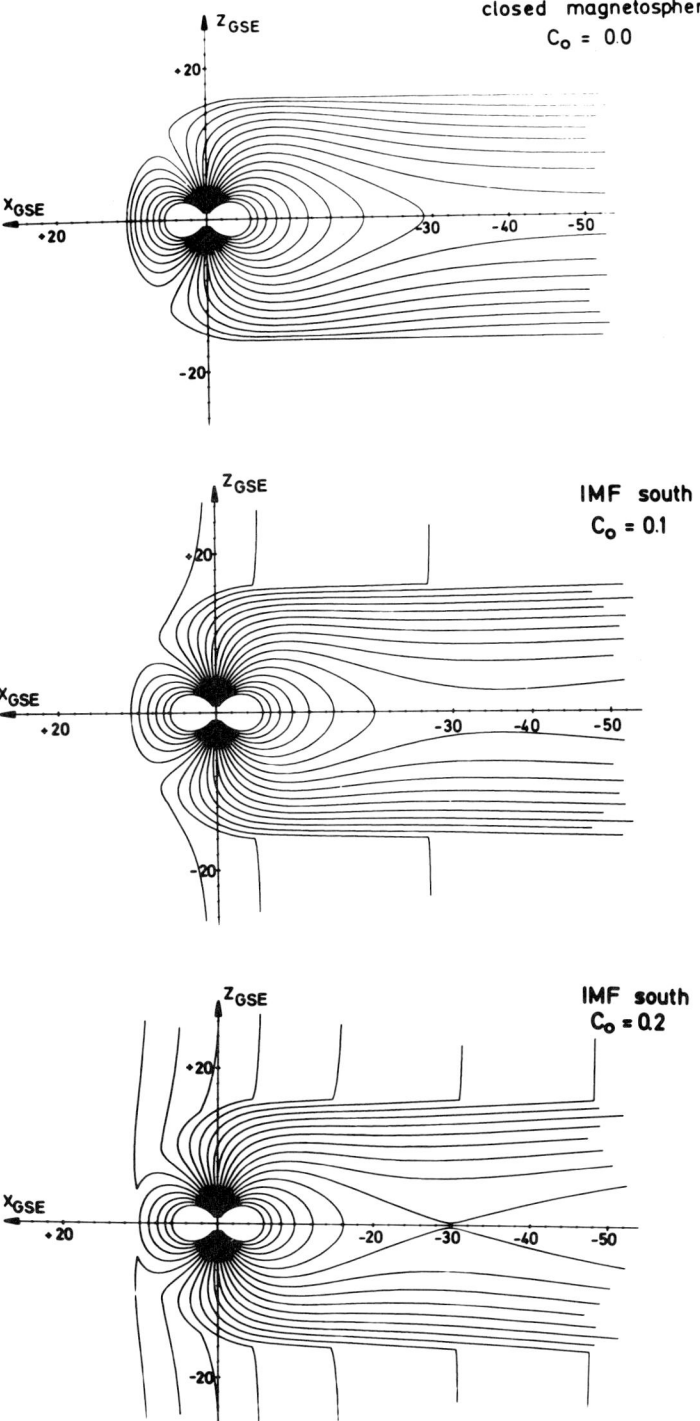

Fig. 1

can be computed in a closed model only by variations of the dipole tilt angle [Voigt, 1974]. The open model allows us to decide whether an observed north-south asymmetry of the polar oval and the polar cusp is due to seasonal variations or to changes of the IMF orientation.

The different field line patterns in Figures 1,2, and 3 show clearly that a minimal variation of the normal component $\underline{B}_\perp$ on the magnetopause changes greatly the field topology in the tail. We find far extended tail field lines when the IMF is southward directed. On the other hand, the field lines are still dipole like beyond the distance of the neutral line when the IMF points northward.

In the next step of our calculations, we are looking for realistic values of $C_d$ and $C_{imf}$. Therefore, we try to reproduce shifts of the poleward and equatorward boundaries of the polar cusp in dependence on the interplanetary $B_z$-component.

We take the position of the polar cusp from data reported by Burch [1973] who identified the polar cusp by the geomagnetic latitudes, $\Lambda_{pol}$ and $\Lambda_{equ}$, of the poleward and equatorward boundaries of dayside cusp electron precipitation. The equatorward boundary was assumed to be coincident with the last closed field line on the dayside.

A second order polynomial fit to the data yields the following expressions for the geomagnetic latitudes of both boundaries in dependence on the IMF $B_z$-component [Burch, 1973]:

$$\Lambda_{equ} = 75.5 + 0.53\, B_z - 0.05\, B_z^2 \qquad (10)$$

$$\Lambda_{pol} = 80.3 + 0.76\, B_z - 0.03\, B_z^2 \quad . \qquad (11)$$

First of all, we choose a set of the model parameters noted in Section 3 by means of which $\Lambda_{equ} = 75.5°$ can be reproduced in the case of $B_z = 0$. Throughout the following computations, these values of the model parameters remain unchanged in order to calculate the influence of the IMF only.

Then we follow the cusp field line which starts at $\Lambda_{pol} = 80.3°$ geomagnetic latitude up to the position $x_{mp}$ on the tail magnetopause where this field line meets the boundary.

The model polar cusp is consequently defined by the latitude $\Lambda_{equ}$ where the last closed dayside field line intersects the earth, and the latitude $\Lambda_{pol}$ where the field line started from $x_{mp}$ intersects the earth.

Finally, we vary both constants of interconnection, $C_d$ and $C_{imf}$, and compute $\Lambda_{equ}$ and $\Lambda_{pol}$ for any value of the IMF $B_z$-component in

---

Fig. 1. Comparison between the closed and open magnetosphere. The field lines are plotted in the noon-midnight meridian plane. They are spaced every 2° in latitude, starting at 66°. The field line topology of the closed model is independent of the IMF strength and direction. The open magnetosphere allows high latitude field lines to connect with the IMF. The IMF points southward in this figure.

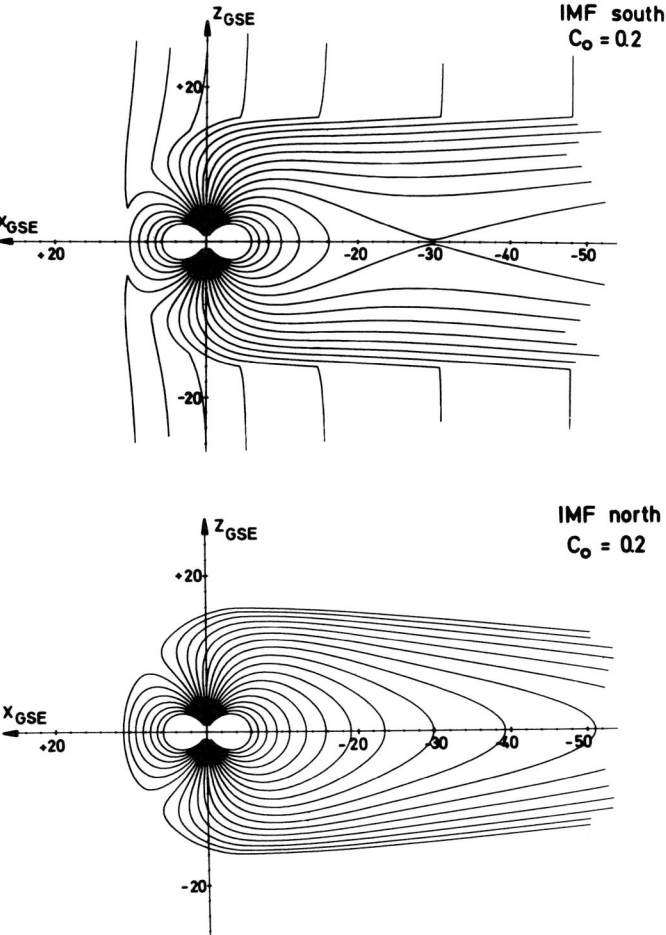

Fig. 2. Comparison between the southward and northward directed IMF. The model parameters are the same as in Figure 1. Note the different topology of the tail field lines in both cases. The southward directed IMF forms a X-type neutral line in the tail. The northward directed IMF forms dipole like field lines beyond the distance of the neutral line. Northward IMF field lines penetrate the magnetopause even in this case; they have not been plotted since they are not connected to the earth's dipole.

the range of $-6$ nT $\leq B_z \leq +6$ nT. The procedure yields a certain combination of $C_d$ and $C_{imf}$ whereby the $B_z$-dependence of $\Lambda_{equ}$ and $\Lambda_{pol}$ in the model comes near to the functions (10) and (11) which are derived from measurements.

Figure 4 shows the most important result: the comparison between experimental data and our model calculations. Data points indicate the location of the poleward and the equatorward boundaries of dayside cusp electron precipitation for 45-minute IMF-$B_z$ averages be-

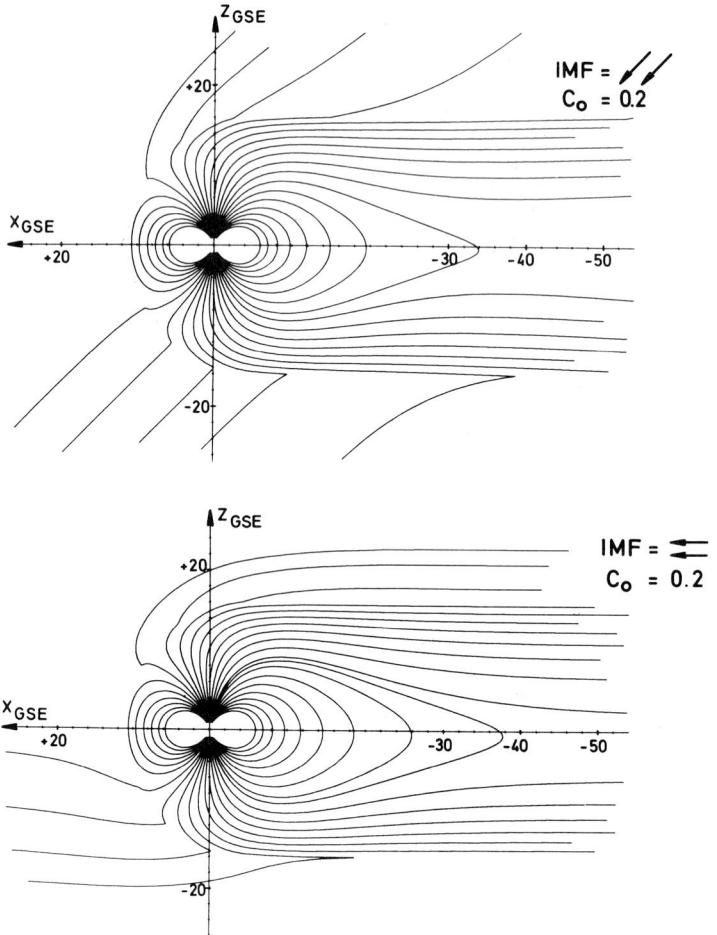

Fig. 3. Effect of other IMF orientations. The model parameters are the same as in Figures 1 and 2. The IMF direction is indicated by the arrows. Note the north-south asymmetries of the polar cusp topologies in both pictures. Asymmetries between both hemispheres depend on the magnitude of the constants of interconnection and on the IMF strengh and orientation.

tween $-6$ nT and $+6$ nT. Solid curves represent second order polynomial fits to the data [Burch, 1973].

The dashed lines are the results of the model calculations: The best agreement with the experimental data could be attained with the values $C_d = 0.12$ and $C_{imf} = 0.93$. In other words, 12% of the magnetospheric magnetic flux, and 93% of the IMF penetrate the magnetopause during average magnetospheric and interplanetary conditions. The great difference between $C_d$ and $C_{imf}$ may be interpreted in the light of the different plasma streaming velocities on both sides of the magnetopause.

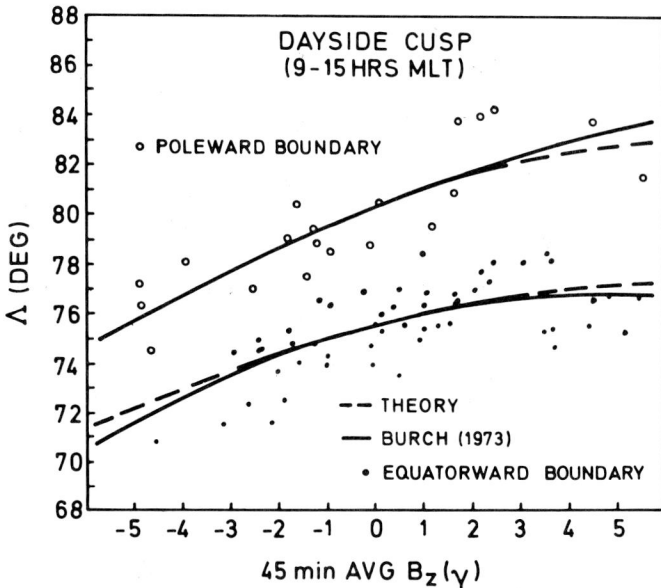

Fig. 4. Comparison between experimental data and model calculations. Data points indicate the location of the polward and the equatorward boundaries of cusp electron precipitation. Solid curves represent second order polynomial fits to the data [Burch, 1973]. The dashed lines are the result of the present model calculations. The best agreement with the data could be attained with $C_d = 0.12$ and $C_{imf} = 0.93$.

The deviation between the theoretical (dashed) and the experimental (solid) curves in the cases of a nearly northward and southward directed IMF (i.e. $|B_z| > 3$ nT ) is due to the fact that the model parameters remained unchanged for all IMF orientations in the present calculations.

It should be noted that the stand-off distance is determined on the one hand by the solar wind plasma parameters and on the other hand by the magnetic pressure on both sides of the magnetopause. Therefore, the stand-off distance increases with an increasing northward IMF component and decreases with an increasing southward IMF component. However, these variations of the subsolar distance depend on the values of both constants $C_d$ and $C_{imf}$. In order to investigate the sensitivity of the constants of interconnection to variations in the stand-off distance, we need simultaneous information on the solar wind parameters, the IMF $B_z$-component, and the magnetospheric subsolar distance. This information was unavailable for our present calculation.

The numerical values of $C_d = 0.12$ and $C_{imf} = 0.93$ obtained from our semi-empirical procedure are a first quantitative approach for calculating the magnetic flux which penetrates the magnetospheric boundary during steady-state conditions.

## 5. Discussion and Conclusion

We have developed a three-dimensional magnetospheric $\underline{B}$-field model which allows us to describe field line interconnection under static conditions. Using this model, we can predict the macroscopic influence of the interplanetary magnetic field (IMF) on the global magnetospheric field topology.

The amount of magnetic flux which penetrates the magnetopause is on the one hand given by the sources located within the magnetospheric cavity and in the interplanetary space which contribute all together to the distribution of the normal $\underline{B}$-field component on the magnetopause. On the other hand, the flux through the boundary is determined by two 'constants of interconnection', $C_d$ and $C_{imf}$. A systematic variation of these two constants leads to a correspondence of the model polar cusp with the cusp in nature derived from particle measurements. The variation yields realistic values for $C_d$ and $C_{imf}$ under magnetospheric quiet and moderately disturbed conditions.

The resulting field line topologies indicate clearly that the main region for field line interconnection is not the subsolar point at the frontside of the magnetopause, but rather the polar cusp region.

We should point out that the results of our model do not take the place of the explanation of field line reconnection (merging) in the light of the MHD equations or the kinetic theory. If we had a global theory for field line reconnection, it would become evident that both constants of interconnection must be replaced by functions $C_d(\underline{x}_{mp})$ and $C_{imf}(\underline{x}_{mp})$ which would depend on specific locations $\underline{x}_{mp}$ on the magnetopause. In an ideal case, these functions would reflect the different plasma processes on both sides of the magnetopause.

The method outlined in Section 2 allows, in principle, the solution of the boundary value problem even for the functions $C_d(\underline{x}_{mp})$ and $C_{imf}(\underline{x}_{mp})$. However, the Chapman-Ferraro potentials specified in equation (3) and (5) can no longer be obtained directly from the closed model in this case. But it is no problem to integrate Laplace's equation numerically with the extended boundary conditions.

In the present state, our model reproduces quasi-static changes of the magnetospheric field topology which are due to changes of the boundary conditions on the magnetopause. The 'closed' magnetosphere remains as a singular case in a generalized class of 'open' models.

## References

Aubry, M.P., C.T. Russell, and M.G. Kivelson, Inward motion of the magnetopause before a substorm, *J. Geophys. Res.*, 75, 7018-7031, 1970.

Burch, J.L., Rate of erosion of dayside magnetic flux based on a quantative study of the dependence of polar cusp latitude on the interplanetary magnetic field, *Radio Sci.*, 8, 955-961, 1973.

Fairfield, D.H., Average and unusual locations of the earth's magnetopause and bow shock, *J. Geophys. Res.*, 76, 6700-6716, 1971.

Kamide, Y., J.L. Burch, J.D. Winningham, and S.I. Akasofu, Dependence of the latitude of the cleft on the interplanetary magnetic field and substorm activity, *J. Geophys. Res.*, 81, 698-704, 1976.

Kamide, Y., and J.D. Winningham, A statistical study of the instantaneous nightside auroral oval: the equatorward boundary of electron precipitation as observed by the Isis 1 and 2 satellites, *J. Geophys. Res.*, *82*, 5573-5588, 1977.

Kivelson, M.G., C.T. Russell, M. Neugebauer, F.L. Scarf, and R.W. Fredericks, Dependence of the polar cusp on the north-south component of the interplanetary magnetic field, *J. Geophys. Res.*, *78*, 3761-3772, 1973.

Roederer, J.G., Global problems in magnetospheric plasma physics and prospects for their solution, *Space Sci. Rev.*, *21*, 23-70, 1977.

Voigt, G.H., A three dimensional analytical magnetospheric model with defined magnetopause, *Z. Geophys.*, *38*, 319-346, 1972.

Voigt, G.H., Calculation of the shape and position of the last closed field line boundary and the coordinates of the magnetopause neutral points in a theoretical magnetospheric field model, *Z. Geophys.*, *40*, 213-228, 1974.

Voigt, G.H., Influence of magnetospheric parameters on geosynchronous field characteristics, last closed field lines and dayside neutral points, in *The Scientific Satellite Programme During the International Magnetospheric Study,* edited by K. Knott and B. Battrick, pp. 381-396, Reidel Publishing Company, Dordrecht-Holland, 1976.

Voigt, G.H., A static-state field-line reconnection model for the earth's magnetosphere, *J. Atmosph. Terr. Phys.*, *40*, 355-365, 1978.

Walker, R.J., An evaluation of recent quantitative magnetospheric magnetic field models, *Rev. Geophys. Space Phys.*, *14*, 411-427, 1976.

Yasuhara, F., S.I. Akasofu, J.D. Winningham, and W.J. Heikkila, Equatorward shift of the cleft during magnetospheric substorms as observed by Isis 1, *J. Geophys. Res.*, *78*, 7286-7291, 1973.

OVERVIEW - MODELING TECHNIQUES

W. P. Olson

McDonnell Douglas Astronautics Company, 5301 Bolsa Avenue, Huntington Beach, California  92647

R. A. Wolf

Rice University, Houston, Texas  77001

In the 1974 meeting on quantitative modeling almost the entire meeting was devoted to quantitative magnetic field models with little attention paid to the electric field and almost no discussion of particle models. Increased emphasis on electric-field modeling was apparent in the present meeting. Also, because of the spacecraft arcing and charging problem, more attention has gone to the modeling of the low energy particle environment, especially in the vicinity of geosynchronous orbit. The availability of these models has generated interest in the possibility of quantitatively modeling the interaction of low energy charged particles with the time varying magnetic and electric fields resident in the inner magnetosphere.

A general examination of the interaction of the solar wind with the geomagnetic field was presented by Leboeuf, et al. They have used an MHD formalism and their analysis is restricted to two dimensions. As such their work does not now provide much new insight into the behavior of a real magnetosphere. However, such analyses, when developed more fully and extended to three dimensions, should give important insights into the self-consistent interactions between magnetospheric particles and fields. Smith, et al., examined the motions of individual charged particles in the equatorial region in a time varying magnetic and electric field environment. They showed a movie demonstrating how computed particle motions of a period of a storm main phase imply formation of an asymmetric ring current. Harel, et al., developed a complex set of computer codes which they are using to model actual substorm events. Their subroutines include time varying magnetic and electric fields where the coupling between ionosphere and the magnetosphere has been taken into account in the determination of the electric fields. These three papers suggest the direction the quantitative modeling of magnetospheric processes will take in the next few years as more accurate descriptions of the time varying behavior of individual magnetospheric features become available. With such models and the appropriate input parameters to them it should be possible to <u>predict</u> the time variations in some magnetospheric features.

Nopper and Carovillano attempted to describe electric fields in the ionosphere that are driven by field aligned currents, as estimated from observations made by the TRIAD satellite. Their work is in some ways similar to the work of Harel, et al., but does not attempt to self consistently describe plasma-field interactions. Hays and Roble showed results of quantitatively modeling the effect of magnetospheric processes on the earth's upper atmosphere. They have found that heating of the upper atmosphere in the auroral regions can be substantial during substorms, enough to cause appreciable variations in the thermosphere.

These papers have all dealt with interactions of the various regions of the magnetosphere and typically also involve both plasmas and fields. There are other papers in the remainder of this section that dealt more specifically with the techniques involving a particular magnetospheric process or procedure. Kisabeth reported on the use of the concept of magnetic charge in order to accurately represent various magnetospheric current systems. The technique is especially well suited for the modeling of field aligned currents. He has also examined the influence of induction currents in the earth's crust on the total magnetospheric magnetic field and induced electric fields at satellite altitudes. Whipple emphasized the usefulness of a rigorous kinetic-theory approach to large-scale magnetospheric modeling, and he demonstrated the approach by applying it to the problem of the interaction between a slowly flowing, collisionless plasma and a magnetic dipole. It is probable that this kinetic approach will shed light on the nature of important magnetospheric features. Luhmann and Schulz described a simplified magnetic shell tracing procedure. The accurate definition and determination of drift shells is important in the study of the motions of low energy particles in the magnetosphere. The availability of a computer-efficient routine for the determination of drift shells will be of great help in the study of low energy particle energization. In a somewhat similar paper Kosik discussed the motion of energetic electrons in terms of electric and magnetic fields and their asymmetries in the region of geosynchronous orbit. This work is analytic but model dependent. Finally, Michel discussed a new method for determining the shape of the magnetosphere. In his shape integral method more complicated geometries than the classical "neutral point" closed magnetosphere may be considered. He suggests that the shape integral method can be used to simultaneously consider the magnetopause and the cusp region current systems. The procedures for such an analysis have been outlined.

These papers indicate that the magnetospheric community is pursuing several activities which relate directly to the quantitative modeling of specific magnetospheric features. In particular papers which discuss the interaction of plasmas and fields have been presented. Also, several papers were given which describe specific techniques that can be used generally to better model magnetospheric behavior.

A KINETIC APPROACH TO MAGNETOSPHERIC MODELING

E. C. Whipple, Jr.

Department of Physics, University of California, San Diego
La Jolla, California 92093

Abstract. The earth's magnetosphere is caused by the interaction between the flowing solar wind and the earth's magnetic dipole, with the distorted magnetic field in the outer parts of the magnetosphere due to the current systems resulting from this interaction. It is surprising that even the conceptually simple problem of the collisionless interaction of a flowing plasma with a dipole magnetic field has not been solved. A kinetic approach is essential if one is to take into account the dispersion of particles with different energies and pitch angles and the fact that particles on different trajectories have different histories and may come from different sources. Solving the interaction problem involves finding the various types of possible trajectories, populating them with particles appropriately, and then treating the electric and magnetic fields self-consistently with the resulting particle densities and currents. This approach is illustrated by formulating a procedure for solving the collisionless interaction problem on open field lines in the case of a slowly flowing magnetized plasma interacting with a magnetic dipole.

## Introduction

The earth's magnetosphere is basically the result of the interaction between the magnetic dipole field of the earth and the flowing solar wind plasma. The distorted magnetic field in the outer parts of the magnetosphere are caused by current systems resulting from this interaction. In some respects, the problem of the earth with its magnetic field immersed in the solar wind plasma is similar to the problem of the interaction of a probe in a plasma in the context of Langmuir probe theory, except, of course, that in the former we are dealing with a much larger physical scale. The top of the earth's atmosphere can be considered to be an electrode with the possibility of charged particle emission--both as a result of the presence of ionospheric plasma, and also as a result of secondary effects from particle impact and photoemission. The neutral particles in the upper atmosphere also act as scattering centers for the charged particles. The external plasma in which the probe (earth) is immersed is an additional source of particles. Solving the interaction problem involves finding the various types of trajectories, populating them with particles appropriately, and then treating the electric and magnetic fields self-consistently with the resulting particle densities and currents.

It is apparent that a kinetic treatment of this problem is required if one is to distinguish between various types of particle trajectories, since fluid descriptions do not allow for this possibility. A kinetic approach is admittedly complicated because of the fact that the trajectories of a great number of particles must be followed. If the velocity distribution functions of the magnetospheric particles were simple Maxwellians, then it would not be necessary to use a kinetic approach. However, observations indicate that velocity distributions are in general not Maxwellian. In particular, there are frequently relatively sharp changes in the character of velocity distributions as a function of velocity or direction. These sharp changes, which at times are almost discontinuities, suggest that the plasmas in these different regions of velocity space (i.e. in different energy ranges or in different directions) have different histories and probably different origins. Collisionless trajectories must be calculated to ascertain the reasons for these boundaries in velocity space, even when the presence of collisions tends to smear these discontinuities.

It is surprising that even the conceptually simple problem of the collisionless interaction of a flowing plasma with a dipole magnetic field has not been solved. A treatment of this idealized problem would be a logical first step in formulating a kinetic approach to the earth's magnetosphere. In this paper we discuss the interaction of a slowly flowing collisionless magnetized plasma with a dipole field and indicate how a solution to this problem could be obtained.

## The Idealized Dipole Problem

Figure 1 illustrates the simple dipole interaction problem. A flowing plasma with velocity $\vec{V}$ containing a uniform magnetic field $\vec{B}_o$ is incident on a magnetic dipole. We treat the steady state problem where the dipole is immersed in the flowing plasma. The total magnetic field $\vec{B}$ is given by the linear superposition of the dipole component, the uniform component, $\vec{B}_o$, and the interaction component $\vec{B}_1$ caused by the induced currents in the system:

$$\vec{B} = \vec{B}_D + \vec{B}_o + \vec{B}_1 \tag{1}$$

The problem is to find the induced currents and from them the interaction component of the magnetic field.

Our treatment of this idealized problem contains a number of assumptions and approximations. As progress is made, hopefully, on this simple problem, we expect to relax these assumptions so as to begin to include more realistic effects. At least, we anticipate that the errors introduced by these approximations can be estimated. The most serious assumptions as we see them are:

1) The assumption of a steady state. The real magnetosphere is certainly not quiescent, and there may not even be in principle a steady state. Nevertheless we feel that this is a useful starting point.

2) Neglect of collisions and other scattering processes. The mean free path for most collision processes in the outer magnetosphere is much longer than the scale of the system. However, scattering is still important in certain regions, such as the lower magnetosphere,

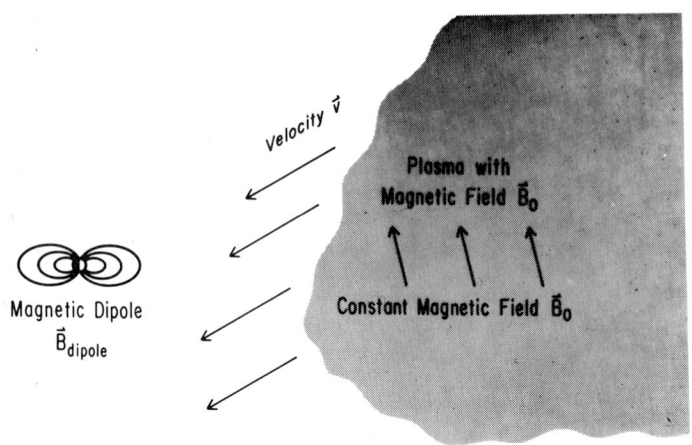

$$\vec{B} = \vec{B}_{dipole} + \vec{B}_0 + \vec{B}_{interaction}$$

where: $\vec{B}_{interaction}$ is due to induced current systems

Fig. 1. The simple dipole problem. The magnetic dipole is immersed in a plasma moving with velocity $\vec{V}$ and containing a uniform magnetic field $\vec{B}_0$.

and for particular effects, such as populating closed orbits for example. We expect to eventually include some kinds of scattering processes, but we neglect them in this paper.

3) The assumption of the constancy of the first two adiabatic invariants. This means that we are neglecting the effects of such things as steep gradients where the scale of the change may be on a smaller scale than the gyroradius of a particle. Such effects may be important at the magnetopause, for example.

Most of the discussion in this paper will in addition be restricted to the geometry illustrated in Figure 2, where the direction of the uniform field $B_0$ is taken to be aligned with the dipole axis, such that in the equatorial plane, the dipole field and the uniform field are either parallel or anti-parallel. The advantage of this geometry is that for small flow velocities one can assume azimuthal symmetry for the problem. With a parallel orientation for the magnetic field directions, the surface of minimum magnetic field does not remain in the magnetic equator. The dashed line in Figure 2 shows the position of the minimum B surface: it stays in the equator from the earth's surface out to about 75% of the distance to the magnetopause (taken as the last closed field line). From there the surface of minimum B goes to higher latitudes, reaching the axis at the "magnetopause" location, and then going out into the "solar wind" plasma.

The magnetic field configuration shown in Figure 2 is the superposition of the dipole and uniform components only. If the density of the flowing plasma is small, then the perturbing effects will be small and the configuration will be only slightly changed by the interaction. It should be possible to start with the treatment of such a case, and then the plasma density can be increased gradually to see how the perturbing field changes.

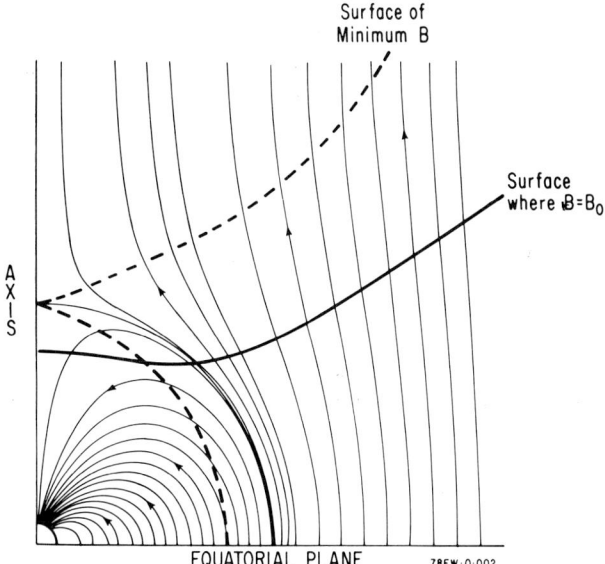

Fig. 2. Magnetic field line configuration for the dipole plus constant magnetic fields ($\vec{B}_D + \vec{B}_o$). In this figure the directions of the two magnetic field components are parallel in the magnetic equator.

## Velocity Distributions

Because of the directions we have chosen for the dipole and uniform magnetic fields, the magnitude of the field is a maximum for any open field line as it crosses the equatorial plane. This is illustrated in Figure 3 where the typical behavior of the magnitude of B is shown as a function of distance along the field line (The drawing is schematic and does not represent an actual calculation.) Particles coming down the field line from the distant solar wind regions where $B = B_o$ can only mirror in regions where $B > B_o$. Consequently, any particles which mirror in regions where $B < B_o$ must be particles which are drifting across field lines. Three kinds of particle trajectories can be identified in the open field line regions: (1) particles coming down field lines from the distant solar wind. These may be regarded as being in a "source cone". At any point, all particles with pitch angles less than the critical pitch angle $\alpha_c$, given by $\sin^2 \alpha_c = B/B_o$ are in this category; (2) particles drifting across field lines which have also come from the solar wind region. These particles will have pitch angles between the critical pitch angle $\alpha_c$ and $90°$; (3) Particles which are trapped in drift orbits such that they have not come directly from the distant solar wind. These particles also have pitch angles between $\alpha_c$ and $90°$. In general, these trapped particles will have higher energies than those in category (2). The population of these trapped particles is determined primarily by the scattering mechanisms which constitute production or loss processes for these trajectories, or by time-dependent effects. For now, we make the assumption that

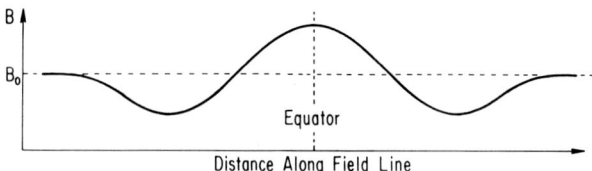

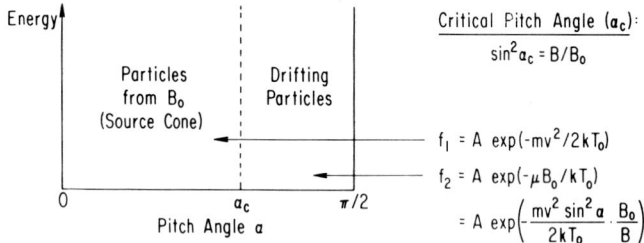

Fig. 3. An illustration of how the magnitude of the magnetic field varies for an open field line, showing how this can be used to infer the particle distribution functions in the perturbed region.

particles in categories (2) and (3) have the same distribution functions. This assumption will be discussed again later in the paper.

The distribution functions of the particles in the source cone and in the drift cone are obtained by using Liouville's theorem together with the known properties of the incident flowing plasma. For small flow velocities, defined by the condition that in the distant undisturbed plasma the drift velocity $v_D$ is small compared to the thermal velocity, $v_t = (2kT_o/m)^{1/2}$, where $T_o$ is the temperature in the undisturbed plasma, we may take the undisturbed velocity distribution functions to be Maxwellian. At a point in the disturbed region, particles in category 1 which are in the source cone will also have a Maxwellian distribution. The velocity distribution for particles in the drift cone is obtained by noting that in the undisturbed plasma such particles must have pitch angles of 90°. Therefore their distribution function may be expressed in terms of the magnetic moment: $f = A \exp(-\mu B_o/kT_o)$. Since the argument of the exponential is an invariant of the motion, the same expression may also be used for the distribution function after the particles have reached the disturbed region. Hence the distribution functions for the two categories may be written as follows:

$$f_1 = n_o (m/2\pi kT_o)^{3/2} \exp(-mv^2/2kT_o) \text{ for } \alpha < \alpha_c \qquad (2)$$

$$f_2 = n_o (m/2\pi kT_o)^{3/2} \exp(-\mu B_o/kT_o) \text{ for } \alpha > \alpha_c \qquad (3)$$

$$= n_o (m/2\pi kT_o)^{3/2} \exp\left(\frac{-mv^2 \sin^2\alpha}{2kT_o} \cdot \frac{B_o}{B}\right) \qquad (4)$$

as shown in Figure 3. Note that in regions where $B > B_o$, there are no particles in category 2.

There are two more assumptions which have been made implicitly in this treatment. The first is related to the condition that the drift velocity is small compared to the thermal velocity. This implies that we have also neglected the drift energy compared to the thermal energy. Second, we have assumed that the electric field has negligible components parallel to the magnetic field. This means that the potential at any point can be obtained by merely tracing the local magnetic field line out into the distant plasma where the potential has a uniform gradient given by

$$\vec{\nabla}\phi = -\vec{E}_o = -\vec{v}_D \times \vec{B}_o \qquad (5)$$

Current Densities

In a neutral plasma in a magnetic field there are three possible contributions to the current density in a steady state: the magnetization current, $\vec{J}_M$; the curvature current, $\vec{J}_C$; and the gradient current, $\vec{J}_G$. The expressions for these are (e.g. Longmire, 1967):

$$\vec{J}_M = -\nabla \times \left(\frac{w_\perp \vec{B}}{B^2}\right) \qquad (6)$$

$$\vec{J}_C = \frac{2w_\parallel}{B^2} \vec{B} \times \left(\frac{\vec{B}}{B} \cdot \vec{\nabla}\right)\left(\frac{\vec{B}}{B}\right) \qquad (7)$$

$$\vec{J}_G = -w_\perp \frac{\vec{B}}{B} \times \vec{\nabla}\left(\frac{1}{B}\right) \qquad (8)$$

In these expressions, the quantity w is the kinetic energy density; we must calculate the parallel ($w_\parallel$) and perpendicular ($w_\perp$) components of the kinetic energy density from the distribution functions given in expressions (2) and (4). The results are:

$$w_\parallel = \frac{n_o kT_o}{2} \text{ for } B > B_o \qquad (9)$$

$$w_\parallel = \frac{n_o kT_o}{2}\left[1 - \left(1 - \frac{B}{B_o}\right)^{\frac{5}{2}}\right] \text{ for } B < B_o \qquad (10)$$

$$w_\perp = n_o kT_o \text{ for } B > B_o \qquad (11)$$

$$w_\perp = n_o kT_o \left[1 - \left(1 - \frac{B}{B_o}\right)^{\frac{3}{2}} \left(1 + \frac{3}{2}\frac{B}{B_o}\right)\right] \text{ for } B < B_o \quad (12)$$

Note that neither the energy densities given in expressions (9) through (12) nor the current densities given in expressions (6) through (8) depend explicitly on the charge or mass of the particle. This independence of $\vec{J}$ on the type of particle is due partly to the assumption of invariance of the magnetic moment which led to the same critical pitch angle, $\alpha_c$, for both ions and electrons, and partly to the neglect of scattering effects which would, of course, affect ions differently than electrons. This means that quasi-neutrality for the plasma is satisfied, and in addition, the only unknown quantity appearing in the expression for the total current density is the interaction component of the magnetic field, $\vec{B}_1$. Consequently, it is possible to use Maxwell's equation to solve for this unknown quantity.

Since
$$\vec{J} = \vec{J}(\vec{B}) = \vec{J}(\vec{B}_1) \quad (13)$$

we have
$$\vec{\nabla} \times \vec{B}_1 = \mu \vec{J}(\vec{B}_1) \quad (14)$$

where
$$\vec{B} = \vec{B}_D + \vec{B}_o + \vec{B}_1 \quad (15)$$

In obtaining equation (14) we have made use of the fact that the curl of both the dipole component ($\vec{B}_D$) and the constant component ($\vec{B}_o$) vanish.

## Method of Solution

The method we have used for the solution of equation (14) for the perturbed field component $\vec{B}_1$ is to work with the vector potential. All of the components of the current density and therefore the total current density are in the azimuthal direction, and we need only to deal with the one component. We define the vector potential $\vec{A}_1$ by

$$\vec{\nabla} \times \vec{A}_1 = \vec{B}_1, \quad \vec{A}_1 = A_1 \vec{n}_\phi \quad (16)$$

It is convenient to define new variables $\vec{b} = \vec{B}/B_o$, $\vec{b}_1 = \vec{B}_1/B_o$, $\vec{a}_1 = \vec{A}_1/B_o$, etc. We also define the dimensionless energy density parameter $P$ by $P = \mu n_o kT_o / B_o^2$. Then equation (14) may be written in the following scalar form (using MKS units):

$$\nabla^2 a_1 = \frac{a_1 \csc^2 \theta}{r^2} - j \quad (17)$$

where the quantity j is given by

$$j = \frac{\frac{5}{2} P (1-b)^{\frac{1}{2}} \left(1 + \frac{b}{2}\right) \left(\frac{\cos\beta}{r} \frac{\partial b}{\partial \theta} - \sin\beta \frac{\partial b}{\partial r}\right)}{\left[b - \frac{5}{2} P (1-b)^{3/2}\right]}. \quad (18)$$

The first term on the right hand side of (17) comes from the expansion of $\vec{\nabla} \times \vec{\nabla} \times \vec{a}_1$ in spherical coordinates $(r, \theta, \phi)$. The angle $\beta$ is the angle between the local radius vector and the total magnetic field vector $\vec{B}$ (or $\vec{b}$). Note that the quantity (5P) is the product of the usual ratio of specific heats (5/3) and the total kinetic energy density in the undisturbed region $(3/2\ n_0 k T_0)$, divided by the magnetic field energy density $(B_0^2/2\mu)$.

The form of equation (17) is the same as Poisson's equation and consequently techniques for solving Poisson's equation may be used here. One method we are trying is a numerical approach, defining a grid of points in the $(r,\theta)$ plane and solving numerically for $a_1$ at each of these grid points. An iteration procedure is involved with the following steps:

1) Start by computing j at each grid point using the undisturbed values for the magnetic field.

2) Use these values for j to obtain a temporary solution for $a_1$ by solving Poisson's equation numerically.

3) Use these values for $a_1$ to obtain new and better values for $\vec{b}_1$ and hence for b. Compute new and better values for j at each grid point.

4) Repeat the process until the procedure converges.

Another method which we are exploring involves the expansion of the vector potential in appropriate functions of the coordinates r and $\theta$.

## Discussion

We have obtained some preliminary solutions to the magnetic field configuration for small values of the pressure parameter P. However, for values of P larger than about 0.20 the iteration procedure outlined in the previous section did not converge. This difficulty in converging is probably associated with the fact that the denominator in (18) goes to zero for a certain value of the magnetic field. This behavior may be due to a breakdown in the assumed adiabatic behavior of the plasma. The denominator in (18) goes to zero when the curl of $\vec{B}$ due to the magnetization current $J_m$ cancels the curl of $\vec{B}$ on the left-hand-side of Maxwell's equation. Thus if there were no other currents in the problem, the magnetic field would vanish because of the diamagnetic effect of $J_m$ and it would not be safe to assume constancy of the first (or second) adiabatic invariants.

Another way of looking at the vanishing of the denominator in (18) is that this condition defines a surface where the curl of B diverges. This implies a very large current density in this surface. It is likely that the solution to the problem for the magnetic field configuration takes on different forms in the two regions on opposite

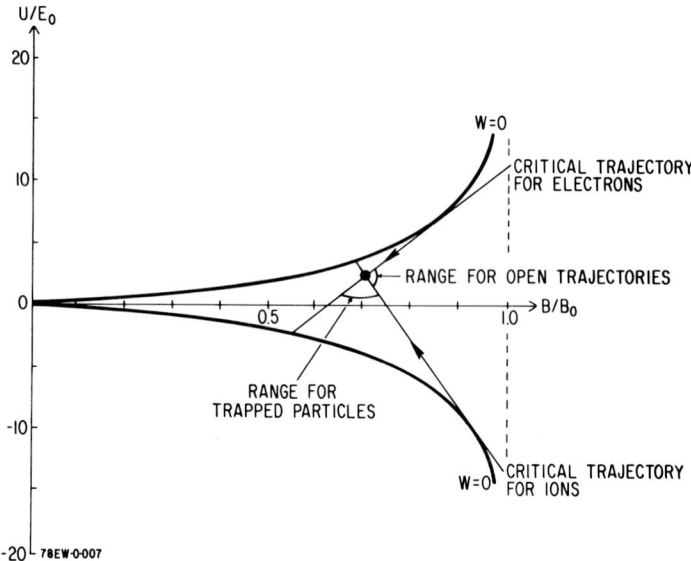

Fig. 4. A (U,B) plot for a point on the minimum B surface of Figure 2 showing how the critical trajectory separating drifting from trapped particles is different for ions as opposed to electrons.

sides of this surface. Across this surface it is necessary to invoke non-adiabatic behavior to get a detailed description of the plasma behavior. The thickness of the surface must then be related to the gyro-radii of the particles. Hence this surface appears to have some of the properties that are usually associated with the magnetopause.

We have assumed that the plasma exists here only on open field lines. Realistically, the plasma on closed field lines must be considered as well as the problem of plasma flow from the open to closed field line regions.

The azimuthal symmetry in the problem as we have discussed it comes from the equal treatment given to the ions and electrons. Two effects will destroy this symmetry: (1) scattering of particles which must necessarily affect ions differently than electrons, whether the scattering is due to collisions, or whether it is due to some kind of wave-particle interaction; (2) a proper treatment of particles in trapped orbits. Even though the drift velocity of the particles in the undisturbed plasma may be small, the electric field associated with this drift as the particles move into the disturbed regions is an important factor in determining the boundary between open and trapped orbits.

The importance of the electric field in determining the boundary between open and trapped orbits may be most easily seen by using the (U,B,K) coordinate system (Whipple, 1978). Figure 4 shows such a plot in the (U,B) plane for a point on the minimum B surface of Figure 2, where the modified longitudinal invariant K is zero. (U is the electric potential, and B the magnitude of the magnetic field.) The two curves which are asymptotic to the vertical line at $B = B_0$ are the boundary curves given by the condition that the generalized velocity

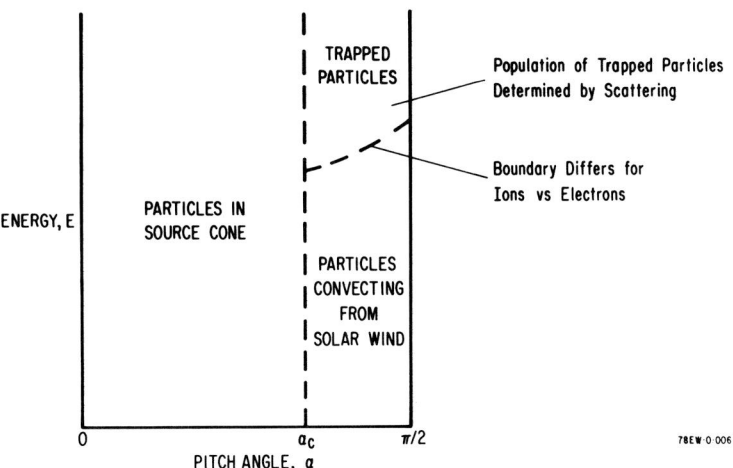

Fig. 5. Boundaries in velocity space showing how different regions in the energy/pitch angle plane are occupied by particles on different kinds of trajectories.

function W vanishes. In this plot, the drift trajectories of particle mirror points follow straight line trajectories. The critical trajectories separating open orbits (particles drifting in from the undisturbed regions) from closed trajectories (trapped particles) are given by the straight lines tangent to these $W = 0$ boundary curves. At an arbitrary point, the critical trajectory for ions will have a different slope than the corresponding critical trajectory for electrons, as shown in the figure. The slope of the line is proportional to the particle magnetic moment and thus its kinetic energy. Consequently, the boundary in velocity space between trapped and drifting particles from the solar wind will differ for ions versus electrons, as shown in Figure 5. Since the electric field is not azimuthally symmetric, it follows that this symmetry will be broken by this effect.

In conclusion, we emphasize that the treatment so far is very preliminary. Our next step must be to obtain a consistent numerical (or other) procedure for solving the magnetic field equation, taking into account the fact that there may be different regions where the solution may have different configurations. Then the restriction to azimuthal symmetry must be removed by both bringing in scattering processes in some way, and also by considering the effects of trapped orbits for electrons and ions. At some point, it will be necessary to consider non-adiabatic processes in regions where B is small or where gradients are large.

Acknowledgments. Part of this work was done while I was a guest worker at the Royal Institute of Technology in Stockholm. I thank Professors Alfvén, Block, and Fälthammar for their invitation and hospitality. I also thank Dr. Lee W. Parker for his work in adapting his computer program for the electrostatic Poisson equation to this problem, and Marian Greenspan for checking the mathematics leading to equations (9) through (12) and (18). This work was supported by the

Atmospheric Research Section of the National Science Foundation under grant ATM 77-04857, by the Air Force Geophysics Laboratory under contract AF-F19628-76-C-0014, and by the National Aeronautics and Space Administration under grant NGL 05-005-007.

## References

Longmire, C. L., Elemetary Plasma Physics, Wiley and Sons, New York, 1963, p. 50.

Whipple, E. C., (U,B,K) coordinates: a natural system for studying magnetospheric convection, J. Geophys. Res., 83, 4318, 1978.

ON CALCULATING MAGNETIC AND VECTOR POTENTIAL FIELDS DUE TO LARGE-SCALE MAGNETOSPHERIC CURRENT SYSTEMS AND INDUCED CURRENTS IN AN INFINITELY CONDUCTING EARTH

J.L. Kisabeth

Institute of Earth and Planetary Physics, Department of Physics, University of Alberta, Edmonton, Alberta, Canada T6G 2J1

Abstract. The use of model current systems in the study of both ground and satellite magnetic field observations is heavily dependent upon the speed and accuracy at which model field calculations can be made. The fact that satellites pass through regions containing currents may further increase the difficulty of the calculations. By utilizing distributions of equivalent magnetic charge and dipoles along with current distributions, both speed and accuracy can be greatly improved even for complex current systems external to an infinitely conducting earth.

Since temporal changes of vector potentials due to time-dependent model current systems are commonly used to infer induction electric fields, the solution for the vector potential associated with an arbitrary distribution of current external to an infinitely conducting earth is given. Preliminary results show that, if induced currents in the earth are neglected, induction electric field estimates may be in serious error, even at satellite altitudes.

## 1. Introduction

The International Magnetospheric Study (IMS) involves the installation of sophisticated arrays of magnetometer stations (e.g. Untiedt et al., 1978) around the world along with ground based experiments measuring ionospheric electric fields and conductivities (Horwitz et al., 1978; Greenwald et al., 1978). Also, continuing and new satellite experiments designed to measure electric and magnetic fields at various positions in the magnetosphere provide an invaluable complement to these ground-based data (McDiarmid et al., 1977; Iijima and Potemra, 1978; Mozer et al., 1978). The excellent data resulting from this international effort should provide the scientific community the unique opportunity of unravelling many of the mysteries of magnetospheric current systems. The fact that both satellite and ground-based measurements are available should help to remove some but not all of the non-uniqueness inherent in the study of magnetic fields and causative current systems.

The proper utilization of this expanded and improved data set has placed heavy demands on the modelling community to develop more complex and realistic models which would also be available for general use. Furthermore, since magnetospheric models have become

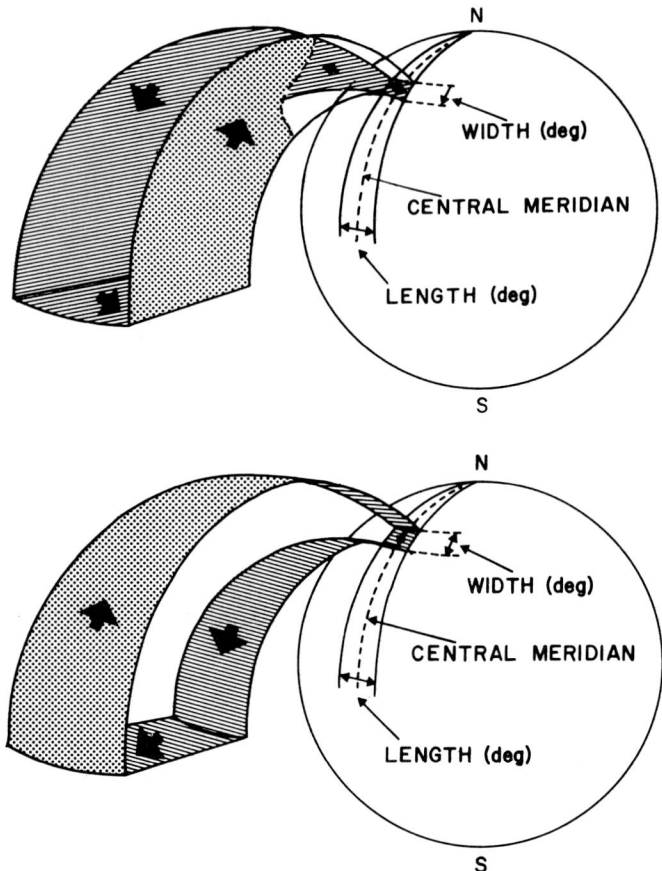

Fig. 1. Model three-dimensional E-W (top) and N-S current systems, each of finite latitudinal and longitudinal extent. These correspond to the Boström no. 1 and Boström no. 2 model respectively.

more accurate in representing the real quiet-time magnetosphere (Olson et al., 1978), there is an urgent need for storm-, substorm-, and quiet-time current system models that would be compatible with these existing magnetospheric models in order to produce time dependent ones valid for all levels of magnetic activity. This time dependence also introduces the added complexity of producing induction electric fields throughout the magnetosphere. In this paper some results are presented that may help in this endeavor.

Sections 2, 3 and 4 utilize the concepts of currents, magnetization and magnetic charge in order to develop fast and accurate methods of calculating magnetic fields associated with the two current systems shown in Fig. 1. In Section 5, the use of these two basic current systems to synthesize extremely complex magnetospheric current systems is demonstrated. This new model termed the 'grid cell' model can be used to model both ground-based and satellite magnetic field observations. In Section 6, the effect of a

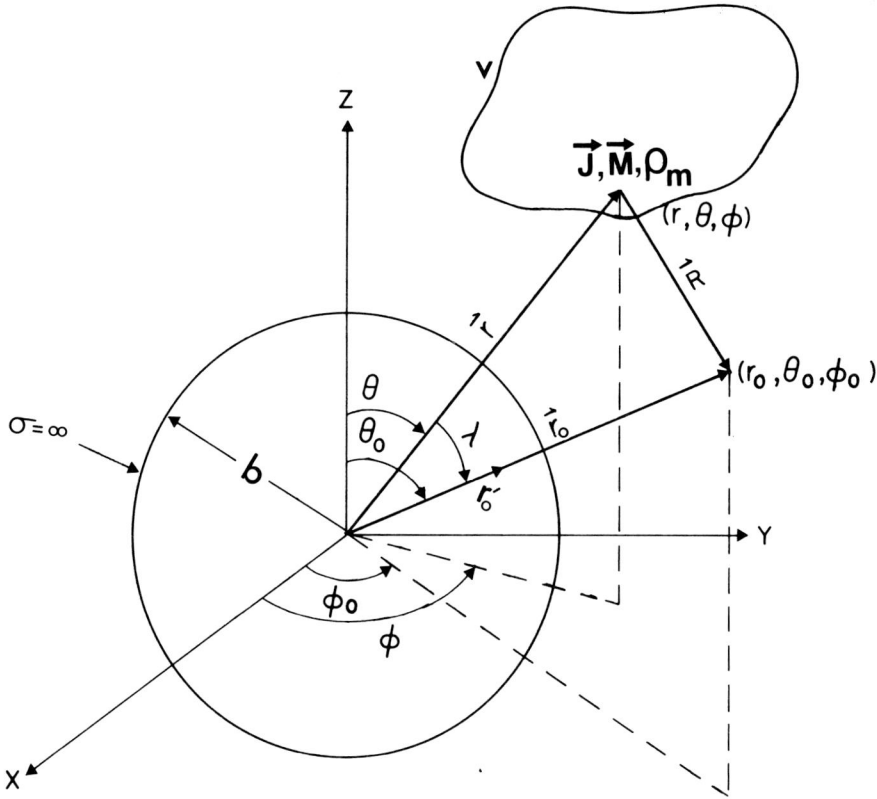

Fig. 2. Diagram defining vectors in spherical coordinates for magnetic field and vector magnetic potential calculations. The conductivity structure of the earth is modelled as an infinitely (superconductive) sphere of radius <u>b</u>.

conducting earth on inductive electric fields derived from time variations of model vector magnetic potential fields is presented. In this particular study, the plasma in the magnetosphere has been neglected (see Mroz et al., 1978). It is, however, believed that the results contained in this section should represent a valid approximation during the first few minutes of a storm or substorm.

## 2. Current Distributions

In this section the method of evaluating the magnetic fields associated with three-dimensional current systems using direct integration over current distributions is presented. Here Biot-Savart's formula is used for the calculation of the magnetic induction

$$\vec{B}(\vec{r}_o) = \frac{\mu_o}{4\pi} \int_v \frac{\vec{J}(\vec{r}) \times (\vec{r}_o - \vec{r})}{|\vec{r}_o - \vec{r}|^3} \, d^3r \quad \text{(MKS)} \tag{1}$$

where $\vec{J}$ is the volume current density vector and v is the volume containing the current system (see Fig. 2). Although it is relatively easy to apply this formula to such distributions as line currents and circular current loops, the problem of applying it to more complex distributions can be very difficult, especially when curvilinear coordinate systems are used. Throughout this paper spherical coordinates are used, but of course, the ideas presented here can be used for various coordinate systems.

In order to simplify the use of (1) in spherical coordinates Kisabeth (1972) (also see Kisabeth and Rostoker, 1977) developed a matrix form of this equation given by

$$B(\vec{r}_o) = \frac{\mu_o}{4\pi} \int_v K(\vec{r}_o,\vec{r}) J(\vec{r}) d^3r \qquad (2)$$

where

$$B(\vec{r}_o) = \begin{pmatrix} B_1 \\ B_2 \\ B_3 \end{pmatrix} \quad (\vec{B}(\vec{r}_o) = B_1 \hat{r}_o + B_2 \hat{\theta}_o + B_3 \hat{\phi}_o)$$

$$J(\vec{r}) = \begin{pmatrix} J_1 \\ J_2 \\ J_3 \end{pmatrix} \quad (\vec{J}(\vec{r}) = J_1 \hat{r} + J_2 \hat{\theta} + J_3 \hat{\phi})$$

and the kernel matrix of the integral is

$$K(\vec{r}_o,\vec{r}) = \frac{1}{R^3} \begin{pmatrix} 0 & ra_{13} & -ra_{12} \\ r_o a_{31} & (ra_{23}+r_o a_{32}) & (r_o a_{33}-ra_{22}) \\ -r_o a_{21} & (ra_{33}-r_o a_{22}) & -(r_o a_{23}+ra_{32}) \end{pmatrix}$$

with elements denoted as $k_{ij}$. The $a_{ij}$'s are elements of the orthogonal transformation matrix relating $(\hat{r}_o,\hat{\theta}_o,\hat{\phi}_o)$ and $(\hat{r},\hat{\theta},\hat{\phi})$ (see Kisabeth and Rostoker, 1977) and

$$R = (r^2 + r_o^2 - 2rr_o a_{11})^{1/2}.$$

It can easily be seen that for a given current density matrix $J(\vec{r})$, expressions for the three components of $\vec{B}$ at $\vec{r}_o$ can be written down immediately.

In this section, (2) is only applied to E-W three-dimensional current systems. Since the use of equivalent magnetic charge for the calculation of N-S three-dimensional current systems is so powerful, this particular system is presented in Section 4.

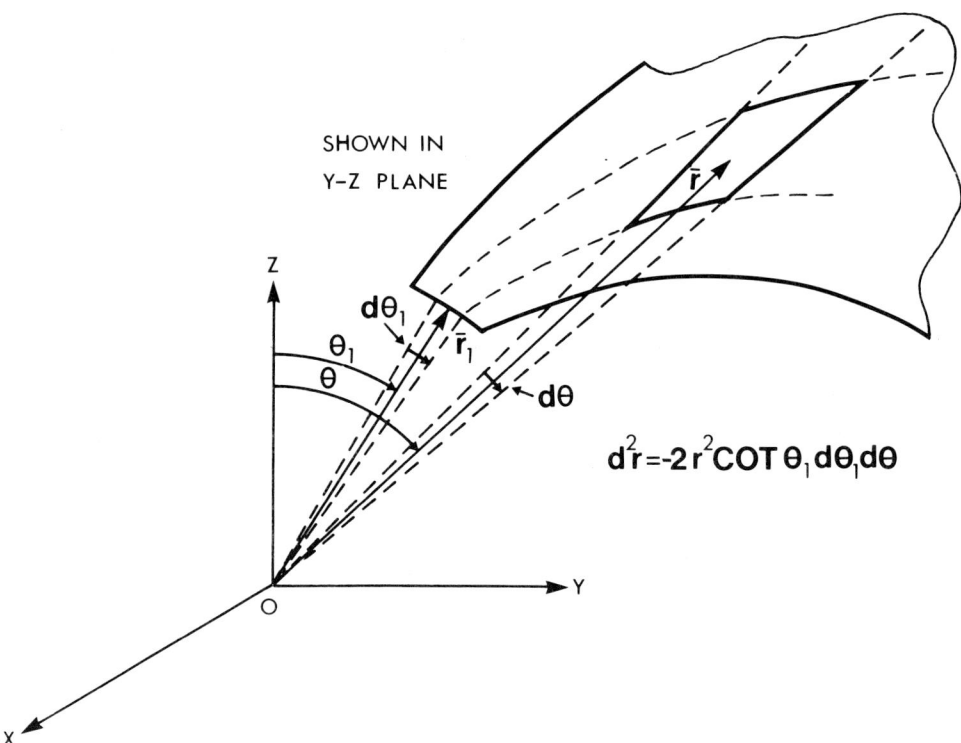

Fig. 3. Diagram showing the change of intergration variables $(r,\theta)$ to $(\theta_1,\theta)$. $d^2r$ represents the differential element of area in the magnetic meridian plane.

The following example is for the system shown in Fig. 1a. It is assumed that the ionospheric current flow is along lines of constant latitude with latitudinal and longitudinal extents of $\theta_N, \theta_S$ and $\phi_W, \phi_E$, respectively and the height-integrated E-W ionospheric current density is $\lambda_\phi$, independent of $\phi$. Thus

$$\vec{J}(\vec{r}) = \lambda_\phi(\theta)\delta(r-r_1)\hat{\phi} \qquad (3)$$

where $r_1$ is the height of the ionospheric current sheet as measured from the center of the earth. Substituting (3) into (2)

$$B(\vec{r}_o) = \frac{\mu_o}{4\pi} \int_0^\infty \int_{\theta_N}^{\theta_S} \int_{\phi_W}^{\phi_E} \begin{pmatrix} k_{31} \\ k_{32} \\ k_{33} \end{pmatrix} \lambda_\phi(\theta)\delta(r-r_1)r^2\sin\theta\, dr\, d\theta\, d\phi \qquad (4)$$

(ionospheric sheet current contribution).

The equatorial plane portion of the current system has a current density vector

$$\vec{J}(\vec{r}) = \lambda_E \frac{\delta(\theta - \pi/2)}{r} \hat{\phi} \qquad (5)$$

where the equatorial surface current density $\lambda_E$ can be expressed in terms of $\lambda_\phi$ for a dipole field (see Section 3, equation 25)

$$\lambda_E = -\frac{\lambda_\phi(\theta_1) r_1}{2r \, \text{ctn}\theta_1}$$

with $r$ and $\theta_1$ given by the equation for a dipole field line intersecting the ionosphere at $r_1$ and $\theta_1$,

$$\frac{r}{\sin^2\theta} = \frac{r_1}{\sin^2\theta_1}. \qquad (6)$$

Before (5) can be used in (2), a change of the variable $r$ to $\theta_1$ at constant $r_1$ must be made using (6) (see Fig. 3),

$$dr = -2r \, \text{ctn}\theta_1 d\theta_1. \qquad (7)$$

Therefore

$$B(\vec{r}_o) = -\frac{\mu_o r_1}{4\pi} \int_{\theta_N}^{\theta_S} \int_o^\pi \int_{\phi_W}^{\phi_E} \begin{pmatrix} k_{31} \\ k_{32} \\ k_{33} \end{pmatrix} \lambda_\phi(\theta_1) \delta(\theta - \pi/2) r \sin\theta d\theta_1 d\theta d\phi \qquad (8)$$

(equatorial sheet current contribution).

The problem of deriving $\vec{J}$ for the field-aligned sheet currents is more difficult. One method is to solve the equation

$$\nabla \cdot \vec{J} = 0$$

and use the given ionospheric surface current density $\lambda_\phi$ as a boundary condition. Another method is to use equivalent magnetization $\vec{M}$ to determine $\vec{J}$. This latter method is used in the next section and yields (see (23) and (24))

$$\vec{J} = \frac{\lambda_\phi(\theta_1) r_1}{2r^2 \sin\theta \, \text{ctn}\theta_1} (2\text{ctn}\theta \hat{r} + \hat{\theta})(\delta(\phi - \phi_E) - \delta(\phi - \phi_W)) \qquad (9)$$

for the eastern and western field-aligned current sheets. Again, using (7) the surface differential (see Fig. 3) can be written

$$d^2 r = r dr d\theta$$
$$= -2r^2 \text{ctn}\theta_1 d\theta_1 d\theta.$$

Thus,

$$\vec{J} \, d^3r = -\lambda_\phi(\theta_1) r_1 r (2\ctn\theta \hat{r} + \hat{\theta})(\delta(\phi-\phi_E) - \delta(\phi-\phi_W)) d\theta_1 d\theta d\phi$$

and

$$B(\vec{r}_o) = \frac{\mu_o r_1}{4\pi} \int_{\theta_N}^{\theta_S} \int_{\theta_1}^{\pi/2} \int_0^{2\pi} \begin{pmatrix} k_{11} & k_{12} \\ k_{21} & k_{22} \\ k_{31} & k_{32} \end{pmatrix} \begin{pmatrix} 2\ctn\theta \\ 1 \end{pmatrix} \lambda_\phi(\theta_1) \quad (10)$$

$$\cdot r(\delta(\phi-\phi_E) - \delta(\phi-\phi_W)) d\theta_1 d\theta d\phi$$

(eastern and western sheet current contributions).

Figure 4 shows an example of a latitude profile beneath an E-W three-dimensional current system along with three latitude profiles illustrating the contributions to the magnetic field from (4) and (10). Since the contribution from the equatorial sheet current is small for this particular system, it is not shown separately. The H, D and Z components are $-B_2$, $B_3$ and $-B_1$ respectively.

Equations (4), (8) and (10) can be combined (after changing the dummy integration variable $\theta$ to $\theta_1$ in (4)) to give

$$B(\vec{r}_o) = \int_{\theta_N}^{\theta_S} L(\vec{r}_o, \theta_1) d\theta_1 \quad (11)$$

where L is the three element column matrix representing the magnetic field due to a three-dimensional 'wire' loop carrying current $(\lambda_\phi(\theta_1) r_1 d\theta_1)$ (see Kisabeth, 1972). Equation (11) was used by Oldenburg (1976) in order to apply Backus-Gilbert linear inversion techniques to polar magnetic substorm fields.

If one wishes to include the effects of an infinitely conducting earth as illustrated in Fig. 2, this can easily be done by using a modified kernel matrix in place of $K(\vec{r}_o, \vec{r})$ in (2). The elements of this modified kernel matrix can be found in Kisabeth (1972) and Kisabeth and Rostoker (1977). The additional computational time necessary to include induction effects is relatively small and thus the modified kernel matrix is used routinely (see Section 5).

## 3. Equivalent Magnetization

The representation of three-dimensional current systems using distributions of infinitesimal dipoles along field lines has been used by Bonnevier et al. (1970). Although the present development yields the same results, its utility lies in providing a relatively simple method of obtaining the current density vectors used in the last section and expressions from which equivalent magnetic charge distributions can be obtained (see Section 4).

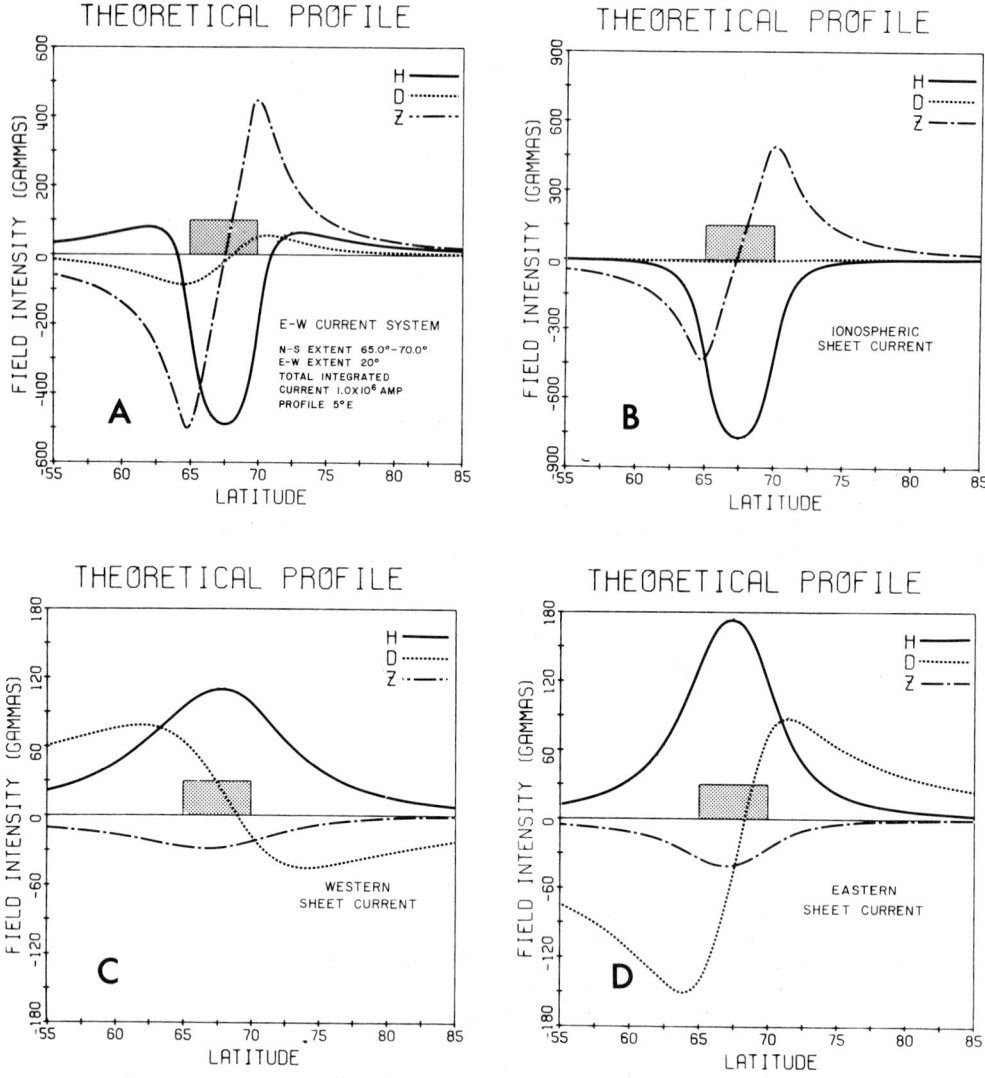

Fig. 4. Latitude profile of the magnetic field components associated with an E-W three-dimensional current system along with profiles attributed to each individual sheet current. Profile A is the sum of profiles B, C and D (after Kisabeth and Rostoker, 1977).

The four pertinent equations from elementary theory of magnetic fields are

$$\psi_m(\vec{r}_o) = \frac{\mu_o}{4\pi} \int_V \frac{\vec{M}(\vec{r}) \cdot (\vec{r}_o - \vec{r}) d^3r}{|\vec{r}_o - \vec{r}|^3} \qquad (12)$$

(scalar magnetic potential)

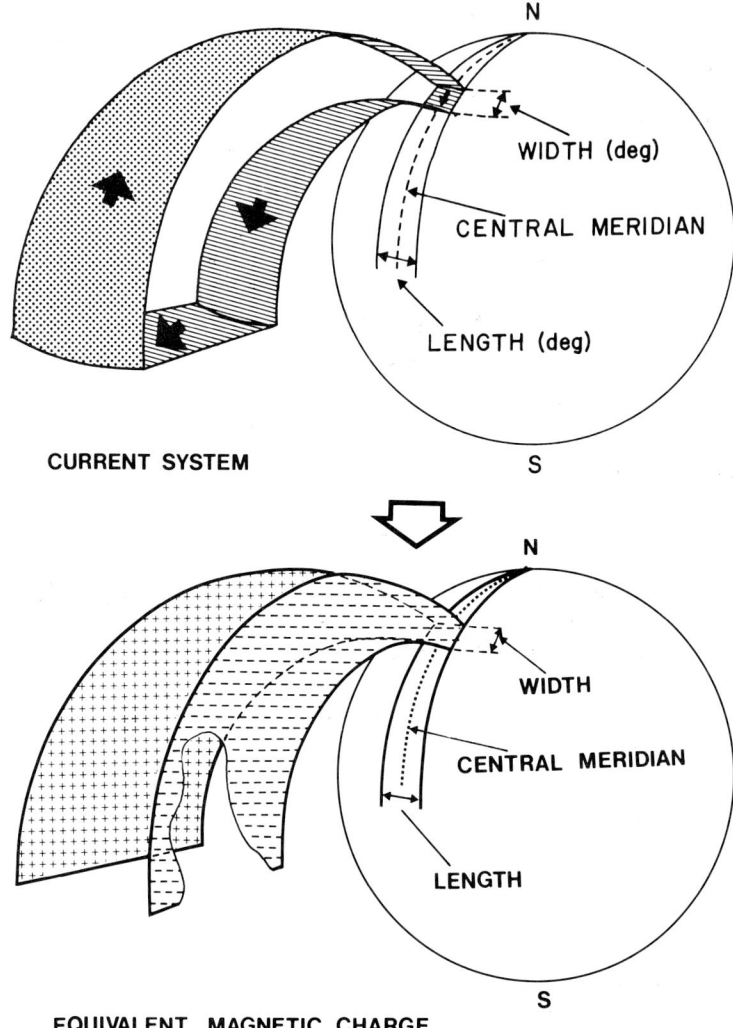

Fig. 5. Model three-dimensional N-S current system along with the equivalent magnetic charge representation (after Kisabeth, 1978a).

$$\vec{B}(\vec{r}_o) = -\nabla_o \psi_m(\vec{r}_o) \quad (13)$$

$$\vec{J} = \nabla \times \vec{M} \quad (14)$$

and

$$\vec{\lambda} = \vec{M} \times \vec{n} \quad (15)$$

(surface current density vector)

where $\vec{M}$ is the volume magnetization vector (see Fig. 2) and $\vec{n}$ the outward normal unit vector to volume $v$. The problem is to find a volume distribution of magnetization $\vec{M}$ which will give rise to the desired current system via (14) and/or (15).

Two conditions which can be imposed on $\vec{M}$ for an E-W three-dimensional current system are:
   i. $\vec{M}$ must be perpendicular to the dipole field lines and must lie in the magnetic meridian plane.
   ii. the volume current density vector must be parallel to the field lines.

These conditions can be expressed mathematically as

$$\vec{M} = M\hat{i}_\perp \quad (16)$$

$$= M \frac{(\sin\theta\hat{r} - 2\cos\theta\,\hat{\theta})}{(3\cos^2\theta+1)^{\frac{1}{2}}}$$

and

$$(\nabla \times \vec{M})_\phi = 0 \quad (\text{i.e., } J_\phi = 0) \quad (17)$$

A solution to (17) is

$$\vec{M} = \frac{K}{r}(\hat{r} - 2\mathrm{ctn}\theta\,\hat{\theta}) \quad (18)$$

where $K$ is a function of $\theta_1$ and $\phi$ to be determined using (15). If the height-integrated ionospheric current density is $\lambda_\phi$ as used in the last section, (15) can be written as

$$\lambda_\phi(\theta_1,\phi)\hat{\phi} = \frac{K}{r_1}(\hat{r} - 2\mathrm{ctn}\theta_1\,\hat{\theta}) \times (-\hat{r}). \quad (19)$$

After solving (19) for $K$, (18) becomes

$$\vec{M} = -\frac{\lambda_\phi(\theta_1,\phi)r_1}{2r\,\mathrm{ctn}\theta_1}(\hat{r} - 2\mathrm{ctn}\theta\,\hat{\theta}). \quad (20)$$

(Note that $\lambda_\phi$ is now a function of both latitude and longitude) and thus the scalar magnetic potential for the E-W system is simply

$$\psi_m(\vec{r}_o) = \frac{\mu_o r_1}{4\pi} \int_{\theta_N}^{\theta_S} \int_{\theta_1}^{\pi/2} \int_{\phi_W}^{\phi_E} \frac{\lambda_\phi(\theta_1,\phi)(\hat{r}-2\mathrm{ctn}\theta\,\hat{\theta})\cdot(\vec{r}_o-\vec{r})r^2\sin\theta d\theta_1 d\theta d\phi}{R^3}. \quad (21)$$

It should be remembered that

$$\vec{r}_o - \vec{r} = (r_o a_{11}-r)\hat{r} + r_o a_{12}\hat{\theta} + r_o a_{13}\hat{\phi}$$

(see Bonnevier et al., 1970 and Kisabeth and Rostoker, 1977) and the integration over $\theta$ must be performed before the $\theta_1$ integration. In order to obtain $\vec{B}$, the gradient of $\psi_m$ is evaluated with respect to the observation coordinates.

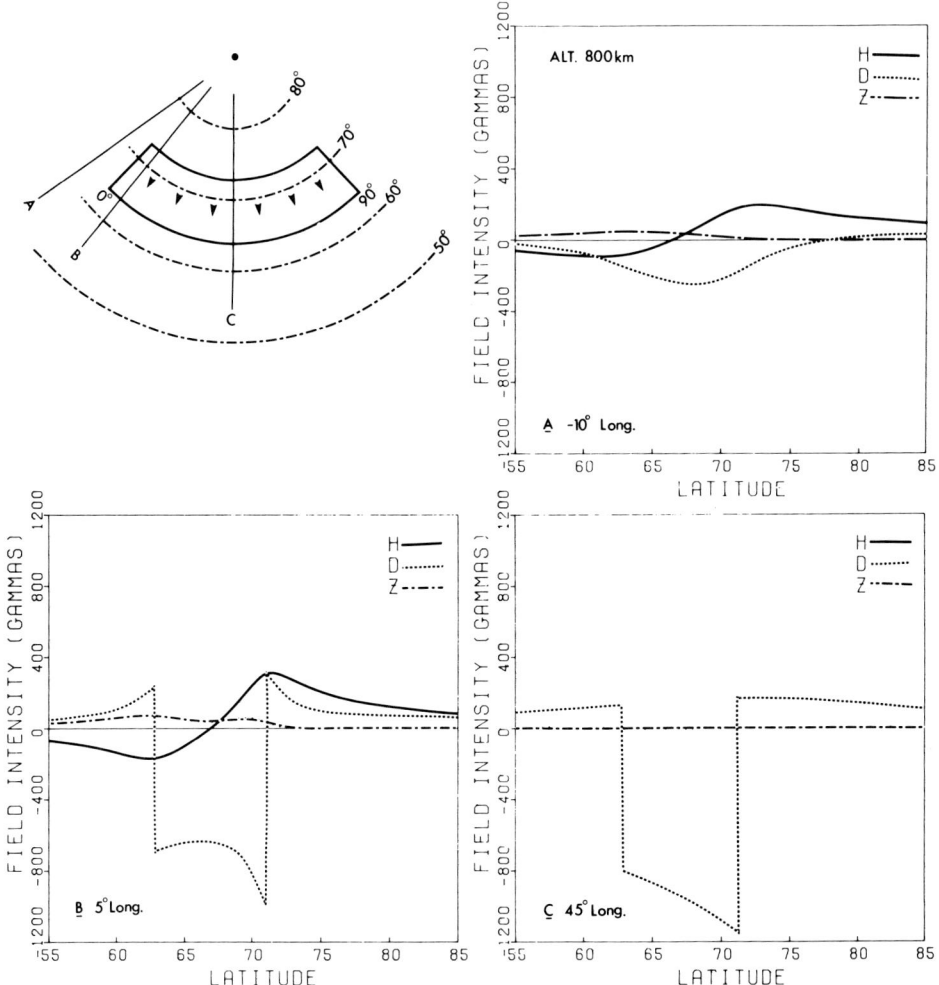

Fig. 6. Simulated satellite passes outside (profile A) and through (profiles B and C) a N-S current system of length 90°. Note that outside the region of current flow (profile A) the D component signature would normally be interpreted as being caused by distributed field-aligned current flow.

Boström (1971) has given a solution for calculating magnetic fields due to dipole distributions external to an infinitely conducting earth as discussed in Section 2 (see Fig. 2).

The equivalent magnetization for the N-S system can be obtained using conditions similar to those applied to the E-W system. The result is

$$\vec{M} = -\lambda_\theta(\theta_1,\phi) \frac{r_1 \sin\theta_1}{r \sin\theta} \hat{\phi} \qquad (22)$$

where $\lambda_\theta$ is the height-integrated N-S ionospheric current density.

Therefore, a combination of the equivalent magnetizations for both the N-S and E-W current systems provides the necessary input to (12) for evaluating current systems with complicated ionospheric current flow resulting from spatial gradients in both the ionospheric electric fields and conductivities. It should be emphasized however, that using this method may consume large amounts of computer time even for relatively simple models, because of the necessity to perform large volume integrations. In Section 5 a method is discussed which circumvents this problem for complex ionospheric current patterns.

In the last section it was noted that the surface current densities used in Biot-Savart's law could easily be evaluated using the equivalent magnetization (see (5) and (9)). For the E-W system and in particular the eastern field-aligned sheet current density

$$\vec{\lambda} = \vec{M} \times \hat{\phi}\Big|_{\phi=\phi_E}$$

$$= \frac{\lambda_\phi(\theta_1)r_1}{2r\ \text{ctn}\theta_1}(2\text{ctn}\theta\hat{r} + \hat{\theta})\Big|_{\phi=\phi_E}$$

or
$$\vec{J}^E = \frac{\lambda_\phi(\theta_1)r_1}{2r^2\text{ctn}\theta_1\sin\theta}(2\text{ctn}\theta\hat{r} + \hat{\theta})\delta(\phi-\phi_E) \qquad (23)$$

and likewise the current density for the western sheet is

$$\vec{J}^W = -\frac{\lambda_\phi(\theta_1)r_1}{2r^2\text{ctn}\theta_1\sin\theta}(2\text{ctn}\theta\hat{r} + \hat{\theta})\delta(\phi-\phi_W). \qquad (24)$$

Also, the equatorial sheet current is

$$\lambda_E\hat{\phi} = \vec{M} \times \hat{\theta}\Big|_{\theta=\pi/2}$$

$$= -\frac{\lambda_\phi(\theta_1)r_1}{2r\ \text{ctn}(\theta_1)}\hat{\phi}$$

and
$$\vec{J}_E = -\frac{\lambda_\phi(\theta_1)r_1}{2r^2\text{ctn}\theta_1}\delta(\theta-\pi/2)\hat{\phi}. \qquad (25)$$

It is, of course, assumed that $\lambda_\phi$ is independent of $\phi$. Similar results may be derived for the N-S three-dimensional system, but it should be remembered that $\lambda_\theta(\theta_1)\sin\theta_1$ must be a constant in order to produce a current system composed of only surface currents as shown in Figure 1b.

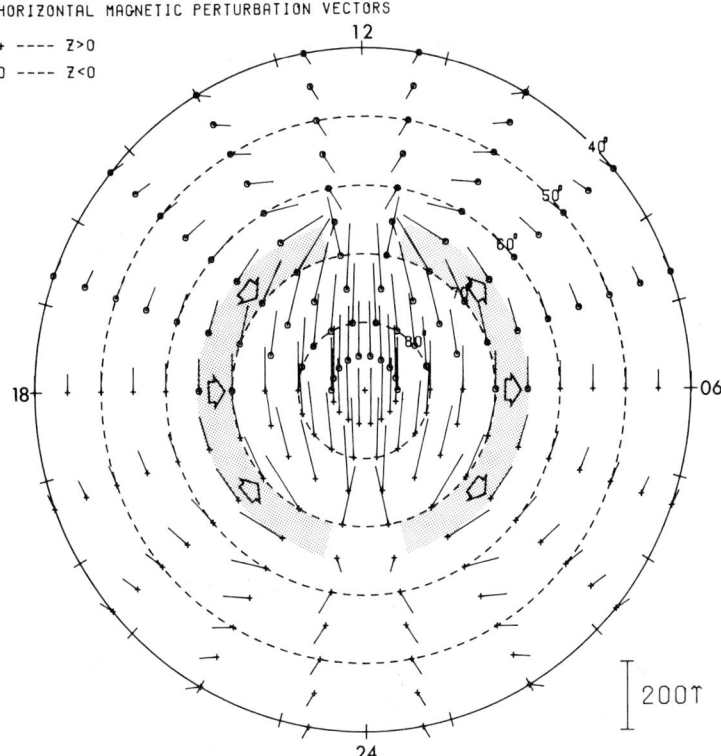

Fig. 7. Magnetic field perturbations associated with two N-S current systems representing the large scale region 1 and 2 field-aligned current distributions. The ionospheric surface current density is 1.0 amp/m at 67.5° latitude.

## 4. Equivalent Magnetic Charge

It is quite evident that the calculation of fields due to large-scale N-S current systems such as those observed by satellites (see Sugiura, 1976; Iijima and Potemra, 1978), would require excessive amounts of computer time if they were derived from (1) or (12). Kisabeth (1978a) has utilized the concept of equivalent magnetic charge in order to alleviate this problem (see Fig. 5). More importantly, this method also allows accurate simulations of satellite magnetic field observations, even when the satellite is in a region containing distributed field-aligned currents (i.e., $\nabla \times \vec{B} \neq 0$).

Again, from elementary theory of magnetic fields,

$$\psi_m(\vec{r}_o) = \frac{\mu_o}{4\pi} \int_v \frac{\rho_m(\vec{r})}{|\vec{r}_o - \vec{r}|} d^3r \qquad (26)$$

and

$$\vec{B}(\vec{r}_o) = \frac{\mu_o}{4\pi} \int_v \frac{\rho_m(\vec{r})(\vec{r}_o - \vec{r})}{|\vec{r}_o - \vec{r}|^3} d^3r \qquad (27)$$

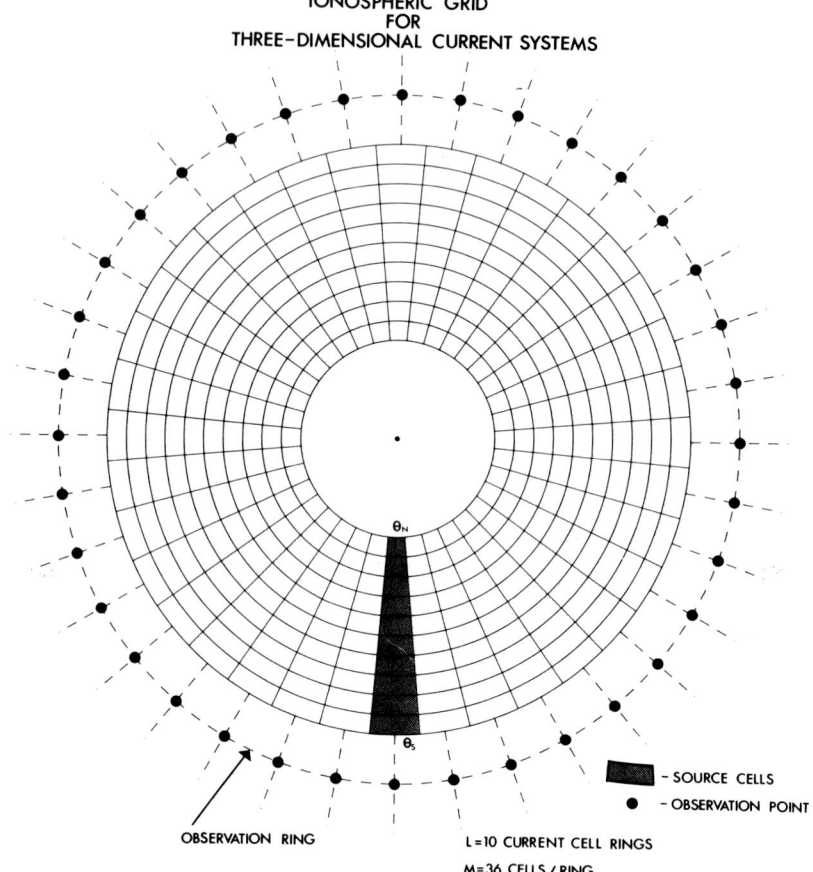

Fig. 8. Ionospheric grid for a 360 cell model three-dimensional current system. The system can be composed of any number of current cell rings, each having an arbitrary latitudinal width and containing M cells. Any number of observation rings can be generated either at the surface of the earth or at satellite altitudes (after Kisabeth, 1978b).

where
$$\rho_m(\vec{r}) = -\nabla \cdot \vec{M} \qquad (28)$$

(volume magnetic charge density)

or
$$\sigma_m(\vec{r}) = \vec{M} \cdot \hat{n} \qquad (29)$$

(surface charge density).

If the equivalent magnetization $\vec{M}$ for a current system is divergence free, the only sources of $\vec{B}$ outside v are given by (29). This is indeed the case for a N-S current system represented by $\vec{M}$ having no

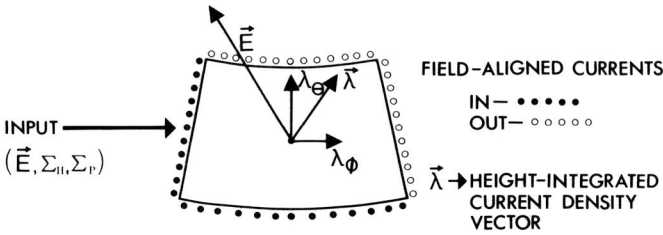

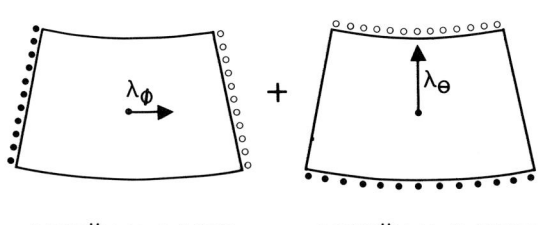

Fig. 9. Composite current system for each ionospheric grid cell shown in Fig. 8. The E-W and N-S current systems for each cell are the same as those illustrated in Fig. 1 (after Kisabeth, 1978b).

$\phi$ dependence. Therefore, using (22) and (29), the surface charge density can be written in terms of a volume density as

$$\rho_m = \lambda_\theta(\theta_1) \frac{r_1 \sin_1}{r^2 \sin^2\theta} \left[ \delta(\phi-\phi_W) - \delta(\phi-\phi_E) \right] \qquad (30)$$

and (27) becomes

$$\vec{B}(\vec{r}_o) = \frac{\mu_o r_1}{2\pi} \int_{\theta_N}^{\theta_S} \int_{\theta_1}^{\pi/2} \int_0^{2\pi} \frac{\lambda_\theta(\theta_1)\cos\theta_1 (\vec{r}_o-\vec{r})(\delta(\phi-\phi_W)-\delta(\phi-\phi_E))d\theta_1 d\theta d\phi}{R^3 \sin\theta}. \qquad (31)$$

Hence, (31) allows rapid calculations of N-S current systems to be performed, regardless of their longitudinal extent. Furthermore, at an observation point inside the volume containing the equivalent magnetization, the magnetic induction is given by

$$\vec{B}(\vec{r}_o) = -\nabla_o \psi_m(\vec{r}_o) + \mu_o \vec{M}(\vec{r}_o). \qquad (32)$$

The term $(\vec{r}_o-\vec{r})/R^3$ has been modified to include the effects of induction currents in an infinitely conducting earth (Kisabeth 1978a).

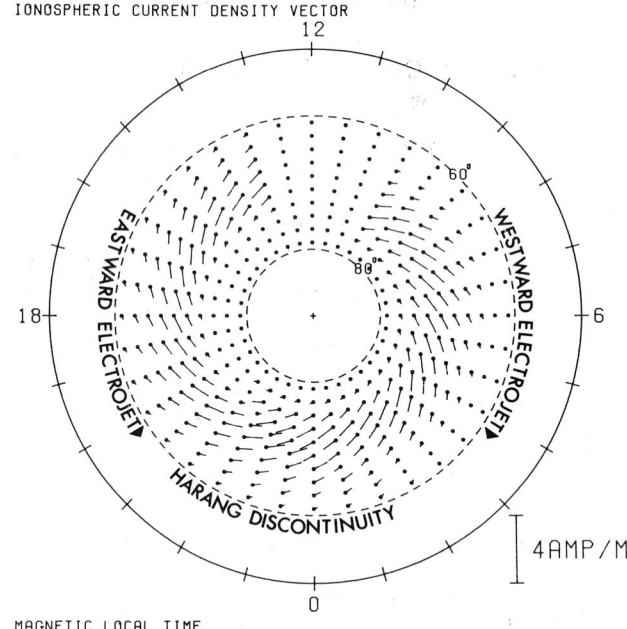

Fig. 10. Distribution of ionospheric current density vectors positioned at the center of each grid cell, which gives rise to eastward and westward electrojets and the Harang discontinuity. The north-south variation of the vectors produces region 1 and 2 field-aligned current distributions.

Figs. 6 and 7 illustrate the usefulness of (31) and (32) for N-S current systems replaced by equivalent charge and dipole distributions. In Fig. 6, three simulated satellite passes are shown, one outside the current system and two penetrating the field-aligned current sheets. Of interest here is the D signature of distributed field-aligned currents (see Zmuda and Armstrong, 1974) in a region void of any such currents (profile A). Also, in profiles B and C, there exist substantial north-south variations of D in regions containing no field-aligned currents. Hence, these profiles show the importance of including edge effects (finite length of current systems) and the proper geometry (i.e. current systems corresponding to dipolar field geometry) when inferring field-aligned current flow from the observed magnetic fields at satellite altitudes.

A common statement made at many conferences and in some of the published literature is that very large N-S current systems (e.g. balanced region 1 and 2 currents) produce large toroidal fields between the field-aligned sheet currents and negligible perturbation fields on the ground. It is hoped that Fig. 7 will remove this misconception. In this figure two N-S current systems are used to represent balanced region 1 and 2 type currents (see Iijima and Potemra, 1978). A surface current density of 1.0 amp/m (defined in the ionosphere at 67.5° latitude) was used to represent very active conditions. For quiet conditions 0.1 amp/m is probably more realistic

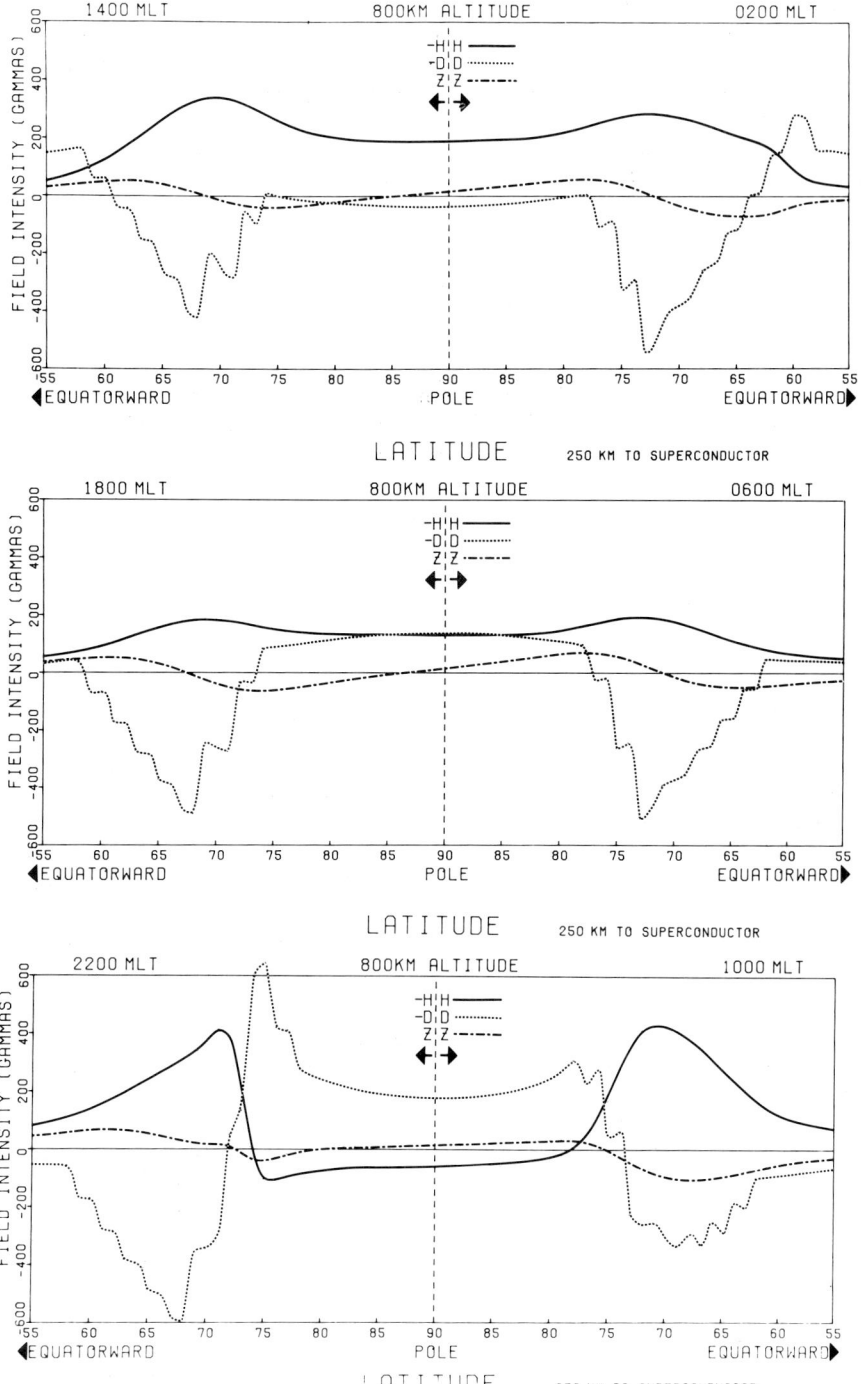

Fig. 11. Simulated satellite passes at an altitude of 800 km through the current system shown in Fig. 10.

and thus the scale in Fig. 7 would be 20 nT instead of 200 nT. Therefore, regardless of whether conditions are quiet or active, the magnetic field perturbations are quite significant at all latitudes. There is no doubt that even Sq fields will contain substantial contributions from these types of current systems.

It should also be noted that (11) in Section 2 can be rewritten to include (31) or (32) as well as (4), (8) and (10), thus giving a compound system represented by

$$B(\vec{r}_o) = \int_{\theta_N}^{\theta_S} N(\vec{r}_o, \theta_1) d\theta_1. \qquad (33)$$

The column matrix N is similar to L in (11), but now contains the contribution to the magnetic field from a field line distribution of equivalent magnetic charge representing a N-S current system of latitudinal width $r_1 d\theta_1$ carrying current $(\lambda_\theta(\theta_1) r_1 \sin\theta_1 (\phi_E - \phi_W))$. Hence, linear inversion techniques can be applied using (33) instead of (11).

### 5. Grid Cell Model Current System

During the past few years and especially during the IMS program, large arrays of magnetometers have contributed greatly to the existing world network of stations (i.e., see Untiedt et al., 1978; Bannister and Gough, 1977, 1978). This enhanced data set has made it necessary to develop more sophisticated model current systems in order to adequately explain magnetic variations observed during quiet and disturbed conditions. One proposed solution is to divide the conducting ionosphere into an array of cells as shown in Fig. 8. Thus each cell would have associated with it a N-S and an E-W three-dimensional current system with height-integrated ionospheric current densities $\lambda_\theta$ and $\lambda_\phi$, respectively. Surface current densities $\lambda_\theta$ and $\lambda_\phi$ could be derived from a knowledge of the electric field $\vec{E}$ and conductivities $\Sigma_H$ and $\Sigma_P$ for each cell (see Fig. 9). The conductivity estimates could possibly be derived using all-sky cameras, photometers, riometers, incoherent radar and other pertinent experiments.

In order to calculate the magnetic fields due to the grid cell model, Z-transforms are taken of various data ensembles representing the N-S and E-W surface current densities ($\lambda_\theta$ and $\lambda_\phi$) in all cells along with ensembles of magnetic field values calculated for the source cells as depicted in Fig. 8. By taking the necessary products of these Z-transforms, the magnetic field on arbitrary observation rings of constant latitude can be evaluated extremely rapidly (Kisabeth, 1978b).

The usefulness of this technique is illustrated in Figures 10 through 13. In Fig. 10, a distribution of current vectors for a typical current system is shown with each vector being positioned at the center of the respective grid cell for a 360 cell system. The surface current densities vary both in latitude and longitude in such

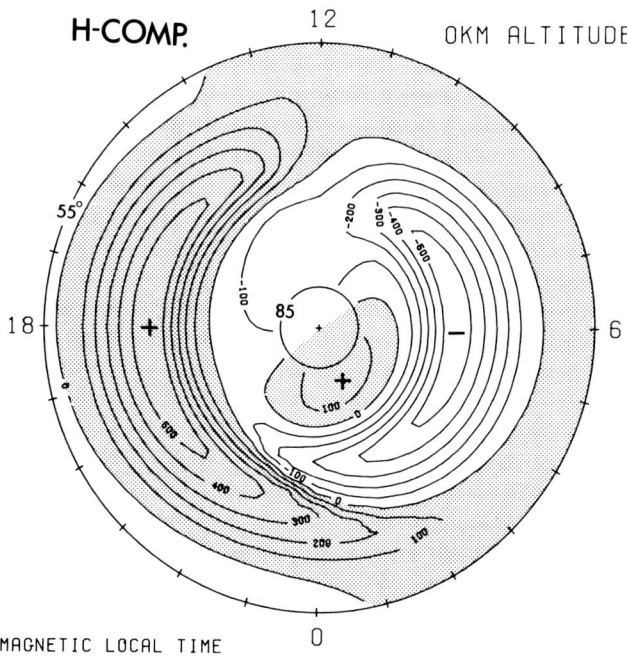

Fig. 12. Contour plot of the H component at the earth's surface due to the current system shown in Fig. 10. The contour interval is 50 nT.

a way as to simulate the eastward and westward electrojets as well as the Harang discontinuity. Examples of latitude profiles showing the predicted magnetic field at satellite altitudes (800 km in this case) are presented in Fig. 11. Note that the H and D component profiles have been inverted in the dusk sector in order to make the dusk-dawn and evening-morning profiles continuous at the pole. It should be pointed out that all three distributions $\vec{J}$, $\vec{M}$ and $\rho_m$ were utilized in the actual calculations of these magnetic field profiles. These profiles are strikingly similar to those observed by Triad and ISIS II satellites (McDiarmid et al., 1977; Iijima and Potemra, 1978).

A contour plot of the H component for this current system is shown in Fig. 12. The large negative and positive regimes associated with the westward and eastward electrojets are clearly defined along with the reverse in sign at lower latitudes. As can easily be seen, this method can allow one to study perturbations along an east-west line of magnetometers at high and low latitudes very easily. Even though the current system for the southern hemisphere would have to be included for low latitude studies, this can readily be done using the cell model.

Fig. 13 demonstrates the power of this technique when studying ΔB variations as reported by Langel (1974a,b). ΔB is defined here as the magnitude of the perturbation in the direction of a pure dipole field. A typical satellite pass as indicated by the arrows in Fig. 13 produces a ΔB profile in almost complete agreement with the typical pass given by Langel (1974a)(his Fig. 3) for OGO 4 data. Of interest

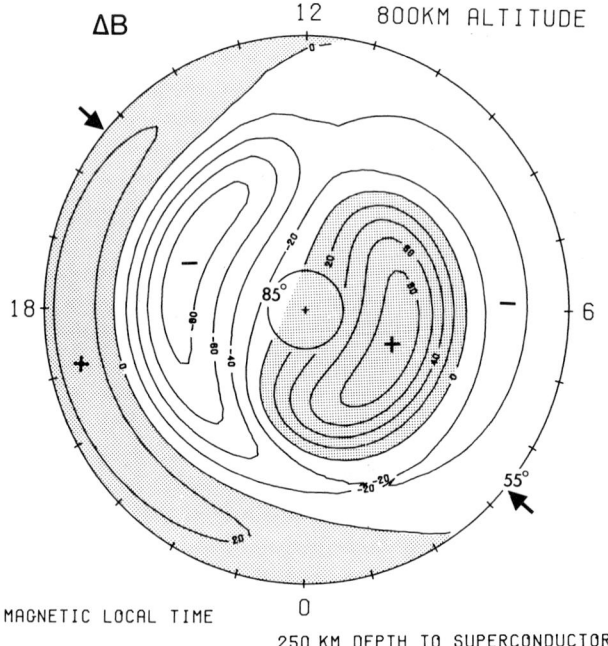

Fig. 13. Contour plot of ΔB at 800 km showing the large regimes of positive and negative ΔB as reported by Langel (1974a,b). The contour interval is 20 nT.

here is the fact that the large offsets of the H component shown in Fig. 11 contribute very significantly to the model ΔB. Also, the induced currents within the earth produce quite large contributions to the overall ΔB pattern. For example, with no induction, the maximum negative and positive ΔB regimes would be ∼100 and -120 nT, respectively.

## 6. Inductive Electric Fields

The importance of inductive electric fields associated with time-dependent magnetospheric current systems has been demonstrated recently (i.e., see Murphy et al, 1974, 1975a,b; Heikkila and Pellinen, 1977; and Mroz et al, 1978). The usual method of obtaining inductive electric fields is by utilizing

$$\vec{E} = -\frac{\partial \vec{A}}{\partial t} \tag{34}$$

where the vector magnetic potential $\vec{A}$ is given by

$$\vec{A}(\vec{r}_o, t) = \frac{\mu_o}{4\pi} \int_v \frac{\vec{J}(\vec{r}, t)}{|\vec{r}_o - \vec{r}|} d^3r . \tag{35}$$

In practice $\vec{A}$ is generally calculated for a given current system

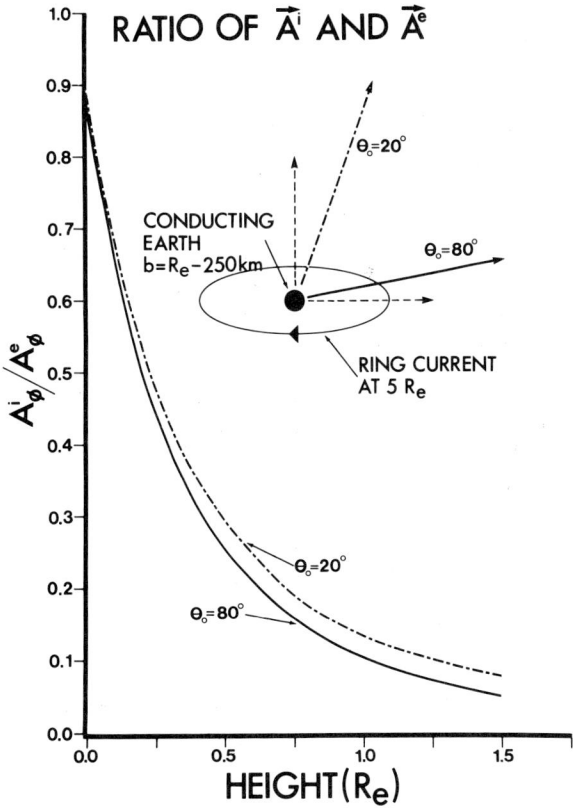

Fig. 14. Graph showing the relative contribution of $\vec{A}^i$ and $\vec{A}^e$ to the total potential for a ring current at 5 Re. The earth is assumed to be infinitely conducting at depths greater than 250 km.

geometry defined by J and v and then a time dependence is added in order that the inductive electric field can be determined.

It is well known that magnetic field variations measured near the earth have significant contributions from induced currents within the earth. Thus an investigation was made to determine what influence a conducting earth might have on inferred inductive electric fields. In order to make this problem mathematically tractable, the infinitely conducting earth approximation was used (see Fig. 2).

By expanding the vector magnetic potential for an arbitrary volume distribution of current in spherical harmonics and incorporating the relevant boundary conditions at the surface of an infinitely conducting sphere of radius b, the total vector magnetic potential (internal + external) was found to be

$$A_i^t(\vec{r}_o) = \frac{\mu_o}{4\pi} \int_V a_{ij} J_j \left( \frac{1}{R} + e_i \frac{b}{r_o R'} - \frac{1}{b} I \delta_{i1} \right) d^3r \quad (36)$$

where
$$\vec{A}^t = A_1^t \hat{r}_o + A_2^t \hat{\theta}_o + A_3^t \hat{\phi}_o$$

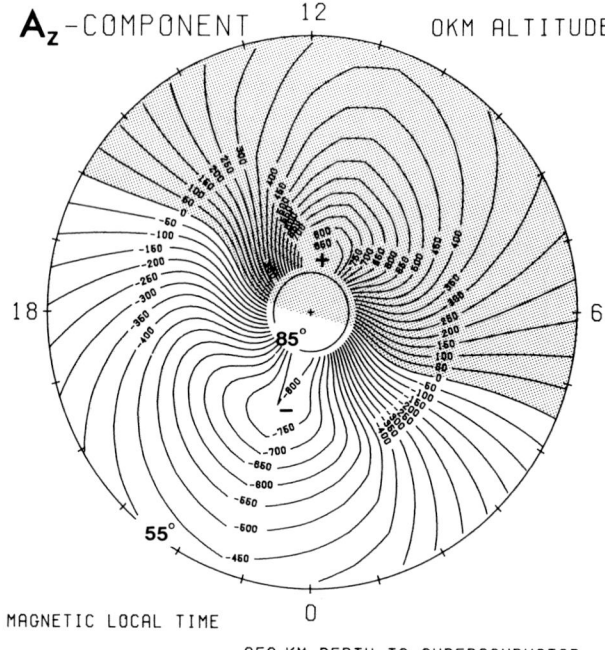

Fig. 15. Contour plot of the $A_z$ component ($A_z = -A_3$) at the surface of the earth due to the current system shown in Fig. 10. The contour interval is 50 volt-sec/km.

$$e_1 = 1; \quad e_{2,3} = -1$$

$$I = \ln\left[\frac{R' + r'_o - ra_{11}}{r(1-a_{11})}\right] \quad (a_{11} \leq 0)$$

$$= \ln\left[\frac{r(1 + a_{11})}{R' - r'_o + ra_{11}}\right] \quad (a_{11} > 0)$$

$$R' = (r^2 + (r'_o)^2 - 2r\, r'_o a_{11})^{\frac{1}{2}}$$

and

$$r'_o = \frac{b^2}{r_o} \quad \text{(see Fig. 2)}.$$

The remaining variables are defined in Section 2. Since the derivation of this formulation is quite lengthy, it will be published separately in a forthcoming paper. It should be pointed out that a similar formulation is available to handle the case where the current system is represented by a volume distribution of equivalent magnetization. It is, however, unfortunate that the concept of magnetic charge cannot be used to evaluate $\vec{A}$.

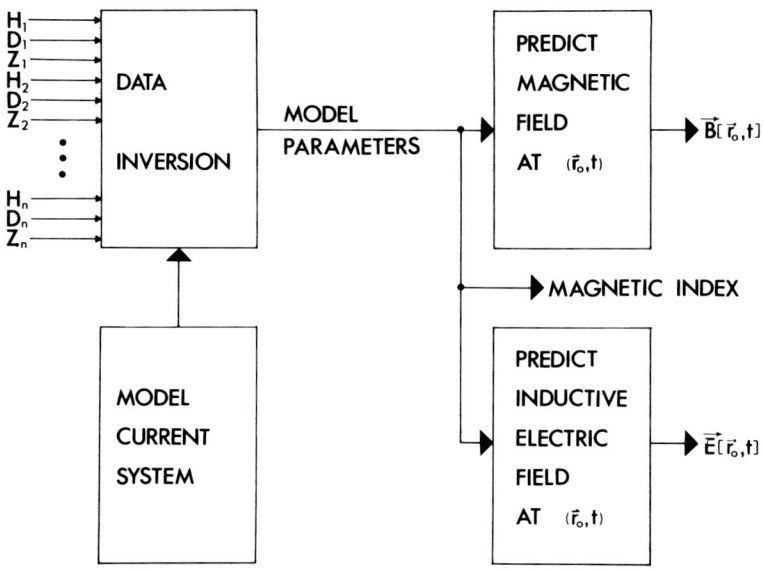

Fig. 16. Block diagram showing a possible magnetic data inversion scheme to predict magnetic and inductive electric fields in various regions of the magnetosphere along with a new storm and substorm magnetic index.

The ratio of $\vec{A}^i$ (internal current system) and $\vec{A}^e$ (external current system) as a function of radial distance above the surface of the earth (the earth is assumed to be infinitely conductive at a depth of 250 km) for a ring current at five earth radii is depicted in Fig. 14. This graph clearly demonstrates the importance of including earth induction effects in vector magnetic potential calculations, even at satellite altitudes. For example, at ~3000 km altitude $A_3^i \approx 0.3\ A_3^e$, thus the horizontal component of $\vec{A}$ is highly attenuated. Although there is no vertical component of $\vec{A}$ for this simple ring current, the radial component of $\vec{A}$ due to an arbitrary current system will in general be enhanced due to the induced current system. This reduction of the horizontal components and enhancement of the vertical component is, of course, just the opposite of how the magnetic field behaves when induced currents are taken into account.

Fig. 15 contains a contour plot of the vertical component of $\vec{A}$ at the surface of the earth ($A_z = -A_1$) for the current system shown in Fig. 10. The contours are labelled in units of volt-sec/km. For example, if this particular current system acquired its present strength linearly over 1000 sec, the vertical component of the induction electric field would be ~0.8 mvolts/m in the polar cap just north of the Harang discontinuity. The horizontal component would, however, be much smaller (~0.04 mvolts/m).

## 7. Conclusions

Probably one of the most important aspects of having three methods (currents, equivalent magnetization and charge) at one's disposal for calculating magnetic fields, is that three independent checks can be made to ascertain whether these field calculations are indeed correct. This is also true for vector magnetic potential calculations where distributions of current and equivalent magnetization may be used. As current system models become more complex, the availability of these consistency checks becomes more valuable. These checks can also be applied to complex magnetotail current systems.

Another benefit derived from utilizing these three methods for magnetic field calculations is that simulated satellite passes can be made through regions of current flow (distributed or sheet currents) without encountering the problems of singularities which would be difficult to handle on the computer. Hence, speed and accuracy of these calculations are greatly enhanced.

The grid cell model presented in Section 5 is a very powerful method of calculating fields in that it is the first available model for generating fields on the ground and in the magnetosphere due to very complex ionospheric electric field and conductivity distributions. It is now possible to check various theoretical and observed ionospheric electric field configurations in terms of what one would expect to, or does observe with magnetometers on the ground and on board satellites. Also, it is believed that a data inversion scheme such as that depicted in Fig. 16 can be realized using the grid cell model. Data from selected ground-based stations and satellites could be used to predict both induction electric and magnetic fields. In addition to these predictions, it may be possible to generate a new magnetic index which could be used in conjunction with AE and Kp indices. This inversion scheme is presently being studied.

Although the pure dipole approximation was used for defining the geometry of the current systems in this paper, the concepts introduced here can be combined with more realistic field configurations such as current flow along distended dipole field lines in the nightside magnetosphere. In fact the magnetic charge formulation is very powerful for calculating magnetic fields due to sheet currents on various L-shell geometries. This is indeed very important when combining ionospheric-magnetospheric current systems with magnetospheric models.

The attempt to study the effects a conducting earth may have on induction electric fields in Section 6 represents only a brief look at what is definitely a very complex problem. However, it does show that the earth must be included as a boundary condition in addition to those associated with various plasma regimes in the ionosphere and magnetosphere, when studying inductive electric fields.

Acknowledgements. The author wishes to thank Dr. G. Rostoker and Dr. K. Kawasaki for many helpful discussions and Muriel Tait for typing the manuscript. This research was supported by the National Research Council of Canada, and in part by a Center for Energy and Mineral Resources (Texas A&M University) grant No. 18761. Part of the computing was done at the National Center for Atmospheric Research Computing Facility, Boulder, Colorado.

## References

Bannister, J.R. and D.I. Gough, Development of a polar magnetic substorm: a two-dimensional array study, Geophys. J.R. astr. Soc., 51, 75, 1977.

Bannister, J.R. and D.I. Gough, A study of two polar magnetic substorms with a two-dimensional magnetometer array, Geophys. J.R. astr. Soc., 51, 1, 1978.

Bonnevier, B., R. Boström and G. Rostoker, A three-dimensional model current system for polar magnetic substorms, J. Geophys. Res., 75, 107, 1970.

Boström, R., The magnetic field of three-dimensional magnetospheric model current systems and currents induced in the ground, Acta. Poly. Scand., 77, 1, 1971.

Greenwald, R.A., W. Weiss, E. Nielson and N.P. Thomson, STARE: A new radar auroral backscatter experiment in Northern Scandinavia, Radio Sci. (in press) 1978.

Heikkila, W.J. and R.J. Pellinen, Localized induced electric field within the magnetotail, J. Geophys. Res., 82, 1610, 1977.

Horwitz, J.L., J.R. Doupnik and P.M. Banks, Chatanika radar observations of the latitudinal distributions of auroral zone electric fields, conductivities and currents, J. Geophys. Res., 83, 1463, 1978.

Iijima, T. and T.A. Potemra, Large-scale characteristics of field-aligned currents associated with substorms, J. Geophys. Res., 83, 599, 1978.

Kisabeth, J.L., The dynamical development of the polar electrojets, Ph.D. Thesis, University of Alberta, Edmonton, Canada, 1972.

Kisabeth, J.L., The use of magnetic charge in the calculation of magnetic fields due to magnetospheric current systems and induced current systems within the earth, (in preparation) 1978a.

Kisabeth, J.L., A new method of calculating magnetic fields due to large-scale magnetospheric current systems using Z-transforms, (submitted to Z. Geophys.) 1978b.

Kisabeth, J.L. and G. Rostoker, Modelling of three-dimensional current systems associated with magnetospheric substorms, Geophys. J.R. astr. Soc., 49, 655, 1977.

Langel, R.A., Near-earth magnetic disturbance in total field at high latitudes 1. Summary of data from Ogo 2, 4 and 6, J. Geophys. Res., 79, 2363, 1974a.

Langel, R.A., Near-earth magnetic disturbance in total field at high latitudes 2. Interpretation of data from Ogo 2, 4 and 6, J. Geophys. Res., 79, 2373, 1974b.

McDiarmid, I.B., E.E. Budzinski, M.D. Wilson and J.R. Burrows, Reverse polarity field-aligned currents at high latitudes, J. Geophys. Res., 82, 1513, 1977.

Mozer, F.S., C.W. Carlson, M.K. Hudson, R.B. Torbert, B. Parady, J. Yatteau and M.C. Kelley, Observations of paired electrostatic shocks in the polar magnetosphere, Phys. Rev. Lett., 38, 292, 1977.

Mroz, G.J., W.P. Olson and K.A. Pfitzer, Magnetospheric fields, Quantitative Modeling of the Magnetospheric Processes, Geophys. Monogr. Ser., vol. 21, edited by W.P. Olson, AGU, Washington, D.C. 1978.

Murphy, C.H., C.S. Wang and J.S. Kim, Inductive electric field of a field-aligned current system, J. Geophys. Res., $\underline{79}$, 2901, 1974.

Murphy, C.H., C.S. Wang and J.S. Kim, Inductive electric field of a time-dependent magnetotail current, Geophys. Res. Lett., $\underline{2}$, 165, 1975a.

Murphy, C.H., C.S. Wang and J.S. Kim, Inductive electric field of a time-dependent ring current, Planet. Space Sci., $\underline{23}$, 1205, 1975b.

Oldenburg, D.W., Ionospheric current structure as determined from ground-based magnetometer data, Geophys. J.R. astr. Soc., $\underline{46}$, 41, 1976.

Olson, W.P., K.A. Pfitzer and G.J. Mroz, Modeling the magnetospheric magnetic field, Quantitative Modeling of the Magnetospheric Processes, Geophys. Monogr. Ser., vol. 21, edited by W.P. Olson, AGU, Washington, D.C. 1978.

Sugiura, M., Field-aligned currents observed by the Ogo 5, and Triad satellites, Ann. Geophys., $\underline{32}$, 267, 1976.

Untiedt, J., R. Pellinen, F. Küppers, H.J. Opgenoorth, W.D. Pelster, W. Baumjohann, H. Ranta, J. Kangas, P. Czechowsky and W.J. Heikkila, Observations of the initial development of an auroral and magnetic substorm at magnetic midnight, Submitted to Z. Geophys., 1978.

Zmuda, A.J. and J.C. Armstrong, The diurnal flow pattern of field-aligned currents, J. Geophys. Res., $\underline{79}$, 4611, 1974.

COMPUTER MODELING OF EVENTS IN THE INNER MAGNETOSPHERE

M. Harel, R. A. Wolf and P. H. Reiff

Space Physics and Astronomy Department, Rice University
Houston, Texas 77001

M. Smiddy

U. S. Air Force Geophysics Laboratory
Hanscom Air Force Base, Massachusetts 01731

Abstract. We have completed a first effort at computer simulating the behavior of the inner magnetosphere during a substorm-type event that occurred on 19 September 1976.

Our computer model simulates many aspects of the behavior of the closed-field-line portion of the earth's magnetosphere, and the auroral and subauroral ionosphere. For these regions, the program self-consistently computes electric fields, electric currents, hot-plasma densities, plasma flow velocities and other parameters.

We present here some highlights of the results of our event simulation. Predicted electric fields for several times during the event agree reasonably well with corresponding data from satellite S3-2. Detailed discussion is presented for a case of rapid subauroral flow that was observed on one S3-2 pass and is predicted by our computer runs. Our computed global distribution of Birkeland current agrees reasonably well with the observations of Iijima and Potemra.

## Introduction

There has been a longstanding effort at Rice aimed at accurate computer modeling of the earth's inner magnetosphere. Our most recent work is aimed at simulating a specific observed event, using some observations as input to the model and using other observations as tests of model predictions.

In this paper, we shall present some results of our first attempts to model an event, specifically the substorm-type event that had its onset at about 1000 UT on 19 September 1976. This particular substorm was chosen for its "clean" character and wealth of data usable both for input and model testing (Harel et al., 1977). Given certain initial and boundary conditions, the program self-consistently computes electric fields and plasma flow velocities in the ionosphere and equatorial plane, horizontal ionospheric and field-aligned (Birkeland) currents, temperatures and densities of magnetospheric plasma-sheet plasma and other parameters. In this brief paper, we cannot discuss the time histories of all of these parameters through the event. Instead, we present here just some highlights of the

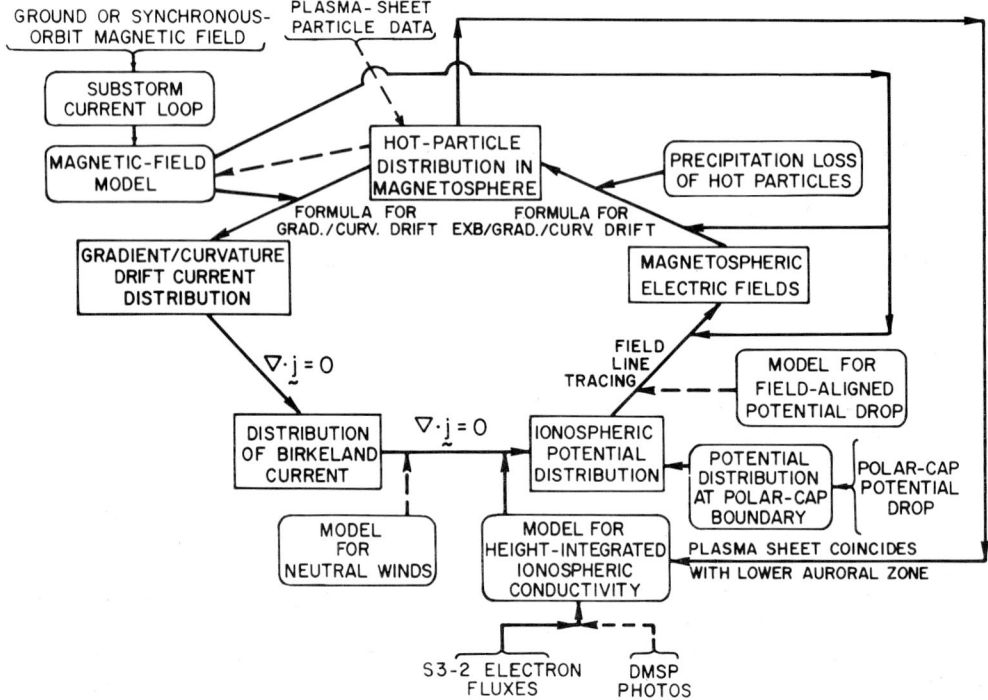

Fig. 1. Overall logic diagram for our program. The central pentagon represents the main computational loop, executed every time step. The rectangles that appear at the corners of the pentagon represent basic parameters computed. Input data are indicated by curly brackets. Subsidiary models, used as input to the main program, are indicated by rectangles with rounded corners. Dashed lines indicate features that we plan to include in the program but have not included yet

results. A much more detailed account will be presented in a future paper (Harel et al., 1979).

The present paper is the latest in a long series of efforts at self-consistent calculation of electric fields and plasma flows in the coupled magnetosphere-ionosphere system (e.g., Karlson, 1963, 1971; Fejer, 1964; Block, 1966; Vasyliunas, 1970, 1972; Swift, 1971; Mal'tsev, 1974; Wolf, 1970; Jaggi and Wolf, 1973; Wolf, 1974; Harel and Wolf, 1976). This work has gradually progressed over the years to include more physical processes and more realistic boundary conditions. In the last few years, some progress has also been made by attacking the ionospheric and magnetospheric portions of the problem separately. Many detailed ionospheric-current and electric-field distributions have been computed assuming, as input, the distributions of ionospheric conductivities and Birkeland currents (Yasuhara and Akasofu, 1977; Nopper and Carovillano, 1978, 1979; Nisbet et al., 1978; Kamide and Matsushita, 1978). Analogously, the injection of ring-current particles has been studied extensively using assumed, though often time-variable, electric fields; these electric fields have been estimated using semiempirical formulas based on data sets of various kinds and for various time periods (e.g., McIlwain, 1974; Roederer and Hones, 1974; Konradi et al., 1976; Cowley, 1976; Kivelson, 1976; Ejiri et al., 1977, 1978). Our approach has the disadvantage of being more complicated and cumbersome than these alternatives, but it has several advantages. Namely, it includes more physics and fewer questionable boundary conditions, and it potentially can provide a comprehensive view of both ionospheric and magnetospheric aspects of an observed event.

## Assumptions and Logic

We attempt to model only the inner magnetosphere, specifically the region where magnetic field lines are certainly closed, available magnetic-field models can be applied with some confidence and plasma-sheet polarization currents are negligible compared with currents due to gradient and curvature drifts. The dynamics of the outer magnetosphere is extremely complicated, and too poorly understood at present for the kind of detailed quantitative modeling that we are attempting. Our choice of modeling region implies an awkward boundary condition, at the boundary between the inner and outer magnetosphere. However, this choice of region allows us to build reasonable models without impractical computing requirements.

Figure 1 shows the basic logic diagram of our model. The basic logical loop (the central pentagon of the figure) is a modification of a diagram given by Vasyliunas (1970).

Let us briefly discuss the diagram, starting with the box labelled "Hot-Particle Distribution." Using a magnetic-field model and assuming isotropic pitch-angle distributions, we can compute gradient-and-curvature drift currents in the magnetosphere. Our magnetic-field model is an Olson-Pfitzer (1974) analytic model, but including, in addition, the effects of a time-dependent substorm-current loop. This current loop, including an eastward perturbation current across the tail, a westward electrojet, and connecting Birkeland currents, is a modification of one proposed by McPherron et al. (1973); its current strength was adjusted as suggested by midlatitude magnetograms for the event.

Continuing counterclockwise around the central logical loop in Figure 1, we compute Birkeland-current strengths from the divergence of the magnetospheric gradient-and-curvature-drift currents, since the magnetization current, while large, is divergence free. Given the Birkeland-current strength, mapped down to the ionosphere, our next step is to derive the potential distribution in the ionosphere. However, to do this, we need two more pieces of input:

(1) the cross-polar-cap potential drop: from S3-2 electric-field data we estimate the potential drop, and assume a simple distribution (basically, a uniform dawn-dusk electric field with a noontime enhancement) at the high-latitude boundary of our calculation. This boundary lies just equatorward of the electric-field-reversal region;

(2) the distribution of ionospheric conductivity: our model Pedersen and Hall conductivities consist of time-dependent terms that include the day-night asymmetry and solar-zenith-angle effect, and a time-dependent term that gives a rough approximation to the auroral conductivity enhancement; in this latter term, the amount of

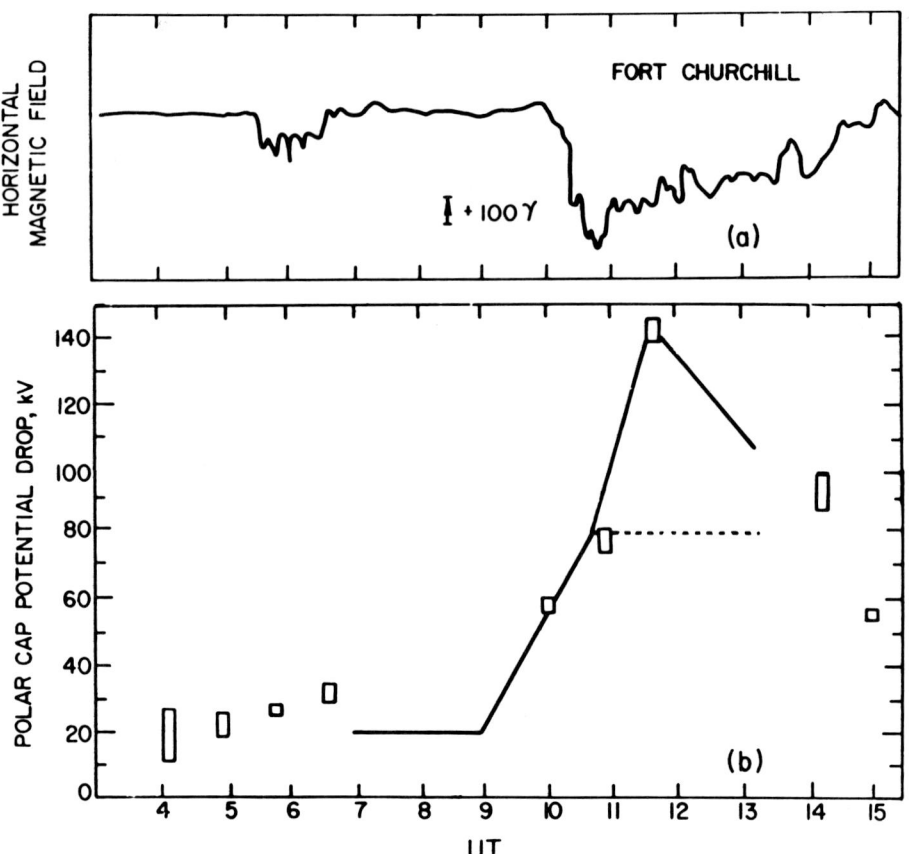

Fig. 2. Fort Churchill H-magnetogram and polar-cap potential drop for 19 September 1976. The lower panel shows polar-cap potential drops estimated from S3-2 electric-field data. Sizes of boxes are indicative of estimated errors. The solid curve shows the potential drop assumed in the simulation. Electric-field data from the 1140 UT pass arrived later than those from the other passes, and as a result some simulation runs followed the dashed line from 1040 to 1300 UT.

enhancement is adjusted as a function of time in an effort to be consistent with electron fluxes observed from S3-2.

The condition of current conservation in the ionosphere then becomes an elliptic equation in two dimensions, which is solved numerically, given the potential distribution at the polar-cap boundary as one boundary condition and a condition of no current across the low-latitude boundary, which is at approximately 21° geomagnetic latitude.

We map the ionospheric potential distribution out along field lines to the equatorial plane, assuming no field-aligned electric fields, but adding on the induction electric field, to get the total magnetospheric electric field.

We now compute the total drift velocities (E x B, gradient and curvature) of magnetospheric ions and electrons of various energies. Specifically, we compute the motions of the inner edges of plasma-sheet electrons of 5 energies, and ions of 11 energies, each inner edge being represented by approximately 18 independently computed points. The boundary positions are advanced by an amount corresponding to multiplying the computed velocities by the time step t. In computing electron boundary positions, we also include, in an approximate way, the effect of loss by precipitation. For simplicity in these initial model calculations, electrons near the inner edge of the plasma sheet are assumed to be lost by simple strong-pitch-angle scattering, as suggested by Vasyliunas (1968) and Kennel (1969).

The program goes completely around this logical loop every time step, which is typically 30 seconds magnetosphere time.

Figure 2 illustrates some aspects of the event being simulated, which might variously be described as a very long substorm, a quick succession of several short substorms or a substorm followed by a "convection-driven negative bay" (Pytte et al., 1978). The lower panel shows cross-polar-cap potential drops as estimated from S3-2 data. Note that this substorm-type event is associated with an increase in the polar-cap potential drop, an association previously suggested by Mozer (1973). Note also that the potential drop continued to rise after substorm onset, which might account for the prolonged negative bay (Pytte et al., 1978).

Results

We briefly present here a few highlights of our results, emphasizing some aspects that have been directly compared with observations.

We must emphasize that we are presenting a comparison of observed data with results of our first tries at computer simulating an observed magnetospheric event. Some data were used as input as described in Section II, to help us determine the polar-cap boundary, the cross-polar-cap potential drop, the conductivity and the magnetic-field model, but data were not used in any other significant way. Given the available input data, there is still some flexibility in the boundary conditions, and we could adjust the boundary conditions in various respects to improve agreement with data, but we have not done that yet. Presented below are our first tries at computer simulating the event, with no effort at optimizing the fit.

We have actually done four computer runs, as indicated in Table 1. Run #3 was done with a time-independent magnetic-field model, to isolate the effects of the induction electric field on ring-current injection. The runs also involved two different degrees of latitudinal smoothing of conductivities. (The reason for smoothing of the conductivities is that the difference equation that we use to conserve current in the ionosphere becomes an inaccurate approximation to the differential equation when there are sharp jumps in conductivity. Run #1 involved about as sharp a conductivity gradient as we can handle accurately with the present 21 x 28 grid and present numerical method.) Except for the dotted and dashed curves in Figure 6, all results presented here are for Run #4.

Table 1. Computer-Simulation Runs

| Computer Run | Peak Polar-Cap Potential Drop | Conductivity Model | Induction Electron Field |
|---|---|---|---|
| 1 | 80 kV | Minimum smoothing | Yes |
| 2 | 80 kV | Greater smoothing | Yes |
| 3 | 80 kV | Greater smoothing | No |
| 4 | 140 kV | Greater smoothing | Yes |

Figures 3-5 show observations vs. theory for the three passes of S3-2 that occurred during the event, before 1300 UT, when the simulation ended. The top two panels of each figure show observed and predicted electric fields. The lower panel shows predicted Birkeland currents. (S3-2 magnetometer data for this date are not reduced yet.) The dotted portion of the top panel represents the

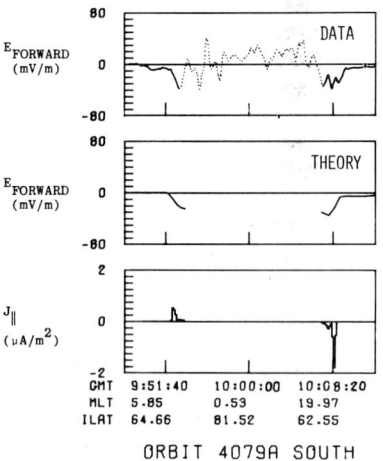

ORBIT 4079A SOUTH

Fig. 3. Data and theory for the 1000 UT pass of satellite S3-2. The top panel shows data from the AFGL electric field instrument. We have plotted the forward component of E, i.e., the component in the direction of satellite motion. The dotted section of the curve is the polar-cap-and-boundary-layer region, which we do not model. The second panel shows the corresponding component of the theoretically predicted electric field at the satellite's location (latitude, longitude and altitude) for the universal times in question. The bottom panel shows predicted Birkeland-current strength (positive values mean upward current). The legend gives Greenwich Mean Time, magnetic local time and invariant latitude. Satellite altitude ranges from 1025 km to 1375 km.

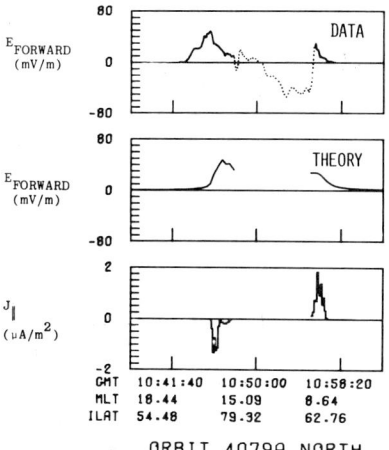

Fig. 4. Data and theory for the 1050 UT pass of S3-2. The format is the same as Figure 3. Satellite altitude ranges from 800 km to 260 km.

polar-cap-and-boundary-layer electric field, which we no not model. However, the input polar-cap potential drop is computed essentially from the area under the dotted curve. The boundary of our calculation (the poleward edge of the computed electric fields), is adjusted in Figures 3-6 (but not Figure 7) to correspond to the observed boundary of the polar-cap-and-boundary-layer region (boundary between dotted and solid observation curves).

We would like to make three general comments concerning the comparison between observed and predicted electric fields in Figures 3-5.

(1) There is little agreement between data and theory with regard to small details, perhaps due to the fact that the model conductivity distribution is smooth and undetailed.

(2) Both data and theory agree that the region below about 60° invariant latitude is rather well shielded from the high-latitude convection field, even in this time-dependent situation. The greatest leakage through the shielding occurred, both in the data and the theory, on the outbound part of pass 4079A South, just after substorm onset. In the model, auroral conductivities were increased suddenly at onset, and the ring current had not had time to rearrange itself completely to restore strong shielding.

(3) Electric fields on the dawn side generally tend to decline smoothly with decreasing latitude, both in the theory and the data, but the same is not true on the dusk side, where, particularly past 1800 local time, the strongest poleward electric field generally tends to occur well equatorward of the polar-cap boundary. Furthermore, in both the model and S3-2 electric-field data, electric fields below the polar-cap boundary tend to be larger on the dusk side than on the dawn side, an effect previously noticed by Kelley (1976). Our models essentially always show greater potential drops at dusk than at dawn -- a result of Hall currents flowing antisunward across the conductivity jumps at dawn and dusk (Wolf, 1970).

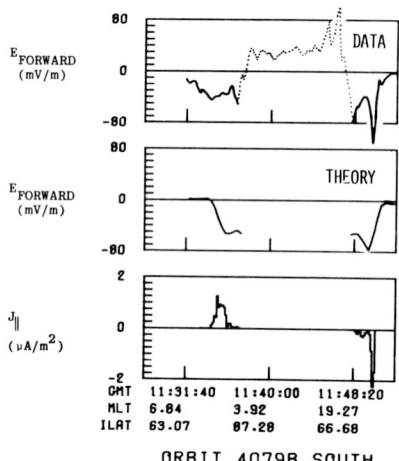

Fig. 5. Data and theory for 1140 UT pass of S3-2. The format is the same as Figure 3. Satellite altitude ranges from 1000 km to 1360 km.

Rapid Subauroral Flow

The most striking feature of the data shown in Figures 3-5 is the sharp electric-field peak observed well below the polar-cap boundary in the last half of orbit 4079B-South (Figure 5). We shall refer to this feature as "rapid subauroral flow." Data from the same auroral-zone pass is shown in more detail in Figure 6 (top panel).

Panels 2-4 display curves for runs 1 (dotted), 2 (dashed), and 4 (solid). Run 3 results are generally similar to run 2 and will not be discussed here.

The second panel of Figure 6 shows calculated electric fields, in a form that displays all the fine structure available in the model. (Our grid spacing is approximately 1.6° in latitude. However, the program employs a special back-correction scheme that allows it to keep track of effects of Birkeland current on a much finer scale. These fine-scale corrections are included in Figure 6, but not Figures 3-5.)

The third panel of Figure 6 shows predicted Birkeland-current strengths along that trajectory.

Panel 4 displays model values of height-integrated Pedersen conductivities. The model global conductivity includes day-night asymmetry, solar-zenith-angle dependence and electron-precipitation effects. The auroral conductivity enhancement is estimated crudely from observed electron fluxes. For a more detailed discussion, see Harel et al. (1977). Our computer model cannot tolerate very large conductivity gradients, so we had to smooth the conductivity profile to some extent (see panel 4 and also Table 1).

The bottom panel of Figure 6 shows height-integrated Pedersen conductivity, estimated directly from measured electron fluxes using the formula

$$\Sigma_p \text{ (mho)} = 0.5 + 5.2 \times \text{(Electron energy flux)}^{\frac{1}{2}} \quad (1)$$

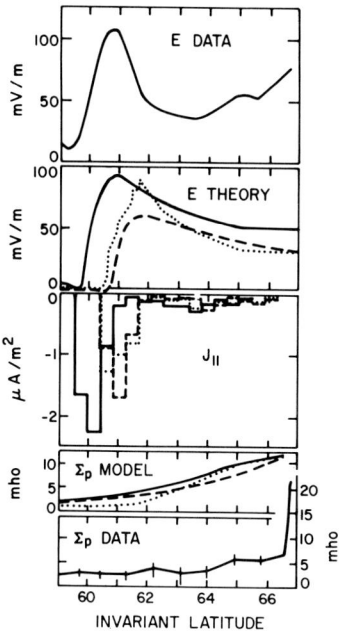

Fig. 6. Detailed view of the dusk-auroral-zone pass for the southern part of orbit 4079B of satellite S3-2. The top panel shows the observed electric-field component opposite to satellite motion (approximately the poleward component). The second panel gives essentially the same component of the theoretical-model electric field. The third panel gives predicted Birkeland currents, and the fourth panel shows the model height-integrated Pedersen conductivities. Solid curves in panels 2-4 pertain to the same computer run as in Figures 3-5; the dotted and dashed curves correspond to lower polar-cap potential drops, the dotted lines one to a less-smeared conductivity model. The bottom panel shows conductivities estimated directly from the data using equation (1).

where the energy flux is in erg $cm^{-2}$ $sec^{-1}$ (Harel et al., 1977). Unfortunately, the geometric factor of the electron detector on S3-2 was too small to allow reliable estimation of the low-latitude edge of the diffuse aurora.

The exciting feature of Figure 6 is, of course, that the computer runs all predicted the observed rapid subauroral flow and at approximately the right location. Similar rapid flows have been observed many times before, often associated with the trough (Heelis et al., 1976; Smiddy et al., 1977; Maynard, 1978; Spiro et al., 1978).

Note that the location of the peak of the rapid subauroral flow computed in run 4 agrees very well with observations, while the other two runs show peaks that lie approximately a degree poleward of the observed one. This difference in model results is easy to understand physically: for runs 1 and 2 we underestimated the polar-cap potential drop; consequently the plasma-sheet ions were not injected as deep into the magnetosphere as was the case for run 4, and the rapid subauroral flow did not extend to as low latitude (Southwood and Wolf, 1978).

An important feature of our predicted Birkeland currents for this pass (panel 3) is that we get only downward currents. This is different from the observation of an upward current sheet at the poleward edge of one rapid subauroral flow (Smiddy et al., 1977). We attribute this difference to the different local time (2100 MLT) of this earlier measurement. Theoretically, only a single current sheet is needed to account for the peak electric field in the trough region, provided that the conductivity gradient there is large enough (Southwood and Wolf, 1978). Approximately a factor-of-two increase in Pedersen conductivity is needed between 61 and 62 to be consistent with the sharp decline of the observed electric field in that region. The data are not inconsistent with such an increase. Also, the conductivity model with the sharpest gradient (dotted curve) gives rise to the sharpest calculated electric-field peak, as expected.

Our computer model often shows rapid subauroral flows in the dusk-to-midnight sector, though not elsewhere, which is consistent with the previously mentioned observations. We should mention that no other clear rapid subauroral flows were observed during this simulated event, and none are predicted by the model for the S3-2 satellite paths, with the following partial exception: one of the computer runs indicated an electric field on pass 4079A South that peaked at about 42 mV/m and had a shape that would classify it as a marginal case of rapid subauroral flow. The observations indicate, for that case, a complicated structure rather than a clear rapid-subauroral-flow signature.

We should also acknowledge a different and conflicting interpretation of rapid subauroral flows (Mozer, 1978), an interpretation in terms of field-aligned potential drops between relevant satellite atltitudes (250-1500 km) and the lower ionosphere.

Birkeland-Current Patterns

Figure 7 compares our computed distribution of Birkeland currents with a summary of active-period observations by Iijima and Potemra (1978). The general pattern of computed currents did not vary much with time through the substorm, although, of course, the low-latitude boundary of the currents moved equatorward during the event, and the current strengths increased. The poleward set of Birkeland currents (region 1 currents in the nomenclature of Iijima and Potemra) are poleward of our modeled region. We estimate their distribution very roughly by calculating the currents into our poleward boundary and assuming that those currents flow directly out along field lines from there. The thickness shown for the computed region 1 currents in Figure 7 is arbitrary.

The observed and predicted patterns agree in their general sense, which is not surprising, since convection theories (Schield et al., 1969; Wolf, 1974) predicted the basic pattern before it was observed, and it has been a consistent feature of our computer models. An encouraging feature of the comparison in Figure 7 is the triple-current-sheet region that exists near midnight in both the

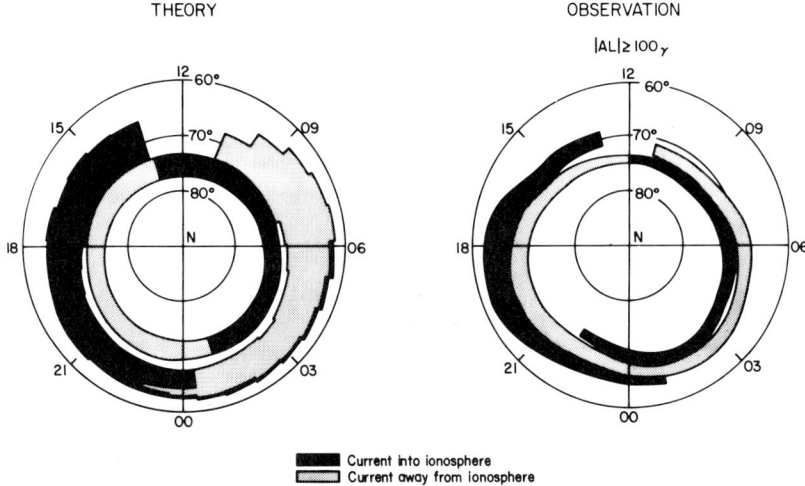

Fig. 7. Comparison of a typical computed Birkeland current pattern (for 1150 UT on 19 September 1976) and an average observed pattern for active times (Iijima and Potemra, 1978).

Triad observations and the theory. (Starting from the poleward boundary and moving toward the equator, we have downward, upward, and then downward currents). The model current structure appears to be rotated about two hours later in local time, as compared to the observations. We suspect that this feature of the model could be brought into agreement with observations by a minor adjustment of the boundary potential at the polar cap.

In the models, we do see multiple reversals of Birkeland currents around local midnight, but we do not see them near dusk. Thus the models always indicate a predominantly downward current in the region of rapid subauroral flow at dusk. On the other hand, Figure 7 predicts Birkeland-current reversals at low latitude near midnight, which may correspond to the effect observed by Smiddy et al. (1977).

## Summary

We have displayed some highlights of results of our first attempt at simulating an observed magnetospheric event. Comparison with observations has come out remarkably well, particularly considering that, in these first tries, we have not adjusted any boundary conditions or assumptions to improve agreement with the data. Of course, much work remains to be done to include more physics in the models and to model more and different events.

Acknowledgments. We are grateful to Ameen Ahmad, H. Kent Hills, Janice Karty, and Robert Spiro for their work in displaying model results, to W. J. Burke, D. A. Hardy and F. J. Rich for their efforts in reducing data from the S3-2 satellite, and to Robert Spiro for several illuminating discussions and for helpful comments on the manuscript. This work has been supported by the National Science

Foundation grant ATM74-21185, by U. S. Air Force Contracts F19628-77-C-0005 and F19628-78-C-0078, and by NASA grants NGR44-006-137 and NGL44-006-012.

References

Block, L. P., On the distribution of electric fields in the magnetosphere, J. Geophys. Res., 71, 855, 1966.

Cowley, S. W. H., Energy transport and diffusion, in Physics of Solar Planetary Environments, ed. D. J. Williams, Amer. Geophys Un., Washington, D.C., p. 582, 1976.

Ejiri, M., R. A. Hoffman and P. H. Smith, Energetic particle penetrations into the inner magnetosphere, Goddard Space Flight Center Report X-625-77-254, 1977

Ejiri, M., R. A. Hoffman and P. H. Smith, The convection electric field model for the magnetosphere based on Explorer 45 observations, J. Geophys. Res., 83, 4811, 1978

Fejer, J.A., Theory of geomagnetic daily disturbance variations, J. Geophys. Res., 69, 123, 1964.

Harel, M., and R. A. Wolf, Convection, in Physics of Solar-Planetary Environments, Vol. II, edited by D. J. Williams, Amer. Geophys. Un., Washington, D. C., p. 617, 1976.

Harel, M., R. A. Wolf, P. H.Reiff and H. K. Hills, Study of plasma flow near the earth's plasmapause, U.S. Air Force Geophysics Laboratory Report AFGL - TR-77-0286, 1977.

Harel, M., R. A. Wolf, P. H. Reiff, R. W. Spiro, H. K. Hills, M. Smiddy W. J. Burke and F. J. Rich, in preparation, 1979.

Heelis, R. A., R. W. Spiro, W. B. Hanson and J. L. Burch, Magnetosphere-ionosphere coupling in the mid-latitude trough, Trans. Am. Geophys. Union, 57, 990, 1976.

Iijima, T., and T. A. Potemra, Large-scale characteristics of field-aligned currents associated with substorms, J. Geophys. Res., 83, 599, 1978.

Jaggi, R. K. and R. A. Wolf, Self-consistent calculation of the motion of a sheet of ions in the magnetosphere, J. Geophys. Res., 78, 2852, 1973.

Kamide, Y. and S. Matsushita, Simulation studies of ionospheric electric fields and currents in relation to field-aligned currents. 1. Quiet periods, submitted to J. Geophys. Res., 1978.

Karlson, E. T., Streaming of plasma through a magnetic dipole field, Phys. Fluids, 6, 708, 1963.

Karlson, E. T., Plasma flow in the magnetosphere. I. A two-dimensional model of stationary flow, Cosm. Electrodyn., 1 474, 1971.

Kelley, M. C., Evidence that auroral-zone electric fields act in opposition to super-rotation of the upper atmosphere, Planet Space Sci., 24, 355, 1976.

Kennel, C. F., Consequences of a magnetospheric plasma, Rev. Geophys. 7, 379, 1969.

Kivelson, M. G., Magnetospheric electric fields and their variation with geomagnetic activity, Rev. Geophys. Space Phys., 14, 189, 1976.

Konradi, A., C. L. Semar, and T. A. Fritz, Injection boundary dynamics during a geomagnetic storm, J. Geophys. Res., 81, 3851, 1976.

Mal'tsev, Yu. P., The effect of ionospheric conductivity on the convection system in the magnetosphere, Geomag. Aeron., 4, 128, 1974.

Maynard, N. C., On large poleward-directed electric fields at sub-auroral latitudes, Geophys. Res. Lett., 5, 617, 1978.

McIlwain, C. E., Substorm injection boundaries, in Magnetospheric Physics, edited by B. M. McCormac, D. Reidel, Dordrecht-Holland, p. 143, 1974.

McPherron, R. L., C. T. Russell and M. P. Aubry, Satellite studies of magnetospheric substorms on August 15, 1968. 9. Phenomenological model of substorms, J. Geophys. Res., 78, 3131, 1973.

Mozer, F. S., On the relationship between the growth and expansion phases of substorms and magnetospheric convection, J. Geophys. Res., 78, 1719, 1973.

Mozer, F. S., Implications of S3-3 and ISEE electric field data on models of magnetospheric electric fields, paper presented at the Chapman Conference on Quantitative Modeling of Magnetospheric Processes, La Jolla, Ca., 1978.

Nisbet, J. S., M. J. Miller and L. A. Carpenter, Currents and electric fields in the ionosphere due to field-aligned auroral currents, J. Geophys Res., 83, 2647, 1978.

Nopper, R. W. Jr., and R. L. Carovillano, Polar equatorial coupling during magnetically active periods, Geophys. Res. Lett., 5, 699, 1978.

Nopper, R. W. Jr., and R. L. Carovillano, Ionospheric electric fields driven by field-aligned currents, in Quantitative Modeling of the Magnetospheric Processes, Geophys. Monogr. Soc., Vol. 21, edited by W. P. Olson, AGU, Washington, D. C., 1979.

Olson, W. P., and K. A. Pfitzer, A quantitative model of the magnetospheric magnetic field, J. Geophys. Res., 79, 3739, 1974.

Pytte, T. R., R. L. McPherron, E. W. Hones, Jr., and H. I. West, Jr., Multiple-satellite studies of magnetospheric substorms: distinction between polar magnetic substorms and convection-driven negative bays, J. Geophys. Res., 83, 663, 1978.

Roederer, J. G., and E. W. Hones, Jr., Motion of magnetospheric particle clouds in a time-dependent electric field model, J. Geophys. Res., 79, 1432, 1974.

Schield, M. A., J. W. Freeman, and A. J. Dessler, A source for field-aligned currents at auroral latitiudes, J. Geophys. Res., 74, 247, 1969.

Smiddy, M., M. Kelley, W. Burke, F. Rich, R. S. Sagalyn, B. Schumann, R. Hays and S. Lai, Intense poleward-directed electric fields near the ionospheric projection of the plasmapause, Geophys. Res. Lett., 4, 543, 1977.

Smith, P. H., H. K. Bewtra and R. A. Hoffman, Motions of charged particles in the magnetosphere under the influence of a time-varying large-scale convection electric field, in Quantitative Modeling of the Magnetospheric Processes, Geophys Monogr. Ser., vol. 21. edited by W. P. Olson, AGU, Washington, D.C., 1979.

Southwood, D. J., and R. A. Wolf, An assessment of the role of precipitation in magnetospheric convection, J. Geophys Res., 83, 5227, 1978.

Spiro, R. W., R. A. Heelis and W. B. Hanson, Ion convection and the formation of the midlatitude F-region ionospheric trough, J. Geophys. Res., 83, 4255, 1978.

Swift, D. W., Possible mechanisms for formation of the ring current belt, J. Geophys. Res., 76, 2276, 1971.

Vasyliunas, V. M., A mathematical model of plasma motions in the magnetosphere, Trans. Am. Geophys. Un., 49, 232, 1968.

Vasyliunas, V. M., Mathematical models of magnetospheric convection and its coupling to the ionosphere, in Particles and Fields in the Magnetosphere, edited by B. M. McCormac, D. Reidel, Dordrecht-Holland, p. 60, 1970.

Vasyliunas, V. M., The interrelationship of magnetospheric processes, in Earth's Magnetospheric Processes, edited by B. M. McCormac, D. Reidel, Dordrecht-Holland, p. 29, 1972.

Wolf, R. A., Effects of ionospheric conductivity on convective flow of plasma in the magnetosphere, J. Geophys. Res., 75, 4677, 1970.

Wolf, R. A., Calculations of magnetospheric electric fields, in Magnetospheric Physics, edited by. B. M. McCormac, D. Reidel, Dordrecht-Holland, p. 167, 1974.

Yasuhara, F. and S. -I. Akasofu, Field-aligned currents and ionospheric electric fields, J. Geophys. Res., 82, 1279, 1977.

# MOTIONS OF CHARGED PARTICLES IN THE MAGNETOSPHERE UNDER THE INFLUENCE OF A TIME-VARYING LARGE SCALE CONVECTION ELECTRIC FIELD

Paul H. Smith

Laboratory for Planetary Atmospheres, NASA/Goddard Space Flight Center
Greenbelt, Maryland 20771

N. K. Bewtra

Computer Sciences Corporation, Silver Spring, Maryland 20910

R. A. Hoffman

Laboratory for Planetary Atmospheres, NASA/Goddard Space Flight Center
Greenbelt, Maryland 20771

Abstract. The motions of charged particles under the influence of the geomagnetic and electric fields are quite complex in the region of the inner magnetosphere. The Volland-Stern type large scale convection electric field ($\vec{E} = -\nabla\Phi$ and $\Phi = AR^\gamma \sin\phi$) with $\gamma = 2$ has been used successfully to predict both the plasmapause location and particle enhancements determined from Explorer 45 ($S^3$-A) measurements. We have recently introduced into the trajectory calculations of Ejiri et al. (1978) a time dependence in this electric field based on the variation in Kp for actual magnetic storm conditions. The particle trajectories are computed as they change in this time-varying electric field. Several storm "fronts" of particles of different magnetic moments are allowed to be injected into the inner magnetosphere from L = 10 in the equatorial plane. The motions of these fronts are presented in a movie format. The local time of injection, the particle magnetic moments and the subsequent temporal history of the magnetospheric electric field play important roles in determining whether the injected particles are trapped within the ring current region or whether they are convected to regions outside the inner magnetosphere.

## Introduction

Particle convection in the magnetosphere has been discussed by a number of authors since the early convection models of Dungey (1961), and Axford and Hines (1961), and its significance in the dynamics of the magnetosphere is well recognized. The review on this subject by Axford (1969) provides quite extensive discussion and references to the many important papers up to that time. Chappell (1974) reviewed the direct and indirect measurements of the convection electric field that were made since Axford's (1969) review, and discussed the way in which the measurements supported the basic convection theory. Two works of special importance in the early quantitative modeling of mag-

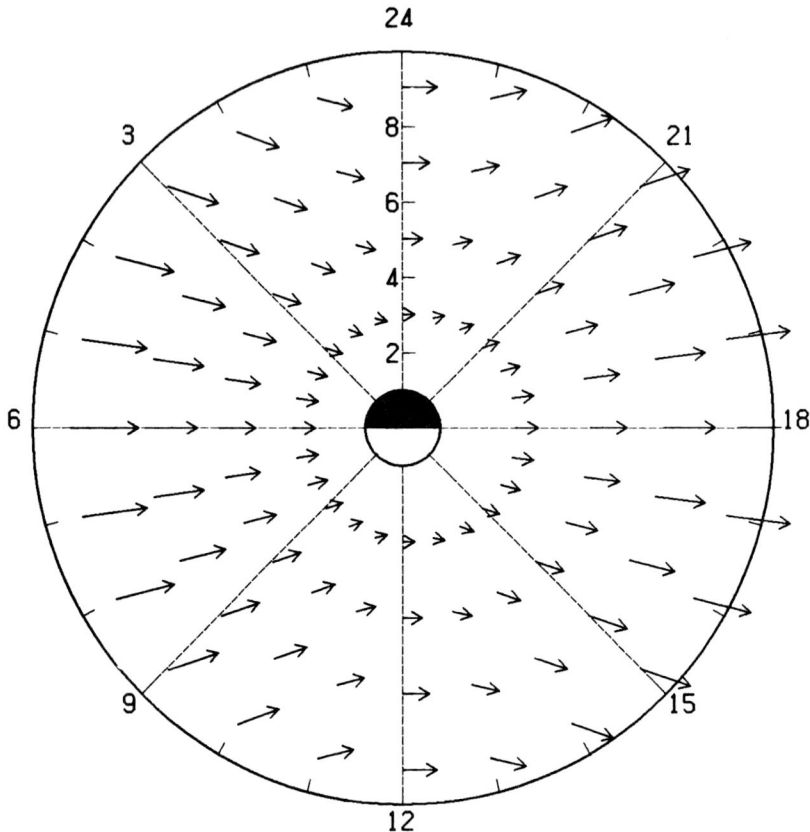

Fig. 1. Normalized large scale convection electric field of the Volland-Stern type with $\gamma = 2$. The vectors are shown in the equatorial plane in L, MLT coordinates.

netospheric processes are those by Kavanagh et al. (1968) and Chen (1970). Kavanagh et al. (1968) developed a simple analytical model which combined a large scale uniform electric field, similar to that given by Brice (1967) with a corotation electric field and the geomagnetic field given by Mead (1964). Chen (1970) presented in detail the motions of low energy protons in a dipole geomagnetic field under the superimposed convection and corotation electric fields and showed that the trajectories of these low-energy protons are topologically quite different from those of other classes of particles. As an extension of these works, principally that of Kavanagh et al. (1968), Wolf (1970) added another dimension by considering the ionospheric conductivity effects on the convective flow patterns of magnetospheric plasma. The theory of convection in the magnetosphere had become so well established based on the various works indicated above and in the re-

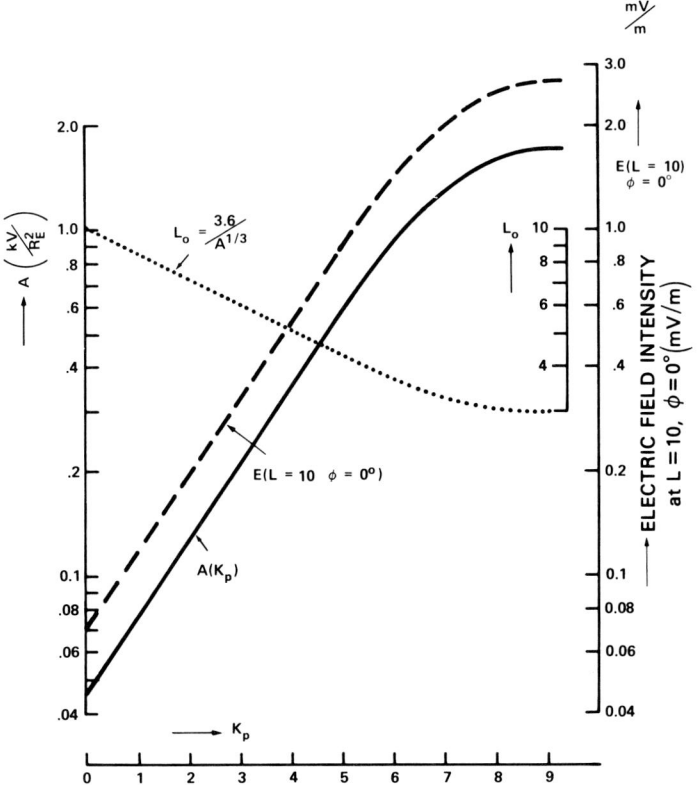

Fig. 2. Parameters of the convection electric field plotted against Kp: (1) A(Kp) given by equation 5, (2) E, the electric field at L = 10 and at midnight ($\phi = 0$), and (3) $L_o$, the radial distance of the duskside stagnation point from the earth's center in the equatorial plane in earth radii (from Fig. 2 Maeda et. al. (1978)).

views by Kivelson (1976) and Stern (1977) that it became generally accepted that the source of the ring current protons associated with the main phase magnetic storms was the convection of plasma sheet protons into low L values (Axford, 1969; Vasyliunas, 1972; Nishida and Obayashi, 1972), even though there had been almost no direct experimental measurements of the characteristics of these storm time ring current particles that would substantiate this concept. The characteristic features of the initial enhancement of the storm time ring current particles in the evening hours became available through the Explorer 45 ($S^3$-A) program and the measurements were found to be qualitatively consistent with the flow patterns resulting from a combination of inward convection, gradient drift, and corotation (Smith and Hoffman, 1974).

In more recent years, one aspect of this convection theory has come into question, that is, the source of the particles. The discoveries

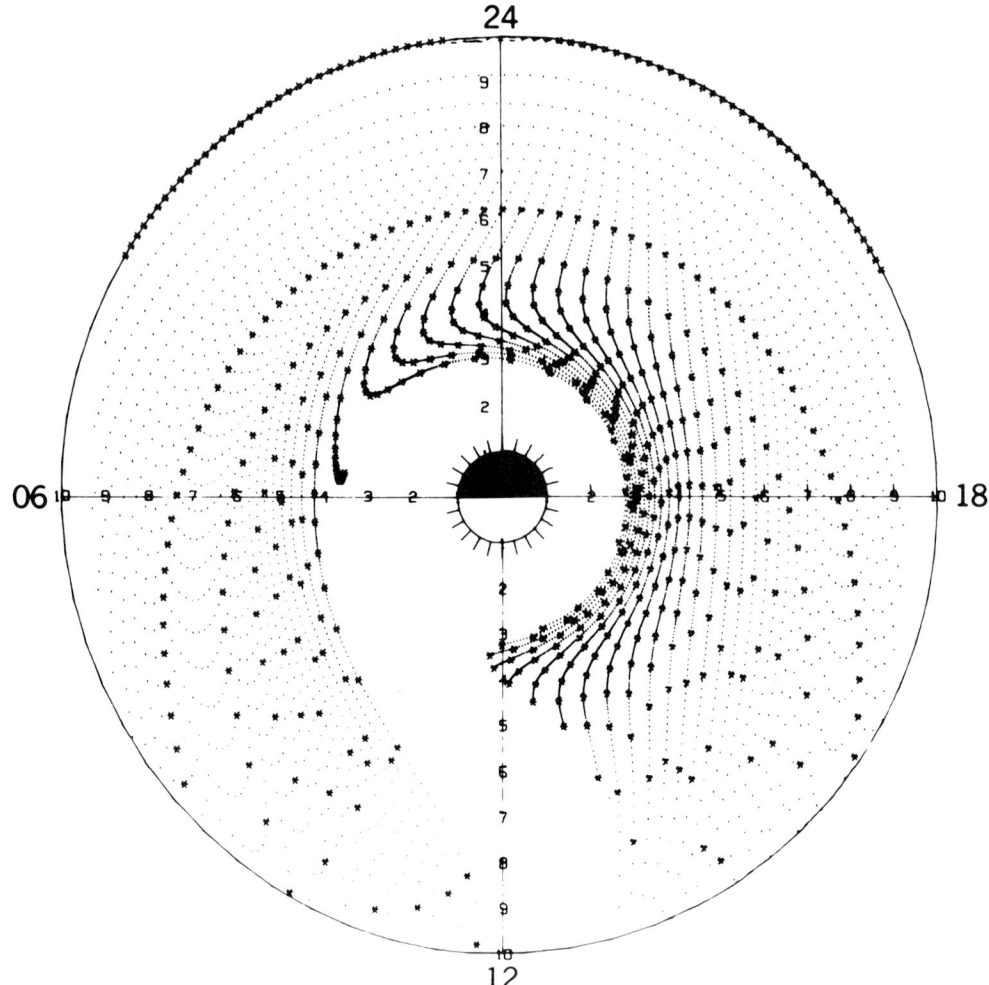

Fig. 3. Particle flow patterns for $\mu = 0.02$ keV/$\gamma$ particles with $90°$ pitch angles under a static electric field from Fig. 9 of <u>Ejiri</u> (1978) for Kp = 4. The "*" are the hour markers along the fixed trajectories. This shows the particle locations after 20 hours of motion from injection.

of ions in the magnetosphere heavier than hydrogen (Shelley et al., 1972) and ion and electron beams directed up along magnetic field lines from the atmosphere (Johnson et al., 1977; McIlwain, 1975) have suggested that the ionosphere may be a source for ring current particles, rather than, or in addition to, the plasma sheet.

This paper briefly traces the evolution of the work of a number of researchers which has lead from the stage of qualitative agreement to a stage of quantitative agreement between the convection theory and measurements of related magnetospheric phenomena. Facets of the evolution have been described in the review by <u>Kivelson</u> (1976) on the vari-

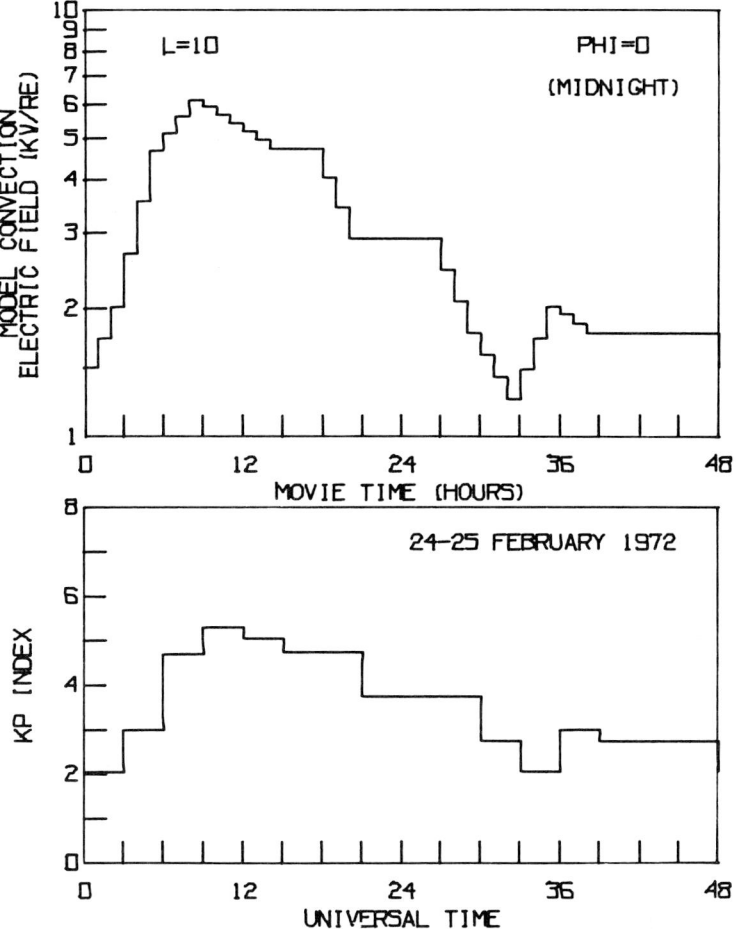

Fig. 4. Kp index for 48 hours during the February 24, 1972, geomagnetic storm. The computed electric field intensity at 10 $R_E$, at midnight ($\phi = 0°$) determined from one hour interpolated values of Kp is given in the upper portion and the time agrees with the value of T given in Figures 5-8.

ation of magnetospheric electric fields with geomagnetic activity and in the extensive review by Stern (1977) on the theoretical concepts of large-scale electric fields. Ejiri (1978) and Ejiri et al. (1978) have developed a quantitative convection model based on Explorer 45 observations. A static electric field is used in this model and the particles move in time (and space) along fixed trajectories. In the situations where the electric field, parameterized by Kp, is fairly constant over a time period of several hours, this is a reasonable approach (Maeda et al., 1978). A time-dependent electric field is needed, however, for the description of the particle trajectories associated with magnetic storms which have a several day duration and decreasing convection electric field (Smith et al., 1978), and for transiently intensified

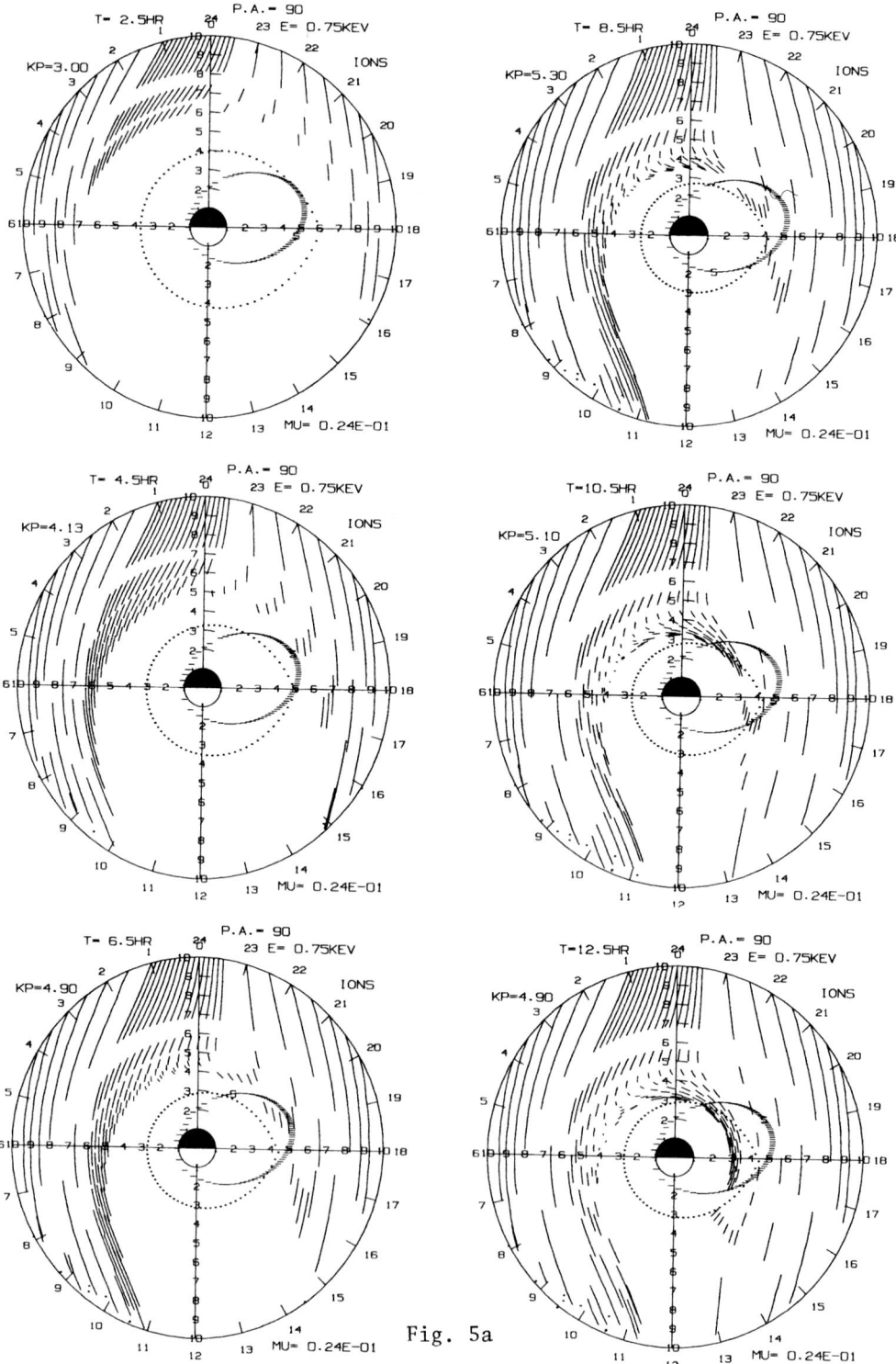

Fig. 5a

electric fields of substorm time scales (Roederer and Hones, 1974).

In the present paper the approach we have employed to add a large-scale time-dependent convection electric field is described. We have also developed along with this approach a technique for the visual description of the complex motions of the charge particles. By using a computer-generated motion picture of storm "fronts" of injected particles of various magnetic moments we have produced both an educational tool to demonstrate the convective motion of particles in the ring current region and an analytical tool for future studies of particle dynamics.

## Formulism and Basic Assumptions

The basic formulism which we use for the computation of the trajectories of charge particles in the magnetosphere is given by Ejiri (1978) and this same computation has been used by Maeda et al. (1978) for electron trajectories. In the equation of motion for the particles it is assumed that non-electromagnetic forces such as gravitational forces, etc. are negligible and that the particles can be represented non-relativistically. It is assumed that the particle motion is adiabatic, i.e. the first and second invariants (magnetic moment and longitudinal invariant) are conserved, and that changes in the electric and magnetic fields are slow and the particle motion can be represented as a motion of its guiding center. The following constraints and conditions are also imposed:
1. The geomagnetic field, B, is an earth centered dipole magnetic field with the same axis as the earth's rotation axis and the magnetic field lines are equipotentials.
2. There are no <u>local</u> energization or loss processes (e.g. charge exchange, wave-particle interactions or collisional loss processes).

The drift velocity, $\vec{u}$, in the equatorial plane is obtained by averaging over a cyclotron motion and a bouncing motion and is given by

$$\vec{u} = \frac{\vec{F} \times \vec{B}}{qB^2} \qquad (1)$$

The force, $\vec{F}$, on the particles is given by

$$\vec{F} = q\vec{E} - q(\vec{\omega} \times \vec{R}) \times \vec{B} - W \cdot G(\alpha_o) \cdot \frac{\nabla B}{B} \qquad (2)$$

where,

$\vec{R}$ is the radial vector from the center of the earth,
$\vec{\omega}$ is the angular velocity of the earth's rotation,
$G(\alpha_o)$ is the term associated with the helical path off the equator

---

Fig. 5a. Selected snapshot frames from the movie "Convection of Magnetospheric Particles in a Time-Varying Electric Field" (Version 780919) for particles of magnetic moment, $\mu = 0.024$ keV/$\gamma$. The configuration at six times, T, are shown. (See text for detailed discussion.)

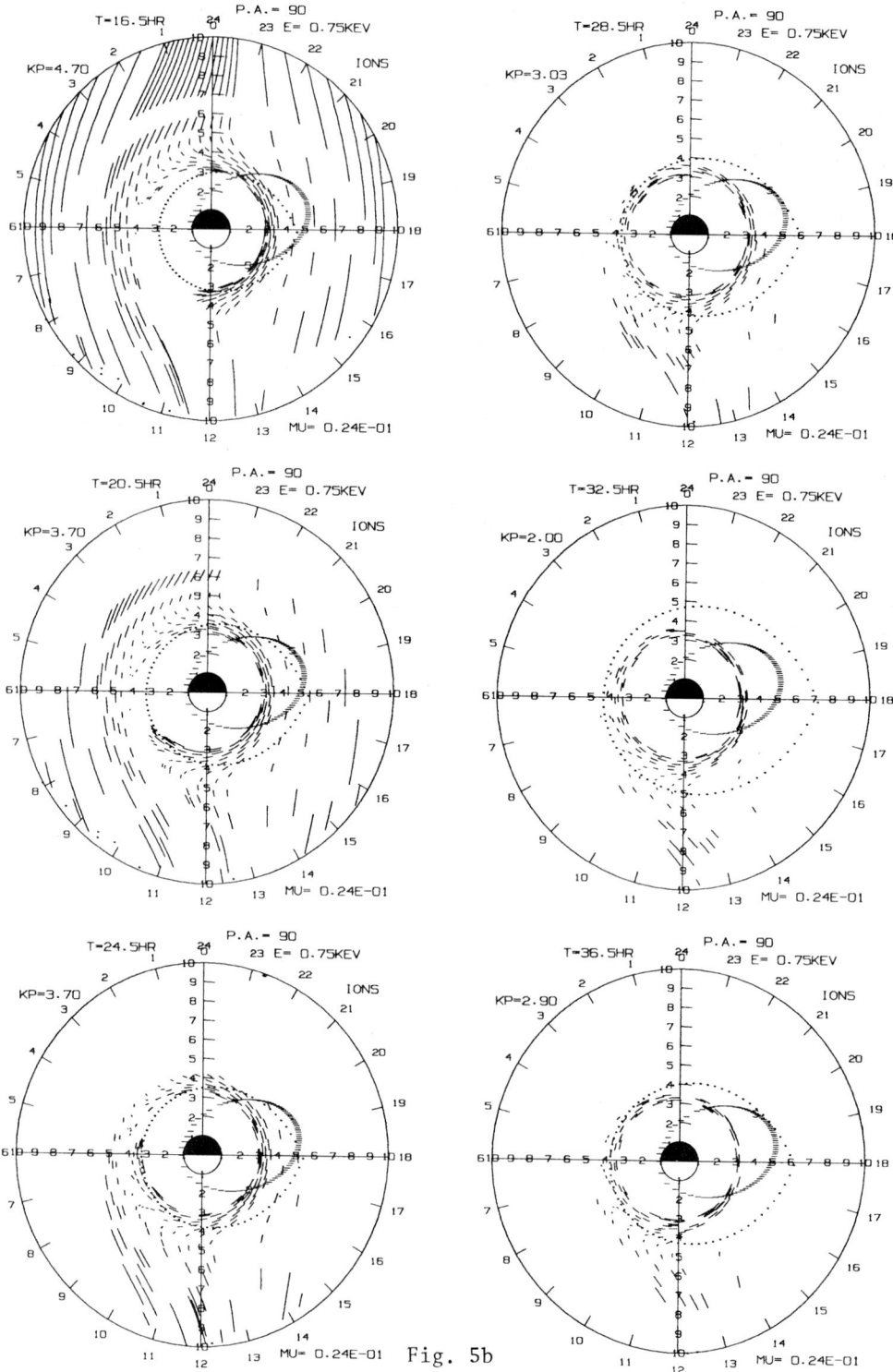

Fig. 5b

over the bouncing motion of the particle. It is a function of the particles' equatorial pitch angle, $\alpha_o$, W is the kinetic energy of the particle and can be written as $W = \frac{\mu B}{\sin^2 \alpha_o}$, $\mu$ is the magnetic moment of the particle. The reader is refered to Roederer (1970) for more detailed discussion on the dynamics of geomagnetically trapped particles.

The first term in Eq (2) is the large-scale convection electric field force which will be discussed below, the second term is the corotational electric force and the third term represents the magnetic curvature and gradient drift forces. When the equatorial pitch angle of the particles is $90^\circ$, the last term in equation (2) becomes $-\mu \nabla B$. Note that only equatorially mirroring particles are considered in this paper. This restriction does not apply to the general formalism of Ejiri (1978).

## Convection Electric Field Model

A simple model of the convection electric field of the magnetosphere has been proposed (Volland, 1973; Stern, 1974, 1975). This Volland-Stern model assumes that the large-scale convection electric field, $\vec{E}_{conv}$, is derivable from a quasi-static electric potential, $\Phi$:

$$\vec{E}_{conv} = -\nabla \Phi_{conv} \qquad (3)$$

and that there is an absence of electric fields parallel to the geomagnetic field lines, $\vec{E} \cdot \vec{B}_o = 0$, where $B_o$ is assumed for convenience to be a coaxial dipole geomagnetic field. This model is also valid only for closed magnetic field lines and within ten earth radii. The scalar potential in the equatorial plane can then be written as

$$\Phi_{conv} = AR^\gamma \sin \phi \qquad (4)$$

where R is the equatorial radial distance in earth radii and A is a coefficient which determines the electric field intensity. This coefficient will be discussed in more detail below. The local time dependence is given by the azimuthal angle, $\phi$, with $\phi = 0$ at midnight. The value of the exponent, $\gamma$, distinguishes this Volland-Stern model from the uniform dawn to dusk electric field. The uniform field used by Chen (1970) is obtained if $\gamma = 1$. Stern (1974) suggested $\gamma = 2$ from a consideration of the OGO-6 measurements of the electric field at high latitudes (Heppner, 1972). Volland (1973) arrived at the same value for $\gamma$ from a best fit to the shape of the plasmapause and Ejiri et al. (1978) in an analysis of the shape and location of the pre-midnight plasmapause measured by Explorer 45, determined an average value of $\gamma$ to be 2.4. The configuration of this large-scale convection electric field, $\vec{E}_{conv}$, is shown in Figure 1 for a normalized value for A and with $\gamma = 2$. The effect of the radial dependence and local time effect are easily seen in this diagram.

In order to account for changes in the convection electric field

---

Fig. 5b. Same as Figure 5a, but for six later times.

522

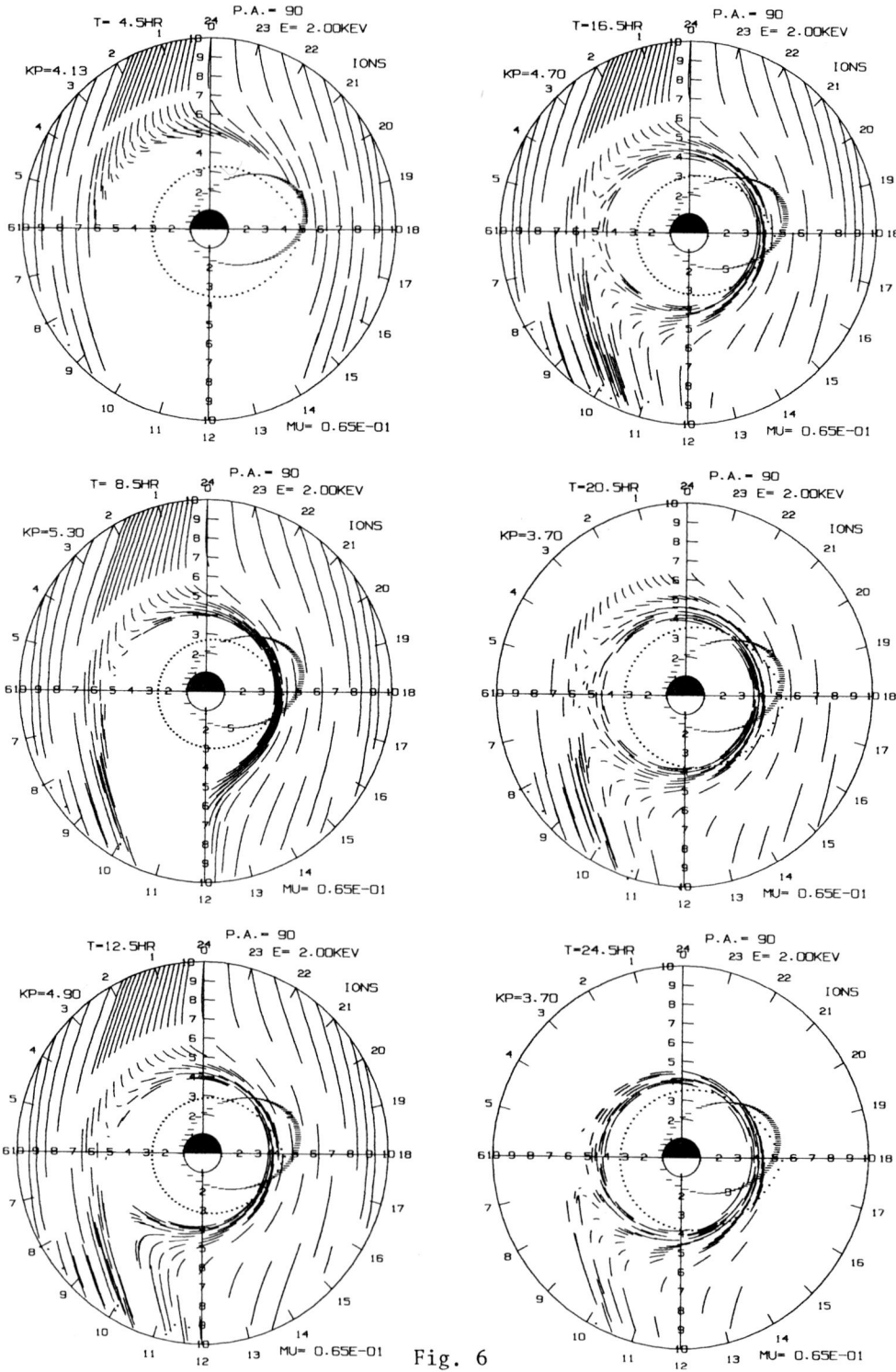

Fig. 6

associated with changes in magnetic activity, Grebowsky and Chen (1975) concluded that the model parameter, A, must be related to some geophysical parameter which describes changes in the large-scale convection. Based on previous correlations with Kp (Carpenter and Park, 1973, Kivelson, 1976), they chose to relate A to the Kp index. By fitting the midnight plasmapause locations measured by the spectrometers on OGO-3 and OGO-5, Grebowsky and Chen (1975) determined the following quadratic dependence on Kp:

$$A = 0.045 / (1 - 0.159 Kp + 0.009 Kp^2)^3 \qquad (5)$$

where the units of A are $kV/R_E^2$. In Figure 2 we show the plot of A vs. Kp as well as the resultant $|\vec{E}|$ and $L_o$, the stagnation distance at dusk, from Figure 4 of Maeda et al. (1978).

The Kp index, which is constructed to indicate global magnetic activity, using the 3 hour values from 12 middle latitude geomagnetic observations to eliminate longitudinal variations, is actually a good indicator of the magnetosphere electric field intensity. It is interesting to note that the intensification of the magnetospheric electric field is the source of the enhancement of geomagnetic activity, in contrast to Dst, which is a sole indicator of ring current intensity.

## Agreement With Explorer 45 Observations

The Volland-Stern type convection electric field model, described in the previous section, with the addition of the dependence of the electric field strength on the Kp index has been used to study various magnetospheric phenomena especially those measured by Explorer 45. Maynard and Chen (1975) were able to interpret with this model their observations from Explorer 45 of regions of isolated cold plasma. Grebowsky and Chen (1975) used this $\gamma = 2$ model to explain the general relation of the observed location of the nose events of Smith and Hoffman (1974) to the observed plasmapause location measured on the same satellite (Maynard and Cauffman, 1973) and predicted the spatial location of the nose by computing the forbidden region boundaries in the same manner as in Stern's (1975) analysis.

The shape of the nose structures in the energy-time spectrograms and its location just inside the plasmapause was studied by Cowley (1976) and Cowley and Ashour-Abdalla (1976) and they concluded that the observed noses cannot be modeled with any uniform dawn-dusk convection electric field since that field ($\gamma = 1$ in equation 5) predicts that the nose should penetrate down to much lower L-values than is observed. They suggested that the convection electric field may be shielded from the plasmasphere region so that the penetration could not occur. Ejiri et al. (1978) have noted that the intensity of the electric field inside the plasmapause is smaller with the Volland-Stern field and in fact the shielding effects can only be introduced if $\gamma > 1$. Cowley

Fig. 6. Same as Figure 5a, but for $\mu = 0.065$ keV/$\gamma$ particles. Six time periods are shown.

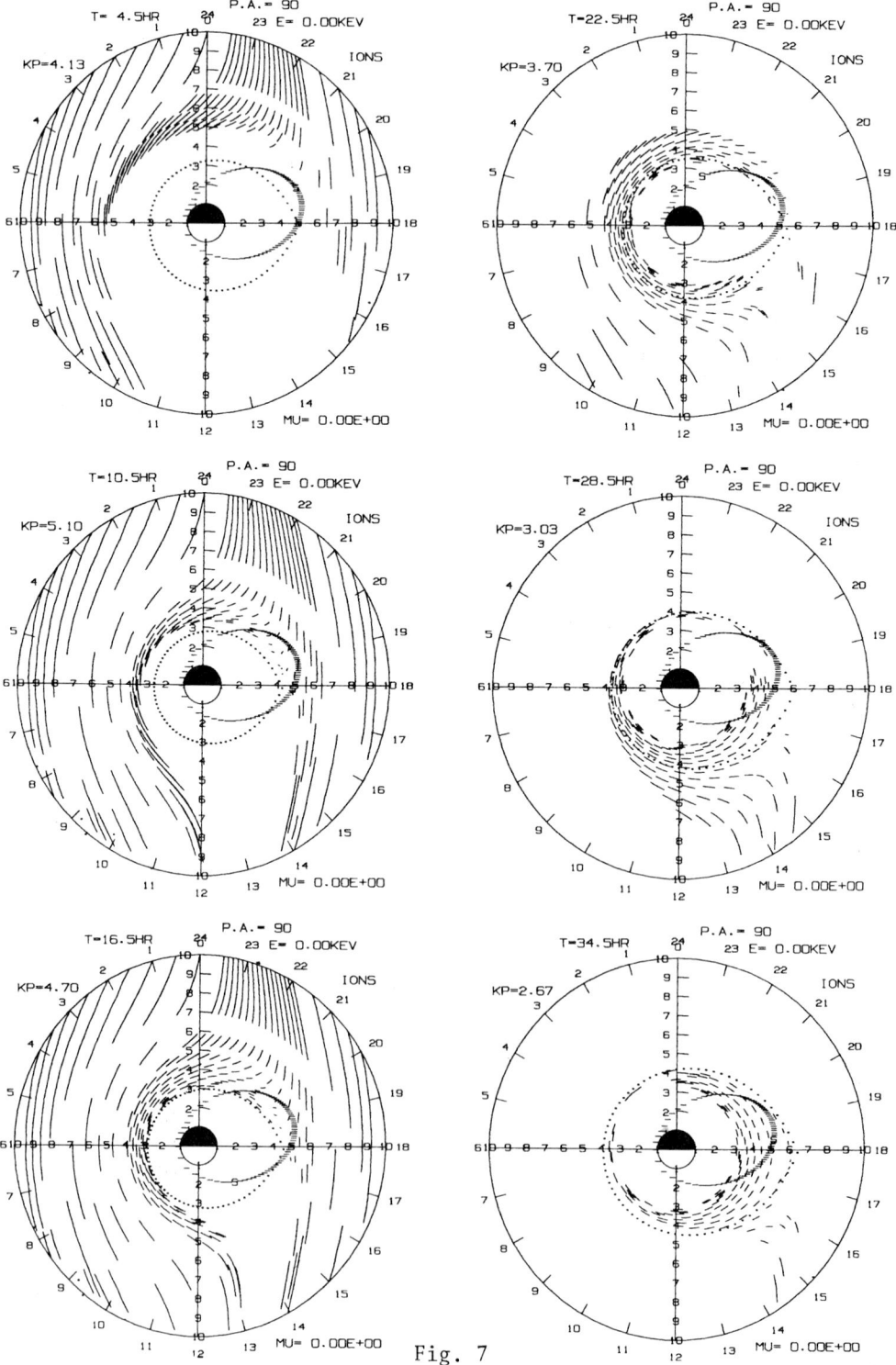

Fig. 7

(1976) also pointed out that the particles may have been "lost" before they could penetrate very deeply, either by charge exchange or strong diffusion. Thus Cowley's conclusions are not inconsistent with this $\gamma = 2$ model of Volland and Stern.

Intensity enhancements of the ring current electrons associated with VLF-emissions during geomagnetic storms and the energy dispersion of these enhancements have been explained by Maeda et al. (1978) using the computed electron trajectories given by this model (Ejiri, 1978). Additional progress was made by Ejiri et al. (1978) in the utilization of trajectories calculated in a static-electric field. They found that to explain the sequence of positions of nose structures during a magnetic storm, the location of particle fronts had to be calculated, taking into account the finite traveling time of newly-injected particles, instead of considering only the particle inner boundaries determined for $t = \infty$. Ejiri et al. (1978) noted that McIlwain's (1974) E3 or E3H electric field models correspond to a weak field and do not predict the plasmapause position inside the Explorer 45 orbit. Ejiri et al. (1979) also used the $\gamma = 2$ convection model for explaining the ion nose structure for both equatorially mirroring ions and those with other pitch angles. They examined the energy spectra and penetration distances for both electrons and ions in the post-midnight to morning hours local time.

## Addition of Time-Dependence to Convection Electric Field

In the previous sections we have shown how the Volland-Stern convection electric field, Eq (3), can be related through Eqs (4) and (5) to the Kp index. The use of this index to parameterize convection boundaries is discussed by Kivelson (1976), principally for the uniform ($\gamma = 1$) dawn-dusk electric field, but also for the model we have been discussing and which Kivelson (1976) refers to as the VSMC (Volland-Stern-Maynard-Chen) model. While this model is more sophisticated than the $\gamma = 1$ model and it provides good agreement with a large data set (see previous section), one of the principal advantages is that it is a simple model and does not require large amounts of computer time to generate the particle trajectories described by Ejiri (1978). In keeping with this approach we have added a time dependence to this static electric field model in a very simple way. As shown by Maeda et al. (1978) the electric field intensity is related to Kp through a very smooth function (Figure 2). We merely use the time history of Kp, which is available to a wide user community, to determine the large scale convection electric field intensity. We interpolate the three-hourly values of Kp to get an hourly quasi-value called Kp' from which the electric field can be determined on an hourly basis. This is done under the gross assumption that the large scale convection electric field must change smoothly over the large areas considered or that at least the particles' response to the changing electric field can be, on average, represented by this type of field. In the previous work with

Fig. 7. Same as Figure 5a, but for $\mu = 0.0$ keV/$\gamma$ particles with six time periods shown.

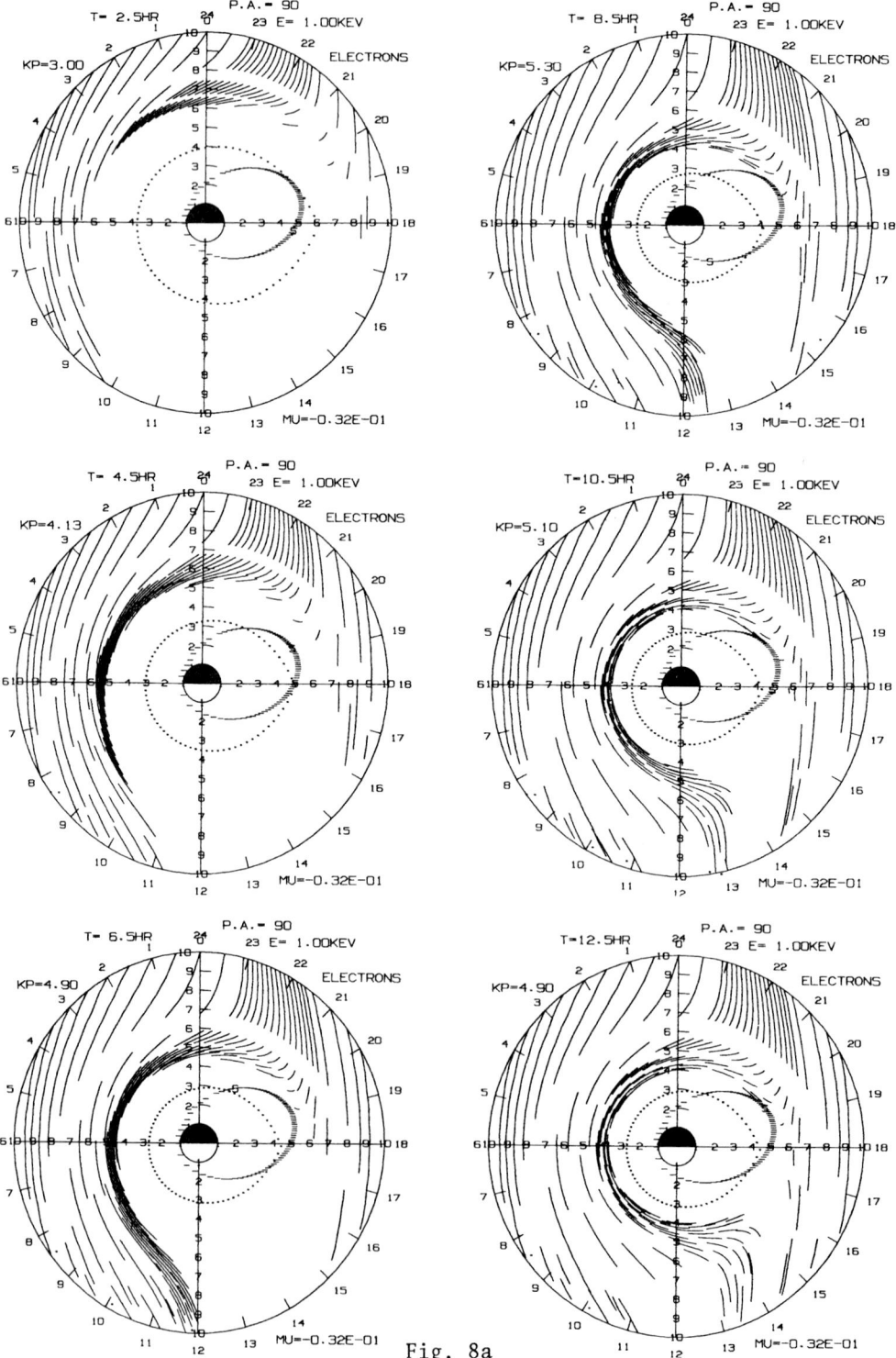

Fig. 8a

this model (Ejiri et al. 1978, 1979) a static electric field was used with the particles moving in time along fixed trajectory paths (Figure 3). With the addition of this time-varying electric field the trajectories themselves change in time as the particles move through the magnetosphere.

The utilization of a time-dependent electric field for particle convection is not a new idea (Kellogg, 1959). Roederer and Hones (1974) devised a time-dependent electric field model, composed of a time-dependent uniform dawn-dusk field component, a static component of the dawn-dusk and corotational fields, and a localized azimuthal field, to simulate the particle injections observed at the geo-synchronous orbit by ATS-5. McIlwain (1972) also pointed out that transient field changes are required to cause the observed particle injections to remain on stably trapped orbits.

In Figure 4 the Kp index is shown for the February 24-25, 1972, geomagnetic storm. The computed electric field intensity at 10 $R_E$, at midnight ($\phi = 0°$) determined from Kp' is shown in the upper portion. Note that after the initial Kp increase, the index decreased during the next 24 hours and then increased slightly from Kp = 2 to Kp = 3. During this period of Kp decrease the computed field weakens by an order of magnitude from a peak value of about 1 mV/m. In the next section we will use this time-varying convection $\vec{E}$ field in the computation of the motions of charged particles of various magnetic moments (both ions and electrons).

## Visual Description of Particle Motion

The addition of the time-varying convection $\vec{E}$ field to the trajectory traces of Ejiri (1978) shown in Figure 3, suggested to us that the most beneficial display technique is motion pictures which would provide the time dimension not only for the variation in electric field strength but also for the movement of the particles themselves. Computer-generated movies of storm "fronts" of injected particles of various magnetic moments have been produced and have been shown at scientific meetings (Smith et al., 1978). In the figures that follow some snap-shot frames of the movie titled "Convection of Magnetospheric Particles in a Time-Varying Electric Field" (Version 780919) are presented. The general characteristics and format of the movie frames (Figures 5-8) are:

1. Singly-charged particles are injected from a distance of 10 $R_E$ and only equatorially mirroring particles are considered.
2. Injection is over a wide local time region around midnight and a uniform spacing of the injection points is assumed, except for the extra trajectories in small local time regions which are added in order to show which trajectories are "trapped". This will be discussed later.
3. The trajectories are not followed after they intersect the 10 $R_E$ boundary.
4. The time history of the electric field intensity is shown in

---

Fig. 8a. Same as Figure 5a, but for $\mu = -0.032$ keV/$\gamma$ particles (electrons). The configuration at six times, T, are shown.

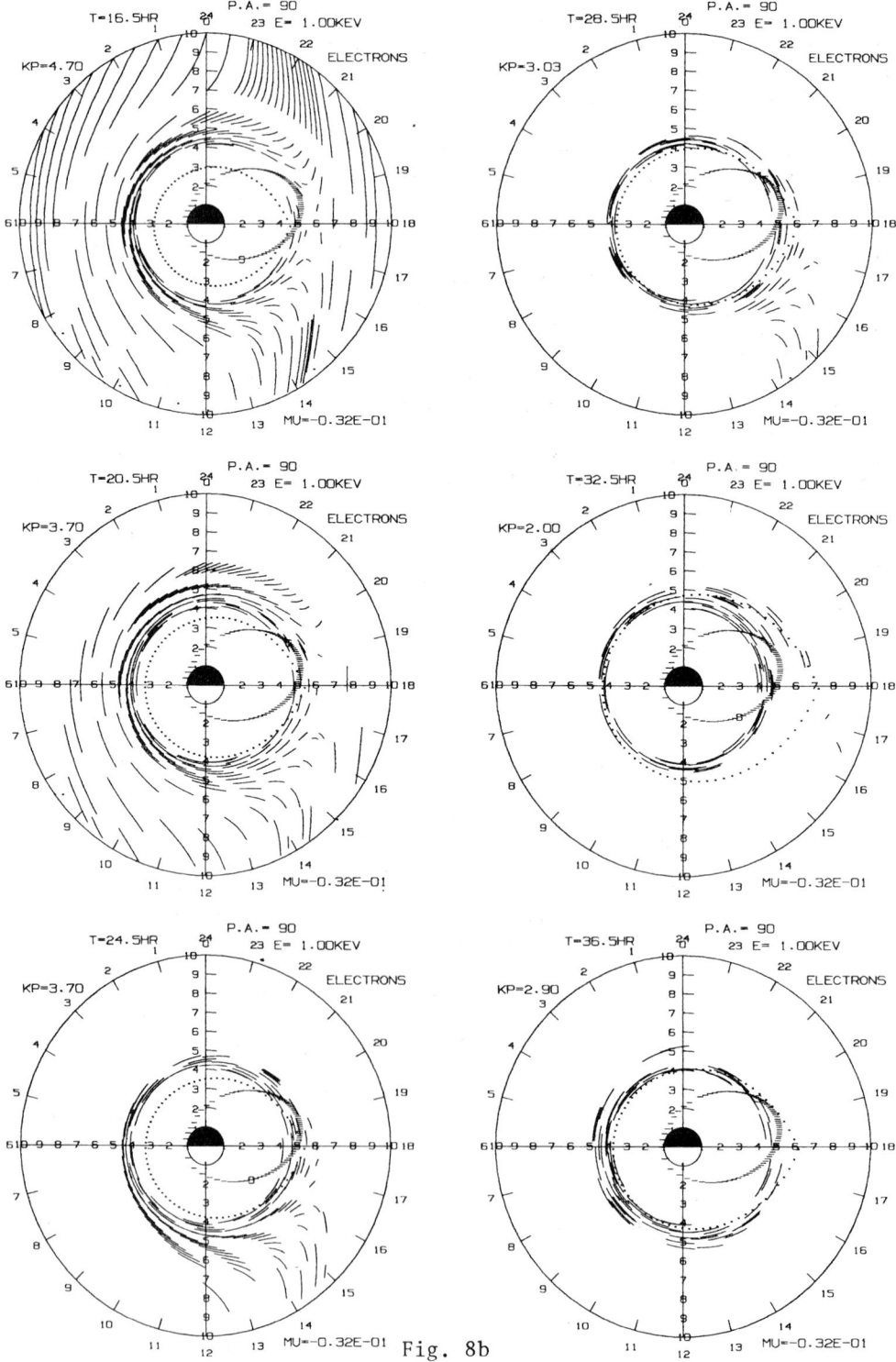

Fig. 8b

Figure 4. The field changes every hour and the particles will move during each hour under the field prescribed by Kp' shown on the individual movie frames.

5. The time, T, changes in the movie every 0.1 hour and a new injection of particles occurs every hour and is arbitrarily stopped after 20 injections.

6. The magnetic moment is conserved in each of the cases and is shown on the frames. The kinetic energy, E, of the particle at the time of injection at L = 10 is shown in the upper right-hand portion of the frame. The particles are energized in this injection process by cross-L drift.

7. The dotted shape is essentially the plasmapause, i.e. the last closed equipotential lines for this convection model. The size varies with Kp'. It should be noted that the shape of McIlwain's injection boundary (Mauk and McIlwain, 1974) normalized by the stagnation point at dusk coincides well with this shape, but that the unnormalized distance is well beyond this "plasmapause".

8. The dashed ellipse is the $S^3$-A (Explorer 45) orbit. The satellite is indicated by the "S" moving around the $\sim$ 8 hour orbit.

9. In this version of the movie only the line segments of the trajectories since the last whole hour are shown. At T = 2.5 hours, for example, line segments of trajectories are shown for three injection fronts; each line shows where the particle has traveled since T = 2.0 hours.

In Figures 5-8 the trajectories for four selected magnetic moment cases are shown and are for the following $\mu$'s: 1) $\mu$ = 0.024, 2) $\mu$ = 0.065, 3) $\mu$ = 0.0 and 4) $\mu$ = -0.032, respectively where $\mu$ is given in units of keV/$\gamma$. Cases 1 and 2 are for ions, case 3 for thermal particles and case 4 for electrons. In Figure 5 the injection energy at L = 10 is 0.75 keV and when the ions have reached L = 5 they have been energized to 6 keV. The six snapshots in Figure 5a are taken at two hour intervals starting at T = 2.5 hours. The snapshots in Figure 5b are shown at four hour intervals. It can be seen that the particles injected at local times within an hour or two of midnight are much slower in moving around the earth than those injected at local times away from midnight. At T = 2.5 hours certain trajectories have intersected L = 10 on the day side before other trajectories have crossed L = 6 near midnight. A fine line separates those trajectories going around the earth on the dawn side from those going around (counter to corotation) through the dusk hours. The trajectories begin to intersect the $S^3$-A orbit after T = 4.5 hours and by T = 8.5 hours have moved to the lower altitude regions traversed by the satellite. Note that, at T = 8.5 hours and corresponding to the high Kp' value of 5.3, the particles at midnight convect from L = 10 to L = 6.5 within 30 minutes. The plasmapause has contracted to L = 2.5 at dawn and L = 4 at dusk. By 12.5 hours the particles are convected into L = 3 on the dusk side and are well inside the plasmapause. Twenty separate injections are sim-

Fig. 8b. Same as Figure 8a, but for six later times.

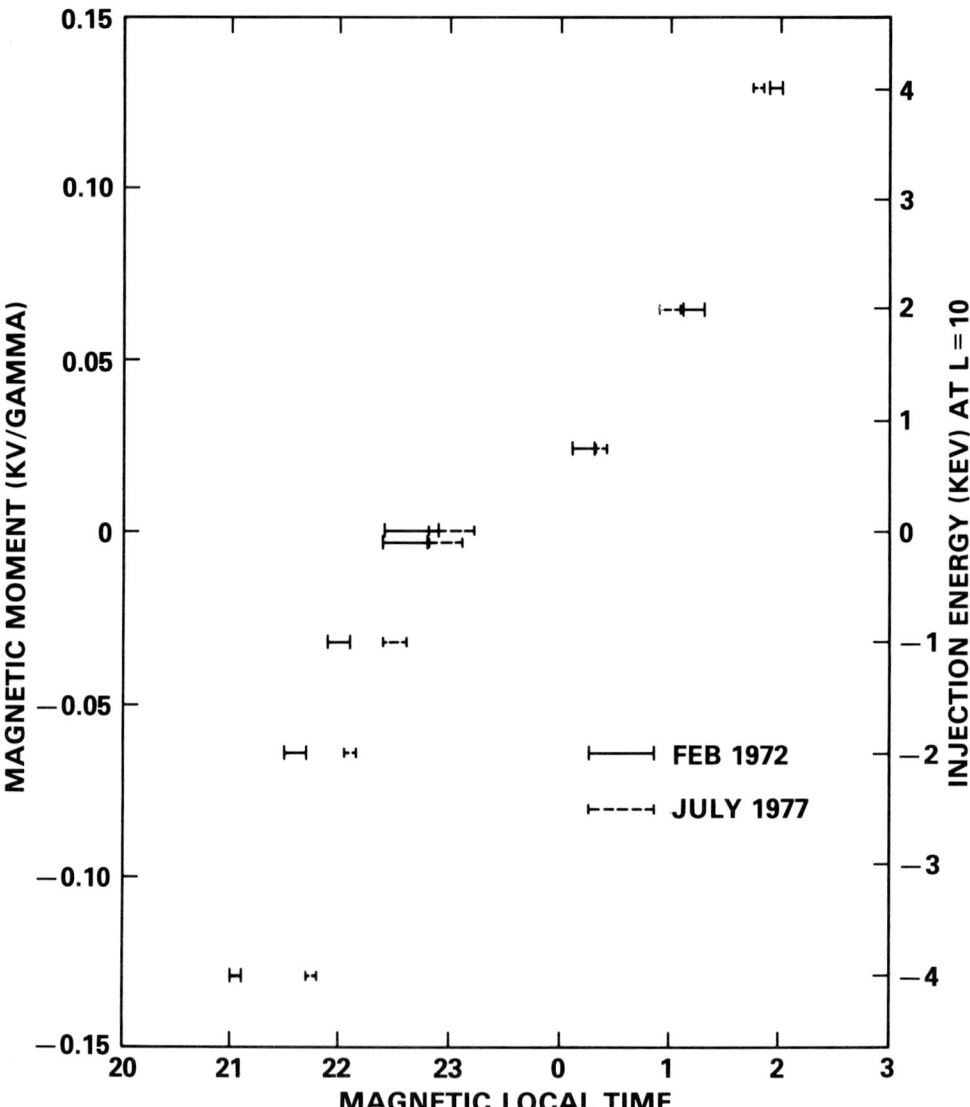

Fig. 9. Magnetic local time of injection at L = 10 for those trajectories which subsequently get "trapped" around the earth using the described convection model. The MLT varies with the magnetic moment of the particles and with the time history of the convection electric field (Kp). Two magnetic storms, February 24, 1972 and July 29, 1977, are shown.

ulated and then arbitrarily stopped. It can be seen in the frames at T = 20.5 and T = 24.5 hours that certain of the trajectories are beginning to appear to be trapped. Even with the large number of simulated trajectories only a very few are actually trapped, and even at T = 24.5 hours those particles have not yet made a full revolution around the earth. As shown in Figure 4 the Kp index continued to decrease to a value of 2, which is shown at T = 32.5 hours in Figure 5b. At this time the "plasmapause" had expanded to 7 $R_E$ at dusk and most of the remaining trajectories were within this plasmasphere. An interesting effect is that even at this relatively low Kp value there exists a stream of particles near noon which are being convected toward the front side L = 10 boundary. It is as if these particles are being kicked out of their trapped orbits by the changing $\vec{E}$ field. The electric field is subsequently enhanced and this effect persists for the remaining time up to the full 48 hours considered in the movie, but not shown in Figure 5. The other interesting and obvious point is the particle diffusion effects produced by this model and most easily seen at the later times.

In Figure 6 the injection energy of the ions is 2.00 keV, and these particles reach 31.25 keV at L = 4. The principal difference is that these particles move much faster. At T = 16.5 hours certain of the trajectories are clearly trapped and the time to make one revolution of the earth was only about 15 hours. These ions, on the other hand, do not penetrate as deep as the lower energy ions. By T = 24.5 hours there is the clearly established ring of particles but also the "noon" convection flow which was described for Figure 5.

The trajectories for the thermal particles ($\mu$ = 0.0) are shown in Figure 7 at a six-hour spacing starting at T = 4.5 hours. These particles convect in further from midnight before they begin their corotational drifts. The trajectories begin to intersect the $S^3$-A orbit near L = 5 in the T = 10.5 frame and it takes well over a day for the injected particles to reach the outbound portion of the orbit at 1600 MLT after they have gone around the earth on the dawn side. The trajectories are well bounded inside the "plasmasphere" after about 30 hours except for the few trajectories which exit and go out the front of the boundary near 1400 MLT.

The one example of electron trajectories is shown in Figure 8 for electrons injected at L = 10 with kinetic energy of 1.0 keV. Figure 8a shows six snapshots at two hour intervals and 8b shows the six at four hour intervals. While a few of the trajectories originate and exit on the dusk side, the vast majority travel through the dawn hours even though they might have originated on the dusk side of midnight. There is a heavy concentration of trajectories in the morning hours starting at about 6 $R_E$ at T = 4.5 hours and moving closer to the earth as the electric field increases. As seen in Figure 8a only a couple of trajectories intersect the $S^3$-A orbit (at L = 5 inbound) during those hours and the trajectories do not begin to intersect the outbound portion of the orbit until about 16.5 hours, after they have nearly encircled the earth. Within about a day there is formed a ring of trapped electrons whose trajectories are at or beyond the "plasmapause" location. It is not until the electric field has decreased (at T = 32.5 hours) to a low level that the electrons are substantially inside the plasmapause and only on the dusk side. The electrons do become trapped and exhibit the "dumping" of particles from the "trapped" orbits into the noon convection stream as described also for the ions.

As shown in the figures above there exist narrow local time regions at L = 10 from which the subsequently trapped particles are injected onto stable orbits. The trapping depends on the electric field time history and hence the local time region of injection varies according to the magnetic storm conditions. Also the region varies in local time according to the magnetic moment of the injected particles. This dependence is shown in Figure 9 for two geomagnetic storms, the February 1972 storm which we have previously discussed and the July 29, 1977 storm shown for comparison. The region from which the trapped particles emerge varies from pre-midnight ($\sim$ 21 MLT) for large negative $\mu$'s (electrons) to post midnight ($\sim$ 02 MLT) for large positive $\mu$'s (ions). The exact dependence is of course coupled directly to the convection model we are using, but this could be a good additional test for this general convection formulism.

## Discussion

The basic convection model described in this paper agrees well with the Explorer 45 results. The comparison of the time-dependent electric field extension of this model to the data has not yet been made, however. The model does provide for the trapping of the particle trajectories which was not possible in the static convection electric field case. It is an analytically simple model to use and can provide the foundation for more refined large-scale convection models especially one which takes into account the tilt and the tail-like geometry of the geomagnetic field. It should also be emphasized that the large scale convection electric field which has been used in this model is a smooth power-law electric field and it does not provide for the rapid plasma flows observed at subauroral latitudes in the topside ionosphere (Shiro et al., 1978), nor does it explicitly include the type of large (20-30 mV/m) transient electric fields which occur during short periods (30-60 sec.) and which have been observed, for example, on GEOS (Pedersen and Grard, 1979). The overall magnetude of the convection electric field, however, does agree with the average field ($\sim$ 1.0 mV/m) measured on GEOS (Pedersen and Grard, 1979) and with the field derived from the particle velocity distributions obtained from ISEE (Frank et al., 1978). It is, of course, not the only electric field model currently in use (Stern, 1977) and other models such as McIlwain's (1974) based on ATS-5 results and the self-consistent modeling described by Wolf (1975) must also be considered. The Wolf model (Harel et al., 1979), for example, is more detailed and takes into account the boundary currents and ionospheric coupling effects.

It is expected that the computer-generated movie display technique, which we have employed, will be increasingly used in the development of the various quantitative magnetospheric models. This technique provides an exceptional educational tool for magnetospheric physics, and future studies of particle dynamics in the magnetosphere will require this technique as an analytical tool due to the increased complexity of the analysis with the recent addition of time-dependent electric fields.

## References

Axford, W. I., Magnetospheric convection, Rev. Geophys. Space Phys. 7, 421-459, 1969.

Axford, W. I., and C. O. Hines, A unifying theory of high-latitude geophysical phenomena and geomagnetic storms, Can. J. Phys., 39, 1433-1464, 1961.

Brice, N. M., Bulk motion in the magnetosphere, J. Geophys. Res., 72, 5193-5211, 1967.

Carpenter, D. C. and C. Park, On what ionospheric workers should know about the plasmapause-plasmasphere, Rev. Geophys. Space Phys., 11, 133-154, 1973.

Chappell, C. R., The convergence of fact and theory on magnetospheric convection, in Correlated Interplanetary and Magnetospheric Observations, edited by D. E. Page, pp 277-295, D. Reidel, Dordrecht, Netherlands, 1974.

Chen, A. J., Penetration of low-energy protons deep into the magnetosphere, J. Geophys. Res., 75, 2458-2467, 1970.

Cowley, S. W. H., Energy transport and diffusion, in Physics of Solar Planetary Environments, edited by D. J. Williams, pp 582-607, Proceedings of the International Symposium on Solar-Terrestrial Physics, American Geophysical Union, Washington, D.C., 1976.

Cowley, S. W. H. and M. Ashour-Abdalla, Adiabatic plasma convection in a dipole field: Proton forbidden-zone effects for a simple electric field model, Planet. Space Sci., 24 821-833, 1976.

Dungey, J. W., Interplanetary magnetic field and the auroral zones, Phys. Rev. Letters, 6, 47-48, 1961.

Ejiri, M., Trajectory traces of charged particles in the magnetosphere, J. Geophys. Res., 83, 4798-4810, 1978.

Ejiri, M., R. A. Hoffman and P. H. Smith, The convection electric field model for the magnetosphere based on Explorer 45 observations, J. Geophys. Res., 83, 4811-4815, 1978.

Ejiri, M., R. A. Hoffman and P. H. Smith, Energetic particle penetrations into the inner magnetosphere, to be published in, J. Geophys. Res., (Preprint NASA/GSFC X-625-77-254), 1979.

Frank, L. A., K. L. Ackerson, R. J. De Coster, and B. G. Burek, Three-dimensional plasma measurements within the earth's magnetosphere, to be published in, Space Sci. Rev., presented at 13th ESLAB Symposium, Innsbruck, Austria, 1978.

Grebowsky, J. M. and A. J. Chen, Effects of convection electric field on the distribution of ring current type protons, Planet. Space Sci., 23, 1045-1052, 1975.

Harel, M., R. A. Wolf, P. H. Reiff, and M. Smiddy, Computer modeling of events in the inner magnetosphere, Quantitative Modeling of the Magnetospheric Processes, Geophys. Monogr. Ser., vol. 21, edited by W. P. Olson, AGU, Washington, D.C. 1979.

Heppner, J. P., Electric field variations during substorms: OGO-6 measurements, Planet. Space Sci., 20, 1475-1498, 1972.

Jaggi, R. K., and R. A. Wolf, Self-consistent calculation of the motion of a sheet of ions in the magnetosphere, J. Geophys. Res., 78, 2852-2866, 1973.

Johnson, R. G., R. D. Sharp, and E. G. Shelley, Observations of ions of ionospheric origin in the storm-time ring current, Geophys. Res. Letters, 4, 403-406, 1977.

Kavanagh, L. D. Jr., J. W. Freeman, Jr., and A. J. Chen, Plasma flow in the magnetosphere, J. Geophys. Res., 73, 5511-5519, 1968.

Kellogg, P. J., Van Allen radiation of solar origin, Nature, 183, 1295-1297, 1959.

Kivelson, M. G., Magnetospheric electric fields and their variation with geomagnetic activity, Rev. Geophys. Space Phys., 14, 189-197, 1976.

Maeda, K., N. K. Bewtra, and P. H. Smith, Ring current electrons trajectories associated with VLF-emissions, J. Geophys. Res., 83, 4339-4346, 1978.

Mauk, B. H. and C. E. McIlwain, Correlation of Kp with the substorm-injected plasma boundary, J. Geophys. Res., 79, 3193-3196, 1974.

Maynard, N. C., and D. P. Cauffman, Double-floating probe measurements on $S^3$-A, J. Geophys. Res., 78, 4745-4750, 1973.

Maynard, N. C., and A. J. Chen, Isolated cold plasma regions: Observations and their relation to possible production mechanisms, J. Geophys. Res., 80, 1009-1013, 1975.

McIwain, C. E., Plasma convection in the vicinity of the geosynchronous orbit, in Earth's Magnetospheric Processes, edited by B. M. McCormac, p 268-279, D. Reidel, Dordrecht, Netherlands, 1972.

McIwain, C. E., Substorm injection boundaries, in Magnetospheric Physics, edited by B. M. McCormac, pp 143-154, D. Reidel, Dordecht, Netherlands, 1974.

McIwain, C. E., Auroral electron beams near the magnetic equator, Nobel Symposium Proceedings, Plenum Publishing Co., London, 1975.

Mead, G. D., Deformation of the geomagnetic field by the solar wind, J. Geophys. Res., 69, 1181-1195, 1964.

Nishida, A., and T. Obayashi, Magnetosphere convection, in Critical Problems of Magnetospheric Physics, edited by E. R. Dyer, p. 179, Inter-Union Commission on Solar-Terrestrial Physics Secretariat, National Academy of Sciences, Washington, D. C., 1972.

Pedersen, A., and R. Grard, A review of dc electric field measurements performed on GEOS, Quantitative Modeling of the Magnetospheric Processes, Geophys. Monogr. Ser., vol. 21, edited by W. P. Olson, AGU, Washington, D.C., 1979.

Roederer, J. G., Dynamics of Geomagnetically Trapped Radiation, Springer, New York, 1970.

Roederer, J. G. and E. W. Hones, Jr., Motion of magnetospheric particle clouds in a time-dependent electric field model, J. Geophys. Res., 79, 1432-1438, 1974.

Shelley, E. G., R. G. Johnson, and R. D. Sharp, Satellite observations of energetic heavy ions during a geomagnetic storm, J. Geophys. Res., 77, 6104-6110, 1972.

Shiro, R. W., R. A. Heelis, and W. B. Hanson, Ion convection and the formation of the mid-latitude F region ionization trough, J. Geophys. Res., 83, 4255-4264, 1978.

Smith, P. H., and R. A. Hoffman, Direct observation in the dusk hours of the characteristics of the storm time ring current particles during the beginning of magnetic storms, J. Geophys. Res., 79, 966-971, 1974.

Smith, P. H., R. A. Hoffman, and N. K. Bewtra, A visual description of the dynamical nature of magnetospheric particle convection in a time-varying electric field, EOS, 59, p. 361, 1978.

Stern, D. P., Models of the earth's electric field, NASA/GSFC X-602-74-159, May 1974.

Stern, D. P., The motion of a proton in the equatorial magnetosphere, J. Geophys. Res., 80, 595-599, 1975.

Stern, D. P., Large-scale electric fields in the earth's magnetosphere, Rev. Geophys. Space Phys., 15, 156-194, 1977.

Vasyliunas, V. M., The interrelationship of magnetospheric processes, in Earth's Magnetospheric Processes, edited by B. M. McCormac, pp 29-38, D. Reidel, Dordrecht, Netherlands, 1972.

Volland, H., A semiempirical model of large-scale magnetospheric electric fields, J. Geophys. Res., 78, 171-180, 1973.

Wolf, R. A., Effect of ionospheric conductivity on convective flow of plasma in the magnetosphere, J. Geophys. Res., 75, 4677-4698, 1970.

Wolf, R. A., Ionosphere-magnetosphere coupling, Space Sci. Rev., 17, 537-562, 1975.


This paper has described in detail the model which was used to generate the movie "Convection of magnetospheric particles in a time-varying electric field," Version 780907. Copies of this ten minute, black and white movie are available for loan to interested scientists and appropriate institutions.

Requests should be addressed to:
    Dr. Paul H. Smith
    NASA/Goddard Space Flight Center
    Code 626
    Greenbelt, Maryland 20771

GLOBAL MAGNETOHYDRODYNAMIC SIMULATION OF THE
TWO-DIMENSIONAL MAGNETOSPHERE

J. N. Leboeuf, T. Tajima, C. F. Kennel, and J. M. Dawson

Department of Physics, University of California at Los Angeles
Los Angeles, California 90024

Abstract. The time-dependent magnetohydrodynamic interaction of the solar wind with a two-dimensional dipole magnetic field has been simulated using a novel Lagrangian particle type of MHD code that can treat local low density or vacuum regions without numerical instability. This enables us to simulate the time-dependent magnetic tail. When the solar wind field is southward, a magnetic field line topology consistent with Dungey's model emerges in steady state. The tail, however, is short, and the x-points are only slightly shifted from their vacuum locations, because of strong numerical resistivity. Different configurations resulting from different relative orientations of the solar wind magnetic field and dipole axis are also presented. While the magnetic field is relatively steady, the density and flow in the magnetosheath are turbulent, as are the bow shock and magnetopause; the Kelvin-Helmholtz instability may account for these phenomena. We also model a "substorm" as the passage of a rotational discontinuity in the solar wind over the dipole. Both 90° and 180° shifts to a southward solar wind field cause a violent readjustment of the magnetic tail (involving emission of a closed magnetic island downstream) which eventually settles down to the Dungey configuration.

Introduction

Global models of the earth's interaction with the solar wind have thus far been constructed in the "Cartoon Approximation" [R. Z. Sagdeev, private communication, 1975] - that is, inspired guesswork based upon synthesis of diverse single-point satellite measurements and partial theoretical calculations. Our lack of the global information characteristic of laboratory experimentation stems from our experimental inability to "photograph" the magnetosphere and our theoretical inability to solve the highly non-linear time dependent three-dimensional MHD and plasma equations describing the magnetosphere. Numerical and laboratory simulation can potentially bridge this gap for the theoretician. The first numerical solutions of a global MHD model of the earth's magnetosphere were carried out by Spreiter and his co-workers [Spreiter and Alksne, 1969]. However, these calculations were time-independent and failed to model the earth's tail. Notwithstanding a very considerable effort, it is fair to say that we have failed to produce a consensus picture of the time dependent tail which includes unsteady reconnection. A few numerical calculations have been devoted to processes involved in the global behavior of the magnetosphere, such as reconnection [Fukao and Tsuda, 1975; Yang and Sonnerup, 1977] or convection in the inner magnetosphere [Wolf, 1974]. Such calculations all ultimately depend on boundary conditions imposed by the global behavior of the magnetosphere.

In this paper, we present results from two-dimensional, time-dependent global MHD simulations of the earth's magnetospheric interaction with the solar wind. Our two-dimensional steady state magnetosphere has the topology proposed by Dungey [1961]; to our knowledge, this is the first time that reconnection has been solved in the complex geometry of a magnetosphere. We also find time-dependences which will be interesting to pursue in later work. Our magnetosheath and magnetopause are turbulent, and we observe the formation and disconnection of a large magnetic bubble in the deep magnetic tail during "substorms", which are modelled by the passage of rotational discontinuities over the two-dimensional dipole. Our simulations are highly dissipative, with typical magnetic Reynolds numbers the order of 3-10. In addition, we have imposed periodic boundary conditions on the magnetic field. We will discuss the implications of these restrictions.

## Numerics

A principle difficulty with numerical simulation codes for global magnetospheric studies has been the unfortunate tendency for such codes to produce grossly unphysical results when the fluid density becomes low. This difficulty has prevented numerical models of the geomagnetic tail from being created. Our MHD particle code has recently overcome this difficulty and has produced numerically stable models of the magnetosphere and its tail [Leboeuf et al., 1978a].

Our newly developed Lagrangian type of MHD code [Leboeuf et al., 1978b] treats elements of the fluid as finite size particles. The particle quantities such as position, mass and momentum are pushed in a Lagrangian way while the magnetic fields are advanced in an Eulerian manner.

The position of a particle is found by

$$d\underline{r}_j(t)/dt = \underline{v}_j(t) \qquad (1)$$

In the ideal, one fluid description of a plasma, the equation of motion for each particle in a momentum conservative form is given by

$$d\underline{v}_j(t)/dt = -1/\rho [\nabla P + 1/8\pi \nabla B^2 - 1/4\pi \, \underline{\nabla} \cdot \underline{BB}] \qquad (2)$$

The left hand side of Eq. (2) is the total derivative of the j-th particle velocity while the right hand side only involves macroscopic quantities defined at the mesh points of the fixed background grid.

The plasma is modeled as an ideal two-dimensional (2-D) MHD fluid with an adiabatic equation of state so that

$$\frac{T}{n^{\gamma-1}} = \text{constant} \qquad (3)$$

The density is in turn expressed as

$$n(\mathbf{r}) = \sum_j f(\mathbf{r} - \mathbf{r}_j)$$

where $f(\mathbf{r} - \mathbf{r}_j)$ is the form factor of the particle, typically a Gaussian one. The pressure gradient calculation ($\nabla P = \nabla(nT)$ in the 2-D pressure expression) is carried out in Fourier space by the use of Fast Fourier Transforms in analogy to the calculation of the electric field from the charge density in conventional particle codes [Kruer et al., 1973]. The terms in Maxwell's stress tensor are evaluated by conventional finite difference methods.

The magnetic fields are updated at the mesh points as follows

$$\partial \mathbf{B}/\partial t = \nabla \times (\mathbf{v}_g \times \mathbf{B}) \tag{5}$$

where the fluid velocity $\mathbf{v}_g$ is defined as

$$\mathbf{v}_g(r) = \sum_{j \in g} \mathbf{v}_j(r) / \sum_{j \in g} 1 \tag{6}$$

where the denominator on the right hand side represents the number of particles in the g-th cell. Equation (5) is not intrinsically time-centered. To achieve numerical stability in Eq. (5), the conservative lax method [Potter, 1973] is employed. The adoption of this algorithm brings in an effective diffusion of magnetic fields numerically; we are effectively solving the approximate differential equation

$$\partial \mathbf{B}/\partial t - \nabla \times (\mathbf{v}_g \times \mathbf{B}) = (\Delta^2/2N\Delta t - \Delta t \, c^2/2)\nabla^2 \mathbf{B} \tag{7}$$

where $\Delta$, $\Delta t$, $N$ and $c$ are respectively the unit grid spacing, the time step, the spatial dimensionally (here $N = 2$) and the fastest propagation velocity on the mesh.

The first term on the right-hand side of (7) originates from the spatial averaging algorithm in the Lax method. The second one arises from the uncentered finite differencing in time. These two terms together represent an effective resistivity which goes to zero when $\Delta t = \Delta/c\sqrt{N}$. A Courant-Friedrich-Levy condition is obtained by demanding that there be no negative diffusion in Eq. (7), i.e.

$$\Delta^2/2N\Delta t - \Delta t c^2/2 > 0 \tag{8}$$

When the equality if taken in Eq. (8), the numerical diffusion is reduced to zero. It is not possible, however, in general to enforce that condition everytime and everywhere in the algorithm, since the system is evolving in space and time and $c^2$ depends on the instantaneous, local plasma density, temperature and magnetic field strength. Since in our simulation $\Delta t < \Delta/c\sqrt{N}$, except at the dipole points where the equality was taken, it means that our magnetosphere is substantially resistive. An additional cause of dissipation to a lesser extent comes from the finite cell size effect in the finite differences. This allows some slippage of the particles across the field lines. The fluid obeys, however, zero ion Larmor radius equations customary in the ideal MHD approximation.

For the simulations reported here, we use a 2-1/2 dimensional version of the code (two spatial and three velocity and field dimensions) with typically 192 × 128 neutral fluid particles on a 64 × 32 uniform x-y spatial grid. The field boundary conditions are periodic. We replace each fluid particle at x = 64 with a new one at x = 0 that has the specified "solar wind" speed and a random y-position; thus the incoming flow pattern is independent of the presence of the obstacle. At t = 0, the density and the velocity of the fluid particles are spatially uniform. The earth's dipole field is modeled by a pair of oppositely flowing z-currents. Our experience thus far indicates that our MHD particle simulation algorithm is stable without adjustment of the time step in the low density region in the "magnetospheric tail". Such a low density region or partial vacuum causes conventional Eulerian codes to become numerically unstable because the Alfven speed becomes infinite locally. Moreover, the strong advection inherent to the solar wind - earth interaction process induces the density to become negative in conventional Eulerian codes, a difficulty which is naturally resolved by the particle nature of our code. With this code, for example, forward MHD shocks created by an obstacle in the flow have also been observed [Tajima et al., 1978a]. The adiabatic gas index is taken to be 2. Finally we estimate the effective magnetic Reynolds number of our flow. Using the first term on the RHS of (7) to estimate the effective conductivity, the Reynolds number $R_m$ is

$$R_m = 2N \left(\frac{v \Delta t}{\Delta}\right) \ell \qquad (9)$$

where v is a typical flow speed and $\ell$ is the system scale length in units of the grid spacing $\Delta$. For the runs to be reported here $\Delta t = 1/20 \Delta/c_s$, N = 2, and we estimate v by the flow speed $v \simeq 2.5 \, c_s$, $c_s$ being the initial solar wind sound speed, whereupon $R_m \simeq \ell/2$. Our system scale length has $\ell \simeq 64$, the magnetospheres produced in our runs have $\ell \simeq 10$-$20$, and the scale of turbulent fluid elements turns out to be $\ell \simeq 6$. To rough approximation, the magnetic Reynolds is thus 3-10. Therefore, our flow is highly dissipative relative to the earth's magnetosphere. We are working on an algorithm to reduce the numerical resistivity.

Zero Solar Wind Magnetic Field

The simplest situation one may consider is the interaction of a supersonic solar wind, with no magnetic field imbedded in it, and a dipole field, which is in many ways similar to the flow of a supersonic fluid past a blunt obstacle, with the dipole field playing here the role of the obstacle. Since the characteristics of such a flow are well established, this experiment will serve as a test for the validity of our model.

In Figure 1, we present three aspects of the data returned from a

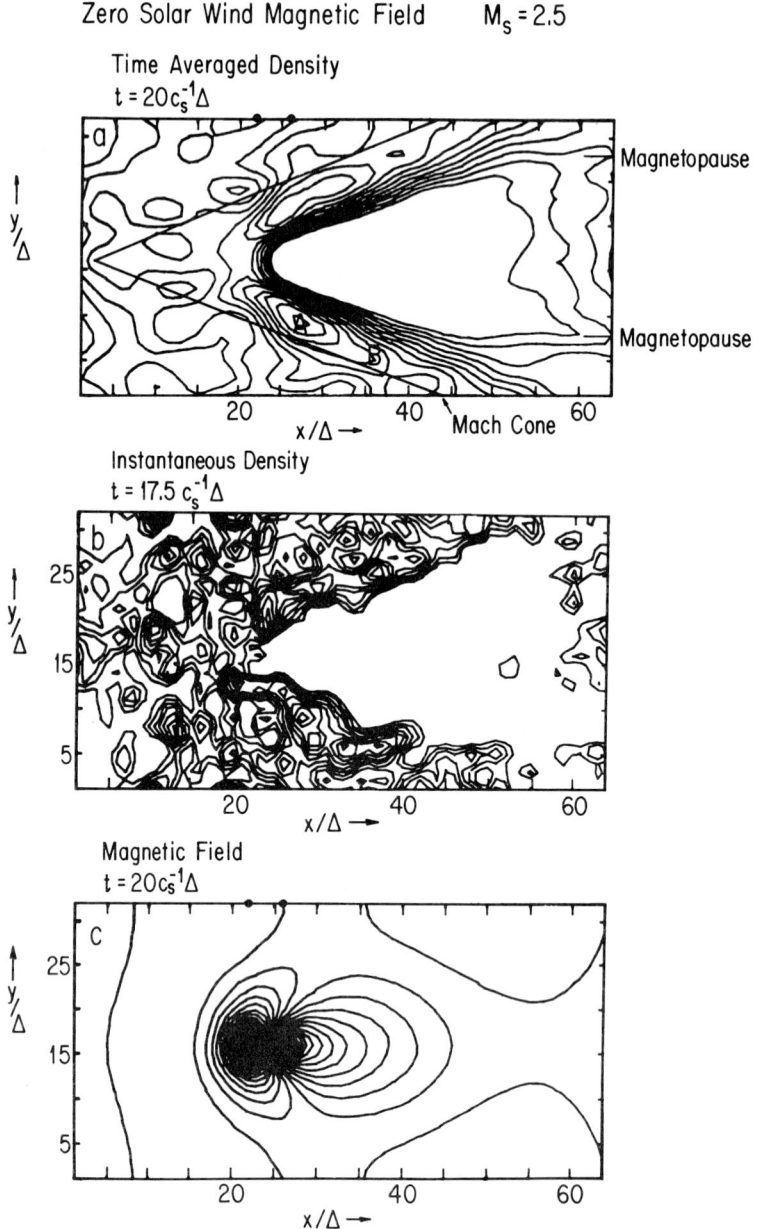

Fig. 1. Magnetosphere with no Solar Wind Magnetic Field. The top inset shows time averaged density contours at $t = 20\Delta/c_S$, the middle inset shows instantaneous contours at $t = 17.5\Delta/c_S$, and the bottom one shows the magnetic field at $t = 20\Delta/c_S$. Contours corresponding to densities less than the initial solar wind density have been suppressed for simplicity. The sonic Mach number was 2.5; the Mach cone is drawn in the top inset.

numerical run which started with a solar wind sonic Mach number of 2.5 and zero solar wind magnetic field. The top inset shows density contours <u>time averaged</u> about the time t = 20 $\Delta/c_s$, where $\Delta$ is the grid spacing and $c_s$ the sound speed. The middle inset shows an instantaneous density profile at t = 17.5 $\Delta/c_s$. The contours obtained for densities less than the initial upstream solar wind density have been suppressed for visual clarification. The bottom inset shows the magnetic field at t = 20 $\Delta/c_s$. The locations of the two dipole currents are indicated by black dots at the top of the diagram. The Mach cone of the bow shock is drawn in to guide the eye. Interesting features of the averaged density contours include (a) the appearance of the density compression associated with the bow shock, (b) the density maximum in the magnetosheath over the polar caps, (c) the formation of a low density cavity in the downstream region, (d) the inhomogeneous high density region near and within the magnetopause.

The measured density jump across the shock can be compared to analytic Rankine-Hugoniot relations for the ordinary fluid [Kellogg, 1962; Axford, 1962; Obayashi, 1964, 1970]. If the shock front is approximated by a hyperbola $(x/x_f)^2 - y^2/x^2_f \tan^2\theta = 1$ with the origin at the crossing of the hyperbola asymptote, the effective Mach number $M'(x,y)$ is given by

$$M' = M \left| \frac{(x/y)^2}{(M^2 - 1)^2 + (x/y)^2} \right| \qquad (10)$$

and the density ratio is

$$\frac{\rho_2}{\rho_1} = \frac{(\gamma+1)M'^2}{(\gamma-1)M'^2 + 2} \qquad (11)$$

The normal components of the fluid velocities across the shock front are, therefore, also obtained as

$$\frac{v_1^n}{v_2^n} = \frac{(\gamma+1)M'^2}{(\gamma-1)M'^2 + 2}, \qquad (12)$$

because the continuity equation $\rho_1 v_1^n = \rho_2 v_2^n$. This pressure for this adiabatic "gas" satisfies

$$\frac{P_2}{P_1} = \frac{2\gamma M'^2}{+1} - \frac{\gamma-1}{\gamma+1}, \qquad (13)$$

where 1 and 2 refer to up and downstream respectively. At the point marked by "A" at x = 28 the measured $\rho_2/\rho_1$ = 1.29, while from (2), $\rho_2/\rho_1$ = 1.31. At the point marked by "B" at x = 36, the measured $\rho_2/\rho_1$ = 1.15, while from (2), $\rho_2/\rho_1$ = 1.10. The measured Mach angle of the asymptotic shock is in good agreement with the theoretical value $\sin^{-1}(1/M) = 23.6°$. Thus there is reasonable agreement between time averaged simulation and the theoretical properties of the bow shock.

Time resolved diagnostics reveal that the shock front, magnetosheath, and magnetopause flutter with a typical period of $2.85 \pm 0.05\ \Delta/c_s$ and scalelength $6.2\Delta$. The instantaneous density profile at $t = 17.5\ \Delta/c_s$ (middle inset). shows a many local density maxima and a wavy magnetopause. We have followed a number of these density maxima and have concluded that they convect with the fluid. One possible explanation for this behavior is a Kelvin-Helmholtz instability generated at the shock, the magnetopause, or both. The Kelvin-Helmholtz period for a scalelength of $6.2\Delta$ and a flow velocity of $2.5\ c_s$ is $2.5\Delta/c_s$, close enough to the simulation fluttering period for the possibility to be retained. Furthermore velocity vector field plots (not shown) reveal that tangential discontinuities, without which the Kelvin-Helmholtz instability could not take place, occur at the shocks, along with normal discontinuities.

The magnetic field (bottom inset) reveals an essentially closed "teardrop" magnetosphere, similar to that predicted for this case [Johnson, 1960]. The outermost magnetic field contours are an artifact of our periodic boundary conditions.

Southward Solar Wind Magnetic Field

A 2-D reconnection magnetosphere differs from its 3-D analog in one significant respect: flux reconnected at the tail cannot convect around the dipole to the nose. To ensure the possibility of a steady reconnection magnetosphere, we have constrained the Z-currents creating the dipole always to be constant, so that flux reconnected at the nose is immediately replaced, and closed flux created at the tail is absorbed. In this and the next sections, we present sample results from runs in which the solar wind magnetic field was assumed everywhere uniform initially, and the system was allowed to evolve in time to a quasi-steady state.

In figure 2, we show steady state magnetic configurations when the solar wind field is southward. The field strength is chosen so that the Alfven speed $c_A$ is one fifth the sound speed $c_s$, the fast Mach number $M_F = 2.45$. The solar wind velocity is along the x-axis, i.e. directed at the earth. When the solar wind field is southward, there are neutral lines in the nose and tail, the topology originally visualized by Dungey [1961]. Because of strong dissipation, the tail is very short. Nonetheless, the neutral lines have been slightly displaced from their vacuum positions by the MHD interaction as is clear from a comparison of vacuum field and total magnetic field in Fig. 2. The induced plus dipole field plot (upper right inset in Fig. 2), obtained from the total magnetic field by subtracting the constant external southward solar wind field, presents a pattern similar to Fig. 1. It indicates the MHD interaction, i.e., compression of the field lines at the nose of the dipole and stretching of the field lines in the tail by the flow. The effect of the MHD interaction shows up equally clearly in the flow induced magnetic field (bottom left inset). Weak induced currents at the x-lines are apparent from a comparison of total magnetic fields and induced fields.

That the solar wind field interacts with a lattice of magneto-

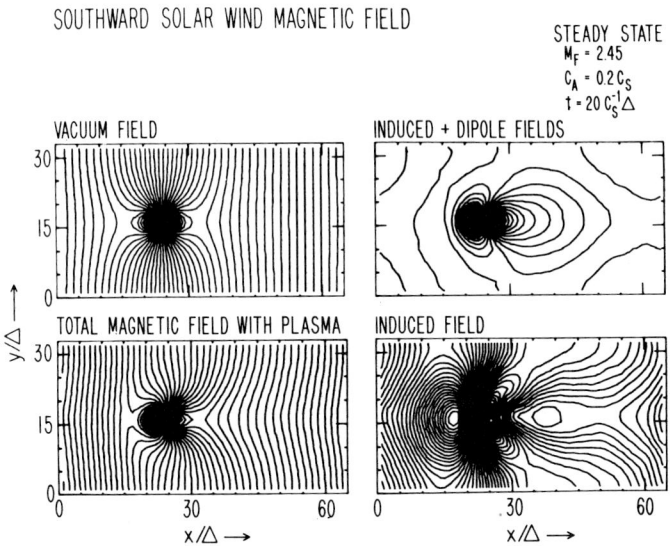

Fig. 2. Magnetosphere with Southward Magnetic Field. In these runs, the solar wind was initially assumed southward everywhere, the wind was turned on at t = 0 and the configuration was allowed to evolve to an apparently steady state. The locations of the x-lines at the nose and tail are slightly displaced from the vacuum locations by the MHD interaction. Weak induced currents are associated with the x-points.

spheres rather than a single magnetosphere is illustrated in Fig. 3 where a 2 × 3 collage of the total magnetic field pattern is displayed. Figure 3 also shows that periodicity effects in the magnetic field are very slight in this case. We have observed the neutral line displacement to vary with different solar wind parameters. Runs with a sonic Mach number of 10 and a strong solar wind magnetic field ($c_A = 1.5\ c_s$) continue to show the Dungey [1971] configuration instead of that of Dubinin et al. [1978].

## Dependence of Magnetic Field Pattern on Angle between Solar Wind Field and Dipole Moment

Because of the 23.5° tilt of the earth's spin axis with respect to the normal to the plane of the ecliptic and the 11.6° angle between the earth's geographic and geomagnetic axis, the angle between the solar wind velocity and the geomagnetic equatorial plane is not always small. Furthermore, the magnetic field imbedded in the solar wind is not restricted to be southward but can assume any direction. It is of interest, therefore, to determine the resulting effect on the shape of the magnetosphere.

In Fig. 4, we show steady-state magnetic configurations corresponding to various angles between solar wind field and dipole moment vector. The solar wind velocity continues to be along the x-axis, i.e., directed at the earth. The solar wind magnetic field is

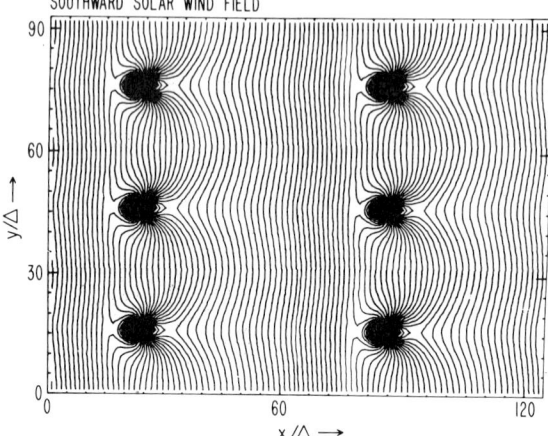

Fig. 3. Dungey Lattice. This 2 × 3 collage of the total magnetic field pattern with southward solar wind field at t = 20 $c_S/\Delta$ shows that the interaction of the solar wind with a lattice of magnetospheres is actually modelled. The effect of the periodic field boundary conditions is however very slight in this case.

chosen so that the Alfven speed $c_A = 0.2\ c_S$, the sonic Mach number $M_S = 2.5$ while the fast Mach number $M_F = 2.45$.

The top inset is for a southward solar wind field but a 34° dipole tilt with respect to the y-axis. The next two diagrams display the magnetic field when there is no dipole tilt but the solar wind field is at 90° with respect to the dipole axis, i.e., directed at the earth, and at 45° with respect to the dipole axis. The locations of both the nose and tail x-lines in the Dungey type magnetospheric magnetic field depend on the angle between the upstream solar wind field and the dipole moment vector. Their locations may be estimated from the vacuum ones when the dissipation is substantial as is the case in these runs. It is not clear whether these patterns are valid for the 3-D magnetosphere. When the solar wind field is northward (bottom inset), the magnetosphere is essentially closed. A teardrop like magnetospheric magnetic field is produced with x-lines over the polar "cusps", which would collapse to points in 3-D. In order for the northward field to penetrate through the lattice of magnetospheres to arrive at the observed steady state, substantial resistive dissipation was necessary.

## 2-D Numerical Substorm and Anti-Substorm - 90° Solar Wind field Switch

We have shown above that the topology of the magnetosphere depends on the solar wind field orientation. Here we postulate that a "substorm" is the time dependent adjustment of magnetospheric topology

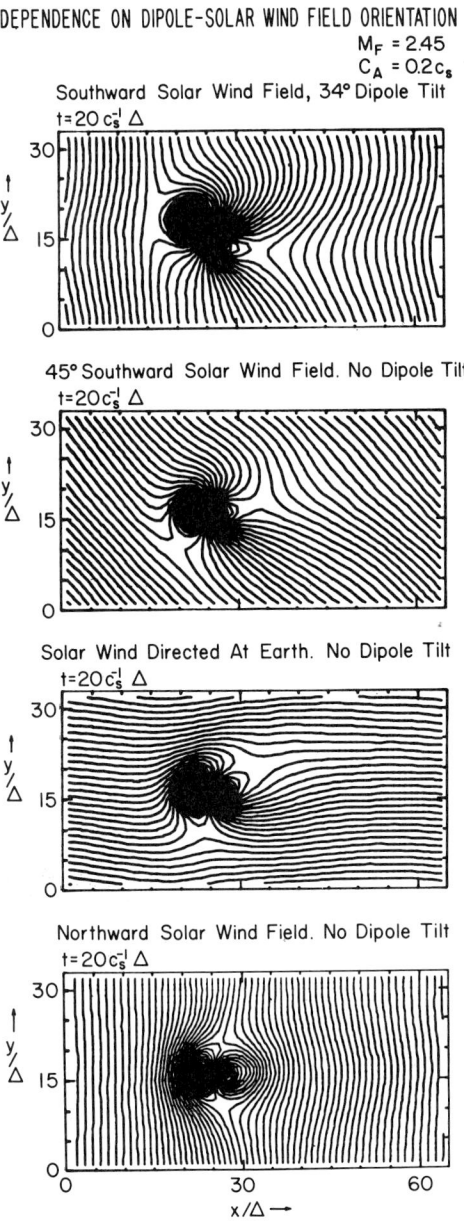

Fig. 4. Dependence of Magnetospheric Topology on Dipole-Solar Wind Field Orientation. These four insets show that the locations of both the nose and tail x-lines depend strongly on the angle between the upstream solar wind field and the dipole moment vector. When the solar wind field is northward, a teardrop-like magnetospheric configuration is produced.

resulting from a sudden rotation of the upstream solar wind magnetic field. A "substorm" corresponds to a southward rotation, and an "anti-substorm", and east-west rotation.

Figure 5 shows our first attempt at a substorm. Here we chose the input solar wind magnetic field to have the form

$$\underset{\sim}{B} = B_o \left\{ \hat{z} H(x - M_s c_s t) - \hat{y} H(M_s c_s t - x) \right\}, \qquad (14)$$

where H is the Heavyside step function. In other words, the initial solar wind field is everywhere uniform and parallel to the geomagnetic equatorial plane. At t = 0, the field suddenly rotates southward at x = 0. $B_0$ is chosen so that the Alfven speed $c_A = 0.2\ c_S$ always, and the density and pressure do not change. We have therefore modelled the passage of a rotational discontinuity over the magnetosphere. Since $c_A$ is small, this discontinuity propagates essentially at the speed of the fluid, which has a fast Mach number $M_F = 2.45$.

At $t = 5\Delta/c_S$, (top inset in Fig. 5) the rotational discontinuity has approached the nose of the magnetosphere and has apparently initiated reconnection there. Note the deformation of the tail. At $t = 10\Delta/c_S$, (middle inset in Fig. 5), the rotational discontinuity is passing over the "polar caps"; the magnetic field is apparently connected to the polar caps. Note that tail reconnection has been initiated and that a bubble of closed flux has formed downstream. Inspection of other figures show that the bubble first moves slowly in the anti-sunward direction; when rapid tail reconnection is initiated, the bubble suddenly disconnects and accelerates to approach the wind velocity. At $t = 30\ c_S/\Delta$, (bottom inset in Fig. 5) the rotational discontinuity has passed beyond the far edge of the diagram, and the magnetopshere has arrived at a Dungey configuration similar to steady state solution of Figure 2. Note that in this run however the tail has been lengthened appreciably from the Fig. 2 case.

We have displayed in Fig. 6 the various contributions to the total magnetic field, in addition to the current density, defined as $\underset{\sim}{J} = (c/4\pi)(\underset{\sim}{\nabla} \times \underset{\sim}{B})$. A weak induced current associated with the bubble downstream is apparent in the current density and flow induced magnetic field. This field is northward upstream and sufficiently so to suppress the external southward solar wind field. It is southward beyond the dipole center.

The current density at various times in this "substorm" run is displayed in Fig. 7. In these contour plots, the rotational discontinuity appears as closely bunched vertical lines, indicative of a sharp current gradient, over the "polar caps" at $t = 10\ c_s^{-1}\Delta$ (top inset) and at the rightmost edge of the picture at $t = 25\ c_s^{-1}\Delta$ (bottom inset) for example. It moves downstream at the speed of the fluid. The flow induced currents are strongest at the dipole. Because of the passage of the rotational discontinuity, fairly strong currents are maintained in the tail region. They have displaced the tail x-line downstream and thus produced temporarily a longer tail than in the steady-state southward solar wind field case of Fig. 2. At $t = 10\ c_s^{-1}\Delta$ (top inset) weak induced currents are also associated with the bubble in the total magnetic field plot of Fig. 5 (middle inset).

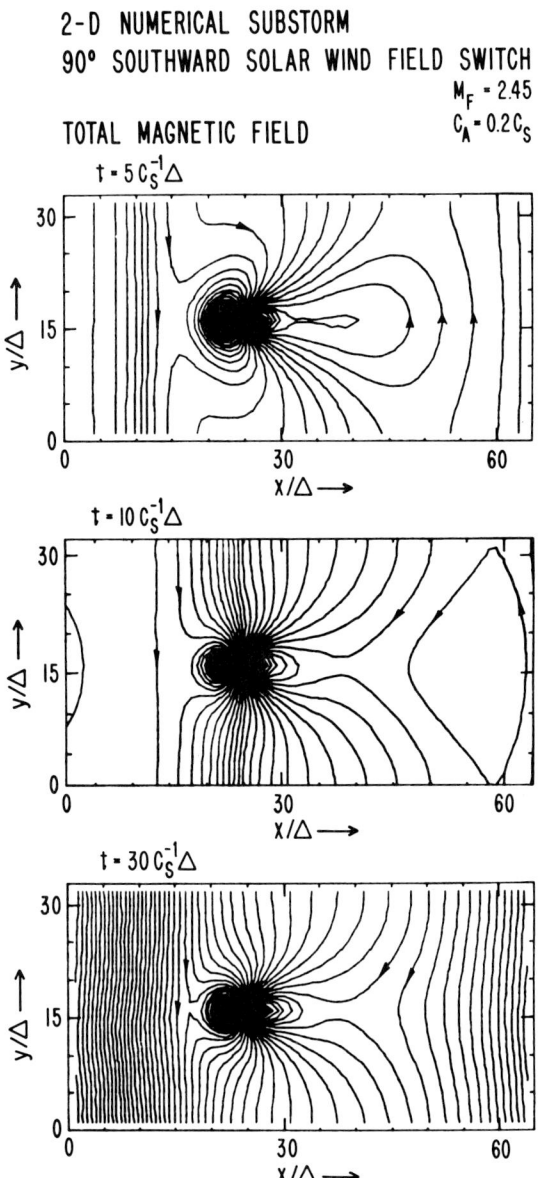

Fig. 5. "Substorm" Simulation: 90° Southward Solar Wind Field Switch. These three insets show the evolution of the magnetospheric topology resulting from the passage over the magnetosphere of a rotational discontinuity which switches the solar wind field from east-west to southward. The magnetosphere evolves from a closed teardrop shape to an open Dungey configuration.

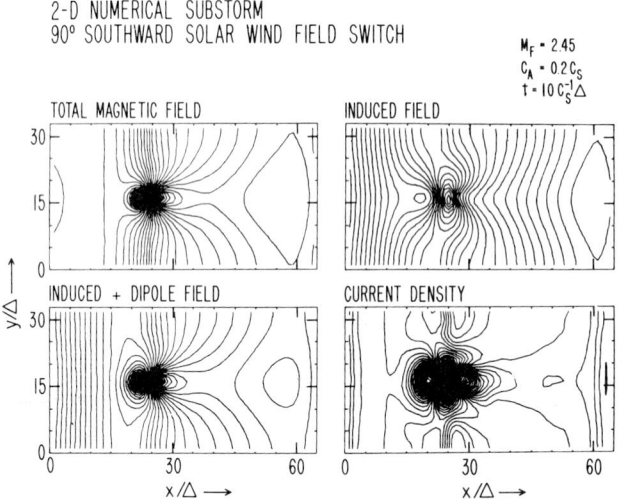

Fig. 6. "Substorm" Simulation: 90° Southward Solar Wind Field Switch. The vacuum contributions to the total magnetic field and the current density are displayed at $t = 10\Delta/c_s$. A weak current is associated with the bubble of closed flux induced downstream by the MHD interaction.

During this substorm, the magnetosphere has gone from a closed tear-drop shaped magnetosphere similar to that of the bottom inset of Fig. 1, when no external field is present, to a Dungey topology when the field has switched southward over the whole magnetosphere. We next present an attempt at an antisubstorm by reversing the field directions in the rotational discontinuity.

Plots of the magnetic field at various times in the run in Fig. 8 illustrate the "antisubstorm" corresponding to the "substorm" of Fig. 5. The input solar wind field now has the form:

$$\underline{B} = B_o \left\{ -\hat{y} H(x - M_s c_s t) + \hat{z} H(M_s c_s t - x) \right\} \quad (15)$$

The initial solar wind field is then everywhere uniform and southward. At $t = 0$, the field suddenly rotates to a direction parallel to the geomagnetic equatorial plane at $x = 0$. $B_o$ is such that $c_A = 0.2\ c_s$.

At $t = 5\ c_s^{-1}\Delta$, the discontinuity has reached the nose of the magnetosphere. The field lines are very sparse upstream because there is no more southward field there. Downstream, the magnetosphere preserves a Dungey configuration because of the remaining southward solar wind field. At $t = 20\ c_s^{-1}\Delta$, the discontinuity has almost reached the rightmost edge of the system. The flow induced field is northward over most of the tail and produces closed field lines there. The x-line at the discontinuity is a result of merging of the induced northward field in the tail region and the remnant external solar wind field. The MHD interaction has apparently induced a southward solar wind field upstream which produces an x-line at the magnetosphere's nose. At $t = 40\ c_s^{-1}\Delta$, the rotational discontinuity has passed be-

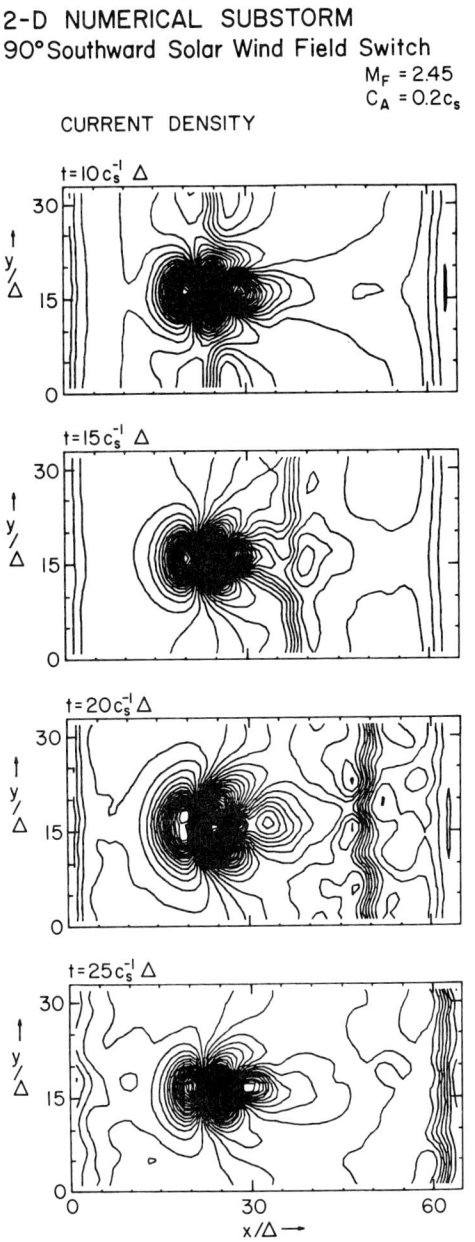

Fig. 7. "Substorm" Simulation: 90° Southward Solar Wind Field Switch. These four insets show the time development of the current density. The passage of the rotational discontinuity over the magnetosphere induces strong currents downstream which result in a slightly longer tail than in Fig. 3.

yond the right edge of the picture. The flow induced northward field over most of the magnetosphere has produced an essentially closed shape with x-lines over the "polar cusps". The x-lines in front and in back are an artifact of our periodic boundary condition. They are induced by the discontinuity which revolves around the magnetosphere at the speed of the fluid and which, without our periodic boundary condition, would have left the system upon reaching the rightmost edge.

Apart from distortions due to periodicity, then, the magnetosphere evolved during the "antisubstorm" from an open magnetosphere of the Dungey type with southward solar wind field to an essentially closed teardrop shape.

## 2-D Numerical Substorm: 180° North-South Solar Wind Field Switch

Magnetospheric substorms are believed to be associated with the efficiency of the solar wind - magnetosphere dynamo. This efficiency is low for a northward solar wind field and it is high for a southward one, in which case magnetic energy is stored in the magnetotail to be converted intermittently to the energy of magnetospheric substorms [Akasofu, 1977]. Thus, when the North-South component of the solar wind field varies in the turbulent solar wind, the efficiency of the dynamo changes and so does the possibility of a substorm. Here we simulate a strong substorm by modeling the passage of a 180° north-south discontinuity over the magnetosphere.

The input solar wind magnetic field has the form

$$\underset{\sim}{B} = B_o [\hat{y} H(x - M c_s t) - \hat{y} H(M c_s t - x)] \qquad (16)$$

The intial solar wind field is then everywhere northward. At $t = 0$, it suddenly switches by 180° to a southward orientation at $x = 0$. $B_o$ is such that $C_A = 0.2 \ C_S$; the sonic Mach number $M_s$ is 2.5 while the fast Mach number $M_F = 2.45$. Fig. 9 shows the time evolution of the total magnetic field as the rotational discontinuity passes over the magnetosphere.

At $t = 5 \ C_s^{-1} \Delta$ (top inset in Fig. 9) the discontinuity has approached the nose of the magnetopshere and reconnection has been initiated there. Downstream, the magnetospheric topology is unchanged from the teardrop shape, with x-lines over the "polar cusps", it assumes with northward solar wind field. At $t = 10 C_s^{-1} \Delta$ (second from the top in Fig. 9), the discontinuity is passing over the "polar caps". Note the hammerhead pattern induced by reconnection between southward solar wind field and dipole field at the nose. The MHD interaction has moved the x-lines from over the "polar cusps" to over the tail. At $t = 30 \ C_s^{-1} \Delta$ (third from the top in Fig. 9), the external solar wind field has switched southward over the whole magnetosphere. There is the expected x-line at the nose but no x-line at the tail. The MHD interaction has induced northward fields near and at the tail. The 2 × 3 collage of the total magnetic field pattern at $t = 30 C_s^{-1} \Delta$ displayed in Fig. 10 indicates that line tying between image magnetospheres due to our periodic field boundary conditions is responsible for the remnant x-lines over the tail and the island like

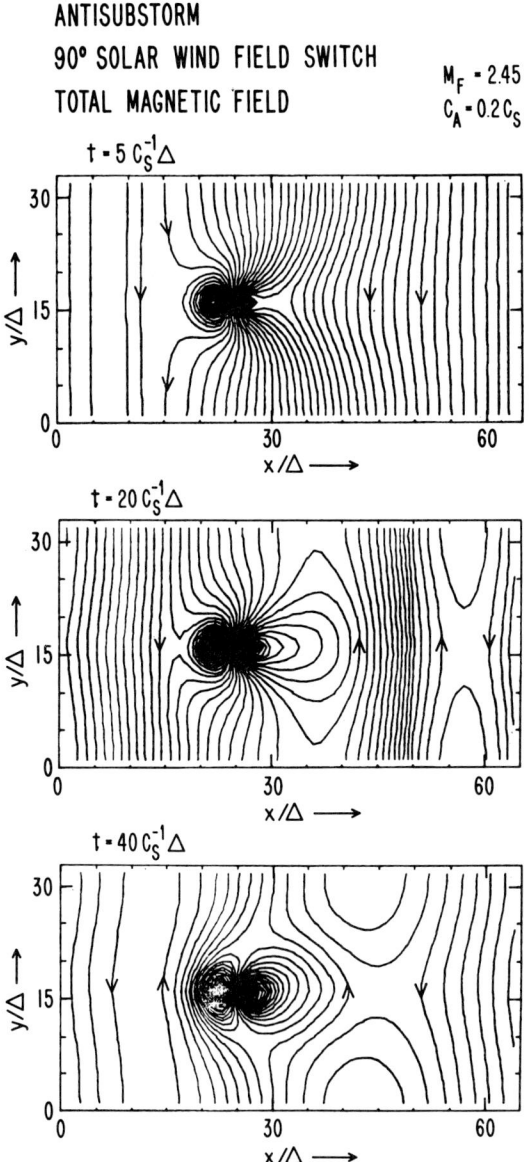

Fig. 8. "Antisubstorm" Simulation: 90° East-West Solar Wind Field Switch. The changes in magnetospheric topology resulting from the passage over the magnetosphere of a rotational discontinuity which switches the solar wind field from southward to east-west are displayed. The magnetosphere gradually evolves from an open Dungey configuration to an essentially closed teardrop shape. The remnant x-point downstream is an artifact of the periodic field boundary conditions.

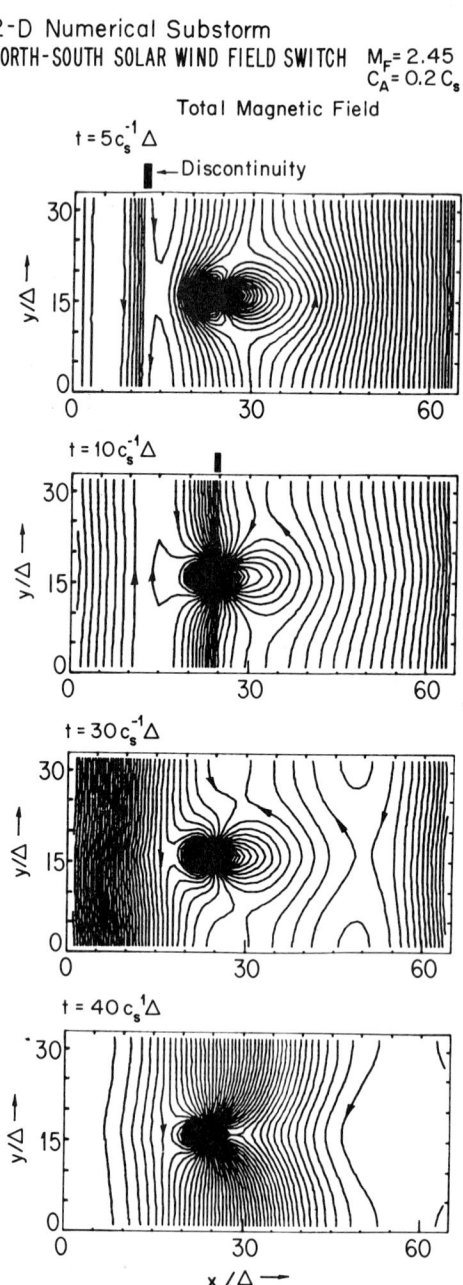

Fig. 9. "Substorm" Simulation: 180° North-South Solar Wind Field Switch. These four insets depict the evolution of magnetospheric topology from an intially teardrop shape with northward solar wind field to an open Dungey configuration with southward solar wind field everywhere. The passage of the rotational discontinuity over the magnetosphere induces rapid and violent topological changes. The resultant tail is however very short in this case.

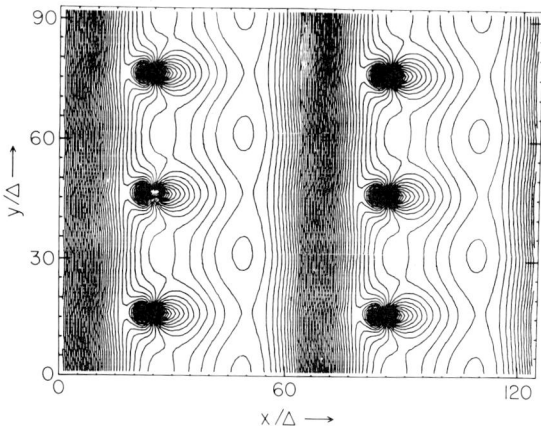

Fig. 10. Lattice of Magnetospheres for 180° North-South Solar Wind Field Switch. This 2 × 3 collage of the total magnetic field pattern at t = 30Δ/$c_s$ illustrates the adverse effects of the periodic field boundary conditions: the island-like structure downstream is an artifact of these boundary conditions.

structure in back of it. In Fig. 11, the various contributions to the magnetic field and the current density are displayed at t = $30C_s^{-1}\Delta$. From the current density, it appears that the strongest currents are induced around the dipole center at x = 24Δ and at the tail. Weaker currents are associated with the island-like structure in the total magnetic field. The flow induced magnetic field is southward up to the dipole center and northward beyond for the most part. The induced plus dipole field configuration is again northward for the most part except at the nose where it is southward up to the x-line at x = 15Δ. A superposition of the external southward solar wind field over the previous pattern gives rise to the spider-like pattern in the total field of Fig. 9 and Fig. 10. At t = 40 $C_s^{-1}\Delta$ (bottom inset of Fig. 9), the magnetosphere has finally settled into a Dungey type configuration, after many rapid and violent directional changes of the total magnetic field. The induced magnetic fields upon passage of the discontinuity are so large that the magnetic field lines experience totally different topologies in time. The tail is however very short in this case.

During this "substorm", the magnetosphere has then gone from a teardrop shape with x-lines over the "polar-cusps" to a quasi-steady reconnected configuration, with x-lines at the nose and tail, similar to the total magnetic field plot of Fig. 2.

The numerical experiments of this section and the previous one indicate that rotational magnetic field discontinuities embedded in the solar wind can indeed produce rapid changes in the topology of the magnetosphere from a closed shape unfavorable to the occurrence

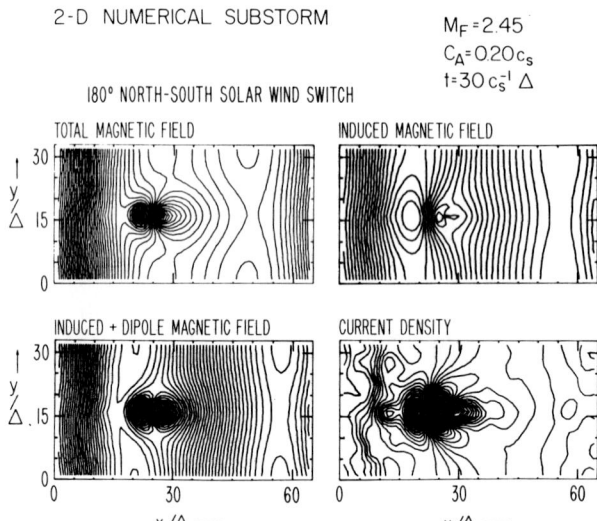

Fig. 11. "Substorm" Simulation: 180° North-South Solar Wind Field Switch. The various contributions to the total magnetic field and the current density are displayed at t = $30\Delta/c_S$.

of substorms to an open one of the Dungey type where substorms are more likely to occur.

## Discussion

The apparent limitations of our model can be classified as follows. First, our magnetosphere is highly dissipative with the numerics controlling the dissipation. The effective magnetic Reynolds number is 3-10. This might account for the short tail because of the difficulty in sustaining a current there. Second, we have so far modeled the interaction between the solar wind and a lattice of magnetospheres because of the periodicity of the magnetic fields. In some of our substorm simulations, this resulted in the discontinuity created by the solar wind field switch in the induced field to revolve around our lattice of magnetospheres instead of being absorbed upon reaching the right edge of the system. Third, our simulation is limited to two spatial dimensions and it is not clear if the results are applicable to the real 3-D magnetosphere. Indeed, magnetic merging is inevitable in 2-D because there can be no exchange of flux between nose and tail via flow in the third dimension. Our restriction to 2-D might also account for the short tail since plasma cannot enter it from the sides.

As far as improvements of the numerics are concerned, going to 3-D is straightforward and a periodic code exists. The much higher cost of these simulations is however the main reason why we have not pursued this yet. Moreover, some more realistic modeling of the 2-D magnetosphere remains to be done. Removal of periodicity in the field boundary conditions would be one major improvement of the model. We could, for instance, enforce that any flux reaching the edges of the system be absorbed. Such a model is presently under consideration. The other undesirable characteristic of our model, namely its high dissipation rate, can be alleviated by using a different algorithm to advance the magnetic field. A leapfrog scheme, which makes the field pushing equation time centered and the absence of spatial averaging inherent to the conservative Lax method, insures that the model is virtually dissipationless. A controllable resistivity can then be reintroduced. The changes in magnetic field topology with varying magnetic Reynolds number will then be studied in a 2-D context.

So far the physics contained in the model has been limited to "ideal" one-fluid magnetohydrodynamics. We have recently developed [Tajima et al., 1978b] an algorithm whereby the addition of the Hall term in the magnetic field pushing equation means that the model contains some of the electron dynamics without explicitly following the electrons. Our particles would then truly represent an ion fluid.

Notwithstanding their limitations and without the above improvements, these preliminary two-dimensional runs have graphically illustrated a turbulent shock, magnetosheath, and magnetopause, and bubble formation during unsteady tail reconnection, as well as support through direct calculation Dungey's classic model of the steady magnetopshere. We have learned that time dependent calculations are necessary even to determine the average steady state  configuration of the magnetosphere; first, time dependent calculations converge to steady state efficiently; second because of the unsteady interaction, the time-averaged steady state may differ from that found from calculations which assume no time-dependence ab initio. As the art of time-dependent global magnetospheric simulation develops, we expect it to be at first a source of new ideas - particularly concerning the deep tail where we have no measurements - and thereafter a source of quantitatively reliable information suitable for comparison with experiment.

We believe that this approach will provide us with a rich variety of possible nonlinear MHD effects in complicated geometry, which can then be sought out in space data. Strong interaction between simulation, experiment and theory has characterized recent research in laboratory plasma physics, and we therefore have reason to believe it may also be effective in space plasma physics.

Acknowledgements. The work was supported in part by NSF 76-83686-4-444024-21461 and by NASA NGL-05-007-190.

## References

Akasofu, S. I., Physics of Magnetospheric Substorms, D. Reidel, Hingham, Mass. 1977.
Axford, W. I., The interaction between the solar wind and the earth's magnetosphere, J. Geophys. Res., 67, 3791, 1962.
Dubinin, E. M., I. M. Podgorny and Yu. N. Potanin, Experimental evidence of opened and closed magnetospheres existence, I and II, Kosmicheskiye Issledovaniya, 1978 (to be published).
Dungey, J. W., Interplanetary magnetic field and the auroral zones, Phys. Rev. Lett., 6, 47, 1961.
Fukao, S., and T. Tsuda, Re-connection of magnetic lines of force:

evolution in incompressible MHD fluids, Planet. Space Sci., 21, 1151, 1973.

Johnson, F. S., The gross character of the geomagnetic field in the solar wind, J. Geophys. Res., 65, 3049, 1960.

Kellogg, P. J., Flow of plasma around the earth, J. Geophys. Res., 67, 3805, 1962.

Kruer, W. L., J. M. Dawson and B. Rosen, The dipole expansion method in plasma simulation, J. Comp. Phys., 13, 114, 1973.

Leboeuf, J. M., T. Tajima, C. F. Kennel and J. M. Dawson, Global simulation of the time-dependent magnetosphere, Geophys. Res. Lett., 5, 609, 1978a.

Leboeuf, J. N., T. Tajima and J. M. Dawson, A magnetohydrodynamic particle code for fluid simulation of plasmas, accepted by J. of Comp. Phys., 1978b.

Levy, R. H., H. E. Petscheck and G. L. Siscoe, Aerodynamic aspect of the magnetospheric flow, AIAA J., 2, 2065, 1966.

Obayashi, T., Interaction of solar plasma streams with outer geomagnetic field, J. Geophys. Res., 60, 861, 1964.

Obayashi, T., Space Science - Solar Terrestrial Physics, Syokabo, Tokyo, 195, 1970.

Potter, D. E., Computational Physics, Academic Press, New York, 110, 1973.

Spreiter, J. R. and A. Y. Alksne, Plasma flow around the magnetosphere, Rev. Geophysics, 7, 11, 1969.

Tajima, T., J. N. Leboeuf and J. M. Dawson, Double-layer forward shocks in a magnetohydrodynamic fluid, Phys. Rev. Letts., 40, 652, 1978a.

Tajima, T., J. N. Leboeuf and J. M. Dawson, A magnetohydrodynamic particle code with force free electrons, paper OA-2, Proceedings of the Eight Conference on Numerical Simulation of Plasmas, U.S. Department of Energy, Monterey, California, June 1978.

Tverskoy, B. A., Dynamics of the earth's radiation belts, Science Publishing House, Physico-Mathematical Literature, Moscow, 1968.

Wolf, R. A., Calculations of magnetospheric electric fields, Magnetospheric Physics, ed. B. M. McCormac, D. Reidel, Dordrecht-Holland, 167, 1974.

Yang, C.-K., and B. U. O. Sonnerup, Compressible magnetopause reconnection, J. Geophys. Res., 82, 699, 1977.

IONOSPHERIC ELECTRIC FIELDS DRIVEN BY FIELD-ALIGNED CURRENTS

R. W. Nopper, Jr. and R. L. Carovillano

Department of Physics, Boston College
Chestnut Hill, Massachusetts 02167

Abstract. Field-aligned currents provide an important coupling between magnetospheric and ionospheric processes, driving ionospheric electric fields on a global scale. We have developed a model in which global distributions of electric fields and currents are calculated using mathematical boundary conditions on field-aligned currents that are chosen to correspond to observations made on the Triad satellite. The ionospheric conductivity tensor is represented empirically and includes spatial inhomogeneities resulting from the solar-zenith angle variation and a local auroral zone enhancement. Numerical results are accurate over the entire latitude range, pole to equator.
  One application made was to examine the role of the ionospheric conductivity in coupling electric fields from high to low latitude. Ground magnetic observations and direct measurements of equatorial ionospheric phenomena have indicated that magnetospheric coupling effects penetrate at least to L = 3.2 during quiet times and to the equator during disturbed times. Calculations using quiet-time sources as observed by Triad yield electric fields at all latitudes which are at least comparable in strength to those produced by dynamo wind models and which have non-diurnal morphologies. In our calculations for disturbed times, imposed fluctuations in the strength of the field-aligned (Region 2) currents are found to produce large variations in the mid- and low-latitude electric field and account for reversal from its quiet time configuration, as is often observed. The electric field in the polar cap is shown to vary significantly in response to substorm conditions when Region 2 currents are appreciable. We conclude that under both quiet and disturbed conditions magnetospheric sources produce ionospheric effects on a global scale.

## Introduction

  Most past efforts at modeling the coupling between the magnetosphere and ionosphere have excluded proper consideration of the low-latitude ionosphere. The importance of auroral zone effects on the coupling has recently been emphasized with the observations that have established Birkeland currents to be a permanent feature at high latitudes. Important low-latitude effects occur essentially by leakage of current from high latitudes. Thus, calculated ionospheric results of Wolf [1970], Jaggi and Wolf [1973], Maeda and Maekawa [1973], Lyatsky et al. [1974], Yasuhara et al. [1975], and others are generally adequate at high latitudes but are not designed to apply at middle and low latitudes.

Ionospheric evidence is now available that suggests a direct relationship between high- and low-latitude phenomena. Thus, during disturbed times, rapid magnetic variations closely correlated in time are observed at auroral zone and equatorial stations (e.g., Akasofu and Chapman [1963]; Onwumechili et al. [1973]). Ionospheric electric fields and currents sometimes fluctuate and reverse in concert with this class of magnetic fluctuations as shown by correlative studies of magnetograms and radar measurements of mid- and low-latitude ionospheric plasma drift episodes (e.g., Balsley [1966], Van Zandt et al. [1971], Carter et al. [1976], Fejer et al. [1976]), and other ionospheric phenomena, such as the sudden disappearance of equatorial sporadic-E and vhf radar echoes (Cohen and Bowles [1963], Rastogi [1977], Rastogi et al. [1977], Patel [1978]). Thus the existence of a polar-equatorial ionospheric coupling during disturbed times is documented observationally.

At quiet times, Matsushita [1971] and Leont'yev and Lyatskiy [1975] proposed that the Sq equivalent current system might be driven in part by magnetospheric sources, the conventional generating mechanism being the E-region dynamo. Observationally, F-region plasma drift measurements at Millstone Hill by Carpenter and Kirchhoff [1974, 1975] corroborate a magnetospheric origin for the quiet-time electric field down to $L = 3.2$. More recently, Gonzales et al. [1978] have shown that only in extremely quiet times is the classical dynamo effect dominant over magnetospheric sources in plasma drifts seen at Millstone Hill. It remains an open question how far equatorward the quiet-time configuration of magnetospheric sources exerts a measurable effect.

Our motivation to calculate the distribution of ionospheric electric fields was to evaluate the degree of coupling between the high- and low-latitude ionospheres. As detailed in the next section, our modeling procedure utilizes a realistic model of the ionospheric conductivity and the observed field-aligned current distribution; careful numerical treatment of dip-angle effects in the equatorial regions renders the calculation accurate globally. These inputs (conductivity and Birkeland current) can be modified to conform to improved empirical requirements or special physical circumstances. Model results are presented both for quiet and for disturbed periods. We find that magnetospheric sources contribute on an equal footing with calculated dynamo sources to the quiet-time electric field morphologies at all latitudes; on the other hand, magnetospheric sources alone produce large enough electric field variations at middle and low latitudes to account for observed disturbed-time fluctuations and reversals. Thus effects of magnetospheric coupling to the ionosphere are significant at all latitudes during both quiet and disturbed times.

## Definition of the Model

Our approach is to solve the equation of continuity of the height-integrated ionospheric current, given the ionospheric conductivity distribution $\underline{\underline{\Sigma}}(\theta,\phi)$ and the driving field-aligned currents J(source). The electric fields we calculate may not be the total ionospheric field; for example, we do not include neutral wind effects. We impose boundary conditions that require north-south symmetry in the conductivities and sources. If $(\theta,\phi)$ denote the co-latitude and azimuth and $\nabla_2$ is the transverse gradient, we obtain the ionospheric potential $\Phi(\theta,\phi)$ from the equation of continuity

$$J(\text{source}) = \underline{\nabla}_2 \cdot \underline{\underline{\Sigma}}(\theta,\phi) \cdot \underline{\nabla}_2\, \Phi(\theta,\phi)$$

by numerically relaxing its finite difference representation on a 30 x 30 grid of nodal points covering the northern hemisphere ionosphere. Southern hemisphere results are determined by the symmetry condition of zero current crossing the equator. The numerical accuracy is verified by checking that the calculated current is locally conserved among adjacent nodal points to a good approximation. The calculated electric fields are accurate solutions to the continuity equation (with assumed source and conductivity models) within a few percent everywhere on the grid.

The ionospheric height-integrated conductivities are expressed by simple functions multiplied by adjustable coefficients. These expressions give a good fit to the principal structures observed, namely, a base level constant conductivity (denoted by subscript "K") a solar-zenith angle variation (subscript "X"), and an auroral enhancement (subscript "A"):

Pedersen
$$\Sigma_P(\theta,\phi) = \Sigma_{PK} + \Sigma_{PX}\exp\{-[\cos^{-1}(\sin\theta\cos\phi)]^2/1.804\}$$
$$+ \Sigma_{PA}\exp\{-[\theta - 0.35]^2/0.01\}\ \text{mhos},$$

Hall
$$\Sigma_H(\theta,\phi) = \Sigma_{HK} + \Sigma_{HX}\exp\{-[\cos^{-1}(\sin\theta\cos\phi)]^2/1.804\}$$
$$+ \Sigma_{HA}\exp\{-[\theta - 0.35]^2/0.01\}\ \text{mhos},$$

direct
$$\Sigma_0(\theta,\phi) = 31.62\ \Sigma_P(\theta,\phi)\ \text{mhos}.$$

Here, $\theta$ and $\phi$ (measured eastward from noon) are in radians. The coefficients are chosen to reproduce the most characteristic or typical values from Kennel and Rees [1972], Rowe and Mathews [1973], Brekke et al. [1974], and others. Those employed in the present modeling are as follows:

|          | $\Sigma_{PK}$ | $\Sigma_{PX}$ | $\Sigma_{PA}$ | $\Sigma_{HK}$ | $\Sigma_{HX}$ | $\Sigma_{HA}$ |
|----------|------|------|------|------|------|------|
| Quiet    | 0.3  | 5.0  | 3.0  | 0.6  | 10.0 | 6.0  |
| Disturbed| 0.3  | 5.0  | 6.0  | 0.6  | 10.0 | 18.0 |

These conductivities are then adjusted for dip angle effects in the usual way [Rishbeth and Garriott, 1969] to yield the elements of the tensor $\underline{\underline{\Sigma}}(\theta,\phi)$ used in the continuity equation. Figure 1 illustrates the variation in $\Sigma_{\theta\theta}$ as a function of $\theta$ along the noon-midnight meridian. Note the different auroral enhancements, peaking at 69° latitude, for quiet and disturbed times, and the 20 to 1 contrast between day and night values.

For driving currents we utilize the Triad results reported by Iijima and Potemra [1976, 1978], shown in Figure 2. These consist of two large-scale current structures: Region 1, often associated with magnetospheric convection, which flows into the ionosphere on the dawnside and out on the duskside; Region 2, often correlated with auroral electrojet activity, which has the opposite polarity. In our idealiza-

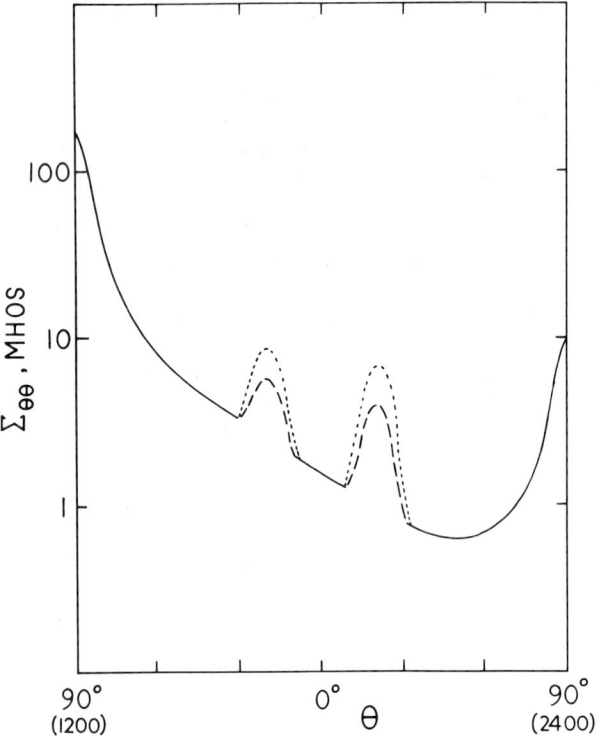

Fig. 1. Variation of conductivity tensor element $\Sigma_{\theta\theta}$ along the noon-midnight meridian from the dayside to the pole to the nightside equator. The dashed (dotted) portions give the model auroral conductivity values during quiet (disturbed) times.

tions, Region 1 currents are located at latitude 72°, and Region 2 at 66°. We abstract the quiet- and disturbed-time current strengths from the Triad results to be:

|  | Region 1 | Region 2 |
| --- | --- | --- |
| Quiet | $10^6$ A | $5 \times 10^5$ A |
| Disturbed | $10^6$ A | $10^6$ A |

It is characteristic of quiet times that the Region 1 currents are roughly twice as strong as the Region 2, whereas during more active periods the two sets carry equal currents, from $10^6$ A (as tabulated here) to several times $10^6$ A.

At present there is no definitive empirical evidence available that specifies the origin or the time-dependent behavior of the field-aligned currents. Likewise, theoretical simulation efforts of the dynamical

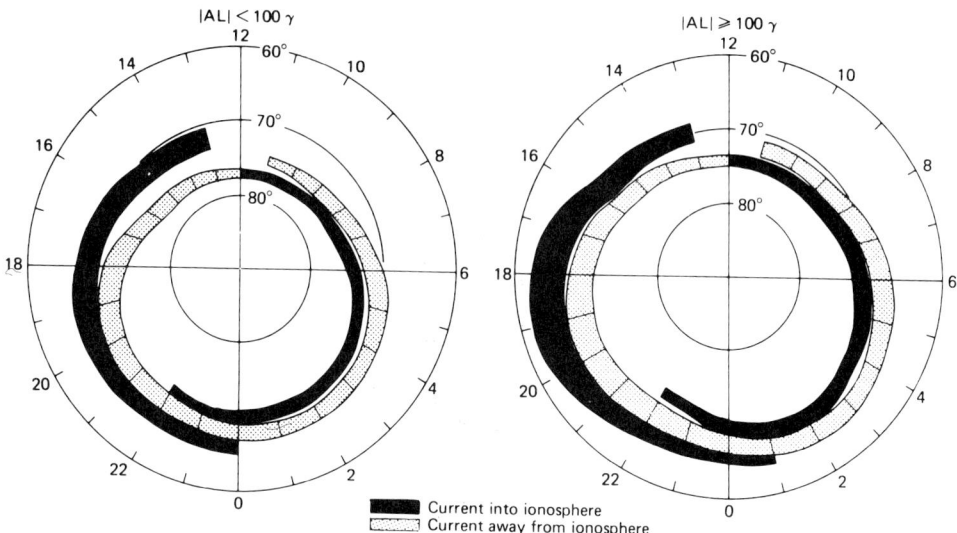

Fig. 2. The field-aligned current distributions determined using the Triad satellite [Iijima and Potemra, 1978].

processes that couple the magnetosphere and ionosphere are in their early stages of development. (See the paper by Harel et al., this volume.) The Triad results adopted in our modeling scheme have only a statistical significance; thus our calculated results would only have statistical significance and would not necessarily be representative of particular events on a case by case basis. Rather, our results should characterize typical distributions of ionospheric electric fields and currents that are generated by typically observed field-aligned currents.

## Quiet-Time Results

Our main interest in this first application is to determine the contribution made by the field-aligned currents to the quiet time electric field morphology at middle and low latitudes. In Figure 3 the equipotential curves are given for the quiet-time modeling parameters presented above. It is seen that a two-cell pattern results from the quiet time pattern of Region 1 and Region 2 currents. This agrees with the two-cell convection patterns of Heppner [1977]. The polar-cap field is directed roughly dawn to dusk with magnitude 23 mV/m. There is a striking dusk-dawn asymmetry, in which duskside equipotentials are pushed toward higher latitudes and those on the dawnside are pulled to lower latitudes. This effect was explained by Wolf [1970] as due to the difference in conductivity between day and night. It is interesting to note, since the ionospheric currents are likewise asymmetrical, that even for a symmetrical disposition of field-aligned currents (as taken here) the resulting ground magnetic effect will be asymmetrical dusk to dawn.

Figure 4 shows the meridional (or $\theta$) component of the electric field as a function of local time at the latitudes of Millstone Hill (54°)

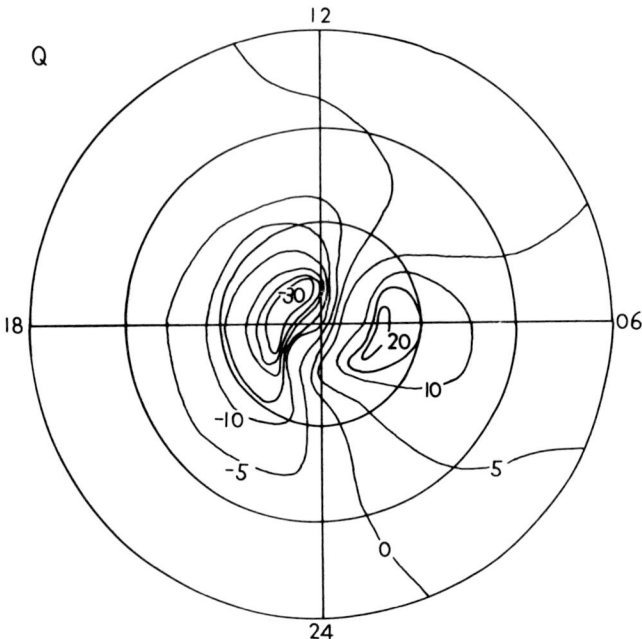

Fig. 3. Polar plot of the equipotential curves, in kV, for quiet times. Here, the Region 1 currents are twice as strong as the Region 2, and the auroral conductivity is moderate. Circles are centered at the pole at colatitude 30°, 60°, and 90° (the equator). The total cross-polar-cap potential drop is about 50 kV in this model.

and Arecibo (30°); $E_\theta$ at the equator is identically zero in these models. This field component would correspond to an eastward component of drift of ionospheric plasma. The calculated magnitudes, 5 mV/m at 54° and 1.5 mV/m at 30°, are approximately as large as those generated in neutral wind models [Matsushita and Tarpley, 1970]. Thus, magnetospheric sources and neutral winds must both be included and play comparable roles in modeling quiet time effects at low latitudes. Figure 4 displays a predominantly diurnal morphology at 54°, where maximum and minimum electric field values occur about twelve hours apart. In our model, the premidnight field is northwards and this would generate a westward F-region plasma drift at Millstone Hill; this drift is observed by Gonzales et al. [1978] on all but the quietest days. The classical dynamo effect, which gives southward fields in this sector, would be dominant over magnetospheric driving effects, according to our results, only during very quiet times.

In Figure 5 the azimuthal (or $\phi$, positive eastwards) quiet-time electric fields are given at 54°, 30°, and the equator, corresponding to northward plasma drifts. Again the calculated magnitudes are com-

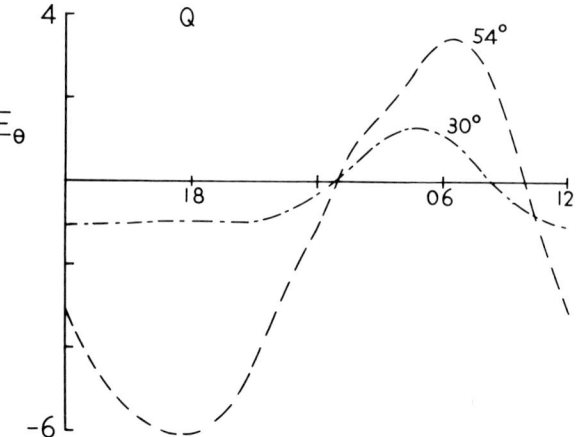

Fig. 4. Meridional component of the electric field, in mV/m, at 54° and 30° latitude, as a function of local time, for the quiet-time model (Q). Positive $E_\theta$ corresponds to a southward directed electric field.

parable to those from neutral wind models. A new result is that this component has strong non-diurnal Fourier components, with principal maxima and minima occurring about six hours apart. This results as a beating effect between the diurnal field-aligned source structure and the diurnal but out-of-phase conductivity distribution. To see the magnitude of the non-diurnal components, we idealize the morphologies of Figure 5 and show the Fourier amplitude spectrum of this signal in Figure 6. We find that the 12 hour component dominates over the diurnal, and that the 8 hour component is also quite strong. Thus the subdiurnal harmonics in observed quiet-time plasma drift morphologies (e.g., Blanc et al. [1977]) may have significant contributions from magnetospheric sources (as well as from the atmospheric dynamo).

## Disturbed-Time Results

Our goal in modeling active times is to investigate the fluctuations and reversals observed in the otherwise quiet-time electric field at low latitudes. In order to do this it is necessary to impose a time-dependence in our modeling parameters. According to observations, Region 2 currents tend to be variable and roughly equal in intensity to the relatively fixed Region 1 currents during disturbed times [Iijima and Potemra, 1976, 1978]. In addition, the auroral conductivities increase, due to enhanced particle precipitation, during times when the auroral electrojets are active. Thus we adopt the basic disturbed-time model parameters as tabulated above in section 11. Fluctuations are considered to be alternating sequences of the quiet and disturbed model-results.

The disturbed time equipotential pattern is given in Figure 7. The

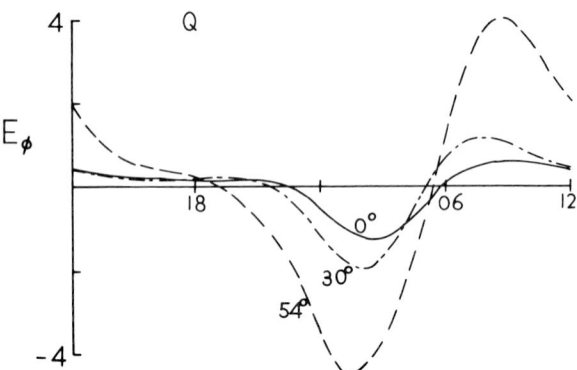

Fig. 5. Azimuthal component of the electric field (mV/m) at 54°, 30°, and 0°, as a function of local time, for quiet times (Q). Positive $E_\phi$ corresponds to an eastward directed field.

enhanced Region 2 source has moved the formerly prominent duskside vortex to the noon sector and reduced it in size. The equipotential curves tend to be distorted poleward (equatorward) on the dusk (dawn) side, as in the quiet-time model. An interesting feature of this case is the sunward-directed electric field over the polar cap. The magnitude and intensity of the polar cap electric field is a sensitive function of characteristics of the Region 2 current system. The sunward direction of the electric field shown in Figure 7 results because in this case the Region 1 and Region 2 currents have equal intensities. The direction would rotate towards dusk if the Region 2 intensity were decreased, and towards dawn if increased. The polar cap electric field is weak (8 mV/m here) because of the opposite polarity and equal strength of the Region 1 and Region 2 currents. Polar cap convection in this case is directed dawn-to-dusk. Such deviations from quiet-time antisunward convection are occasionally seen during disturbed periods (e.g., Heppner [1977]).

In Figure 8 a comparison is made of the azimuthal electric field components at 30°, between quiet- and disturbed-time results, curves Q and D'. Over most of local times, the strengthened Region 2 currents have reversed the electric field at this latitude. (The added auroral conductivity during disturbed times has little effect upon the reversal, as can be seen from curve D, which uses the sources of D' but the conductivity of Q.) The equatorial electric field is similar to that at 30°, but of roughly half the magnitude, as shown in Figure 9. Since the characteristic Triad active field-aligned current strengths often tend to be several times the value ($10^6$ A) used in obtaining Figures 7 through 9, it follows that the equatorial electric field can be several times that shown in these figures. Thus strengths up to 2 mV/m can sometimes be obtained and these will, for most local times, be directed oppositely to the quiet-time electric fields. Hence, variations in the Region 2 currents may be responsible for the fluctuations and reversals

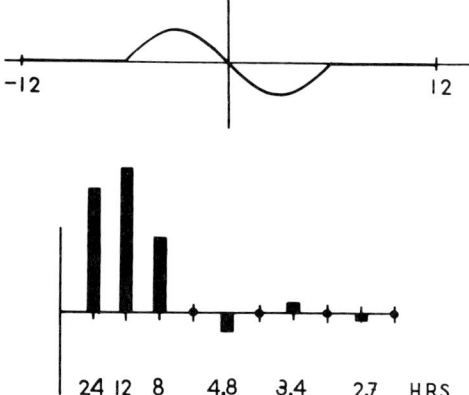

Fig. 6. Idealized morphology of the eastward electric fields in Figure 5, and the Fourier spectrum of this signal.

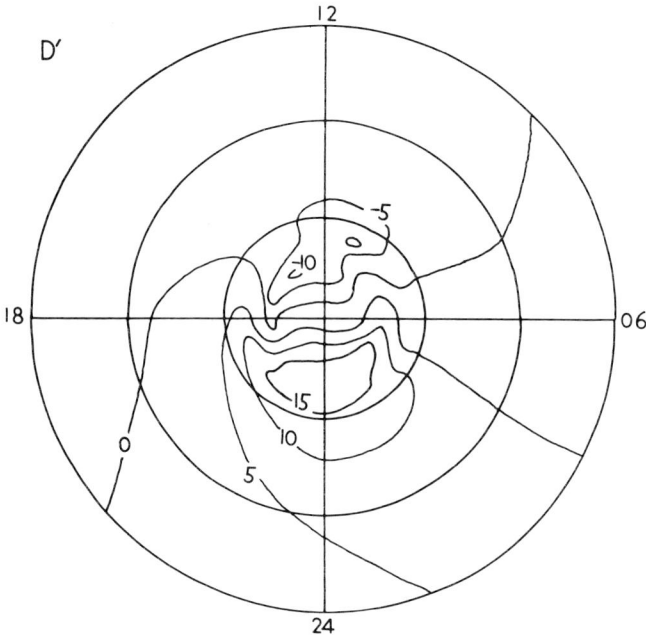

Fig. 7. Equipotential pattern corresponding to the disturbed-time model parameters: equal Region 1 and Region 2 currents and a greatly enhanced auroral conductivity.

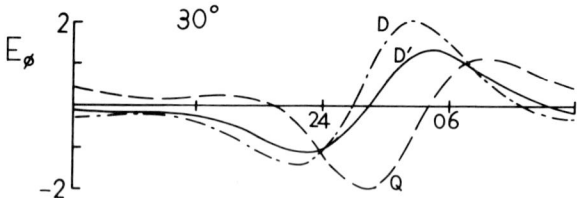

Fig. 8. Azimuthal electric field components at 30° for quiet (Q) and disturbed (D') models. (Model D utilizes sources of D' and conductivities of Q.) Fluctuations in the strength of the Region 2 currents and auroral precipitation would cause this electric field component to vary between Q and D'.

in the equatorial electric field and associated ionospheric phenomena, as referenced earlier. The observed rapid variations in the equatorial field cannot be accounted for by the neutral wind system with its large inertia and slower response time. Further discussion of this case is given in Nopper and Carovillano [1978].

Summary and Conclusions

The principal outcome of our modeling study is the calculation of relatively large electric fields due to magnetospheric sources at all ionospheric latitudes. During quiet times, the observed configuration of field-aligned currents contributes large magnitudes and non-diurnal spectral terms to mid- and low-latitude electric fields and, consequently, to plasma drifts. These effects of magnetospheric origin should not be excluded from models of the quiet ionosphere since they compete with the traditional source of quiet-time variations, dynamo winds. Adopting a plausible model for time variations in the Region 2 currents leads to an explanation of the fluctuations and reversals of the low-latitude electric field observed in disturbed times. Region 2 currents also affect the direction and magnitude of polar cap electric fields in an important way. In summary, the dynamics of the magnetosphere/ionosphere coupled system, as manifest in field-aligned currents,

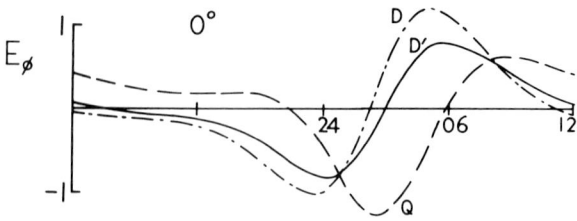

Fig. 9. Same as Figure 8 for the equator.

exerts a global influence on the ionosphere for all but the quietest levels of geomagnetic activity.

Acknowledgements. The authors are grateful to N. U. Crooker, G. L. Siscoe, and M. A. Heinemann for discussions on various aspects of this research. This research was supported by the Atmospheric Sciences Section, National Science Foundation, under Grant ATM75-22873, and by the Air Force Geophysics Laboratory under Contract F19628-77-C-0085.

## References

Akasofu, S. -I. and S. Chapman, The enhancement of the equatorial electrojet during polar magnetic storms, J. Geophys. Res., 68, 2375-2382, 1963.

Balsley, B. B., Evidence of the nighttime current reversal in the equatorial electrojet, Ann. Geophys., 22, 460-462, 1966.

Blanc, M., P. Amayenc, P. Bauer, and C. Taieb, Electric field induced drifts from the French incoherent scatter facility, J. Geophys. Res., 82, 87-97, 1977.

Brekke, A., J. R. Doupnik, and P. M. Banks, Incoherent scatter measurements of E region conductivities and currents in the auroral zone, J. Geophys. Res., 79, 3773-3790, 1974.

Carpenter, L. A., and V. W. J. H. Kirchhoff, Daytime three-dimensional drifts at Millstone Hill Observatory, Radio Sci., 9, 217-222, 1974.

Carpenter, L. A., and V. W. J. H. Kirchhoff, Comparison of high-latitude and mid-latitude ionospheric electric fields, J. Geophys. Res., 80, 1810-1814, 1975.

Carter, D. A., B. B. Balsley, and W. L. Ecklund, VHF Doppler radar observations of the African equatorial electrojet, J. Geophys. Res., 81, 2786-2794, 1976.

Cohen, R., and K. L. Bowles, Ionospheric VHF scattering near the magnetic equator during the International Geophysical Year, J. Res. NBS, sect. D., 67D, 459-480, 1963.

Fejer, B. G., D. T. Farley, B. B. Balsley, and R. F. Woodman, Radar studies of anomalous velocity reversals in the equatorial ionosphere, J. Geophys. Res., 81, 4621-4626, 1976.

Gonzales, C. A., M. C. Kelley, L. A. Carpenter, and R. H. Holzworth, Evidence for a magnetospheric effect on mid-latitude electric fields, J. Geophys. Res., 83, 4397-4399, 1978.

Harel, M., R. A. Wolf, and P. H. Reiff, Computer modeling of events in the inner magnetosphere, Quantitative Modeling of Magnetospheric Processes, Geophys. Monogr. Ser., vol. 21, ed. by W. P. Olson, AGU, Washington, D.C., 1979.

Heppner, J. P., Empirical models of high-latitude electric fields, J. Geophys. Res., 82, 1115-1125, 1977.

Iijima, T., and T. A. Potemra, The amplitude distribution of field-aligned currents at northern high latitudes observed by Triad, J. Geophys. Res., 81, 2165-2174, 1976.

Iijima, T., and T. A. Potemra, Large-scale characteristics of field-aligned currents associated with substorms, J. Geophys. Res., 83, 599-615, 1978.

Jaggi, R. K., and R. A. Wolf, Self-consistent calculation of the motion of a sheet of ions in the magnetosphere, J. Geophys. Res., 78, 2852-2866, 1973.

Kennel, C. F., and M. H. Rees, Dayside auroral-oval plasma density and conductivity enhancements due to magnetosheath electron precipitation, J. Geophys. Res., 77, 2294-2302, 1972.

Leont'yev, S. V., and V. B. Lyatskiy, Three-dimensional current system of a partial ring current; relation between DP2 and Sq variations, Geomagn. Aeron., 15, 91-94, 1975.

Lyatsky, W. B., Yu. P. Maltsev, and S. V. Leontyev, Three-dimensional current system in different phases of a substorm, Planet. Space Sci., 22, 1231-1247, 1974.

Maeda, H., and K. Maekawa, A numerical study of polar ionospheric currents, Planet. Space Sci., 21, 1287-1300, 1973.

Matsushita, S., Interactions between the ionosphere and magnetosphere for Sq and L variations, Radio Sci., 6, 279-294, 1971.

Matsushita, S., and J. D. Tarpley, Effects of dynamo-region electric fields on the magnetosphere, J. Geophys. Res., 75, 5433-5443, 1970.

Nopper, R. W., Jr., and R. L. Carovillano, Polar-equatorial coupling during magnetically active periods, Geophys. Res. Lett., 5, 699-702, 1978.

Onwumechili, A., K. Kawasaki, and S. -I. Akasofu, Relationships between the equatorial electrojet and polar magnetic variations, Planet. Space Sci., 21, 1-16, 1973.

Patel, V. L., Interplanetary magnetic field variations and the electromagnetic state of the equatorial ionosphere, J. Geophys. Res., 83, 2137-2144, 1978.

Rastogi, R. G., Coupling between equatorial and auroral ionospheres during polar substorms, Proc. Indian Acad. Sci., 86A, 409-416, 1977.

Rastogi, R. G., B. G. Fejer, and R. F. Woodman, Sudden disappearance of VHF radar echoes from equatorial E-region irregularities, Indian J. Radio and Sp. Phys., 6, 39-43, 1977.

Rishbeth, H., and O. K. Garriott, Introduction to Ionospheric Physics, Academic Press, New York, 1969.

Rowe, J. F., Jr., and J. D. Mathews, Low-latitude nighttime E-region conductivities, J. Geophys. Res., 78, 7461-7470, 1973.

Van Zandt, T. E., V. L. Peterson, and A. R. Laird, Electromagnetic drift of the mid-latitude $F_2$ layer during a storm, J. Geophys. Res., 76, 278-281, 1971.

Wolf, R. A., Effects of ionospheric conductivity on convective flow of plasma in the magnetosphere, J. Geophys. Res., 75, 4677-4698, 1970.

Yasuhara, F., Y. Kamide, and S. -I. Akasofu, Field-aligned and ionospheric currents, Planet. Space Sci., 23, 1355-1368, 1975.

# INFLUENCE OF ELECTRIC FIELDS ON CHARGED PARTICLE MOTION AND ELECTRON FLUXES AT SYNCHRONOUS ALTITUDES

Jean Claude Kosik

Division Mathématiques, Centre Spatial de Toulouse

Av. E. Belin - 31400 Toulouse

<u>Abstract</u>. The motion of energetic electrons in an asymmetric magnetosphere including a convection electric field is obtained through perturbation methods. The drift shells, as well as the changes in kinetic energy and pitch-angle, are strongly dependent on the relative ratio of kinetic energy to electric field intensity. As a consequence the noon-midnight meridian is no longer a plane of symmetry for drift shells. The azimuthal variation of fluxes of particles that mirror at the equator during quiet magnetic periods is obtained and compared to ATS I results. Tthe observational results can be explained only by a low convection electric field intensity.

## Introduction

In a series of recent papers different authors have studied the motion of charged particles in a dipole field in presence of convection and corotation electric fields [Roederer and Schulz, 1971 ; Stern, 1971, 1975 ; Chen, 1970 ; Stern, 1977]. The consequences of the electric fields for particle fluxes were obtained in the dipole field by Kivelson and Southwood [1975], Cowley and Ashour Abdalla [1975, 1976] and Solomon [1976]. The dipole representation of the magnetosphere is justified for low energy particles (E < 1 KeV) and for altitudes of the order of synchronous distances. However, with increasing particle energy, the particle motion becomes sensitive to any deformation of the magnetic field, and this is particularly true when we study such motion in the regions where L is greater than 5 or 6. In an early work [Kosik, 1971b], we made a very crude analysis of the problem, neglecting the change in kinetic energy during the motion of the particle. This error has been corrected in the present study and our results apply down to 40 KeV particles. The accuracy of the model is checked by using particles that mirror at the magnetic equator and invoking the conservation of the first invariant. In the first part of this work we show that electric fields can introduce a strong modification of the classical properties encountered in an asymmetric magnetosphere : the noon-midnight meridian plane is no longer a plane of symmetry. Different planes of symmetry appear for drift shells corresponding to different kinetic energies. Other consequences of the presence of electric fields are the change in kinetic energy during the drift and a modified variation of the equatorial pitch-angle with longitude. In the second part of this work we apply the results to synchronous altitudes observations made in different energy channels. As can be expected, the minimum flux is no longer detected at midnight but at another longitude which depends on the kinetic energy of the particles. At high energies ($E_e >$ 500 KeV) the deviation of the longitude of the flux minimum from midnight is negligible.

## MOTION OF PARTICLES

The Mead [1964] magnetospheric model provides a very simple asymmetric magnetosphere. Let us recall some features of its topology obtained elsewhere [Kosik, 1971a]. The field lines are given to lowest order in $(L/10)^4$ by the following equations :

$$r = L \sin^2 \theta \left\{ 1 - \frac{1}{2} \frac{\overline{g_1^0}}{g_1^0} L^3 \sin^6 \theta + 2\sqrt{3} L^4 \frac{\overline{g_2^1}}{g_1^0} \left( \frac{\sin^7 \theta}{7} - \frac{\sin^9 \theta}{3} \right) \cos \varphi_0 \right\} \quad (1)$$

$$\varphi = \varphi_0 + \frac{\sqrt{3}}{7} L^4 \frac{\overline{g_2^1}}{g_1^0} \sin^4 \theta \sin \varphi_0$$

where L is equal to $1/\sin^2 \theta_0$, r is the radial distance expressed in Earth radii, $\theta_0$ is the colatitude of the "foot" of the field line on the Earth and $\varphi_0$ is the longitude of the "foot", counted from the noon meridian. More precisely we should define

$$L \equiv \lim_{r \to 0} (r/\sin^2 \theta) \quad \text{and} \quad \varphi_0 \equiv \lim_{r \to 0} \varphi$$

In the following part of the text we will omit the subscript for the longitude of the foot of the field line. From Mead's scalar potential one gets the magnetic field intensity along a magnetic field line (L, $\varphi_0$) :

$$B = |g_1^0| \frac{(1 + 3\cos^2 \theta)^{1/2}}{L^3 \sin^6 \theta} \left\{ 1 + \frac{1}{2} \frac{\overline{g_1^0}}{g_1^0} L^3 \sin^6 \theta \left( \frac{5 + 3\cos^2 \theta}{1 + 3\cos^2 \theta} \right) \right.$$
$$\left. + \frac{3\sqrt{3}}{7} \frac{\overline{g_2^1}}{g_1^0} L^4 \sin^7 \theta \left( \frac{5 - 13\cos^2 \theta}{1 + 3\cos^2 \theta} \right) \cos \varphi \right\} \quad (2)$$

In a steady convection pattern, field lines are also equipotentials [Axford and Hines, 1961]. We define the corotation electric potential as

$$V_1 = - C_1 / L \quad \text{with} \quad C_1 = \Omega a^2 g_1^0 = 92 \text{ kilovolt}/R_e \quad (3)$$

where a is the radius of the Earth. It is possible to express L as a function of (r, $\theta$, $\varphi$) by using equation (1). Taking the partial derivatives of (3), we obtain the three components of the electric field :

$$E_r = \frac{C_1}{r^2} \sin^2 \theta \left\{ 1 + \frac{\overline{g_1^0}}{g_1^0} r^3 - 6\sqrt{3} \frac{\overline{g_2^1}}{g_1^0} \frac{r^4}{\sin \theta} \left( \frac{1}{7} - \frac{\sin^2 \theta}{3} \right) \cos \varphi \right\}$$

$$E_\theta = \frac{C_1}{r^2} \sin \theta \cos \theta \left\{ 1 + \frac{\overline{g_1^0}}{g_1^0} r^3 + 2\sqrt{3} \frac{\overline{g_2^1}}{g_1^0} \frac{r^4}{\sin \theta} \left( \frac{1}{7} - \sin^2 \theta \right) \cos \varphi \right\} \quad (4)$$

$$E_\varphi = - 2\sqrt{3} C_1 \frac{\overline{g_2^1}}{g_1^0} r^2 \left( \frac{1}{7} - \frac{\sin^2 \theta}{3} \right) \sin \varphi$$

The equations correspond exactly to the formulae obtained by Schulz [1971] for a corotative model of the Mead type, using the equations $\vec{\nabla} \times \vec{E} = 0$ and $\vec{E} \cdot \vec{B} = 0$. To this corotation electric field we add

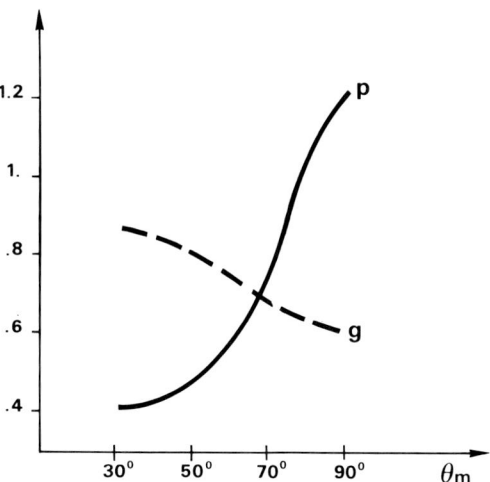

Fig. 1. The solid and dashed curves give the mirror-point colatitude dependence of magnetic and electric shell-splitting as needed in (11).

a convection electric field in the equatorial plane and directed from dawn to dusk. It is derived from the following potential :

$$V_2 = - C_2 L \left\{ 1 - \frac{1}{2} \frac{\overline{g}_1^0}{g_1^0} L^3 - \frac{5}{21} \sqrt{3} L^4 \frac{\overline{g}_2^1}{g_1^0} \cos \varphi \right\} \sin \varphi \qquad (5)$$

where $C_2 = 2$ kilovolt/$R_e$. This reduces to $V_2 = C_2 \, r \sin \varphi$ in the equatorial plane.

E = 100 keV

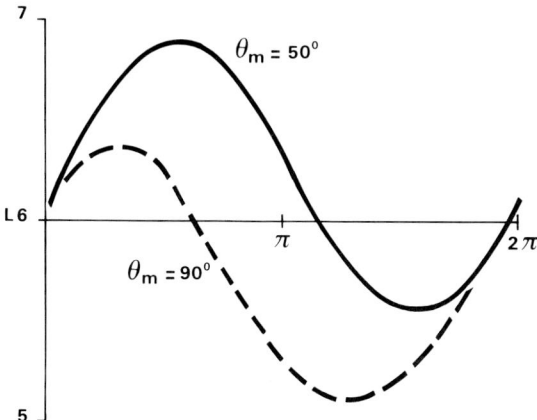

Fig. 2. Azimuthal variation of L for particles having different mirror latitudes.

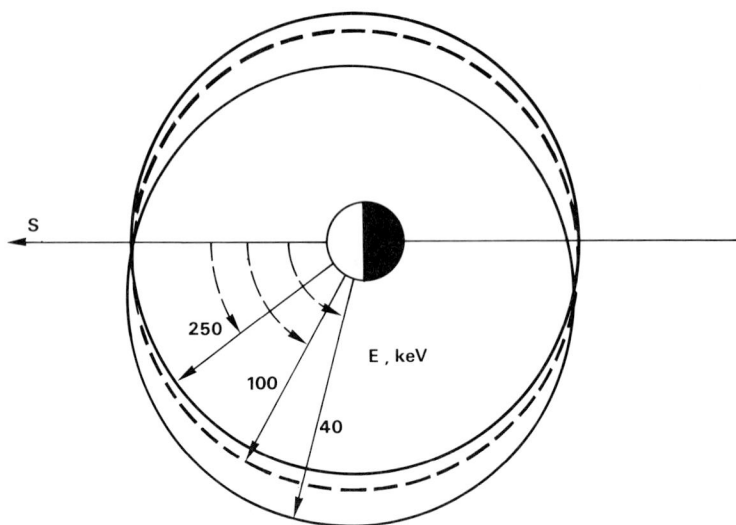

Fig. 3. Drift shells of particles having J = 0. One notices the increasing dawn-dusk asymmetry with decreasing kinetic energy, as evaluated at $\varphi = 0$ and $\varphi = \pi$

Drift Shells of Electrons

The motion of nonrelativistic charged particles is characterized by three constants :

$$\mu(L, \theta_m, E, \varphi) = \text{Cst} \; ; \; J(L, \theta_m, E, \varphi) = \text{Cst} \; ; \; W(L, \theta_m, E, \varphi) = \text{Cst} \quad (6)$$

where $\mu \; (= E/B_m)$ is the first invariant, $J = \oint v \, ds$ is the second invariant $(= 2p \, I)$, $W \; (= E + qV_1 + qV_2)$ is the total energy, E is the kinetic energy, $B_m$ is the mirror-point field and $\theta_m$ the mirror-point colatitude. In order to solve these equations we use a perturbation technique of the type described by Kosik [1971a] or Schulz [1972] : in equations (1) and (2) the terms involving the external field are considered small compared to the main field terms. We decompose I and $B_m$ in the folowing way :

$$I = I_0 + I_1 + I_2 \cos\varphi + \cdots \cdots$$
$$B_m = B_{m0} + B_{m1} + B_{m2} \cos\varphi + \cdots \cdots \quad (7)$$

The three invariants depend on four variables, and we can eliminate E and $\theta_m$. By taking the partial derivatives with respect to $\varphi$ and L, keeping only the first order terms, we obtain the following equations :

$$\left[\left(\frac{\partial I_0}{\partial \theta_m}\frac{\partial B_{0m}}{\partial L} - \frac{\partial I_0}{\partial L}\frac{\partial B_m}{\partial \theta_m}\right) + \left(\frac{\partial B_{0m}}{\partial \theta_m}\frac{I_0}{2} - \frac{\partial I_0}{\partial \theta_m}B_{0m}\right)\left(\frac{C_1 q}{EL^2} - \frac{qC_2}{E}\sin\varphi\right)\right]dL$$
$$= \left[-\left(I_2\frac{\partial B_{0m}}{\partial \theta_m} - \frac{\partial I_0}{\partial \theta_m}B_{2m}\right)\sin\varphi + q\frac{C_2 L}{E}\cos\varphi\left(\frac{\partial B_{0m}}{\partial \theta_m}\frac{I_0}{2} + \frac{\partial I_0}{\partial \theta_m}B_{0m}\right)\right]d\varphi \quad (8)$$

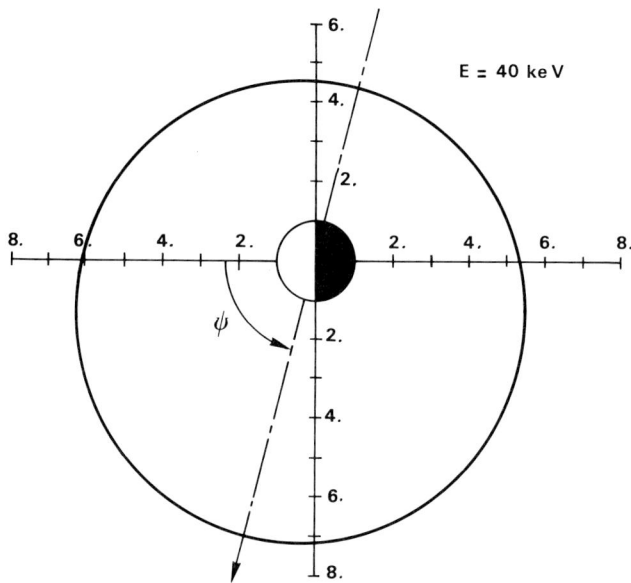

Fig. 4. Drift shell of a particle having 40 keV at $\theta = 0$ and $\theta = \pi$ with $J = 0$. The drift shell is pushed out in the noon-dusk quadrant.

To solve this equation we choose a power series solution for L :

$$L = L_0 + L_1 + L_2 \cos\varphi + \cdots \cdots \quad (9)$$

To the first order we get $L_0$ = constant, which corresponds to motion in an azimuthally symmetric B field in absence of electric fields. To the next order we get :

$$L = L_0 \left[ 1 + \frac{\overline{g}_2^1}{g_1^0} L_0^4 M \left\{ \cos(\varphi - U) - \cos(\varphi_0 - U) \right\} \right] \quad (10)$$

where the subscript o indicates that the values correspond to a given reference meridian, usually the noon meridian ( $\varphi_0 = 0$ )

where
$$M^2 = p^2(\theta_m) + \frac{C_2^2 \, q^2}{E^2} \frac{g_1^0}{\overline{g}_2^1} \frac{g^2(\theta_m)}{L^6} \quad (11a)$$

and
$$\tan U = -\frac{C_2 \, q}{E \, L^3} \frac{g_1^0}{\overline{g}_2^1} \frac{g(\theta_m)}{p(\theta_m)} \quad (11b)$$

The functions p ($\theta$m) and g ($\theta$m) are plotted in Figure 1. Equation (10) shows that L attains the minimum at an angle $\pi$ + U (different from $\Pi$ ) and that this longitude changes with the mirror point colatitude of the particle, as is shown in Figure 2. If we use (10) and (1) we obtain the radial distance of the particle during its drift :

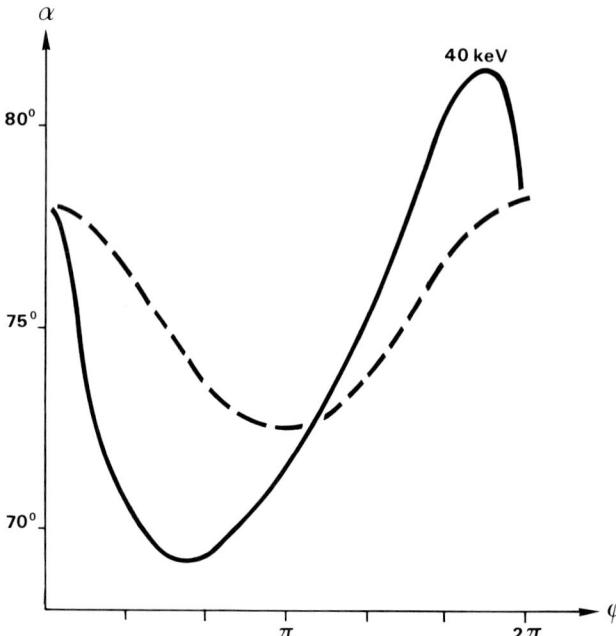

Fig. 5. Relative variation of energy during azimuthal drift.

$$r = L_0 \sin^2 \theta \left[ 1 - \frac{1}{2} \frac{\overline{g}_1^0}{g_1^0} L_0^3 \sin^6 \theta + \frac{\overline{g}_2^1}{g_1^0} L_0^4 \left\{ 2\sqrt{3} \left( \frac{\sin^7 \theta}{7} - \frac{\sin^9 \theta}{3} \right) \cos \varphi + M \left( \cos(\varphi - U) - \cos U \right) \right\} \right] \quad (12)$$

It is possible to find the longitude of the minimum radial distance when particles cross the equator and thus locate the "plane of symmetry" of the drift shell. The critical longitude $\Psi$ is given by :

$$\tan \Psi = - \frac{M \sin U}{(8\sqrt{3}/21) - M \cos U} \quad (13)$$

Figure 3 shows drift paths of electrons havin J = 0 and having reference energies E at noon of 40 keV, 100 keV, 250 keV, respectively. one notices that the deformation of the drift shell increases as the particle kinetic energy decreases. The drift "shell" is pushed out in the noon-dusk quadrant, as is clearly shown in Figure 4 for a particle having 40 keV at $\varphi = 0$ and $\varphi = \Pi$.

Change in Kinetic Energy and Pitch-angle

The total energy W being constant during the drift, the change in kinetic energy is easily obtained if we take into account the change in L given by (10). We obtain, for $\varphi_0 = 0$.

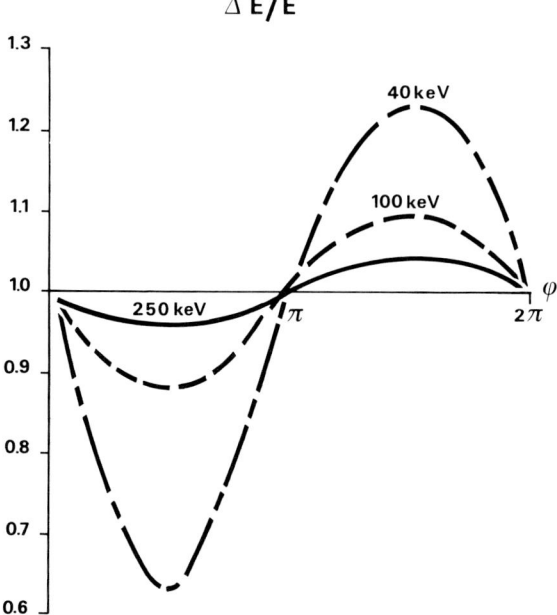

Fig. 6. Variation of equatorial pitch-angle of a particle in presence of electric fields (solid curve) or in absence of electric fields (dashed curve).

$$\Delta E = C_2 q L_0 \sin \varphi + \frac{\overline{g}_2^{-1}}{g_1^0} M L_0^4 \left\{ \cos(\varphi - U) - \cos U \right\} \sin \varphi$$
$$- q C_1 \frac{\overline{g}_2^{-1}}{g_1^0} M L_0^3 \left\{ \cos(\varphi - U) - \cos U \right\} \tag{14}$$

The change in kinetic energy must be taken into account together with the change in the equatorial magnetic field to obtain the pitch-angle evolution by using the first invariant $\mu$. After some calculations, we obtain:

$$\sin^2 \alpha_a = \sin^2 \alpha_b \left[ 1 + L_a^4 \frac{\overline{g}_2^{-1}}{g_1^0} \left\{ \frac{15 \sqrt{3}}{7} (\cos \varphi_b - \cos \varphi_a) - 3 M \left\{ \cos(\varphi_b - U) - \cos(\varphi_a - U) \right\} - \left( \frac{C_2 q L_a}{E_a} \sin \varphi_a - \frac{C_1 q}{E_a L_a} \right) M \left\{ \cos(\varphi_b - U) - \cos(\varphi_a - U) \right\} - q \frac{C_2 L_a}{E_a} (\sin \varphi_2 - \sin \varphi_1) \right] \tag{15}$$

where the subscripts a and b refer to different values of $\varphi$. Figure 5 shows the relative variation in energy $\Delta E/E$ with longitude for particles of 40, 100, 250 keV : as expected, the variation is the greatest for 40 keV particles. Figure 6 shows the variation in equatorial pitch-

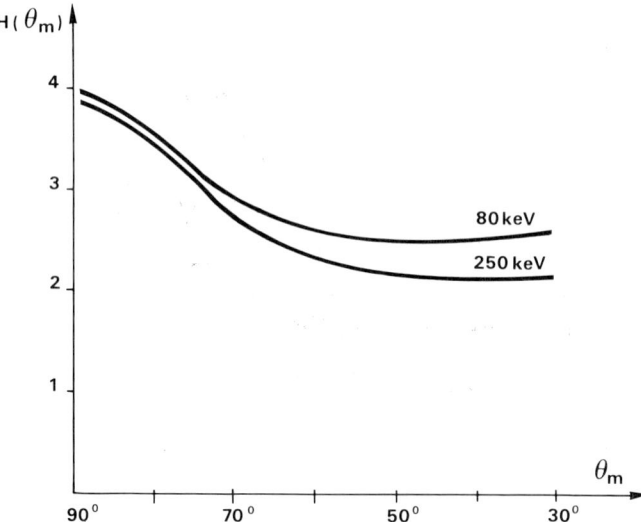

Fig. 7. Functional dependence of the drift velocity with mirror latitude at selected kinetic energies.

angle for a 40 keV particles : it departs greatly from the usual variation in the absence of electric fields which is plotted as a dashed curve having a minimum at midnight [Roederer, 1970].

Longitudinal Change in the Drift Velocity

To obtain the azimuthal variation of the drift velocity, we make use of a theorem derived by Northrop and Teller [1960] : "Suppose we observe a collection of particles with the same invariants $\mu$ and J distributed on a bundle of magnetic field lines which form a finite flux tube. These particles will have different Euler potentials $\alpha$ and $\beta$, and each will drift according to the equations of motion. At any later time, the particles will be bound within a new flux tube. The flux of B through the tube is the same at a later time".

To construct the bundle of magnetic field lines we choose two drift shells $L_o$ and $L_o + \Delta L_o$ and take a certain extent in longitude $\Delta \varphi$. The conservation of $\mu$ and J for particles that belong to the same drift shell is easily expressed. To express this conservation for particles that belong to different shells, we must assume some spread $\Delta \theta_m$ in the colatitude $\theta_m$ of the mirror-points. Conservation of $\mu$ and J implies conservation of $K = (B_m)^{-1/2} I$. Thus, any change $\Delta L_o$ in $L_o$ that keeps K constant leads to a change $\Delta \theta_m$ given by :

$$\Delta \theta_m = \frac{\partial K / \partial L_o}{\partial K / \partial \theta_m} \Delta L_o = F(\theta_m) \frac{\Delta L_o}{L_o} \qquad (16)$$

The magnetic flux imbedded in the bundle of field lines is more easily obtained in the limit $r \to 0$, in which limit the normal component of B

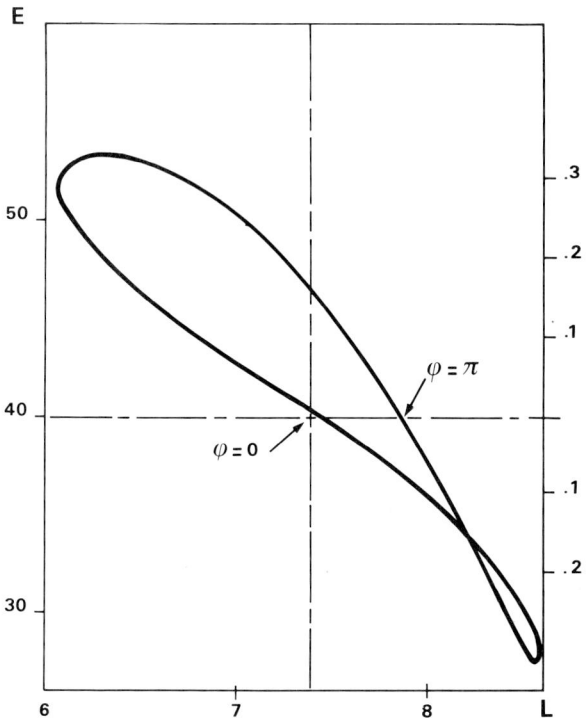

Fig. 8. As the satellite drifts in longitude, its detector measures particles having different energies and L origins at noon. The sampling in (E, L) space at noon is described by the solid curve.

is given by $B_r = 2 g_1^0 / r^3 \cos \theta_0$. However, it is a good approximation to say that $B_r \to 2 g_1 \cos \theta_0$ as $r \to 1$. Thus, we obtain:

$$\Delta \Phi = 2 g_1^0 \cos \theta_0 \sin \theta_0 \, \Delta \theta_0 \, \Delta \varphi \tag{17}$$

Using the fact that $L_0 = 1/\sin^2 \theta_0$ we can express (17) in the form:

$$\Delta \Phi = g_1^0 \frac{\Delta L_0}{L_0} \Delta \varphi \tag{18}$$

Now, as the particles drift, $\Delta \Phi$ must remain constant, and the change in L is given by (10). Taking into account this equation and the fact that there is some spread in the mirror point colatitude $\Delta \Phi_m$, we get the flux $\Delta \Phi$ for any other longitude:

$$\Delta \Phi = \frac{g_1^0}{L_0^2} \left[ 1 + L_0^4 \frac{\overline{g}_2^1}{g_1^0} H(\theta_m) \left\{ \cos(\varphi - U) - \cos(\varphi_0 - U) \right\} \right] \Delta L_0 \, \Delta \varphi \tag{19}$$

where $H(\theta_m) = 3 M(\theta_m) + \frac{\partial M}{\partial \theta_m} F(\theta_m)$. Newkirk and Walt [1964], have

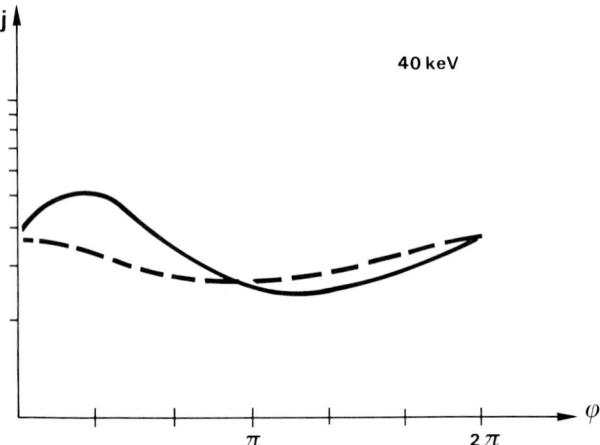

Fig. 9. Diurnal variation of electron fluxes as would be measured by a synchronous satellite ($E_d \simeq 40$ keV). The dashed curve corresponds to the absence of electric fields.

shown that the extent in longitude is proportional to the averaged drift velocity $<\dot\varphi>$. Using the constancy of $\Delta\Phi$ we obtain :

$$<\dot\varphi> = <\dot\varphi_0> \left[ 1 - L_0^4 \frac{\overline{g}_2^{-1}}{g_1^0} H(\theta_m) \left\{ \cos(\varphi - U) - \cos(\varphi_0 - U) \right\} \right] \quad (20)$$

where $<\dot\varphi_0>$ is the bounce averaged drift velocity at longitude $\varphi_0$. Figure 7 gives the values of $H(\theta_m)$ for different mirror-point colatitudes.

## Application to Synchronous measurements

Let us now examine the consequences on the energetic electron environment and on measurements by synchronous satellites. During its orbit a synchronous satellite samples different L shells : if $L_s$ is the L-value when the satellite is at noon, then at any other longitude $\varphi_d$ the satellite will be located on a field line whose L value is given by :

$$L_d = L_s \left[ 1 + \frac{8\sqrt{3}}{21} \frac{\overline{g}_2^{-1}}{g_1^0} L_s^4 (\cos\varphi_d - 1) \right] \quad (21)$$

Suppose now that the satellite detector measures particles at a given pitch-angle $\alpha_d$, above a given threshold energy $E_d$, and in a finite width $\Delta E_d$. The detector will count only those particles (subscript p) which satisfy simultaneously the three conditions $L_p = L_d$, $\alpha_p = \alpha_d$ and $E_d \leqslant E_p \leqslant E_d + \Delta E_d$ for $\varphi = \varphi_d$. It is thus possible to evaluate the flux measured at longitude $\varphi_d = 0$, if we are given a map of directional fluxes at longitude $\varphi = 0$, provided that we know the origin in $(L, E, \alpha)$ at noon of those particles which are detected at longitude $\varphi_d$. From Liouville's theorem we have :

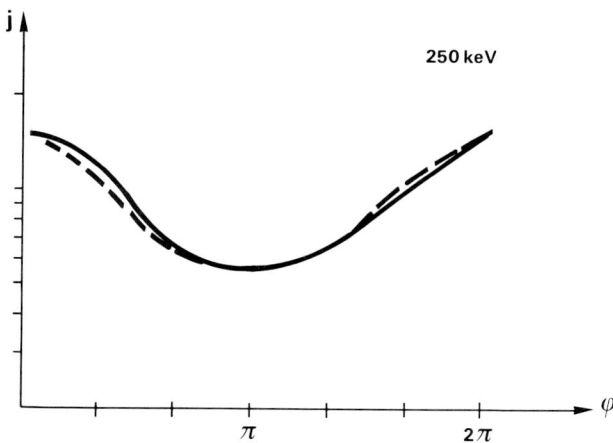

Fig. 10. Diurnal variation of electron fluxes as would be measured by a synchronous satellite ( $E_d \simeq 250$ keV ). The dashed curve is obtained in absence of electric fields.

$$j / E = j_o / E_o \quad (22)$$

where the subscript o refers to longitude 0. The sampling made by the detector in the (E, L) space at noon is shown on figure 8 for particles mirroring at the equator (40 keV threshold energy detector). We use the following spectrum :

$$J_o = A\, e^{-.15(E/E_o)(r/r_o)^{2.4}}\, e^{-.5(r-r_o)}\, (\sin \alpha)^{1.2} \quad (23)$$

where $r_o$, $E_o$ are reference values at longitude 0 ( $r_o = 5$ , $E_o = 40$ keV )
Taking into account equations (22) and (23) we obtain the results shown in figures 9 and 10 for particles of 40 and 250 keV respectively. In these figures are also plotted as dashed curves the fluxes expected in a purely asymmetric magnetosphere, whithout electric fields. As expected, the electric fields have a negligible effect for particle measurements at energies ⩾ 250 keV : a minimum in the count rate is observed near midnight. The situation is completely different for low energy measurements : the minimum is shifted toward the dawn meridian,
while the maximum appears in the afternoon sector. Another feature is that we have a rather broad minimum and a rather sharp maximum. The shift of the minimum comes simply from the duskwards displacement of the drift shells : when the satellite is at dawn it measures particles that originate far out in the outer zone at noon, a region where fluxes are quite low, and we have the opposite situation at dusk.

## Discussion of the Results-Conclusion

The two primary effects of the electric fields are the deformation of drift shells and the change in energy along these drift shells. Formulae (11) show that the "tilt" U can reach 80° for a 40 keV electron in a 2 kilovolt/$R_e$ convection electric field. If one reduces the intensity of this electric field, the "tilt" decreases. Under the same conditions the relative variation in energy reaches 30 %. On the other hand, if $g_2^1$ vanishes, we get $U = \pi/2$ : we just have the well-known electric field effect in an azimuthally symmetric magnetosphere, and the dawn-dusk meridian is a plane of symmetry for the shells. The change of energy along the shell is more than a consequence of a drift trajectory in a potential : for particles of vanishing second invariant J the conservation of the first invariant $\mu$ implies that any change in E exactly compensates the change in equatorial B. Figure 3 perfectly illustrates this fact : when the kinetic energy of the particle decreases, the drift path is located at higher radial distance. As a consequence particles having J = 0 are a good check for the accuracy of the model.

We have tried to compare the azimuthal variation of particle intensities given by our model with observational results obtained by Winckler [1970] from the ATS-I satellite. We have selected the data corresponding to magnetically quiet days [Coleman and Mc Pherron, 1970]. It was not possible to discern any electric field effect on low energy particles. The discrepancy between our results ant the observation can be explained by a convection electric field lower than the one assumed in our calculations. A convection electric field of 0.8 or 1.0 kilovolt/$R_e$ acting on 100 keV particles would produce effects similar to those shown in Figure 8 for 250 keV particles with a 2 kilovolt/$R_e$ field. Such effects are probably small during the quiet periods and hardly detectable among the other uncertainties of the model. However, Paulikas et al. [1968] have observed fluctuations in the location of maximum and minimum fluxes during quiet periods and during periods of moderate magnetic activity : these fluctuations could arise from fluctuations in the strength of the electric field. The other effect of the electric field would be that the region of minimum fluxes around midnight is larger than the region of maximum fluxes around noon (Figure 9), and this is observed by Paulikas et al. [1968].

From the present work we can draw the two following conclusions : (a) During quiet magnetically periods, the magnetic field asymmetry produces the dominant effect because the convection electric field is so weak (less than 1 kilovolt/$R_e$). When the electric field increases, a competition between magnetic and electric field effects distorts the azimuthal variation of fluxes, and the distortion increases with decreasing kinetic energy. (b) During substorms, in which important modifications of the electric field and magnetic field occur, the use of a simple dipole field and magnetic field occur, the use of a simple dipole field is probably inaccurate for particle energies $\leq$ 10 keV.

Acknowledgements. The electron spectrometer data of Prof. J.R. Winckler was kindly provided by J. Vette (NSSDC and World Data Center A for Rockets and Satellites). We are also indebted to Mike Schulz for a careful reading of the manuscript and helpful suggestions.

## References

Axford, W.I., and C.O. Hines, An unifying theory of high latitude geophysical phenomena and geomagnetic storms, Can. J. Physics, 39, 1433, 1961

Chen, A.J., Penetration of low energy protons deep into the magnetoshere, J. Geophys. Res., 75, 2458, 1970

Coleman, P.J., and R.L. Mc Pherron, Fluctuations in the distant geomagnetic field during substorms : ATS I, in Particles and Fields in the Magnetosphere, edited by B. Mc Cormac, D. Reidel, Dordrecht, Holland, 1970

Cowley, S.W.H., and M. Ashour Abdalla, Adiabatic plasma convection in a dipole field : Variation of plasma bulk parameters with L, Planet. Sp. Sc., 23, 1527, 1975

Cowley, S.W.H., and M. Ashour Abdalla, Adiabatic plasma convection in a dipole field : Electron forbidden-zone effects for a simple electric field model, Planet. Sp. Sc., 24, 805, 1976

Kivelson, M.G., and D.J. Southwood, Local time variations of particle flux produced by an electrostatic field in the magnetosphere, J. Geophys. Res., 80, 56, 1975

Kosik, J.Cl., Mouvement des particules chargées dans un modèle analytique approche de la magnétosphère de Mead, Ann. Geophys., 27, 11, 1971a

Kosik, J.Cl., Motion of energetic particles in a magnetospheric model including a convection electric field, Planet. Sp. Sc., 19, 1209, 1971b

Newkirk, L.L., and M. Walt, Longitudinal drift velocity of geomagnetically trapped particles, J. Geophys. Res., 69, 1759, 1964

Paulikas, G.A., J.B. Blake, S.C. Freden, and S.S. Imamoto, Observations of energetic electrons at synchronous altitudes (1), J. Geophys. Res., 73, 4915, 1968

Roederer, J.G., Dynamics of geomagnetically trapped radiation, Springer, New York, 1970

Roederer, J.G., and M. Schulz, Splitting of drift shells by the magnetospheric electric field, J. Geophys. Res., 76, 1055, 1971

Schulz, M., Compressible corotation of a model magnetosphere, J. Geophys. Res. 75, 6329, 1970

Schulz, M., Drift shell-splitting at arbitrary pitch-angle, J. Geophys. Res., 72, 624, 1972

Solomon, J., On the azimuthal drift of substorm-injected protons, J. Geophys. Res., 81, 3452, 1976

Stern, D.P., Shell-splitting due to electric fields, J. Geophys. Res., 76, 7787, 1971

Stern, D.P., The motion of a proton in the equatorial magnetosphere, J. Geophys. Res., 80, 595, 1975

Stern, D.P., Large scale electric fields in the Earth's magnetosphere, Rev. Geophys., 15, 2, 1977

Vette, J.I. and A.B. Lucero, Models of trapped radiation environment, Vol. 3, "Electrons at synchronous altitudes", NASA SP 3024, 1967

Winckler, J.R., Energetic electrons in the Van Allen radiation belts, in Particles and Fields in the Magnetosphere, Edited by B. Mc Cormac, D. Reidel Dordrecht, Holland, 1970

# MAGNETIC SHELL TRACING: A SIMPLIFIED APPROACH

J. G. Luhmann and Michael Schulz

Space Sciences Laboratory, The Aerospace Corporation
El Segundo, California 90245

**Abstract.** We consider the adiabatic motion of electrons trapped in the model geomagnetic field derived from the simplified (three-coefficient) scalar potential proposed by Mead [J. Geophys. Res., 69, 1181, 1964]. The scalar potential contains contributions from the geomagnetic dipole (with coefficient $g_1^0$) and from external currents (with coefficient $\bar{g}_1^0$ for the azimuthally symmetric part and coefficient $\bar{g}_2^1$ for the antisymmetric part). Numerical values for the external coefficients $\bar{g}_1^0$ and $\bar{g}_2^1$ are deduced from the diurnal variation of $\underset{\sim}{B} = -\underset{\sim}{\nabla} V$ observed at the satellite ATS-1 (geocentric distance $r = 6.6a$, magnetic colatitude $\theta = \pi/2$, and local time $\varphi$, where $a$ is the radius of the earth). The drift shell is specified analytically (i.e., by an approximation given in closed form) in terms of the equatorial pitch angle $\alpha_0 = \sin^{-1} y$ attained at the noon meridian ($\varphi = \pi$). The "radial gradient" $\partial \ln j/\partial \ln B$ at $\alpha_0 = \pi/2$ is inferred from the diurnal variation of particle flux $j(\alpha_0 = \pi/2, \varphi)$ at ATS-1. This procedure offers an internal test for the absence of time-variation in the expansion coefficients contained in V. The pitch-angle distribution $j(\alpha_0, \varphi = \pi)$ is directly observed at the noon meridian ($\varphi = \pi$). The parameters thus determined allow the pitch-angle distributions at other longitudes to be obtained from Liouville's theorem. The results, at least for a well-studied test case, are comparable in quality to those obtained by Pfitzer et al. [J. Geophys. Res., 74, 4687, 1969], who used a more complicated field model, a purely numerical tracing of drift shells, and the data from two spacecraft. In other words, our results are in similarly good qualitative agreement with the observational data obtained at other longitudes. For example, we find that the particle flux at any pitch angle $\alpha_0 \gtrsim 46°$ has its diurnal maximum at noon ($\varphi = \pi$) and that on the midnight meridian the maximum in the pitch-angle distribution occurs at $\alpha_0 \neq \pi/2$.

## Introduction

It is well known [Serlemitsos, 1966; Hills, 1967; Haskell, 1969; Pfitzer et al., 1969; West et al., 1973] that energetic radiation-belt electrons have a qualitatively different pitch-angle anisotropy on the day side than on the night side in the earth's outer magnetosphere. The dayside anisotropy is such that the maximum flux at a given energy is attained at an equatorial pitch angle $\alpha_0 = 90°$.

The nightside anisotropy is such that the maximum flux is attained at a pair of equatorial pitch angles ($\alpha_0$, $\pi - \alpha_0$) different from $90°$; the detector sees a relative minimum in the flux when looking at right angles to the equatorial magnetic field $\underline{B}$. This initially surprising observation has been shown to be a natural consequence of Liouville's theorem [Pfitzer et al., 1969].

The peculiar nightside anisotropy arises from the adiabatic motion of charged particles in an azimuthally asymmetric $\underline{B}$ field. In order to conserve the first two adiabatic invariants, a particle having $\alpha_0 \approx 90°$ must trace a drift shell consisting of field lines that have very nearly the same equatorial magnitude of $\underline{B}$. A particle having $\alpha_0 \approx 0°$ must trace a drift shell consisting of field lines that have very nearly the same arc length. If the magnetic field is not azimuthally symmetric, there is no reason why the two drift shells described above must coincide at all longitudes, although they may intersect at two or more longitudes. Indeed they do not coincide; drift shells generated by particles that mirror at different latitudes along a given field line are not degenerate, but are dispersed (at other longitudes) with respect to $\alpha_0$ [Northrop and Teller, 1960; Hones, 1963; Fairfield, 1964; Mead, 1966; Roederer, 1967; Stern, 1968]. In fact, of electrons or protons impacting the same synchronous satellite at midnight, those particles having $\alpha_0 = 90°$ belong to a drift shell that has a larger diameter than the drift shell executed by those particles having equatorial pitch angles of (for example) $65°$ and $115°$ relative to $\underline{B}$. Moreover, the particle intensity at fixed energy $E$ and equatorial pitch angle $\alpha_0$ is a decreasing function of drift-shell diameter in the vicinity of the synchronous orbit [e.g., Vette et al., 1966]. This "negative radial gradient" thus conspires with the $\alpha_0$-dependent asymmetry of the drift shells to produce the peculiar nightside anisotropy that is observed [Pfitzer et al., 1969].

The pioneering work of Pfitzer et al. [1969] in this area represented a truly formidable exercise in numerical analysis. Pfitzer et al. [1969] used an elaborate program developed by Roederer [1967] in order to trace the drift shells executed by particles trapped in the complicated model $\underline{B}$ field of Williams and Mead [1965]. Moreover, they found that the predicted diurnal variations of particle fluxes at various values of $\alpha_0$ were sometimes unduly sensitive to parameters in the field model that were poorly determined by the available observations of $\underline{B}$ at synchronous altitude.

In an effort to alleviate these difficulties, we have taken a second look at the data of Pfitzer et al. [1969] and have attempted to recover their essential conclusions from a much simpler numerical analysis. Our efforts have been quite successful. We have adopted a simple $\underline{B}$-field model that is derived from a scalar potential containing three spherical-harmonic terms. We have traced the relevant drift shells in perturbation theory [cf. Stern, 1968; Kosik, 1971a] through the use of simple analytic functions of $\alpha_0$ and the field-line parameters [Schulz, 1972; Schulz and Lanzerotti, 1974]. Finally, we have deduced the "radial gradient" from an internally consistent treatment of the synchronous-satellite data rather than from extraneous sources. A by-product of this treatment (see below) is our ability to diagnose genuine temporal variations of the

$\underset{\sim}{B}$ field that would otherwise contaminate intensity predictions based on the postulate of a static $\underset{\sim}{B}$ field. Thus, we can identify objectively certain time periods to which our method of mapping fluxes, as well as the method of Pfitzer et al. [1969], should not be applied. Our model is well suited to use on programmable pocket calculators, such as the HP-67 or the TI-59, for investigators who lack access to major computing facilities. Moreover, our model yields diurnal predictions that agree with the observational data of Pfitzer et al. [1969] about as well as their significantly more complicated model did.

## Approach

We follow Pfitzer et al. [1969] in assuming that the normalized equatorial pitch-angle distribution $g(\alpha_0^*)$ on the noon meridian is independent of geocentric distance r for drift shells accessible to the synchronous orbit. Thus, we express the count rate $j(\alpha_0, L_0, \varphi_0)$ as

$$j(\alpha_0, L_0, \varphi_0) = g(\alpha_0^*) h(L_0^*), \qquad (1)$$

where $h(L_0^*)$ represents the "radial" dependence of the particle flux at the equator in the noon meridian and $L_0$ is the field-line parameter introduced by Stone [1963]. In other words, we identify the value of $|\underset{\sim}{B}|$ observed at the synchronous satellite ATS-1 with $B_0$, the minimum value of $|\underset{\sim}{B}|$ along the field line. Moreover, we label the field line by the parameter $\varphi_0$, i.e., by the local time at which its minimum $|\underset{\sim}{B}|$ is found, and by the parameter

$$L_0 \equiv (-g_1^0/B_0)^{1/3}, \qquad (2)$$

where $g_1^0 a^3$ is the earth's magnetic dipole moment and $\underline{a}$ is the earth's radius (= 6371.2 km, by convention, for geomagnetic purposes). The asterisks in (1) denote evaluation at the reference longitude $\varphi_0 = \varphi_0^* \equiv \pi$, i.e., at noon. Thus, the parameters $(L_0, \varphi_0)$ label a field line; the parameters $(L_0^*, \alpha_0^*)$ label the adiabatic drift shell executed by a particle that starts out with equatorial pitch angle $\alpha_0^*$ on the field line $(L_0, \pi)$. The correspondence between $(\alpha_0, L_0)$ at $\varphi_0$ and $(\alpha_0^*, L_0^*)$ at $\varphi_0^*$ is supposed to be established by tracing the equatorial intercept of the drift shell. The equality of $j(\alpha_0, L_0, \varphi_0)$ with $j(\alpha_0^*, L_0^*, \varphi_0^*)$ follows from Liouville's theorem.

Pfitzer et al. [1969] deduced $g(\alpha_0^*)$ from ATS-1 data acquired at the noon meridian. The results are shown in their Figure 3. Pfitzer et al. [1969] deduced a radial profile equivalent to $h(L_0^*)$ from OGO-3 data under the assumption that $h(L_0^*)$ is independent of $\alpha_0^*$. We make the same assumption, but we deduce $h(L_0^*)$ directly from the ATS-1 data. In order to accomplish this, we note that $\alpha_0 = 90°$ at all longitudes for particles having $\alpha_0^* = 90°$. Moreover, the value of $L_0$ for any such particle is equal to its value of $L_0^*$, since particles that mirror on the equator are known to follow a path of constant $B_0$ there (in order to conserve the first adiabatic

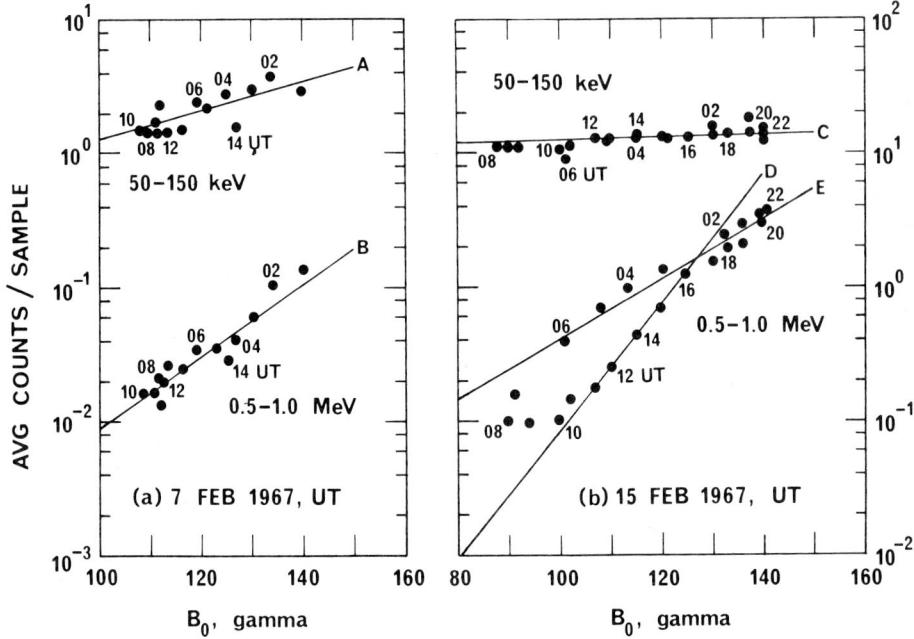

Fig. 1. Observed variation of counting rate ($\alpha_0 = 90°$) with $B_0$ in two energy channels on ATS-1 at indicated hours of universal time. Best fits: (A) $0.138 \exp(B_0/43.7 \gamma)$; (B) $1.62 \times 10^{-5} \exp(B_0/15.8 \gamma)$; (C) $10.10 \exp(B_0/486.4 \gamma)$; (D) $1.42 \times 10^{-6} \exp(B_0/9.12 \gamma)$; (E) $2.41 \times 10^{-3} \exp(B_0/19.5 \gamma)$.

invariant). Thus, the diurnal variation of the ATS-1 count rate at $\alpha_0 = 90°$, when plotted against the diurnal variation of $B_0$ at the satellite, generates the function $h(L_0^*)$ in a static magnetosphere if electric-field effects can be ignored. We have plotted the count rate against $B_0$ in Figure 1 for the cases treated by Pfitzer et al. [1969]. The data from 7 February 1967 (UT) in Figure 1a appear to define $h(L_0^*)$ rather well in each energy channel. The linear fit between $\log_{10}[h(L_0^*)]$ and $B_0$ defines $\log_{10}[h(L_0^*)]$ as the sum of a constant and a quantity proportional to $(L_0^*)^{-3}$ in each case. It would have been much better to plot at least 24 hours' worth of data preceding the magnetic storm which began at 1310 UT [Pfitzer et al., 1969], so as to show that the data truly define $h(L_0^*)$ as a single-valued function of $L_0^*$. In order to proceed, we must assume that the magnetosphere was indeed static for at least the 13 hours that preceded the storm of 7 February 1967 (UT).

However, the higher-energy data for 15 February 1967 (UT) obviously fail to define $h(L_0^*)$ as a single-valued function of $L_0^*$ (see Figure 1b). This failure is evidence of an intervening temporal variation in the state of the magnetosphere. Thus, it would be unrealistic to model the diurnal variation of $j(\alpha_0, L_0, \varphi_0)$ on 15 February 1967 (UT) in terms of a static magnetosphere, and we shall not attempt to do so. The foregoing exercise illustrates the benefit of using the available ATS-1 data instead of OGO-3 results for determining the "radial" profile $h(L_0^*)$. The present procedure automatically warns us against trying to fit genuine temporal variations with a static model of the magnetosphere. We know that in a static magnetosphere (with electric-field effects ignored), the $\alpha_0 = \pi/2$ flux must be a single-valued function of $B_0$, plotted with universal time (UT) as a parameter. However, we cannot determine just *when* the temporal variation occurred if the data happen to fail this test. Of course, it is possible for $h(L_0^*)$, as determined here, to be single-valued even if the magnetospheric $\underline{B}$ field is not static. This could happen if the energy spectrum and "radial" profile were fortuitously related (as perhaps they were for the lower-energy data in Figure 1b), or if the magnetic disturbance were fortuitously symmetric about the UT at which the satellite passes through noon or midnight. It might be possible [cf. Kosik, 1973] to develop a model for mapping fluxes in a non-static magnetosphere, by arranging to conserve all three adiabatic invariants, but this lies beyond the scope of the present work. The present procedure assures a reasonable fit to the observed diurnal variation at $\alpha_0 = 90°$. We need not contend with possible discrepancies between OGO-3 and ATS-1 in the present work.

The above determination of $h(L_0^*)$ is independent of the model that specifies the global form of the magnetic field $\underline{B}$. However, we do need such a model in order to trace drift shells and thus map particle fluxes at $\alpha_0 \neq 90°$. For this purpose we adopt the model scalar potential

$$V = [(a^3/r^2)g_1^0 + r\bar{g}_1^0]\cos\theta$$
$$+ \sqrt{3}\,\bar{g}_2^1(r^2/a)\cos\theta\sin\theta\cos\varphi, \qquad (3)$$

where $r$, $\theta$, and $\varphi$ are the geocentric distance, colatitude (measured from north pole), and local time (longitude from midnight), respectively, and $a$ is the radius of the earth. The form of (3) is motivated by a suggestion of Mead [1964], but we treat the expansion coefficients $\bar{g}_1^0$ and $\bar{g}_2^1$ as independently adjustable parameters of the field model. Thus, we view (3) as the representation of a scalar potential attributable to any and all sources. It leads to an equatorial $\underline{B}$ field that points in the $-\underline{\hat{\theta}}$ direction and bears a magnitude

$$B_0 = -(a/r)^3 g_1^0 - \bar{g}_1^0 - \sqrt{3}\,\bar{g}_2^1(r/a)\cos\varphi_0 \qquad (4)$$

that varies sinusoidally with the local time of an observer at fixed $r$. The parameter $g_1^0 = -0.31$ G is fixed by the magnitude of the earth's dipole moment, but the parameters $\bar{g}_1^0$ and $\bar{g}_2^1$ are to be adjusted so as to fit the quiescent diurnal variation of $B_0$ observed at ATS-1 ($r = 6.6a$). The results for the ATS-1 data of 7 February 1967 (UT) are $\bar{g}_1^0 = -19.7\,\gamma$ and $\bar{g}_2^1 = 1.66\,\gamma$; the good fit of $B_0$ to local time ($\varphi$) in Figure 2 tends to support our assumption that $\underline{B}$ was static prior to 1310 UT on 7 February 1967. However,

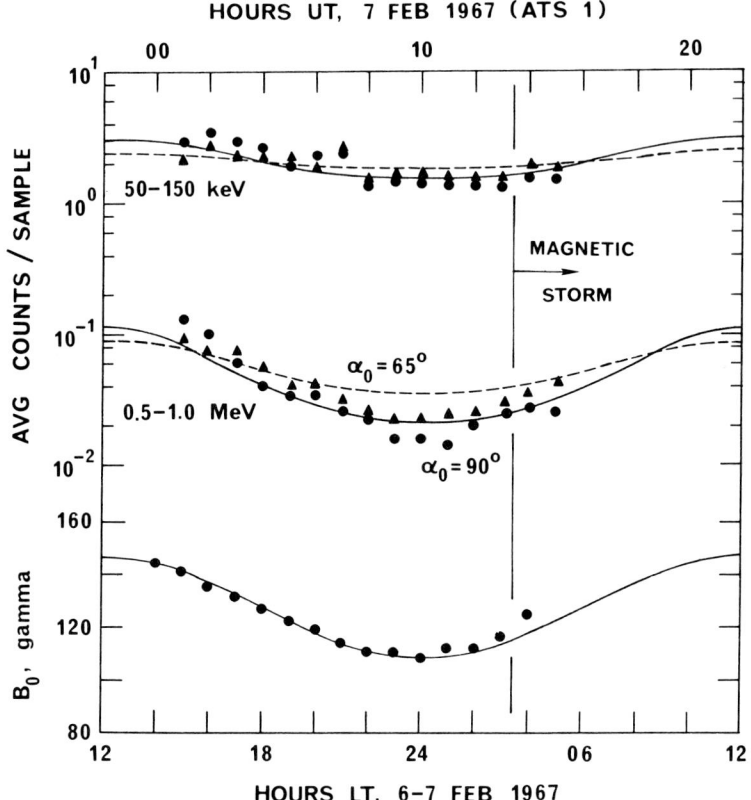

Fig. 2. Comparison of ATS-1 data with expectations based on our model for 7 February 1967 (UT). Solid curves represent either direct fits ($B_0$) or indirect fits ($\alpha_0 = 90°$) to the corresponding data (●). Dashed curves ($\alpha_0 = 65°$) represent genuine extrapolations based on drift-shell tracing, and comparison with corresponding data (▲) determines the success of the model.

it would be futile (in view of Figure 1b) to fit the ATS-1 data for 15 February 1967 in terms of a static model of the magnetosphere. Indeed, the diurnal minimum in $B_0$ at ATS-1 occurred at about 2200 LT (14 February 1967) rather than at midnight [Pfitzer et al., 1969], and so any static fit based on (4) would necessarily have been a poor fit.

It is interesting to compare the parameters $\bar{g}_1^0$ and $\bar{g}_2^1$ obtained from the ATS-1 data of 7 February 1967 (UT) with the parameters $\bar{g}_1^0 = -25.15(10a/b)^3 \gamma$ and $\bar{g}_2^1 = 1.215(10a/b)^4 \gamma$ originally suggested by Mead [1964] for the average magnetospheric condition, i.e., for a magnetopause having a nominal stand-off distance $b \sim 10a$ along the earth-sun line. Mead [1964] intended this model to account for the effects of magnetopause currents only. The appearance of a tail-like structure on the night side was an inadvertent consequence of truncation after three terms. However, it is clear that one could simulate the effects of a ring current beyond the observer's position by reducing the magnitude of Mead's $\bar{g}_1^0$, and that one could simulate the effects of a tail current by increasing the magnitude of Mead's $\bar{g}_2^1$. We see from the above comparison (for b ~ 10a) that these are precisely the modifications required to bring about agreement between (4) and the ATS-1 data.

The tracing of drift shells in such a B-field model has been described by Stern [1968], Kosik [1971a], and Schulz [1972]. The summary provided by Schulz and Lanzerotti [1974] represents a good starting point for the present work. We seek to trace the value of $L_0$, as defined by (2), around the drift shell of a high-energy charged particle in the static B-field model specified by (3). If $L_0^*$ is the value of $L_0$ at some reference longitude $\varphi_0^*$ on the drift shell, then we regard the azimuthal deviation of $L_0/L_0^*$ from unity as a "small" quantity of order $(L_0/10)^4$ that is adequately approximated in lowest order. In other words, we consider that $(L_0/10)^4 \ll 1$, even in the vicinity of the synchronous orbit. Thus, given a particle of equatorial pitch angle $\alpha_0$, we determine that the particle would reach the reference longitude $\varphi_0^*$ on a field line bearing the label

$$L_0^* \approx L_0 + \sqrt{3}\,(5\bar{g}_2^1/7g_1^0)L_0^5\{1 + [Q(y)/D(y)]\}(\cos\varphi_0 - \cos\varphi_0^*), \quad (5)$$

and that the particle's equatorial pitch angle there would be given by

$$\alpha_0^* \equiv \sin^{-1} y^* = \sin^{-1}[(L_0/L_0^*)^{3/2} y], \quad (6)$$

where $y \equiv \sin\alpha_0$. The function $[Q(y)/D(y)]$ had been tabulated by Stern [1968] and plotted graphically by Kosik [1971a]. A good analytical approximation for it is given [Schulz, 1972; Schulz and Lanzerotti, 1974; Davidson, 1976] by

$$Q(y)/D(y) \approx -(27.12667 + 45.39913y^4 - 5.88256y^8)$$
$$\div (82.81038 + 19.19078y^{3/4} - 35.35791y). \quad (7)$$

Since $Q(1)/D(1) = -1$, we recover from (5) the familiar result that a particle mirroring at the equator will trace a path of constant $L_0$ (i.e., constant $B_0$) on the equatorial surface.

In order to evaluate the expected $j(\alpha_0, L_0, \varphi_0)$ as a function of longitude along the orbit of ATS-1, we first calculate $B_0$ from (4) at r = 6.6a, using the coefficients $g_1^0$, $\bar{g}_1^0$, and $\bar{g}_2^1$ given above. We next calculate $L_0$ by means of (2), $L_0^*$ by means of (5), and $\alpha_0^*$ by means of (6), all in terms of the desired equatorial pitch angle $\alpha_0$ and longitude $\varphi_0$. Finally, we evaluate $g(\alpha_0^*)$ with the help of Pfitzer et al. [1969] and $h(L_0^*)$ with the help of Figure 1. For example, we find $g(\alpha_0^*)/g(90°) \approx 0.335 \exp(\alpha_0^*/82.3°)$ for the higher-energy channel on 7 February 1967. It follows from Liouville's theorem that $j(\alpha_0, L_0, \varphi_0) = g(\alpha_0^*)h(L_0^*)$, just as in (1) above.

Following Pfitzer et al. [1969], we have chosen $\varphi_0^* = \pi$ (i.e., the noon meridian) as our reference longitude.

## Results

Our expectations for the diurnal variation of particle flux at $\alpha_0 = 90°$ (solid curves) and $\alpha_0 = 65°$ (dashed curves), as determined by means of the analysis outlined above, are shown in Figure 2 for 7 February 1967 (UT) together with the diurnal ATS-1 data used by us and by Pfitzer et al. [1969]. We obtain only qualitative agreement between our diurnal curves and the ATS-1 particle data. However, our model does predict that $j(65°)/j(90°) > 1$ in the midnight sector of the magnetosphere, i.e., at $|\varphi_0| \lesssim \pi/4$, as the observations clearly require. Moreover, the midnight value of $j(65°)/j(90°)$ is predicted (mainly on the basis of Figure 1a) to increase with energy, and the ATS-1 diurnal observations confirm this prediction; we note from Pfitzer et al. [1969] that $g(\alpha_0^*)/g(90°)$ shows at most a weak dependence on particle energy.

The results of Pfitzer et al. [1969], as shown in their Figure 4 for 7 February 1967, reveal a similarly qualitative agreement between their diurnal curves and the ATS-1 particle data. However, Pfitzer et al. [1969] were using a much more complicated B-field model [Williams and Mead, 1965] and needed a complicated numerical code [Roederer, 1967] in order to trace drift shells. Moreover, they needed a second satellite (OGO-3) in order to identify the "radial gradient." It would be fair to say that Pfitzer et al. [1969] described a very formidable procedure that does not readily lend itself to routine application.

Our scheme is much simpler. We do not claim that it organizes the data any better than the method of Pfitzer et al. [1969], but we do believe that the advantages of our approach are well documented in the above examples. Specifically, we have adopted a simple B-field model, based on (3), whose adjustable parameters $\bar{g}_1^0$ and $\bar{g}_2^1$ are easy to identify from the diurnal variation of B at ATS-1. Moreover, we have identified the radial profile $h(L_0^*)$ in (1) from the diurnal variation of $j(\alpha_0 = \pi/2, L_0, \varphi_0)$ at ATS-1 and have thus circumvented the need for a second spacecraft such as OGO-3. Finally, we have traced the adiabatic drift shell of any charged particle from equatorial pitch angle $\alpha_0$ on field line $L_0$ at longitude $\varphi_0$ to equatorial pitch angle $\alpha_0^*$ and field line $L_0^*$ at the reference longitude $\varphi_0^* = \pi$, and we have done this by simple algebra rather than with an elaborate numerical code. In summary, we have described a procedure that is simple enough to be used routinely for organizing charged-particle data acquired in the earth's outer radiation belt.

The present scheme requires certain refinements if the particle energy is too low or too high. At low energies ($E \lesssim 500$ keV) one should take account of magnetospheric electric fields when mapping drift shells [Pfitzer et al., 1969; Roederer and Hones, 1970; Kosik, 1971b; McIlwain, 1974]. A superficial reading of the paper by Kosik [1971b] might suggest that our failure (in Figure 1b) to obtain $j(\alpha_0 = \pi/2, L_0, \varphi_0)$ as a single-valued function of $B_0$ in the higher energy channel ($E = 0.5$-$1.0$ MeV) on 15 February 1967 (UT) is a consequence not of a time-varying $\underset{\sim}{B}$ (as we have assumed above) but rather of a static electric field associated with magnetospheric convection. However, Kosik [1971b] appears to have neglected, in his application of Liouville's theorem, the fact that a particle's energy varies with longitude ($\varphi_0$) around the drift shell in the presence of a static electric field. When invoking electric-field effects, one must model the energy spectrum at the reference longitude ($\varphi_0^*$) in order to predict the counting rate elsewhere. At very high particle energies (perhaps at $E \gtrsim 500$ MeV) an entirely different refinement is required. At such energies one must take account of a particle's radius of gyration and (thus) of the possible offset between the drift shell of the guiding center and the location of the satellite [Kosik, 1971c; Blake et al., 1974]. A significant "radial gradient" in $h(L_0^*)$ will then produce a significant modulation of the observed particle intensity as the orientation of the detector collimator rotates from east to west.

We conclude that the present work forms the basis for a simplified mapping of the outer radiation belt, but that certain further refinements (which should be easy to implement) will usually be necessary for the construction of a truly accurate radiation-belt model.

Acknowledgments. The authors thank L. J. Lanzerotti and C. G. Maclennan for past discussions of the present method of extracting a "radial gradient" from the diurnal variations of particle flux and magnetic field at ATS-1. The present work was conducted under U. S. Air Force Space and Missile Systems Organization (SAMSO) contract F04701-77-C-0078.

## References

Blake, J. B., E. F. Martina, and G. A. Paulikas, On the access of solar protons to the synchronous altitude region, J. Geophys. Res., 79, 1345, 1974.
Davidson, G. T., An improved empirical description of the bounce motion of trapped particles, J. Geophys. Res., 81, 4029, 1976.
Fairfield, D. H., Trapped particles in a distorted dipole field, J. Geophys. Res., 69, 3919, 1964.
Haskell, G. P., Anisotropic fluxes of energetic particles in the outer magnetosphere, J. Geophys. Res., 74, 1740, 1969.
Hills, H. K., Observations of energetic electron intensities in the earth's outer radiation zone with OGO-1, Univ. of Iowa Res. Rept. 67-72, Iowa City, 1967.
Hones, E. W., Jr., Motion of charged particles trapped in the earth's magnetosphere, J. Geophys. Res., 68, 1209, 1963.
Kosik, J. Cl., Mouvement des particules chargées dans un modèle analytique approché de la magnétosphère de Mead, Ann. Géophys. 27, 11, 1971a.
Kosik, J. Cl., Motion of energetic particles in a magnetospheric model including a convection electric field, Planet. Space Sci., 19, 1209, 1971b.
Kosik, J. C., L'environnement en particules énergétiques aux altitudes synchrones au cours de périodes géomagnétiques calmes, Ann. Géophys., 27, 175, 1971c.

Kosik, J. Cl., Influence of a sudden compression of the magnetosphere on outer zone electron fluxes measured at arbitrary pitch angle, Planet. Space Sci., 21, 1345, 1973.

McIlwain, C. E., Substorm injection boundaries, in Magnetospheric Physics, edited by B. M. McCormac, pp. 143-154, Reidel, Dordrecht, 1974.

Mead, G. D., Deformation of the geomagnetic field by the solar wind, J. Geophys. Res., 69, 1181, 1964.

Mead, G. D., The motion of trapped particles in a distorted magnetosphere, in Radiation Trapped in the Earth's Magnetic Field, edited by B. M. McCormac, pp. 481-490, Reidel, Dordrecht, 1966.

Northrop, T. G., and E. Teller, Stability of the adiabatic motion of charged particles in the earth's field, Phys. Rev., 117, 215, 1960.

Pfitzer, K. A., T. W. Lezniak, and J. R. Winckler, Experimental verification of drift-shell splitting in the distorted magnetosphere, J. Geophys. Res., 74, 4687, 1969.

Roederer, J. G., On the adiabatic motion of energetic particles in a model magnetosphere, J. Geophys. Res., 72, 981, 1967.

Roederer, J. G., and E. W. Hones, Jr., Electric field in the magnetosphere as deduced from asymmetries in the trapped particle flux, J. Geophys. Res., 75, 3923, 1970.

Schulz, M., Drift-shell splitting at arbitrary pitch angle, J. Geophys. Res., 77, 624, 1972.

Schulz, M., and L. J. Lanzerotti, Particle Diffusion in the Radiation Belts, pp. 30, 41-45, Springer, Heidelberg, 1974.

Serlemitsos, P., Low energy electrons in the dark magnetosphere, J. Geophys. Res., 71, 61, 1966.

Stern, D., Euler potentials and geomagnetic drift shells, J. Geophys. Res., 73, 4373, 1968.

Stone, E. C., The physical significance and application of L, $B_0$, and $R_0$ to geomagnetically trapped particles, J. Geophys. Res., 68, 4157, 1963.

Vette, J. I., A. B. Lucero, and J. A. Wright, Models of the Trapped Radiation Environment, Volume II: Inner and Outer Zone Electrons, p. 20, NASA SP-3024, Washington, D. C., 1966.

West, H. I., Jr., R. M. Buck, and J. R. Walton, Electron pitch angle distributions throughout the magnetosphere as observed on Ogo 5, J. Geophys. Res., 78, 1064, 1973.

Williams, D. J., and G. D. Mead, Nightside magnetosphere configuration as obtained from trapped electrons at 1100 kilometers, J. Geophys. Res., 70, 3017, 1965.

# SHAPE INTEGRAL METHOD FOR MAGNETOSPHERIC SHAPES

F. Curtis Michel

Space Physics and Astronomy Department, Rice University
Houston, Texas 77001

**Abstract.** We discuss the application of a new method for calculating the shape of a magnetospheric boundary. The method uses an integral equation which is evaluated for a trial shape. The resultant values of the integral equation as a function of auxiliary variables then tells one how close one is to the desired solution. A variational method can then be used to improve the trial shape.

## Introduction

An important feature of physical theories is the ability to predict and test. Only by comparing the actual shape of the magnetopause with a prediction can we confirm or deny our assessment of the overall physical situation. Only by such comparison can we be surprised. Certainly we know the rough shape of the magnetopause, thanks largely to direct observation, but that only allows us to draw cartoons of the shape. Certainly pressure balance arguments allow us to estimate correctly the scale of the magnetopause, but neither is enough to tell us if a specific auroral field line maps to the nose or to the tail. Knowing how the shape changes as a function of dipole tilt, ring current contributions, etc. would give us a physical feeling of the magnetospheric response that is virtually out of the question to derive from direct observation.

We have developed a new method for calculating the shape of any magnetopause to arbitrarily high precision. In any magnetopause calculation, one must be given the external pressure difference as a function of position that must be balanced by the internal magnetic field. At any instant, the total pressure

$$P(\text{plasma}) + P(\text{magnetic}) + P(\text{dynamic})$$

balances across the magnetopause, thus the internal magnetic field satisfies

$$B^2/8\pi = P(\text{total}) - P(\text{internal plasma}) - P(\text{internal dynamic}). \quad (1)$$

In principle one can obtain the right hand side either from observational measurements or from models for the external flow hydrodynamics and internal magnetospheric dynamics. In practice one may have to make do at first with rather idealized boundary conditions at the surface, such as a ballistic flow model where the pressure is given by Midgley and Davis (1963)

$$P = P_0\cos^2\psi \qquad (2)$$

and

$$\cos\psi = \vec{n}\cdot\vec{v}/v. \qquad (3)$$

Here n is the outward normal to the surface and v is the incident flow velocity. It does not follow however that it is unimportant to solve such idealized models accurately, even though they may not represent the full complexity of the actual magnetosphere. In virtually every other field of physics, the solving of representative models is regarded to be of central and fundamental importance. Too often in magnetospheric physics simple models are discounted because they do not immediately represent the complex time-dependent and turbulent processes so frequently observed. However the implicit assumption that such complex situations can be understood without first understanding the simple situations has yet to be supported by experience.

## Solving the Axisymmetric Magnetopause

Davis and Midgley (1962) first treated the problem of a dipole moment confined by a constant external pressure by using a multipole expansion. The basis for their method was the fact that the external pressure (P) gave the current (J) that would have to flow on the magnetopause to drop the internal field (B) to zero outside, since

$$P \propto B^2; \quad B \propto J \qquad (4)$$

and if one knew the shape of the magnetopause one could calculate the magnetic field outside the magnetopause in terms of an (infinite) series of magnetic moments, the dipole, octopole, etc. moments. By superposition the total field outside then consists of the source dipole plus the magnetospheric dipole and then plus all the higher magnetospheric multipole moments. Since the outside field is zero, all the higher moments must vanish and the magnetospheric dipole moment must be equal and opposite to the source dipole. One then simply parameterizes the magnetospheric shape and adjusts the parameters to achieve the above conditions.

Unfortunately, one has then an infinite number of multipoles to equal to zero, while in practice one must be content with setting only a finite number of higher multipoles equal to zero . Now one would hope that the precision of the model shape would increase with the number of such conditions satisfied, and therefore that "in principle" one could calculate the shape to whatever precision was desired. Unfortunately the correct magnetosphere has a cusp above each magnetic pole and the multipolar representation of such a cusp is an asymptotically divergent series. That is, there is a length beyond which a truncated series actually becomes a worse, not a better, approximation to the correct answer.

We have shown (Michel, 1977a) that the series can be summed analytically. Consequently the asymptotic divergence is eliminated. As a result, the magnetic moment representation of the magnetopause can be written as a single integral of the current over the surface. The form of this equation is

$$\int \frac{\text{(current)(weight function) d(surface)}}{\text{(Kernel with an auxiliary parameter, h)}} \quad (5)$$

= (a function of h representing the source multipoles).

Let us write out the latter function; it is just

$$I(h) = \text{(dipole moment)} + \text{(quadrupole)} + \text{(octopole)} \; h^2 + \ldots \quad (6)$$

In other words, the multipole moments of the magnetopause have been "repackaged" in a function I(h) of which they are the power-series coefficients.

If, as if so often the case, one simply assumes a dipole, then

$$I(h) = \text{constant} \quad (7)$$

Thus if one guesses a shape, it can be immediately tested to <u>all orders</u>. Indeed, an objective "goodness-of-fit" parameter can be defined as

$$\varepsilon \equiv \int \left(\frac{dI}{dh}\right)^2 dh, \quad (8)$$

since by definition $\varepsilon$ is zero for the correct shape and is greater than zero otherwise, becoming larger as I(h) deviates more from constancy. One simply minimizes $\varepsilon$ and one has the best possible fit given any finite-parameter shape. Figure 1 shows one such shape parameterization. The cusp height has been chosen to be unit height above the poles, the second term gives the correct shape at the cusp, and the third term is the first fourier expansion correction to describe the residual corrections to the shape.

If one wished to include a "ring current," the first nontrivial term beyond the dipole moment would be the octopole moment and therefore one would have

$$I(h) = A + Bh^2 \quad (9)$$

Then if we define

$$J(h) = I - \frac{1}{2} h^2 \frac{d^2 I}{dh^2} = A \quad (10)$$

we simply redefine $\varepsilon$ to be

$$\varepsilon \equiv \int \left(\frac{dJ}{dh}\right)^2 dh \quad (11)$$

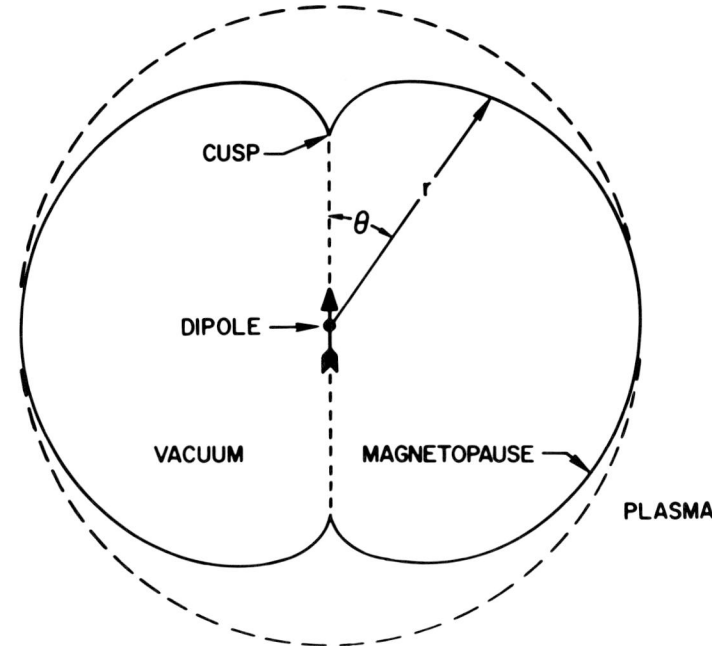

Fig. 1. Calculated shape for constant external pressure magnetopause using approximate shape $r = 1 + 0.69(\sin \theta) - 0.0.5 \sin \theta$.

and the same procedure follows as before.

We envision $\varepsilon$ as being calculated numerically, so including a complicated source really does not require much more effort than the simple dipole. In particular, consider the case where a finite magnetic flux escapes from the magnetic poles, as shown in Figure 2. Here one has

$$I(h) = A + Bh^{-3} \qquad (12)$$

The negative power of h may seem surprising, but the power series expansion of h fits smoothly as one goes from the multipole expansion of the external field to that of the internal field. Thus the series stops with the dipole moment (constant) because an $h^{-1}$ term would correspond to an external monopole (nonexistent). Next begin the internal multipole moments with $h^{-2}$ corresponding to an internal monopole (also nonexsitent) and then $h^{-3}$ as the internal "dipole," i.e., uniform field. The significance of such possible solutions should become clear in the next section.

One way to handle the cusp problem is to choose a shape parameterization that automatically gives the correct shape of the cusp, as was done for Figure 1. A constant pressure cusp opens so that

$$r = a + b \, (\sin \theta)^{2/3} \qquad (13)$$

in the immediate vicinity of $r = a$, the height of the cusp above the polar cusps. Other orthogonal function series expansions would have

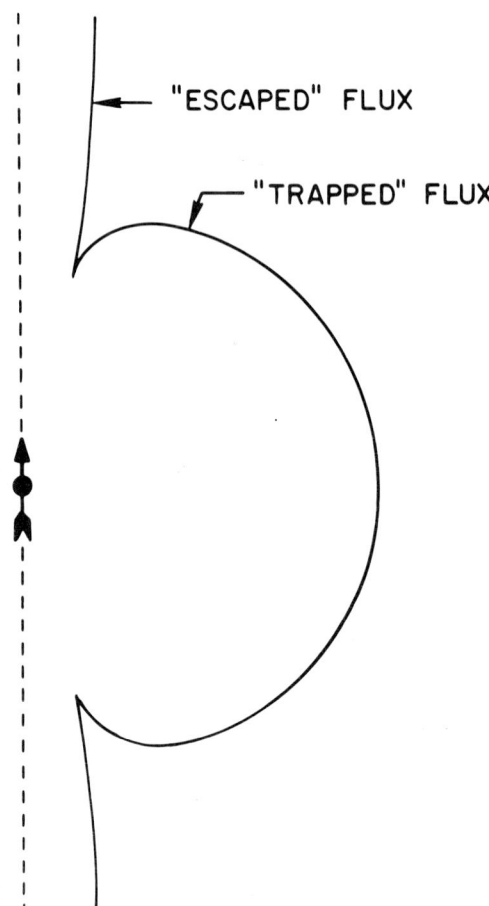

Fig. 2. Another possible magnetopause shape for the constant pressure case, if some of the field lines are allowed to "escape" to infinity along the polar axis. These flux tubes have the same topology as the geomagnetic tail, where they have been wrapped around the night-time magnetopause and pushed together

to produce such a cusp but convergence near $r = a$ is typically poor, (for a power series, say) which introduces unnecessary computational difficulties.

### General Shape (No Symmetry Conditions)

Often simplifications obtained by imposing symmetry conditions cannot be generalized to problems lacking that symmetry. Thus it is not obvious that the method in the previous section could work without the axisymmetry. However, we have shown (Michel, 1977b) that in fact the method does generalize. Now we obtain an integral equation of the form

$$\int \frac{\text{(vector current)(weighting function) d(surface)}}{\text{(Kernel with auxiliary variables, h and k)}} \quad (14)$$

=(vector function of h and k representing the source multipoles).

Naturally this latter equation is more complicated but basically h again represents the multipolarity of the source and k represents the direction. (We cannot repeat here the technical detail that one would need to actually do these calculations. We can only convey the general ideas behind them.) The terms in k (for given h) tell us how much the dipole, for example, is tilted with respect to a wind direction.

This is the equation one needs for a magnetosphere simulating that of the earth. In the geomagnetic case one can see more clearly the problems with a multipole expansion per se. The long geomagnetic tail would require higher and higher order moments to represent its extent, yet these multipoles drop off at high powers of radial distance and therefore their coefficients must be larger and larger, again leading to an asymptotic series if one tries to do better than model the near-earth almost-spherically-symmetric environment. Again the integral equation does not suffer from these limitations.

The method has been tested for $\cos^2 \psi$ flow past a perpendicular dipole and works in the sense that the appropriate version of $\varepsilon$ is minimized as one approaches a realistic magnetopause shape. We have not yet calculated a precision shape owing to lack of support for either the computational requirements (small, but not negligible) or the surface parametrization (the efficiency of such calculations is enormously enhanced by cleverness in choosing the most nearly appropriate shapes; brute-force approaches can compromise any advantages gained method). In this latter regard, the nonsymmetric case is complicated by the need to specify the direction of the current on the magnetopause surface. Since the integral equation is solved by what amounts to a numerical variational principle (minimizing $\varepsilon$), it seems likely that the correct current direction can be obtained automatically as part of minimizing $\varepsilon$. (Thus while we recognize the value of being clever in choosing our shape parameterization, as noted above, it is also useful to minimize the sensitivity to lack of cleverness.)

## Some Applications

1. Representation of internal magnetopause field by magnetic moments. The perturbation field at the earth, say, can easily be derived simultaneously with the shape. Thus a uniform field contribution at the earth (plus quadrupole etc.) from the magnetopause is an immediate practical output of such a calculation. The sensitivity of such terms to $\varepsilon$ moreover allows one to estimate the error from using an approximate shape.
2. First principles representation of the geomagnetic tail analogous to the escaped flux problem illustrated in Figure 2.
3. Inclusion of realistic effects such as the ring current. Note that even a partial ring current can be modeled. In the case of

Jupiter, such effects may be quite important.
4. Fast computations for use in attempts to simulate dynamic time-varying problems self-consistently.

Acknowledgments. The research was supported in part by the National Aeronautics and Space Administration under grant NGL44-006-012.

## References

Midgley, J. E., and L. Davis, Jr., Computation of the bounding surface of a dipole field in a plasma by a moment technique, J. Geophys Res. 67, 499-504, 1962.

Midgley, J. E., and L. Davis, Jr., Calculation by a moment technique field by the solar wind, J. Geophys. Res., 68, 5111-5123, 1963.

Michel, F. C., Accretion magnetospheres: general solutions, Astrophys. J., 213., 836-839, 1977.

Michel, F. C., Magnetopause shapes: general solutions, J. Geophys Res. 82, 5181-5186, 1977.

OVERVIEW - MODEL APPLICATIONS

H. B. Garrett,

Air Force Geophysic Laboratory, Hanscom Air Force Base, Massachusetts 01731

In the Modeling Applications Session, four invited papers were presented which covered various practical applications of magnetospheric models. They ranged from the long-standing problem of the effects of the radiation environment on spacecraft systems to the new problems associated with plasma interactions with high voltage devices in low earth orbit. The concluding paper, in contrast, treated the philosophical issue of what constitutes a model and how models are applied to practical solutions. A central theme of all these papers was the need for and importance of practical, quantitative magnetospheric models. Another important consideration to emerge was the need to provide proper interfacing between models and user needs. The Model Applications Session, the first of its kind at an AGU symposium, went far in outlining the importance of the consideration of possible user needs in designing quantitative models of the magnetosphere.

There has never been a session at an AGU symposium devoted to model applications. It is, therefore, important to establish the rationale for such a session before summarizing the Model Applications Section. Simply put, as quantitative models of the magnetosphere mature, it is encumbent upon us as researchers, particularly considering the amount of manpower and computer time devoted to model development, to plan for the use of our codes by others so that our efforts may be applicable to a much larger community. We must concern ourselves with potential users and their needs. Hence, the need for a Model Applications Session was foreseen and incorporated into this Conference.

Four invited papers were selected for the Model Applications Session. These papers were chosen to give a broad overview of the user community and, at the same time, provide some indepth consideration of a particularly critical long-standing problem - the effects of radiation on spacecraft systems - and of a very new problem - the interaction of high voltage structures with the magnetospheric plasma.

The paper by Snyder dealt with ionospheric and magnetospheric modeling for Air Force applications. After reviewing current ionospheric and magnetospheric research aimed at specific Air Force needs, such as predicting geomagnetic activity, Snyder made comments related to future work. In particular, he called attention to the obvious maturity that the field of quantitative magnetospheric modeling is slowing approaching: .."increased emphasis (is needed) on weaving theory with well-planned problem oriented experiments as opposed to survey data acquisition." He further underscores the importance to the applications community of realistic initialization and parameterization of inputs that can only come from a comprehensive physical understanding of how the magnetosphere operates. Finally, he discussed the importance of on-going research to possible practical applications 10 to 20 years from now - an issue that continues to plague research work that may have no apparent immediate practical application.

Janni and Radke, in their discussion of the effects of the radiation environment on spacecraft, draw attention to important, practical gaps that we have in our knowledge of probably the most basic magnetospheric problem - knowledge of the high energy particle environment. Somewhat surprisingly, following an initial overriding interest in the radiation environment by early space researchers, there has been a relative lack of interest in the high energy particle population above 2 MeV in the case of electrons and 30 MeV for protons. Janni and Radke present convincing evidence that the electrons and ions above these energies are precisely the particles which have the greatest potential threat. As this is an area in which there is a clearly defined user need and one that is also of interest to the modeling community, it should serve as a test case for the relationship between users and modelers in the future.

Konradi et al., discussed a new potential problem area for satellite designers. If, as current plans indicate, large voltages are built up on solar panels by photovoltaic cells connected in series, current leakage and arcing can be enhanced by interactive coupling with the ambient plasma. This is especially true at low earth orbit where it is planned to assemble the structure. Although the power loss through the plasma is relatively unimportant, the sheath and its focusing effect on the ions are potentially significant effects. The determination of such satellite effects is not only dependent upon the detailed physics of the plasma sheath surrounding the spacecraft, but also of the ambient plasma itself which determines the sheath - thus, the very real need for accurate quantitative magnetospheric models.

DeForest, although addressing the specific problems associated with modeling charge buildup on satellites, discussed the general philosophy of adapting models to a user's needs. As he indicated, the definition of modeling is somewhat confusing and must, itself, be adapted to the user. He suggests that a model is basically a means of linking cause and effect. As it is impossible to meet all users' needs, the model should contain as much relevant information as possible so that it can be readily modified for specific user needs. Finally, DeForest defines what he feels are four principal functions for quantitative models. These are design (for which historical data are needed), diagnostic (for the purpose of simulation and physical understanding), safe operations (for which are required simultaneous environment measurements, physical understanding, and real time response capabilities), and modeling of other heavenly bodies (or scientific study).

In this last decade during which there has been increasing pressure for space research to yield "practical" benefits, it has become incumbent on the individual researcher to tailor models to the needs of potential users. For example, the recent growth of interest in low energy plasma models has come about in part through the apparent effects of the plasma on spacecraft charging. With the advent of a myriad of complex commercial and government communications and meteorological satellites, their susceptibility to spacecraft charging has become a subject of great interest, enlarging the potential market for low energy models. Intimate knowledge, in the form of quantitative magnetospheric models, of the entire near-earth plasma environment will be needed to properly design the solar power stations or space-based radars planned for the 1990's and beyond as they will, in all likelihood, be built in low-earth orbit and moved slowly to geosynchronous orbit. In turn, the potential impact of such extensive operations on the space environment can only be discussed in the light of accurate models. Finally, the low-energy plasma environment is a rich plasma and astrophysics laboratory allowing in situ experiments.

To date much of the effort in modeling has and is going into the static and time-dependent models, whereas users often require no more than adequate statistical or analytic models. As can be traced in the proceedings of this conference, however, there is a growing concern with this factor. Several recent models have been specifically developed in response to user needs. Considering the potential benefits of quantitative magnetospheric models, it would be of great value to each investigator to weigh the needs of possible users in developing his models (especially in the areas of data input and output) as has been done in the recent models of Vette, Olson, and their co-workers.

Generally it is concluded that model makers must address the following issues:
1. A need for consideration of availability of input parameters.
2. Consideration of ease of use.
3. Consideration of output parameterization.
4. The need to match the sophistication of the model to level of sophisitcation required by the user.

To conclude, if work in quantitative modeling is to be of lasting value by having practical application, then these issues must be addressed. Hopefully, this, the first session on model applications, will help us progress in this direction.

ENVIRONMENTAL MODEL NEEDS FOR SPACECRAFT INTERACTIONS

S. E. DeForest

Department of Physics,
University of California, San Diego
La Jolla, California 92093

**Abstract.** An environmental model suitable for supporting studies in spacecraft interactions might differ greatly from a model derived from the same sources but used primarily for study of magnetospheric dynamics. A useful interaction model must be as simple as possible while retaining all of the information relevant to the situation. For most calculations, the minimum needed is the number flux of electrons and ions as a function of energy (perhaps in a compressed form such as a best fit two-maxwellian plasma). More sophisticated calculations require the particle pitch angle dependence and energy dependence both as a function of time. These extra dimensions complicate the model considerably, but they are required to make accurate predictions. Some calculations for some orbits also require knowledge of total magnetic field, the ion composition, and the plasma wave field. Present efforts to model the environment at geosynchronous orbit are concentrating on the reduction of previously collected plasma data. With the launch of SCATHA in January, 1979, a new body of data tailored to the needs of modeling spacecraft interactions (including active control of spacecraft potentials) will become available. The modeling techniques now being developed and used will aid in disseminating these new data in a timely fashion. A final consideration in constructing an interaction model is that almost all users will wish to modify it slightly to fit their special needs or to incorporate new information. Therefore, the model should not try to satisfy everyone, but rather it should contain the useful information in a way that can be easily modified.

## Introduction

Other papers in this meeting have commented on the nature of environmental models and their uses. In this paper we will consider the specific problem of preparing a model which is appropriate for the study of interactions of spacecraft with their environments. By the nature of the problem, this means the bulk of the discussion will concern the charged particle populations.

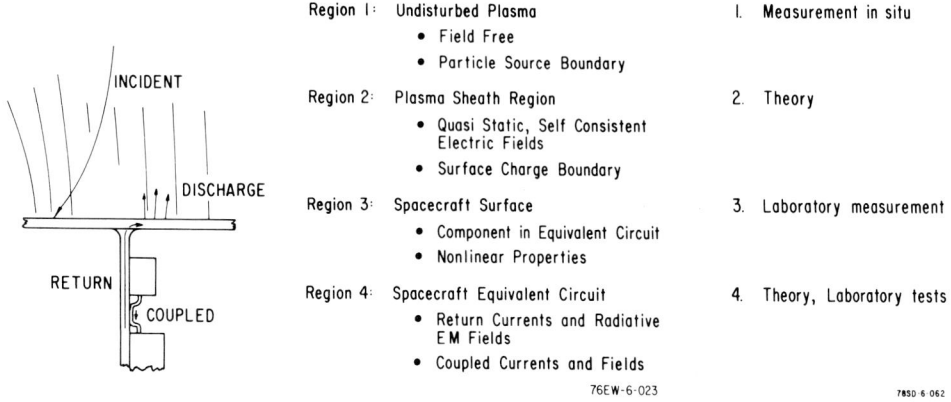

Fig. 1. Schematic representation of regions to be considered in studying spacecraft interactions with natural plasma (after Whipple, 1977).

## Nature of Model

In considering a similar problem, Whipple (1977) emphasized the point that the function of a model is to relate a given cause to an effect. In this case, one supposes based on past observations that the environment about a spacecraft can produce electrostatic charging which can adversely affect operations. With an appropriate model, one hopes to be able to predict the nature of these interactions before they happen. In that

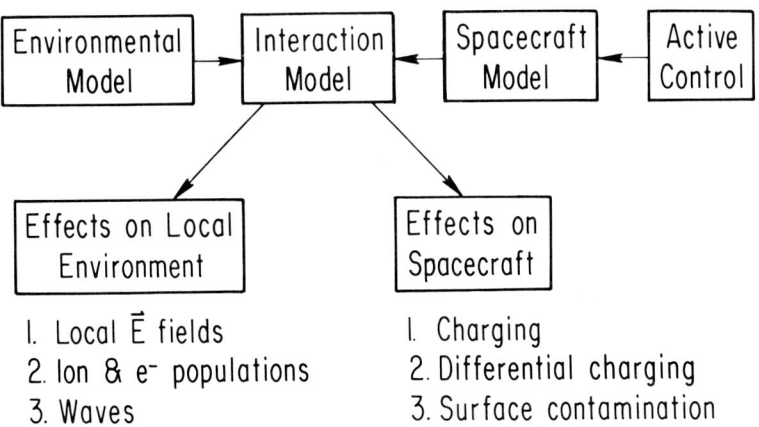

Fig. 2. Relation between the available inputs to an interaction model and the expected outputs.

Table 1. List of some relevant plasma parameters to demonstrate the difficulty of presenting an accurate, but concise representation of natural plasma.

## SUBSET OF RELEVANT PLASMA PARAMETERS

Particles

| Density, Number Flux, Energy Flux, Isotropy | Electrons, Protons, Ions | 12 |

Fields

| Magnetic, Electric | 3 components each | 6 |

Position

| Longitude, Orientation (2 angles) | | 3 |

| Total | | 21 |

76SD-6-033A

way, design changes can be implemented or operational configuration changed before damage is done. Since the input to such a model must be derived from the plasma environment itself, one might ask how to summarize this complex plasma. Whipple lists three types of models. The statistical is the simplest and is based directly on sensor input. The parametric model is characterised by a partial understanding of underlying physics. The complete physical model is a goal which is sought by scientists, but might not be necessary for satisfactory treatment of the interaction problem.

In Figure 1, taken from Whipple, the plasma and spacecraft surface have been separated into four regions of interest. We have added to Whipple's original figure by including the prime source of descriptive data. In the rest of this paper, we will be concerned primarily with region 1. However, it is important to realize that in the preparation of a useful interaction model the ultimate user must be considered. If the environmental model is not in a form suitable to aid the flow of calculations from region 1 to 4, then it is not going to be useful.

Table 2. Partial list of types of models that have been used or proposed in approximate order of increasing sophistication and accuracy.

## TYPES OF MODELS

1. Summary of Integrals

2. Samples of Actual Data

3. Parameter Reduction by Correlation

4. Parameter Reduction by Fitting
    (Bi-Maxwellian fits)

    ⋮

n. Physical Understanding

Much effort has already gone into modeling the sheath about the spacecraft (region 2) (see other papers in the conference). In particular, the NASCAP program (Stevens, 1978) has the ability to model the complete three-dimensional field around arbitrary spacecraft. The region 4 must be determined for each spacecraft configuration. Frequently it must be verified by measurements on the spacecraft itself or on a suitable mockup.

The nature of a useful environmental model will be determined not only by the character of the modeling of the other regions, but also by the intended use. Typical uses are to:

1. Design spacecraft (need environmental history).
2. Diagnose spacecraft anomalies (need simultaneous environmental measurements and some physical understanding).
3. Operate spacecraft safely (need same conditions as above plus real time response).
4. Study cosmic plasma (need some physical understanding).

Obviously the type of model used to monitor spacecraft operation is more involved and expensive to operate than a model to design spacecraft. At the present time, we can use environmental models with reasonable success for parts of goals 1, 3 and 4.

The role of environmental models is shown in a different way in Figure 2. Here we have added the possibility of active control and explicitly shown that

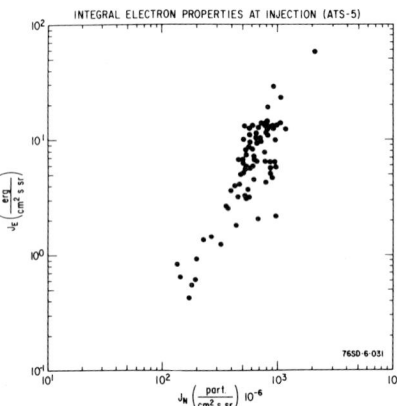

Fig. 3. Example of parameter reduction by empirical correlation without the necessity of detailed physical understanding: if $J_E$ can be expressed as a function $J_N$, then the plasma might have a uniform temperature for a given particle flux.

two classes of possible predictions might be of use. In order to simplify the link between the environment and interaction, one wishes to specify the minimum possible set of parameters consistent with accurate predictions. This has led some investigators to use only two numbers; the maximum flux of particles, and the equivalent temperature. In this approach time variations are either neglected or treated by having the plasma "turn on" at a specified time and watching the spacecraft respond.

To see how much of a simplification this type of assumption is, consider Table 1 which lists a subset of plasma parameters which are necessary to calculate the charge state of a geosynchronous spacecraft even for the static case, use of a single temperature and flux for the plasma can result in large error in the predicted spacecraft charging. Recent results using NASCAP (Katz, Private Communication) have shown a factor of four difference between the predicted potentials using typical one and two Maxwellian fits.

## Current Status

In Table 2 we show four types of models that have been published. The bottom entry is an indication that continued study in this area will lead to better results. The summary of integrals is convenient because it is compact, but accuracy and completeness is sacrificed. Samples of actual data when used with an appropriate computer program, can in principle produce the most realistic results. However, a considerable effort is needed to set up such a program. In practice this has not been popular. Parameter reduction by correlation is

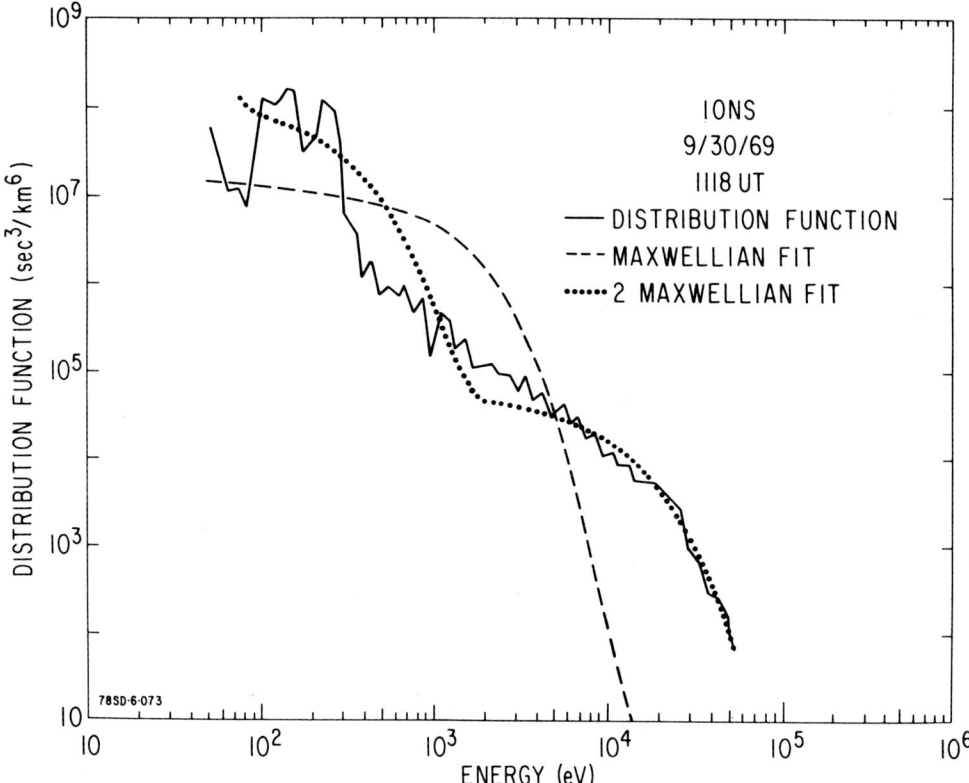

Fig. 4. Example of the utility of a two Maxwellian fit to a measured plasma distribution function (courtesy of H. Garrett).

illustrated by Figure 3 which shows a scatter plot of parameters measured by ATS-5 when the spacecraft was situated at the inferred point of plasma injection. Without understanding why these two parameters are correlated, we can specify only one and use an empirical fit for the other with about a factor of two accuracy.

The problem of parameter reduction by fitting to standard forms is how to trade between complexity and accuracy. Since one would like to have some physical reasoning for choosing a particular type of fit, a Maxwellian choice is common. However, the natural plasma is not Maxwellian. Therefore, investigators at UCSD in conjunction with AFGL have distributed information derived from ATS-6 which has been fit to a bi-Maxwellian. This choice has no real physical motivation other than to empirically fit the data while also matching results which have been calculated using the complete spectra. A sample of this approach is shown in Figure 4. Note that this fit is much better for the high energy particles which ultimately determine the maximum potential.

We do not predict how many entries will have to be made in Table 2 before a completely adequate particle model is developed. Referring to Figure 2, we might even predict that several standard models might develop to statisfy certain special needs.

## Conclusion

Because of the difficult and highly variable nature of the plasma parameters, satisfactory models are just now being developed even though very sophisticated magnetic field models and high energy particle models have been known for some time.

A satisfactory compromise between complexity of models and reliability of results has not yet been reached. However, we do know that simple Maxwellian theory will give predictions sufficiently in error to be of little use. A bi-Maxwellian does much better, and makes a better interim model.

In the future, models with time dependence and pitch angle variations will probably be developed. Measurements of ion composition are just now becoming available for modeling. Their importance on the interaction problem has not yet been determined.

**Acknowledgments.** This work was supported by Air Force Contract SCATHA F04701-77-C-0062.

## References

Stevens, N. John, J. L. Roche, and M. J. Brandt, NASA Charging Analyzer Program - a computer tool that can evaluate electrostatic contamination, NASA Technical Memorandum, NASA TM-73889, 1978.

Whipple, E. C., Modeling of spacecraft charging, Proceedings of the Spacecraft Charging Technology Conference, NASA TMX-73537, 1977.

IONOSPHERIC AND MAGNETOSPHERIC MODELING FOR AIR FORCE APPLICATIONS

A.L. Snyder, Jr.

Ionospheric Dynamics Branch, Air Force Geophysics Laboratory
Hanscom AFB, MA 01731

Abstract. General user requirements are reviewed relative to five present interest areas:
  Ionospheric Synoptic Pattern Modeling
  Electron Density Profile Modeling
  Ionospheric Scintillation Modeling
  Monitoring Techniques
  Magnetospheric Substorm Prediction
Trends are reviewed toward future emphasis on quantitative modeling. The USAF Scientific Advisory Board recommends future emphasis on:
  Remote Sensing
  Numerical Models of the Solar Terrestrial Environment
  Solar Disturbance Modeling
Comments are presented regarding study approaches and considerations that are important to the community of science users.

## Introduction

Since the May 1975 Conference on Quantitative Magnetospheric Models, my organizational affiliation has changed from the community of science users to the community of scientists. However, my association with the users of science continues to influence my concern for the pragmatic application of space science. With resource needs often exceeding monetary restrictions, I recall a 1975 statement by V.E. Hildebrand (Olson, 1975):
  "...the scientific community (should) keep in mind the needs
  of the practical users because they can exert considerable
  influence on the determination of the... funding levels in
  several government agencies."
As I was preparing my remarks, I found that my concerns are divided between the science needed for immediate applications and science goals the achievement of which will advance understanding and provide the basis for future, not yet defined applications. My remarks will attempt to address both issues.

While the "space age" began just over 20 years ago with the launch of Soviet and American satellites, literature documents the early influence of space phenomena on defense. Svyatskiy (1934) dramatically recounts Yaroslav's 12th Century battle between the Russians and the Polovets. At a critical point in the conflict, the Polovets were frightened by the aurora and surrendered to the Russians. The latter concluded that they could actually frighten the Polovets with the aurora and influence the battle.

Over eight centuries later, we are likewise concerned with the environmental impact on defense systems and operations. This concern extends from the depths of the terrestrial sciences to the horizons of astrophysics. My comments are restricted to the ionospheric and magnetospheric regimes and will purposely avoid mention of those areas, of obvious Air Force concern, treated by other authors. I will briefly review five present interest areas:

1. Ionospheric Synoptic Pattern Modeling
2. Electron Density Profile Modeling
3. Ionospheric Scintillation Modeling
4. Monitoring Techniques
5. Magnetospheric Substorm Prediction

Problems of Present Interest

Ionospheric Synoptic Pattern Modeling

The high frequency (HF) radio wave user community requires advice for frequency management on time scales from minutes to months (Thompson, 1978). The ITS-78 (Barghausen, et al., 1968) provides the present operational HF propagation model. However this model does not represent the instantaneous polar ionosphere nor the ionospheric variability associated with world-wide disturbances. The Air Weather Service, of the United States Air Force, developed spectral analysis techniques to incorporate real-time data and realistic representations of the polar ionospheric patterns (Flattery and Ramsay, 1975; Flattery and Davenport, 1977). The latter are important for assessing HF system performance through the main F-layer trough and for determining the effects of horizontal electron density gradients on HF propagation paths (Snyder and Buchau, 1977). Pike (1976) and Halcrow and Nisbet (1977) developed models of the horizontal gradients through the F-layer trough. Gassmann (1973) suggested ordering real-time polar ionospheric patterns relative to auroral boundaries routinely observed by the USAF Defense Meteorological Satellite Program (DMSP). These operational techniques are limited to phenomenological models tied to observational anchor points and mathematically pieced together through spectral analysis. Other example efforts that are recognized as having potential for operational use include the International Reference Ionosphere (Rawer, 1978) and Watkins (1978). The latter is a recent numerical computer model of the polar F-region that represents many features of the quiet ionosphere (cf. Wagner et al, 1973; Pike, 1970). Watkins (1978) emphasizes the importance of the convective electric field for quantitative polar ionospheric models. The operational impact of this point will be discussed in a later section.

Electron Density Profile Modeling

Operational radio wave systems that need ray path and propagation mode diagnostics require vertical electron density profiles. These profiles may then be used in ray tracing calculations. The present operational techniques center on a modified version of the Damon and Hartranft (1970) profile model. This simple model uses three Chapman layers for the E, F1, and F2 regions and is combined with a variable scale height topside that is fit to match total electron content measurements. Recent AWS work (Flattery and Davenport, 1977) represents the vertical profiles through a combination of four empirical orthogonal functions derived from Millstone Hill data. Data are needed from other latitude regimes (e.g. Alaska and Puerto Rico) before one can confidently model the auroral and low latitude electron density profiles with the empirical orthogonal functions. A basic approach has been taken by Jasperse and his colleagues to model the E and lower F-layer profiles (Jasperse, 1976, 1977; Jasperse and Smith, 1978). These techniques are presently restricted to the sunlit ionosphere; however work is underway to calculate E-layer profiles in the continuous aurora from precipitating electron data (Jasperse and Strickland, 1978). These quantitative techniques provide a partial alternative to the need for additional profile data from other latitude regimes. Work is needed to extend quantitative profiling techniques into the topside ionosphere where the local approximation breaks down.

## Ionospheric Scintillation Modeling

Ionospheric irregularities produce amplitude and phase scintillations that affect transionospheric (VHF-UHF) propagation systems. Amplitude scintillations may reduce the signal level below the fade margin for satellite communications; phase scintillations may limit the coherent integration times that are needed for deep space surveillance. Transionospheric propagation system users need advice on the occurrence of significant scintillations. Aarons and his colleagues (Aarons, 1978; Aarons, et al, 1977; Mullen, et al, 1977; Basu et al, 1976; Basu and Kelley, 1978) developed primarily statistical models for auroral and equatorial amplitude scintillations. These pragmatic models use readily available input parameters and provide average scintillation conditions. These results are derived from extensive monitoring programs. Current studies are aimed at increased understanding of longitudinal variations in equatorial amplitude scintillations and in extending knowledge of the polar cap scintillations. Unified knowledge of the disturbed equatorial ionosphere and its impact on transionospheric propagation systems advanced dramatically with the development of an airborne all-sky imaging photometer (cf. Weber et al, 1978). This work combined with recent quantitative studies (Scannapieco and Ossakow, 1976) of the Rayleigh-Taylor instability show promise for a comprehensive understanding of the equatorial ionosphere. More observational and theoretical work is needed to reach a similar understanding of high latitude irregularities.

## Monitoring Techniques

Obviously data are needed for magnetospheric and ionospheric specifications as well as for initialization, adjustment, and scaling of models. In the past four years (cf. Snyder, 1974;

Thompson, 1978) the emphasis has been toward the development of
DMSP satellite-borne sensors for operational data acquisition
(Smiddy, et al, 1978; Rush, et al, 1978; Huber, et al., 1977).
Data from orbiting satellites circumvent certain logistical and
geo-political problems, but introduce problems associated with
incomplete local-time coverage and a revisit time of 100 minutes
to a similar latitude/local-time band. For these reasons, ground-
based data sources will continue to serve a useful purpose for
operational applications; however, spiraling costs may limit
labor intensive data gathering. Efforts are underway (Buchau,
et al, 1978; Hall, 1978) to develop automated techniques for
remote ground-station data acquisition and relay to a central
analysis site. Such techniques are feasible because of the
rapid development of the microprocessor technology field. It is
a continuing problem to acquire a near real-time, comprehensive
data set, for disturbance monitoring, environmental analyses, and
model uses, that meets logistical and geopolitical constraints. This
problem will continue to evolve as science develops and pragmatic
concerns change. For these reasons, it is important that close
liaison exist between the science and science-user communities.
It should also be recognized that operational networks often
yield unique data that are essential to increased understanding -
witness the DMSP auroral photography and the AWS solar optical and
radio observing networks.

Magnetospheric Substorm Prediction

The substorm is likely the magnetosphere's primary mode of energy
dissipation. The energy release has important effects on Air Force
systems that operate in or through the ionosphere and plasmasphere.
These include radiation and spacecraft charging effects, discussed
by other speakers, increased high latitude ionospheric irregularities, changes in the ionospheric synoptic patterns, and auroral
absorption of HF radio waves. Monitoring techniques exist that
detect the majority of the substorms within minutes to an hour of
onset. Scientific community interest is high in defining what
triggers a substorm; however the lack of comprehensive unambiguous
signatures of a substorm have led to generally inconclusive results.
Perreault and Akasofu (1978) have recently shown a good correlation
between an interplanetary quantity $\varepsilon(t) = V|B|^2 \sin^4(\theta/2) \ell o^2$ and the
occurrence of individual substorms.

where: $V$ = Solar Wind Speed
$|B|$ = Magnitude of the Interplanetary Magnetic Field (IMF)
$\theta$ = Polar Angle of the IMF
$\ell o$ = 7 Earth Radii

The quantity $\varepsilon(t)$ is an approximation of that part of the interplanetary energy or Poynting flux which enters the magnetosphere
as a function of time (cf. Perreault and Akasofu, 1978, pp. 548-549).
If this relationship proves substantial, a satellite such as ISEE-C
could provide sufficient lead time (30 to 60 minutes) for meaning-

ful operational forecasts of magnetospheric substorms. The forecast would include timing and magnitude of the impending substorms. While this interplanetary parameter provides promise, other techniques should be pursued that may lead to reliable substorm predictions.

## The Future

When one considers application of science, for example five years downstream, the future is now for the scientific community. The future may be now for periods of 10 to 20 years downstream necessary for the solutions of complex problems. Recognizing this, I welcome the opportunity to comment on the present direction of space science as it may impact future applications. The USAF Scientific Advisory Board Executive Committee on Basic Research (1977) stated that:

"The ultimate goal of such research programs in space science is a sufficiently quantitative understanding of the space environment to be able to predict, in some detail, its properties and the variations of these properties in space and time with high resolution."

The Space Science Board of the National Research Council (1978) also concluded that theory must play a central role in the development of the field.

Of special interest to this conference is the emphasis on quantitative understanding. The conferees, and especially Dr. W.P. Olson, are to be commended for their continuing unselfish interest in advancing quantitative modeling throughout the space science community.

It is obvious that the applications community needs, objective, quantitative techniques that produce numerical results. Without the quantitative models, the operational space scientist is left to seat-of-pants techniques that do not match the 20th Century hardware technology.

The obvious question is where can the modeling efforts be most beneficial for the future? The Scientific Advisory Board (1977) concludes:

a. Remote Sensing.
b. Numerical models of the interplanetary-magnetosphere-ionosphere-atmosphere system.
c. Solar disturbance modeling.

Electromagnetic remote sensing from X-rays to beyond the microwave region coupled with parameterized theories may enable one to map the ionosphere at a number of altitudes. Recent work toward solution of the electromagnetic inverse source problem (cf. Bleistein and Cohen, 1977) shows potential for remote passive monitoring. Such techniques could possibly replace the point (ground-based and geostationary satellite-borne) or line (orbiting satellite-borne) sensor systems of today by an areal, high time resolution system.

Numerical models of the solar-terrestrial environment are needed to accurately portray the mean and extreme conditions that systems will encounter as well as to predict the space and time variations.

It appears that quantitative understanding of the solar flare and the auroral substorm are integral to many transient phenomena that affect Air Force space systems.

## Comments Regarding Study Approaches

It is apparent that the third decade of space science must see an increased emphasis on combining theory and data analysis. The NRC report (1978) calls for closure-the activity of seeking agreement between theory and experiment. We need increased emphasis on weaving theory with well-planned problem oriented experiments as opposed to survey data acquisition.

Of importance to the applications community is the need for realistic model initialization and parameterization. If for example, quantitative polar ionosphere models require electric field measurements, these data must be available in real-time and be reliable. If the model complexity requires extensive computational time, the models must be parameterized to ease the calculations routinely required. Such questions as model initialization and parameterization require not only the initial model solution but also deep insight into the problem and creativity in experimental and computational techniques.

As space science matures, the future is bright for the quantitative physicist. However, I recall one of Jacob Bronowski's Silliman Foundation Lectures on Natural Philosophy (1978). Bronowski reminds us "...that the world is totally connected; and that when we practice science, we decode a part of nature that is not complete. We are limited by our own finiteness." The ultimate goal of science is truth and consistency which is the heart of good quantitative models.

## Acknowledgements

I extend my thanks, appreciation and gratitude especially to several AFGL Space Physics Division colleagues: H.B. Garrett, J. Buchau, J. Aarons, and C.P. Pike who skillfully and enthusiastically provided advice and encouragement in the preparation of this paper. I also acknowledge my continuing ties to the USAF AWS and to their challenging work of merging science and reality with pragmatism to provide real-time advice to DoD space systems.

## References

Aarons, J., Ionospheric Scintillations, A Review, URSI General Assembly, Helsinki, Finland, 1978.

Aarons, J., J. Mullen, H. Whitney, E. Martin, K. Bhavnani, and L. Whelan, A High Latitude Empirical Model of Scintillation Excursions: Phase I, AFGL-TR-76-0210, 1976.

Barghausen, A.F., J.W. Finney, L.L. Proctor, L.D. Schultz, Predicting Long-Term Operational Parameters of High Frequency Sky-Wave Telecommunication Systems, Institute for Telecommunication Sciences, ESSA TR ERL 110-ITS-78, 1969.

Basu, S., S. Basu, and B.K. Khan, Model of Equatorial Scintillations from In Situ Measurements, AFGL-TR-76-0080, 1976.

Basu, S. and M.C. Kelley, A Review of Recent Studies of Equatorial F-region Irregularities and Their Impact on Scintillation Modelling, 1978 Symposium on the Effect of the Ionosphere on Space and Terrestrial Systems, Arlington, VA, 1978.

Bleistein, N. and J.K. Cohen, Non-Uniqueness to the Inverse Source Problem in Acoustics and Electromagnetics, J. of Mathematical Physics, 18, 194, 1977.

Bronowski, J., The Origins of Knowledge and Imagination, Yale University Press, New Haven, 1978.

Buchau, J., W.N. Hall, B.W. Reinisch, and S. Smith, Remote Ionospheric Monitoring, 1978 Symposium on the Effect of the Ionosphere on Space and Terrestrial Systems, Arlington, VA, 1978.

Damon, T.D. and F.R. Hartranft, Ionospheric Electron Density Profile Model, Aerospace Environmental Support Center TM-70-3, 1970.

Flattery, T.W. and A.C. Ramsay, Derivation of Total Electron Content for Real-Time Global Applications, in Effects of the Ionosphere on Space Systems and Communication, Edited by J.M. Goodman, 1975.

Flattery, T.W. and G.R. Davenport, Four Dimensional Ionosphere Model, URSI Meeting, Stanford, CA, 1977.

Gassmann, G.J., Analog Model 1972 of the Arctic Ionosphere, AFCRL-TR-73-0151, 1973.

Halcrow, B.W. and J.S. Nisbet, A Model of F2 Peak Electron Densities in the Main Trough of the Ionosphere, Radio Science, 12, 815, 1977.

Hall, W.N., Private Communication, AFGL/PHI, Hanscom AFB, MA 01731, 1978.

Huber, A., J. Pantazis, A.L. Besse, and P.L. Rothwell, Calibration of the SSJ/3 Sensor on the DMSP Satellites, AFGL-TR-77-0202, 1977.

Jasperse, J.R., Boltzmann-Fokker-Planck Model for the Electron Distribution Function in the Earth's Ionosphere, Planet. Space Sci., 24, 33, 1976.

Jasperse, J.R., Electron Distribution Function and Ion Concentrations in the Earth's Lower Ionosphere from Boltzmann-Fokker-Planck Theory, Planet. Space Sci., 25, 743, 1977.

Jasperse, J.R. and E.R. Smith, The Photoelectron Flux in the Earth's Ionosphere at Energies in the Vicinity of Photoionization Peaks, Geophys. Res. Letters, 5, 843, 1978.

Jasperse, J.R. and Strickland, D.J., Private Communication, AFGL/PHI, Hanscom AFB, MA 01731, 1978.

Jones, W.B. and F.G. Stewart, A Numerical Method for Global Mapping of Plasma Frequency, Radio Science, 5, 773, 1970.

Mullen, J.P., H.E. Whitney, S. Basu, A. Bushby, J. Lanat, and J. Pantoja, Statistics of VHF and L-Band Scintillation at Huancayo, Peru, J. Atmos. Terr. Phys., 39, 1243, 1977.

Olson, W.P., Summary of the La Jolla Conference on Quantitative Magnetospheric Models, EOS, 56, 607, 1975.

Perreault, P. and S.-I. Akasofu, A Study of Geomagnetic Storms, Geophys. J. Roy. Astron. Soc., 54, 547, 1978.

Pike, C.P., Universal Time Control of the South Polar F Layer During the IGY, JGR, 75, 4871, 1970.

Pike, C.P., An Analytical Model of the Main F-layer Trough, AFGL-TR-76-0098, 1976.

Rawer, K., International Reference Ionosphere (in preparation) 1978.

Rush, C.M., A.L. Snyder, E. Ziemba, V. Patterson, T. Tascione, and D. Nelson, Global Behavior of HF Radio Noise in the Topside Ionosphere, (submitted) Radio Science, 1978.

Scannapieco, A.J. And S.L. Ossakow, Nonlinear Equatorial Spread F, Geophys. Res. Letters, 3, 451, 1976.

Smiddy, M., R.C. Sagalyn, and P.J.L. Wildman, The Topside Ionosphere Plasma Monitor (SSIE) on the DMSP Block 5D F2 Satellite, AFGL Memo, 1978.

Snyder, A.L., Real-Time Magnetospheric and Ionospheric Monitoring, 6th Conference on Aerospace and Aeronautical Meteorology, El Paso, TX, 1974.

Snyder, A.L. and J. Buchau, Environmental Studies Relating to Ionospheric Effects on Department of Defense Systems, AIAA 15th Meeting, Los Angeles, CA, 1977.

Space Science Board, National Research Council, Space Plasma Physics, The Study of Solar-System Plasmas, National Academy of Sciences, Washington, D.C., 1978

Svyatskiy, D.O., Aurora Borealis in Russian Literature and Science from the X to the XVIII Century, Arkhiv Istorii Nanki I Tephniki, Leningrad, Seriya 1, Issue 4, 47, 1934.

Thompson, R.L., User Requirements of Aerospace Propagation-Environment Modeling and Forecasting, AGARD Symposium, Ottawa, Canada, 1978.

USAF Scientific Advisory Board, Special Report of the USAF Scientific Advisory Board, Executive Committee on Basic Research, Hq USAF (NB), Washington, D.C., 20330, 1977.

Wagner, R.A., A.L. Snyder, and S.-I. Akasofu, The Structure of the Polar Ionosphere During Exceptionally Quiet Periods, Planet. Space Sci., 21, 1911, 1973.

Watkins, B.J., A Numerical Computer Investigation on the Polar F-region Ionosphere, Planet. Space Sci., 26, 559, 1978.

Weber, E.J., J. Buchau, J.G. Moore and R.H. Eather, Airborne All-Sky Imaging of Equatorial Airglow, AFGL Technical Report (in press), 1978.

CURRENT LEAKAGE FOR LOW ALTITIDE SATELLITES: MODELING APPLICATIONS

A. Konradi, J.E. McCoy and O. K. Garriott

NASA-Johnson Space Center
Houston, Texas 77058

Abstract. To simulate the behavior of a high voltage solar cell array in the ionospheric plasma environment we used the large (90 ft x 55 ft diameter) vacuum chamber to measure the high-voltage plasma interactions of a 3 ft x 30 ft conductive panel. The chamber was filled with Nitrogen and Argon plasma at electron densities of up to $10^6 cm^{-3}$. Since for a solar array parasitic current losses depend almost exclusively on the magnitude of the ion current all measurements reported here were done with a negative bias. When biased at voltages up to -2500V the panel developed a negative sheath clearly visible on a low-light television monitor. Measurements of current flow to the plasma were made in three configurations: (a) with one end of the panel grounded, (b) with the whole panel floating while a high bias was applied between the ends of the panel, and (c) with the whole panel at high negative voltage with respect to the chamber walls. The results indicate that a simple model with a constant panel conductivity and plasma resistance can adequately describe the voltage distribution along the panel and the plasma current flow. As expected, when a high potential difference is applied to the panel ends more than 95% of the panel floats negative with respect to the plasma. For voltages in excess of -500V the current-voltage characteristic becomes linear. Extrapolations to tens of kilovolts based on measurements up to -2000V indicate that in the F2 region current losses from a high voltage solar array could become a problem if not taken into account during the design phase. No definite conclusions, however, can be drawn at this point until actual measurements in the tens of kilovolt range are made.

## Introduction

In recent years the study of problems due to the interaction between the plasma environment and spacecraft have assumed great importance. These problems fall into two basic categories: One in which a spacecraft is immersed in a very hot plasma ($T_e \sim$ 10-20 keV) and builds up a charge as a result of the initial excess electron current impinging on its surface, the other in which the spacecraft is immersed in a relatively cool plasma ($T_e \lesssim 1$ eV) while a high voltage is applied to the whole or part of the spacecraft by such means as firing of high voltage electron or ion guns or series connection of photovoltaic cells to build up high voltage power sources. The first case is well known and goes under the name of spacecraft charging. This charging occurs frequently in

618   CURRENT LEAKAGE FOR LOW ALTITUDE SATELLITES

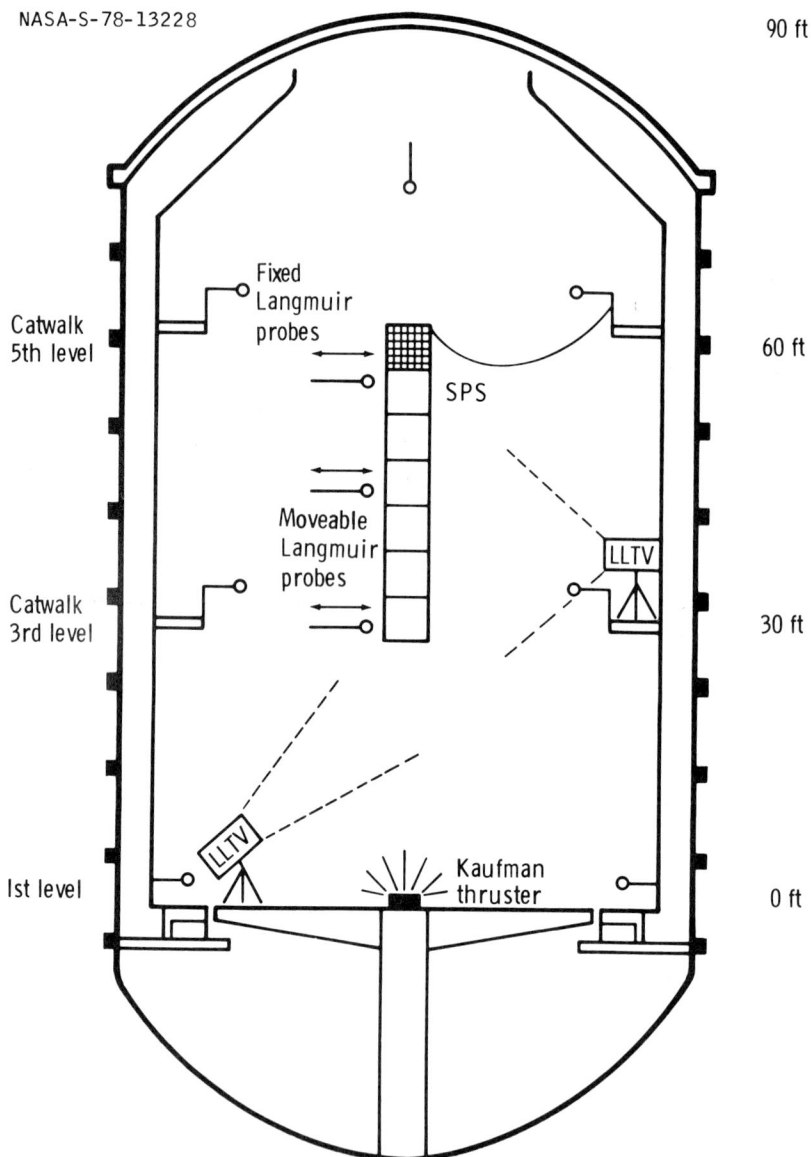

Fig. 1. Set-up of experimental equipment used for current leakage studies in Chamber "A".

the geosynchronous orbit when spacecraft are enveloped by hot magnetospheric substorm plasma. As long as there are sufficient photoelectrons emitted, the excess charge can be carried off. If, however, the spacecraft is in the Earth's shadow or rotating slowly so that one of its sides does not receive sunlight for sufficiently long periods, a very high voltage may be built up, which under adverse conditions, may lead to arcing and can cause anomalies in the operation of the spacecraft (Pike and Lovell, 1977 and references

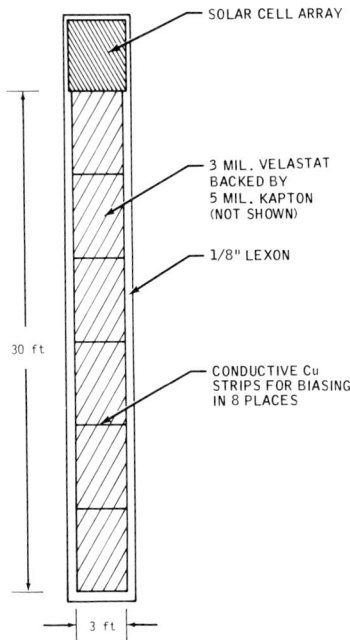

Fig. 2. Panel of conductive plastic used to measure plasma currents and observe sheath formation and ion focusing.

therein). The other case is less well known because in the past, spacecraft power systems have been deliberately designed for low voltages. With one exception electron accelerators to date have been flown only on rockets and no problems due to the charging of the rocket bodies have been encountered. Provisions were made to collect sufficient electrons, or neutralizing plasma generators were activated, or the electron beam due to rather high ambient neutral densities caused enough ions to be produced in the vicinity of the rocket to assure sufficient return current collection.

However, in the future, when large structures such as a Solar Power Station, will be built in space the need to go to high voltage systems may be dictated by economic considerations just as on Earth. During accelerator operation vehicle charging is at least a nuisance and may even prohibit beam emission and must be avoided through plasma current collection. On the other hand, for high voltage power systems any plasma current short circuiting the system lead to parasitic power loss and therefore need to be eliminated. The present work was motivated by the desire to understand the processes responsible for, and the extent of the current leakage when large, high voltage structures are immersed in plasma of various densities. The results should be considered as preliminary since the measurements were not exhaustive.

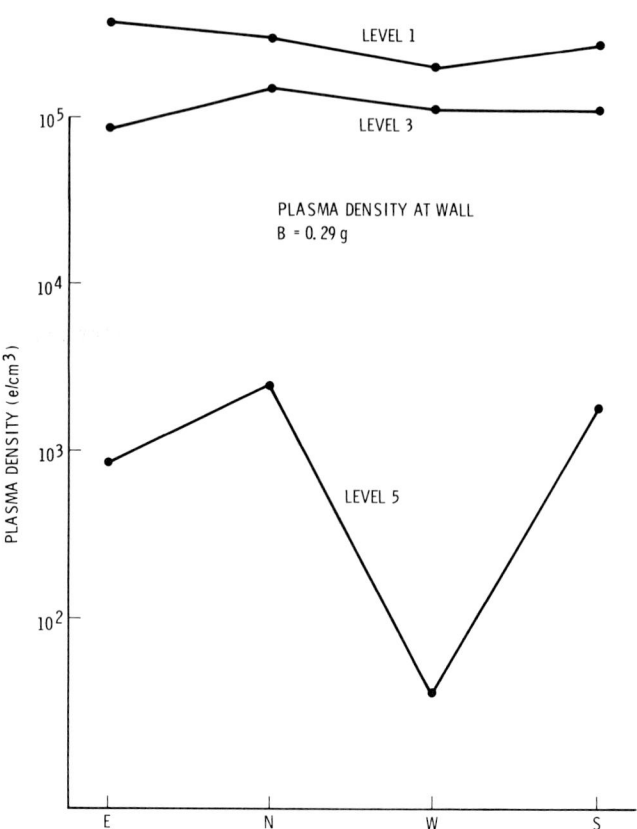

Fig. 3. Plasma densities measured at 1st level (0 ft), 3rd level (30 ft), and 5th level (60 ft). The ion generator was positioned at the center bottom of the chamber.

Experimental Instrumentation and Environment

All experiments described here were performed in a large vacuum chamber, Chamber "A", located at the Johnson Space Center in Houston. Figure 1 shows a schematic diagram of the chamber. The outside dimensions of the chamber are 120 ft height and 65 ft diameter. The effective working volume has a 90 ft height and a 55 ft diameter. The chamber can be pumped down to about $1 \times 10^{-6}$ torr pressure and the thermal environment maintained between 90° and 312°K. No thermal requirements were imposed during this experimental run, which means that all measurements were made at roughly room temperature. Key to the success of the experiments were several instruments which allowed production of plasma, monitoring plasma parameters, and visually observing plasma effects.

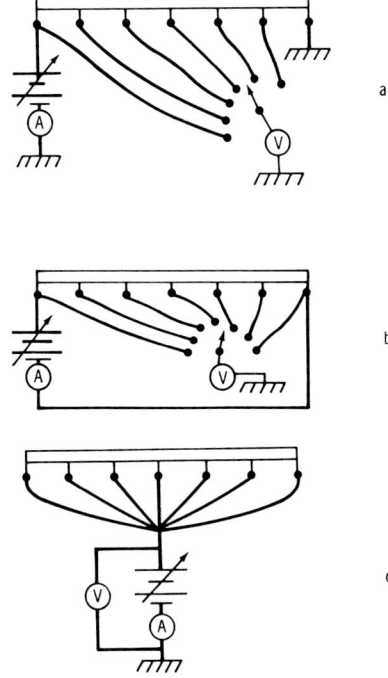

Fig. 4. Electrical interconnections used to measure plasma currents from a negatively biased array with one end grounded (a), floating with respect to the plasma (b), and with all sections at the same potential (c). (a) and (b) simulate a high voltage photovoltaic array with a monotonically changing potential distribution. (c) is used to study sheath formation and ion focusing.

Solar Panel Simulator (SPS)

The key device was a Solar Panel Simulator (SPS) shown in Figure 2. The 3 foot by 30 foot panel was constructed in seven sections with thin transverse copper strip electrodes for biasing and potential measurements at the ends of each section. Six sections were made of a conductive plastic material (Velostat), and the seventh section was composed of actual solar cells. The resistance from one end of the conductive plastic to the other was 140 k$\Omega$. The back side of the panel was made of a rigid dielectric material (Lexan). The objective was to simulate a large solar array with all of its cells series connected. This was done by producing either a monotonic potential gradient or allowing the whole panel to be at the same potential.

Plasma Generator

A Kaufman thruster operating on Argon produced a plasma flow of an estimated 0.5-1 ampere. The electron temperature of the plasma

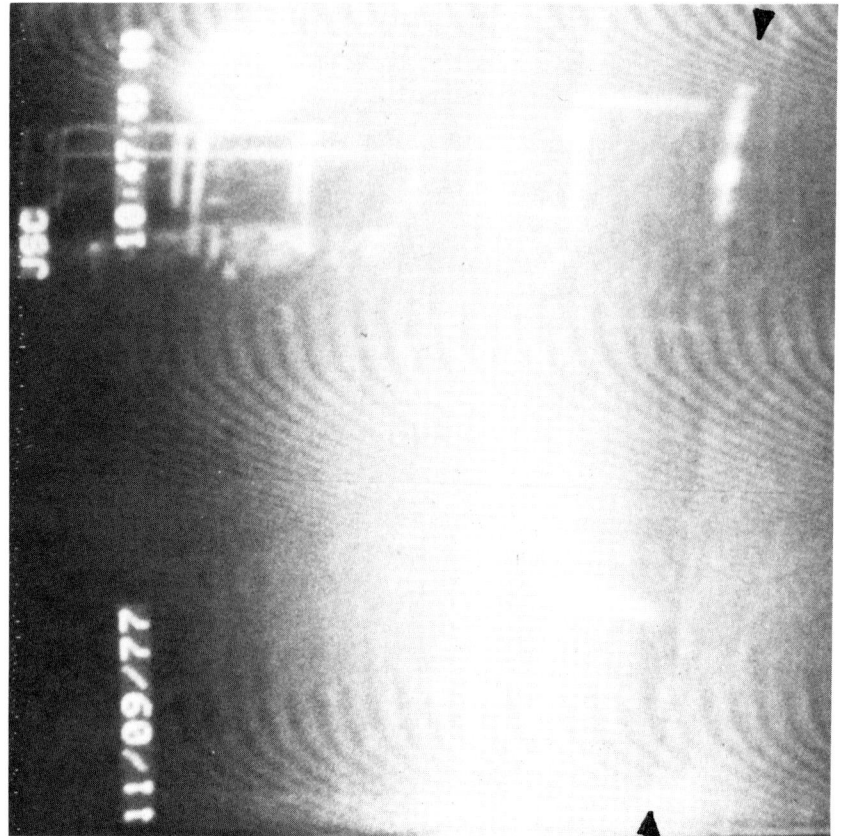

Fig. 5. Plasma sheath formed around the SPS at -100V (a) and -1500V (b) (See opposite page) bias. The bulbous object at the bottom of the SPS in (b) is due to image blooming caused by a glow produced by focused impinging ions. The SPS is seen edge-on and is located between the arrows.

was typically 1-3 eV. At the bottom of the chamber the measured electron density could be made as high as $10^6 cm^{-3}$.

## Langmuir Probes

The plasma parameters were measured at various locations in the chamber with a number of ½ inch diameter spherical Langmuir probes. The probes were placed at the East, North, West, and South sides of the chamber wall at levels 1, 3 and 5 for a total of 12 probes with an additional probe at the top of the chamber and three movable probes near the center of the chamber. The movable probes were used to study the extent of the SPS plasma sheath when the panel was subjected to high voltages.

## Low Light Level Television (LLTV)

Two very low light level television cameras, normally used for the study of the polar Aurora, were provided by the University of Alaska.

These cameras proved to be of great value because they furnished visual evidence for a variety of plasma phenomena which might otherwise not have been recognized.

Plasma

With the thruster at the bottom of the chamber the plasma density was not uniform. Figure 3 shows typical plasma density measurements at levels 1, 3 and 5. The differences in the density measurements between any two levels is probably due to the variation in local plasma conditions. The effects of surface contamination of the probes was also an unkonwn. Unfortunately no instrumentation was available to measure ion temperature and streaming velocity. From considerations of the potential distributions internal to the thruster we estimate an ion energy of 10-30 eV, mostly as a result of bulk flow. When the Kaufman thruster is operating a glow emanates from the plasma. This glow, estimated at about 10 kR, can be seen by the dark adapted eye and the LLTV. Preliminary calculations indicate that the glow may be caused through direct excitation of the background gas by plasma electrons, or through a series of charge transfer reactions and dissociative recombination yielding electronically excited OH and $H_2O$ molecules.

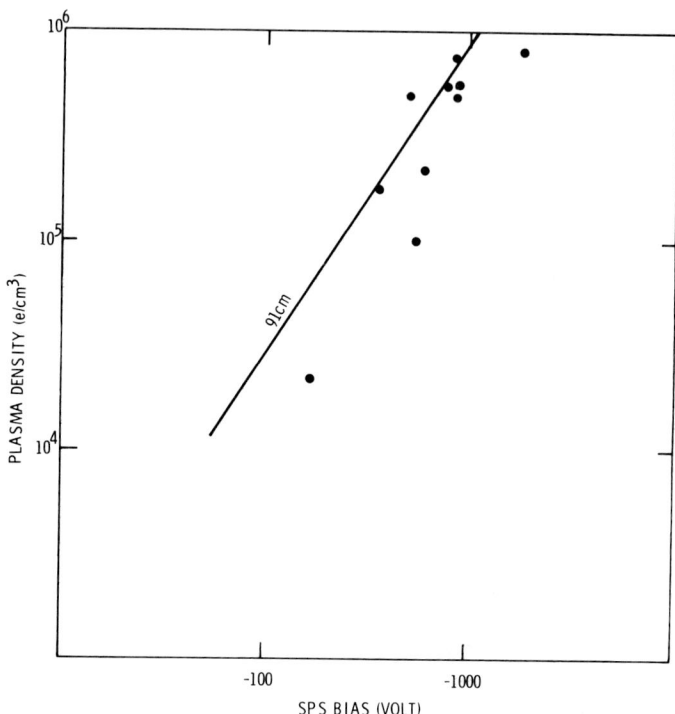

Fig. 6. Experimental measurements of the plasma density and the SPS bias for a sheath boundary stand-off distance of 3 feet (91 cm). The measurement was done with a Langmuir probe located at the center of the SPS.

The complete set-up of the equipment is shown in Figure 1. The SPS was suspended roughly in the middle of the chamber in a vertical position. The plasma generator (Kaufman thruster) was placed either at the bottom center of the chamber or on the side at the third level (same level as the TV camera). The Langmuir probes are shown near the walls at levels 1, 3 and 5, as well as at the center top and near the center of the chamber. The LLTV cameras are shown on level 1 and 3.

## Measurements

For experimental purposes the SPS was connected in three different configurations shown in Figure 4. The first configuration shown in Figure 4a has one end of the SPS grounded while the other end is at a high potential. Current flowing through both the SPS and the plasma to ground as well as voltages at six points are measured. The second configuration shown in Figure 4b is similar to the first, except that both the SPS and the high voltage power supply remain floating and

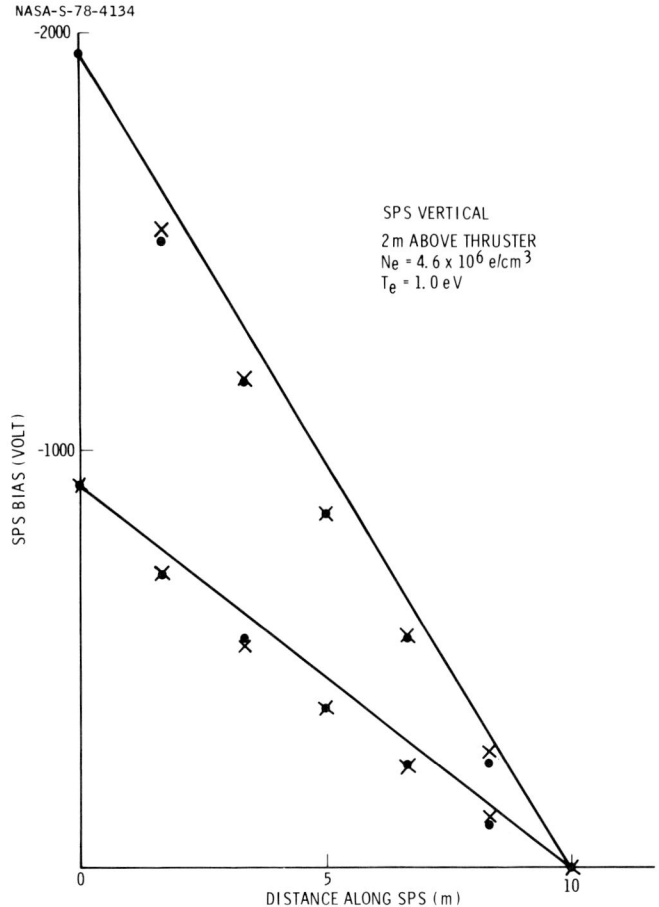

Fig. 7. Experimental (·) and calculated (x) voltage distribution along the SPS with the top grounded.

the voltages at seven points are measured with respect to the ground. In the third configuration shown in Figure 4c all seven copper strips are tied together and to the HV power supply. The SPS-to-plasma current is monitored as a function of voltage.

Sheath Dimension

It is well known that when a body is immersed in a plasma, a sheath will form around the body. The thickness of the sheath and the potential distribution inside will depend on the body-to-plasma potential. We have produced such sheaths by biasing the SPS to high voltages. The magnitude of the attainable positive voltage is limited by the large electron current that is drawn by the SPS. On the other hand with plasma densities of about $10^6 cm^{-3}$ biases of up to -2500V have been applied to the SPS before spontaneous arcing limited the magnitude of the applied voltage. Figure 5 shows the sheath formation

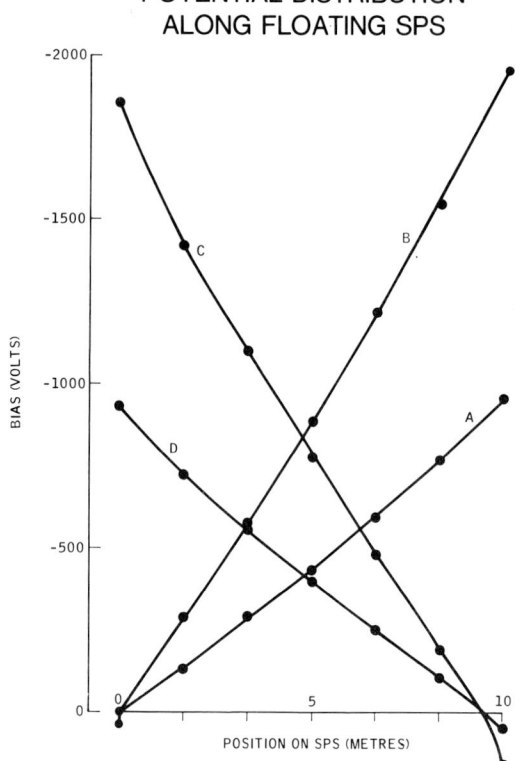

Fig. 8. Four measurements of the voltage distribution along the SPS with respect to ground with the whole SPS floating. In A and B the bottom of the SPS was positive with respect to the top. In C and D the bias was reversed.

for −100V and −1500V biases on the SPS when connected as shown in figure 4c. The figure is an edge-on view as seen by the LLTV. The position of the SPS is indicated between the arrows. The dark region to the left of the SPS is the sheath. At low voltages, when the sheath is thin, it has a well defined wedge-like appearance. At higher voltages it becomes more rounded and the edge is less clearly visible in the TV image. The bulbous, overexposed glow visible at the lower end of the SPS is a phenomenon described later in the section on ion focusing. To do quantitative comparision between theory and experiment we used the movable probes to determine the sheath edge. The edge was defined as the location at which the probe began to measure a potential appreciably different from the plasma potential. Figure 6 shows one example of the bias voltages and plasma densities for a measured sheath thickness of 3 feet. The straight line indicates the biases and plasma densities expected for the 3 foot sheath thickness based on a simple Child-Langmuir law calculation for a one dimensional geometry. The scatter of the points is probably due to uncertainties in the measurement of the plasma density.

## Voltage Distribution Along the SPS with One End Grounded

The simplest potential distribution of a high voltage power array is represented by a uniform gradient from one end of the array to the other. As will be shown in the next section such an array will float with respect to the plasma with over 90% of the panel at negative bias. Thus an SPS with one end grounded and the other at high negative voltage is a reasonable approximation of a solar array and represents a simple means for studying the potential distribution and the flow of parasitic plasma currents. To simulate a series connected high voltage solar cell array we used the electrical configuration shown in Figure 4b. The top end of the SPS was grounded while the bottom end was biased at a high negative voltage. The power supply current and the voltage distribution along the SPS were measured. The results for 2 values of the potential are shown in Figure 7. The dots represent experimental values. The shape of the I-V curve can be explained in the following way. The SPS has a resistance per unit length of $\lambda$ $\Omega$/m. Similarly one can assign a conductance to the plasma sheath of $\rho$ mho/m. The current from the power supply will partially flow along the SPS to ground and partially from the SPS through the plasma sheath to the plasma and thence to ground. The resulting voltage drop along the SPS can be described with a second order differential equation.

$$\frac{d^2V}{dx^2} = \lambda \rho V.$$

The solution of this equation is in terms of $\sinh\sqrt{\rho\lambda}x$ and $\cosh\sqrt{\rho\lambda}x$. Applying boundary conditions and using the midpoint of the SPS to find $\rho$ we get a set of theoretical points represented by crosses. The good agreement between the experimental and theoretical values suggests the usefulness of this simple model to calculate potential distributions and plasma current flows. The straight line represents the potential distribution due to the SPS ohmic resistance and holds true for the no plasma limit ($\rho = 0$).

## Voltage Distribution Along the SPS with Both Ends Floating

A high voltage photovoltaic power system operating in an outer space plasma environment will float electrically in such a way that by far the major part of the structure will be negative with respect to the plasma. This effect has been observed in experiments with the SPS and is predicted by several models with various degrees of realism. To illustrate this phenomenon we shall use a very simplified model. Let us assume that the plasma sheath is thin with respect to the dimensions of the photovoltaic array, then we shall make the approximation that the area of the sheath edge equals the area of the photovoltaic array. To keep charge neutrality the total plasma ion and electron currents to the array must be the same.

$$A_i j_i = A_e j_e; \quad A_i + A_e = A$$

where $A_i$ = area over which ions are collected, $A_e$ = area over which electrons are collected, $j_i$ = ion current density, $j_i$ = electron current density, and $A$ = total area of the photovoltaic array.

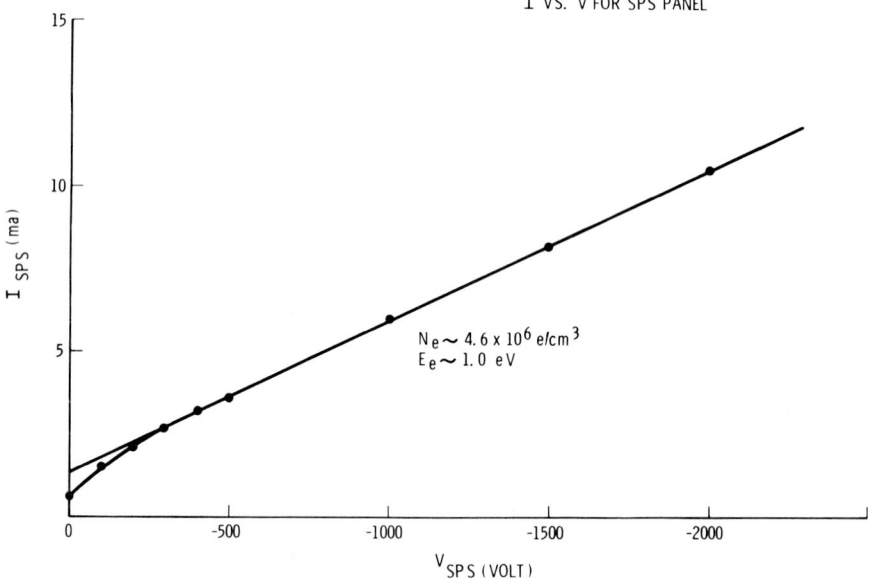

Fig. 9. I-V characteristic of the SPS. For V between -100V and -500V I is proportional to $\sqrt{V}$ while for V greater than -500V I increases linearly with V.

We shall equate the incident electron current density to the thermal current density.

$$j_e = n_e q_e V_e = -n_e e \sqrt{\frac{kT_e}{m_e}}$$

For ions we shall write

$$j_i = n_i q_i V_i = n_i e \sqrt{\frac{2 E_i}{m_i}}$$

where $E_i$ is determined by the Bohm criterion (Bohm, 1949) to be $E_i \geq \tfrac{1}{2} kT_e$. Here k is the Boltzman constant and V, T, n, and m are the velocity, the temperature, concentration, and mass of the particle with e or i referring to electrons or ions. Assuming equality we get

$$\frac{j_e}{j_i} = \frac{A_i}{A_e} = \frac{m_i}{m_e} = 270$$

for an Argon plasma.

In our experimental set-up the ions did have an initial flow velocity equivalent to about 20 eV of energy. Depending on the angle of their impingement the inequality is probably more correct and the ratio of the ion collecting area to the electron collecting area is smaller but still large enough to suggest that more than 95% of the SPS is ion collecting. In any case it means that for an electrical configuration shown in Figure 4b only a small portion of the array will be positive and collecting electron current. Figure 8 shows four cases of voltage measurement when the SPS was allowed to float in a plasma. In cases A and B the bottom of the array was positive, while for C and D the reverse was true. Clearly the measurements are consistent with the above considerations. The relatively larger positive part of the array in cases C and D can be explained as being a consequence of the uneven plasma distribution shown in Figure 2. Since the plasma density is lower at the top of the array, a larger than prediction portion is needed to collect the electron current. The significance of this experiment is to demonstrate that the important factor in the determination of parasitic plasma currents is the negatively biased portion of the array collection ion currents which is almost as large as the whole array.

Total Plasma Current Measurement

To get some preliminary values for the dependence of the plasma current on biasing voltage we used the electrical configuration shown in Figure 4c. The whole SPS was given the same bias and the plasma current was measured. A typical result is shown in Figure 9. For biases greater than -500V the relationship between current and voltage is linear. For voltages between -100 and -500 the current increases very much as the square root of the voltage. Unfortunately little data is available at voltages in excess of -2000V for plasma densities of the order of $10^6 e/cm^3$. When this critical voltage was approached arcing occurred at several places on the SPS. The cause of the arcing is not yet understood and will be the subject of further study. Quantitatively the functional dependence of the current on the voltage is similar to what one sees in the case of a Langmuir probe operating in an orbit limited regime: (Langmuir and Mott-Smith, 1926). For moderate biasing voltages the long, narrow, SPS develops a sheath similar to a cylindrical probe and the current increases as the square root of the voltage. For higher voltages the sheath continues to grow and the SPS begins to look like a spherical probe, i.e., the current increases linearly with voltage. As will be seen in the next section, however, the SPS draws current in a space charge limited mode and the orbit limited theory is strictly speaking not applicable. At this point the question arises whether we are indeed operating in an environment in which the sheath at its maximum extent does not interact with the walls of the chamber. Recordings of the sheath done with the LLTV at plasma densities $n_e \sim 10^6 cm^{-3}$ indicate that the sheath edge is far from the wall. The sheath grows approximately as $n^{-\frac{1}{2}}$ and for low enough densities can touch the chamber walls. Because of this possibility measurements were performed only in plasma density regimes where this effect did not take place. On the other hand the simple fact is that the SPS is a large object and therefore through ion current absorption and electron current repulsion alters the plasma density at significant distances beyond the sheath edge. The importance of this fact will have to be assessed through future scaling experiments.

Fig. 10. Glow pattern produced on the SPS by ion focuaing (a) -500V bias and (b) -1500V bias (See opposite page).

<u>Ion Focusing</u>

When a negative bias is applied to the panel the LLTV shows the appearance of a bright plume extending the length of the panel. Figure 10 shows this plume for bias voltages of -500V and -1500V. While the cause of the plume formation is not definitely known, after considering various alternatives we believe that the plume outlines the locus where ions accelerated in the sheath impinge upon the surface of the panel. Presumably the ions produce secondary electrons in the vicinity of the panel and these interact with the ambient neutral gas to form N* which leads to visible light emission in the plasma. By comparing the shape of the plume for the two biases it is clear that at higher voltages the plume narrows indicating a focusing of the ion flux. This focusing is caused by the electric field of the negative sheath. To show this focusing analytically one has to find a self-consistent solution of the Poisson equation, the continuity equation, and the Vlasov equation. For simple geometries where the trajectories of accelerated ions coincide with the field lines such solutions can be obtained, however, for a more complex

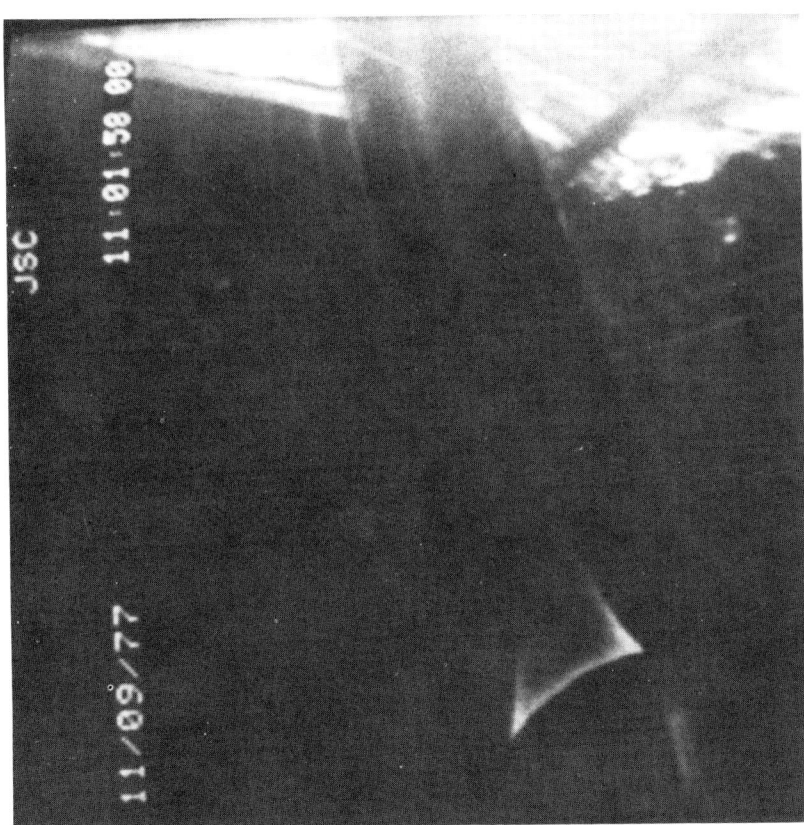

geometry the solution can be obtained only numerically. At this time we do not have a full solution of this problem. The appearance of the well defined region where ions are impinging on the surface and a dark region surrounding it indicates that we are dealing with a space charge limited ion flow and not an orbit limited case. To check our assumption of a focusing effect we have reduced the model to two dimensions representing a cut through the SPS at its mid-point.

We used an elliptic cylindrical coordinate system and represented the sheath with a Laplacian field as shown in Figure 11. Two cases were examined: A sheath thickness at the mid point of the SPS equal to one and two width of the SPS. The sheath boundaries are indicated by dashed lines while the equipotentials are represented by solid ellipses. The particles were started at the sheath boundary with zero initial velocity and allowed to be accelerated toward the surface of the SPS. The left side of the figure represents trajectories of particles accelerated from the sheath with a mid-point thickness of one SPS width. The right side shows trajectories from a sheath with a mid-point thickness of two SPS width. Clearly one can notice a trajectory focusing toward the middle of the SPS, especially at the larger sheath dimension corresponding to higher negative panel potential.

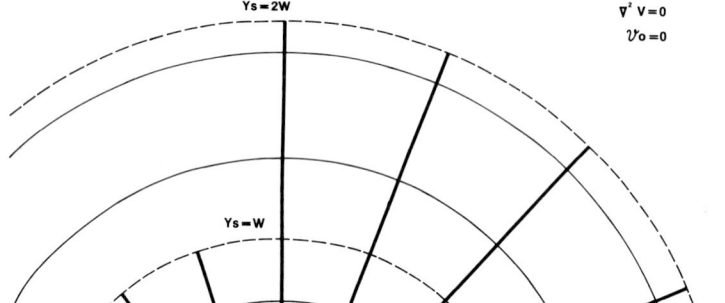

Fig. 11. Sheath focusing of ions. The diagram represents a cross sectional cut of the SPS at its midpoint. The thick lines represent trajectories of ions accelerated in a two dimensional Laplacian field with the SPS (W) at a constant -2000V bias. The dashed lines represent a sheath edge with a stand-off distance $Y_s = W$ and $Y_s = 2W$.

## Conclusions

We have presented some initial results dealing with the study of the behavior of a high voltage panel exposed to an ionosphere-like plasma in a laboratory. The study was motivated by the requirement to understand the interaction of high voltages with ambient plasma and the role of parasitic plasma currents which may short circuit high voltage photoelectric power arrays. The work was restricted to studies in the simulated F2 region where power drainage is expected to be most severe. Since plasma currents to high voltage arrays are determined by the magnitude of the ion current alone the work reported here is solely concerned with the negative sheath and ion currents.

Our findings indicate that qualitatively the behavior of the high voltage panel may be understood in terms of electrostatic probe theories. However, the geometry of the test article prevents the direct application of known analytic solutions and requires finding a numerical solution. Experimentally we have found several interesting and potentially important effects. One of these indicates that at high negative voltages incident ions are strongly focused onto a narrow region of the high voltage panel. Another is that for biases in excess of -500V and up to -2500V, the maximum reached in the test, the current-voltage relationship becomes almost linear.

The focusing effect is important in that it leads to local increases in the current density of accelerated ions. While it is difficult to predict what effect this may have on the structure it can possibly lead to ion sputtering and implantation. In view of the previously discussed arcing observed at high voltages we have firm data for parasitic current power loss only up to about -2500V. At -2000V bias our power loss is $2.5 \times 10^{-4}$ W/cm$^2$. Assuming we manage to circumvent the problem of arcing and breakdown and that the I-V characteristics extrapolate linearly to much higher voltages; at 10kV and 20kV we can expect power losses of 5.5 mW/cm$^2$ and 22mW/cm$^2$ respectively. Typical power outputs of solar cells are about 15mW/cm$^2$. We must stress that the above considerations are based on measurements done on a 3 ft by 30 ft panel and we currently do not know how the power losses scale with array size. However, if we assume that the SPS represents a high voltage solar array, parasitic power losses in the ionization maximum of the ionosphere could become a problem. The fact that solar cells are covered by dielectric material most likely will not affect the total current loss since for any voltages above a few hundred volts the sheath stand-off distance is much greater than the inter-cell distance at which the current carrying conductors of the individual cells are located, and from the point of view of the sheath edge the solar array would look like a single, continuous, high voltage panel. Clearly our assumptions and extrapolations must first be justified before any meaningful predictions of power losses are made. So far we have mainly strong evidence suggesting need for further experimental work. There is also no doubt in our minds that all the potential problems can be designed around; on the other hand these problems must be fully determined before appropriate remedial measures are taken.

Acknowledgements. We would like to thank Dr. R. S. Clark and Mr. D. R. White for their help with the experiment. The assistance of the Space Environment Test Division under Mr. J. C. McLane, Jr. was invaluable.

## References

Bohm, D., The Characteristics of Electrical Discharges in Magnetic Fields, Chapter 3. Edited by Futhrie, A., and Wakerling, R.K., McGraw-Hill Book Co. Inc. (1949).

Langmuir, I. and H.M. Mott-Smith, The Theory of Collectors in gaseous discharges, Phys. Rev. 28, 727, 1926.

Pike, C. P. and R. R. Lovell, ed. Proceedings of the Spacecraft Charging Technology Conference, Air Force Geophysics Laboratory, Hanscomb AFB, Mass., 1977, AFGL-TR-77-0051, also NASA TMX-73537.

# THE RADIATION ENVIRONMENT AND ITS EFFECTS ON SPACECRAFT

Joseph Janni and George Radke

Applied Physics Division
Air Force Weapons Laboratory
Kirtland Air Force Base, NM 87117

**Abstract.** A parametric study has been performed to evaluate the effects of different electron and proton spectral shapes on the dose deposited behind realistic satellite shielding. The high energy component above 1-MeV of geomagnetically trapped electrons is not well known, introducing a considerable uncertainty in shield design. Only recently has emphasis been placed on accurately measuring the spectrum of electrons above 1-MeV. However, these are the electrons that produce most of the dose within typical electronics subsystems. Lower energy electrons usually do not penetrate the shielding provided by satellite structure and electronics housings. The sensitivity of the dose has been parametrically evaluated for electron spectra having moderately different high energy components. This evaluation has been performed for simple spherical shields and for the actual shielding provided by an operational USAF satellite. Moderate changes in shielding can provide drastic reductions in the accumulated electron dose.
  The inner radiation zone contains protons with very high energies that extend well beyond 30-Mev. The shielding provided by most contemporary satellites is insufficient to stop these protons. Unlike the situation for electrons, it is very difficult to easily shield these protons.

## Introduction

  The radiation environment to be discussed in this report will be limited to the geomagnetically trapped electron and proton components. This environment consists of electrons whose energy extends as high as several MeV and protons having energies out to hundreds of MeV. These environments can produce high background levels in satellite sensors, atomic displacements in solar cells, spacecraft charging effects of several types, and accumulated radiation dose failure of electronic components. It is the dose rate and accumulated dose effects that constitute the main emphasis of this report, and for which component hardening and satellite shielding are used to decrease satellite vulnerability.
  In order to determine satellite vulnerability to total dose effects, the space radiation environment to be encountered by the satellite must be specified. Uncertainties in this environment translate into related uncertainties in the dose. The vulnerabilities of individual or aggregate electronic components in the

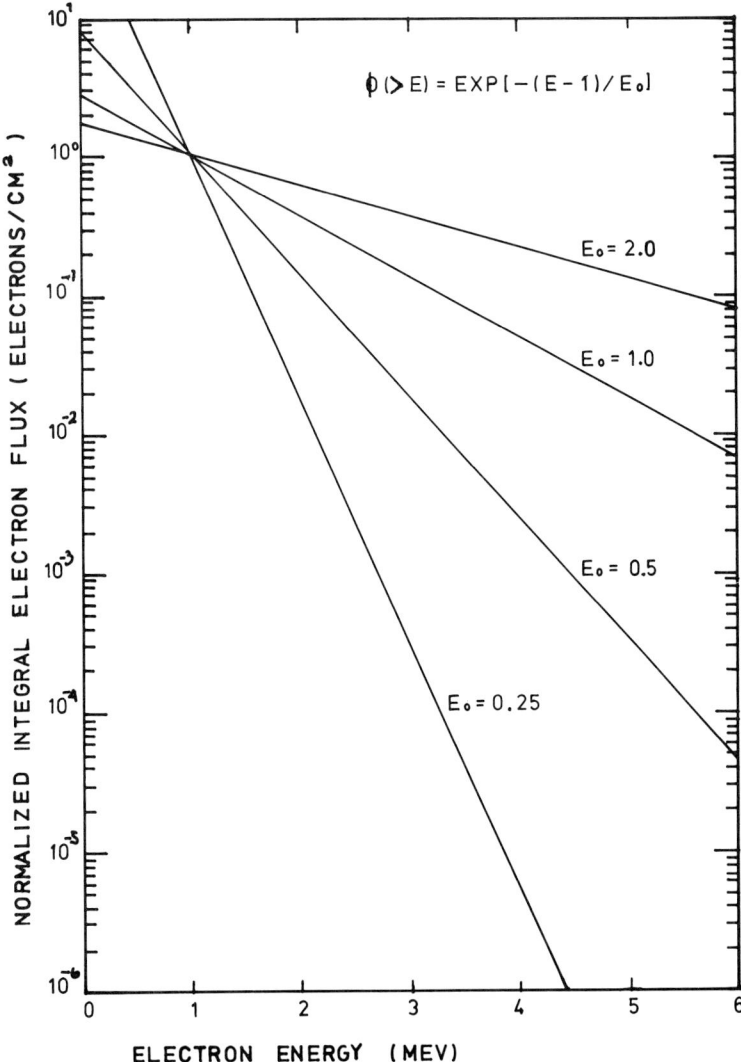

Fig. 1. Normalized integral electron flux as a function of electron energy.

satellite must be known. Errors in the understanding of component vulnerabilities are a second source of uncertainty in a satellite vulnerability analysis. In order to determine radiation doses at locations throughout a satellite, the shielding provided by the satellite structure must be accurately known so that the exterior radiation environment can be transported into the satellite to the locations of interest. The accuracy of radiation transport techniques also affects the overall uncertainty of the analysis. Uncertainties in the knowledge of the high energy electron environment above 1-MeV and of the high energy proton environment above 30-MeV are the dominant sources of error in performing dose related

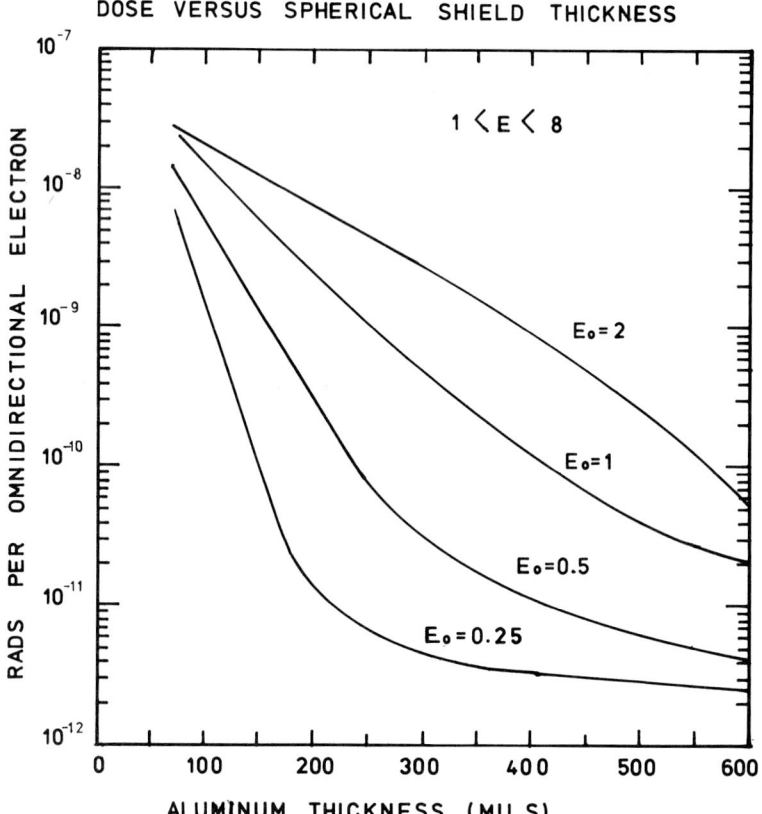

Fig. 2. Depth-dose per omnidirectional electron as a function of spherical aluminum thickness.

vulnerability analyses for naturally occurring geomagnetically trapped radiations.

## Parametric Study for Solid Aluminum Spheres Using Incident Electrons

The variation in spectral shape of incident electrons has been used as the variable of interest. The different spectral shapes that have been used are presented in Figure 1. All spectra were purely exponential in shape, with only the e-folding parameter to be varied. The range of electron e-folding parameters goes from 0.25 to 2.0 because this spans the value of spectral shapes actually found in the geomagnetically trapped electron population. All radiation was treated as omnidirectionally incident. The integral electron flux has been normalized to unity at 1-MeV in order to make the comparisons easier. The functional form of the spectrum is displayed near the top of the figure. Each electron spectrum extends only to 8-MeV.

The depth-dose profiles produced by the different electron

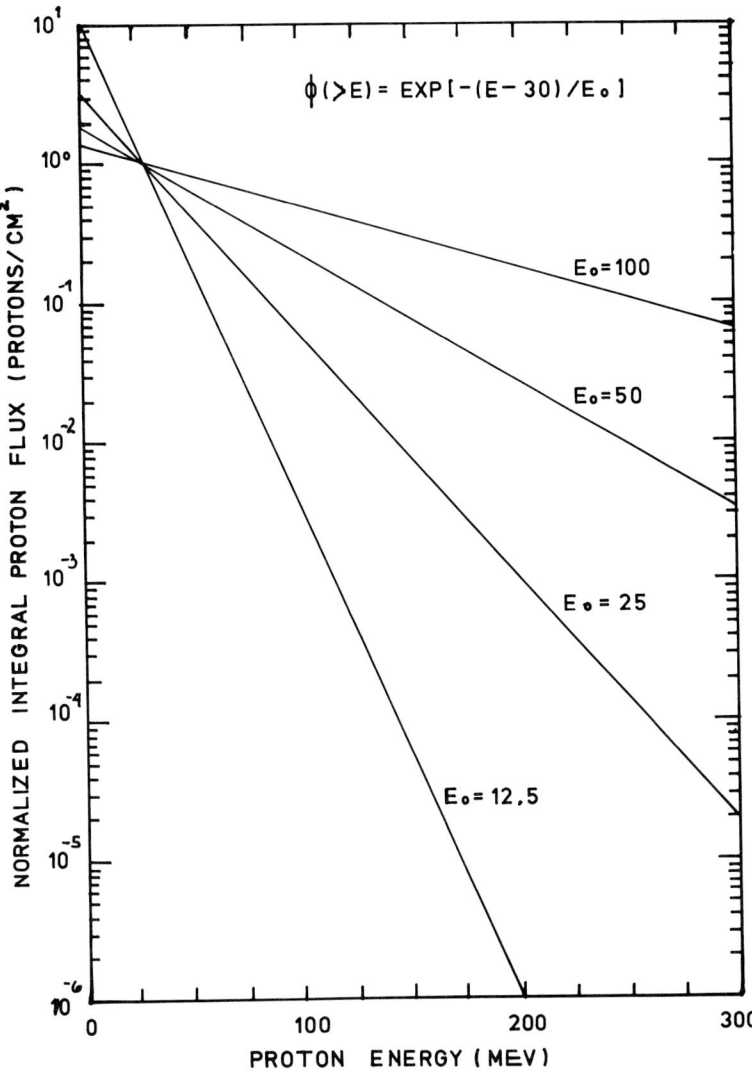

Fig. 3. Normalized integral proton flux as a function of proton energy.

spectra are illustrated in Figure 2, which presents rad dose as a function of aluminum thickness normalized to one incident electron. Only the high energy component between 1-MeV and 8-MeV have been considered. The range of aluminum thickness covers the regime usually found on modern spacecraft. The dose behind shielding greater than about 75-mils is completely dominated by the free space electron population above 1-MeV. Lower energy electrons do not produce energy deposition behind the structure used to house the typical electronic subsystems in modern satellites, unless a component is exposed to the free space environment.

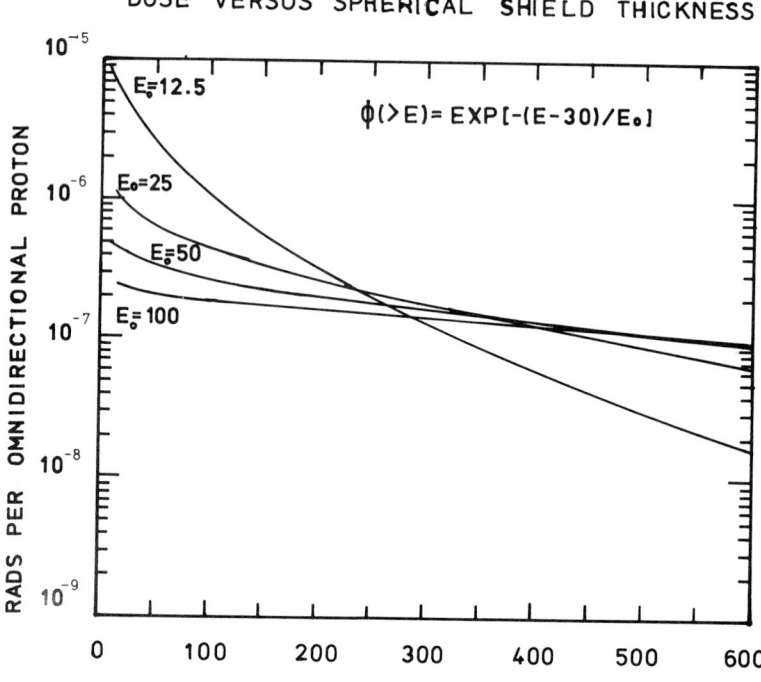

Fig. 4. Depth-dose per omnidirectional proton as a function of spherical aluminum thickness.

## Parametric Study for Solid Aluminum Spheres Using Incident Protons

Geomagnetically trapped protons have been represented in Figure 3, and are assumed to have an exponential shape whose e-folding values extend from 12.5 to 100. These extremes span the region of geomagnetically trapped proton spectra. The integral proton flux has been normalized to unity at 30-MeV, exponentially extends to 1000-MeV, and is assumed to be omnidirectional in free space.

The variation of dose deposited behind each different proton spectrum is shown in Figure 4 as a function of spherical aluminum thickness. There is a strong dependence of depth-dose on the spectrum e-folding value for thicknesses less than about 250-mils, while at greater thicknesses the dose is dominated by very high energy protons which are extremely difficult to shield.

## Parametric Study for a USAF Satellite Using Incident Electrons

A detailed radiation shielding analysis has been performed of an advanced design U. S. Air Force spacecraft, which contains several thousand complementary metal oxide semiconductors (CMOS). In order to precisely determine the on-orbit radiation dose to these semiconductors, a very extensive computer model of the entire satellite was prepared using engineering drawings, photographs,

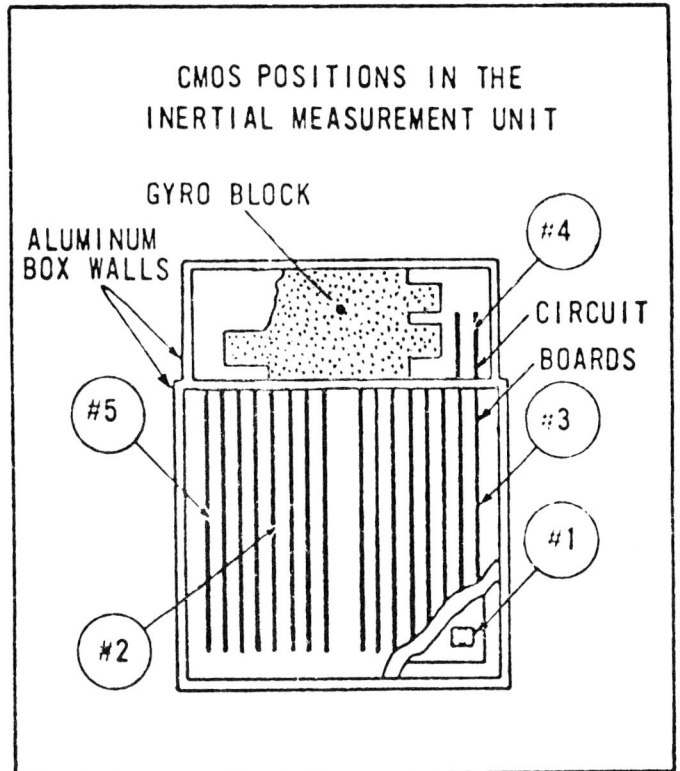

Fig. 5. Cross-sectional diagram of the IMU geometry with five representative CMOS locations.

and direct measurements. Radiation transport calculations were then performed using this model to determine the radiation doses for a carefully selected sample of representative CMOS in each electronics subsystem. Shielding was then designed for the more sensitive components, with considerable care given to mass minimization. This study was performed to improve the on-orbit lifetime of the satellite in geomagnetically trapped radiation.

One important difference between geomagnetically active and quiet environments is the higher flux of electrons above 1-MeV during active times. These spectral differences become particularly important in computing doses due to primary electrons penetrating moderate to heavy shielding. The doses from the geomagnetically active environment can be about 4 times higher than those from the quiet time environment. For very heavily shielded components, primary electrons cannot penetrate and bremsstrahlung becomes the dominant contributor to the dose.

In order to calculate accurate doses to the CMOS components, it was necessary to develop a sophisticated computer model of the entire spacecraft geometry. The model was complete in all important aspects and contained considerable physical detail; however, the subsystems associated with the primary spacecraft mission

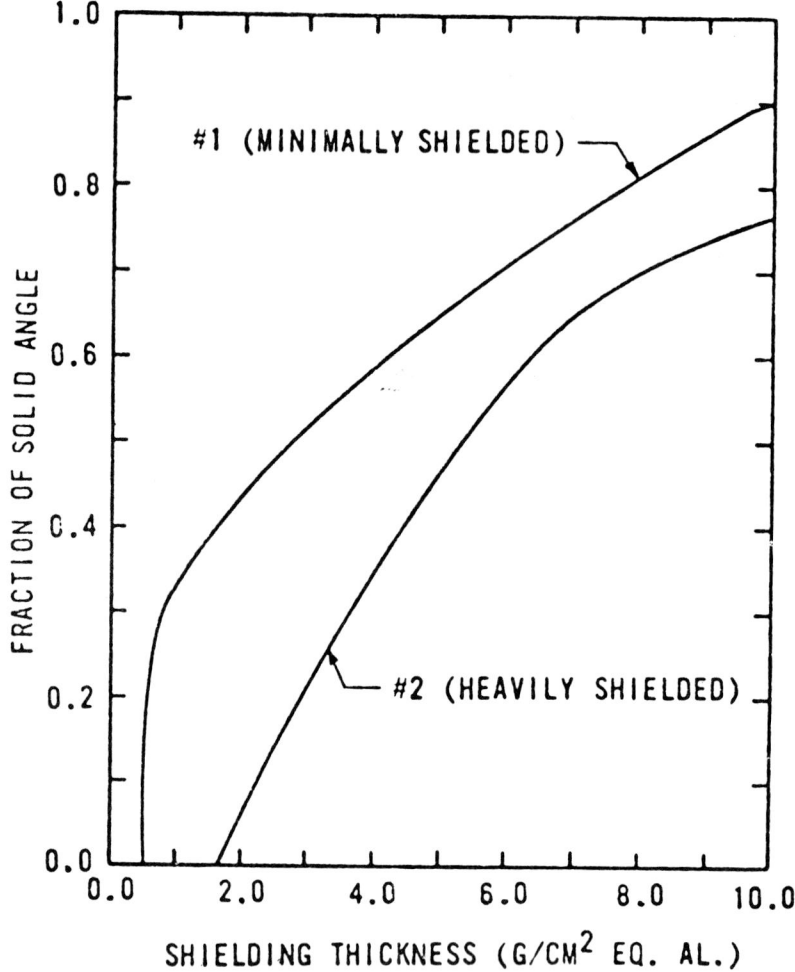

Fig. 6. Mass distribution for CMOS locations 1 and 2 in the IMU.

were given priority. The secondary subsystems were not modeled exactly but were included as homogeneous units of the proper size, mass, and orientation. The remainder of the spacecraft was modeled down to the circuit board level with individual CMOS components positioned in the correct locations. Small bolts, nuts, washers, and wires were not modeled. Most circuit board components other than CMOS were not included directly in the physical mock-up, but were conservatively accounted for by homogenizing the circuit boards.

After the spacecraft model was completed, it was checked for accuracy by developing computer-generated cross-sectional plots of the spacecraft through selected planes. Figure 5 was drawn directly from these cross-sectional plots of the Inertial Measurement Unit (IMU), and illustrates 5 CMOS locations that were selected as examples. The cross-sectional plots were then carefully com-

TABLE 1. Unshielded Electron Dose Ratios

| Electron Spectral Parameter | Position Number | | |
|---|---|---|---|
| $E_o$ | 1 | 2 | 5 |
| .25 | 80. | 1.5 | 2.5 |
| .50 | 164. | 1.0 | 11. |
| 1.0 | 294. | 5.9 | 48. |
| 2.0 | 433. | 28. | 119. |

pared to the engineering drawings to locate any misplaced or omitted parts and to make appropriate corrections.

The shielding for any component can be converted to equivalent aluminum and plotted to give a visual representation of the mass distribution around the dose point. Figure 6 shows two typical examples of these mass distributions. The ordinate is the fraction of the total solid angle around the CMOS location which is shielded by a lesser number of $gm/cm^2$ than indicated on the abscissa. The curve for CMOS position No. 1 (see Fig. 5) is representative of a minimally shielded CMOS component which is facing an outside wall near the corner of an electronics box. This configuration allows substantial amounts of radiation to enter. The curve for CMOS position No. 2 is typical of a component buried deep within the electronics box and substantially shielded by nearby circuit boards. The large range in shielding around different components in the same box as illustrated in Figure 6 is not unusual.

The detailed model also permitted the investigation of the effects of the CMOS packaging and circuit board orientation on the overall life of the spacecraft. A parametric study of the effect of the different electron and proton spectral parameters from

TABLE 2. Unshielded Proton Dose Ratios

| Proton Spectral Parameter | Position Number | | |
|---|---|---|---|
| $E_o$ | 1 | 2 | 5 |
| 12.5 | 10. | .53 | 1.8 |
| 25. | 6.9 | 1.0 | 2.2 |
| 50. | 4.8 | 1.8 | 2.6 |
| 100. | 3.7 | 2.2 | 2.7 |

TABLE 3. Unshielded/Shielded Electron Doses

| $E_o$ | \multicolumn{3}{c}{Position Number} | | |
|---|---|---|---|
| | 1 | 2 | 5 |
| .25 | 41.9 | 1.00 | 1.00 |
| .50 | 15.1 | 1.00 | 1.05 |
| 1.0 | 6.14 | 1.02 | 1.26 |
| 2.0 | 3.24 | 1.04 | 1.30 |

Figures 1 and 3 was performed as a function of the dose point position shown in Figure 5. The electron and proton dose ratios for 3 IMU positions are in Tables 1 and 2 for the satellite prior to the addition of the extra shielding that was eventually added. Three extremes in geometry are given in Figure 6. CMOS position No. 3 faces an outer box wall and is shielded from behind by several other circuit boards. In this geometry almost all of the dose is caused by electrons entering through the box wall. The back is already adequately shielded by other circuit boards. The geometry for position No. 5 faces away from the outside wall. The primary direction from which the dose is received is through the adjacent exterior wall. In position No. 1, the CMOS chip is facing an outside wall, but it is in the corner of the box.

In order to compare the effects of the different spectral shapes for the positions 1, 3, and 5 of Figure 5, all results were normalized to unity using the second e-folding parameter and the second position. The results for the satellite prior to the addition of shielding are in Tables 1 and 2. Position No. 1 is lightly shielded, and is very sensitive to the e-folding values. Positions 2 and 5 are more heavily shielded, with position 2 being the most

TABLE 4. Unshielded/Shielded Proton Doses

| $E_o$ | Position Number | | |
|---|---|---|---|
| | 1 | 2 | 5 |
| 12.5 | 2.48 | 1.18 | 1.98 |
| 25. | 1.96 | 1.04 | 1.09 |
| 50. | 1.46 | 1.03 | 1.04 |
| 100. | 1.24 | 1.01 | 1.02 |

heavily shielded. All positions exhibit a strong dose dependence on the electron spectral parameter, but only position 1 exhibits a strong dependence on the proton spectral parameter. Positions 2 and 5 are minimally sensitive to the proton spectral parameter because they are relatively well shielded.

The ratios of the unshielded to shielded satellite doses are presented in Tables 3 and 4 for electron and proton dose to the added shielding to be evaluated. The added shielding is extremely effective in reducing the electron dose, but only minimally reduces the proton dose except for the most thinly shielded position for the softest spectrum.

## Conclusion

An accurate knowledge of the flux of electrons above 1-MeV and of protons above 30-MeV is very important in the assessment of satellite vulnerability to geomagnetically trapped radiations. Modern satellites usually have sufficient shielding surrounding their electronic subsystems to stop most of the electrons below 1-MeV and the protons below about 30-MeV. Therefore, it is the high energy component of the geomagnetically trapped space radiation environment that is most important in determining satellite vulnerability to total dose effects.

OVERVIEW - MODELS AND DATA - DISCUSSION OF THE JULY 29, 1977 SUBSTORM

R. H. Manka

Atmospheric Science Section, National Science Foundation, Washington, DC 20550

The discussion of the July 29, 1977 substorm accomplished two purposes. First, the event was used as an example event from the modeling standpoint in order to discuss what can be calculated for such an event given the present modeling state of the art. Second, the modeling discussion contributed to the overall analysis of this event and it is hoped that several members of the modeling community will participate in the further analysis of this event. This event has also been the topic of preliminary data-oriented discussions at the April 1978 Spring AGU meeting in Miami (see IMS Newsletter 78-5) and the Innsbruck meeting in June 1978.

It provides an excellent data set for the study of magnetospheric processes. Especially of note is that there is an excellent set of coordinate spacecraft and ground-based data for the event, and these data are being further reduced and studied by the investigators. It is likely that this event will be the topic of an intensive workshop to be held in May 1979. The spacecraft data set includes data from interplanetary spacecraft (IMP-7,8), four geosynchronous satellites, GEOS-1, and several low altitude satellites such as S3-3, and other polar orbiters.

The substorm event is relatively clearly defined. It was preceded by a quiet period on July 28 and initiated by a strong shock wave in the solar wind. The resulting density and velocity increases in the solar wind were sufficient to push the magnetopause in to $\sim 7~R_E$ and past GEOS-1 which was near apogee. During the twelve hour period following the shock there were four distinct substorms, with well-defined westward and eastward electrojets.

Figures 1 and 2 show the stacked magnetograms for the July 29, 1977 substorms. These magnetograms, based upon magnetometer data in the World Data Center, were quickly digitized by Dr. Akasofu in time for the Innsbruck discussion. Several key features of the event can be seen in the figures. A large shock wave arrived at the earth at about 0027 UT on July 29, with a resulting compression of the magnetopause into the GEOS position ($\sim 7~R_E$). This solar wind pressure gradually relaxed, but then at least four distinct substorms were seen with maximum field depressions at 0430, 0630, 0900, and 1230 UT. As can be seen from the auroral zone latitude magnetograms, the westward electrojet of the 1230 substorms extends from local dawn through midnight to local dusk. These two magnetogram figures, as provided by Dr. Akasofu, were published in IMS Newsletter 78-9, and a summary of the Miami AGU discussion was published in IMS Newsletter 78-5.

In addition to the ground based data shown in the figures, there is a large, comprehensive data set from other ground-based arrays and a

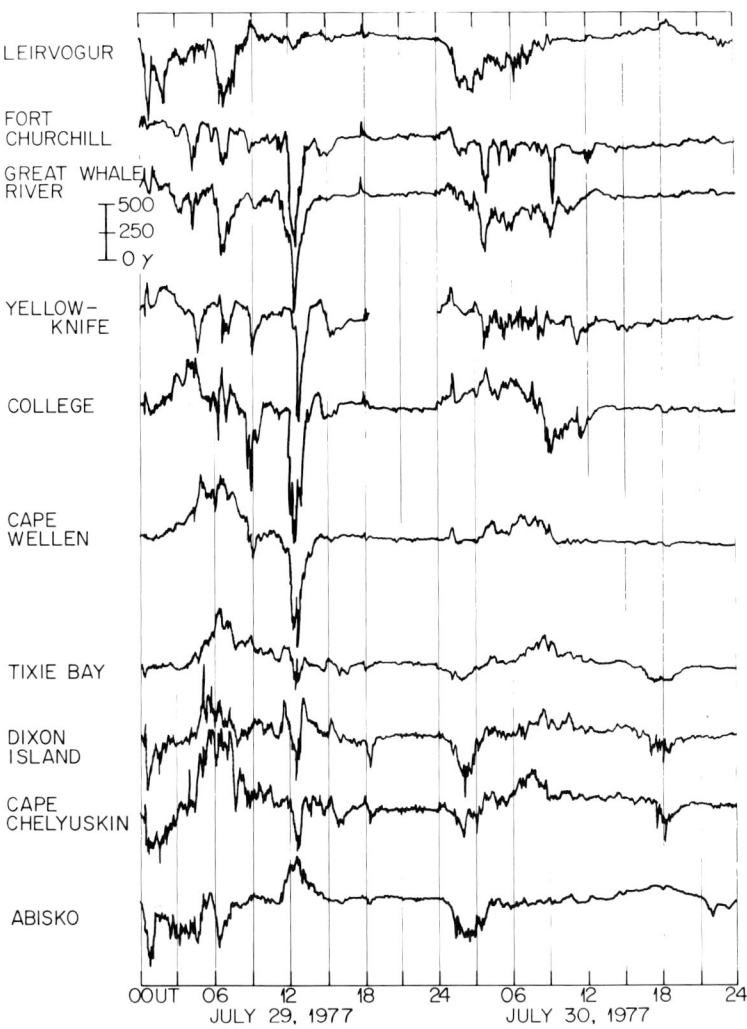

Figure 1

group of high altitude, geosynchronous, and low altitude satellites. This includes solar wind plasma and magnetic fields from IMP-7,8; energetic particles and magnetic fields from ATS-6, the NOAA SMS and GOES satellites and from GEOS 1 which was at apogee at the time of the initial shock (0027 UT) and again during the final substorm (1230 UT); and ion composition, particles, electric fields, etc., from low altitude spacecraft such as S3-2 and S3-3. Figure 3 shows the relative positions of the geosynchronous and high altitude spacecraft during the first three hours of July 29 during the time that the initial shock wave was observed. The magnetopause was pushed in past GEOS-1, but apparently not past ATS-6.

Further participation by the modeling community in the analysis of

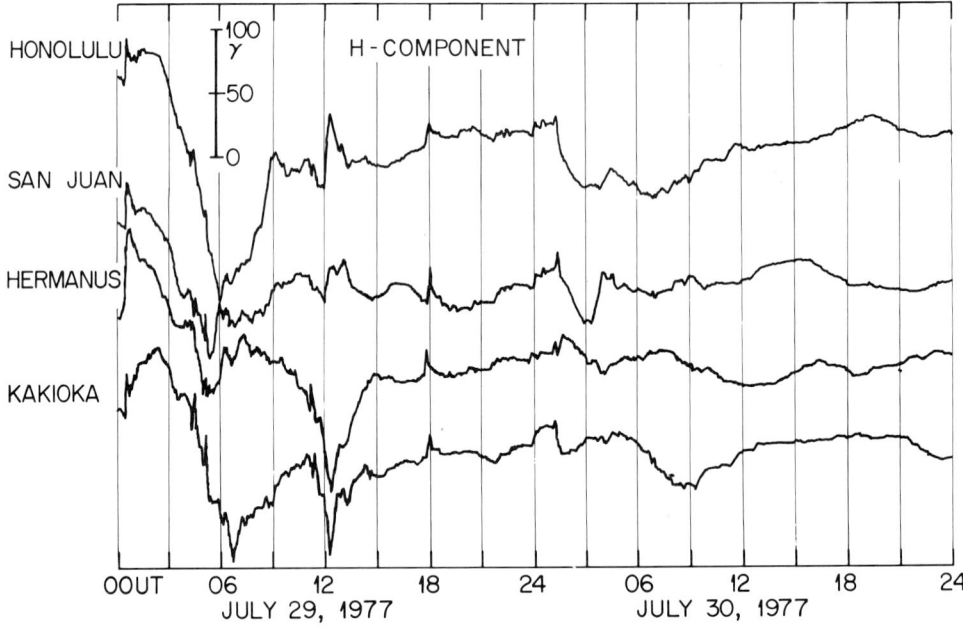

Figure 2

this interesting event is welcomed. In particular, it is desirable to calculate as many quantities as possible which would complement the data analysis and theoretical work on the event which has already taken place.

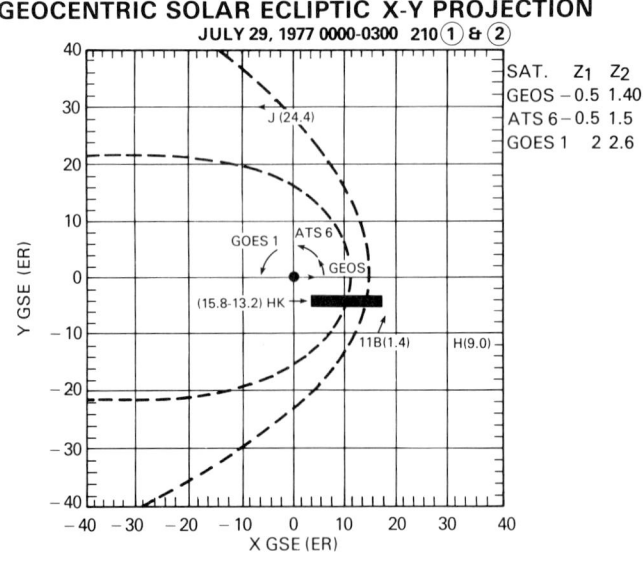

Figure 3

DESCRIPTIONS OF AVAILABLE QUANTITATIVE MAGNETOSPHERIC MODELS

The following lists of models describe their regions of validity, the type of model, authorship and availability. These models all provide a quantitative description of some magnetospheric feature or process. Most of them are "representative" as opposed to "explanatory" (see Cauffman electric field model summary).

Most of these models are available from the authors or the National Space Science Data Center. Users of these models are requested to communicate back to the authors their comments and criticisms. The users can make their experiences known to their colleagues generally by inputing them to the International Association of Geomagnetism and Aeronomy's Working Group on Quantitative Magnetospheric Models whose charter is to continuously monitor modeling activities. Address such correspondence to W. P. Olson, McDonnell Douglas Astronautics Company, A3-204-AATO, Building 13-3, Huntington Beach, California 92647.

## MAGNETIC FIELD MODELS

The magnetospheric magnetic field is usually considered to have contributions from the magnetopause, ring, and tail currents. More localized contributions from field-aligned and ionospheric currents are just now being discussed quantitatively and no explicit models of them are contained in the following list.

These magnetospheric field models are usually used together with a dipole representation of the main field. When a more accurate description of the main field is required, a spherical harmonic expansion is used. Several such expansions are described in Mead's paper in this volume. All of the main field models are available from the National Space Sciences Data Center.

## MAGNETIC FIELD MODELS

| Model | Description | Valid Region | Primary Reference | Availability |
|---|---|---|---|---|
| Antanova and Shabansky | A tilted image dipole representation of the magnetopause currents and a infinitesimally thin set current in the magnetotail. No ring current representation is included. Inputs similar to Hones-Taylor model with the addition of tilt angle of image dipole. | Earthward of inner edge of tail currents | Geomagnetism and Aeronomy, 8, p 891. 1968. | From National Space Science Data Center |
| Choe-Beard | Magnetopole shape described by pressure balance. A description of the tail current system is included but the model contains no ring current representation. Model valid for all tilt angles. | Earthward of inner edge of tail currents. | Haldeson, et al., Planetary and Space Science, 23. p 887. 1975. Kosik, J.-C. Planetary and Space Science. 25. p 457. 1977. | This model is available as a spherical harmonic expansion, Halderson, et al. Not available as deck. It is also available as an analytic function, Kosik. |
| Hedgecock-Thomas | A tilt dependent empirical model. Model is entailed to represent only quiet magnetic conditions as defined by Kp less than 2.0 or Bst greater than -10. | Entire inner magnetosphere and tail to 20 R$_E$. | Hedgecock, P.C., B.T.Thomas, A.M. Cornwall and C.W. Davis. Imperial College Preprint. | Contact authors. |
| Hones-Taylor | An image dipole representation of the magnetopause currents plus a thin tail current set. The location of the inner edge of the tail currents and tail intensity and image dipole location and moment. Available as a similarity expression. No ring current representation is included. | Earthward of current set | Taylor and Hones, J. of Geophys. Res. 70. p 3605. 1965 | Available as an analytical expression (see reference) |

## MAGNETIC FIELD MODELS

| Model | Description | Valid Region | Primary Reference | Availability |
|---|---|---|---|---|
| Mead-Fairfield | An empirical model based on data from 5 to 17 $R_E$. Model is tilt dependent and available for four ranges of magnetic activity as parameterized by Kp. Model is expressed as a second order power series expansion in solar magnetic coordinates, quadratic in position and linear in tilt. Coefficients are given in reference. | Use is out to 17 $R_E$. | Mead, D.C. and D.H.Fairfield. J. Geophys. Res. 80. p 523. 1975. | Deck can be constructed by user from information in reference. |
| Mead-Williams | Magnetopause currents determined from pressure balance. Includes infinitesimally thin tail current sets extending infinitely in east west direction. Inner edge of current set is input parameters. Subsolar point distance is also an input parameter. No ring current is included. Tailward edge of current set and field intensity to the current set are also input parameters. Tile dependent. | | J. Geophys. Res. 70. p 3017. 1965. | Model is represented as spherical harmonic expansion. It is valid earthward of the tail current systems and generally to geocentric distances not larger than 10 $R_E$. A computer deck is available from National Space Sciences Data Center. |
| Olson-Pfitzer | Magnetopause, ring and tail currents represented for quiet magnetic conditions. Perpendicular incidence of solar wind on geomagnetic dipole axis (zero tilt) is assumed. Magnetopause currents determined from pressure balance relm. | Throughout magnetosphere to linear orbit in tail. | Olson and Pfitzer. J. Geophys. Res. 79. p 3739. 1974. | Model available as power series expansion from authors. |
| Olson-Pfitzer | Includes boundary, ring, and tail currents. | Valid for all tilt and to lunar orbit in tail. | Documentation included with deck in form of comment cards. Reference to model in this volume. | Model available as power series expansion with exponential terms from authors. |
| Olson-Pfitzer | Includes boundary, ring and tail contributions. The strength and location of each current system can be varied individually and used to represent time varying fields. | Throughout magnetosphere to lunar orbit in tail. For zero tilt only. | See article by Olson-Pfitzer-Mroz in this volume. | Available in mid 1979. |
| Sugiura-Poros | Using Mead's expansion for magnetopause field. Includes ring and tail currents contribution. Only valid for zero-tilt. | Valid throughout magnetosphere except in region of currents systems boundary. | Sugiura, M. and D.J.Poros. Planetary and Space Science. 21. p 1763. 1973. | Magnetic field data available in tabular form only. |
| Tsyganenko | Magnetopause shape is defined using Fairfield data. Infinitesimally thin ring and tail currents are included. Model is valid for all tilt angles. | Entire inner magnetosphere to inner edge of tail currents. | Annals Geophys. 32. p 1. 1976. | Not available. |

## MAGNETIC FIELD MODELS

| Model | Description | Valid Region | Primary Reference | Availability |
|---|---|---|---|---|
| Voigt | Model uses a predetermined magnetopause shape, hemispherical on day side with cylindrical tail geometry. The model includes infinitesimally thin tail and equatorial ring currents. The model is tilt dependent. Inputs include the standoff distance to the subsolar point, the flair of the tail, tail field intensity, and the ring current strength (determined by the magnetic index $D_{st}$). The "tilt angle" is also input. | Throughout magnetosphere | The 1976 Scientific Satellite Programme during the International Magnetospheric Study. Edited by K. Knott and B. Pattrick, Reidel Publishing Company, Dordrecht-Holland, pp 381-396. 1976. | Not available. |
| Willis-Pratt | A zero-tilt engine dipole model with described currents whose strength decreases with distance down the tilt. The model contains no ring current representation. | Model valid earthward of tail currents | J. Atm and Terrest. Phys. 34. p 1955. 1972 | |
| Alekseev and Shabansky | Magnetopause shape is predefined as paraboloid of revolution. It has thin tail current but no ring current. The model is tilt dependent. Input parameters unknown. | Valid earthward of tail current inner edge. | Planetary and Space Sciences. 20. p 117. 1972. | Not available. |

## ENERGETIC PARTICLE RADIATION MODELS

Current models of trapped radiation are summarized in the following tables. These models and supporting documentation are all available from the National Space Science Data Center (NSSDC), Goddard Space Flight Center, Greenbelt, Maryland 20771. Users are encouraged to discuss their requirements with the NSSDC staff to assure compatibility between the software and card decks provided and computer capabilities available to users.

## ENERGETIC PARTICLE RADIATION MODELS

| Model | Description | Valid Region Energy Range | L Range | Availability |
|---|---|---|---|---|
| AE-6 (electrons) | Flux values for Oct. 1967, but Starfish residue subtracted. | 0.04 - 4 MeV | 1.2 - 2.8 | (see introductory statement) |
| AP-8 (protons) | Model has both solar minimum and maximum values; these changes only occur at low altitudes. Data used in making model are shown in Table 1 of reference (pages 19 through 22). | 0.1 - 400 MeV | 1.17 - 6.6 | NSSDC-/WDC-A-R&S-76-06. 1976 |
| AEI-7 (electrons) | Interim model with upper and lower limit for energies above 1.5 MeV to account for ATS 6 data at 3.9 MeV and OV1-19 data up to 5 MeV. Model is interim until discrepancies | 0.04 - 5 MeV | 2.8 - 11.0 | (see introductory statement) |

## ENERGETIC PARTICLE RADIATION MODELS

| Model | Description | Valid Region Energy Range | L Range | Availability |
|---|---|---|---|---|
| | between Azur data at 4.5 MeV and OV1-19 data can be properly studied. At energies below 1.5 MeV, this model is same as AE-4 in the outer zone for solar maximum. | | | |
| AE-4 (electrons) | Flux values for epoch 1964 and 1967 corresponding to solar minimum and maximum conditions. Solar cycle effects only occur between $2.8 \leq L \leq 5$. Energies above 4 MeV are strictly extrapolations, and data above 1.9 MeV did not have proper calibration. Data used in making model are shown in Table 1 of the reference (page 31.) | 0.04 - 4 MeV | 2.8 - 11.0 | NSSDC 72-13. 1972 |
| AE-5 (1975 projected) (electrons) | Flux values for Oct. 1967, but Starfish residue subtracted and solar cycle effects used so projected valid epoch is 1975, corresponding to solar minimum conditions. Data sets used in making model are shown in Table 3 of reference (page 57) with Starfish model of second reference used in subtraction. | 0.04 - 4 MeV | 1.2 - 2.8 | NSSDC 74-03. 1974<br><br>GSFC-X-601-72-487. 1974 |

## ELECTRIC-FIELD MODELS

Models of the magnetospheric electric-field are gradually becoming more sophisticated, but they still must be used with caution for the following reasons: (1) the electric-field pattern varies substantially with geomagnetic activity, but this variation is not well understood yet; (2) short time and/or length-scale fluctucations are typically as large as (often larger than) the large-scale convection electric-field, but these fluctuations are not included in the models; (3) magnetic-field-aligned electric-fields are not included in any of the models, because their pattern is still in the process of being sorted out observationally; (4) induction electric-fields are clearly important during substorms or other active periods, but they are neglected in most of the models; (5) there are still very few direct measurements of electric-fields out in the magnetosphere beyond 8000 km altitude.

## ELECTRIC FIELD MODELS

| Model | Description | Valid Region | Primary Reference | Availability |
|---|---|---|---|---|
| Harel-Wolf | Theoretical model that results from self-consistent calculation of ionospheric and magnetospheric electric fields and currents. Time-dependent electric fields and currents have been computed through a substrom type event that occured 19 Sep 76. Other parameters computed: joule heating, plasma distribution in the magnetosphere. Crucial input data: polar-cap potential drop and boundary location as functions of time: electron fluxes used to estimate conductivities; midlatitude magnetograms, to estimate strength | 1) Only valid for $L \leq 10$; 2) Neglects parallel electric fields and neutral winds; 3) Magnetic-field model is Olson-Pfitzer (1974) analytic model, perturbed by time-dependent substorm current loop; the magnetic field is not computed self-consistently; 4) Only one | Harel, M., R.A.Wolf, P.H.Reiff, and M. Smiddy. Computer Modeling of Events in the Inner Magnetosphere, This Volume, 1979.<br><br>Harel, M., R.A.Wolf, P.H.Reiff, and H.K.Hills. Study of Plasma Flow Near the Earth's Plasmapause. U.S.A.F. Geophysics Labora- | Data decks give computed potentials at grid points, and locations of the grid points in ionosphere and equatorial plane, for various times during event. FORTRAN interpolation routine gives model electric-field at an arbitrary |

ELECTRIC FIELD MODELS

| Model | Description | Valid Region | Primary Reference | Availability |
|---|---|---|---|---|
| | of substorm current loop, for estimation of induction electric-field. Tests: Good, but not precise, agreement with 53-2 measurements of subpolar-cap electric-fields in the 19 Set 76 event. Computed pattern of Birkeland currents resemble Iijima-Potemra pattern. Model ring current injection is similar to usual observations by McIlwain and co-workers. Limitations: 1) Only valid for $L \leq 10$. 2) Neglects parallel electric fields and nuetral winds. 3) Magnetic-field model is Olson-Pfitzer (1974) analytic model, | event has been modeled so far. | tory. Report AFGL-TR-77-0286. 1977. | point in the ionosphere or equatorial plane. Contact M.Harel or R.A.Wolf, Dept. of Space Physics and Astronomy, Rice University, Houston, Texas 77001 |
| Harel-Wolf (Continued) | perturbed by time-dependent substorm current loop; the magnetic field is not computed self-consistently. 4) Only one event has been modeled so far. | | | |
| McIlwain | A static semi-empirical potential-electric-field model is established by detailed comparison of computed drift trajectories with particle fluxes observed at synchronous orbit after substorms. The electrostatic potential is given by an expansion of approximately 120 terms. Particles are assumed to originate at an injection boundary, the location of which can be determined from the particle data. The model agrees with geosynchronous-orbit data from a great number of events. | Although model is presumed to be reliable near geosynchronous orbit, validity becomes unclear far from that orbit. No system is suggested for mapping to low altitudes. Induction electric fields are ignored. Coefficients are derived for post substorm conditions with $Kp \gtrsim 1$, although coefficients can be scaled to correspond to other Kp values. (see reference) | McIlwain,C.E. Plasma convection in the vicinity of geosynchronous orbit, in Earth's Magnetospheric Processes, ed. B.M. McCormac, R.Reidel, Dordrecht-Holland, p. 268, 1972. | A computer deck that generates electric-field values is available from Prof. C.E. McIlwain, Physics Dept. U. of California, San Diego, La Jolla, California 92037 |
| Kamide-Matsushita | The ionospheric electric potential contours, latitudinal and longitudinal distributions of the electric fields, ionospheric currents, and the equivalent ionospheric current vectors with their equivalent current systems are calculated. The model is essentially similar to that of Nopper and Carovillano, but uses a denser grid point and ionospheric conductivities and field-aligned currents based on most recent observations including seasonal changes. | Currently valid only for steady state away from equatorial region. Cannot distinguish various origins of electric field. | Kamide, Y. and S.Matsushita. Simulation Studies of Ionospheric Electric Fields and Currents in Relation to Field-aligned Currents, Quiet Periods. submitted to J. Geophys. Res. 1978. | PLIB (Programmers will be available from the National Center for Atmospheric Research, Boulder, Colorado 80307, after refinements are made around October 1979. |
| Nopper-Carovillano | Global ionospheric potential pattern is computed from current conservation, given models for height-integrated conductivities and Birkeland currents. A semi-empirical conductivity model represents the gross structures of solar-zeneth | 1) Does not extend out into magnetosphere. 2) Computed electric-field patterns vary substantially with empirical input | Nopper, Jr., R.W. Ionospheric Electric Fields and Currents of Magnetospheric Origin. Ph.D. Thesis, Boston College. June 1978. | Card decks containing calculated results of quiet-time model, disturbed-time model, etc., are available from |

## ELECTRIC FIELD MODELS

| Model | Description | Valid Region | Primary Reference | Availability |
|---|---|---|---|---|
| Nopper-Carovillano (Continued) | angle variation and auroral enhancement; idealizations of the field-aligned currents seen by Triad serve to drive the ionospheric potential. The auroral conductivity and configuration of field-aligned currents are adjustable to correspond to "quiet times" and "disturbed times"; other situations can also be modeled and solved. The model is versatile and can accommodate a wide range of input parameters. Tests: Interal consistency checks based on the physical and mathematical requirement of conservation of ionospheric current have been run to verify the reliability of the final results at all latitudes. Numerical errors in current density are limited to a few percent. Limitations: 1) Has not been extended out into magnetospher. 2) Computed electric-field patterns vary substantially with input parameters, which are difficult to estimate accurately. 3) Neutral winds are neglected. | parameters, which are difficult to estimate accurately. 3) Neutral winds are neglected. | Nopper, Jr., R.W. and R.L. Carovillano. Polar-Equatorial Coupling During Magnetically Active Periods. Geophys. Res. Lett. 5. p. 699. 1978. Nopper, JR., R.W. and R.L. Carovillano. Ionospheric Electric Fields Driven by Field-Aligned Currents. This Volume. 1979. | R.W.Nopper, Jr., or R.L. Carovillano, Dept. of Physics, Boston College, Chestnut Hill, Massachusetts 10267 |
| Richmond | An empirical model has been constructed of the quiet-time, electric-field in the inner magnetosphere and subauroral ionosphere. Observed electric field data from incoherent scatter radar and from whistler duct drift observations have been fitted with a spherical harmonic series representation of the electrostatic potential. Coordinates used are magnetic latitude at 300 km altitude and local time (assumed interchangeable with longitude). Tests: Direct comparison of the model $\vec{E} \times \vec{B}$ drifts with the data base shows reasonable, but not excellent, agreement. | The model is valid only for magnetically quiet days within L=4 (below 60° magnetic latitude) | Details are given in Electric Field in the Ionosphere and Plasmasphere on Quiet Days by A.D. Richmond. J. Geophys. Res. 81. pp 1447-1450, 1976. | A computer-card deck is available on request from A.D.Richmond, NOAA/ERL/SEL, R43, Boulder, Colorado 80303. The sub-routine of this deck yields the electrostatic potential and the components of $\vec{E} \times \vec{B}$ drift velocity at 300km altitude at any point on the earth. |
| Schulz | The corotational electric field associated with a magnetopause plus dipole magnetic field geometry (as given by Mead) is presented. | Valid for untilted magnetic dipole only. Does not include convection or induction electric fields, or effects of currents in the earth's surface. | Schulz, M. J. Geophys. Res. 75. p 6329. 1970. | The expression and derivation for $\phi$ are given in the reference. |
| Volland | Semi-empirical analytic expressions are given for the horizontal components of the ionospheric electric field at all latitudes and local times. They are arranged to fit most general features of electric fields observed by polar-orbiting satellites. Magnetospheric electric fields are derived assuming a static magnetic-dipole field and no parallel electric field. Tests: Gives reasonable plasmapause, $Sq^p$ currents, Birleland currents, Harange discontinuity. | 1) Has only been used with a dipole magnetic-field model for mapping out into the magnetosphere. 2) Includes many adjustable parameters. Author has suggested parameter values for a "low" and "moderate" geomagnetic activity. | Volland, H. J. Geophys. Res. 83. p 2695. 1978. Volland, H. This Volume. 1979 | Since the published analytic expressions (see references) are simple, users are encouraged to create their own little programs. |

# KEY WORD LIST INDEX

adiabatic motion, 582
anisotropy, 203
artificially injected ions, 340
atmosphere, 326
    magnetosphere, 297
atmospheric electricity, 326
ATS-1, -5, -6, 35
auroral
    currents, 499
    images, 96
AWS/75, 121
Birkeland currents, 499, 557
boundary layer, 389, 401, 412
bow shock, 389
charge leakage, 617
charged particles, 499
charging electrostatic, 602
CMOS (Complementary Metal
    Oxide Semiconductors), 634
composition, plasma, 340
computer simulation, 499, 513
conductivity, 326, 557
    Hall, Pederson, 261
contours
    $\alpha$, 48
    $\Delta B$, 9, 48
    $\Delta I$, 48
convection, 297, 423, 513, 557, 569
    electric field, 261
corotation, electric field, 261
cosmic ray
    cutoffs, 242
    pitch angle dependence, 242
    tilt dependence, 242
currents
    auroral, 499
    Birkeland, 499
    boundary, 9, 77
    distributed, 77, 86
    field aligned, 261, 297
    ionospheric, 473
    parallel, 499
    polarization, 389
    ring, 64, 77, 592
    systems, 462, 582
        magnetospheric, 77
    tail, 9, 77
cusp shape, 592
discontinuities, rotational, 536

dissipation, 536
distortion, 64
distribution functions, 203, 364
disturbed times, 64
double layer, 389
drift shells, 569, 582
$D_{ST}$, 121
dynamic model, 77, 513
electric fields, 281, 297, 499, 557, 569
    convection, 513
    fluctuation, 281
    induced, 316, 412
    parallel, 389
electrons, 150, 203, 634
    density profiles, 609
    energetic, 180, 401
    low energy, 96
    trapped, 121
energization, 316
equivalent magnetic charge, 473
field aligned currents, 557
field lines, 9, 77, 86, 96, 150, 261
    closed, 401
    foot points, 96
    IMF, 423
    open, 401
    reconnection, 448
flux
    averages, 121
    erosion, 423
    transfer, 423
    tubes, 592
geomagnetic storm, 121
geometric activity, 436
GEOS-1, -2, 35, 281
geosynchronous orbit, 35, 364
GOES-1, -2, -3, 35
IGRF/75, 110
IGS/75, 121
index
    $A_L$, 423
    $K_p$, 242
indices, 436
induced
    earth currents, 473
    electric fields, 473, 536
injection plasma, 364
interplanetary magnetic field, 423, 448

ion beams, 340
ions, magnetospheric, 9
ionosphere, 281, 326, 557
ionospheric
    dynamo, 473
    electrojet charge, 609
    scintellations, 609
Jupiter, 9
kinetics, 462
low altitude, 64
magnetic
    charge, 473
    effects, 436
    field, 35, 64, 462
        disturbed, 77
        internal, 110
        interplanetary, 297
        lines, 64
        quiet, 77
        time varying, 316
    force, 389
    index, $K_p$, 48
    lines, 64
    merging, 448
    pressure, 401
    tail, 592
    vector potential, 473
magnetohydrodynamics, 536
magnetometer, 35
magnetopause, 389, 401, 412
    shadowing, 150
    shape, 86, 592
magnetosheath, 389
magnetosphere
    asymmetric, 569
    boundary, 412
    closed, 9, 448
    inner, 297, 499
    open, 9, 448
    outer, 150
magnetospheric
    current systems, 77
    tail, 297
mass spectrometer, 340
Mercury, 9
merging, 297
meteorology, 326
MHD Simulation, 401
model
    comparisons, 9
    statistical, 364
Momentum transfer
    diffusive, 297

wave, 297
neutral sheet, 242
observations, vector, 9
particles, high energy, 180
PC5 waves, 281
pitch angle
    anisotropy, 582
    distribution, 150, 340, 364, 582
plasma
    drifts, 364, 401
    hot, 281, 557
    injection, 340
    mantle, 389
    parameter, 448
    pause, 261, 513
    sheath, 401
    sheet, 389
    solar wind, 462
    spacecraft interaction, 602
    temperature, 364
polar
    cap, 242
        electric field, 261
    cusp, 448
POGO/71, 121
potential scalar, 582
protons, 634
    trapped, 121
    vector, 77, 316, 462
radiation
    dose, 634
    shielding, 634
reconnection, 423
ring current, 64, 77, 592
satellite
    ATS-1, -5, -6, 35, 96
    DMSP-32, 96
    Explorer-33, 401
        -35, 423
    GEOS-1, -2, 35, 340
    GOES-1, -2, -3, 35
    HEOS, 401
    IMP6, 401
    ISEE, 340, 401, 412
    SMS-1, -2, 35
    TRIAD, 557
    VELA, 401
sector
    boundary crossing, 326
    structure, 180
secular variation, 110
self consistent, 86, 499

sheath, 401
shell splitting, 150, 536
simultaneous observations, 96
solar
    flares, 326
    terrestrial coulping, 326
    wind, 261, 389, 412, 473
        magnetospheric interaction, 297
        parameter, 436
        plasma, 462
        velocity, 150, 436
spacecraft, 602
statistical study, 436
substorm, 150, 180, 203, 281, 536
    predictions, 609
synchronous orbit, 180
tail, 242
vacuum field model, 86
velocity distributions, 462
Venus, 9
zone, 9
    inner, 121
    outer, 121

QC
809
M35
Q37

OCT 15 1979